Macromolecular Engineering

Edited by
Krzysztof Matyjaszewski,
Yves Gnanou,
and Ludwik Leibler

1807–2007 Knowledge for Generations

Each generation has its unique needs and aspirations. When Charles Wiley first opened his small printing shop in lower Manhattan in 1807, it was a generation of boundless potential searching for an identity. And we were there, helping to define a new American literary tradition. Over half a century later, in the midst of the Second Industrial Revolution, it was a generation focused on building the future. Once again, we were there, supplying the critical scientific, technical, and engineering knowledge that helped frame the world. Throughout the 20th Century, and into the new millennium, nations began to reach out beyond their own borders and a new international community was born. Wiley was there, expanding its operations around the world to enable a global exchange of ideas, opinions, and know-how.

For 200 years, Wiley has been an integral part of each generation's journey, enabling the flow of information and understanding necessary to meet their needs and fulfill their aspirations. Today, bold new technologies are changing the way we live and learn. Wiley will be there, providing you the must-have knowledge you need to imagine new worlds, new possibilities, and new opportunities.

Generations come and go, but you can always count on Wiley to provide you the knowledge you need, when and where you need it!

William J. Pesce
President and Chief Executive Officer

Peter Booth Wiley
Chairman of the Board

Macromolecular Engineering

Precise Synthesis, Materials Properties, Applications

Edited by
Krzysztof Matyjaszewski, Yves Gnanou, and Ludwik Leibler

Volume 3
Structure-Property Correlation
and Characterization Techniques

WILEY-VCH Verlag GmbH & Co. KGaA

The Editors

Prof. Dr. Krzysztof Matyjaszewski
Carnegie Mellon University
Department of Chemistry
4400 Fifth Ave
Pittsburgh, PA 15213
USA

Prof. Dr. Yves Gnanou
Laboratoire de Chimie des Polymères Organiques
16, ave Pey-Berland
33607 Pessac
France

Prof. Dr. Ludwik Leibler
UMR 167 CNRS-ESPCI
École Supérieure de Physique
et Chimie Industrielles
10 rue Vauquelin
75231 Paris Cedex 05
France

■ All books published by Wiley-VCH are carefully produced. Nevertheless, authors, editors, and publisher do not warrant the information contained in these books, including this book, to be free of errors. Readers are advised to keep in mind that statements, data, illustrations, procedural details or other items may inadvertently be inaccurate.

Library of Congress Card No.: applied for

British Library Cataloguing-in-Publication Data
A catalogue record for this book is available from the British Library.

Bibliographic information published by the Deutsche Nationalbibliothek
The Deutsche Nationalbibliothek lists this publication in the Deutsche Nationalbibliografie; detailed bibliographic data are available in the Internet at http://dnb.d-nb.de.

© 2007 WILEY-VCH Verlag GmbH & Co. KGaA, Weinheim, Germany

All rights reserved (including those of translation into other languages). No part of this book may be reproduced in any form – by photoprinting, microfilm, or any other means – nor transmitted or translated into a machine language without written permission from the publishers. Registered names, trademarks, etc. used in this book, even when not specifically marked as such, are not to be considered unprotected by law.

Composition   K+V Fotosatz GmbH, Beerfelden
Printing   betz-druck GmbH, Darmstadt
Bookbinding   Litges & Dopf GmbH, Heppenheim
Cover   Grafik-Design Schulz, Fußgönheim
Wiley Bicentennial Logo   Richard J. Pacifico

Printed in the Federal Republic of Germany
Printed on acid-free paper

ISBN 978-3-527-31446-1

Contents

Preface *XIX*

List of Contributors *XXI*

Volume 1
Synthetic Techniques

1   **Macromolecular Engineering** *1*
          Krzysztof Matyjaszewski, Yves Gnanou, and Ludwik Leibler

2   **Anionic Polymerization of Vinyl and Related Monomers** *7*
          Michel Fontanille and Yves Gnanou

3   **Carbocationic Polymerization** *57*
          Priyadarsi De and Rudolf Faust

4   **Ionic and Coordination Ring-opening Polymerization** *103*
          Stanislaw Penczek, Andrzej Duda, Przemyslaw Kubisa,
          and Stanislaw Slomkowski

5   **Radical Polymerization** *161*
          Krzysztof Matyjaszewski and Wade A. Braunecker

6   **Coordination Polymerization: Synthesis of New Homo- and Copolymer Architectures from Ethylene and Propylene using Homogeneous Ziegler–Natta Polymerization Catalysts** *217*
          Andrew F. Mason and Geoffrey W. Coates

7   **Recent Trends in Macromolecular Engineering** *249*
          Damien Quémener, Valérie Héroguez, and Yves Gnanou

8   **Polycondensation** *295*
          Tsutomu Yokozawa

Macromolecular Engineering. Precise Synthesis, Materials Properties, Applications.
Edited by K. Matyjaszewski, Y. Gnanou, and L. Leibler
Copyright © 2007 WILEY-VCH Verlag GmbH & Co. KGaA, Weinheim
ISBN: 978-3-527-31446-1

9	**Supramolecular Polymer Engineering** *351* G. B. W. L. Ligthart, Oren A. Scherman, Rint P. Sijbesma, and E. W. Meijer
10	**Polymer Synthesis and Modification by Enzymatic Catalysis** *401* Shiro Kobayashi and Masashi Ohmae
11	**Biosynthesis of Protein-based Polymeric Materials** *479* Robin S. Farmer, Manoj B. Charati, and Kristi L. Kiick
12	**Macromolecular Engineering of Polypeptides Using the Ring-opening Polymerization-Amino Acid *N*-Carboxyanhydrides** *519* Harm-Anton Klok and Timothy J. Deming
13	**Segmented Copolymers by Mechanistic Transformations** *541* M. Atilla Tasdelen and Yusuf Yagci
14	**Polymerizations in Aqueous Dispersed Media** *605* Bernadette Charleux and François Ganachaud
15	**Polymerization Under Light and Other External Stimuli** *643* Jean Pierre Fouassier, Xavier Allonas, and Jacques Lalevée
16	**Inorganic Polymers with Precise Structures** *673* David A. Rider and Ian Manners

Volume 2
Elements of Macromolecular Structural Control

1	**Tacticity** *731* Tatsuki Kitayama
2	**Synthesis of Macromonomers and Telechelic Oligomers by Living Polymerizations** *775* Bernard Boutevin, Cyrille Boyer, Ghislain David, and Pierre Lutz
3	**Statistical, Alternating and Gradient Copolymers** *813* Bert Klumperman
4	**Multisegmental Block/Graft Copolymers** *839* Constantinos Tsitsilianis
5	**Controlled Synthesis and Properties of Cyclic Polymers** *875* Alain Deffieux and Redouane Borsali

6	**Polymers with Star-related Structures** *909* *Nikos Hadjichristidis, Marinos Pitsikalis, and Hermis Iatrou*	
7	**Linear Versus (Hyper)branched Polymers** *973* *Hideharu Mori, Axel H. E. Müller, and Peter F. W. Simon*	
8	**From Stars to Microgels** *1007* *Daniel Taton*	
9	**Molecular Design and Self-assembly of Functional Dendrimers** *1057* *Wei-Shi Li, Woo-Dong Jang, and Takuzo Aida*	
10	**Molecular Brushes – Densely Grafted Copolymers** *1103* *Brent S. Sumerlin and Krzysztof Matyjaszewski*	
11	**Grafting and Polymer Brushes on Solid Surfaces** *1137* *Takeshi Fukuda, Yoshinobu Tsujii, and Kohji Ohno*	
12	**Hybrid Organic Inorganic Objects** *1179* *Stefanie M. Gravano and Timothy E. Patten*	
13	**Core–Shell Particles** *1209* *Anna Musyanovych and Katharina Landfester*	
14	**Polyelectrolyte Multilayer Films –** **A General Approach to (Bio)functional Coatings** *1249* *Nadia Benkirane-Jessel, Philippe Lavalle, Vincent Ball, Joëlle Ogier,* *Bernard Senger, Catherine Picart, Pierre Schaaf, Jean-Claude Voegel,* *and Gero Decher*	
15	**Bio-inspired Complex Block Copolymers/Polymer Conjugates** **and Their Assembly** *1307* *Markus Antonietti, Hans G. Börner, and Helmut Schlaad*	
16	**Complex Functional Macromolecules** *1341* *Zhiyun Chen, Chong Cheng, David S. Germack, Padma Gopalan,* *Brooke A. van Horn, Shrinivas Venkataraman, and Karen L. Wooley*	

Volume 3
Structure-Property Correlation and Characterization Techniques

1	**Self-assembly and Morphology Diagrams for Solution and Bulk Materials: Experimental Aspects** 1387	
	Vahik Krikorian, Youngjong Kang, and Edwin L. Thomas	
1.1	Introduction: Bulk Phase 1388	
1.1.1	Block Copolymer Blends 1390	
1.1.1.1	AB/hA Blends 1391	
1.1.1.2	ABC/hA Blends 1392	
1.1.1.3	BCP/Nanoparticle Blends 1394	
1.1.2	Self-assembly of Rod–Coil BCPs 1399	
1.2	Introduction: Solution Phase 1405	
1.2.1	General Features of Block Copolymer Micelles 1407	
1.2.2	Artifacts from Sample Preparation (Impurities, Drying, Cross-linking, Staining) 1409	
1.2.2.1	Impurities 1410	
1.2.2.2	Staining and Drying 1410	
1.2.2.3	Crosslinking 1410	
1.2.3	Coil–Coil Block Copolymers in Solvents 1411	
1.2.3.1	Theory 1412	
1.2.3.2	Concentrated Systems 1413	
1.2.3.3	Diluted Systems 1416	
1.2.3.4	ABC Triblock Polymers 1418	
1.2.3.5	Micelles in Homopolymer Solvent 1420	
1.2.4	Rod–Coil Block Copolymer in Solvents 1421	
1.2.5	Inorganic Hybrid Micelles 1424	
1.3	Summary and Outlook 1425	
	Acknowledgments 1426	
	References 1426	
2	**Simulations** 1431	
	Denis Andrienko and Kurt Kremer	
2.1	Introduction 1431	
2.2	Basic Techniques: Monte Carlo and Molecular Dynamics 1435	
2.2.1	Molecular Dynamics (MD) 1435	
2.2.2	Monte Carlo (MC) 1436	
2.3	Polymer-specific Properties 1438	
2.3.1	Static Observables 1438	
2.3.2	Elementary Dynamic Observables 1440	
2.3.3	Collective Observables 1441	
2.4	Scaling Relations for Dynamics 1441	
2.4.1	Rouse Model 1441	
2.4.2	Reptation Model 1443	

2.4.3	Zimm Model	*1445*
2.5	Molecular Interactions	*1446*
2.5.1	Potential Energy Functions	*1447*
2.5.2	Software	*1451*
2.6	Coarse-graining and Simple Generic Models	*1452*
2.6.1	Simple Generic Models	*1453*
2.6.2	Coarse-graining Techniques	*1455*
2.6.2.1	Penalty Function	*1455*
2.6.2.2	Force-matching Scheme	*1456*
2.6.2.3	Inverse Boltzmann Method	*1457*
2.6.3	Mapping Points	*1458*
2.6.4	Reintroduction of Chemical Details: Inverse Mapping	*1459*
2.6.5	Application of Coarse-graining	*1460*
2.7	Other Illustrative Examples	*1462*
2.7.1	Diffusion of Small Molecules	*1462*
2.7.2	Osmotic Coefficient of Polyelectrolytes	*1463*
2.7.3	Phase Diagrams	*1464*
2.8	Outlook	*1464*
	Acknowledgments	*1465*
	References	*1465*

3   Transport and Electro-optical Properties in Polymeric Self-assembled Systems   *1471*
Olli Ikkala and Gerrit ten Brinke

3.1	Introduction	*1471*
3.1.1	Background on Polymeric Self-assembly and Supramolecular Concepts	*1471*
3.1.2	General Remarks on Polymeric Self-assembly in Relation to Electrical and Optical Properties	*1475*
3.2	Transport of Ions and Protons within Self-assembled Polymer Systems	*1476*
3.2.1	Self-assembled Protonically Conducting Polymers	*1476*
3.2.2	Self-assembled Ionically Conducting Polymers	*1483*
3.3	Self-assembly of Conjugated Polymers	*1487*
3.4	Optical Properties of Self-assembled Polymer Systems	*1496*
3.5	Conclusion	*1502*
	Acknowledgments	*1502*
	References	*1503*

4   Atomic Force Microscopy of Polymers: Imaging, Probing and Lithography   *1515*
Sergei S. Sheiko and Martin Moller

4.1	Introduction	*1515*
4.2	General Principles of AFM	*1517*
4.2.1	Resolution	*1518*

4.2.2	Probe Calibration	1520
4.2.3	Imaging and Operation Modes	1520
4.2.3.1	Contact Mode	1521
4.2.3.2	Intermittent Contact Mode	1521
4.2.3.3	Chemical Force Microscopy	1522
4.2.3.4	Electric Force Microscopy	1523
4.2.3.5	Spectroscopic Imaging	1524
4.2.3.6	Rapid Scanning AFM	1525
4.3	Visualization of Polymer Morphologies	1526
4.3.1	Molecular Visualization	1526
4.3.1.1	Molecular Resolution	1527
4.3.1.2	Using Molecular Visualization to Prove Synthetic Strategies	1530
4.3.1.3	Molecular Characterization	1532
4.3.1.4	Visualization Molecular Processes	1534
4.3.2	Crystallization of Polymer Chains	1538
4.3.3	Wetting and Dewetting Phenomena	1544
4.3.3.1	Polymer Blends	1545
4.3.3.2	Spreading	1546
4.3.3.3	Dewetting	1548
4.3.4	Block Copolymers	1549
4.3.4.1	Micelles	1550
4.3.4.2	Alignment of Microdomains	1551
4.4	Measuring Properties	1554
4.4.1	Adhesion and Colloidal Forces	1554
4.4.2	Friction	1556
4.4.3	Mechanical Properties	1557
4.4.4	Molecular Force Spectroscopy	1558
4.4.5	Thermal Properties	1561
4.5	Scanning Probe Lithography	1562
4.5.1	Writing with Molecular Inks	1562
4.5.2	Thermomechanical Surface Patterning	1563
	References	1565
5	**Scattering from Polymer Systems**	1575
	Megan L. Ruegg and Nitash P. Balsara	
5.1	Introduction	1575
5.2	Scattering from Binary Mixtures	1577
5.3	Homogeneous Multicomponent Polymer Mixtures	1584
5.4	Mixtures Containing Non-overlapping Hard Objects	1588
5.5	Disordered Isotropic Systems	1589
5.6	Ordered Arrays	1590
5.7	Examples	1591
5.7.1	Scattering from Homogeneous Binary Homopolymer Blends	1591
5.7.2	Homogeneous Multicomponent Blends	1594
5.7.3	Microphase Separated Multicomponent Phases	1596

5.8	Conclusion *1603*	
	References *1603*	

6	**From Linear to (Hyper) Branched Polymers:**	
	Dynamics and Rheology *1605*	
	Thomas C. B. McLeish	
6.1	Introduction *1605*	
6.2	Synthesis and Rheology of Controlled Topology Polymers *1610*	
6.2.1	Linear Rheology *1613*	
6.2.2	Non-linear Rheology *1615*	
6.3	Entangled Branched Chains in Linear Response *1618*	
6.3.1	A Tube Model for Star Polymers and Critical Tests *1619*	
6.3.2	H-Polymers and Combs *1624*	
6.3.3	Complex Topologies – the Seniority Distribution *1630*	
6.4	Long-chain Branching in Nonlinear Response *1635*	
6.4.1	Stretch and Branch-point Withdrawal (BPW):	
	the Priority Distribution *1635*	
6.4.2	Branched Polymers – a Minimal Model *1639*	
6.5	Application to Other Topologies and Challenges *1643*	
	References *1646*	

7	**Determination of Bulk and Solution Morphologies**	
	by Transmission Electron Microscopy *1649*	
	Volker Abetz, Richard J. Spontak, and Yeshayahu Talmon	
7.1	Introduction *1649*	
7.2	Background of Electron Microscopy *1650*	
7.3	Conventional TEM of Bulk Materials *1654*	
7.3.1	Sectioning of Samples *1655*	
7.3.2	Staining of Samples *1657*	
7.4	Cryo-TEM and Freeze-fracture TEM of Solutions *1662*	
7.4.1	Direct Imaging Cryo-TEM *1663*	
7.4.2	Freeze-fracture Replication *1665*	
7.4.3	Limitations, Precautions, Artefacts and Extensions	
	of the Technique *1666*	
7.5	Transmission Electron Microtomography *1670*	
7.5.1	Background *1670*	
7.5.2	Methodology *1671*	
7.5.3	Reconstruction Fidelity *1673*	
7.5.4	Quantitative Analysis *1676*	
7.5.5	Emerging Opportunities *1677*	
7.6	Analytical Electron Microscopy *1679*	
7.6.1	Energy-dispersive X-Ray Mapping *1679*	
7.6.2	Energy-filtered Transmission Electron Microscopy *1679*	
7.6.2.1	Zero-loss Imaging *1680*	
7.6.2.2	Structure-sensitive Imaging *1681*	

	7.6.2.3	Element-specific Imaging *1682*
		Acknowledgments *1683*
		References *1683*
8		**Polymer Networks** *1687*
		Karel Dušek and Miroslava Dušková-Smrčková
8.1		Introduction *1687*
8.1.1		Polymer Networks as Materials *1688*
8.2		Network Formation *1689*
8.2.1		Chemical Reactions Most Frequently Used for Preparation of Polymer Networks *1690*
8.2.1.1		Epoxide Group *1691*
8.2.1.2		Isocyanate Group *1693*
8.2.2		What Do We Need to Know about Chemistry of Network-forming Reactions and Why? *1696*
8.3		Polymer Networks Precursors *1696*
8.4		Crosslinking Kinetics *1699*
8.5		Buildup of Polymer Networks *1702*
8.5.1		Network Formation Theories and their Application *1704*
8.5.1.1		Classification of Network Formation Theories *1704*
8.5.2		Application of Network Formation Theories *1709*
8.6		Properties of Polymer Networks *1711*
8.6.1		Structural Features *1711*
8.6.2		Equilibrium Rubber Elasticity *1711*
8.6.3		Dynamic Mechanical Properties *1713*
8.6.4		Equilibrium Swelling *1714*
8.6.5		Phase Separation *1718*
8.6.6		Volume Phase Transition *1720*
8.7		More Complex Polymer Networks *1722*
8.7.1		Interpenetrating Polymer Networks (IPN) *1724*
8.8		Concluding Remarks *1727*
		References *1727*
9		**Block Copolymers for Adhesive Applications** *1731*
		Costantino Creton
9.1		Introduction *1731*
9.2		Rheological Properties and Structure *1732*
9.3		Large Strain and Adhesive Properties *1736*
9.3.1		Effect of the Diblock Content on Adhesive and Deformation Properties *1738*
9.3.2		Understanding the Structure of the Extended Foam *1741*
9.3.3		Interfacial Fracture *1746*
9.4		Conclusions *1749*
		References *1750*

10	**Reactive Blending** *1753*	
	Robert Jerome	
10.1	Polymer Blending: Benefits, Challenges and Compatibilization	*1753*
10.2	Preparation of Reactive Polymers *1755*	
10.3	Compatibilization Reactions *1758*	
10.4	Molecular Architecture of the Compatibilizer *1758*	
10.5	Phase Morphology Generation and Stability *1760*	
10.6	Kinetics of the Interfacial Reaction *1763*	
10.7	Molecular Weight of the Reactive Chains *1765*	
10.8	Content of Reactive Group of the Reactive Polymers *1769*	
10.9	Nanostructured Polyblends *1771*	
10.10	Conclusions *1779*	
	Acknowledgments *1780*	
	References *1780*	
11	**Predicting Mechanical Performance of Polymers** *1783*	
	Han E. H. Meijer, Leon E. Govaert, and Tom A. P. Engels	
11.1	Introduction *1783*	
11.2	Modelling Intrinsic Behavior: the Rejuvenated State *1789*	
11.3	The State Parameter S that Captures Physical Ageing *1792*	
11.4	Evolution of S, Physical Ageing During Processing *1796*	
11.5	Why Heterogeneity is Always Needed: Craze-initiation by Surpassing a Critical Cavitation Stress and Its Consequences for the Microstructure *1799*	
11.6	Semi-crystalline Polymers, Intrinsic Behavior and Geometrical Softening *1803*	
11.7	Semi-crystalline Polymers: the Influence of Flow *1808*	
11.8	From Heterogeneous Structures to Anisotropic Mechanical Properties *1814*	
11.9	Conclusions *1816*	
	Acknowledgments *1822*	
	References *1823*	
12	**Scanning Calorimetry**	
	René Androsch and Bernhard Wunderlich	
12.1	Introduction *1827*	
12.2	Heat Capacity and Thermodynamic Functions of State *1830*	
12.3	Measurement of the Heat Capacity and Separation of Thermal Events by Temperature-modulated DSC *1835*	
12.4	Contributions to the Apparent Heat Capacity of Semicrystalline Polymers *1845*	
12.5	Recent Advances in Thermal Analysis of Polymers *1849*	
12.5.1	Reversible Melting of Polymers *1850*	
12.5.2	The Rigid Amorphous Fraction *1861*	
12.5.3	High-speed Calorimetry *1868*	

12.6	Experimental Tools for Further Thermal Analysis of Polymers	*1872*
	References *1876*	

13	**Chromatography of Polymers** *1881*
	Wolfgang Radke
13.1	Introduction *1881*
13.1.1	Heterogeneity of Polymers *1881*
13.2	Modes of Liquid Chromatography of Polymers *1882*
13.2.1	Size-Exclusion Chromatography (SEC, GPC) *1884*
13.2.1.1	Basic Size-Exclusion Chromatography *1884*
13.2.1.2	Calibration in SEC *1888*
13.2.1.3	SEC with Molar Mass-sensitive Detection *1892*
13.2.2	Interaction Chromatography of Polymers *1906*
13.2.2.1	Liquid Adsorption Chromatography (LAC) of Polymers *1907*
13.2.2.2	Chromatography at the Critical Point of Adsorption (LCCC) *1911*
13.2.2.3	Gradient Chromatography of Polymers *1918*
13.2.3	Two-dimensional Chromatography *1922*
13.3	Coupling Liquid Chromatography with Spectrometric and Spectroscopic Methods *1926*
13.3.1	Coupling LC with FTIR Spectroscopy *1926*
13.3.2	Coupling LC with NMR Spectroscopy *1928*
13.3.3	Coupling LC with MALDI-TOF-MS *1931*
13.4	Conclusions *1934*
	References *1934*

14	**NMR Spectroscopy** *1937*
	Hans Wolfgang Spiess
14.1	Introduction *1937*
14.2	Background *1938*
14.2.1	Anisotropic Spin Interactions *1939*
14.2.2	Manipulation of Spin Interactions *1941*
14.2.3	Double Quantum NMR *1942*
14.2.4	Two-dimensional NMR Spectroscopy *1944*
14.3	Applications *1945*
14.3.1	Chain Microstructure *1946*
14.3.2	Heterogeneous Polymer Melts *1947*
14.3.3	Micellar Aggregates from Block Copolymers *1948*
14.3.4	Elastomers *1950*
14.3.5	Melts Composed of Stiff Macromolecules *1951*
14.3.6	Conformational Memory in Poly(n-Alkyl Methacrylates) *1952*
14.3.7	Applications in Supramolecular Chemistry *1954*
14.3.7.1	Hydrogen Bonds in Supramolecular Polymers *1954*
14.3.7.2	Proton Conductors *1955*
14.3.7.3	Supramolecular Assembly of Dendritic Polymers *1956*

14.3.7.4	Discotic Photoconductors Based on Hexabenzocoronene (HBC)	1957
14.3.7.5	Polyphenylene Dendrimers as Shape-persistent Nanoparticles	1958
14.3.7.6	Organic–Inorganic Hybrid Materials	1959
14.4	Conclusion and Outlook	1961
	Acknowledgments	1963
	References	1963

15	**High-throughput Screening in Combinatorial Polymer Research**	*1967*
	Michael A. R. Meier, Richard Hoogenboom, and Ulrich S. Schubert	
15.1	Introduction	1967
15.2	Automated and Parallel Polymer Synthesis	1969
15.3	High-throughput Screening	1971
15.4	Screening for Molecular Weight and Polydispersity Index	1972
15.5	Screening for Polymer Composition and Polymerization Kinetics	1983
15.6	Polymer Property Screening	1987
15.7	Conclusion	1995
	Acknowledgments	1995
	References	1996

Volume 4
Applications

1	**Applications of Thermoplastic Elastomers Based on Styrenic Block Copolymers** *2001*	
	Dale L. Handlin, Jr., Scott Trenor, and Kathryn Wright	
2	**Nanocomposites** *2033*	
	Michaël Alexandre and Philippe Dubois	
3	**Polymer/Layered Filler Nanocomposites: An Overview from Science to Technology** *2071*	
	Masami Okamoto	
4	**Polymeric Dispersants** *2135*	
	Frank Pirrung and Clemens Auschra	
5	**Polymeric Surfactants** *2181*	
	Henri Cramail, Eric Cloutet, and Karunakaran Radhakrishnan	
6	**Molecular and Supramolecular Conjugated Polymers for Electronic Applications** *2225*	
	Andrew C. Grimsdale and Klaus Müllen	

7 **Polymers for Microelectronics** *2263*
Christopher W. Bielawski and C. Grant Willson

8 **Applications of Controlled Macromolecular Architectures to Lithography** *2295*
Daniel Bratton, Ramakrishnan Ayothi, Nelson Felix, and Christopher K. Ober

9 **Microelectronic Materials with Hierarchical Organization** *2331*
G. Dubois, R. D. Miller and James L. Hedrick

10 **Semiconducting Polymers and their Optoelectronic Applications** *2369*
Nicolas Leclerc, Thomas Heiser, Cyril Brochon, and Georges Hadziioannou

11 **Polymer Encapsulation of Metallic and Semiconductor Nanoparticles: Multifunctional Materials with Novel Optical, Electronic and Magnetic Properties** *2409*
Jeffrey Pyun and Todd Emrick

12 **Polymeric Membranes for Gas Separation, Water Purification and Fuel Cell Technology** *2451*
Kazukiyo Nagai, Young Moo Lee, and Toshio Masuda

13 **Utilization of Polymers in Sensor Devices** *2493*
Basudam Adhikari and Alok Kumar Sen

14 **Polymeric Drugs** *2541*
Tamara Minko, Jayant J. Khandare, and Sreeja Jayant

15 **From Biomineralization Polymers to Double Hydrophilic Block and Graft Copolymers** *2597*
Helmut Cölfen

16 **Applications of Polymer Bioconjugates** *2645*
Joost A. Opsteen and Jan C. M. van Hest

17 **Gel: a Potential Material as Artificial Soft Tissue** *2689*
Yong Mei Chen, Jian Ping Gong, and Yoshihito Osada

18 **Polymers in Tissue Engineering** *2719*
Jeffrey A. Hubbell

**IUPAC Polymer Terminology
and Macromolecular Nomenclature** *2743*
R. F. T. Stepto

Subject Index *2747*

Preface

Macromolecular Engineering: From Precise Macromolecular Synthesis to Macroscopic Materials Properties and Applications aims to provide a broad overview of recent developments in precision macromolecular synthesis and in the design and applications of complex polymeric assemblies of controlled sizes, morphologies and properties. The contents of this interdisciplinary book are organized in four volumes so as to capture and chronicle best, on the one hand, the rapid advances made in the control of polymerization processes and the design of macromolecular architectures (Volumes I and II) and, on the other, the noteworthy progress witnessed in the processing methods – including self-assembly and formulation – to generate new practical applications (Volumes III and IV).

Each chapter in this book is a well-documented and yet concise contribution written by noted experts and authorities in their field. We are extremely grateful to all of them for taking time to share their knowledge and popularize it in a way understandable to a broad readership. We are also indebted to all the reviewers whose comments and remarks helped us very much in our editing work. Finally, Wiley-VCH deserves our sincere acknowledgements for striving to keep the entire project on time.

We expect that specialist readers will find *Macromolecular Engineering: From Precise Macromolecular Synthesis to Macroscopic Materials Properties and Applications* an indispensable book to update their knowledge, and non-specialists will use it as a valuable companion to stay informed about the newest trends in polymer and materials science.

November 2006
Pittsburgh, USA   Krzysztof Matyjaszewski
Bordeaux, France   Yves Gnanou
Paris, France   Ludwik Leibler

Macromolecular Engineering. Precise Synthesis, Materials Properties, Applications.
Edited by K. Matyjaszewski, Y. Gnanou, and L. Leibler
Copyright © 2007 WILEY-VCH Verlag GmbH & Co. KGaA, Weinheim
ISBN: 978-3-527-31446-1

List of Contributors

Volker Abetz
GKSS Research Centre Geesthacht GmbH
Institute of Polymer Research
Max-Planck-Straße 1
21502 Geesthacht
Germany

Basudam Adhikari
Indian Institute of Technology
Materials Science Centre
Polymer Division
Kharagpur 721302
India

Takuzo Aida
University of Tokyo
School of Engineering
Department of Chemistry
and Biotechnology
7-3-1 Hongo, Bunkyo-ku
Tokyo 113-8656
Japan

Michaël Alexandre
Materia Nova Research Centre asbl
Parc Initialis
1 avenue Nicolas Copernic
7000 Mons
Belgium

Xavier Allonas
University of Haute Alsace, ENSCMu
Department of General
Photochemistry, UMR 7525 CNRS
3 Alfred Werner street
68093 Mulhouse Cedex
France

Denis Andrienko
Max Planck Institute for Polymer Research
Ackermannweg 10
55128 Mainz
Germany

René Androsch
Martin Luther University
Halle-Wittenberg
Institute of Materials Science
06099 Halle
Germany

Markus Antonietti
Max Planck Institute of Colloids and Interfaces
Colloid Department
Research Campus Golm
14424 Potsdam
Germany

Macromolecular Engineering. Precise Synthesis, Materials Properties, Applications.
Edited by K. Matyjaszewski, Y. Gnanou, and L. Leibler
Copyright © 2007 WILEY-VCH Verlag GmbH & Co. KGaA, Weinheim
ISBN: 978-3-527-31446-1

List of Contributors

Clemens Auschra
CIBA Specialty Chemicals, Inc.
Research and Development
Coating Effects
Schwarzwaldallee 215
4002 Basel
Switzerland

Ramakrishnan Ayothi
Cornell University
Department of Materials Science
and Engineering
Bard Hall
Ithaca, NY 14853-1501
USA

Vincent Ball
Institut National de la Santé
et de la Recherche Médicale
INSERM Unité 595
11 rue Humann
67085 Strasbourg Cedex
France
and
Université Louis Pasteur
Faculté de Chirurgie Dentaire
1 place de l'Hôpital
67000 Strasbourg
France

Nitash P. Balsara
University of California
Department of Chemical Engineering
and Materials Sciences
and Environmental Energy
Technologies Divisions
Lawrence Berkeley National
Laboratory
Berkeley, CA 94720
USA

Nadia Benkirane-Jessel
Institut National de la Santé
et de la Recherche Médicale
INSERM Unité 595
11 rue Humann
67085 Strasbourg Cedex
France
and
Université Louis Pasteur
Faculté de Chirurgie Dentaire
1 place de l'Hôpital
67000 Strasbourg
France

Christopher W. Bielawski
The University of Texas at Austin
Department of Chemistry
and Biochemistry
Austin, TX 78712
USA

Hans G. Börner
Max Planck Institute of Colloids
and Interfaces
Colloid Department
Research Campus Golm
14424 Potsdam
Germany

Redouane Borsali
Université Bordeaux 1
CNRS, ENSCPB
Laboratoire de Chimie des Polymères
Organiques
16 avenue Pey-Berland
33607 Pessac
France

Bernard Boutevin
Ingénierie et Architectures
Macromoléculaires
Institut Gerhardt, UMR 5253
Ecole Nationale Supérieure Chimie
de Montpellier
8 rue de l'Ecole Normale
34296 Montpellier
France

Cyrille Boyer
Ingénierie et Architectures
Macromoléculaires
Institut Gerhardt, UMR 5253
Ecole Nationale Supérieure Chimie
de Montpellier
8 rue de l'Ecole Normale
34296 Montpellier
France

Daniel Bratton
Cornell University
Department of Materials Science
and Engineering
Bard Hall
Ithaca, NY 14853-1501
USA

Wade A. Braunecker
Carnegie Mellon University
Department of Chemistry
4400 Fifth Avenue
Pittsburgh, PA 15213
USA

Cyril Brochon
Laboratoire d'Ingénierie des
Polymères pour les Hautes
Technologies
UMR 7165 CNRS
Ecole Européenne Chimie Polymères
Matériaux
Université Louis Pasteur
25 rue Becquerel
67087 Strasbourg
France

Manoj B. Charati
University of Delaware
Department of Materials Science
and Engineering
201 DuPont Hall
and
Delaware Biotechnology Institute
15 Innovation Way
Newark, DE 19716
USA

Bernadette Charleux
Université Pierre et Marie Curie
Laboratoire de Chimie des Polymères
4, Place Jussieu, Tour 44, 1er étage
75252 Paris Cedex 5
France

Yong Mei Chen
Hokkaido University
Section of Biological Sciences
Faculty of Science
Laboratory of soft & wet matter
North 10, West 8
060-0810 Sapporo
Japan

List of Contributors

Zhiyun Chen
Washington University in Saint Louis
Center for Materials Innovation and
Department of Chemistry
One Brookings Drive
St. Louis, MO 63130-4899
USA

Chong Cheng
Washington University in Saint Louis
Center for Materials Innovation and
Department of Chemistry
One Brookings Drive
St. Louis, MO 63130-4899
USA

Eric Cloutet
Université Bordeaux 1
Laboratoire de Chimie des Polymères
Organiques
Unité Mixte de Recherche
(UMR 5629) CNRS
ENSCPB
16 avenue Pey-Berland
33607 Pessac Cedex
France

Geoffrey W. Coates
Cornell University
Department of Chemistry
and Chemical Biology
Baker Laboratory
Ithaca, NY 14853
USA

Helmut Cölfen
Max Planck Institute of Colloids
and Interfaces
Colloid Chemistry
Research Campus Golm
Am Mühlenberg 1
14476 Potsdam-Golm
Germany

Henri Cramail
Université Bordeaux 1
Laboratoire de Chimie des Polymères
Organiques
Unité Mixte de Recherche
(UMR 5629) CNRS
ENSCPB
16 avenue Pey-Berland
33607 Pessac Cedex
France

Costantino Creton
Laboratoire PPMD
ESPCI
10 rue Vauquelin
75231 Paris
France

Ghislain David
Ingénierie et Architectures
Macromoléculaires
Institut Gerhardt, UMR 5253
Ecole Nationale Supérieure Chimie
de Montpellier
8 rue de l'Ecole Normale
34296 Montpellier
France

Priyadarsi De
University of Massachusetts Lowell
Polymer Science Program
Department of Chemistry
One University Avenue
Lowell, MA 01854
USA

Gero Decher
Institut Charles Sadron
(C.N.R.S. UPR 022)
6 rue Boussingault
67083 Strasbourg Cedex
France
and
Université Louis Pasteur
Faculté de Chimie
1 rue Blaise Pascal
67008 Strasbourg Cedex
France

Alain Deffieux
Université Bordeaux 1
CNRS, ENSCPB
Laboratoire de Chimie des Polymères
Organiques
16 avenue Pey-Berland
33607 Pessac
France

Timothy J. Deming
University of California, Los Angeles
Department of Bioengineering
420 Westwood Plaza
7523 Boelter Hall
Los Angeles, CA 90095
USA

G. Dubois
IBM Almaden Research Center
650 Harry Road
San Jose, CA 95120
USA

Philippe Dubois
Université de Mons-Hainaut
Matériaux Polymères et Composites
Place du Parc 20
7000 Mons
Belgium

Andrzej Duda
Center of Molecular and
Macromolecular Studies
Polish Academy of Sciences
Department of Polymer Chemistry
Sienkiewicza 112
90-363 Łodz
Poland

Karel Dušek
Academy of Sciences
of the Czech Republic
Institute of Macromolecular
Chemistry
Heyrovského nám. 2
162 06 Praha
Czech Republic

Miroslava Dušková-Smrčková
Academy of Sciences
of the Czech Republic
Institute of Macromolecular
Chemistry
Heyrovského nám. 2
162 06 Praha
Czech Republic

Todd Emrick
University of Massachusetts Amherst
Department of Polymer Science
and Engineering
120 Governors Drive
Amherst, MA 01003
USA

Tom A. P. Engels
Eindhoven University of Technology
Department of Mechanical
Engineering
P.O. Box 513
5600 MB Eindhoven
The Netherlands

Robin S. Farmer
University of Delaware
Department of Materials Science
and Engineering
201 DuPont Hall
and
Delaware Biotechnology Institute
15 Innovation Way
Newark, DE 19716
USA

Rudolf Faust
University of Massachusetts Lowell
Polymer Science Program
Department of Chemistry
One University Avenue
Lowell, MA 01854
USA

Nelson Felix
Cornell University
Department of Materials Science
and Engineering
Bard Hall
Ithaca, NY 14853-1501
USA

Michel Fontanille
Université Bordeaux 1
Laboratoire de Chimie des Polymères
Organiques
ENSCPB
16 avenue Pey-Berland
33607 Pessac
France

Jean Pierre Fouassier
University of Haute Alsace,
ENSCMu
Department of General
Photochemistry, UMR 7525 CNRS
3 Alfred Werner street
68093 Mulhouse Cedex
France

Takeshi Fukuda
Kyoto University
Institute for Chemical Research
Uji, Kyoto 611-0011
Japan

François Ganachaud
Ecole Nationale Supérieure de Chimie
de Montpellier
Laboratoire de Chimie
Macromoléculaire
8 rue de l'Ecole Normale
34296 Montpellier Cedex 5
France

David S. Germack
Washington University in Saint Louis
Center for Materials Innovation and
Department of Chemistry
One Brookings Drive
St. Louis, MO 63130-4899
USA

Yves Gnanou
Université Bordeaux 1
Laboratoire de Chimie des Polymères
Organiques
ENSCPB
16 avenue Pey-Berland
33607 Pessac
France

Jian Ping Gong
Hokkaido University
Section of Biological Sciences
Faculty of Science
Laboratory of soft & wet matter
North 10, West 8
060-0810 Sapporo
Japan

Padma Gopalan
University of Wisconsin – Madison
Department of Materials Science
and Engineering
1117 Engineering Research Building
1500 Engineering Drive
Madison, WI 53706
USA

Leon E. Govaert
Eindhoven University of Technology
Department of Mechanical
Engineering
P.O. Box 513
5600 MB Eindhoven
The Netherlands

Stefanie M. Gravano
University of California, Davis
Department of Chemistry
One Shields Avenue
Davis, CA 95616-5295
USA

Andrew C. Grimsdale
Nanyang Technological University
School of Materials Science and
Engineering
50 Nanyang Avenue
Singapore 639798

Nikos Hadjichristidis
University of Athens
Department of Chemistry
Panepistimiopolis Zografou
15771 Athens
Greece

Georges Hadziioannou
Laboratoire d'Ingénierie des
Polymères pour les Hautes
Technologies
UMR 7165 CNRS
Ecole Européenne Chimie Polymères
Matériaux
Université Louis Pasteur
25 rue Becquerel
67087 Strasbourg
France

Dale L. Handlin, Jr.
Kraton Polymers
700 Milam
North Tower
Houston, TX 77002
USA

James L. Hedrick
IBM Almaden Research Center
650 Harry Road
San Jose, CA 95120
USA

Thomas Heiser
Université Louis Pasteur
Institut d'Electronique du Solide
et des Systèmes
UMR 7163, CNRS
23 rue du Loess
67087 Strasbourg
France

Valérie Héroguez
Université Bordeaux 1
Laboratoire de Chimie des Polymères
Organiques
ENSCPB
16 avenue Pey-Berland
33607 Pessac
France

Richard Hoogenboom
Eindhoven University of Technology
and Dutch Polymer Institute (DPI)
Laboratory of Macromolecular
Chemistry and Nanoscience
P.O. Box 513
5600 MB Eindhoven
The Netherlands

Jeffrey A. Hubbell
Ecole Polytechnique Fédérale
de Lausanne
Institute of Bioengineering
1015 Lausanne
Switzerland

Hermis Iatrou
University of Athens
Department of Chemistry
Panepistimiopolis Zografou
15771 Athens
Greece

Olli Ikkala
Helsinki University of Technology
Department of Engineering Physics
and Mathematics
and Center for New Materials
P.O. Box 2200
02015 Hut
Espoo
Finland

Woo-Dong Jang
The University of Tokyo
School of Engineering
Department of Chemistry
and Biotechnology
7-3-1 Hongo, Bunkyo-ku
Tokyo 113-8656
Japan

Sreeja Jayant
Rutgers, The State University
of New Jersey
Department of Pharmaceutics
160 Frelinghuysen Road
Piscataway, NJ 08854-8020
USA

Robert Jerome
University of Liège
Center for Education and Research
on Macromolecules (CERM)
Sart-Tilman, B6a
4000 Liège
Belgium

Youngjong Kang
Massachusetts Institute of Technology
Department of Materials Science
and Engineering and
Institute for Soldier Nanotechnologies
Cambridge, MA 02139
USA

Jayant J. Khandare
Rutgers, The State University
of New Jersey
Department of Pharmaceutics
160 Frelinghuysen Road
Piscataway, NJ 08854-8020
USA

Kristi L. Kiick
University of Delaware
Department of Materials Science
and Engineering
201 DuPont Hall
and
Delaware Biotechnology Institute
15 Innovation Way
Newark, DE 19716
USA

Tatsuki Kitayama
Osaka University
Department of Chemistry
Graduate School of Engineering
Toyonaka, Osaka 560-8531
Japan

Harm-Anton Klok
Ecole Polytechnique Fédérale
de Lausanne (EPFL)
Institut des Matériaux
Laboratoire des Polymères
STI – IMX – LP, MXD 112
(Bâtiment MXD), Station 12
1015 Lausanne
Switzerland

Bert Klumperman
Eindhoven University of Technology
Laboratory of Polymer Chemistry
P.O. Box 513
5600 MB Eindhoven
The Netherlands

Shiro Kobayashi
Kyoto Institute of Technology
R&D Center for Bio-based Materials
Matsugasaki, Sakyo-ku
Kyoto 606-8585
Japan

Kurt Kremer
Max Planck Institute for Polymer
Research
Ackermannweg 10
55128 Mainz
Germany

Vahik Krikorian
Massachusetts Institute of Technology
Department of Materials Science
and Engineering and
Institute for Soldier Nanotechnologies
Cambridge, MA 02139
USA

Przemyslaw Kubisa
Center of Molecular and
Macromolecular Studies
Polish Academy of Sciences
Department of Polymer Chemistry
Sienkiewicza 112
90-363 Łodz
Poland

Alok Kumar Sen
Indian Institute of Technology
Materials Science Centre
Polymer Division
Kharagpur 721302
India

Jacques Lalevée
University of Haute Alsace, ENSCMu
Department of General
Photochemistry, UMR 7525 CNRS
3 Alfred Werner street
68093 Mulhouse Cedex
France

Katharina Landfester
University of Ulm
Department of Organic Chemistry III
– Macromolecular Chemistry
and Organic Materials
Albert-Einstein-Allee 11
89081 Ulm
Germany

Philippe Lavalle
Institut National de la Santé
et de la Recherche Médicale
INSERM Unité 595
11 rue Humann
67085 Strasbourg Cedex
France
and
Université Louis Pasteur
Faculté de Chirurgie Dentaire
1 place de l'Hôpital
67000 Strasbourg
France

Nicolas Leclerc
Université Louis Pasteur
Laboratoire d'Ingénierie des
Polymères pour les Hautes
Technologies
UMR 7165 CNRS
Ecole Européenne Chimie Polymères
Matériaux
25 rue Becquerel
67087 Strasbourg
France

Young Moo Lee
Hanyang University
School of Chemical Engineering
College of Engineering
Seoul 133-791
Korea

Ludwik Leibler
Matière Molle et Chimie
UMR 167 CNRS
ESPCI
10 rue Vauquelin
75005 Paris
France

G. B. W. L. Ligthart
DSM Campus Geleen
Performance Materials
PO Box 18
6160 MD Geleen
The Netherlands

Wei-Shi Li
ERATO-SORST Nanospace Project
Japan Science and
Technology Agency (JST)
National Museum of Emerging
Science and Innovation
2-41 Aomi, Koto-ku
Tokyo 135-0064
Japan

Pierre Lutz
Institut Charles Sadron
6 rue Boussingault
67083 Strasbourg Cedex
France

Ian Manners
University of Bristol
Department of Chemistry
Cantock's Close
Bristol BS8 1TS
UK

Andrew F. Mason
IBM Almaden Research Center
650 Harry Road
San Jose, CA 95120
USA

Toshio Masuda
Kyoto University
Department of Polymer Chemistry
Graduate School of Engineering
Katsura Campus
Kyoto 615-8510
Japan

Krzysztof Matyjaszewski
Carnegie Mellon University
Department of Chemistry
4400 Fifth Avenue
Pittsburgh, PA 15213
USA

Thomas C. B. McLeish
University of Leeds
IRC in Polymer Science
and Technology
Polymers and Complex Fluids
Department of Physics
and Astronomy
Leeds LS2 9JT
UK

Michael A. R. Meier
Eindhoven University of Technology
and Dutch Polymer Institute (DPI)
Laboratory of Macromolecular
Chemistry and Nanoscience
P.O. Box 513
5600 MB Eindhoven
The Netherlands

E. W. Meijer
Eindhoven University of Technology
Laboratory of Macromolecular
and Organic Chemistry
P.O. Box 513
5600 MB Eindhoven
The Netherlands

Han E. H. Meijer
Eindhoven University of Technology
Department of Mechanical
Engineering
P.O. Box 513
5600 MB Eindhoven
The Netherlands

R. D. Miller
IBM Almaden Research Center
650 Harry Road
San Jose, CA 95120
USA

Tamara Minko
Rutgers, The State University
of New Jersey
Department of Pharmaceutics
160 Frelinghuysen Road
Piscataway, NJ 08854-8020
USA

Martin Moller
RWTH Aachen
Institut für Technische
und Makromolekulare Chemie
Pauwelsstraße 8
52056 Aachen
Germany

Hideharu Mori
Yamagata University
Faculty of Engineering
Department of Polymer Science
and Engineering
4-3-16, Jonan
Yonezawa 992-8510
Japan

Klaus Müllen
Max Planck Institute
for Polymer Research
Ackermannweg 10
55128 Mainz
Germany

Axel H. E. Müller
University of Bayreuth
Macromolecular Chemistry II
95440 Bayreuth
Germany

Anna Musyanovych
University of Ulm
Department of Organic Chemistry III
– Macromolecular Chemistry
and Organic Materials
Albert-Einstein-Allee 11
89081 Ulm
Germany

Kazukiyo Nagai
Meiji University
Department of Applied Chemistry
1-1-1 Higashi-mita, Tama-ku
Kawasaki 214-8571
Japan

Christopher K. Ober
Cornell University
Department of Materials Science
and Engineering
310 Bard Hall
Ithaca, NY 14853-1501
USA

Joëlle Ogier
Institut National de la Santé
et de la Recherche Médicale
INSERM Unité 595
11 rue Humann
67085 Strasbourg Cedex
France
and
Université Louis Pasteur
Faculté de Chirurgie Dentaire
1 place de l'Hôpital
67000 Strasbourg
France

Masashi Ohmae
Kyoto University
Department of Materials Chemistry
Graduate School of Engineering
Katsura, Nishikyo-ku
Kyoto 615-8510
Japan

Kohji Ohno
Kyoto University
Institute for Chemical Research
Uji, Kyoto 611-0011
Japan

Masami Okamoto
Advanced Polymeric Materials
Engineering
Graduate School of Engineering
Toyota Technological Institute
2-12-1 Hisakata, Tempaku
Nagoya 468-8511
Japan

Joost A. Opsteen
Radboud University Nijmegen
Institute for Molecules and Materials
Toernooiveld 1
6525 ED Nijmegen
The Netherlands

Yoshihito Osada
Hokkaido University
Section of Biological Sciences
Faculty of Science
Laboratory of soft & wet matter
North 10, West 8
Sapporo 060-0810
Japan

Timothy E. Patten
University of California, Davis
Department of Chemistry
One Shields Avenue
Davis, CA 95616-5295
USA

Stanislaw Penczek
Centre of Molecular
and Macromolecular Studies
Polish Academy of Sciences
Department of Polymer Chemistry
Sienkiewicza 112
90-363 Łodz
Poland

Catherine Picart
Université de Montpellier II
Laboratoire de Dynamique
des Interactions Membranaires
Normales et Pathologiques
(C.N.R.S. UMR 5235)
Place Eugène Bataillon
34095 Montpellier Cedex
France

Frank Pirrung
CIBA Specialty Chemicals, Inc.
Research and Development
Coating Effects
Schwarzwaldallee 215
4002 Basel
Switzerland

Marinos Pitsikalis
University of Athens
Department of Chemistry
Panepistimiopolis Zografou
15771 Athens
Greece

Jeffrey Pyun
University of Arizona
Department of Chemistry
1306 E. University Boulevard
Tucson, AZ 85721
USA

Damien Quémener
Université Bordeaux 1
Laboratoire de Chimie des Polymères
Organiques
ENSCPB
16 avenue Pey-Berland
33607 Pessac
France

Karunakaran Radhakrishnan
University of Akron
Institute of Polymer Science
170, University Avenue
Akron, OH 44325
USA

Wolfgang Radke
Deutsches Kunststoff-Institut
Darmstadt
Schlossgartenstraße 6
65289 Darmstadt
Germany

David A. Rider
University of Toronto
Department of Chemistry
80 St. George Street
M5S 3H6 Toronto, Ontario
Canada

Megan L. Ruegg
University of California
Department of Chemical Engineering
Berkeley, CA 94720
USA

Pierre Schaaf
Institut Charles Sadron
(C.N.R.S. UPR 022)
6 rue Boussingault
67083 Strasbourg Cedex
France
and
Ecole Européenne de Chimie
Polymères et Materiaux
25 rue Bequerel
67087 Strasbourg Cedex 2
France

Oren A. Scherman
University of Cambridge
Department of Chemistry
Lensfield Road
Cambridge CB2 1EW
UK

Helmut Schlaad
Max Planck Institute of Colloids
and Interfaces
Colloid Department
Research Campus Golm
14424 Potsdam
Germany

Ulrich S. Schubert
Eindhoven University of Technology
and Dutch Polymer Institute (DPI)
Laboratory of Macromolecular
Chemistry and Nanoscience
P.O. Box 513
5600 MB Eindhoven
The Netherlands

List of Contributors

Bernard Senger
Institut National de la Santé
et de la Recherche Médicale
INSERM Unité 595
11 rue Humann
67085 Strasbourg Cedex
France
and
Université Louis Pasteur
Faculté de Chirurgie Dentaire
1 place de l'Hôpital
67000 Strasbourg
France

Sergei S. Sheiko
University of North Carolina
at Chapel Hill
Department of Chemistry
Chapel Hill, NC 27599-3290
USA

Rint P. Sijbesma
Eindhoven University of Technology
Laboratory of Macromolecular
and Organic Chemistry
P.O. Box 513
5600 MB Eindhoven
The Netherlands

Peter F. W. Simon
GKSS Research Centre
Geesthacht GmbH
Institute of Polymer Research
Max-Planck-Straße
21502 Geesthacht
Germany

Stanislaw Slomkowski
Center of Molecular and
Macromolecular Studies
Polish Academy of Sciences
Department of Polymer Chemistry
Sienkiewicza 112
90-363 Łódź
Poland

Hans Wolfgang Spiess
Max-Planck-Institute for Polymer
Research
Spectroscopy
P.O. Box 3148
55021 Mainz
Germany

Richard J. Spontak
North Carolina State University
Departments of Chemical and
Biomolecular Engineering and
Materials Science and Engineering
Raleigh, NC 27695
USA

R. F. T. Stepto
The University of Manchester
School of Materials
Materials Science Centre
Polymer Science
and Technology Group
Grosvenor Street
Manchester M1 7HS
UK

Brent S. Sumerlin
Southern Methodist University
Department of Chemistry
3215 Daniel Avenue
Dallas, TX 75240-0314
USA

Yeshayahu Talmon
Technion – Israel Institute
of Technology
Department of Chemical Engineering
32000 Haifa
Israel

M. Atilla Tasdelen
Istanbul University
Department of Chemistry
Maslak
Istanbul 34469
Turkey

Daniel Taton
Université Bordeaux 1
Laboratoire de Chimie des Polymères
Organiques
LCPO CNRS
ENSCPB
16 avenue Pey-Berland
33607 Pessac
France

Gerrit ten Brinke
University of Groningen
Laboratory of Polymer Chemistry
Materials Science Centre
Nijenborgh 4
9747 AG Groningen
The Netherlands

Edwin L. Thomas
Massachusetts Institute of Technology
Department of Materials Science
and Engineering
and
Institute for Soldier Nanotechnologies
Cambridge, MA 02139
USA

Scott Trenor
Kraton Polymers
Houston, TX 77002
USA

Constantinos Tsitsilianis
University of Patras
and FORTH/ICEHT
Department of Chemical Engineering
Karatheodori 1
26504 Patras
Greece

Yoshinobu Tsujii
Kyoto University
Institute for Chemical Research
Uji, Kyoto 611-0011
Japan

Jan C. M. van Hest
Radboud University Nijmegen
Institute for Molecules and Materials
Toernooiveld 1
6525 ED Nijmegen
The Netherlands

Brooke A. van Horn
Washington University in Saint Louis
Center for Materials Innovation and
Department of Chemistry
One Brookings Drive
St. Louis, MO 63130-4899
USA

Shrinivas Venkataraman
Washington University in Saint Louis
Center for Materials Innovation and
Department of Chemistry
One Brookings Drive
St. Louis, MO 63130-4899
USA

Jean-Claude Voegel
Institut National de la Santé
et de la Recherche Médicale
INSERM Unité 595
11 rue Humann
67085 Strasbourg Cedex
France
and
Université Louis Pasteur
Faculté de Chirurgie Dentaire
1 place de l'Hôpital
67000 Strasbourg
France

C. Grant Willson
The University of Texas at Austin
Departments of Chemistry
and Biochemistry
and Chemical Engineering
Austin, TX 78712
USA

Karen L. Wooley
Washington University in Saint Louis
Center for Materials Innovation and
Department of Chemistry
One Brookings Drive
St. Louis, MO 63130-4899
USA

Kathryn Wright
Kraton Polymers
700 Milam
North Tower
Houston, TX 77002
USA

Bernhard Wunderlich
University of Tennessee
Department of Chemistry
200 Baltusrol Road
Knoxville, TN 37992-3707
USA

Yusuf Yagci
Istanbul University
Department of Chemistry
Maslak
Istanbul 34469
Turkey

Tsutomu Yokozawa
Kanagawa University
Department of Material
and Life Chemistry
Rokkakubashi, Kanagawa-ku
Yokohama 221-8686
Japan

1
Self-assembly and Morphology Diagrams for Solution and Bulk Materials: Experimental Aspects

Vahik Krikorian, Youngjong Kang, and Edwin L. Thomas

Block copolymers (BCPs) are composed of different polymer chains linked together with covalent bonds. By changing the chemical identity of each block, the relative volume fraction, and the molecular architecture, a variety of different ordered microstructures with the characteristic dimensions in the meso-regime can be formed. The spontaneous self-assembly of BCPs is of academic and industrial interest. BCPs possess a rich phase behavior displaying a large variety of differently shaped and connected microdomains in the bulk and thin films. In addition, the variety of these microstructures can be further enhanced and controlled by blending with homopolymers, selective solvents and with other types of BCPs or by the addition of inorganic particles. Recently, emphasis has been focused on utilizing BCPs as templates for inorganic phase growth or for the organization of nanoparticles within the microphase separated domains with the aim of fabricating advanced functional materials for applications in photonics, catalysis, separation, high capacity magnetic storage media, solar cells etc. [1, 2].

This chapter is divided into two major sections. The first part is devoted to self-assembly in the bulk and in the second part the focus is on the self-assembly within liquid phases. This chapter is not a thorough review of the different aspects of self-assembly in the BCP systems, since a number of excellent reviews have already been published and are cited where appropriate. Here, our goal is to familiarize the reader with the fascinating world of BCP self-assembly and at the same time give a flavor of recent advances in this field. The focus will be primarily on the experimental aspects and to clarify some of the concepts, brief theoretical aspects are presented. We will start with the blends of BCPs with organic as well as inorganic species, then focus on the self-assembly of rod–coil BCP systems which have recently attracted significant attention. In the second part, the self-assembly of the coil–coil and rod–coil BCPs from aqueous and/or organic solvents will be explored.

Macromolecular Engineering. Precise Synthesis, Materials Properties, Applications.
Edited by K. Matyjaszewski, Y. Gnanou, and L. Leibler
Copyright © 2007 WILEY-VCH Verlag GmbH & Co. KGaA, Weinheim
ISBN: 978-3-527-31446-1

1.1
Introduction: Bulk Phase

Since the advent of anionic polymerization [3] BCP chemistry has reached a point where synthesis of copolymers with complex architectures can be readily achieved. In addition to the simplest linear AB diblock, linear ABA or ABC triblock, 3-arm or 4-arm miktoarm star copolymers, cyclic, H-shaped, tetrablocks, pentablocks and $(AB)_n$ multiblocks all with well defined compositions and narrow molecular weight distributions have been synthesized (Fig. 1.1). Depending on the physical architecture of the BCP and the degree of segregation, denoted by the value of χN (where χ is the temperature dependent Flory–Huggins interaction parameter[1]) and N is the degree of polymerization) and the processing conditions, the self-assembly pathway can be altered significantly and a wide variety of periodic microdomain structures are possible.

Similar to surfactant systems, AB diblock copolymers can form a variety of morphologies in the melt as well as in solution. In the case of self-assembly from the melt, at low B volume fractions (ϕ_B), spheres of B are formed within the matrix A and are typically arranged in an ordered body centered cubic (BCC) lattice (Fig. 1.2a). Increasing the volume fraction of the B block results in the formation of hexagonally packed B cylinders within the matrix of A (Fig. 1.2b). In the case of symmetric diblocks (ϕ_B close to 0.5) alternating lamellae of A and B are formed (Fig. 1.2c). Moreover, in the compositions between cylinders and lamellae a complex bicontiuous cubic double gyroid is formed (Fig. 1.2d). A representative theoretical phase diagram for conformationally symmetric diblock melt is shown in Fig. 1.2e.

Initial theories for self-assembly of block copolymers were developed for the cases where the χN values were large ($\chi N > \sim 100$) also known as the strong segregation limit (SSL). In this regime the segregation between each of the blocks is such that the domains are nearly pure and the boundary between them, the intermaterial dividing surface (IMDS), is very narrow compared to

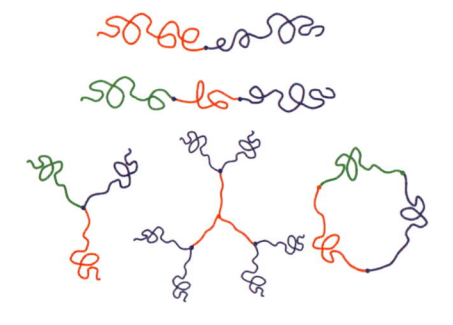

Fig. 1.1 Representative architectures for BCPs. Chain stretching at the junction points is exaggerated for the sake of illustrating the strong segregation limit (SSL) and less conformational entropy.

---

1) $\chi = A + B/T$; where A and B are constants and T is the absolute temperature.

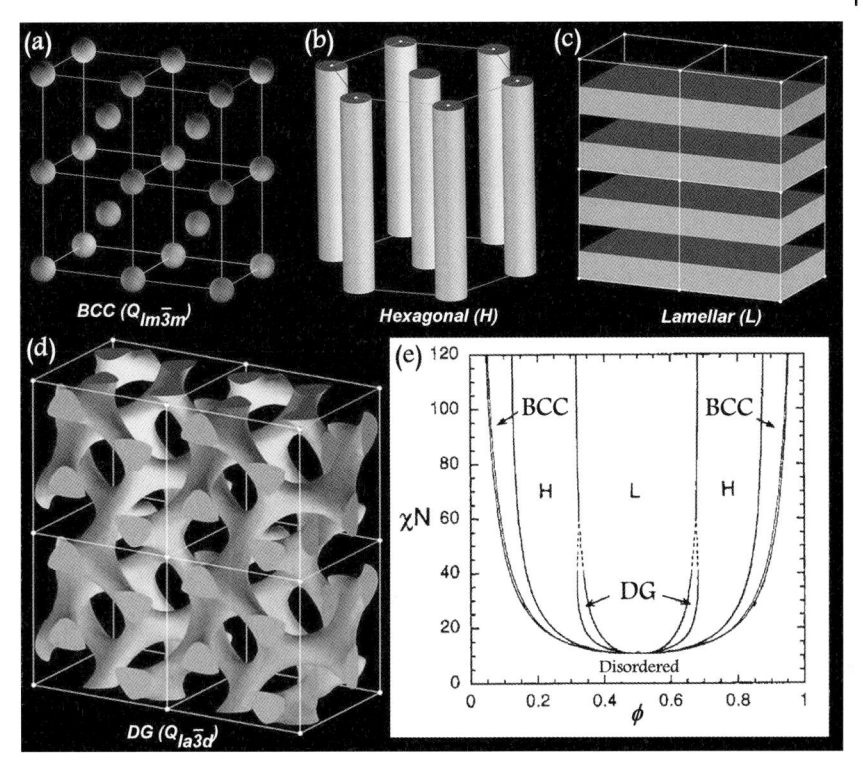

Fig. 1.2 Schematics of microphase separated morphologies in BCP systems.
(a) Body centered cubic (BCC), (b) hexagonally packed cylinders, (c) lamellae, (d) bicontinous cubic double gyroid and (e) theoretical phase diagram for conformationally symmetric diblock copolymer melts. (e) is adapted from Ref. [162].

the domain size. In contrast to the SSL, at $\chi N \leq \sim 12$ the morphology is mainly dominated by the entropic terms, this regime is referred to as the weak segregation limit (WSL). In this regime full segregation between the constituent blocks is not present, the IMDS is broad, and there are significant density fluctuations. A variety of theoretical approaches have been undertaken in order to predict the melt phase behavior of BCPs [4–7]. Leibler [8] described the microphase separation, i.e. the transition from disordered to ordered state, and demonstrated the composition dependent phase behavior within the WSL using a mean-field theory. Later, Matsen and Shick [9] theoretically predicted the phase diagram of an AB diblock copolymer. For a thorough review of the theoretical work behind equilibrium states and the phase behavior of these systems the reader is referred to the recent review by Fredrickson [10].

1.1.1
Block Copolymer Blends

Similar to alloying in metallurgy, blending in polymers has proved to be a versatile means to further extend the space of microstructures. One of the obvious advantages of blending is the ease of tailoring the final properties of the materials without undertaking extensive (and costly) chemical synthesis routes. Blending BCPs with homopolymers or other types of BCPs further expands the functionality of the BCP toolbox for fabricating custom-made materials with unique properties and functionalities. For instance, in order to access domain spacing large enough for applications in visible light photonics, the block molecular weights must be greater than about 5×10^5 g mol^{-1}. It has been shown [11, 12] that by adding selective solvents or homopolymers, one can controllably swell a specific domain and tailor the morphology without the need for synthesizing new molecules. One of the main advantages of using BCPs in blends was first realized in homopolymer/homopolymer mixtures. Due to the typical poor miscibility between polymeric compounds, the resulting blends are often coarse-grained, resulting in structures with inadequate mechanical properties. Using BCPs with blocks having favorable interaction parameters with respect to each phase remedies this issue by having the BCP act as macromolecular nonionic surfactant thus forming a monolayer at the interface and compatibilizing the blend. This BCP monolayer results in a significant increase in the radius of curvature and consequently reduces the average stable domain size and increases the fracture toughness [13, 14].

Winey and Thomas [15–18] reported the systematic modification of the phase behavior of blends of poly(styrene)-b-poly(isoprene) (PS-PI) with various amounts and molecular weights of PS homopolymers. A number of different blends involving BCPs have been reported in the literature. These include binary blends of AB with the homopolymer of A (hA) or B (hB), a blend of two AB diblocks with different degrees of polymerization $(AB)_\alpha/(AB)_\beta$ [19], blends of AB and BC diblocks [20], non-centrosymmetric microstructure forming blocks of ABC and AC [21], blends of ABC blocks with different degrees of polymerization, AB/ABA [22] and AB/ABC [23]. The corresponding theoretical efforts in predicting the phase diagrams of simpler systems such as AB/hA/hB have been investigated [24–28] however the prediction of phase diagrams for more complex structures has not yet been rigorously studied.

For instance, incorporation of small amounts of low molecular weight hA or hB into a simple AB diblock copolymer can be readily accommodated. As the concentration of the homopolymer is increased, the solubility limit can be ultimately reached, which results in macrophase separation of the homopolymer. The degree of solubility of the homopolymer within the BCP is governed by the χ parameter between the BCP segments and the homopolymer, the length of the homopolymer chain, the degree of polymerization of the BCP segments as well as the particular microphase separated structure. Due to the temperature dependence of χ and the associated temperature-induced structural transforma-

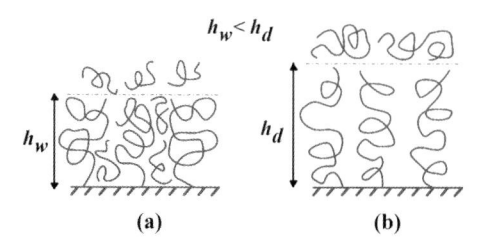

Fig. 1.3 Schematic view of (a) wet brush and (b) dry brush.

tions in BCPs, the phase diagram of the homopolymer/BCP systems is also temperature dependent. From the theoretical models and the experimental studies, it has been shown that, in the case of neutral interactions between the homopolymer and one of the BCP segments, "dry"-brush formation and macrophase separation can be avoided using lower molecular weight homopolymers, Fig. 1.3. Moreover, for shorter homopolymers, higher volume fractions of the homopolymer can be accommodated within the BCP domains. Taking advantage of enthalpic interactions between the homopolymer and the BCP would avoid the entropy-driven phase separation associated with blends of the higher molecular weight chains. This further expands the possibilities of fabricating blends with certain physical properties where higher molecular weights are desired. For instance, exothermic mixing occurs by the addition of poly(vinylidene floride) (PVDF) or poly(vinyl chloride) (PVC) to poly(methyl methacrylate) (PMMA) blocks and results in no macrophase separation at high loadings or larger molecular weights [29–31]. The extreme case for the exothermic mixing would be when the affinity between the homopolymer and the BCP is up to an extent that would result in a chemical reaction under certain conditions.

1.1.1.1 AB/hA Blends

Blending of homopolymers can induce phase transitions. This has been the topic of extensive theoretical and experimental research that will be briefly touched upon in this chapter.

The experimental phase diagram together with the mean-field calculations for a symmetric PS-PI BCP blended with short hPS [32] (molecular weight of the hPS to PS block in the BCP is about 0.55) is shown in Fig. 1.4. As the concentration of the hPS is increased, a phase transformation from lamellae to double gyroid to cylinders occurs. By further increase of the hPS volume fraction the cylinders transform into a disordered sphere morphology (micelles). An increase in the volume fraction of hPS also shifts the order–disorder transition (ODT) to lower temperatures. Moreover, an order–order transition (OOT) can be observed between the double gyroid and cylindrical morphologies. Upon rapid cooling from the higher temperature homogeneous disordered phase to temperatures below the OOT a stable cylindrical microstructure results. Heating this morphology leads to a double gyroid phase while cooling the gyroid phase back to below the OOT temperature does not result in phase transition to initial

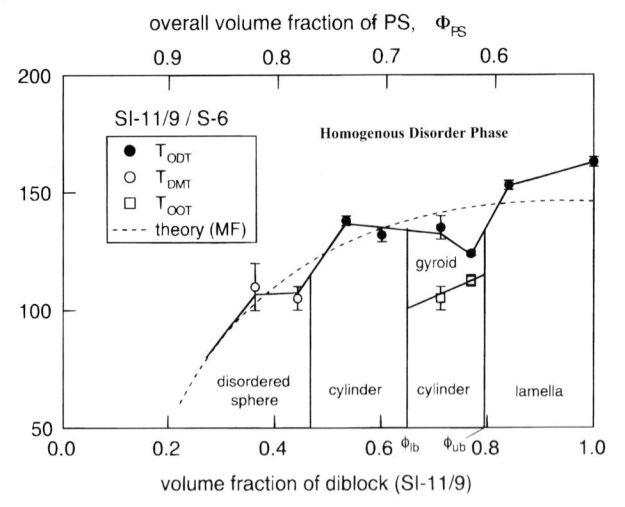

Fig. 1.4 TODT, TDMT, and TOOT vs. volume fraction diblock in Sl-11/9/S-6 blend. Vertical lines separating microdomain structures are obtained from the total volume fraction of PS in the system and other reports on the effect of volume fraction of PS on the microdomain structure. The dashed line indicates the results of a mean field calculation for ODT. Adapted from Ref. [32].

cylindrical morphology. Similar behavior was observed by Hajduk et al. [33] for the transition from lamella to gyroid.

1.1.1.2 ABC/hA Blends

Because of the additional interfacial constraint created in linear ABC triblock copolymers their ABC/hA phase transformations do not exactly follow those of the AB/hA type blends. For example, Lescanec et al. [34] studied the phase transition of a compositionally symmetric poly(2-vinylpyridine)-b-polyisoprene-b-polystyrene (P2VP-PI-PS) system upon addition of 10 wt.% hPS with different molecular weights. The system formed a microphase separated hexagonal lattice of P2VP cylinders, surrounded by a PI annulus within a PS matrix, Fig. 1.5. Because of the non-uniform degree of chain deformation, the PI/PS IMDS is strongly coupled to the shape of the Wigner–Seitz cell of the microdomain lattice, see Fig. 1.6. This is a unique system in which the IMDS is not of constant mean curvature (CMC), which is usually the approximate case in the strongly segregated block copolymer systems. The modification of the PI/PS IMDS was probed by the addition of 10 wt.% hPS, using different molecular weights. It was observed that up to a certain molecular weight of hPS, the PS blocks behave as "wet" brushes preserving the non-CMC nature of the IMDS. Upon further increase in the molecular weight of the hPS, the PS blocks start to behave as "dry" brushes, the hexagonal shaped PI/PS IMDS becomes more circular, accompanied by an increase in the domain spacing relative to that of the

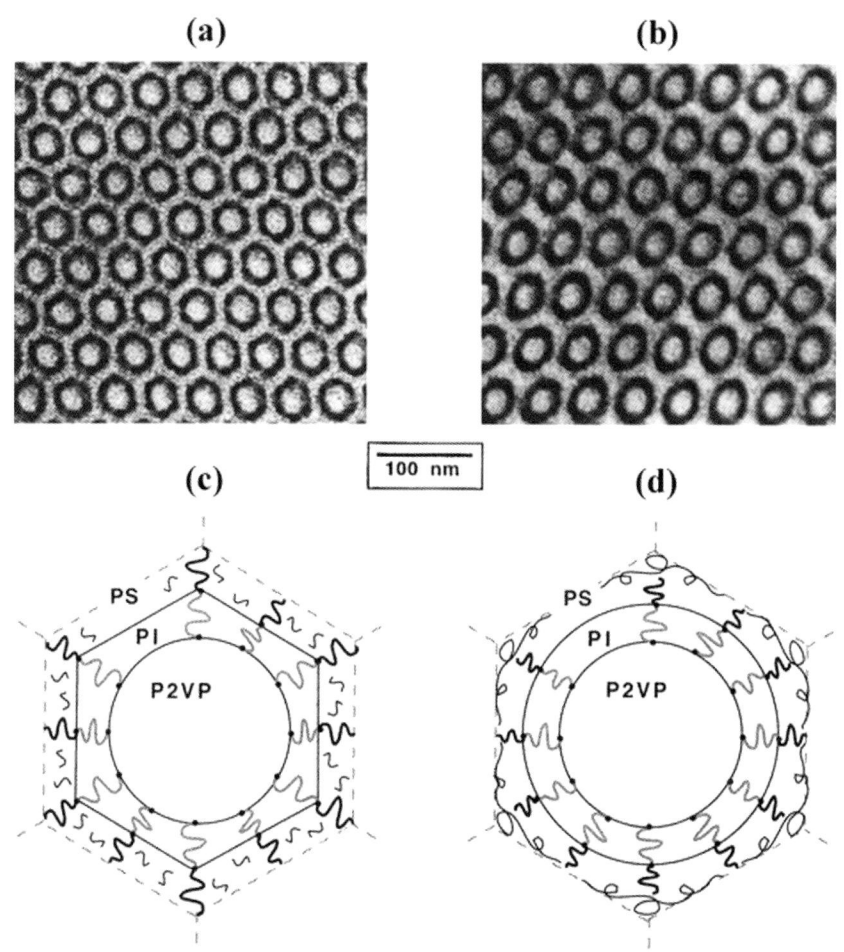

Fig. 1.5 Bright-field TEM micrographs of P2VP-PI-PS/hS (4000 g mol^{-1}) (a) and P2VP-PI-PS/hS (50000 g mol^{-1}) (b). Nearly axial projections of hexagonally ordered microdomains consisting of a P2VP core surrounded by a PI matrix are observed. Schematic representations (c and d) of the block copolymer chain conformations in a single microdomain observed in parts (a) and (b), respectively. In each schematic, the boundary of the Wigner–Seitz cell is indicated as dashed gray lines. The PI and PS blocks are shown by thick gray and dark curves, respectively. hPS is shown by thin dark lines. Note that the polymer chain trajectories are artificially constrained to lie within one Wigner–Seitz cell. Adapted from Ref. [34].

neat triblock. By further increase in the molecular weight of the hPS, macrophase separation occurs and the triblock returns back to its non-CMC form. This system shows subtle differences with the simple AB/hA system in which shorter molecules have more pronounced effect on the interfacial curvature than their higher molecular weight counterparts. This can be explained in terms of the additional interfacial constraint that is present on the mid B block in the

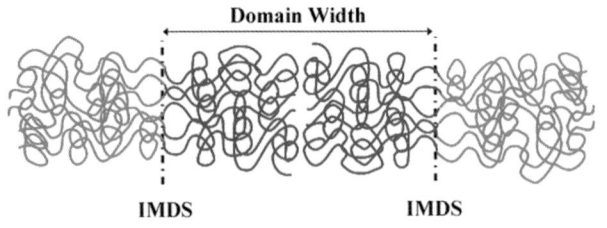

Fig. 1.6 Schematic representation of the inter-material dividing surface (IMDS) in a symmetric diblock copolymer. Chain stretching at the junction points is exaggerated for the sake of illustrating the strong segregation limit (SSL) and less conformational entropy.

case of ABC as opposed to a simple AB block. This observation was exhibited in a similar system [35] by addition of B homopolymers or AC diblocks.

1.1.1.3 BCP/Nanoparticle Blends

The incorporation of inorganic nanoparticles into BCPs is an elegant way of controlling the spatial location and the mean interparticle distances with the aim of tuning the optical, electronic or magnetic properties of the composite material. The spatial arrangement of the nanoparticles is governed by the BCP architecture, the effective nanoparticle effective size, its surface functionality (corona), surface grafting density, volume fraction, and its shape/aspect ratio, Fig. 1.7. Recent developments in the nano-patterning and the nano-lithography of BCPs have shown considerable promise for technological applications in on-chip photonic, magnetic and or electronic devices.

Research pathways in the area of BCP/inorganic phase systems have been undertaken in a number of different directions. In one approach, the microphase

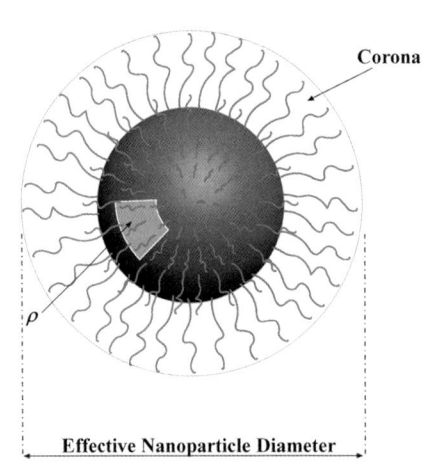

Fig. 1.7 Schematic representation of a spherical nanoparticle functionalized with a polymeric ligand. The effective nanoparticle diameter and the surface grafting density (ρ = number of chains per unit area) are shown.

separated BCPs act as templates for binding of the inorganic species onto a specific domain with the aim of fabricating ordered nanoparticle arrays. The nature of the bonding can be covalent [36], electrostatic [37], hydrogen bonding [38] or solely hydrophobic/hydrophilic interactions.

BCPs have also been used as templates for *in situ* synthesis of the inorganic phases, as in the sol–gel processes or reduction of metal ions to form metallic nanoparticles in a selected domain [39, 40]. The BCP in this case might be used as a sacrificial component or it can remain in the material for a specific function in the overall composite.

In another approach, which is primarily used for nanoparticle systems, the inorganic phase is synthesized *ex situ*, its size/shape and surface functionality is tuned and then the nanoparticles are incorporated within the microphase separated BCP domains by co-assembly with the BCP from a common solvent. The BCP acts as a scaffold for controlled distribution of the nanoparticles in three dimensions. Recent developments in this approach will be further expanded in this chapter. For a thorough review of sol–gel and metal ion reduction approaches to BCP/nanoparticle composites, the reader is referred elsewhere [1, 2, 19, 41].

Interfacial interactions in the nanoparticle-filled BCPs play a critical role in the degree of dispersion and in the particle spatial distribution within the targeted BCP domain. For instance, in an A-B diblock copolymer, depending on the functionalization of the nanoparticles by A or B homopolymers, the particles can preferentially segregate in the corresponding A or B block domain due to the enthalpic interactions. In contrast, if the particles are functionalized with a random mixture of A and B blocks they will phase separate to the IMDS. In addition to the enthalpic interactions between the nanoparticle- corona and the BCP chains, it was theoretically shown by Balazs et al. [20, 42] that entropic interactions can play an important role in the location determination of the nanoparticles. As the diameter of the nanoparticles increases, the nanoparticles tend to sequester at the center of the domains while smaller particles are mainly located at the IMDS. Bockstaller et al. [43] have demonstrated experimentally the effect of particle size in their spatial distribution within the symmetric poly(styrene)-b-poly(ethylene propylene) (PS-PEP) domains (Fig. 1.8). In the case of gold nanoparticles having a diameter of $d=3.5$ nm in a diblock with lamellar spacing of 58 nm ($d/L=0.06$), the nanoparticles were segregated to the IMDS, while in the case of larger silica particles with the same surface functionality $d=21.5$ nm ($d/L=0.26$), the particles were found near the center of the domains. Neglecting any differences in the enthalpic interactions between the nanoparticles and the polymer matrix, the observed locations of the nanoparticles suggest a dominant entropic contribution to the free energy. This is because, in the case of larger nanoparticles, the decrease in the conformational entropy of the corresponding block is dominant, while for smaller particles the decrease in the entropy is balanced by the translational entropy of the particles.

Kramer et al. [44] have demonstrated experimentally that for a symmetric PS-P2VP BCP, addition of gold nanoparticles functionalized with a short brush of homopolymers of each block can preferentially sequester the nanoparticles at

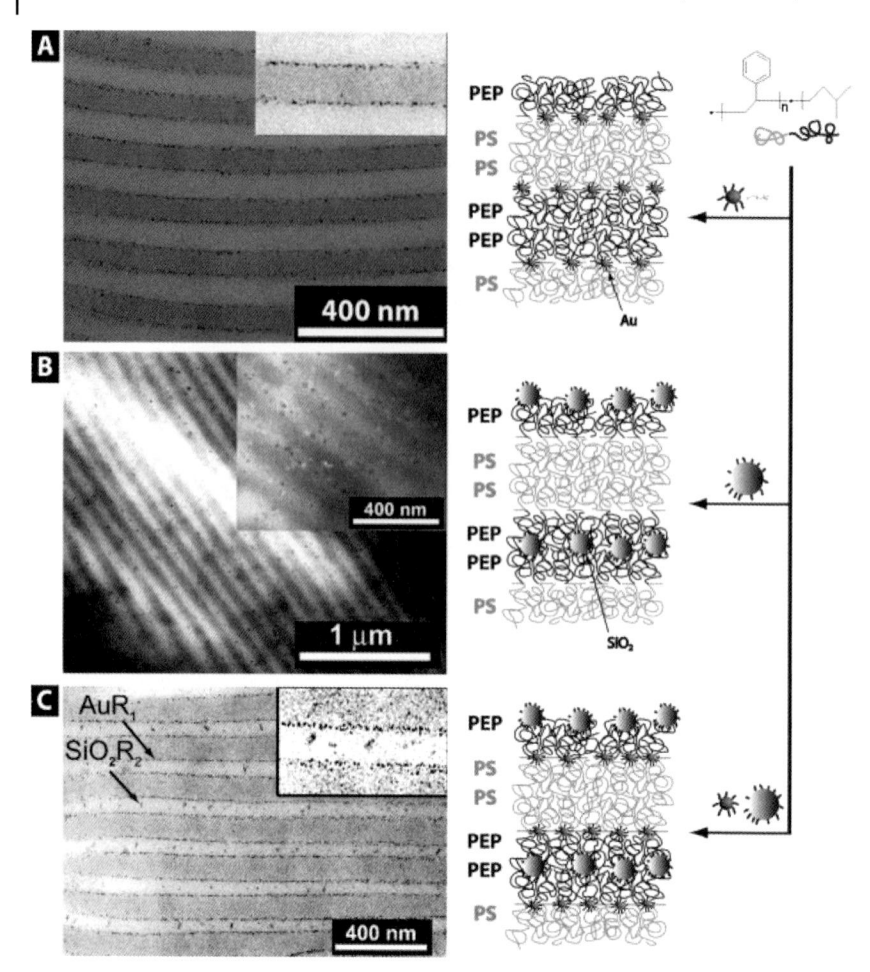

Fig. 1.8 Schematics of composite morphology and respective bright-field transmission electron micrographs revealing (A) interfacially segregated particle morphology (PS-PEP/AuSC12H25), (B) center morphology (PS-PEP/SiO2-(C$_3$H$_9$)), (C) Microstructure of a ternary blend PS-PEP/AuSC$_{12}$H$_{25}$/SiO$_2$-(C$_3$H$_9$) exhibiting combined morphological characteristics: gold particles at the IMDS, silica particles in the center of the PEP domains. Adapted from Ref. [1].

the center of the corresponding block. The particles organize at the center of the domains since the translational entropy loss of the particles, by segregating in the center of the blocks, is by far less than the chain stretching penalty in order to accommodate the nanoparticles throughout the corresponding domain. Functionalizing the particles with a mixture of homopolymers from both components resulted in particle location to the IMDS (Fig. 1.9).

Very recently Kramer et al. [45] have investigated the effect of the chain density of PS or P2VP on the placement of the nanoparticles within the PS-P2VP

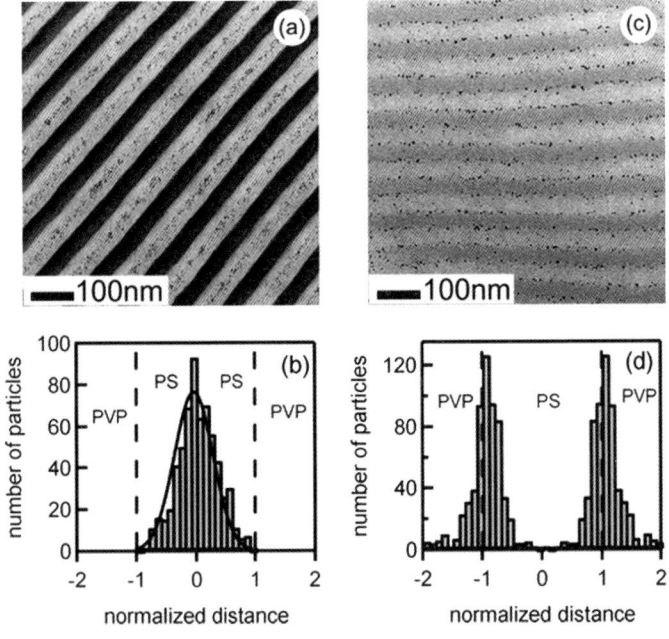

Fig. 1.9 Cross-sectional TEM images of gold/block copolymer (PS-P2VP) diblock with $M_n \sim 196\,500$ g mol^{-1}) composite films using gold particles coated with (a) 100% PS thiol with a grafting density of ~ 0.14 chain Å^{-2}, and (c) a 1:1 mixture of thiols that produces a particle coating that is 20% P2VP with a grafting density of ~ 0.11 chain Å^{-2}. Graphs (b) and (d) show the corresponding histograms of particle locations for samples (a) and (c), respectively. Adapted from Ref. [44].

system. Particles with higher chain density tend to segregate at the center of the domains while those with the same ligand but lower chain density segregate at the IMDS. Further experiments need to be performed in order to better understand this effect in detail.

There has been no comprehensive study on the phase behavior of nanoparticle-filled BCP systems to date. However, there have been some scattered reports on the effect of nanoparticles in phase transformation within BCP. Kramer et al. [46] have observed a transition from lamellar to spherical morphology in a symmetric poly(styrene)-b-poly(2-vinylpyridine)/gold nanoparticle system when the nanoparticle volume fraction exceeds a critical value. The phase transition was observed in thick films cast from solution in which a gradient in the nanoparticle concentration was observed throughout the film thickness (Fig. 1.10). However, the role of the gold nanoparticle in this transition was not critical since addition of the same volume fraction of PS used in the gold functionalization also resulted in conversion to fully spherical morphology throughout the films. In a different system where poly(styrene)-b-poly(4-vinyl pyridine) (PS-P4VP) was loaded with different volume fractions of mercaptoacetic acid modified CdS nanocrystals, the nanoparticles were preferentially segregated in the

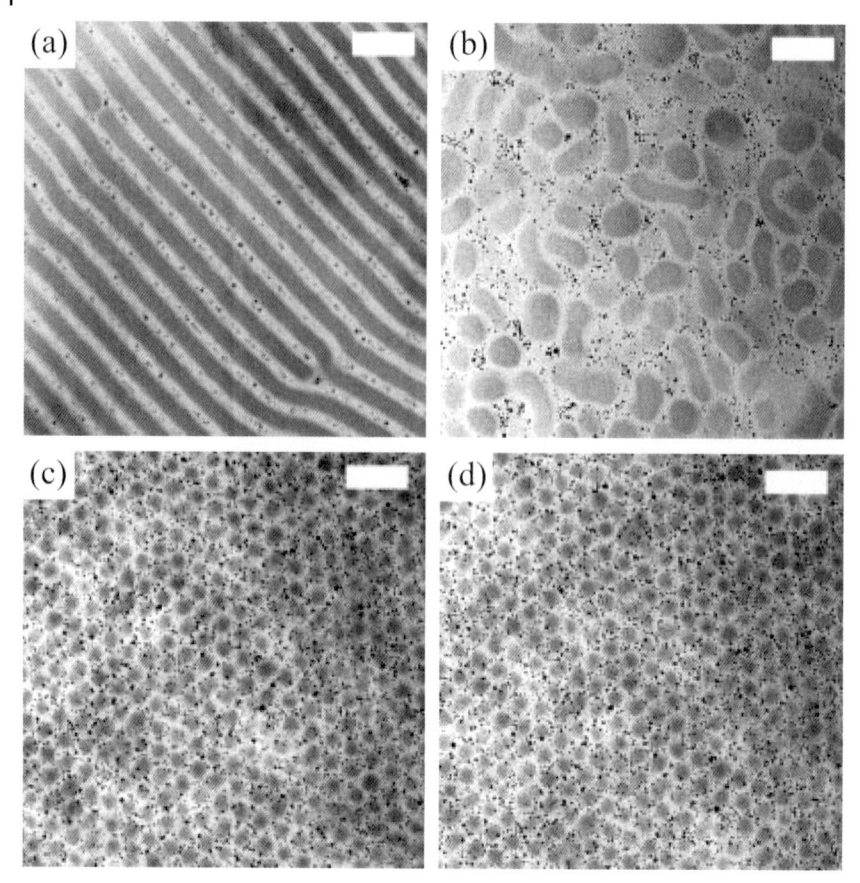

Fig. 1.10 Cross-sectional TEM images of a film of gold particles/PS-b-P2VP ($Mn=59$ kg mol^{-1}). The overall volume fraction of PS-coated gold particles is 0.5. The distance from the top of the film (L) is (a) 27, (b) 36, (c) 52, and (d) 95 lm. All scale bars are 100 nm. Adapted from Ref. 46.

poly(4-vinyl pyridine) phase due to the hydrogen bonding affinity between the nitrogen lone-pair electrons present in the P4VP block and the ligands on the CdS nanoparticles [47]. Increasing the volume fraction of CdS resulted in the transformation of hexagonally packed P4VP cylinders to the lamellar morphology. The same trend was observed in work by Ho et al. [48] where gold nanoparticles where hybridized by *in situ* reduction of hydrogen tetrachloroaurate trihydrate (HAuCl$_4$) within poly(4-vinylpyridine)-b-poly(ε-caprolactone). Increasing the gold nanoparticle concentration resulted in a phase transformation from cylindrical microdomains into the lamellar morphology.

Theoretical treatments of the phase behavior of BCP/nanoparticle systems are still preliminary at this point. Balazs et al. [20, 42] demonstrated the synergistic interaction between hard particles and a symmetric diblock copolymer and their cooperative final morphology and the thermodynamic behavior. In their approach, they used a self-consistent field (SCF) theory for the BCP part and a density functional theory for the nanoparticle system. It was suggested that, depending on the volume fraction and the size of the nanoparticles, the particles can localize at the center or at the IMDS. A number of other studies including the Monte Carlo simulation of Wang et al. [49], the discontinuous molecular dynamics simulation of Schultz et al. [19] or the SCF approach of Lee et al. [50] on the nanoparticle-filled BCPs under confinement have been studied that will not be further elaborated in this chapter.

1.1.2
Self-assembly of Rod–Coil BCPs

Rod-coil BCPs are composed of a rigid rod segment joined to a flexible coil block; therefore they exhibit certain general characteristics of liquid-crystalline rods and the microphase separation of the coil–coil BCPs. In contrast to the flexible homopolymers that follow Gaussian chain statistics (persistence lengths ~ 1 nm), rod-like homopolymers possess very large persistence lengths, of the order of 10 to 100 nm. This conformational asymmetry induces significant packing differences for the rod segment compared to the coil block. Depending on the aspect ratio (length to diameter) of the rod, the amount of IMDS required for the rod block is much smaller than that for the coil block, leading to severe chain stretching in the coil block. The parallel packing of the rods in the stiff coil block results in a variety of microphase-separated morphologies not observed in the common coil–coil BCPs. In addition, unlike random coil non-crystalline homopolymers where the final structure is not significantly affected by the local conformation of the chains, the geometric constraints present in rod-like copolymers result in anisotropic liquid crystal (LC) and crystalline rod phases which significantly affect the structures formed by self-assembly. The self-assembly of the rod-like polymers can be initiated from the isotropic melt (for low $N\chi$ systems) or from solution and triggered by either temperature (thermotropic) or solvent concentration (lyotropic).

Recently rod–coil BCPs have attracted considerable attention due to their unique phase behavior, nanostructured morphologies and the numerous possibilities for incorporating various functionalities in the rod segment. For instance, electronically active polymers as multi-component active layers with applications in LEDs or photovoltaics [51, 52] have been incorporated into rod–coil BCP. Integrating such active functional segments in the framework of BCPs provides the possibility of controlling the spatial arrangement and the mean effective distances between domains having a given property functionality. This considerably enhances the fundamental study of organic electronic materials as well giving control over their final optical/electronic properties [53].

To date, a number of different types of LC-based BCPs have been synthesized. This includes simple AB diblocks (A rod and B coil), ABA or BAB triblocks, BCPs with rod-like side chains etc. (Fig. 1.11). For the sake of brevity, we will

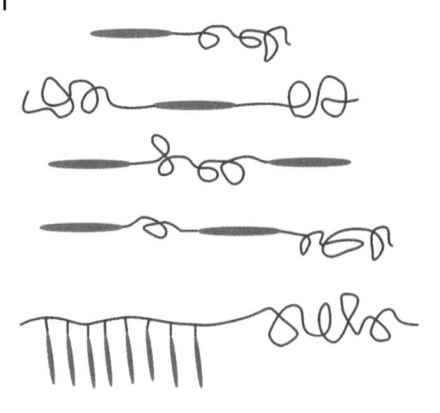

Fig. 1.11 Representative structures of rod–coil BCPs illustrating the rigid rod and the flexible coil.

primarily focus on the simple rod–coil diblock copolymers in this chapter. However, for thorough review of this field the reader is referred elsewhere [54–56].

In rod–coil BCPs, the tendency of the block segment to form liquid-crystalline phases greatly affects the self-assembly pathway and ultimately the final microphase separated morphology. Early theoretical work of Semenov et al. [57], Halperin [58] and Fredrickson et al. [59] predicted transitions between nematic, smectic A, smectic C, bilayer and "puck" morphologies at higher coil volume fractions and low χ_{AB}. Different possible packing schemes for rod–coil BCPs are illustrated in Fig. 1.12. More recently, zig-zag lamellae, elliptical cross-sectional cylinders and hexagonally packed cylinders have also been discovered [58, 60]. Based on these studies it was shown that lowering the temperature or evaporating the solvent (i.e. increasing the polymer concentration) from the isotropic state, first results in a homogenously nematic phase. Later, the incompatibility between the constituent blocks drives the system to microphase separated ordered smectic-like lamellar morphologies. For a thorough review of the theoretical treatment of such systems the reader is referred elsewhere [10].

In the rod–coil BCPs three major thermal transitions govern the overall self-assembly process. Depending on the relationship between the ODT temperature for the BCP, the isotropization temperature ($T_{LC \rightarrow i}$) for the rod segment and the T_g of the coil block, the self-assembly process can follow different pathways (Fig. 1.13). Because of the normally high incompatibility between the coil segment and the rod segment (i.e. very high χ_{AB}) strong segregation is the usual situation, which results in ODT temperatures in most cases above the degradation temperature of the BCP. There have been some attempts [61] to lower the incompatibility between the rod segment and the coil without sacrificing the microphase separation. This is mainly aimed at lowering the ODT temperature and allowing processing from the melt. However, most of the rod–coil BCPs studied to date have microphase separated structures arising from solution casting (lyotropic) as opposed to melt cooling (thermotropic).

In the case of self-assembly from the isotropic melt, as the temperature is lowered to below ODT temperatures a phase transition from a homogenous dis-

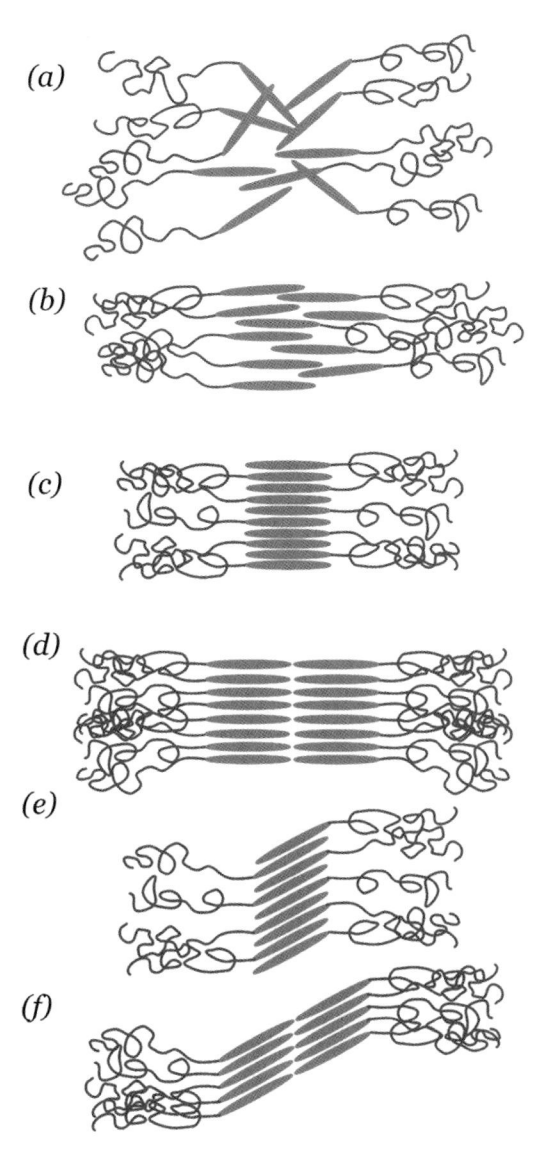

Fig. 1.12 Illustration of possible molecular arrangements/packing in rod–coil BCPs (a) isotropic (b) nematic (c) smectic A interdigitated (d) smectic A bilayer (e) smectic C interdigitated (f) smectic C bilayer. For the sake of simplicity the BCPs are assumed to be monodisperse.

ordered state to a microphase separated ordered state takes place (Fig. 1.14). If T_g for the coil block is higher than the LC clearing transition temperature ($T_{LC \rightarrow i}$) for the rod segment, lowering the temperature results in vitrification of the already microphase-separated coil block, presenting a confined environment for the subsequent LC phase transition. On the other hand, if the $T_{LC \rightarrow i}$ for the rod segment is larger than the T_g of the coil block, the LC phase forms without significant hindrance from the coil blocks.

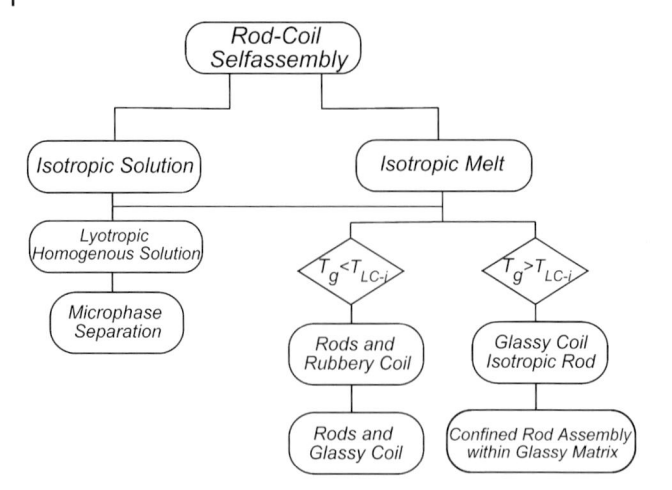

Fig. 1.13 Self-assembly pathways from solution or melt in rod–coil BCPs.

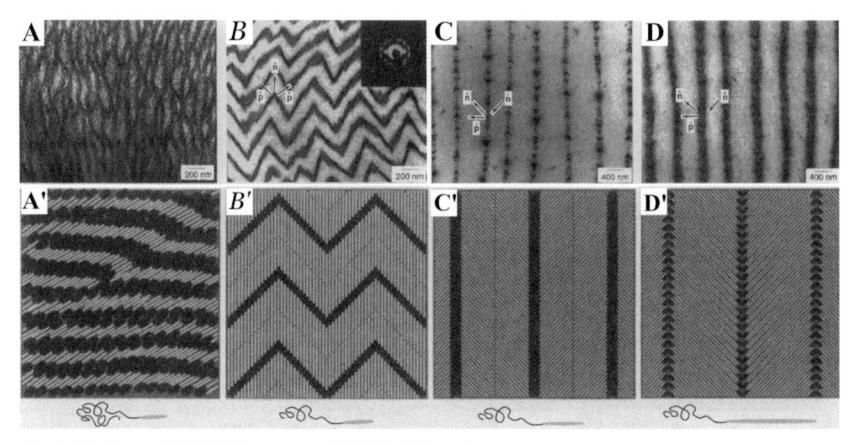

Fig. 1.14 Bright field TEM images of (PS-PHIC) and corresponding packing models (a, a′) $f_{PHIC}=0.42$, (b, b′) $f_{PHIC}=0.90$, (c, c′) $f_{PHIC}=0.96$, (d, d′) $f_{PHIC}=0.98$ (adapted from Ref. 69).

In the case of self-assembly from solution, if a good and neutral solvent for both constituent blocks is used, the final morphology will depend on the interplay between competing liquid-crystalline inter-rod interactions and that of the microphase separation. However, if the solvent used is selective to one of the blocks, the scenario would be different. The self-assembly will be initiated by segregation of the block unfavorable to the solvent and at a relative volume fraction that is biased by the presence of the selective solvent, which drives the self-assembly of the other block upon further solvent removal.

The structure of the rod segment can be helical, mesogenic or π-conjugated. Stiff helical rod-like structures have advantages over their other synthetic counterparts. They possess a secondary structure stabilized by inter-chain interactions. The most widely known group of this family are the polypeptides. Depending on their microenvironments, they can form a-helical, β-sheet or random coil conformations. These structural transitions can be triggered by altering the pH, ionic strength, applied magnetic/electric fields or simply by varying the temperature. The first synthesized rod–coil BCPs were based on polypeptide rods. These included BCPs with rod segments of poly(γ-benzyl-L-glutamate) (PBLG), poly(benzyl-L–leucine), poly(benzyl-L–lyisne), poly(valine), etc. while the coil blocks were poly(styrene), poly(butadiene), poly(dimethylsiloxane), poly(ethylene oxide), etc. [62].

PBLG is a model synthetic polypeptide, capable of forming both a-helical and β-sheet conformations. In the a-helical form it is stabilized by the intramolecular hydrogen bonding while in the β-sheet form the intermolecular hydrogen bonding interactions are the main stabilizing factor. The crystallization of PBLG and its conformational states, both in solution and in the bulk, have been the focus of intensive studies. Due to its molecular rigidity it can be treated as a rod-like structure. However as the rod length is increased it tends to fold rather than tilt, which further complicates the already complex rod–coil phase behavior. Recently Schlaad et al. have demonstrated that, depending on the type of the solvent, the intra-chain hydrogen bonding can be promoted, resulting in fully stretched rigid helices [63], while changing the solvent results in folded helices.

Gallot et al. studied diblock BCPs with high molecular weight poly(styrene) (PS) or poly(butadiene) (PB) as the A block with a peptidic rod block [64, 65]. All of these studies revealed predominantly lamellar microphase separation except one case where the amphiphilic diblock copolymer, with 40% coil volume fraction, formed tilted rods. Within polypeptide-rich domains, the PBLG helices were aligned perpendicular to the IMDS and were not completely stretched but folded. By using shorter rod blocks, Klok et al. have observed hexagonal morphologies in addition to the lamellar microphase separation [66, 67].

In contrast to polypeptides where rod-like behavior is due to the intramolecular hydrogen bonding, the inherent stiff backbone structure of poly(hexyl isocyanate) (PHIC) is responsible for its rigidity. PHIC has an 8_3 helical conformation with a large persistence length of 50 to 60 nm. PHIC can retain its stable rod-like behavior in a variety of organic solvents, while it adopts random coil conformation in other solvents such as pentafluorophenol [56]. Thomas et al. [68, 69] have studied the phase behavior of a poly(hexyl isocyanate)-b-poly(styrene) (PHIC-b-PS) system over a wide range of molecular weights and different volume fractions of the rod segment. Casting films from a good solvent for both blocks (toluene) resulted in unprecedented new morphologies. At lower volume fractions of the rod segment ($f_{PHIC}=0.42$) lenticular aggregates or wavy lamellar (WL) morphology consisting of discontinuous lens-shaped PHIC domains surrounded by a continuous PS matrix (Fig. 1.14 (A,A′)) was observed. Increasing the volume fraction of the rod segment ($f_{PHIC}=0.73$ or 0.90) resulted in zigzag-shaped lamellar domains of PS embedded within the PHIC matrix with a high degree of smectic C-like long-range order (Fig. 1.14 (B,B′)). Based on the electron diffraction studies, it was observed that the PHIC domains were highly crystalline and their chain axis was tilted ~45° with respect to the lamellar normals. At very high volume fractions of the rod segment (f_{PHIC}=0.96 or 0.98), lamellar morphologies were still observed (Fig. 1.14 (C,C′,D,D′)). The PS block formed "arrowhead"-shaped domains whose orientation flips by 180° between adjacent layers. Based on the electron diffraction observations, the PHIC chain axes were alternating between ~+45° and −45° in adjacent domains. This type of morphology is reminiscent of smectic O mesophase formation observed in short LC or racemic mixtures of chiral molecules. As the solvent is evaporated and the concentration of the BCP increases and forms a homogenous lyotropic nematic LC phase, that latter microphase separates into the smectic-like lamellar morphology. A tentative morphology diagram for this system is shown in Fig. 1.15. This system is a good illustration of the competition between the onset of microphase separation and the crystallization of the rod segment during solvent evaporation.

A number of BCPs with different constituent rod blocks have been reported in the literature. Depending on the length of the rod block and its volume fraction, a variety of different microphase-separated morphologies are observed. For instance, poly(styrene-b-(2,5-bis[4-nmethoxyphenyl]oxycarbonyl)-styrene) (PS-b-PMPCS), in which the PS is the coil block while PMPCS forms the rod block, forms a tetragonal (non-hexagonal) perforated layer structure over a wide rod volume fraction ($f_{PMPCS} \approx 0.37$ to 0.52) [70, 71].

As mentioned earlier, in the rod–coil BCPs, typically $\chi_{RC} > \chi_{CC}$ for coil–coil BCPs. This results in strong segregation between the blocks at much smaller chain lengths for each block, which can lead to domain periodicities significantly smaller than what could be achieved in coil–coil BCPs. For instance, Stupp et al. [72] found that in a PI-based rod–coil BCP with very short rod seg-

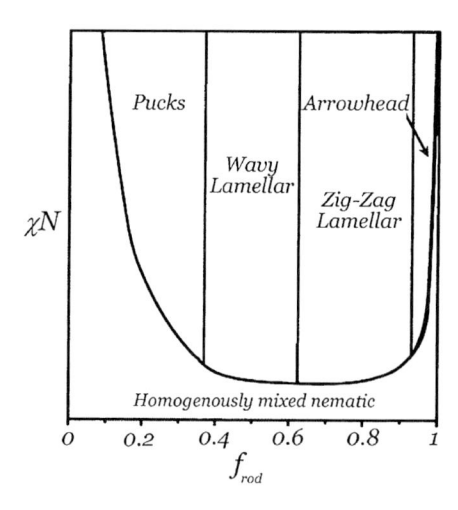

Fig. 1.15 Morphology diagram of rod–coil diblock copolymers (adapted from Ref. [163]).

ment, the morphology varies from alternating strips of rod- and coil-rich domains to discrete aggregates of rods arranged in a hexagonal superlattice as the rod volume fraction is decreased. At intermediate volume fractions a coexistence of strips and aggregates was observed.

Gyroid and spherical phases have only been observed in the short oligomer rod–coil systems. Typically, increasing the length of the rod segment drives the tendency of the system to lamellar phases where interdigitation or tilting of the rod segment lowers the overall free energy of the system.

A variety of rod–coil molecules with different poly(propylene oxide) lengths as the coil block were synthesized and characterized by Lee et al. [73]. Based on the X-ray scattering data, it was demonstrated that by fixing the rod segment and increasing the coil length the microphase-separated morphology could be transformed from lamellar to bicontinous cubic with Ia3d symmetry (double gyroid) and ultimately to a hexagonally packed cylindrical phase. Complex morphologies such as the double gyroid have not yet been observed in rod–coil BCPs for higher molecular weight rod blocks, suggesting that packing constraints in microdomains with high curvature of the IMDS are inconsistent with packing of high aspect ratio rod blocks.

1.2
Introduction: Solution Phase

When block copolymers (BCP) are diluted in a liquid that selectively solvates one of the blocks, the insoluble blocks tend to spontaneously form aggregates. These BCP micelles consist of a desolvated core surrounded by a swollen corona (Fig. 1.16). Similar to the bulk phase transitions described in the previous section, copolymer micelles may show a variety of morphologies in solution, including spheres, worm-like cylinders and vesicles. In the formation of micelles, the overall degree of polymerization (N), the interaction parameter between monomers in the blocks (χ_{AB}), and the volume fraction of each block (f) remain key parameters, in addition to the characteristics of each block with respect to the solvent (χ_{AS}, χ_{BS}). Hence the extra degrees of freedom due to the interplay between a solvent and the copolymer enable structural polymorphism of BCP micelles, and thus allow tailoring of the physical properties available from a given copolymer, without the need for synthesizing new copolymers.

While copolymer micelles are often compared with those formed from small molecule surfactants [74], they differ in several significant aspects. BCP micelles exhibit a much lower critical micelle concentration (cmc) compared with small molecule surfactants [75]. For example, a typical cmc for polystyrene-b-poly(acrylic acid) copolymer micelles in water is about six orders of magnitude lower than that of sodium dodecylsulfate ($\sim 10^{-2}$ M), a common surfactant [76, 77]. Importantly, the kinetic exchange between soluble single chains and aggregates is quite slow in the case of BCP micelles, compared with small molecule surfactants [78]. This is due to both the relatively high viscosities in the cores of BCP

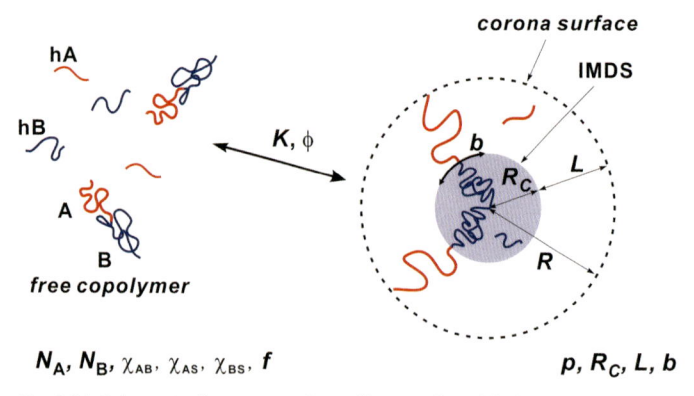

Fig. 1.16 Schematic illustration of micellization from block copolymers and key parameters determining micellar structures. Homopolymers (hA, hB) represent typical impurities formed by incomplete polymerization.

micelles and the large thermodynamic barrier to dissociation of a free chain into the surrounding solvent: for example, the enthalpic penalty for dissociation into the solvent is $N_B \chi_{BS}$ where N_B is the number of monomer units of the less soluble B block. This leads to long residence times for single chains within the aggregates, and makes polymer micelle structures kinetically very stable [75, 79]. On account of their morphological variety and structural stability, BCP micelles provide numerous applications in the fields of biotechnology and nanotechnology: e.g. cosmetics, drug delivery vehicles, compatibilizers and nanotemplates.

Much theoretical and experimental research has been directed towards predicting and controlling the characteristics of micellar structures: a search on this research subject from the CAS database gives about 500 references, indicative of the wide range in morphological and physical properties of the micelles that can be manipulated by subtle variation in the BCP molecular architecture and composition, particularly in the dilute limit.

Many excellent comprehensive review articles on BCP micelles have been published over the past two decades [7, 75, 78, 80–87]. A main goal of this chapter is to illustrate the relationship between the molecular architecture of BCPs and the micellar morphology from both theoretical and experimental points of view. In the following discussion, we review how the molecular architecture of BCP chains (unimers), preparation methods and the properties of solvents dictate the characteristics of the subsequent micellar structures and the micellization kinetics. We then give an overview of the control of the micelle morphology with typical examples of micellar systems composed of coil–coil, rod–coil and copolymers containing crystallizable blocks in both aqueous and organic solvents. We also cover recent research on organic–inorganic hybrid micelle systems, and some examples of promising applications.

1.2.1
General Features of Block Copolymer Micelles

When the concentration of BCP is higher than a certain value, the so-called critical micelle concentration (cmc), BCP chains dissolved in a selective solvent associate into micelles which are in equilibrium with free copolymer chains (i.e. closed association mechanism). In this case, the structural parameters of copolymer micelles (e.g. the aggregation number (p), the radius of core (R_C), the corona thickness (L), the hydrodynamic radius (R_h), and the cmc) can be predicted simply as function of the characteristics of copolymers such as the composition (f), the degree of polymerization (N) and the concentration (ϕ) (Fig. 1.17). Leibler and coworkers investigated the critical micelle concentration behavior for solutions of BCPs using mean field theory [88]. It was found that there was the abrupt onset of an increase in the fraction of chains involved in the formation of micelles around a certain concentration of unimers (ϕ_0 = cmc) in the system, and the average aggregation number (p) was dependent on the molecular weight of copolymer and on the interaction parameter (χ_{AB}), but only weakly dependent on the concentration of copolymers. Simple scaling models describing the structural parameters of the micelles were derived for the crew-cut where R_C is much larger than L, the star-like micelle where R_C is much smaller than L and the intermediate regimes (Fig. 1.17) [89–92] and these scaling relations are summarized in Table 1.1.

Usually the cmc is considered as an empirical value of a certain system rather than a thermodynamic property, because it is highly dependent on the characterization method. Experimentally, the cmc is often defined as the concentration

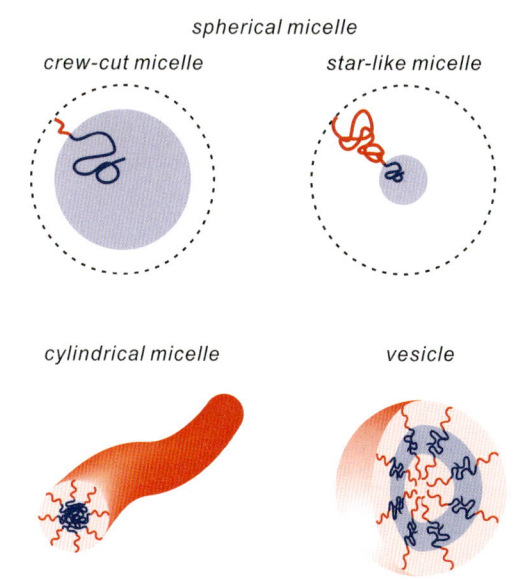

Fig. 1.17 Schematic illustration of various morphologies of micelles including spherical, cylindrical and vesicle.

Table 1.1 The scaling relations for spherical A-B block copolymer micelles, predicted by Zhulina and Birshtein [92].

	Polymer composition	p	R_C	L	b
Crew-cut	$N_A < N_B^{\nu/6}$	N_B	$N_B^{2/3}$	N_A^ν	$N_B^{1/3}$
	$N_B^{\nu/6} < N_A < N_B^{(1+2\nu)/6\nu}$	N_B	$N_B^{2/3}$	$N_A N_B^{(\nu-1)/6\nu}$	$N_B^{1/3}$
	$N_B^{(1+2\nu)/6\nu} < N_A < N_B^{(1+2\nu)/5\nu}$	$N_B^2 N_A^{-6\nu/(1+2\nu)}$	$N_B^2 N_A^{-2\nu/(1+2\nu)}$	$N_A^{3\nu/(3\nu+1)}$	$N_A^{2\nu/(1+2\nu)}$
Star-like	$N_A > N_B^{(1+2\nu)/5\nu}$	$N_B^{4/5}$	$N_B^{3/5}$	$N_A^\nu N_B^{2(1-\nu)/5}$	$N_B^{2/5}$

A parameter ν is the scaling exponent for the radius of gyration of linear polymers, $R_g \sim M^\nu$.

to form a sufficient number of micelles which can be detectable by a certain method (e.g. a surface tension vs. a composition measurement). The influence of chain architecture on micellization was investigated for various structures of copolymers (Figs. 1.18 and 1.19) [93–97]. For example, due to the additional entropic penalty for end block aggregation in triblock copolymers, the cmc of triblock copolymers is much higher due to unfavorable formation of loops.

Addition of salts can affect the critical micelle temperature (cmt) which is the lowest temperature to form micelles, and the tendency follows the Hofmeister series of anions [98–101]. The Hofmeister series orders anions with increasing salting-out tendency (i.e. decreasing polymer solubility), and is as follows:

$$SO_4^{2-} > HPO_4^{2-} > OH^- > F^- HCOO^- > CH_3COO^- > Cl^- > Br^- > NO_3^-$$
$$> I^- > SCN^- > ClO^-$$

Hence when copolymers are dissolved in aqueous solution in the presence of NaCl or LiCl, the micelles can be formed at a lower temperature than when they are formed with NaSCN [102, 103]. Typically the anions impact much more strongly on the salting-out effect than do the cations, but Alexandridis et al. found that the cations also favorably affect the micellizations with the order $Na+ > K+ > Li^+$ [101].

Since the kinetic exchange between soluble single chains and micelles is usually very slow or even frozen, especially for micelles with a high T_g core

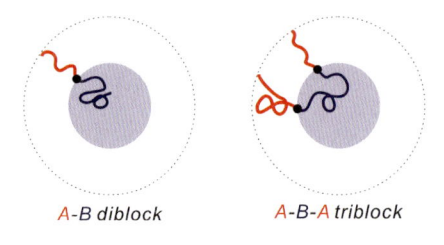

Fig. 1.18 Schematic illustration of chain conformation in micelles formed from various copolymer architectures.

A-B diblock   A-B-A triblock

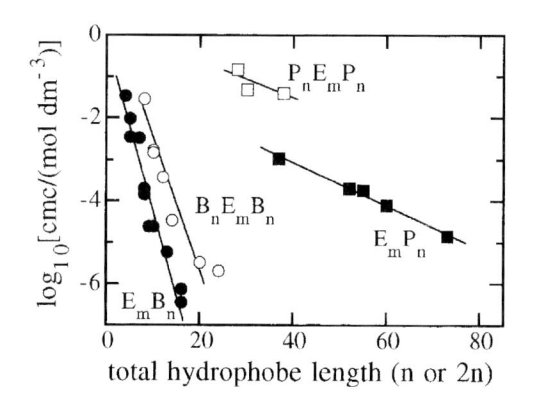

Fig. 1.19 Critical micelle concentration versus hydrophobe length (n) for diblock and triblock copolymer in aqueous solution at 30 °C. E, P and B denote poly(ethylene oxide), poly(propylene oxide) and poly(oxybutylene) chains respectively. Adapted from Ref. [93].

block, the experimentally observed structure of the BCP micelles is quite dependent on the preparation history. Hence it is very important to find a suitable micellization condition to allow the system to reach thermodynamic equilibrium. Detailed preparation methods have been well reviewed in several articles [82, 85, 86]. The simplest way to form BCP micelles in solution is by directly dissolving the solid copolymers into a selective solvent, and letting the solution reach equilibrium by waiting a sufficiently long aging time. For example, PEO-b-PB BCP micelles in aqueous solution were prepared by dissolving low molecular weight PEO-b-PB in water, followed by aging for about 1 week [104]. However this method is restricted to low molecular weight and low T_g core BCPs, otherwise the structures may be in a nonergodic state. Another method is based on a gradual change of solvent quality by adding a selective solvent to the copolymer dispersed in a nonselective solvent. This method can be applied to most copolymer systems including high M_w and high T_g copolymers. The molecular processes during the solvent exchange were investigated by Munk [82]. During the early stage, the micellar solution was in a dynamic equilibrium and the aggregation number of the micelles changed with varying solvent quality. However the dynamics could be effectively quenched when the solution was exchanged to a strongly selective solvent. One advantage of the slow kinetics is the possibility to study metastable or intermediate structures by freezing out a structure by appropriate choice of the T_g of the aggregating block [85].

1.2.2
Artifacts from Sample Preparation (Impurities, Drying, Cross-linking, Staining)

In the characterization of BCP micelles, the role of the particular sample preparation method and analytical technique on the observed structure of the micelles has not always been fully appreciated. Various artifactual micellar structures can be caused in many different ways including insufficient sample quality (high PDI or impurities), improper micellization, drying, cross-linking, and staining processes.

1.2.2.1 Impurities

Anomalous micellization, a phenomenon forming unusually large micelles, is one of the most common problems encountered in dilute copolymer solutions [105–110]. There are some explanations attributing the origins of anomalous micellization to the intrinsic processes forming large hollow spherical or wormlike micelles [105, 106], but many recent results argue strongly against intrinsic process [107–109, 111]. Chu et al. showed that the anomalous micellization can be suppressed by filtering out the impurities, and this was further confirmed by others. A common impurity in BCPs is trace amounts of homopolymer caused by incomplete sequential polymerization: e.g. some GPC analyses showed that copolymers even having very low PDI (≈ 1.02) still contained $\sim 1\%$ of homopolymers [111].

1.2.2.2 Staining and Drying

Many artifacts may arise during the process of staining and drying the micellar solution for electron microscopy imaging. In the process of staining and drying, the system undergoes composition changes in two ways: one is by adding stain which may increase the salt content of the solution, and the other is by increasing the total concentration as evaporation proceeds. Talmon et al. carefully investigated the effects of the staining with heavy atom salts, and the subsequent drying process on the micelle morphology [112, 113]. In a comparison of air-dried with cryo-TEM (believed to be a nearly artifact-free method), they found that the staining process alone could change the size of the micelles, and the change was even more drastic after drying. The drying process could induce change in the structural parameters, or induce phase transformations, so that the final structure corresponds to a very different point on the phase diagram or is even unrelated to the copolymer–solvent phase diagram. As described above, cryo-electron microscopy provides a useful way to characterize the structure of micelle solutions without artifacts, but it is worth noting that other types of artifact may occur from the TEM instrumentation. For example, the false fringes around the periphery of micelles are observed when the microscope is defocused, while enhancing contrast can confuse measurement of micellar dimensions due to the granularity of the phase contrast in the background of the support film [114].

1.2.2.3 Crosslinking

If a method other than cryo microscopy is to be used, one is faced with fixating the micellar structures, but one must take care to avoid introducing artifacts due to the altered chemical potential of the system. For example, crosslinking of the core or shell can fix the solution structure [115–119], so that the subsequent drying of the solvent is not able to induce a compositionally driven shape change. The general strategy is to create a situation when the kinetics of "freezing" the micelles (e.g. by cryo, by crosslinking etc.) is much faster than the kinetics of rearrangement which is driven by the modification of the chemical po-

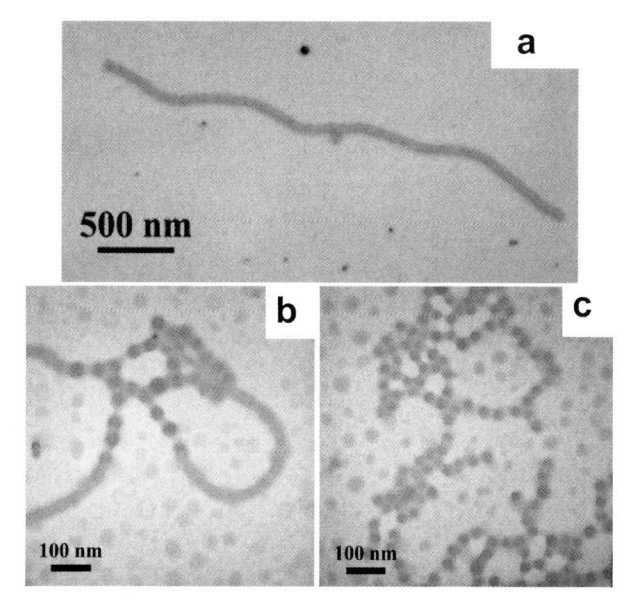

Fig. 1.20 TEM images of the micelles prepared from PAA$_{90}$-b-PMA$_{80}$-b-PS$_{100}$ in THF/water (vol/vol = 1:2): (a) the cylindrical micelles before cross-linking, (b) and (c) the intermediates of the rod-to-sphere transition induced by cross-linking of the shell by using carbodiimide activator/diamino crosslinker. Adapted from Ref. 120.

tential of the system components due to external influences (crosslinking/staining, drying). However, crosslinking can sometimes lead to significant changes in the micelle structure [120]. Wooley et al. investigated the effect of the shell crosslinking on the morphological behavior of cylindrical micelles (Fig. 1.20) [120]. They observed rapid transformation of the cylindrical micelles to spherical micelles upon introduction of 1-[3-(dimethylamino)propyl]-3-ethylcarbodiimide methiodide (ETC), which is an activator for the crosslinking reaction. This phase-induced phase transiton was the result of the chemical modification of PAA by reaction with ETC, and the consequent frustration of morphology. The cross-linking can be suppressed by introduction of a diamino crosslinker before the addition of ETC. Micelles have been crosslinked by UV, X-ray or γ-ray radiation, without any significant morphological changes [86].

1.2.3
Coil–Coil Block Copolymers in Solvents

BCP domain morphologies are basically the result of the competition of the enthalpic and entropic contributions to the system free energy. Swelling copolymers with solvents can sometimes alter the phase stability of the system by decreasing the number of chemical contacts between monomers (i.e. the $-\Delta H$ contribution), and increasing the degree of chain stretching (i.e. the $-T\Delta S$ contribution).

1.2.3.1 Theory

self-consistent mean-field theory (SCFT) has been widely used for prediction of the stable morphologies of BCPs in a solvent [121–124]. With extension of the theory for a copolymer melt in the weak segregation limit [8], Fredrickson and Leibler predicted the a phase diagram for diblock copolymer in a nonselective solvent [121]. Similarly, Huang and Lodge investigated the phase behavior as a function of solvent selectivity, temperature, copolymer concentration, composition, and molecular weight employing self-consistent mean-field theory (Fig. 1.21) [122]. Comparing with the melt calculations described in the previous section, a noticeable difference is that the solution phase diagram does not include the gyroid phase, although such bicontinuous morphologies were demonstrated experimentally [125]. For nonselective solvents, the phase diagram was equivalent to that in the melt, when the copolymer volume fraction (ϕ) was taken into account. In contrast, an asymmetric phase diagram was obtained in a slightly selective solvent. When the solvent is not selective, the stable phase separation region was reduced as the polymer concentration decreased, whereas decreasing polymer concentration broadened the stable region for very selective solvents. In the dilute selective solvent systems, theoretical investigations of phase behavior are limited. Rubinstein et al. performed theoretical investigations on the association of diblock copolymers into micelles in a dilute solution with selective solvent, and compared with the experimental results, within both the crew-cut and the star-like micelle regimes (Fig. 1.22) [126]. In a selective solvent, the total free energy of a micelle is the sum of the surface free energy of the core, the corona free energy, and the elastic stretching of core blocks. Since the free energy of the corona always prefers spherical morphology and because of the large difference in the free energy of the corona between spherical and other morphologies when the corona thickness is much larger than the core radius, the star-like micelles are always spherical. However the corona free energy is weakly dependent on the morphology in the crew-cut micelles because the IMDS in a crew-cut corona is almost planar: i.e. the free energy of a spherical corona is slightly less than that of a cylindrical corona, which is in turn slightly less than that of a lamellar corona. For the transition from spheres, to cylinders, and to vesicles, the small increment in the corona free energy can be compensated by the decrease in the elastic free energy of the core. As a result, the polymorphism of micelles in a dilute solution with selective solvent could be possible only in the crew-cut regime.

The more complicated systems such as ABC triblock copolymers are of great interest due to their large number of different morphologies. Wang and coworkers investigated the aggregate morphology of ABC triblock copolymers in dilute solution using SCFT in two dimensions [127]. Depending on the hydropobicity differences between the blocks and the degrees of segregation, the various shapes of micelles which are not usually observed from diblock copolymers (e.g. torous and peanut-like micelles) were predicted.

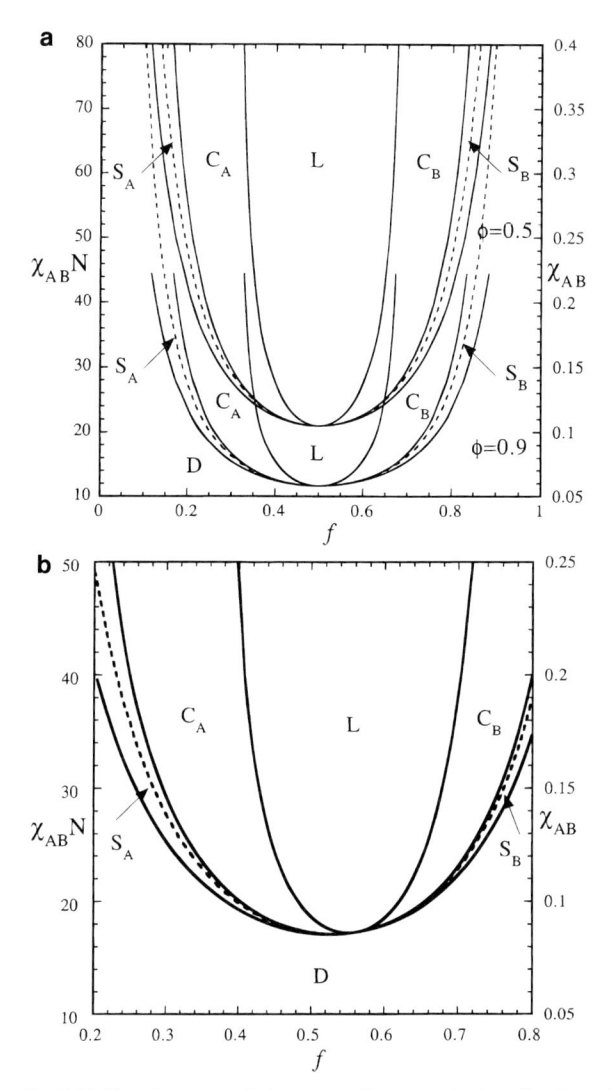

Fig. 1.21 Two-dimensional phase map for a diblock copolymer with $N=200$. (a) In a non-selective solvent ($\chi_{AS}=\chi_{BS}=0.4$) for the constant polymer volume fraction (ϕ), and (b) in a selective solvent with $\chi_{AS}=0.6$ and $\chi_{BS}=0.4$ for $\phi=0.5$. The phases represent L=lamella, C_A and C_B=hexagonally-packed cylinders, D=disordered. The dashed curves (S_A and S_B) correspond to the spinodal instability of the disordered state. Adapted from Ref. [122].

1.2.3.2 Concentrated Systems

Solvent selectivity and the concentration of polymers both play a crucial role in the assembly of copolymers in solution. Many investigations have been conducted for characterizing the morphological transition from the melt-state to dilute solution in solvents having various selectivity (i.e. neutral, slightly selective or strongly se-

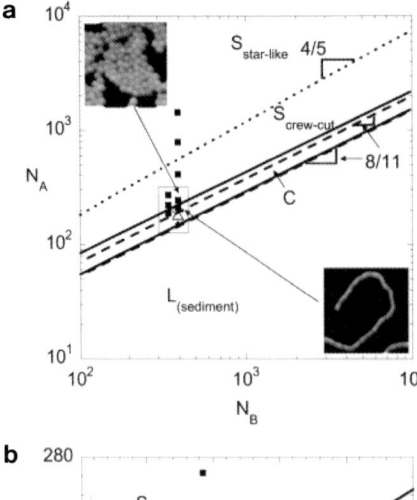

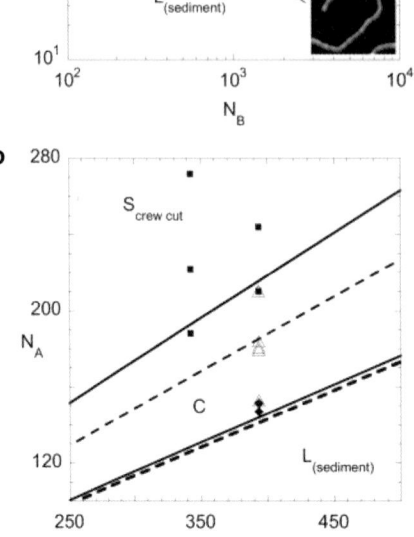

Fig. 1.22 (a) Theoretical diagram of states in logarithmic N_A, N_B coordinates. The crossover boundary between star-like and crew-cut spherical micelles is indicated by the dotted line. The morphological boundaries between spherical (S) and cylindrical (C) and between cylindrical (C) and lamellar (L) aggregates are shown by solid lines. Asymptotic boundaries between spherical and cylindrical and between cylindrical and lamellar aggregates are indicated by the dashed lines. (b) Experimental diagram of states in N_A, N_B coordinates. Samples with spherical micelles are marked by (■), samples with cylindrical micelles are indicated by (Δ), and insoluble samples are shown by (◆). Adapted from Ref. [126].

lective). If selectivity is highly temperature dependent, small changes in temperature could result in large changes in the solution properties. Lodge et al. explored the phase behavior of PS-*b*-PI in four different solvents, a neutral solvent bis(2-ethylhexyl) phthalate (DOP), a slightly styrene-selective solvent di-*n*-butyl phthalate (DBP), a strongly styrene-selective solvent diethyl phthalate (DEP), and an isoprene-selective solvent *n*-tetradecane (C14) (Fig. 1.23) [128]. The phase behavior was examined using a combination of SAXS, static birefringence, and dynamic light scattering. When the solvent selectivity was increased, several novel features were observed. For example, a relatively large window of coexisting lamellar (L) and inverted cylinder (C_2) appears in a strong selective solvent (DEP) between the respective single phases, which is attributed to a destabilization of the inverted gyroid (G_2) phase as a result of a heightened free energy penalty for packing frustration. The spherical phase adopted a face-centered-cubic lattice in DEP and not the body-centered-cubic lattice found in the neat DBP [129] and C14 solutions and

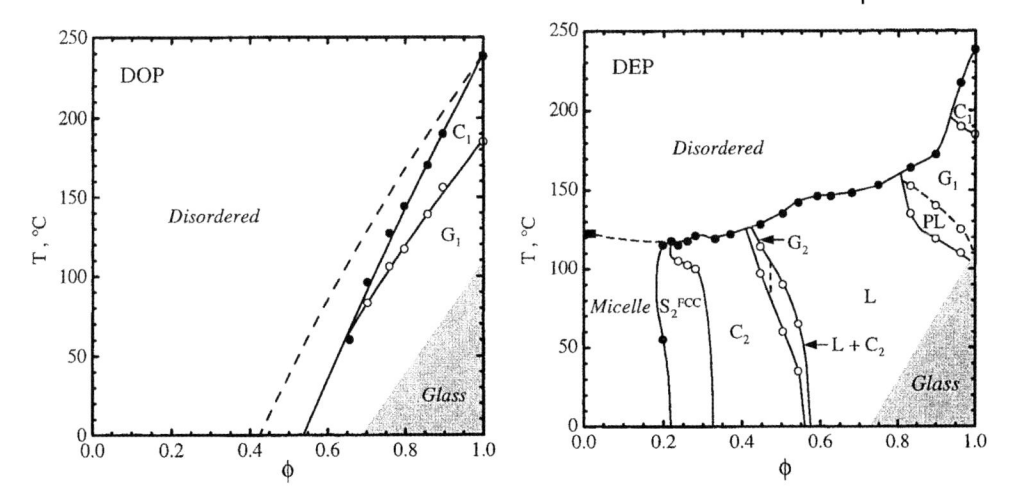

Fig. 1.23 Phase diagrams for PS-b-PI as a function of temperature (T) and polymer volume fraction (ϕ) for solutions in DOP (a neutral solvent) and DEP (a strongly-PS-selective solvent). Filled and open circles identify ODTs and OOTs, respectively. The dilute solution critical micelle temperature (cmt) is indicated by a filled square. The ordered phases are denoted by: C, hexagonal-packed cylinders; G, gyroid; PL, perforated lamellae; L, lamellae; S, cubic packed spheres. The subscript 1 identifies the phase as "normal" (PS chains reside in the minor domains) and the subscript 2 as "inverted" (PS chains located in the major domains). Adapted from Ref. [128].

neat BCPs. The scaling of the domain period with concentration depends strongly on the solvent selectivity.

Alexandridis et al. investigated a ternary system consisting of a PEO-b-PPO-b-PEO Pluronic copolymer, D_2O ("water") and p-xylene ("oil") (Fig. 1.24) [130]. In this system, the solvent quality was varied by changing the ratio of D_2O and p-xylene, which are selective solvents for PEO and PPO respectively. The self-assembly of this ternary system exhibited significant microstructural polymorphism, with as many as nine independent morphologies including seven ordered phases: lamellae, water-continuous cylindrical, oil-continuous cylindrical, water-continuous and oil-continuous micellar cubic, two bicontinuous cubic; as well as two disordered phases water-continuous micellar and oil-continuous micellar. This rich polymorphism is considered to arise partially from the crystallizability of PEO within the reverse micelles. The self-assembly of the block copolymer/water/oil system is of course fundamentally similar to self-assembly of the small molecular surfactant/water/oil system. The main difference between those two systems is that the phase behavior of surfactants is limited by the surfactant molecular geometry so that the isothermal phase diagram exhibits either normal or reverse phases, but not both, while that of the more conformationally shape-adaptive block copolymers can be readily tuned by the varying degree of swelling of the different blocks by the respective selective solvent. The extra de-

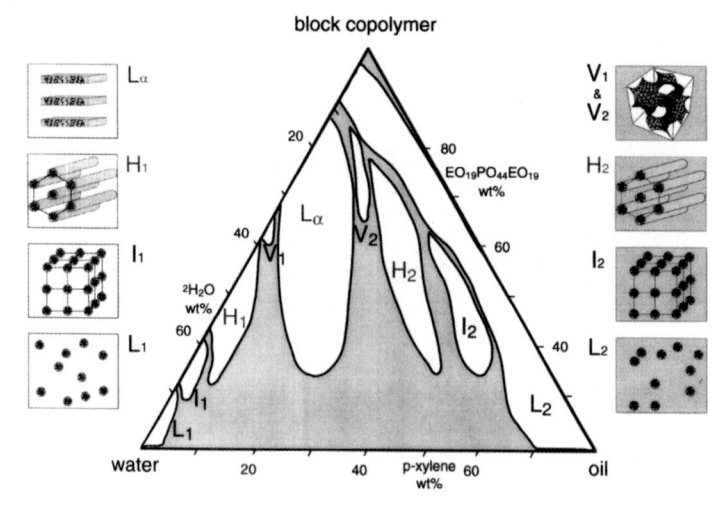

Fig. 1.24 Phase diagram of the $(EO)_{19}(PO)_{43}(EO)_{19}$-$D_2O$ ("water")-p-xylene ("oil") ternary system at 25 °C. Adapted Ref. [130].

gree of freedom affords a much nicer polymorphism, showing both normal and reverse structures at the same temperature with the same block copolymer.

1.2.3.3 Diluted Systems

Numerous morphologies have been observed for BCPs in aqueous solution, but many of them may not be in thermodynamic equilibrium due to the sample preparation method used, as described in the previous section. Bates et al. reported giant worm-like micelles and novel structural feature such as Y-shaped junctions and network structures for PB-b-PEO BCPs in aqueous solution [104, 131]. Branched worm-like micelles having Y-juctions and network structures were found in the intermediate regions of lamellae and cylinders for high molecular weight PB-b-PEO (Fig. 1.25). Increasing the molecular weight of the copolymer results in a sharp decrease in the equilibrium solubility of the copolymer in water, and consequently slows the exchange dynamics between micelles. In this case, the network structures can be fragmented into a nonergodic dispersion of particles by simple stirring or sonication. The worm-like cylindrical micelles can be spontaneously aligned into a nematic phase and then packed into a hexagonal phase as the polymer concentration is increased [104].

Crew-cut amphiphilic BCPs in aqueous solution can also exhibit various polymorphism, with variation of the micellar shape induced by the solvent quality or ionic strength as well as polymer compositions [132, 133]. Eisenberg et al. demonstrated a variety of morphological transitions for PS-b-PAA crew-cut micelles (Fig. 1.26) [134, 135]. In this study, the micelle morphology can be varied from spheres to cylinders, and cylinders to lamellae by increasing the water/di-

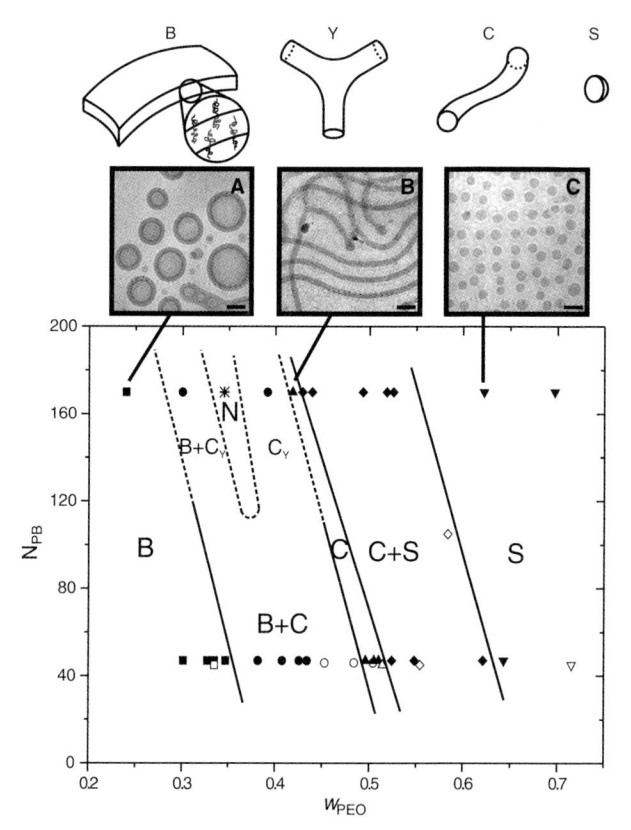

Fig. 1.25 Morphology diagram for PB-b-PEO (1 wt.%) in water as a function of molecular size and composition, where N_{PB} and w_{PEO} are the degree of polymerization and weight fraction of the PB and PEO blocks, respectively. Four basic structural motifs – bilayers (B), Y-junctions (Y), cylinders (C), and spheres (S) – have been identified by cryo-TEM, as illustrated in the micrographs, (A to C). Bars (A to C), 100 nm. Adapted from Ref. [131].

oxane ratio. The morphological transitions were also reversible: i.e. the morphology was varied from lamellae to cylinders, and cylinders to spheres by decreasing the water/dioxane ratios. This reversibility of morphology suggests that each morphological structure is a thermodynamically stable structure at each condition. By the same token, the size of vesicles was also able to be controlled (Fig. 1.27) [136, 137]. Unlike vesicles formed from low molecular weight surfactants, which are typically kinetically trapped, nonequilibirum structures [138], these vesicles prepared from diblock copolymers were proposed as a thermodynamically stable structure, and the stability may be increased with the larger polydisperisty of copolymer because the copolymers having different chain lengths can effectively relax the lateral strain which arises from a mismatched number of molecules between the inside and outside of the vesicular layers

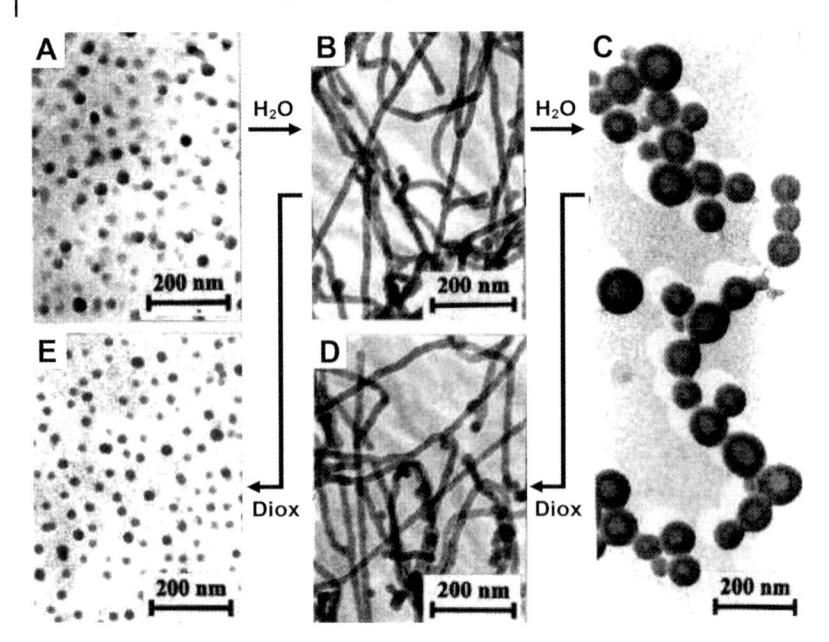

Fig. 1.26 Reversibility of various morphological transitions for PS$_{310}$-b-PAA$_{52}$ copolymer solution. Adapted from Ref. [134].

[136, 137]. The addition of salt results in a change in the equilibrium solubility of the copolymer chain following the Hofmeister series, and also alters the chain–chain interactions. Especially when copolymers containing an ionizable block like PAA are in an aqueous solution, the interfacial curvature of the micelle can be dramatically modulated by the salt concentration: i.e. the effective volume of the ionizable block decreases with increasing salt concentration by neutralizing the charge repulsions between chains via the screening or salt-bridge effect, and thus the preferred interfacial curvature decreases. Salt-induced phase transitions were demonstrated for PS-b-PAA and PS-b-PEO BCPs in aqueous solutions [135, 139]. Depending on the initial salt concentration, the morphology varied from spheres to cylinders, and then to vesicles. As expected, multivalent salts affect the phase transition much more strongly than monovalent salts.

1.2.3.4 ABC Triblock Polymers

As might be anticipated, a more complex system employing an ABC triblock terpolymer can result in several unusual micellar morphologies (Fig. 1.28). Pochan, Wooley and coworkers reported toroidal micelles prepared from PAA-b-PMA-b-PS triblock copolymers in water/THF and diamine mixture (Fig. 1.28a) [140]. Toroidal micelles are believed to be formed by the collapse of negatively charged cylindrical micelles, driven by interaction with a divalent organic cation

1.2 Introduction: Solution Phase | 1419

Water content (%) / Size (nm)

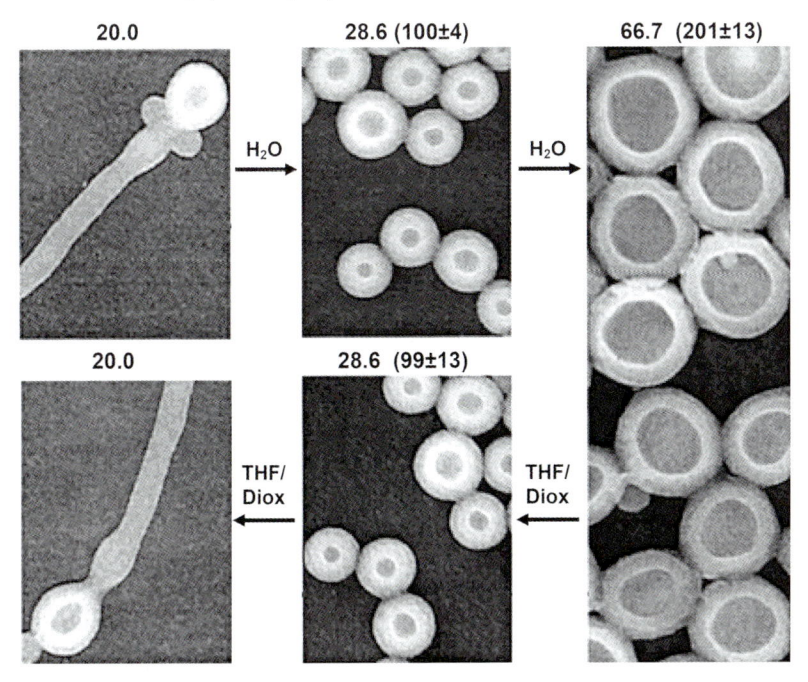

Fig. 1.27 Reversibility of vesicle sizes in response to increasing or decreasing water contents for PS_{300}-b-PAA_{44} vesicles in a THF/Dioxane (44.4/55.6) solvent mixture. Adapted from Ref. [136].

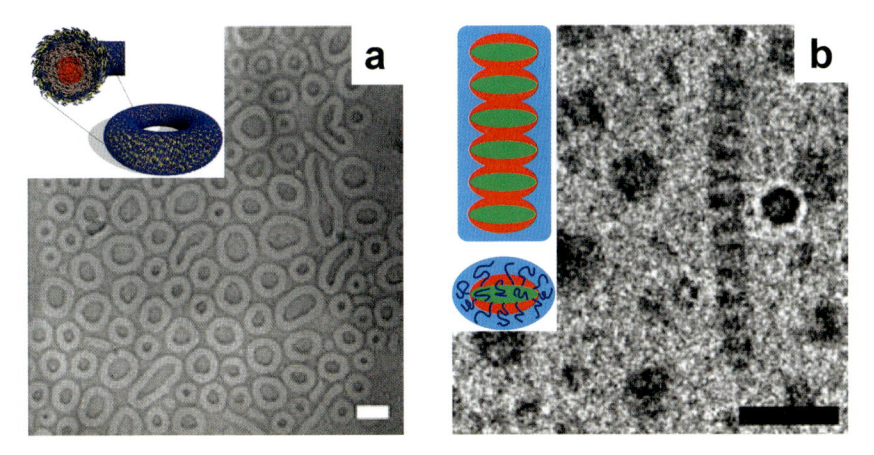

Fig. 1.28 TEM images of (a) toroid micelles prepared from PAA-b-PMA-b-PS triblock copolymers in THF/water (1:2 volume ratio) [140] and (b) segmented rod micelles prepared from PEO-PEE-PFPO miktoarm copolymers [141].

during the evaporation of THF. The initial volume fraction of THF and the evaporation speed strongly affected the final micelle structures, which implies that the toroidal micelles are likely a kinetically trapped structure. Lodge and co-workers investigated the morphological behavior of micelles in the super strong segregation limit for ABC miktoarm copolymers containing a fluorinated polymer block (Fig. 1.28 b) [141]. The segmented rod micelles were observed for miktoarm copolymers consisting of poly(ethylene oxide) (PEO), polyethylethylene (PEE) and poly(perfluoropropylene oxide) (PFPO) in aqueous solution. Formation of this unusual structure was explained from the mutual contributions of the super-hydrophobic block and the structural restriction imposed by a 3 arm star. The super-hydrophobicity of the fluorinated polymer block imparts a large amount of interfacial tension at the junction with other blocks, which may overwhelm the entropic penalties for stretching the shorter blocks. When the corona block is much longer than the inner core and shell blocks, the tendency to form a flat surface by the high interfacial tension induced by super-hydrophobicity will be offset by the fact that the one junction point for three blocks in a miktoarm architecture tends to increase the interfacial curvature.

1.2.3.5 Micelles in Homopolymer Solvent

In the formation of BCP micelles, solid homopolymers can be used as solvents instead of conventional low molar mass solvents. The first theory for the critical micelle concentration in diblock-homopolymer blends was presented by Leibler, Orland and Wheeler [88] and later extended by other researchers [142].

The Thomas group investigated the morphological behavior of micelles as function of copolymer concentration as well as the dependence of the phase on the molecular weight of homopolymer in PS-b-PB/PS blends [143–145]. Similar to the case of a low molar mass solvent, PS-b-PB copolymer dissolved in a PS homopolymer formed micelles beyond the critical micelle concentration of copolymer with a PB core surrounded by a PS corona swollen with the PS homopolymer. The cmc was observed to increase with decreasing homopolymer molecular weight, increasing content of polystyrene in the copolymer, or decreasing overall copolymer molecular weight. Depending on the relative molecular weights of the homopolymer and copolymer, various morphologies were also found including spherical, cylindrical micelles and vesicles (Fig. 1.29). The transitions in micelle geometry were explained in terms of the relative volume fractions of core and corona regions of the micelle. The micelle core volume fraction can be augmented either by directly increasing the core block molecular weight or by increasing the homopolymer molecular weight which decreases the corona thickness as a result of a lower degree of mixing.

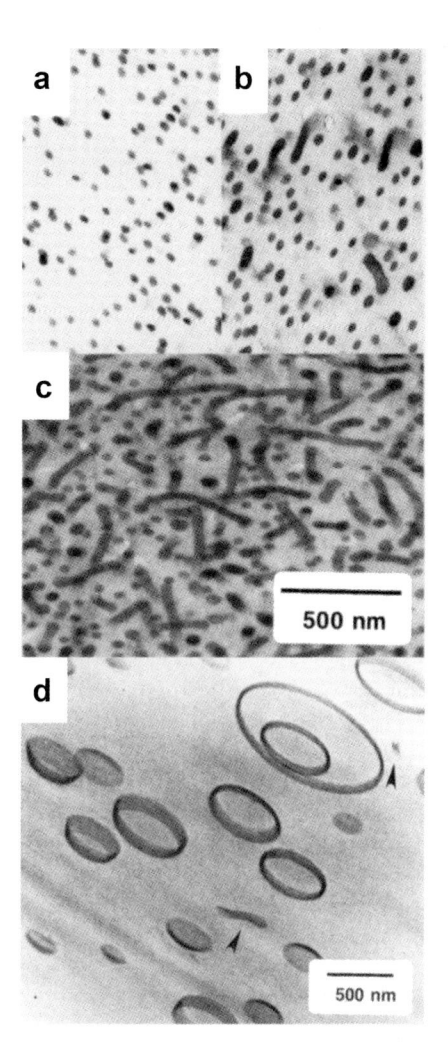

Fig. 1.29 Morphological transition of PS-b-PB micelles from spheres to cylinder, and then to lamella bilayers (vesicles) with increasing homopolymer molecular weight. Adapted from Ref. [143].

1.2.4
Rod–Coil Block Copolymer in Solvents

In the context of controlling the micelle morphology, rod–coil BCP, are of interest. In this case, the rigid rod blocks tend to aggregate into domains possessing liquid-crystalline orders (Fig. 1.30) [146, 147]. Compared to coil–coil copolymer systems, the solution phase behavior of rod–coil copolymers is less established. Some significant theoretical studies on the rod–coil BCPs in a selective solvent have been done by Halperin using a scaling method [91, 148]. Two thermodynamically stable structures of micelles are the extended star-like coronas (i.e. very long coil blocks compared with rod blocks), and the tilted sematic-C lamellar phase.

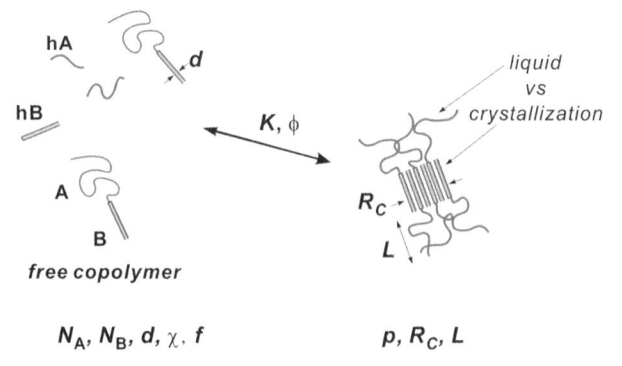

Fig. 1.30 Schematic illustration of micellization from rod–coil block copolymers and key parameters determining micellar structures.

In experimental investigations of rod–coil BCP/solvent systems, strong phase separation was observed, and the resulting structures are very stable, even for low molecular weight copolymers having oligomeric rod blocks, due to strong interactions between rod blocks. Kilbinger et al. found clear evidence for the hockey-puck micelles prepared from highly asymmetric oligo(p-benzamide)-b-poly(ethylene glycol)monomethyl ether ($OPBA_6$-b-$MPEG_{110}$) in chloroform as a selective solvent for the MPEG coil blocks (Fig. 1.31) [149]. The structures were characterized by using dynamic and static light scattering, and AFM techniques. Consistent with the theoretical expectation, bilayer hockey-puck micelles were

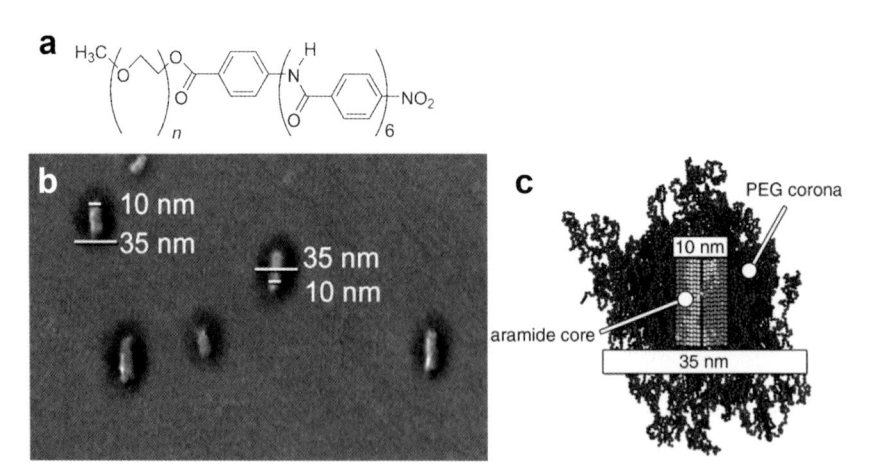

Fig. 1.31 (a) The molecular structure of oligo(p-benzamide)-b-MPEG ($OPBA_6$-b-$MPEG_{110}$) rod–coil block copolymer; (b) AFM phase-contrast tapping-mode image of hockey-puck micelles prepared by depositing a drop of solution on mica; (c) a simulation model of a bilayer hockey-puck micelle. Adapted from Ref. [149].

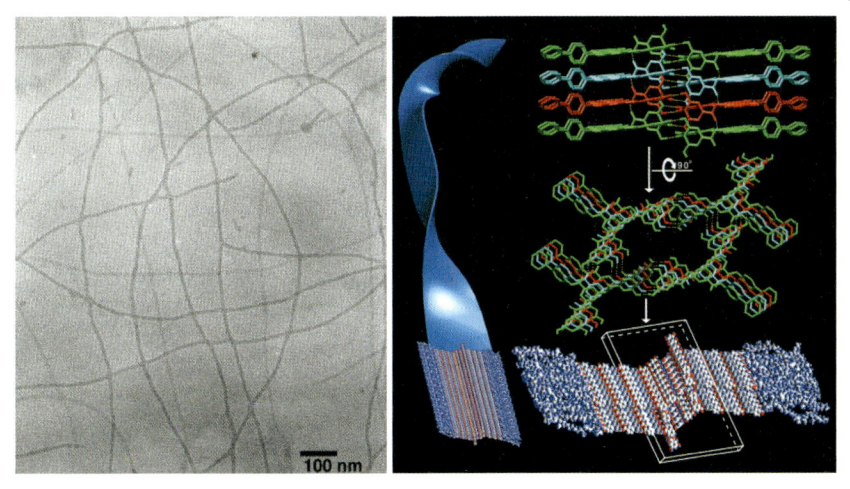

Fig. 1.32 TEM image and schematic illustration of dendron rod–coil supramolecular assembly. Adapted from Ref. [150].

formed due to the strong interactions between the OPBA chains via hydrogen bonding and π–π stacking. Phase-contrast AFM revealed that the shape anisotropy of the micellar core is more pronounced that that of the overall micelle, which is most likely a result of the stretching of the MPEG at the rod–coil boundary.

Another interesting structure was reported by Stupp [150]. In contrast to simple rod–coil molecules, the molecule referred to as a dendron rod–coil has a rod block which has dendritic segments (Fig. 1.32). These dendron rod–coil molecules form a well-defined ribbon-like 1D-nanostructure in CH_2Cl_2. In this case, the bulky geometry of the dendron could frustrate the formation of two-dimensional assemblies. Nonetheless, the identical aromatic rod-dendron segments of the molecules should be strongly driven to aggregate in one dimension through noncovalent interactions. π–π stacking interactions between aromatic segments and directional hydrogen bonding in the dendron-like rod blocks play important roles in the aggregation.

The interplay of the strong interactions of rod blocks and solvents may lead many rod–coil copolymers to form the kinetically trapped structures. For example, coil–rod–coil triblock copolymers consisting of PEO-*b*-PPP-*b*-PEO in water showed a very slow evolution, even at very low molecular weights: the oligomeric PEO-*b*-PPP-*b*-PEO formed spherical micelles at the initial stage, and then these micelles gradually changed into the cylindrical morphologies on the time scale of a week [151]. The kinetics of the transformations could be very slow when the copolymer concentrations are very low [152].

1.2.5
Inorganic Hybrid Micelles

BCP micelles have been widely used both to synthesize inorganic nanoparticles within the micelle core regions and to stabilize colloidal particles. Numerous approaches for synthesizing metallic, semiconducting or magnetic nanoparticles with various shapes have been reported, and this field has been recently reviewed [83, 153]. Nanoparticles are synthesized by reducing metal salt precursors which were stabilized by ligand-like micelle core blocks such like PAA, P4VP, P2VP, PVA and PEO [154–156]. For example, due to the ability of the pyridine moiety to form complexes with various metal ions, BCPs containing polyvinylpyridine block have been used to make gold, platinum, rhodium and cobalt nanoparticles from organic solvents [154, 157–159]. In this case, the resulting nanoparticle shapes were dictated by the micelle morphologies. Recent results, however, showed that metal nanoparticles nucleated by slow reduction may cause a shape transitions of micelles [156, 160]. Sidorov et al. investigated

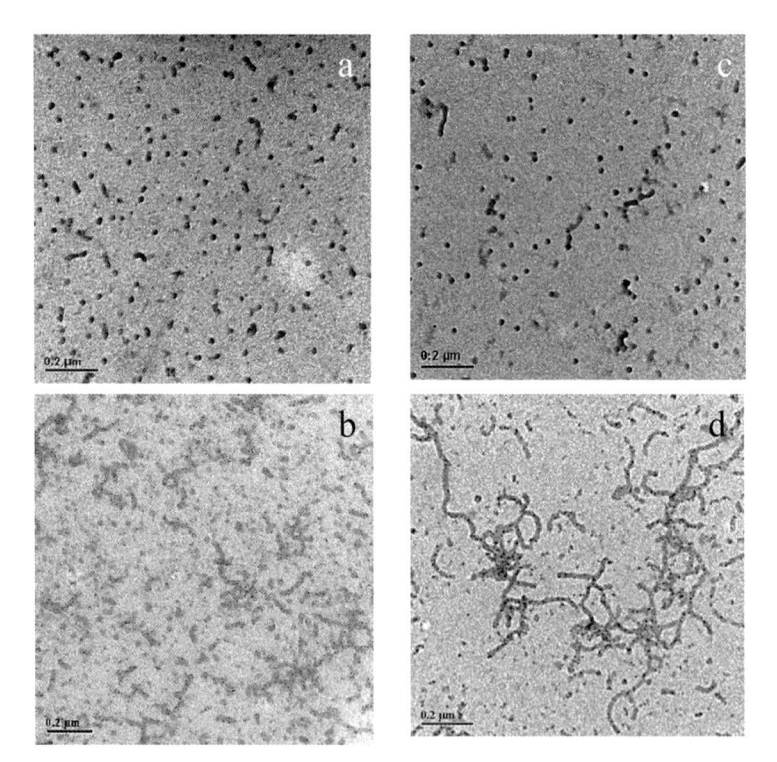

Fig. 1.33 TEM images of metalated block copolymer micelles based on as-prepared PEO45-b-P4VP28 after incorporation of HAuCl$_4$·H$_2$O (a, c) and after Au nanoparticle formation (b, d). In images a and b, molar ratio 4VP:Au=8:1, while in images c and d 4VP:Au=4:1. Adapted from the Ref. [160].

the effect of metallization on the morphology of micelles in the PEO-*b*-P4VP/ water system [160]. Micelles were compared before and after metallization of the gold salts within the micelles (Fig. 1.33). While spherical micelles were observed before the metallization, cylindrical micelles were observed after the metallization, and this phase transition was more prominent at the higher concentration of gold salt.

Similarly, Gerardin et al. also investigated the effect of the lanthanum ion concentration on the morphology of micelles [156]. In the synthesis of lanthanum hydroxide nanoparticles from poly(acrylic acid)-*b*-poly(acrylamide) (PAA-*b*-PAM) micelles in aqueous solution, the acrylic acid/lanthanum molar ratio ([AA]/[La^{3+}]) ranged from 0.1 to 1. Before reducing the lanthanum ions, all micelles were characterized as spherical micelles by cryo-TEM and light scattering techniques. However rod-like structures were observed after reducing lanthanum hydroxide nanoparticles, especially when the concentration of lanthanum ions was high. These phase transitions were attributed to the change in salt concentration or pH while the nanoparticles were formed as well as the polymer chain deformation by the nanoparticles [160]. Crosslinking of micelles before formation of nanoparticles could prevent the phase transition during the formation of nanoparticles [161].

1.3
Summary and Outlook

Self-assembly plays a key role in current research in chemistry, physics, biology and materials. BCP self-assembly bridges between different research fields. Block copolymers have long been studied as model building blocks to obtain a fundamental understanding of the self-assembly process and the factors affecting the morphology of assemblage in both melt and solution phases. Moreover self-assembled BCPs are also becoming technologically important materials beyond their conventional applications. Current understanding of the phase behavior of block copolymers has been established with parallel advances in synthetic methods for precise control of copolymer composition and architecture. Characterization tools, including TEM and SAXS, are able to identify domain structural information in the nanoscopic regime. As reviewed here, the phase behavior of simple systems consisting of diblock copolymers are currently very well understood in both melt and solution phases, and has been mapped as a function of polymer composition, molecular weight or concentration. However, many more complicated systems have not yet been fully investigated. For example, ABC triblock copolymers considerably enrich the phase library with new morphologies which are not observed for diblock copolymer systems. Because of the larger number of variables (block sequence, topology, composition etc.), it will be a challenge to map the phase behavior as a function of these variables. Blending block copolymers with inorganic nanoparticles provides an even more complex quaternary system consisting of AB diblock copolymers, nanoparticles, surfactants and solvent. As reviewed here, the surface density of surfactants on nanoparticles as well as the interactions among particles, surfactants and copolymers must be carefully considered in order to understand phase transitions in the system. In addition, particle–particle interactions and packing for dimensionally and morphologically mismatched particles should lead to novel structures and properties.

Block copolymer micelles have been of great interest due to their many potential applications. The initial emphasis has been mainly on spherical micelles, with an increased emphasis recently on worm-like, vesicles, toroidal structures. The experimental investigation of micellar morphology, requires considerable effort to examine genuine structures and avoid the many artifacts. Understanding how to form particular structures and manipulate them via phase transitions remains an important challenge. Most work on micelles has been in aqueous media; examples in organic media are still very limited. In this context, development of explicit characterization techniques such as cryo-TEM for micelles in organic solvents would be advantageous. Due to their nature as discrete nanocages with high stability, block copolymer micelles have been employed in many applications: for example, nanoparticle synthesis as templates and drug delivery vehicles. Further detailed applications will be extensively handled in the next volume (Volume IV) of this series.

Acknowledgments

The authors would like to thank Professor Ludwik Leibler for his kind invitation to contribute this chapter. This work was supported by the National Science Foundation (NSF) under grant No. 0308133 and the U.S. Army through the Institute for Soldier Nanotechnologies (ISN) under contract number DAAD-19-02-D0002. The content of the information does not necessarily reflect the position or the policy of the Government, and no official endorsement should be inferred.

References

1 M. R. Bockstaller, R. A. Mickiewicz, E. L. Thomas, *Adv. Mater.* **17**, 1331 (2005).
2 C. Park, J. Yoon, E. L. Thomas, *Polymer* **44**, 6725 (2003).
3 M. L. M. Szwarc, R. Milkovich, *J. Am. Chem. Soc.* **78** (1956).
4 E. Helfand, Z. R. Wasserman, *Macromolecules* **11**, 960 (1978).
5 E. Helfand, *J. Chem. Phys.* **62**, 999 (1975).
6 F. S. Bates, G. H. Fredrickson, *Phys. Today* **52**, 32 (Feb.1999).
7 I. W. Hamley, *The Physics of Block Copolymers*, Oxford science publications, Oxford University Press, Oxford, 1998, 432 pp.
8 L. Leibler, *Macromolecules* **13**, 1602 (1980).
9 M. W. Matsen, M. Schick, *Phys. Rev. Lett.* **72**, 2660 (1994).
10 G. Fredrickson, *The Equilibrium Theory of Inhomogeneous Polymers*, The International Series of Monographs on Physics, Oxford University Press, New York, 2006.

11 A. Urbas et al., *Adv. Mater.* **12**, 812 (2000).
12 Y. Fink, A.M. Urbas, M.G. Bawendi, J.D. Joannopoulos, E.L. Thomas, *J. Lightwave Technol.* **17**, 1963 (1999).
13 H. Watanabe, T. Kotaka, *Macromolecules* **16**, 769 (1983).
14 R.E. Cohen, D.E. Wilfong, *Macromolecules* **15**, 370 (1982).
15 K.I. Winey, E.L. Thomas, L.J. Fetters, *J. Chem. Phys.* **95**, 9367 (1991).
16 K.I. Winey, E.L. Thomas, L.J. Fetters, *Macromolecules* **24**, 6182 (1991).
17 K.I. Winey, E.L. Thomas, L.J. Fetters, *Macromolecules* **25**, 2645 (1992).
18 K.I. Winey, E.L. Thomas, L.J. Fetters, *Macromolecules* **25**, 422 (1992).
19 A. Haryono, W.H. Binder, *Small* **2**, 600 (2006).
20 J. Huh, V.V. Ginzburg, A.C. Balazs, *Macromolecules* **33**, 8085 (2000).
21 T. Goldacker, V. Abetz, R. Stadler, I. Erukhimovich, L. Leibler, *Nature* **398**, 137 (1999).
22 D.A. Norman, L. Kane, S.A. White, S.D. Smith, R.J. Spontak, *J. Mater. Sci. Lett.* **17**, 545 (1998).
23 T.M. Birshtein, E.B. Zhulina, A.A. Polotsky, V. Abetz, R. Stadler, *Macromol. Theory Simul.* **8**, 151 (1999).
24 S.W. Sides, G.H. Fredrickson, *Polymer* **44**, 5859 (2003).
25 P.K. Janert, M. Schick, *Macromolecules* **31**, 1109 (1998).
26 M.W. Matsen, *Macromolecules* **28**, 5765 (1995).
27 A.N. Semenov, *Macromolecules* **26**, 2273 (1993).
28 L. Leibler, H. Orland, J.C. Wheeler, *J. Chem. Phys.* **79**, 3550 (1983).
29 I. Borukhov, L. Leibler, *Phys. Rev. E* **62**, R41 (2000).
30 H.R. Brown, K. Char, V.R. Deline, *Macromolecules* **23**, 3383 (1990).
31 T.A. Vilgis, J. Noolandi, *Macromolecules* **23**, 2941 (1990).
32 J. Bodycomb, D. Yamaguchi, T. Hashimoto, *Macromolecules* **33**, 5187 (2000).
33 D.A. Hajduk, R.M. Ho, M.A. Hillmyer, F.S. Bates, K. Almdal, *J. Phys. Chem. B* **102**, 1356 (1998).
34 R.L. Lescanec, L.J. Fetters, E.L. Thomas, *Macromolecules* **31**, 1680 (1998).
35 J. Suzuki, M. Furuya, M. Inuma, A. Takano, Y. Matsushita, *J. Polym. Sci. Part B-Polym. Phys.* **40**, 1135 (2002).
36 W.H. Binder, C. Kluger, C.J. Straif, G. Friedbacher, *Macromolecules* **38**, 9405 (2005).
37 R.R. Bhat, J. Genzer, B.N. Chaney, H.W. Sugg, A. Liebmann-Vinson, *Nanotechnology* **14**, 1145 (2003).
38 A.K. Boal et al., *Nature* **404**, 746 (2000).
39 R. Saito, S. Okamura, K. Ishizu, *Polymer* **33**, 1099 (1992).
40 Y.N.C. Chan, G.S.W. Craig, R.R. Schrock, R.E. Cohen, *Chem. Mater.* **4**, 885 (1992).
41 P.F.W. Simon, R. Ulrich, H.W. Spiess, U. Wiesner, *Chem. Mater.* **13**, 3464 (2001).
42 R.B. Thompson, V.V. Ginzburg, M.W. Matsen, A.C. Balazs, *Science* **292**, 2469 (2001).
43 M.R. Bockstaller, Y. Lapetnikov, S. Margel, E.L. Thomas, *J. Am. Chem. Soc.* **125**, 5276 (2003).
44 J.J. Chiu, B.J. Kim, E.J. Kramer, D.J. Pine, *J. Am. Chem. Soc.* **127**, 5036 (2005).
45 B.J. Kim, J. Bang, C.J. Hawker, E.J. Kramer, *Macromolecules* **39**, 4108 (2006).
46 B.J. Kim, J.J. Chiu, G.R. Yi, D.J. Pine, E.J. Kramer, *Adv. Mater.* **17**, 2618 (2005).
47 S.W. Yeh, K.H. Wei, Y.S. Sun, U.S. Jeng, K.S. Liang, *Macromolecules* **38**, 6559 (2005).
48 R.M. Ho et al., *Macromolecules* **38**, 8607 (2005).
49 Q. Wang, P.F. Nealey, J.J. de Pablo, *J. Chem. Phys.* **118**, 11278 (2003).
50 J.Y. Lee, Z.Y. Shou, A.C. Balazs, *Macromolecules* **36**, 7730 (2003).
51 B. de Boer et al., *Polymer* **42**, 9097 (2001).
52 J.A. Osaheni, S.A. Jenekhe, *J. Am. Chem. Soc.* **117**, 7389 (1995).
53 M. Lee et al., *Adv. Mater.* **13**, 1363 (2001).
54 H.A. Klok, S. Lecommandoux, *Adv. Mater.* **13**, 1217 (2001).
55 M. Lee, B.K. Cho, W.C. Zin, *Chem. Rev.* **101**, 3869 (2001).
56 G. Mao, C.K. Ober, *Acta Polym.* **48**, 405 (1997).
57 A.N. Semenov, *Mol. Cryst. Liq. Cryst.* **209**, 191 (1991).

58 A. Halperin, *Macromolecules* **23**, 2724 (1990).
59 D. R. M. Williams, G. H. Fredrickson, *Macromolecules* **25**, 3561 (1992).
60 V. Pryamitsyn, V. Ganesan, *J. Chem. Phys.* **120**, 5824 (2004).
61 B. D. Olsen, R. A. Segalman, *Macromolecules* **38**, 10127 (2005).
62 B. Gallot, *Prog. Polym. Sci.* **21**, 1035 (1996).
63 H. Schlaad, B. Smarsly, I. Below, *Macromolecules* **39**, 4631 (2006).
64 J. P. Billot, A. Douy, B. Gallot, *Makromol. Chem.-Macromol. Chem. Phys.* **177**, 1889 (1976).
65 A. Douy, B. Gallot, *Polymer* **28**, 147 (1987).
66 H. A. Klok, J. F. Langenwalter, S. Lecommandoux, *Macromolecules* **33**, 7819 (2000).
67 S. Lecommandoux, M. F. Achard, J. F. Langenwalter, H. A. Klok, *Macromolecules* **34**, 9100 (2001).
68 J. T. Chen, E. L. Thomas, C. K. Ober, S. S. Hwang, *Macromolecules* **28**, 1688 (1995).
69 J. T. Chen, E. L. Thomas, C. K. Ober, G. P. Mao, *Science* **273**, 343 (1996).
70 K. K. Tenneti et al., *J. Am. Chem. Soc.* **127**, 15481 (2005).
71 Y. F. Tu et al., *Macromolecules* **36**, 6565 (2003).
72 L. H. Radzilowski, S. I. Stupp, *Macromolecules* **27**, 7747 (1994).
73 M. Lee, B. K. Cho, H. Kim, J. Y. Yoon, W. C. Zin, *J. Am. Chem. Soc.* **120**, 9168 (1998).
74 J. N. Israelachvili, *Intermolecular and Surface Forces*, Academic Press, London, 1992.
75 B. Lindman, P. Alexandridis, *Amphiphilic Block Copolymers : Self-assembly and Applications*, Elsevier, Amsterdam, 1st edn., 2000, pp. xii, 435.
76 I. W. Hamley, *The Physics of Block Copolymers*, Oxford science publications, Oxford University Press, Oxford, 1998, 432 pp.
77 K. Khougaz, X. F. Zhong, A. Eisenberg, *Macromolecules*, 3937 (1996).
78 Z. Tuzar et al. *Surface and Colloid Science*, E. Matijevic (ed.), Plenum Press, New York, 1993, vol 15.
79 Y.-Y. Won, H. T. Davis, F. S. Bates, *Macromolecules* **36**, 953 (2003).
80 C. Price, in *Developments in Block Copolymers-1*, I. Goodman (Ed.), Applied Science, London, 1982, vol. 1, pp. 39–80.
81 I. Piirma, *Surfactant Science Series, Vol. 42: Polymeric Surfactants*, 1992, 289 pp.
82 P. Munk, in *NATO ASI Series, Series E: Applied Sciences* S. E. Webber, P. Munk, Z. Tuzar, (Eds.), Kluwer Academic Publishers, London, 1996, vol. 327, pp. 19–32.
83 S. Foerster, M. Antonietti, *Adv. Mater.* **10**, 195 (1998).
84 S. Forster, T. Plantenberg, *Angew. Chem., Int. Ed.* **41**, 688 (2002).
85 G. Riess, *Prog. Polym. Sci.* **28**, 1107 (2003).
86 J.-F. Gohy, *Adv. Polym. Sci.* **190**, 65 (2005).
87 I. W. Hamley, *Block Copolymers in Solutions: Fundamentals and Applications*, J. Wiley & Sons Ltd., Chichester, 2005.
88 L. Leibler, H. Orland, J. C. Wheeler, *J. Chem. Phys.* **79**, 3550 (1983).
89 P. G. de Gennes, in *Solid State Physics*, L. Leibler, (Ed.), Academic Press, New York, 1978, vol. 14, pp. 1–18
90 M. Daoud, J. P. Cotton, *J. Phys. (Paris)* **43**, 531 (1982).
91 A. Halperin, *Europhys. Lett.* **10**, 549 (1989).
92 E. B. Zhulina, T. M. Birshtein, *Vysokomolekulyarnye Soedineniya, Ser. A* **27**, 511 (1985).
93 C. Booth, D. Attwood, *Macromol. Rapid Commun.* **21**, 501 (2000).
94 S. Pispas, N. Hadjichristidis, I. Potemkin, A. Khokhlov, *Macromolecules* **33**, 1741 (2000).
95 H. Iatrou, L. Willner, N. Hadjichristidis, A. Halperin, D. Richter, *Macromolecules* **29**, 581 (1996).
96 E. Minatti et al., *Macromol. Rapid Commun.* **23**, 978 (2002).
97 R. Borsali et al., *Langmuir* **19**, 6 (2003).
98 E. Leontidis, *Curr. Opin. Colloid Interface Sci.* **7**, 81 (2002).
99 P. Bahadur, K. Pandya, M. Almgren, P. Li, P. Stilbs, *Colloid. Polym. Sci.* **271**, 657 (1993).
100 P. Bahadur, P. Li, M. Almgren, W. Brown, *Langmuir* **8**, 1903 (1992).
101 P. Alexandridis, J. F. Holzwarth, *Langmuir* **13**, 6074 (1997).

102 J. P. Mata, P. R. Majhi, C. Guo, H. Z. Liu, P. Bahadur, *J. Colloid Interface Sci.* **292**, 548 (2005).

103 H. Ankleshwaria, J. Mata, P. Bahadur, *Tenside, Surfactants, Detergents* **40**, 303 (2003).

104 Y.-Y. Won, H. T. Davis, F. S. Bates, *Science* **283**, 960 (1999).

105 K. Iyama, T. Nose, *Polymer* **39**, 651 (1998).

106 C. Price, N. Briggs, J. R. Quintana, R. B. Stubbersfield, I. Robb, *Polym. Commun.* **27**, 292 (1986).

107 Z. Tuzar, P. Kratochvil, K. Prochazka, P. Munk, *Coll. Czech. Chem. Commun.* **58**, 2362 (1993).

108 Z. Zhou, B. Chu, *Macromolecules* **20**, 3089 (1987).

109 Z. Zhou, B. Chu, *Macromolecules* **21**, 2548 (1988).

110 T. P. Lodge, *Macromol. Chem. Phys.* **204**, 265 (2003).

111 T. P. Lodge et al., *Langmuir* **19**, 2103 (2003).

112 Y. Talmon, *J. Colloid Interface Sci.* **93**, 366 (1983).

113 P. K. Vinson, Y. Talmon, *J. Colloid Interface Sci.* **133**, 288 (1989).

114 E. J. Roche, E. L. Thomas, *Polymer* **22**, 333 (1981).

115 K. Prochazka, M. K. Baloch, Z. Tuzar, *Makromol. Chem.* **180**, 2521 (1979).

116 Z. Tuzar et al., *Makromol. Chem.* **183**, 399 (1982).

117 D. J. Wilson, G. Riess, *Eur. Polym. J.* **24**, 617 (1988).

118 A. Guo, G. Liu, J. Tao, *Macromolecules* **29**, 2487 (1996).

119 K. L. Wooley, *J. Polym. Sci., Part A: Polym. Chem.* **38**, 1397 (2000).

120 Q. Ma, E. E. Remsen, C. G. Clark, Jr., T. Kowalewski, K. L. Wooley, *Proc. Natl. Acad. Sci. U. S. A.* **99**, 5058 (2002).

121 G. H. Fredrickson, L. Leibler, *Macromolecules* **22**, 1238 (1989).

122 C.-I. Huang, T. P. Lodge, *Macromolecules* **31**, 3556 (1998).

123 M. D. Whitmore, J. Noolandi, *J. Chem. Phys.* **93**, 2946 (1990).

124 K. M. Hong, J. Noolandi, *Macromolecules* **16**, 1083 (1983).

125 P. Alexandridis, *Macromolecules* **31**, 6935 (1998).

126 E. B. Zhulina, M. Adam, I. LaRue, S. S. Sheiko, M. Rubinstein, *Macromolecules* **38**, 5330 (2005).

127 R. Wang, P. Tang, F. Qiu, Y. Yang, *J. Phys. Chem. B* **109**, 17120 (2005).

128 K. J. Hanley, T. P. Lodge, C.-I. Huang, *Macromolecules* **33**, 5918 (2000).

129 E. L. Thomas, D. J. Kinning, D. B. Alward, C. S. Henkee, *Macromolecules* **20**, 2934 (1987).

130 P. Alexandridis, R. J. Spontak, *Curr. Opin. Colloid Interface Sci.* **4**, 130 (1999).

131 S. Jain, F. S. Bates, *Science* **300**, 460 (2003).

132 D. E. Discher, A. Eisenberg, *Science* **297**, 967 (2002).

133 K. Yu, A. Eisenberg, *Macromolecules* **29**, 6359 (1996).

134 H. Shen, A. Eisenberg, *J. Phys. Chem. B* **103**, 9473 (1999).

135 L. Zhang, A. Eisenberg, *J. Am. Chem. Soc.* **118**, 3168 (1996).

136 L. Luo, A. Eisenberg, *Langmuir* **17**, 6804 (2001).

137 L. Luo, A. Eisenberg, *J. Am. Chem. Soc.* **123**, 1012 (2001).

138 I. W. Hamley, *The Physics of Block Copolymers*, Oxford science publications, Oxford University Press, Oxford, 1998, 432 pp.

139 L. Zhang, A. Eisenberg, *Macromolecules* **29**, 8805 (1996).

140 D. J. Pochan et al., *Science* **306**, 94 (2004).

141 Z. Li, E. Kesselman, Y. Talmon, M. A. Hillmyer, T. P. Lodge, *Science* **306**, 98 (2004).

142 A. M. Mayes, M. Olvera de la Cruz, *Macromolecules* **21**, 2543 (1988).

143 D. J. Kinning, K. I. Winey, E. L. Thomas, *Macromolecules* **21**, 3502 (1988).

144 D. J. Kinning, E. L. Thomas, L. J. Fetters, *J. Chem. Phys.* **90**, 5806 (1989).

145 D. J. Kinning, E. L. Thomas, L. J. Fetters, *Macromolecules* **24**, 3893 (1991).

146 L. Onsager, *Ann. N. Y. Acad. Sci.* **51**, 627 (1949).

147 P. J. Flory, *J. Chem. Phys.* **17**, 303 (1949).

148 A. Halperin, *Macromolecules* **23**, 2724 (1990).

149 T. W. Schleuss et al., *Angew. Chem., Int. Ed.* **45**, 2969 (2006).

150 E. R. Zubarev, M. U. Pralle, E. D. Sone, S. I. Stupp, *J. Am. Chem. Soc.* **123**, 4105 (2001).
151 J.-H. Ryu, M. Lee, *J. Am. Chem. Soc.* **127**, 14170 (2005).
152 S. Svenson, P. B. Messersmith, *Langmuir* **15**, 4464 (1999).
153 I. W. Hamley, *Nanotechnology* **14**, R39 (2003).
154 F. S. Diana, S.-H. Lee, P. M. Petroff, E. J. Kramer, *Nano Lett.* **3**, 891 (2003).
155 J. P. Spatz, A. Roescher, M. Moeller, *Adv. Mater.* **8**, 337 (1996).
156 F. Bouyer, C. Gerardin, F. Fajula, J. L. Putaux, T. Chopin, *Colloid. Surf., A: Physicochem. Eng. Aspects* **217**, 179 (2003).
157 M. V. Seregina et al., *Chem. Mater.* **9**, 923 (1997).
158 O. A. Platonova et al., *Colloid Polym. Sci.* **275**, 426 (1997).
159 M. Antonietti, E. Wenz, L. Bronstein, M. Seregina, *Adv. Mater.* **7**, 1000 (1995).
160 S. N. Sidorov et al., *Langmuir* **20**, 3543 (2004).
161 Z. Li, G. Liu, *Langmuir* **19**, 10480 (2003).
162 M. W. Matsen, F. S. Bates, *Macromolecules* **29**, 1091 (1996).
163 E. L. Thomas, J. T. Chen, M. J. E. Orourke, C. K. Ober, G. P. Mao, *Macromol. Symp.* **117**, 241 (1997).

2
Simulations

Denis Andrienko and Kurt Kremer

2.1
Introduction

This chapter will give a short introduction to computer simulations of polymers, or soft matter in general. It should serve as a guide to the literature and also provide a first impression of the opportunities but also of the current limitations.

Polymeric materials can be crystalline, amorphous (glasses, rubbers, melts, gels) or even solutions. Especially polymer melts in the glassy state are standard materials for many applications (yoghurt cups, compact discs, housings of technical equipment etc.). They often combine relatively low specific weight and ductibility with processing at moderate temperatures. In the melt state, polymers are viscoelastic liquids where the crossover from elastic to viscous behavior is determined by the local chemical structure and the chain length. Added to a solvent, polymers can be used as viscosity modifiers and, depending on parameters, lead to either shear thickening or shear thinning, as used for drag reduction. Cross-linking chains into a disordered network results in gels or rubber. Applications range from gels in (low-fat) food, hydrogels in modern body care (diapers/nappies) via biological systems (cytoskeleton) all the way to classical elastomers (car tires), to name just a few. Here, the interplay of connectivity, chain length and local structure parameters, such as stiffness, determines the properties.

This versatility of physical properties is based on the many different chemical molecular building blocks as well as on various molecular architectures and huge differences in molecular weights of polymers. It is the combination and the rather delicate interplay of local chemical with more global architectural and size properties, which makes macromolecules so versatile and interesting. This means that many different length and time scales are relevant, and that understanding the properties on one scale is not at all sufficient to understand material properties. The consequence of this combination of generic and system-specific properties together with the huge variation in chemistry makes it impossi-

Macromolecular Engineering. Precise Synthesis, Materials Properties, Applications.
Edited by K. Matyjaszewski, Y. Gnanou, and L. Leibler
Copyright © 2007 WILEY-VCH Verlag GmbH & Co. KGaA, Weinheim
ISBN: 978-3-527-31446-1

ble to generate a thorough theoretical understanding based just on analytical theory. While this typically treats some idealized or limiting cases, computer simulations are supposed to bridge the gap between the (usually) approximate solutions of idealized models and their "realistic" properties and also to establish a link to the chemistry-dependent details. Of course, one still has to go a long way to reach this goal in general, however, over the last years remarkable success was achieved due to an increase in computer power and especially due to improved simulation methods. It is the aim of the present chapter to provide a short introduction to this field.

Before going into the technical details, it is important to have a general overview and understanding of the advantages and disadvantages of the simulation methods in general. Hence we will address in more detail the question: Why do we need to carry out computer simulations at all?

First of all, computer simulations are carried out with the aim of understanding the assemblies of macromolecules in terms of their structure. Second, for this we eventually have to be able to bridge microscopic length and time scales all the way to the macroscopic world: starting from the interaction between the molecules, which can be an *ad hoc* ansatz for simple models or a parameterized force field (see below), predict the bulk properties of the material. We may also test a theory by conducting a simulation using the same model; we may carry out simulations under extreme conditions that are difficult or impossible in the laboratory [1].

All this becomes possible because simulations can provide the ultimate detail concerning individual conformational statistics of the macromolecules as well as in many cases particle motions as a function of time. That is, the properties of a model system can often be addressed more easily in much more detail than by doing experiments on the actual system. Of course, the experiments still test the calculated results and provide criteria for improving the model and methodology [2].

Roughly speaking there are two important types of simulation methods. The first uses simulations to sample conformation space. To do this, a variety of Monte Carlo and molecular dynamics methods are available, often combined with simulated annealing methods, to determine or refine structures and morphology of the system. This leads to the description of the system at equilibrium, including structural and thermodynamic parameters, as well as the typical soft matter fluctuations. Typical polymeric systems display strong thermal conformational fluctuations. The relevant scale of energy is the thermal energy $k_B T$, which at room temperature is only about 4.1×10^{-21} J or 2.5 kJ mol^{-1} and the intramolecular conformational entropy is proportional to the overall chain length. For such calculations, it is important to sample conformation space very carefully. The second area uses simulations to study actual dynamics. Here not only the sampling of the configuration space is required, but also it is important to have the correct development of the system over time.

For the first area, a variety of Monte Carlo (MC) and molecular dynamics (MD) simulations can be used. By contrast, in the second area molecular dy-

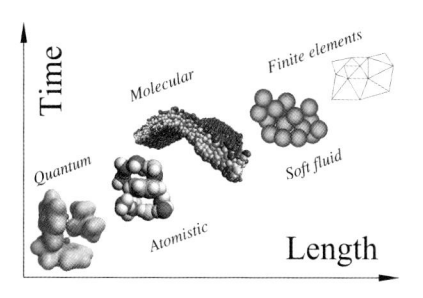

Fig. 2.1 A general multiscale picture of time and length scales involved in computer simulation of macromolecules.

namics is the first natural choice to provide the necessary information. However, properly adapted MC methods can also be used in cases where the motion is well characterized as a Brownian motion. The advantage of MD over MC is that it gives a more straightforward route to dynamic properties of the system such as transport coefficients, time-dependent responses to perturbations, rheological properties, etc.

Computer simulations are done on rather different levels of description (Fig. 2.1). The most rigorous tool involving utmost degrees of freedom is quantum chemistry (electrons); reduction of the degrees of freedom results in atomistic models (potential energy functions, force fields); further reduction leads to mesoscopic models (monomers serve as building blocks); entire polymer chains (soft fluid models), and finally self-consistent field theories, with the density as a degree of freedom (finite elements, etc.). The various levels of description are developed to different degrees and have been applied for simulation of solid-state and soft condensed matter systems for many years.

In the case of polymers we must deal with systems containing several to many thousands of atoms. It is computationally prohibitive to simulate such systems either on the quantum chemical *ab initio* level, or at the atomistic level to the extent required. Even coarse-grained, simplified, models with a significantly reduced number of degrees of freedom cannot treat systems of a realistic (i.e. used by experimentalists) size. However, with the further downsizing of experimental parameters and increase in simulation capabilities the two fields are coming together. Therefore, currently the main problem in many polymer simulations is to bridge different time and length scales, that is, to provide links which would connect polymer structure to macroscopic properties or parameters entering macroscopic theories. Equally important of course is to understand which features are being ignored at which level of coarse-graining.

The key to success in linking the different scales is to combine the knowledge of the universal scaling relations with the computer simulation methods in order to relate chemically specific structure of the polymer to its prefactor in a scaling law predicting universal behavior. In particular, in case of a single-chain behavior one focuses on the atomic-scale mechanisms of *flexibility*, which distinguish one polymer from another. In dense systems, *packing* is an important aspect of specificity, since it is partially responsible for the glass transition temperature and crystallizability. In the case of polymer melts, *entanglement* molecular weights, whose existence is a *universal* property, but whose value is specific, sets characteristic relaxation times and, therefore, defines viscosities and elastic moduli of the systems. Universality tells us that important aspects of polymer structure and dynamics do not depend on atomic-level detail, or at least can be adsorbed in a set of specific properties, such as Kuhn length, entanglement length, etc. If we understand how to apply this, universality can reduce the complexity needed to model specific polymers.

The concept of universality can be explained by considering such an important property as the size of a single polymer chain. Imagine an "ideal" chain of N segments linked together by more or less flexible bonds. The typical size of such a chain (for example, its average end-to-end distance or radius of gyration) scales as some power of N [3, 4]:

$$R = AN^\nu \tag{1}$$

where ν is a scaling exponent. This is a universal property of polymers, and the exponent ν depends only on the environment. For an ideal chain or a polymer melt (random walk) $\nu = 1/2$. For a more realistic chain with excluded volume (self-avoiding walk) $\nu \approx 3/5$. The prefactor A, however, can vary strongly, depending on the size of the segments and the stiffness of the bonds.

A second example concerns the shear viscosity η of a polymer melt. If one changes the process temperature of a bisphenol A polycarbonate (BPA-PC) melt from 500 to 470 K, the viscosity rises by a factor of 10 (the glass transition temperature is $T_G \approx 420$ K). On the other hand, increasing the chain length by a factor of 2 also shifts the viscosity by a factor of 10, since for melts of long chains one observes $\eta \sim N^{3.4}$. This power law is again a universal property of linear polymers, and holds for all known polymers independent of the chemical structure of the backbone. Thus in the simplest case of linear polymers one can write

$$\eta = BN^{3.4} \tag{2}$$

These two examples demonstrate that scaling relations separate the problem into a generic contribution (scaling function) and a material specific contribution, that is prefactors A, B in Eqs. (1) and (2) which are based on local properties determined on a nanometer or Ångstrom length scale. Both contributions can be the origin of viscosity or chain size variations over many decades. Hence the notion of specificity lies in understanding the prefactors in Eqs. (1) and (2) and in many other scaling relations. Our task is to relate the chemically specific structure to its prefactor in a scaling law. In many cases a direct comparison to an experiment allows to fix the prefactor (e.g. for the chain extension). In more complicated cases we need to understand atomic-scale mechanisms of flexibility, packing, etc. This holds especially for modern nanoscopic molecular assemblies and many biological molecules, where this clear separation of length scales and thus of generic and system-specific aspects is often missing.

In the following sections we will briefly review the background coming from the generic models and then, later, go back to the question of the extent to which chemistry-specific aspects play a role.

2.2
Basic Techniques: Monte Carlo and Molecular Dynamics

There are two basic concepts that are used in computer simulations of materials. A conceptually direct approach is the molecular dynamics (MD) method. One numerically solves Newton's equations of motion for a collection of particles, which interact via a suitable interaction potential. Through the equations of motion, a natural time scale is built in. The simulation samples phase space (in principle) *deterministically*. The second approach, the Monte Carlo method, samples phase space *stochastically*. Monte Carlo is intrinsically stable but has no inbuilt time scale. In certain cases there are ways to rectify this. These two very different basic approaches are both used to explore the statistical properties of systems and materials. Nowadays many applications employ hybrid methods, where combinations of both are used.

2.2.1
Molecular Dynamics (MD)

MD simulations date back to the early 1950s. For a detailed overview see [5]. Consider a cubic box of volume $L^3 = V$ containing N identical particles of mass m. In order to avoid surface effects and (as much as possible) finite size effects, one typically uses periodic boundary conditions; the particle number density is $\rho = N/L^3$. The first simulations employed hard spheres, but we here consider a potential $U(\mathbf{r}_{ij})$, with $\mathbf{r}_{ij}$ being the distance vector between particles i and j. The straightforward MD approach is now to integrate Newton's equations of motion:

$$m_i \ddot{\mathbf{r}}_i = -\mathbf{\nabla}_i \sum_{j \neq i} U(\mathbf{r}_{ij}) \tag{3}$$

Since energy in such a simulation is conserved, we generate the microcanonical ensemble. The average kinetic energy defines the temperature T via $\sum_i m_i |\dot{\mathbf{r}}_i|^2 / 2 = 3Nk_B T/2$, with m_i being the particle mass and N the total number of particles.

A simple but very efficient and stable integration scheme for Eq. (3) is the Verlet algorithm. With a simulation time step $\delta t \ll 2\pi/\omega_{max}$ where ω_{max} is the highest characteristic frequency of the system (e.g. the Einstein frequency of a crystal), we have ($m_i = 1$, in one dimension)

$$r_i(t+\delta t) = r_i(t) + \delta t \dot{r}_i(t) + \frac{\delta t^2}{2}\ddot{r}_i(t) + \frac{\delta t^3}{6}\dddot{r}_i(t) \tag{4}$$

$$r_i(t-\delta t) = r_i(t) - \delta t \dot{r}_i(t) + \frac{\delta t^2}{2}\ddot{r}_i(t) - \frac{\delta t^3}{6}\dddot{r}_i(t) \tag{5}$$

Addition and subtraction of these two equations yields

$$r_i(t+\delta t) = 2r_i(t) - r_i(t-\delta t) + \delta t^2 \ddot{r}_i(t) + \mathcal{O}(\delta t^4) \tag{6}$$

$$v_i(t) = \dot{r}_i(t) = \frac{1}{2\delta t}[r_i(t+\delta t) - r_i(t-\delta t)] + \mathcal{O}(\delta t^3) \tag{7}$$

Thus, the position and velocity calculations have an algorithmic error of $\mathcal{O}(\delta t^4)$ and $\mathcal{O}(\delta t^3)$, respectively. There are many variants of this basic method used throughout the literature. If the system is ergodic, one can equate ensemble averages to averages over time steps:

$$\langle A \rangle = \frac{1}{M}\sum_{i=1}^{M} A(t_i) \tag{8}$$

for any physical quantity A of interest. This is the basis for simulating a microcanonical ensemble [1], also called the NVE ensemble, where all extensive thermodynamics variables of the system, namely N, V and E, are kept constant. Nowadays, most applications employ other ensembles such as the canonical (NVT), the isobaric-isothermal (NPT) or even the grand canonical (μPT) ensemble. As a general rule, in order to avoid two-phase coexistence and equilibration problems, one should choose an ensemble which has as many intensive variables kept constant as possible. In addition to this equilibrium simulation approach there is a huge literature on different nonequilibrium simulations to study the shear viscosity of liquids [6, 7] or other transport properties such as heat conductivity [8].

2.2.2
Monte Carlo (MC)

The classical version of MD simulation, just outlined, is a fully deterministic simulation technique. While there are many variants of the classical MD which add stochastic terms (for instance to keep the temperature constant), the other extreme, namely purely stochastic sampling, corresponds to the classical Monte Carlo (MC) approach.

In an MC simulation of the above example the ballistic motion of particles is replaced by a stochastic one. Starting from a particular configuration, randomly a particle is selected and displaced by a random jump. If, in the trivial case of hard spheres, this new configuration does not comply with excluded volume

2.2 Basic Techniques: Monte Carlo and Molecular Dynamics

constraints, the move is rejected; if it does, it is accepted. Then one starts the whole procedure again. Once every particle had a chance to move once (on average), our Monte Carlo step is complete. This is the most basic Monte Carlo simulation (see e.g. [5]). Since for hard spheres there is no energy involved, all states have the very same probability P_{eq}. Also if the system is in state x, the probability $W(x \rightarrow y)$ of it jumping into y is the same as that from y to x and the procedure trivially fulfills the detailed balance condition of statistical mechanics:

$$W(x \rightarrow y) P_{eq}(x) = W(y \rightarrow x) P_{eq}(y) \tag{9}$$

For the case when the interaction energy is not a hard core potential, but the potential $U(r_{ij})$, leading to a total potential energy function (Hamiltonian)

$$H(\{r_i\}) = \sum_{\langle i,j \rangle} U(r_{ij})$$

where $\langle i,j \rangle$ denotes all interacting pairs, P_{eq} is given by

$$P_{eq}(\{r_i\}) = \exp(-H/k_B T)/Z$$
$$Z = \sum_{\{r_i\}} \exp(-H/k_B T) \tag{10}$$

where Z is the partition function.

With $P(x)$ the Boltzmann probability of the original state and $P(y)$ that of the new state, detailed balance is obeyed if

$$\frac{W(x \rightarrow y)}{W(y \rightarrow x)} = \exp\{-[H(x) - H(y)]/k_B T\} \tag{11}$$

This is a sufficient condition for a MC simulation to relax into thermal equilibrium, although this may take a very long time. In some cases improper choice of "moves" can lead to a non-ergodic behavior with equilibrium only attained on a subspace. Algorithms without detailed balance will not be discussed here.

In this context the *Metropolis criterion* is the one most frequently used to accept or reject a move:

$$W(x \rightarrow y) = \Gamma \begin{cases} \exp\{-[H(x) - H(y)]/k_B T\}, & \Delta H > 0 \\ 1, & \Delta H < 0 \end{cases} \tag{12}$$

Since only the ratio of the rates W is relevant, Γ is an arbitrary constant between zero and one, $\Delta H = H(x) - H(y)$, a random number r, equally distributed between 0 and 1, is used to decide upon the acceptance of a move. If $r < W(x \rightarrow y)$ the move is accepted, otherwise rejected. Usually $\Gamma = 1$ is chosen so that any move that lowers the energy is accepted. This is the basic MC procedure to sample phase space.

In many cases, however, one also would like to gain information on the dynamics of a (model) system. How can one use MC simulations, without an intrinsic time scale, to obtain information on the dynamics? The method described before evolves a system from one state to another by a *local* particle move. These *local* moves can also be interpreted as a dynamic MC move based on a stochastic Markov process, where subsequent configurations $x \to x' \to x'' \to \ldots$ are generated with transition probabilities $W(x \to x')$, $W(x' \to x''), \ldots$. To a large extent the choice of the move is arbitrary, as long as one can interpret it as a *local* basic unit of motion. As seen before, the actual choice of W is somewhat arbitrary as well: the prefactor Γ in Eq. (12) can vary. Γ actually can be interpreted as fixing an attempt rate, $\Gamma = \tau_0^{-1}$, for the moves and so introduces a time scale. Thus one reinterprets the transition probability $W(x \to y)$ as a transition probability per unit time [9, 10]. To compare the simulated dynamics with an experiment, the basic task is to determine τ_0 properly (e.g. from the diffusion of chains in a polymer melt). It is obvious, however, that this (overdamped) simulation does *not* include any hydrodynamic effects since there is no momentum involved.

So far, the two extreme cases for classical, particle-based computer simulations were discussed: microcanonical MD and MC. There are many approaches in between.

2.3
Polymer-specific Properties

Before introducing the scaling concepts and linking them to particular polymers, we will briefly outline the terminology and the polymer-specific quantities (observables) which are most commonly investigated by computer simulations. Apart from standard observables such as density, pressure, temperature, stress tensor, radial distribution function, which are not polymer-specific, there are few which are reported in almost every theoretical or simulation paper on polymers.

2.3.1
Static Observables

Static observables are calculated from averaging over simulation snapshots which are assumed (or known) to span or better exceed the observable longest relaxation time or which are by construction of the (MC) algorithm statistically independent.

The simplest polymer-specific static observables reduce the observed chain conformations onto a single length, which also serves as a typical characteristic of the size of the chain. These are the mean-squared end-to-end distance R_{ee}^2, the gyration tensor $R_{\alpha\beta}^2$, and the hydrodynamic radius R_H.

For a sequence of monomers labelled from 1 to N the mean-squared end-to-end distance is defined as

$$R_{ee}^2 = \langle |\mathbf{r}_N - \mathbf{r}_1|^2 \rangle \tag{13}$$

where $\mathbf{r}_i$ is the coordinate of monomer i, $\langle \cdots \rangle$ denotes the ensemble average. Elements of the 3×3 gyration tensor are defined as

$$R_{\alpha\beta}^2 = \frac{1}{2N^2} \left\langle \sum_{ij} (r_{i\alpha} - r_{j\beta})^2 \right\rangle \tag{14}$$

The scalar mean-squared radius of gyration R_g^2 is the trace of $R_{\alpha\beta}^2$ and is invariant under coordinate frame of rotation. The mean-square radius of gyration can also be defined as

$$R_g^2 = \frac{1}{2N^2} \sum_{ij} |\mathbf{r}_i - \mathbf{r}_j|^2 \tag{15}$$

where $\mathbf{r}_{cm}$ is a center of mass vector of the chain:

$$\mathbf{r}_{cm} = \frac{1}{N} \sum_{i=1}^{N_b} \mathbf{r}_i \tag{16}$$

The hydrodynamic radius, R_H, which is important in solutions, has the definition

$$\left\langle \frac{1}{R_H} \right\rangle = \frac{1}{N^2} \sum_{ij, i \neq j} \left\langle \frac{1}{r_{ij}} \right\rangle \tag{17}$$

where r_{ij} is the scalar distance between any two monomers i and j.

Another common intramolecular static quantity is the structure factor:

$$S(q) = \frac{1}{N} \sum_{ij} \langle \exp(i\mathbf{q} \cdot \mathbf{r}_{ij}) \rangle = \frac{1}{N} \sum_{ij} \left\langle \frac{\sin(qr_{ij})}{qr_{ij}} \right\rangle \tag{18}$$

The structure factor is an important quantity which connects simulation predictions to experimental observations and allows a characterization of the average chain conformation on all relevant length scales. In a regime $R_g^{-1} \ll q \ll a^{-1}$, where a is the monomer size, $S(q)$ displays a scaling relationship $S(q) \sim q^{-1/\nu}$. The above equation assumes a constant form factor of one for all scatterers. In experiments, however, this is not always the case and depends on the scattering method (i.e. X-rays or neutrons).

2.3.2
Elementary Dynamic Observables

Mean-squared displacements are typical quantities to track the dynamic evolution of a system. Typically, three mean-squared displacements are considered:

$$g_1(t) = \left\langle |r_i(t-t_0) - r_i(t_0)|^2 \right\rangle_{t_0} \tag{19}$$

$$g_2(t) = \left\langle |r_i(t-t_0) - r_i(t_0) - r_{cm}(t-t_0) + r_{cm}(t_0)|^2 \right\rangle_{t_0} \tag{20}$$

$$g_3(t) = \left\langle |r_{cm}(t-t_0) - r_{cm}(t_0)|^2 \right\rangle_{t_0} \tag{21}$$

where the center-of-mass position r_{cm} is given by Eq. (16), $\langle \cdots \rangle_{t_0}$ denotes an average over all time origins, t is time interval.

$g_1(t)$ is the mean-squared displacement of individual monomers, and is often computed only for a few "inner" monomers of a chain to minimize the effects of the relatively "more mobile" chain ends and thus to mimic better the behavior of asymptotically long chains. $g_2(t)$ is the mean-squared displacement of individual monomers relative to the position of the center of mass of their respective chains. It is clear from the definition of g_2 that it saturates at a value close to R_g^2 as $t \to \infty$. Finally, g_3 is the mean-squared displacement of chain center-of-mass.

The diffusion constant for a chain can be determined from the behavior of the g_3 mean-squared displacement for long times:

$$D = \lim_{t \to \infty} \frac{g_3(t)}{6t} \tag{22}$$

when the mean-squared displacement of chain centers of mass is in the diffusive regime, i.e. becomes a linear function of time.

The three mean-squared displacements also play an important role of providing means to assess, for example, the equilibration of the system. The intersection of g_2 and g_3 is a standard definition of the equilibration time for a melt, as it is the time required for the average chain center of mass to diffuse a distance of the order of the size of an average chain.

In analogy with the static structure factor, for comparison with experiment often the dynamic coherent and incoherent structure functions of the polymers are calculated. They offer the opportunity for a direct comparison to neutron scattering experiments, while the mean square displacements averaged over all beads are quantities which can be obtained by pulsed field NMR. In both cases simulations and experiments on highly entangled polymer melts yield comparable results.

2.3.3
Collective Observables

While the present chapter mostly focuses on single polymer properties, simulations are equally important to study static and dynamic collective properties. Typical static properties are the various density correlation functions, which characterize the morphology of the system under investigation. In q-space the corresponding property is the structure function of the whole system. In both cases also the corresponding dynamic quantities are of great theoretical and experimental interest. On short length scales local packing and dynamics are analyzed. On larger length and time scales collective diffusion and e.g. domain formation in a mixture are investigated.

2.4
Scaling Relations for Dynamics

Here we briefly discuss the three models that are most frequently used to interpret dynamic properties for polymeric systems. In current research usually more complicated situations are investigated; here we mostly concentrate however on isolated ideal chains and short chains in melts (Rouse model), isolated ideal chains in solution (Zimm model) and entangled polymer melts (reptation model). All three models describe the actual dynamics also in experimental terms fairly well.

2.4.1
Rouse Model

In the Rouse model, all the complicated interactions are absorbed into a monomeric friction and a coupling to a heat bath. It was originally proposed to model an isolated chain in solution, though it actually works well for short chains in a melt. The polymer is modeled as a freely jointed chain of N beads connected by $N - 1$ springs, immersed in a continuum, as depicted in Fig. 2.2. Each bead experiences a friction, with friction coefficient ζ. The beads are connected by a

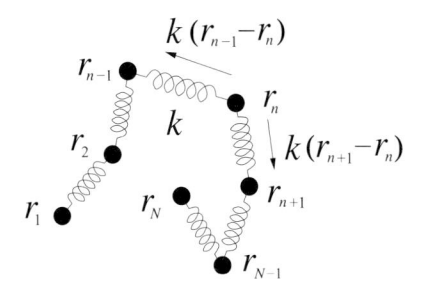

Fig. 2.2 The polymer is modeled as a freely jointed chain of N beads connected by Hookean springs, immersed in a continuum. Each bead experiences a friction ζ. Excluded volume interaction and chain entanglement are not taken into account.

Hookean spring with a force constant $k = 3k_B T/b^2$, where $b^2 = ll_K$. Each bead-spring unit is intended to model a short subchain, not just a monomer. The equation of motion of the beads is given by a Langevin equation. For the monomer i ($i \neq 1, N$) it reads

$$\zeta \dot{r}_i = k(2r_i - r_{i-1} - r_{i+1}) + f_i \tag{23}$$

The distribution of the random forces f_i is Gaussian with zero mean and the second moment

$$\langle f(t)_i f(t')_j \rangle = 6\zeta k_B T \delta_{ij} \delta(t - t') \tag{24}$$

This model does not contain any specific interactions between monomers except those due to the chain connectivity. Since in a melt the long-range hydrodynamic interactions are screened, it was suggested that this model could describe the motion of those chains, except that ζ arises from other chains rather than the solvent.

The advantage of this model is that the Langevin dynamics of a single chain of point masses connected by harmonic springs can be solved exactly by transforming to normal coordinates of the chain [11]. This was first done in a seminal paper by Rouse [12]. It leads to a longest relaxation time (Rouse time)

$$\tau_R = \frac{\zeta b^2 N^2}{3\pi^2 k_B T} \tag{25}$$

and a melt viscosity

$$\eta = \frac{\rho \zeta}{36} R_{ee}^2 \sim N\zeta \tag{26}$$

The self-diffusion constant D can be determined from the mean-square displacement of r_{cm}, using Eq. (22)

$$D = \frac{k_B T}{N\zeta} \tag{27}$$

Within the Rouse model the mean-square displacements of monomers, $g_1(t)$, reads

$$g_1(t) \sim \begin{cases} t^1, & t < \tau_0, & g_1(t) < b^2 \\ t^{\frac{1}{2}}, & \tau_0 < t < \tau_R, & b^2 < g_1(t) < R_{ee}^2 \\ t^1, & t > \tau_R, & g_1(t) > R_{ee}^2. \end{cases} \tag{28}$$

For very short times, when a monomer has moved less than its own diameter, it is affected little by its neighbors along the chain. This short time regime, $t < \tau_0$, is governed by the local chemical properties of the chains. For intermediate times, the motion of a monomer is slowed down because it is connected to other monomers. This can be viewed as the diffusion of a particle with increasing distance-dependent mass.

It turns out experimentally that this extremely simple model provides an excellent description of the dynamics of a melt of relatively short chains, meaning molecular weights of e.g. $M < 20\,000$ for polystyrene [PS, $M_{mon} = 104$] or $M < 2000$ for polyethylene [PE, $M_{mon} = 14$], both qualitatively and quantitatively almost perfectly, although the reason is still not well understood. Only recently some deviations have been observed [13]. The effects are rather subtle and would require a detailed discussion beyond the scope of this chapter. For longer chains, the motion of the chains are observed to be significantly slower.

Measurements of η [14], NMR, neutron spin-echo scattering experiments [15] and results for molecular dynamics simulations also agree surprisingly well with that. For short chains, it turns out that the noncrossability of the chains in addition to the chains' chemical structure mostly affect the prefactors in the diffusion coefficient through the monomeric friction coefficient ζ.

2.4.2
Reptation Model

For chains which significantly exceed the entanglement length N_e, i.e. are much longer than the above examples (e.g. polystyrene with the entanglement molecular weight of $M_e \sim 18\,000$), the motion is slowed down drastically. Experiments show a dramatic decrease in the diffusion constant D [16–18]. For $N \geq N_e$:

$$D \sim N^{-2} \ldots N^{-2.4} \tag{29}$$

Similarly the melt viscosity increases from $\eta \sim N$ for short chains to [11]

$$\eta \sim N^{3.4} \tag{30}$$

The time-dependent elastic modulus $G(t)$ exhibits a solid or rubber-like plateau at intermediate times before decaying completely. Since the properties for all systems start to change at a chemistry- and temperature-dependent chain length N_e or molecular weight M_e in the same way, one is led to the idea that this can only originate from properties common to all chains, namely the chain connectivity and the fact that the chains cannot pass through each other.

Reptation theory can qualitatively and in its original version semiquantitatively explain this behavior [19]. We here restrict ourselves to the basic model. New extensions [20, 21] allow for a detailed quantitative prediction of melt properties. As before, we consider polymer melts where the chains are all identical. They can be characterized by an overall particle density ρ and a number of monomers N per chain. As explained in previous sections, the overall extension of chains in a melt is well characterized by the properties of random walks.

With l being the average bond length we have for the mean square end-to-end distance ($N \gg 1$)

$$R_{ee}^2 = l_K l(N-1) \approx l_K l N \tag{31}$$

Here l_K is the Kuhn length and a measure of the stiffness of the chain. This gives an average volume per chain of

$$V \sim R_{ee}^3 \sim N^{\frac{3}{2}} \tag{32}$$

leading to a vanishing self density of the chains in the melt. To pack beads to the monomer density $\mathcal{O}(N^{\frac{1}{2}})$ other chains share the volume of the same chain. These other chains effectively screen long range excluded volume interaction, since the individual monomer cannot distinguish whether a nonbonded neighbor monomer belongs to the same chain or not.

In reptation theory the motion of the chains is viewed as Rouse motion of chains in a tube of diameter d_T. In the original treatment the tube is fixed and a chain has to move completely out of the tube to relax its conformation and any stress linked to the conformation. All other means of relaxation, such as constraint release due to moving chain ends of those chains which form the tube or fluctuation effects such as contour length fluctuations of the tube modify this scheme only somewhat quantitatively, but do not alter the qualitative picture.

For short time scales the motion of the monomers cannot be distinguished from that of the Rouse model, the motion of the monomer is isotropic and $g_1(t) \sim t^{\frac{1}{2}}$. Only after the motion reaches a distance of $d_T^2 \sim R_{ee}^2(N_e)$ do the constraints from the tube show up. The corresponding time is the Rouse time of a subchain of N_e beads, namely $\tau_e \sim N_e^2$.

The contour length of the tube L_T can be estimated as $L_T \sim d_T N/N_e \sim N/N_e^{\frac{1}{2}}$. For $t > \tau_e$, the chain performs essentially a one dimensional Rouse motion along the random walk like tube turning the $t^{\frac{1}{2}}$ power law for $g_1(t)$ into a $t^{\frac{1}{4}}$ power law.

However, after the Rouse relaxation time τ_R, the monomers have only moved a distance of order $L_T^{\frac{1}{2}} \ll L_T$ for $N \gg N_e$ along the tube. Following this regime the overall diffusion along the tube dominates and results in a second $t^{\frac{1}{2}}$ regime for the motion in space. The initial tube will be destroyed when one of the segments has visited of the order of N different contiguous sites. This requires a time $\tau_d \sim N^3/N_e$. In this time the chain has moved a distance comparable to its own size, therefore the diffusion constant D is expected to scale as N^{-2}.

In summary the theory predicts the following general power-law sequence for the mean-square displacement $g_1(t)$:

$$g_1(t) \sim \begin{cases} t^1, & t < \tau_0 \\ t^{\frac{1}{2}}, & \tau_0 < t < \tau_e \sim N_e^2 \\ t^{\frac{1}{4}}, & \tau_e < t < \tau_R \sim N^2 \\ t^{\frac{1}{2}}, & \tau_R < t < \tau_d \sim N^3/N_e \\ t^1, & t > \tau_d \end{cases} \quad (33)$$

For the motion of the center-of-mass $g_3(t)$ one expects

$$g_3(t) \sim \begin{cases} t^1, & t < \tau_e \sim N_e^2 \\ t^{\frac{1}{2}}, & \tau_e < t < \tau_R \sim N^2 \\ t^1, & t > \tau_R \end{cases} \quad (34)$$

Direct experimental evidence for these intermediate time regimes has been found by NMR [22–24], from diffusion experiments of polymers at an interface, and very recently n-scattering [25], however, simulations were the first to observe the crossover into the $t^{1/4}$ regime [26].

2.4.3
Zimm Model

So far we have only discussed the free draining regime, namely the situation where hydrodynamic interactions do not play a role. The Zimm model [27], a very early extension of the Rouse model, takes hydrodynamic interactions between the beads into account. The equation of motion of the beads is still given by a Langevin equation. The only difference is that the inverse bead friction is replaced by the mobility tensor H_{ij}, which is a second-rank tensor representing how the velocity of the jth bead affects the velocity of the ith bead through the solvent between them:

$$\dot{r}_i = \sum_{j=1}^{N} H_{ij}[k(2r_i - r_{i-1} - r_{i+1}) + f_i] \quad (35)$$

For the Rouse model $H_{ij} = (1/\zeta)\delta_{ij}$. Zimm used the Oseen tensor for H_{ij}, decoupled it from the rest of the equation and replaced it with the average at equilibrium (preaveraging approximation):

$$\langle H_{ij} \rangle = \frac{1}{6\pi\eta} \left\langle \frac{1}{|r_i - r_j|} \right\rangle_{ij} \quad (36)$$

where the statistical average is taken with respect to i and j.

In a θ-solvent, the Gaussian chain model conveniently gives an analytical expression for the Oseen tensor and leads to the diffusion coefficient:

Table 2.1 Summary of the Rouse, reptation and Zimm models. τ is the longest relaxation time and η is the viscosity (intrinsic viscosity in case of the Zimm model, i.e. an increase in the solution viscosity when the concentration of the polymer is raised to a certain level).

Model	Chain statistic	Hydrodynamic interaction	D	τ	η
Rouse	Ideal	Screened	N^{-1}	N^2	N^1
Reptation	Ideal	Screened	N^{-2}	N^3	N^3
Zimm	Ideal	Present	$N^{-\frac{1}{2}}$	$N^{\frac{3}{2}}$	$N^{\frac{1}{2}}$
Zimm	Real	Present	$N^{-\nu}$	$N^{3\nu}$	$N^{3\nu-1}$

$$D_0^{Zimm} = \frac{K_B T}{\eta_s b N^{\frac{1}{2}}} \tag{37}$$

That is the hydrodynamic interactions increase the diffusivity of the chain, compared with a group of N independently moving beads (Rouse model). Zimm model for θ-chains can be also modified for the case of chains in a good solvent. Of course, one neglects numerical coefficients, but it is possible to obtain the scaling exponents. A summary of all discussed models is given in Table 2.1. The three models outlined above give a rough idea about generic dynamic aspects of linear polymers, and introduce several universal parameters, such as entanglement length, Kuhn length, Rouse relaxation time, etc. It is clear that in order to determine these parameters one has to increase the resolution of the model, since most of the generic models abstract from the chemical structure of polymers. The next level in the hierarchy of complexity is an atomistic level of detail. At this level atoms are treated as point masses interacting via simple potentials. The notion of "different" atoms allows one to distinguish between polymer chemistries. At the same time it helps to avoid computationally consuming quantum-chemical or density-functional based calculations to evolve the system.

In the next section we discuss the molecular dynamics approach and various force-fields used in these studies.

2.5
Molecular Interactions

As discussed before, molecular dynamics in a nutshell is a numerical, step-by-step solution of the classical equations of motion written for the point masses representing the atoms (or united atoms when hydrogen atoms are not included explicitly) of the studied substance:

$$m_i \ddot{r}_i = f_i, \quad f_i = -\frac{\partial U(r_1, r_2, \ldots, r_N)}{\partial r_i} \tag{38}$$

Here f_i is the force exerted on the atom i by the rest of the atoms, m_i is the mass of the ith atom and U is a potential energy of the system.

The equations of motion (38) imply that all chemically relevant information about the system is hidden in its potential energy. Therefore, central to the success to the study of chemical systems is the quality of the model used to calculate the energy of the system as a function of its structure.

Polymers involve macromolecules each containing thousands or many more atoms plus their condensed phase environment. This can lead to systems containing millions of atoms. In addition, the inherent dynamic nature of polymers and the mobility of their environments require that a large number of conformations, generated via various methods, be subjected to energy and force calculations. Thus, typically about 10^6–10^7 energy calculations are required per time step for a system containing of the order of 10^5 atoms and optimized algorithms and potential functions are crucial for successful simulations. Hence the mathematical equations in empirical potential energy functions will include relatively simple terms to describe the physical interactions that dictate the structure and dynamic properties of molecules. This simplification allows for the computational speed needed to perform the required number of steps.

2.5.1
Potential Energy Functions

A potential energy function (PEF) allows one to calculate the potential energy, $U(\{r_i\})$, of a chemical system as a function of its three-dimensional structure, which is given by the positions of the atoms. The structure is specified by the set of all degrees of freedom, $\{r_i\}$. The terms of PEF describe the various physical interactions that dictate the structure and properties of a chemical system.

The total potential energy of a typical molecular system, U_{total}, can be separated into terms for the internal, U_{int}, and external, U_{ext}, potential energy:

$$U_{total} = U_{int} + U_{ext} \tag{39}$$

$$U_{int} = \sum_{bonds} \frac{1}{2} k_{ij}^b (r_{ij} - b_{ij})^2 \tag{40}$$

$$+ \sum_{angles} \frac{1}{2} k_{ijk}^\theta (\theta_{ijk} - \theta_{ijk}^0)^2 \tag{41}$$

$$+ \sum_{dihedrals} k_\phi [1 + \cos(n\phi - \phi_0)] \tag{42}$$

$$U_{ext} = \sum_{nonbonded} 4\varepsilon_{ij} \left[\left(\frac{\sigma_{ij}}{r_{ij}}\right)^{12} - \left(\frac{\sigma_{ij}}{r_{ij}}\right)^6 \right] \tag{43}$$

$$+ \sum_{charges} \frac{q_i q_j}{4\pi\varepsilon\varepsilon_0 r_{ij}} \tag{44}$$

The internal terms are associated with covalently connected atoms, and the external terms represent the noncovalent or nonbonded interactions between them.

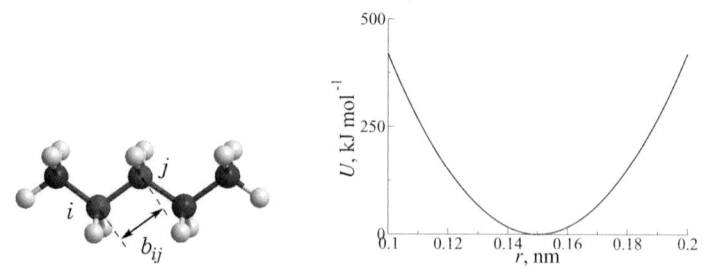

Fig. 2.3 The bond stretching potential between two covalently-bonded atoms i and j is characterized by the bond length b_{ij} and the force constant k_{ij}^b (harmonic bond).

The *nonbonded* interactions are often assumed to be pair-additive. They contain a repulsion term, and a dispersion term, which are normally combined in the Lennard–Jones potential (Eq. 43). If atoms have partial charges then the Coulomb term (Eq. 44) is added to the nonbonded interactions.

The internal terms, or *bonded* interactions, are not exclusively pair interactions, but also include 3- and 4-body interactions. The bond stretching between two covalently bonded atoms, depicted in Fig. 2.3, is usually represented by a harmonic potential (Eq. 40). However, in some force fields different forms of the bond potential are used. For example a fourth power potential is used in GROMOS 96 force field

$$U_b(r_{ij}) = \frac{1}{4} k_{ij}^b \left(r_{ij}^2 - b_{ij}^2 \right)^2 \tag{45}$$

or an anharmonic Morse bond stretching potential can be used which differs from the harmonic potential in having an asymmetric potential well:

$$U_{\text{morse}}(r_{ij}) = D_{ij} \left\{ 1 - \exp\left[-\beta_{ij}(r_{ij} - b_{ij}) \right] \right\}^2 \tag{46}$$

Another anharmonic bond stretching potential that is slightly simpler than the Morse potential adds a cubic term to the simple harmonic form:

$$U_b(r_{ij}) = k_{ij}^b (r_{ij} - b_{ij})^2 + k_{ij}^b k_{ij}^{\text{cub}} (r_{ij} - b_{ij})^3, \tag{47}$$

and can be used to improve the infrared spectrum of an SPC water model.

The bond angle vibration between a triplet of atoms (Fig. 2.4) is represented either by a harmonic potential on the angle θ_{ijk} (Eq. 41) or (as, for example, in GROMOS 96 force field) cosine-based angle potential

$$U_a(\theta_{ijk}) = \frac{1}{2} k_{ijk} \left[\cos(\theta_{ijk}) - \cos(\theta_{ijk}^0) \right]^2 \tag{48}$$

b_{ij} and the force constant k^b_{ij} (harmonic bond).

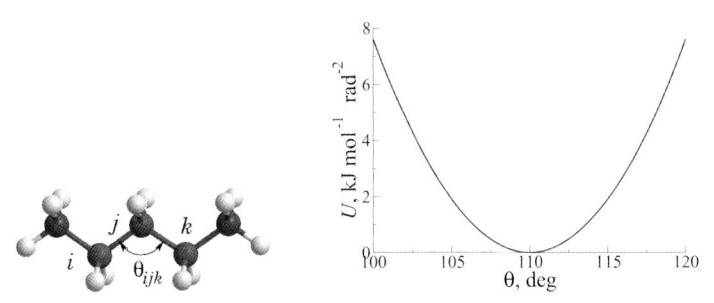

Fig. 2.4 Harmonic bond angle potential between a triplet of atoms $i-j-k$ is represented by the equilibrium bond angle and the force constant k^θ_{ijk}. Atom j is in the middle; atoms i and k are at the ends.

where

$$\cos(\theta_{ijk}) = \frac{\mathbf{r}_{ij} \cdot \mathbf{r}_{kj}}{r_{ij} r_{kj}} \tag{49}$$

Dihedrals (4-body potentials) can be of two types: proper dihedrals, as shown in Fig. 2.5, and improper ones (see Fig. 2.6).

For the proper dihedral interaction the periodic function (Eq. 42) is used. The dihedral angle is normally defined as the angle between the ijk and jkl planes, with zero corresponding to the cis configuration, with i and l on the same sides.

For alkanes, a function based on expansion in powers of $\cos\phi$ is often used (Ryckaert–Bellemans function):

$$U_d(\phi_{ijkl}) = \sum_{n=0}^{5} C_n (\cos\psi)^n \tag{50}$$

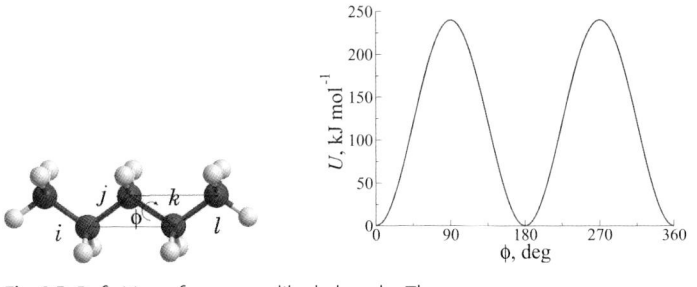

Fig. 2.5 Definition of a proper dihedral angle. The configuration shown is in the trans form, which corresponds to $\phi = \pi$. Zero dihedral angle corresponds to a cis configuration, when i and l are on the same side.

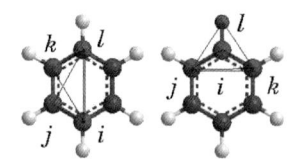

Fig. 2.6 Principle of improper dihedral angles. Out-of-plane bending for rings (left), substituents of rings (right). The improper dihedral angle ϕ is defined as the angle between planes (i,j,k) and (j,k,l) in all cases.

where $\psi = \phi - \pi$. The use of this potential implies exclusion of Lennard–Jones interactions between the first and the last atoms of the dihedral, and ψ is defined according to the convention $\psi_{\text{trans}} = 0$.

The Ryckaert–Bellemans dihedral function can also be used to include the OPLS force-field dihedral potential. The OPLS potential function is given as the first four terms of a Fourier series:

$$U_d(\phi_{ijkl}) = V_0 + \frac{1}{2}[V_1(1 + \cos\phi) + V_2(1 - \cos 2\phi) + V_3(1 + \cos 3\phi)] \tag{51}$$

Note that this potential also needs exclusion of Lennard–Jones interactions between the first and the last atom of the dihedral.

Improper dihedrals (Fig. 2.6) are used to keep planar groups planar (that is, aromatic rings, or hydrogens in benzene) or to prevent molecules from flipping over to their mirror images. This potential has also a harmonic form:

$$U_{id}(\theta_{ijkl}) = k_\theta(\theta_{ijkl} - \theta_0)^2 \tag{52}$$

Although the potentials used in simulations are inherently approximate, they are completely under the control of the user, so that by changing specific contributions their role in determining a specific property can be examined.

In addition to defining the functional form of the interaction potentials, one has to supply the corresponding parameters which enter these potentials. There are several popular parameter sets used. They can be obtained empirically, for example by matching properties of a liquid benzene, as is done in OPLS force-field, or they can be derived from higher-level modeling methods, such as quantum chemical calculations or density functional approaches. Since the main purpose is the correct description of physical properties of a particular compound, these sets can even be heuristic. Below we briefly list some popular force-fields:

- Classical force fields:
  - AMBER (Assisted Model Building and Energy Refinement) – widely used for proteins and DNA [28]. http://amber.scripps.edu/.
  - CHARMM (Chemistry at HARvard Macromolecular Mechanics) – originally developed at Harvard, widely used for both small molecules and macromolecules [29, 30]. http://www.charmm.org/.
  - GROMOS (GROningen MOlecular Simulation package) – a force field that comes as part of the GROMOS, a general-purpose molecular dynamics computer simulation package for the study of biomolecular systems. GRO-

MOS force field (A-version) has been developed for application to aqueous or apolar solutions of proteins, nucleotides and sugars. However, a gas phase version (B-version) for simulation of isolated molecules is also available. http://www.igc.ethz.ch/gromos-docs/.
- OPLS-aa, OPLS-ua, OPLS-2001 – Members of the OPLS family of force fields developed by William L. Jorgensen and coworkers at Yale Department of Chemistry [31].
- ECEPP/2 – free energy force field [32].

- Second-generation force fields:
  - MMFF94 (Merck Molecular Force Field) – developed at Merck, for a broad range of chemicals. They are based on MM3 force field. MMFF is not optimized for a single use, but tries to perform well for a wide range of organic chemistry calculations [33–37].
  - MM2, MM3, MM4 – developed by Norman L. Allinger and coworkers, for a broad range of chemicals. It is mainly used for hydrocarbons and other small organic molecules [38–43].

- Reactive force fields:
  - ReaxFF – reactive force field developed by William Goddard and coworkers. It is fast, transferable and is currently the computational method of choice for atomistic-scale dynamic simulations of chemical reactions [44].

2.5.2
Software

The increase in the number of studies using molecular dynamics has been fuelled by the availability of programs and the computational power required for meaningful studies. This, however, also bears the danger of quick shots, which one frequently finds in the literature and which should be treated with great caution! It is not possible to cover all the programs available on the market, so we will limit ourselves to a few. The main difference between the programs is flexibility (how easy it is to learn and modify the code), scalability (important for running big systems on parallel computers) and speed (amount of computer time needed for a fixed number of MD steps).

NAMD [45] is a parallel molecular dynamics code designed for high-performance simulation of large systems. It scales up to hundreds of processors on high-end parallel platforms, as well as tens processors on low-cost clusters. It works with AMBER, CHARMM potential functions, parameters and file formats. NAMD is implemented in C++ and is combined with the graphical/analysis software VMD and the grid computing software BioCoRE. NAMD is distributed free of charge with the source code available from www.ks.uiuc.edu.

GROMACS [46] (GROningen MAchine for Chemical Simulation) is a very fast program for molecular dynamics simulation. It supports GROMOS, OPLS, AMBER and ENCAD force fields. It can also handle polarizable shell models and flexible constraints. It is written in ANSI C, and distributed together with the source code and documentation under the GNU General Public License from www.gromacs.org. It also has interfaces with popular quantum-chemical packages (MOPAC, GAMES-UK, GAUSSIAN) to perform mixed MM/QM simulations. GROMACS works with the particle and domain decomposition, and is developed and optimized for use on PC clusters.

ESPResSo [47] (Extensible Simulation Package for Research on Soft matter) is a program package designed to perform MD and MC simulations for a broad class of soft matter systems in a parallel computing environment. The main concept is to provide a user-friendly and fast simulation tool which serves at the same time as a research platform capable of rapidly incorporating the latest algorithmic developments in the field of soft matter sciences. A particular strength of ESPResSo is its efficient treatment of long-range interactions for various geometries using sophisticated algorithms such as P3M, MMM2D, MMM1D and ELC. It is equipped with a variety of interaction potentials, thermostats and ensemble integrators; it offers the usage of constraints, masses and rotational degrees of freedom; it allows one to move between different ensembles on-the-fly. An efficient MPI parallelization allows the usage of multi-processor architectures. Strict usage of ANSI-C for the core functions and a Tcl-script driven user interface makes ESPResSo platform independent. This also ensures easily modifiable interfaces to communicate with other MD/MC packages, real-time visualization and other graphic programs. ESPResSo is implemented as an open source project with the goal to stimulate researchers to contribute to the package: www.espresso.mpg.de.

DL_POLY [48] is a general-purpose MD simulation package. It uses domain decomposition parallelization, the particle mesh Ewald method for calculating long-range forces and incorporates a novel three-dimensional fast Fourier transform (Daresbury Laboratory). It is written in a uniform style in FORTRAN 77 and provides a Java-based graphical user interface for viewing the molecular system, job submission and construction and analysis of input and output files. www.cse.clrc.ac.uk/msi/software/DL_POLY/.

2.6
Coarse-graining and Simple Generic Models

Now that the key ingredients of the molecular simulations at the atomistic level are established, it might seem that once we know a reliable potential function and a parameter set, we can, in principle, model any specific polymer we wish.

In practice, of course, this is not true: polymer molecules are complex objects with a large number of internal degrees of freedom. Physical properties of typical systems involve processes occurring on a wide range of length and time scales. For example, bond and angular vibrations occur at an Å length scale, aggregatin and adsorption as well as entanglement processes at a molecular level, which is between 10 and 100 Å. Finally, mechanical properties (e.g. crack propagation) manifest itselft at μm scales. On the other hand, a typical bond vibration time is 10^{-13} s, and a single polymer chain relaxes much slower, on the order of 10^{-4} s plus or minus a few decades depending on chain length and temperature. Therefore, the full atomistic MD simulation is an expensive target, and we shall explore some other means to avoid this problem where possible or to cover simultaneously such wide range of time and length scales.

If one is only interested in universal or generic properties it is useful to employ as simple a model as possible. This leads to very simple bead spring MD simulation or even to lattice models of walks, which can be studied very efficiently by MC simulation techniques. If, however, the link to a specific chemical system is to be maintained, coarse-graining, or systematic reduction of degrees of freedom seems to be one of the most promising ways to link different time and length scales. In what follows we first give a very short account of simple generic simulations and then describe the idea and the principle of coarse-graining, later we will briefly discuss the approximations that are made and a few schemes available for systematic coarse-graining.

2.6.1
Simple Generic Models

Depending on the question of interest, one can reduce the models to the bare essentials in order to reach long effective times and/or huge system sizes. This has been an important research topic over many years, especially when it came to the thorough testing of theoretical concepts. These bare essentials are connectivity of the chains, excluded volume of the beads and (if not automatically resulting from the excluded volume) a noncrossability constraint. Such models allow for a very efficient and fast simulation of single chains but also melts with both MD and MC and have been the first to display unambiguously the slowing down of the bead motion due to reptation by Kremer and Grest [26]. By introduction of different nonbonded interactions (i.e. different excluded volumes) one can also study phase separations of polymer mixtures of structure formation of block copolymer systems. Actually by MC simulations of such (lattice) models, the crossover from the mean field behavior to the Ising critical behavior was demonstrated clearly by Binder [49]. The most basic MD model in this context is a simple freely jointed bead spring chain. All beads in the system interact via a shifted, purely repulsive Lennard–Jones potential:

$$U(r_{ij}) = \begin{cases} 4\varepsilon_{ij}\left[\left(\dfrac{\sigma}{r_{ij}}\right)^{12} - \left(\dfrac{\sigma_{ij}}{r_{ij}}\right)^{6} + \dfrac{1}{4}\right], & r_{ij} < r_c \\ 0, & r_{ij} \geq r_c \end{cases} \quad (53)$$

with a cutoff in the minimum of the potential at $r_c = 2^{\frac{1}{6}}\sigma$. For connected monomers we add an attractive ENE (Finite Extendable Nonlinear Elastic) interaction:

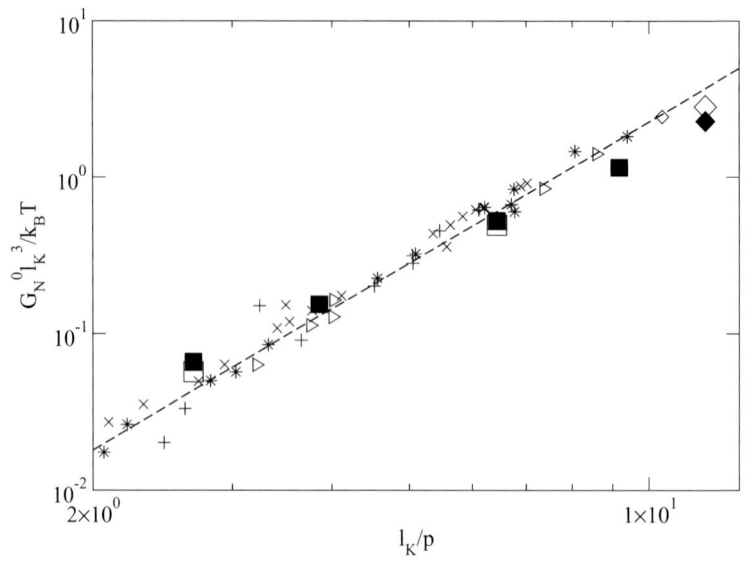

Fig. 2.7 Dimensionless rubber-like plateau moduli $G_N^0 l_K^3/k_B T$ of highly entangled polymer melts as a function of the dimensionless ratio l_K/p of Kuhn length l_K and packing length p. The packing length is a measure of the spatial distance between the neighboring strands. The figure contains (i) experimentally measured plateau moduli for polymer melts [50]: (*) polyolefins, (×) polydienes, (+) polyacrylates and (▷) miscellaneous; (ii) plateau moduli inferred from the normal tensions measured in computer simulation of bead-spring melts [21, 51] (□) and a semi-atomistic polycarbonate melt [52] (◇) under an elongational strain; (iii) predictions of the tube model based on the results of a primitive path analysis for bead-spring melts (■), and the semi-atomistic polycarbonate melt (◆). The line indicates the best fit to the experimental melt data for polymer melts by Fetters et al. [53]. Errors for all the simulation data are smaller than the symbol size. This analysis has also been extended to entangled solutions [21].

$$U^{\text{bond}}(r) = \begin{cases} -0.5\kappa R_0^2 \ln\left[1 - \left(\frac{r}{R_0}\right)^2\right], & r \leq R_0 \\ \infty, & r > R_0 \end{cases} \tag{54}$$

For melts the parameters $\kappa = 30\varepsilon/\sigma^2$ and $R_0 = 1.5\sigma$ are usually chosen [26]. This minimal model has been used in many applications. A first modification is to introduce a chain stiffness potential. By this one can vary the chain extension and also the entanglement molecular weight. Figure 2.7 gives an example of the outcome of a comprehensive analysis of the reptation primitive path and resulting comparison of the moduli to many different experiments. This figure illustrates the strength of suitably chosen most simple models in reproducing and predicting polymer properties.

2.6.2
Coarse-graining Techniques

In recent years, considerable progress has been made in the direction of systematic coarse-graining in different areas of material research, ranging from derivation of the atomistic force fields from quantum-chemical calculations to combining continuum models with atomistic models. Here we will mostly concentrate on linking two levels: atomistic, and the level of superatoms (units of a few atoms). These are probably the most important scales in order to address structure–property relations of polymers.

The original idea of coarse-graining is fairly simple. As a reference system we chose the system with atomistic resolution. Now we represent this system by a coarser model in which we replace the individual atoms by beads, which represent a number of atoms. This ansatz has some similarities to the renormalization group idea in statistical physics. The way in which this is done and the focus of the coarse-grained simulations define certain requirements for the underlying atomistic model. Roughly one can distinguish two classes. If there are no equilibrated short chain systems available, the coarse-grained model have to be derived for individual molecules, not interacting with any other molecule. This can usually be done by quantum chemical methods applied to small chain or monomer fragments. When the closeness to the chemistry is not that important and simulations of all atom systems are available, more pragmatic ways are also employed, as discussed below.

Let us now specify the mapping of our systems on superatoms or beads. Each bead comprises several atoms. This level is denoted a *coarse-grained* level. An example of the 4:1 (four beads per repeat unit) mapping scheme for bisphenol A polycarbonate is shown in Fig. 2.8. We also assume that the atomistic simulation of a small test system has already been performed and the trajectories of the atomistic simulations are available. Our tasks are (i) to find the best mapping of the system, (ii) to calculate a coarse-grained potential function of the system and (iii) to perform back-mapping of the coarse-grained model to the atomistic model (in case some local properties are to be calculated).

We start with point (ii), i.e. for the moment we assume that the mapping scheme is known and that the potential energy function U_a which depends on $r^N = \{r_1, r_2, \ldots, r_N\}$ positions of the particles is known and well parameterized for the problem. There are a few approaches one can follow in order to obtain an approximate potential energy function for the coarse-grained system, U_{cg}.

2.6.2.1 Penalty Function
The method proposed by Müller-Plathe and coworkers [54, 55] defines a penalty function which measures the deviation from the target distribution and which can then be minimized by a suitable numerical procedure. For example, the penalty function can be an integral of a radial distribution function $g(r)$:

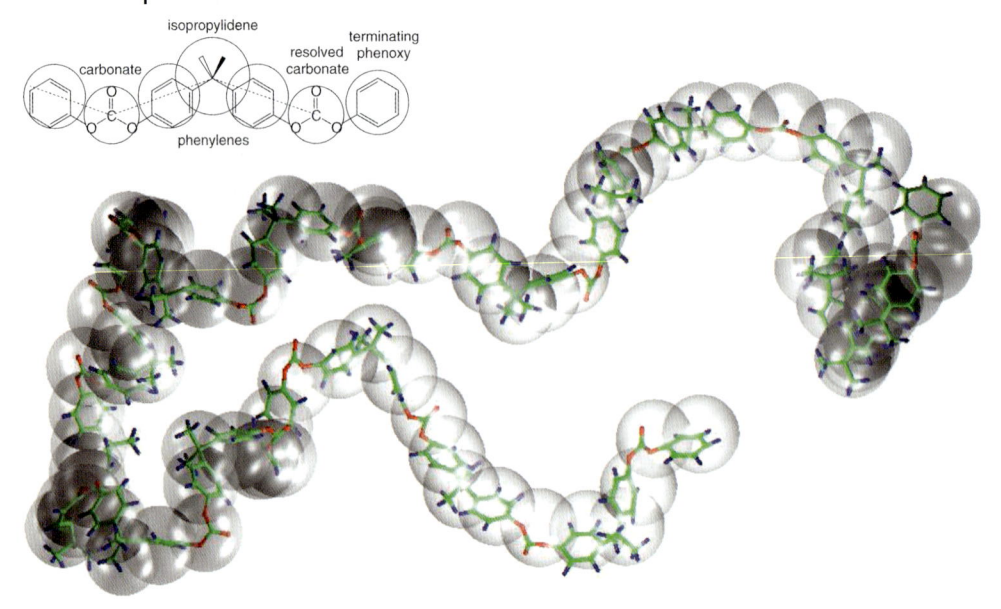

Fig. 2.8 Full atomistic and coarse-grained representation of bisphenol A polycarbonate chain taken out of an equilibrated polymer melt. The inset shows the 4:1 mapping scheme (four beads per repeat unit) used for equilibration.

$$f(p) = \int_0^{\text{cutoff}} w(r)[g^{\text{cg}}(r,p) - g^{\text{ref}}(r)]^2 dr \qquad (55)$$

Both the penalty function and the coarse-grained radial distribution function depend on the set of parameters of the coarse-grained potential p; $w(r)$ is an optional weighting function, to emphasize the desirable parts of the radial distribution function. This approach has been successfully used to study the overall chain extension of polymers (polyacrylic acid) in solution [56, 57].

2.6.2.2 Force-matching Scheme

Ercolessi and Adams [58] suggested fitting the parameters of the potential directly to forces obtained from simulations. The match is achieved by minimizing directly the "penalty function":

$$f(p) = \sum_{l=1}^{L} \sum_{i=1}^{N} |F_{il}^{\text{ref}} - F_{il}^{\text{cg}}(p)|^2 \qquad (56)$$

where F_{il}^{ref}, F_{il}^{cg} are reference (*ab initio* or atomistic) forces which acts on the *i*th atom in the *l*th atomic configuration, N is the number of atoms in the atomic configuration and L is the total number of configurations used in the minimization.

The efficiency of this method can be further improved by presenting the coarse-grained force field as a linear function of fitting parameters p, which allows one to perform the least-squares fit through the solution of an over-determined system of linear equations. To linearize the potential energy function of short-ranged interactions, one can use spline interpolation (e.g. cubic splines), since spline representations depend linearly on its parameters [59, 60].

2.6.2.3 Inverse Boltzmann Method

The use of appropriate distribution functions to construct effective interaction potentials by a Boltzmann inversion of the probability distributions is a well-known procedure for low molecular weight liquids or colloidal systems. There, however, it applies to the nonbonded interactions. Here we follow a different route. We first employ a Boltzmann inversion approach for the bonded short-range interaction along the backbone of the chain and by this separate it from the nonbonded interactions. This, however, relies on the assumption that these two classes of interaction are independent of each other. In addition one usually assumes that the distribution functions of the intramolecular quantities, such as bond lengths, angles and torsion angles, are independent of each other. This is usually not the case and can in principle be taken into account. Here, however, we only discuss the case where this is assumed. For the specific example given later, this is fulfilled almost perfectly [61].

We can, therefore, assume here that the collective variables which are defined by the mapping are uncorrelated. In this case the probability distribution function of the system $P(\{x_1, x_2, \ldots x_N\})$ is a product of the corresponding probabilities:

$$P(\{x_1, x_2, \ldots, x_N\}) = \prod_{i=1}^{N} P(x_i) \tag{57}$$

The coarse-grained potential for each variable can then be calculated as a Boltzmann inversion of the corresponding probability density distribution for this type of variable:

$$P(x_i) \sim \exp\left[-\frac{U(x_i)}{k_B T}\right] \tag{58}$$

Technically, the probability density distributions are computed via Monte Carlo sampling of the atomistic structure of an isolated test chain. This sampling involves random moves of the coarse-grained variables, evaluation of the resulting total energy and acceptance based on a standard Metropolis criterion [61]. By this the temperature is already included in the sampling of the distributions and the resulting effective interaction contains both entropic and enthalpic contributions.

The nonbonded potential of the coarse-grained system is again assumed to be a sum over all pairs:

$$U_{cg}^{nb} = \sum U_{ij}(r_{ij}) \tag{59}$$

This is, again, an approximation: it assumes that the coarse-grained nonbonded potential of a mean force can be approximated by simple pairwise potentials [62].

In most cases parameterization of the nonbonded potentials is done by either giving reasonable guesses about bead excluded volume size or by fitting non-bonded parameters against predicted values of certain thermodynamic observables. For example, in the mapping schemes for polycarbonates, the bead sizes are computed by calculating the gyration radii among atomic positions in each group, and then scaling these radii such that the volume of the repeat unit, accounting for the overlaps of the adjacent beads, is equal to the van der Waals equation-of-state analysis [61, 63].

2.6.3
Mapping Points

As has been mentioned before, the first step for any coarse-graining procedure is to define the mapping points. At the present stage there is no systematic way of choosing the mapping points for an arbitrary molecule; most of the mapping schemes are developed by trial and error. However, it is still possible to formulate several requirements that each mapping scheme satisfies:

1. The topological properties of polymer conformations should be well represented, e.g. if a polymer chain can have, for example, isotactic or syndiotactic conformations, the mapping scheme should be able to distinguish between them. Polystyrene [64] is a typical example where this problem occurs.
2. The coarse-grained degrees of freedom should be independent of each other as much as possible. In practice one should try to minimize the amplitudes of the corresponding correlation functions.
3. The mapping scheme should be chosen in such a way that it simplifies the back-mapping procedure.
4. There is always a certain degree of freedom in how to separate bonded and nonbonded contributions to the coarse-grained potential energy function (where one has to take care to avoid double counting of interactions). This fact, combined with a proper choice of the mapping scheme, can be used to adjust the bead size to the bond length ratio (here the bond length is the distance between the nearest beads connected via bonded potential) that is known to affect the dynamics of the system [65].

A representative example here is bisphenol-A polycarbonate for which two coarse-graining schemes were devised, with two [61, 66] and four [63] beads per repeat unit. The 4:1 scheme turns out to be more efficient even though it has twice as many degrees of freedom, since it avoids large bead frictions and artifacts arising from sphere packing, which dominate the liquid structure and dynamics in the 2:1 scheme.

2.6.4
Reintroduction of Chemical Details: Inverse Mapping

Inverse mapping is a process of reintroduction of chemical details to a coarse-grained configuration. In principle, many atomistically resolved chain conformations give rise to a single coarse-grained configuration. Therefore, inverse mapping must be an optimization process. An implication of this fact is that there cannot be a unique universal protocol for inverse mapping; however, the procedure described below leads to characteristic underlying micro states.

One of the possible schemes is described in [66]. For each chain, an atomistic chain of the same number of chemical repeat units is created with the same center of mass and randomly assigned backbone torsions. All bond lengths and bond angles in the atomistic chains are taken at their equilibrium values. For each coarse-grained–atomistic chain pair, while keeping the coordinates of the beads in the coarse-grained chain fixed, the torsion angles and the Euler angles of the atomic chain are varied according to a conjugate gradient minimization. During the minimization the sum of the squared distances from the coarse-grained beads to the appropriate mapping points on the atomistic chains is minimized. This is done for each chain independently with no energy evaluation. This results in an atomistic replica of the coarse-grained system which still may have many atom–atom overlaps. To resolve these overlaps, a short MD simulation run is performed with the appropriate bonded potentials and "soft" excluded volume potentials. Many difficulties can arise at this stage; for example, in reintroducing aromatic rings, overlaps may occur such that rings entangle one another, forming a physical interlocking structure. However, such pathologies appear rare enough that initiating another inverse mapping run with a different random number generator seed resolves the issue.

The above-described schemes will most probably lead to the configuration with a minimal energy. However, in some cases there could be several possible arrangements of the atoms within each coarse-grained bead, e.g. due to the rotational degree of freedom of the benzene ring [67]. In this case, the distribution should be sampled at the simulation temperature, in order to provide representative starting segments. Then, during the mapping procedure, each chain is reconstructed with the segments randomly taken out of the distribution generated above. After that, usual minimization of the orientations, positions and conformations can be performed. These are two examples of how to reintroduce the chemical details. In both cases the typical atom displacements needed to minimize the energy of the resulting all-atom polymer melt are close to 1 Å and the agreement of the obtained structure functions compared with neutron scattering is excellent. The largest systems studied by such an ansatz is a melt of BPA–PC in a box of about 90^3 nm^3 for times up to 10^{-5} s [67].

2.6.5
Application of Coarse-graining

A first most important application of coarse-graining is the preparation of equilibrated starting configurations. The most direct way would be to set up a system in an arbitrary state and equilibrate it by running for a few relaxation times. This, however, would require prohibitive amounts of CPU time, even for a coarse-grained system. Fortunately coarse-graining can still help here: for moderately dense systems one can run a coarse-grained system by a reptation or slithering-like algorithms, which beat the slow realistic dynamics by slightly more than $\mathcal{O}(N)$, and then reintroduce the atomistic details.

For dense systems, the concept of universality can be used again. This method, however, requires a precise knowledge of the average equilibrium conformation of short chains [68]. In a melt of dense solutions the chains assume random walk statistics for all contour lengths $l \gg l_K$. Once a "sample" which reaches that asymptotic regime is available, one can use this as a reference state even for much longer chains. Values for internal distances, $\langle r_{ij}^2 \rangle$, have to fall on a general master curve. For the internal distances one finds

$$\langle r_{ij}^2 \rangle \sim \begin{cases} |i-j|^2, & |i-j|l < l_K \\ |i-j|, & |i-j|l \geq l_K \end{cases} \tag{60}$$

with additional possible deviations for very small $|i-j|$ due to the chemical structure, if a more detailed model is treated. This function has to be determined in a complete simulation of shorter chains. Thus is a melt a system-specific master plot is $\langle r_{ij}^2 \rangle$ versus $l|i-j|$, the corresponding contour length.

The strategy for the melt equilibration is then as follows:
1. Simulate a melt of many short, but long enough chains ($l \gg l_K$) into equilibrium by a conventional method.
2. Use this melt to construct the master curve $\langle r_{ji}^2 \rangle / (l|i-j|)$ or target function the melts of longer chains.
3. Create nonreversal random walks of the correct bond length, which match the target function closely. They have to have the anticipated large distance conformations. Introduce, if needed beyond the intrinsic stiffness of the bonds, stiffness via a suitable second neighbor excluded volume potential along the chain, since it might be slightly larger than that of the full melt.
4. Place these chains as rigid objects in the system and move them around by a Monte Carlo procedure (shifting, rotating, inverting, etc., but not manipulating the conformation itself) to minimize density fluctuations.
5. Use this state as the starting state for melt simulations.
6. Introduce slowly (but not too slowly) the excluded volume potential by keeping the short-range intra-chain interactions, taking care that in the beginning the chains can easily cross through each other.

7. Run until the excluded volume potential is completely introduced. Control the internal distances master curve permanently to check for possible overshoots and deviations from the master curve.
8. Eventually support long-range relaxation by so-called end bridging [69, 70] or double pivot moves [68].

Starting from equilibrated samples, further applications of coarse-graining include either reintroduction of atomistic details or calculation of dynamic and static properties, surface interactions, studies of the systems under shear, etc. – all in a coarse-grained representation.

Systematic coarse-graining has been applied to several polymers: polyethylene [71], polystyrene [64], polycarbonate [61, 63, 66] and polybutadiene [72]. As an illustrative example we briefly overview the work on polycarbonates.

The coarse-grained model of polycarbonates was applied to measure dynamic and static properties of polycarbonate melts [61]. Diffusivity and the monomeric friction coefficient for the coarse-grained beads were calculated from a Rouse mode analysis. An important result is the prediction of the trend in the Vogel–Fulcher temperature among the family of polycarbonates considered (BPA, BPZ and TMP–PC). This was the first major success for distinguishing quantitative differences among similar polymers. More recently an extensive study of highly entangled polycarbonate was performed [52], where several complementary approaches to dermine N_e were examined and compared with experiment.

A calculation of the structure factor from a sequence of configurations generated by inverse mapping from equilibrated coarse-grained samples showed remarkable agreement between direct simulations of short-chain melts and the results from the inverse mapping [67]. This shows that coarse-graining successfully links the local arrangement and structure of the melts.

Further, coarse-graining has been applied to study the adsorption of a polycarbonate melt [73, 74] on a nickel surface. There a combination of *ab initio* calculations of the interactions of fragments of the repeat units (propane, carbonic

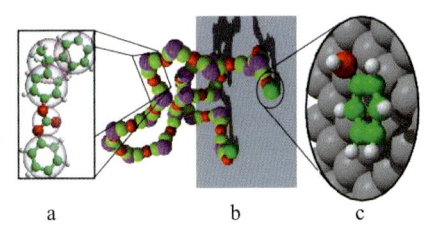

Fig. 2.9 The multiscale model of BPA–PC on nickel. (a) The coarse-grained representation of a BPA–PC segment; the coarse-grained beads are transparent spheres, superimposed on the underlying chemical structure, where the carbon atoms are green, the oxygens red, and the hydrogens white. (b) Coarse-grained model of an $N = 20$ BPA–PC molecule, with ends adsorbed on a flat surface; configuration from a 160-chain liquid simulation. (c) A phenol molecule adsorbed on the bridge site of a (111) nickel surface; configuration computed via density functional simulation.

acid, benzene and phenol) combined with an appropriate dual-scale coarse-grained model was employed to study the effect of different chain ends [75]. Figure 2.9 sketches the whole procedure. Finally, adsorption of polycarbonate blends [76] as well as their shear over the (111) nickel surface [77, 78] were considered.

2.7
Other Illustrative Examples

Apart from already mentioned studies on polycarbonates, computer simulation in general and systematic coarse-graining in particular have been applied to study a number of polymeric systems, only a few of which we will discuss here.

2.7.1
Diffusion of Small Molecules

Industrially manufactured plastics often contain a certain amount of low molecular weight components, e.g. initiators and catalysts from the polymerization process, or additives that provide specific properties such as plasticizers, dyes or antioxidants. Their behavior and mobility often influence the quality of the product because they play a role not only in the mechanical properties but also in aging or toxicity. However, in some cases the necessary key properties are difficult to determine experimentally. Especially under production conditions temperatures can be rather high, so that standard experiments turn out to be very difficult and/or expensive. This is especially true for the measurement of the diffusion coefficient of small molecules in polymer matrices in the low concentration limit.

An illustrative example on how molecular dynamics combined with coarse-graining can help in this situation is given in [79]. Here the diffusion of phenol in a bisphenol-A polycarbonate was studied in detail. The approach used was first to generate coarse-grained melts, as discussed above. These melts served as starting points to reintroduce the all-atom structure of the polymer. By this, well equilibrated polymer melts at the desired high temperatures were created. Within this system the diffusion of a few added phenol molecules could be followed for a long enough time to give accurate results. Without this detour via the coarse-grained model such a study is not possible, because the hole–hole correlations extend over fairly large regions. This study not only provided the characteristic values for the diffusion coefficient, but also revealed the mechanism of diffusion (continuous migration process combined with hopping between cavities that are separated by an average energy barrier of about 42 kJ mol^{-1} about 8 $k_B T$ at typical temperatures), which is right in between the hopping of small atoms (gases) and the continuous creeping of polymers. Analogous studies were done for methane in polyethylene [80] and oxygen in polyisobutene [81]. Figure 2.10 gives an example for the positronium lifetimes in polystyrene from simula-

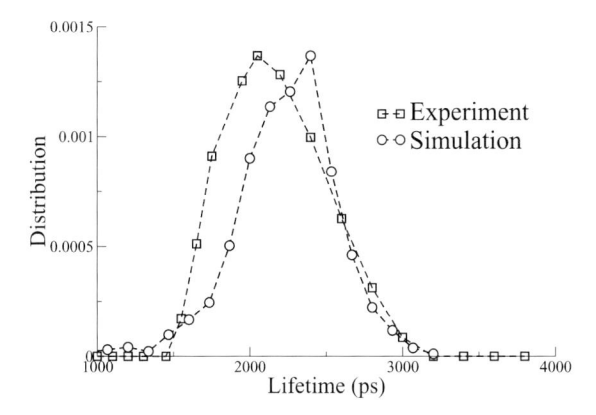

Fig. 2.10 Positronium lifetime spectra in polystyrene at 300 K. Circles, experimental; squares, simulation. Reprinted from [82].

tion and from experiment [82]. Since the positronium samples the whole geometry the coincidence of the lifetimes indicates that the simulations also properly reproduce the free volume structure in the polymer matrix.

2.7.2
Osmotic Coefficient of Polyelectrolytes

Computer simulations have also proven to be useful for studying polyelectrolytes. Here a cell model, combined with MD simulations of simplified models, already predicts qualitatively correct behavior of the osmotic coefficient as a function of counterion concentration for, e.g., poly(*p*-phenylene) [83].

However, it is not very easy to improve the results of simple models by doing more rigorous coarse-graining. The main problem is that the behavior of the system depends strongly on the spatial distribution of the ions. To model the proper ion distribution large length scales are required because of the long-range nature of electrostatic interactions. One has to use solvent-free effective potentials that implicitly take solvation into account. For ionic solutions, however, one cannot directly invert the distribution functions of, for example, three ion pairs, since they are interdependent. One can approximate the effective interactions between the ions with pair potentials with MacMillan–Mayer theory [84]. The potentials can be obtained through a reverse Monte Carlo procedure or the hypernetted-chain approximation [85, 86]. Another approach is to have a concentration-dependent dielectric permittivity, which takes into account multi-body effects and can be used even for inhomogeneous systems [87].

2.7.3
Phase Diagrams

Another illustrative example is the prediction of phase diagrams for solutions of *n*-alkanes in CO_2 [88]. Here thermodynamic perturbation theory (PTP) is first validated against grand-canonical MC simulations for short chains and mixtures (hexadecane and carbon dioxide) and then is extended to solutions of *n*-alkanes in CO_2. This work emphasizes how important it is to combine theoretical methods and simulations, especially when studying such complicated topics as critical behavior of mixtures, bubble nucleation, wetting, etc. Again, the link between exemplary experiments to obtain the "prefactors" and simulations of simplified models, which revealed the general structure of the phase diagram, turned out to be quite powerful.

2.8
Outlook

It was the aim of this contribution to provide a first introduction into the very wide area of computer simulations of soft matter. This necessarily has to stay somewhat superficial and can only give hints on options and possibilities. It also should provide a first guide to further reading and illustrate the power of modern simulation techniques, when properly employed. This is due to a combination of increasing computer speed and, even more importantly, improved modeling schemes. First examples illustrate how modern simulations can turn into a true quantitative tool with predictive power. In spite of all progress made in recent years in this direction a number of challenges remain. The most important task is the systematic improvement of links between different simulation schemes, that is:
- Systematic coarse-graining procedures, including the inverse mapping step. Levels must cover micro (many atoms), meso (many monomers) and macro (many chains) regimes and link quantum simulations at the low end and to self-consistent field calculations and finite element-like approaches at the upper end.
- Quantum simulations of reasonably sized systems (quantum chemistry calculations, density functional methods, path integral quantum Monte Carlo) should be available to couple electronic and conformational degrees of freedom.
- Improved methods to parameterize and validate force fields for atomistic classical molecular simulations, especially suitably parameterized intermolecular interactions.
- New methods for static and dynamic studies on the semimacroscopic to macroscopic level, i.e. for composite materials based on the microstructure of polymers.
- Adaptive multiscale schemes, which allow one to vary systematically the resolution of the simulations locally and temporarily and where the regimes with

the different level of resolution are in true equilibrium with each other and freely exchange molecules.

For further reading on the progress that has been made so far we refer to a number of reviews [5, 9, 49, 88–90], which also provide a fairly comprehensive guide to the literature.

Acknowledgments

It is a pleasure to acknowledge the longstanding collaboration with G. S. Grest and with past and present members of the research group in Mainz, too many to list individually.

References

1 M. P. Allen, D. J. Tildesley. *Computer Simulation of Liquids*. Oxford Science Publications. Clarendon Press, Oxford, **2004**.
2 M. Karplus, J. A. McCammon. Molecular dynamics simulations of biomolecules. *Nat. Struct. Biol.* 9:646–652, **2002**.
3 P. G. de Gennes. *Scaling Concepts in Polymer Physics*. Cornell University Press, Ithaca, NY, **1985**.
4 A. Yu. Grosberg, A. R. Khokhlov. *Statistical Physics of Macromolecules*. AIP Press, New York, **1994**.
5 K. Binder, G. Ciccotti (Eds). *Monte Carlo and Molecular Dynamics of Condensed Matter Systems*, Vol. 49. Italian Physical Society, Bologna, **1995**.
6 D. J. Evans, G. P. Morris. *Statistical Mechanics of Non-equilibrium Liquids*. Academic Press, London, **1990**.
7 J. D. Ferry. *Monte Carlo and Molecular Dynamics Simulation in Polymer Science*. Oxford University Press, New York, **1995**.
8 F. Müller-Plathe. A simple nonequilibrium molecular dynamics method for calculating the thermal conductivity. *J. Chem. Phys.* 106:6082, **1997**.
9 K. Kremer, K. Binder. Monte carlo simulations of lattice models for macromolecules. *Comp. Phys. Rep.* 7:259, **1988**.
10 K. Binder (Ed.). *Monte Carlo in Statistical Physics*. Springer, Berlin, **1979**.
11 M. Doi, S. F. Edwards. *The Theory of Polymer Dynamics*. Clarendon Press, Oxford, **1986**.
12 P. E. Rouse, A theory of the linear viscoelastic properties of dilute solutions of coiling polymers. *J. Chem. Phys.* 21:1272, **1953**.
13 W. Paul, G. D. Smith, D. Y. Yoon, B. Farago, S. Rathgeber, A. Zirkel, L. Willner, D. Richter. Chain motion in an unentangled polyethylene melt: a critical test of the Rouse model by MD simulations and neutron spin echo spectroscopy. *Phys. Rev. Lett.* 80:2346, **1998**.
14 J. D. Ferry. *Viscoelastic Properties of Polymers*. Wiley, New York, **1980**.
15 B. Ewen, D. Richter. Neutron spin echo investigations on the segmental dynamics of polymers in melts, networks and solutions. *Adv. Polym. Sci.* 134:1–129, **1997**.
16 D. S. Pearson. Recent advances in the molecular aspects of polymer viscoelasticity. *Rubber Chem. Technol.* 60:439–496, **1987**.
17 D. S. Pearson, L. J. Fetters, W. W. Graessley, G. V. Strate, E. D. von Meerwall. Viscosity and self-diffusion coefficient of hydrogenated polybutadiene. *Macromolecules* 27:711–719, **1994**.
18 H. Tao, T. P. Lodge, E. D. von Meerwall. Diffusivity and viscosity of concentrated hydrogenated polybutadiene solutions. *Macromolecules* 33:1747–1758, **2000**.

19 N. Attig, K. Binder, H. Grubmüller, K. Kremer (Eds). *Computational Soft Matter: from Synthetic Polymers to Proteins*. NIC Series, Vol. 23, Jülich, **2006**.

20 T.C.B. McLeish. Tube theory of entangled polymer dynamics. *Adv. Phys.* 5:1379–1527, **2002**.

21 R. Everaers, S.K. Sukumaran, G.S. Grest, C. Svaneborg, A. Sivasubramanian, K. Kremer. Rheology and microscopic topology of entangled polymeric liquids. *Science* 303:823–826, **2004**.

22 P.T. Callaghan, A. Coy. Evidence for reptational motion and the entanglement tube in semidilute polymer solutions. *Phys. Rev. Lett.* 68:3176–3179, **1992**.

23 M. Appel, G. Fleischer. Investigation of the chain-length dependence of self-diffusion of poly(dimethylsiloxane) and poly(ethylene oxide) in the melt with pulsed-field gradient NMR. *Macromolecules* 26:5520–5525, **1993**.

24 M. Appel, G. Fleischer, J. Karger, F. Fujara, I. Chang. Anomalous segment diffusion in polymer melt. *Macromolecules* 27:4274–4277, **1994**.

25 A. Wischnewski, D. Richter, M. Monkenbusch, L. Willner, B. Farago, G. Ehlers, P. Schleger. Reptation in polyethylene melts with different molecular weights. *Physica B* 276:337–338, **2000**.

26 K. Kremer, G.S. Grest. Dynamics of entangled linear polymer melts: a molecular-dynamics simulation. *J. Chem. Phys.* 92:5057, **1990**.

27 B.H. Zimm. Dynamics of polymer molecules in dilute solution: viscoelasticity, flow birefringence and dielectric loss. *J. Chem. Phys.* 24:269–278, **1956**.

28 D.A. Case, T.E. Cheatham, III, T. Darden, H. Gohlke, R. Luo, K.M. Merz, Jr., A. Onufriev, C. Simmerling, B. Wang, R. Woods. The amber biomolecular simulation programs. *J. Comput. Chem.* 26: 1668–1688, **2005**.

29 B.R. Brooks, R.E. Bruccoleri, B.D. Olafson, D.J. States, S. Swaminathan, M. Karplus. Charmm: a program for macromolecular energy, minimization, and dynamics calculations. *J. Comput. Chem.* 4: 187–217, **1983**.

30 A.D. MacKerell, B. Brooks, Jr., C.L. Brooks, III, L. Nilsson, B. Roux, Y. Won, M. Karplus. CHARMM: The Energy Function and Its Parameterization with an Overview of the Program. *The Encyclopedia of Computational Chemistry*, Vol. 1. Wiley, Chichester, **1998**.

31 W.L. Jorgensen, D.S. Maxwell, J. Tirado-Rives. Development and testing of the OPLS all-atom force field on conformational energetics and properties of organic liquids. *J. Am. Chem. Soc.* 118:11225–11236, **1996**.

32 G. Nemethy, M.S. Pottle, H.A. Scheraga. Energy parameters in polypeptides, 9. Updating of geometrical parameters, non-bonding interactions and hydrogen bonding interactions for naturally occurring amino acids. *J. Phys. Chem.* 87(11): 1883–1887, **1983**.

33 T.A. Halgren. Merck molecular force field. 1. Basis, form, scope, parameterization, and performance of MMFF94. *J. Comput. Chem.* 17:490–519, **1996**.

34 T.A. Halgren. Merck molecular force field. 2. MMFF94 van der Waals and electrostatic parameters for intermolecular interactions. *J. Comput. Chem.* 17:520–552, **1996**.

35 T.A. Halgren. Merck molecular force field. 3. Molecular geometries and vibrational frequencies for MMFF94. *J. Comput. Chem.* 17:553–586, **1996**.

36 T.A. Halgren, R.B. Nachbar. Merck molecular force field. 4. Conformational energies and geometries for MMFF94. *J. Comput. Chem.* 17:587–615, **1996**.

37 T.A. Halgren. Merck molecular force field. 5. Extension of MMFF94 using experimental data, addition computational data, and empirical rules. *J. Comput. Chem.* 17:616–641, **1996**.

38 N.L. Allinger, K.H. Chen, J.H. Lii, K.A. Durkin. Alcohols, ethers, carbohydrates, and related compounds. I. The MM4 force field for simple compounds. *J. Comput. Chem.* 24:1447–1472, **2003**.

39 J.H. Lii, K.H. Chen, K.A. Durkin, N.L. Allinger. Alcohols, ethers, carbohydrates, and related compounds. II. The anomeric effect. *J. Comput. Chem.* 24:1473–1489, **2003**.

40 J.H. Kii, K.H. Chen. T.B. Grindley, N.L. Allinger. Alcohols, ethers, carbohydrates and related compounds. III. The 1,2-di-

methoxyethane system. *J. Comput. Chem.* 24:1490–1503, **2003**.

41 J. H. Lii, K. H. Chen, N. L. Allinger. Alcohols, ethers, carbohydrates, and related compounds. IV: Carbohydrates. *J. Comput. Chem.* 24:1504–1513, **2003**.

42 N. L. Allinger, Y. H. Yuh, J. H. Lii. Molecular mechanics – the MM3 force-field for hydrocarbons. 1. *J. Am. Chem. Soc.* 111:8551–8566, **1989**.

43 J. H. Lii, N. L. Allinger. Molecular mechanics – the MM3 force-field for hydrocarbons. 2. Vibrational frequencies and thermodynamics. *J. Am. Chem. Soc.* 111:8566–8575, **1989**.

44 A. C. T. van Duin, S. Dasgupta, F. Lorant, W. A. Goddard. ReaxFF: a reactive force field for hydrocarbons. *J. Phys. Chem. A* 105:9396–9409, **2001**.

45 J. C. Phillips, R. Braun, W. Wang, J. Gumbart, E. Tajkhorshid, E. Villa, C. Chipot, R. D. Skeel, L. Kale, K. Schulten. Scalable molecular dynamics with NAMD. *J. Comput. Chem.* 26:1781–1802, **2005**.

46 D. van der Spoel, E. Lindahl, B. Hess, G. Groenhof, A. E. Mark, H. J. C. Berendsen. GROMACS: fast, flexible, and free. *J. Comput. Chem.* 26:1701–1718, **2005**.

47 H. J. Limbach, A. Arnold, B. A. Mann, C. Holm. Espressoan extensible simulation package for research on soft matter systems. *Comput. Phys. Commun.* 174:704–727, **2006**.

48 I. T. Todorov, W. Smith. DL POLY 3: the CCP5 national UK code for molecular-dynamics simulations. *Philos. Trans. R. Soc. Lond. A* 362:1835–1852, **2004**.

49 K. Binder (Ed.). *Monte Carlo and Molecular Dynamics Simulations in Polymer Science*. Oxford University Press, Oxford, **1995**.

50 L. J. Fetters, D. J. Lohse, S. T. Milner, W. W. Graessley. Packing length influence in linear polymer melts on the entanglement, critical, and reptation molecular weights. *Macromolecules* 32:6847–6851, **1999**.

51 M. Pütz, K. Kremer, G. S. Grest. What is the entanglement length in a polymer melt? *Europhys. Lett.* 49:735, **2000**.

52 S. Leon, N. van der Vegt, L. Delle Site, K. Kremer. Bisphenol A polycarbonate: entanglement analysis from coarse-grained MD simulations. *Macromolecules* 38:8078–8092, **2005**.

53 L. J. Fetters, D. J. Lohse, W. W. Graessley. Chain dimensions and entanglement spacings in dense macromolecular systems. *J. Polym. Sci., Part B: Polym. Phys.* 37:1023, **1999**.

54 D. Reith, H. Meyer, F. Müller-Plathe. CG-OPT: a software package for automatic force field design. *Comput. Phys. Commun.* 148:299–313, **2002**.

55 F. Müller-Plathe. Coarse-graining in polymer simulation: from the atomistic to the mesoscopic scale and back. *ChemPhysChem* 3:754–769, **2002**.

56 D. Reith, B. Müller, F. Müller-Plathe, S. Wiegand. How does the chain extension of poly(acrylic acid) scale in aqueous solution? A combined study with light scattering and computer simulation. *J. Chem. Phys.* 116:9100–9106, **2002**.

57 D. Reight, H. Meyer, F. Müller-Plathe. Mapping atomistic to coarse-grained polymer models using automatic simplex optimization to fit structural properties. *Macromolecules* 34:2335–2345, **2001**.

58 F. Ercolessi, J. B. Adams. Interatomic potentials from 1st-principles calculations – the force matching method. *Europhys. Lett.* 26:583–588, **1994**.

59 S. Izvekov, M. Parrinello, C. J. Burnham, G. A. Voth. Effective force fields for condensed phase systems from *ab initio* molecular dynamics simulation: a new method for force-matching. *J. Chem. Phys.* 120:10896–10913, **2004**.

60 S. Izvekov, G. A. Voth. Multiscale coarse graining of liquid-state systems. *J. Chem. Phys.* 123:134105, **2005**.

61 W. Tschop, K. Kremer, J. Batoulis, T. Burger, O. Hahn. Simulation of polymer melts. I. Coarse-graining procedure for polycarbonates. *Acta Polym.* 49:61–74, **1998**.

62 J. D. McCoy, J. G. Curro. Mapping of explicit atom onto united atom potentials. *Macromolecules* 31:9362–9368, **1998**.

63 C. F. Abrams, K. Kremer. Combined coarse-grained and atomistic simulation of liquid bisphenol A-polycarbonate:

liquid packing and intramolecular structure. *Macromolecules* 36:260–267, **2003**.
64 V. A. Harmandaris, N. P. Adhikari, N. F. A. van der Vegt, K. Kremer. Hierarchical modelling of polystyrene: from atomistic to coarse-grained simulations. *Macrom.* 39:6708–6719, **2006**.
65 C. F. Abrams, K. Kremer. Effects of excluded volume and bond length on the dynamics of dense bead-spring polymer melts. *J. Chem. Phys.* 116:3162–3165, **2002**.
66 W. Tschop, K. Kremer, O. Hahn, J. Batoulis, T. Burger. Simulation of polymer melts. II. From coarse-grained models back to atomistic description. *Acta Polym.* 49:75–79, **1998**.
67 B. Hess, S. Leon, N. F. A. van der Vegt, K. Kremer. Long time atomistic polymer trajectories from coarse grained simulations: bisphenol-A polycarbonate. *Soft Matter* 2:409–414, **2006**.
68 R. Auhl, R. Everaers, G. S. Grest, K. Kremer, S. J. Plimpton. Equilibration of long chain polymer melts in computer simulations. *J. Chem. Phys.* 119:12718–12728, **2003**.
69 V. G. Mavrantzas, T. D. Boone, E. Zervopoulou, D. N. Theodorou. End-bridging Monte Carlo: a fast algorithm for atomistic simulation of condensed phases of long polymer chains. *Macromolecules* 32:5072–5096, **1999**.
70 A. Uhlherr, S. J. Leak, N. E. Adam, P. E. Nyberg, M. Doxastakis, V. G. Mavrantzas, D. N. Theodorou. Large scale atomistic polymer simulations using Monte Carlo methods for parallel vector processors. *Comput. Phys. Commun.* 144:1–22, **2002**.
71 H. Fukunagaa, J. Takimoto, M. Doi. A coarse-graining procedure for flexible polymer chains with bonded and non-bonded interactions. *J. Chem. Phys.* 116:8183–8190, **2002**.
72 L. Yelash, M. Muller, P. Wolfgang, K. Binder. How well can coarse-grained models of real polymers describe their structure? The case of polybutadiene. *J. Chem. Theory Comput.* 2:588–597, **2006**.
73 L. Delle Site, C. F. Abrams, A. Alavi, K. Kremer. Polymers near metal surfaces: selective adsorption and global conformations. *Phys. Rev. Lett.* 89:156103, **2002**.

74 C. F. Abrams, L. Delle Site, K. Kremer. Dual-resolution coarse-grained simulation of the bisphenol-A-polycarbonate/nickel interface. *Phys. Rev. E* 67:021807, **2003**.
75 L. Delle Site, S. Leon, K. Kremer. BPA-PC on a Ni(111) surface: the interplay between adsorption energy and conformational entropy for different chain-end modifications. *J. Am. Chem. Soc.* 126:2944–2955, **2004**.
76 D. Andrienko, S. Leon, L. Delle Site, K. Kremer. Adhesion of polycarbonate blends on a nickel surface. *Macromolecules* 38:5810–5816, **2005**.
77 X. Zhou, D. Andrienko, L. Delle Site, K. Kremer. Dynamic surface decoupling in a sheared polymer melt. *Europhys. Lett.* 70:264–270, **2005**.
78 X. Zhou, D. Andrienko, L. Delle Site, K. Kremer. Flow boundary conditions for chain-end adsorbing polymer blends. *J. Chem. Phys.* 123:104904, **2005**.
79 O. Hahn, D. A. Mooney, F. Müller-Plathe, K. Kremer. A new mechanism for penetrant diffusion in amorphous polymers: molecular dynamics simulations of phenol diffusion in bisphenol-A-polycarbonate. *J. Chem. Phys.* 111:6061–6068, **1999**.
80 F. Müller-Plathe. Diffusion of penetrants in amorphous polymers: a molecular dynamics study. *J. Chem. Phys.* 94:3192–3199, **1991**.
81 S. C. Rogers, F. Müller-Plathe, W. F. van Gunsteren. Computational evidence for anomalous diffusion of small molecules in amorphous polymers. *Chem. Phys. Lett.* 199:237–243, **1992**.
82 H. Schmitz, F. Müller-Plathe. Calculation of the lifetime of positronium in polymers via molecular dynamics simulations. *J. Chem. Phys.* 112:1040–1045, **2000**.
83 M. Deserno, C. Holm, J. Blaul, M. Ballauff, M. Rehahn. The osmotic coefficient of rod-like polyelectrolytes: computer simulation, analytical theory, and experiment. *Eur. Phys. J. E* 5:97–103, **2001**.
84 J. Mayer. *Equilibrium Statistical Mechanics*. Pergamon Press, Oxford, **1968**.
85 A. P. Lyubartsev, A. Laaksonen. Calculation of effective interaction potentials

from radial distribution functions: a reverse Monte Carlo approach. *Phys. Rev. E* 52:3730–3737, **1995**.
86 A. P. Lyubartsev, S. Marčelja. Evaluation of effective ion-ion potentials in aqueous electrolytes. *Phys. Rev. E* 65:041202, **2002**.
87 B. Hess, C. Holm, N. van der Vegt. Osmotic coefficients of atomistic NaCl(aq) force fields. *J. Chem. Phys.* 124:164509, **2006**.
88 C. Holm, K. Kremer (Eds). *Advanced Computer Simulation Approaches for Soft Matter Science I. Advances in Polymer Science, Vol. 173*, Springer Verlag, **2005**.
89 C. Holm. K. Kremer (Eds). *Advanced Computer Simulation Approaches for Soft Matter Science II. Advances in Polymer Science Vol. 185*, Springer Verlag, **2005**.
90 J. Baschnagel, K. Binder, P. Doruker, A. A. Gusev, O. Hahn, K. Kremer, W. L. Mattice, F. Müller-Plathe, M. Murat, W. Paul, S. Santos, U. W. Suter, V. Tries. *Advances in Polymer Science, Vol. 152: Viscoelasticity, Atomistic Models, Statistical Chemistry.* Springer, Heidelberg, **2000**.

3
Transport and Electro-optical Properties in Polymeric Self-assembled Systems

Olli Ikkala and Gerrit ten Brinke

3.1
Introduction

3.1.1
Background on Polymeric Self-assembly and Supramolecular Concepts

A central theme in materials science involves efforts to tune the properties of materials according to the requirements of the applications. This usually requires that several properties have to be tailored simultaneously and perhaps even synergistically, which can be nontrivial. For example, the electrical and mechanical properties typically have to be tailored simultaneously in combination with stability, processability and economics. Feasible combinations of polymeric properties have been widely pursued using multicomponent materials and here the controlled periodic nanoscale structures of block copolymers were appreciated at an early stage (e.g. [1–3]). Encouraged by the rapid development of supramolecular chemistry [4, 5], tailored physical interactions have recently been used to allow supramolecular polymers to be obtained [6–11]. Finding and tailoring new functionalities have become one of the main efforts and the biosciences have been a source of inspiration (e.g. [12–15]). Periodic nanoscale structures have been denoted differently depending on the background of the researchers. For example, in block copolymers, periodic nanostructures have been denoted microphase separation, nanophase separation, self-assembly, mesomorphism or self-organization. Shape-persistent mesomorphic groups, such as rigid rods, can also be incorporated to facilitate the structures and in such cases the terms mesomorphism or liquid crystallinity have typically been used [16–22]. Note that the definitions seem not to be strict; for example, the structure formation of aqueous surfactant systems has been denoted lyotropic liquid crystallinity (e.g. [23, 24]). Therefore, in this treatment, the term self-assembly is adopted. It has been discussed previously how nanoscale structures can result from competing attractive and repulsive interactions [25]. There the concept was denoted self-organization. In thermodynamics, however, self-organization some-

Macromolecular Engineering. Precise Synthesis, Materials Properties, Applications.
Edited by K. Matyjaszewski, Y. Gnanou, and L. Leibler
Copyright © 2007 WILEY-VCH Verlag GmbH & Co. KGaA, Weinheim
ISBN: 978-3-527-31446-1

times refers to dissipative structure formation [26, 27] and self-assembly is preferred here. Therefore, in this chapter we understand by self-assembly spontaneous structure formation due to competing attractive and repulsive interactions in the fluid, glassy or crystalline state. Self-assembly can also be static and dynamic [28], and numerous examples have been presented in this and closely related fields (e.g. [19–22, 28–47]).

As mentioned, a classical example of polymeric self-assembly is provided by block copolymers [1, 2, 40, 41, 48–52]. As most polymers do not mix, they are mutually repulsive, taken that the chains are sufficiently long. If they are covalently connected within the same molecules to form block copolymers, i.e. connected by infinite attractive interaction, the free energy is minimized upon forming nanoscale structures. The structures depend on the chemical nature of the blocks and also on the temperature (which are combined within the χ-parameters), the length of the blocks, the architecture and the number and sequence of the blocks. There can be two, three or more repulsive blocks and they can be connected to form different architectures, such as linear, comb-shaped, star-shaped (mictoarm), T-shaped, H-shaped, (hyper)branched or dendritic shapes. The blocks can be flexible, forming coiled conformations, semi-rigid or even completely rod-like, which are prone to mutual aggregation. Figure 3.1 shows some examples of different block architectures. If there are two flexible chains connected in a linear fashion as in a classical diblock copolymer, spherical, cylindrical, lamellar and gyroid and hexagonal perforated structures are obtained (Fig. 3.2). The disordered phase prevails at very high temperatures and there can be a sequence of different phases as a function of temperature. If the

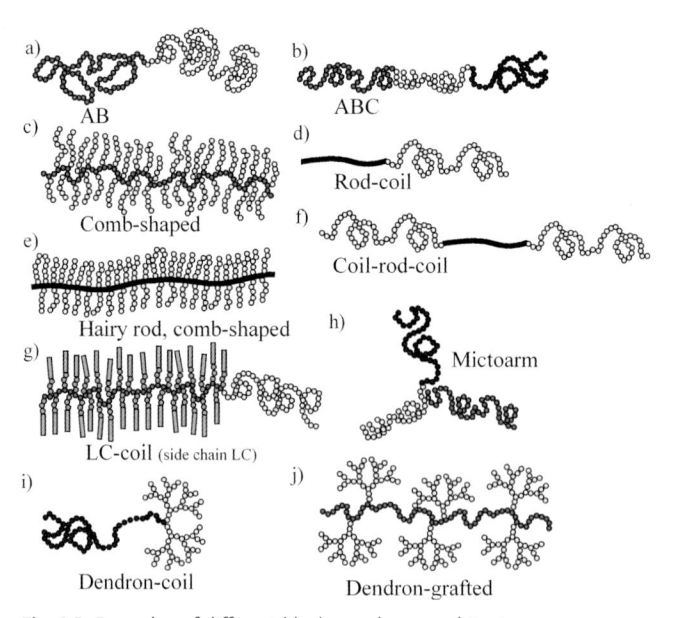

Fig. 3.1 Examples of different block copolymer architectures.

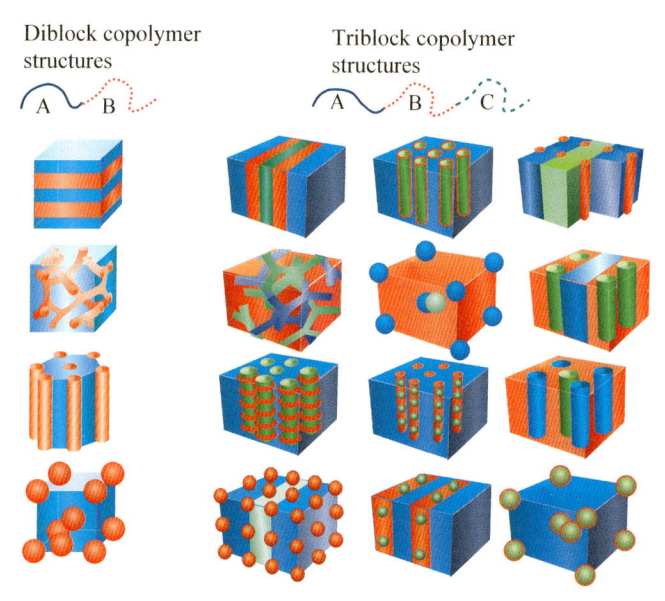

Fig. 3.2 Examples of self-assembled AB diblock copolymer and ABC triblock copolymer structures obtained by connecting flexible polymer chains in a linear architecture. Adapted from [49].

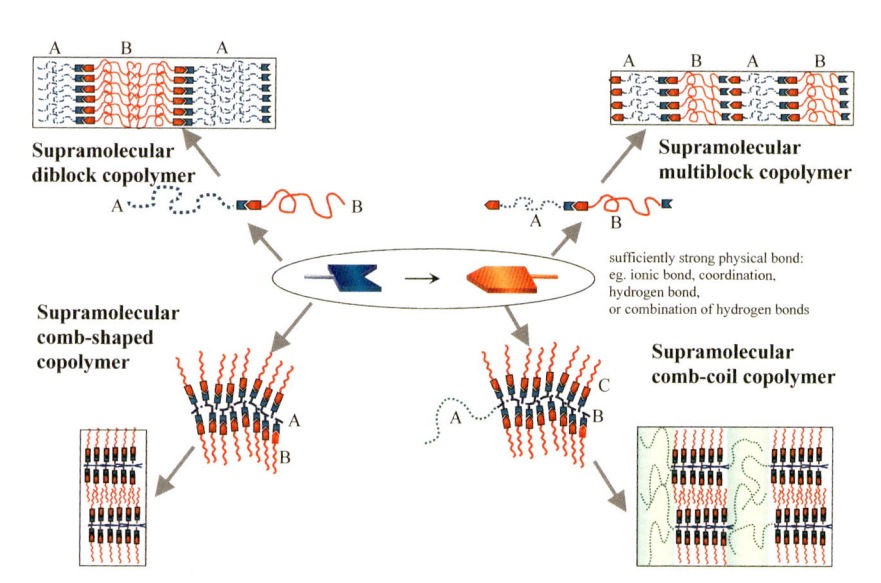

Fig. 3.3 Some examples of various architectures for self-assembly using supramolecular interactions.

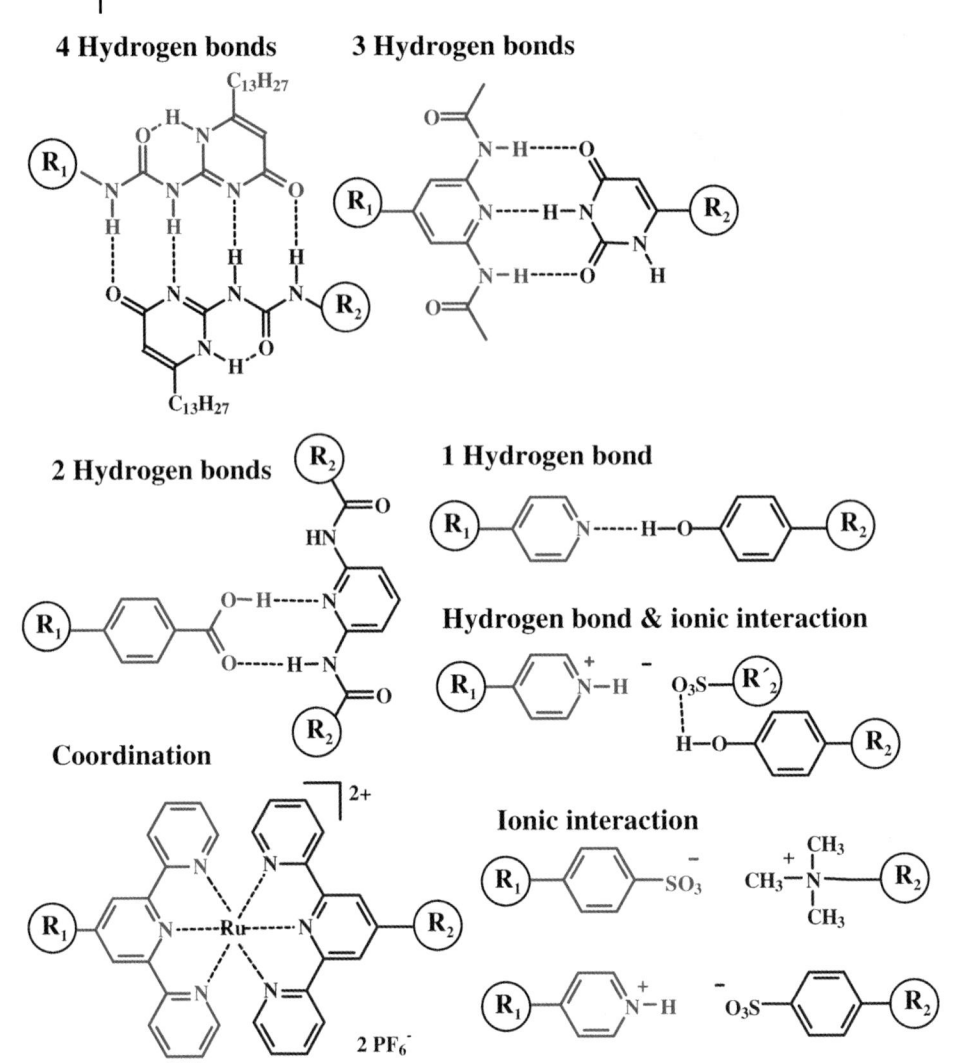

Fig. 3.4 Selected examples of the use of physical interactions to construct self-assembling materials (see also Fig. 3.3) [4, 6, 7, 11, 21, 34, 44, 47, 74, 92, 96]. R_1 and R_2 represent the repulsive moieties to be connected to achieve self-assembly.

number of the blocks is increased to three, more structures are available, for example core–shell cylinders (Fig. 3.2). Further increase in the number of blocks leads to a rapidly increasing number of structures and structural complexity.

Another aspect has become relevant due to supramolecular chemistry [4, 5], i.e. that the attractive interactions can also be physical interactions, taken that they are sufficiently strong to withstand the repulsive interactions between the blocks. The interactions can be ionic [43, 53–66], in which case the term ionic self-assembly can be used [42]; they can be hydrogen bonds or their combinations [10, 11, 21, 22, 34, 67–88]; they can be coordinative [89–93]; or they can in principle be any sufficiently strong attractive interaction or their combination. Over the years, several examples have arisen of physical interactions that enable one to prepare polymer-like materials (Figs. 3.3 and 3.4). For example, ionically interacting sites at the ends of telechelic polystyrene and poly(ethylene oxide) form "ionic multiblock copolymers" [55], the combination of four hydrogen bonds leads to linear polymeric supramolecules [6, 10, 11] where self-assembly can be achieved [94] and coordination bonds can be used to prepare polymers [7, 8, 95] and "coordinative block copolymers" [91, 92].

3.1.2
General Remarks on Polymeric Self-assembly in Relation to Electrical and Optical Properties

This chapter emphasizes selected aspects of tuning the electric and optical functionalities using polymeric self-assembly in the light of selected recent work. Protonic and ionic transport using self-assembled polymers will be discussed first, as they could have relevant applications in energy storage applications, i.e. proton conductors for fuel cell membranes [97] and the Li^+ conductors [98, 99] for batteries, where the aim is to combine high protonic or ion transport with sufficient mechanical properties and stability. Undoped conjugated polymers are semiconducting, they can be doped for conductivity and they have interesting electrical and optical properties [47, 100–102]. Due to the rigidity of the conjugated chains, there is a tendency for aggregation, which can lead to infusibility and poor solubility. Incorporating side-chains improves these aspects (see Fig. 3.1), but simultaneously due to the side-chains they typically also self-assemble. Therefore, self-assembly is intimately but indirectly connected to many of the conjugated polymers.

In general, the electrically conducting multicomponent materials require continuous pathways of conducting channels across the material, i.e. percolative systems. This is a demanding requirement, as self-assembly typically leads only to local order and macroscopically the materials are usually constituted of domains with differently aligned self-assembled domains and domain boundaries between them. Notably, the electrically conducting channels can be discontinuous at the domain boundaries. More generally, aligned and monodomain structures in block copolymers have been pursued using several concepts, for example using thin films where the surface energies play a major role, electric fields, magnetic fields, flow alignment and graphoepitaxy [86, 103–122]. Therefore, in many of the applications requiring enhanced electrical transport, a generic goal in relation to self-assembly is to prepare monodomain structures having e.g. well-aligned cylindrical or lamellar conducting channels or monodomain co-continuous phases, such as the gyroid structure or other networks. Another scheme is to aim at molecular reinforcement due to self-assembly, in which case the matrix is conducting and the material contains self-assembled reinforcements. Still another possibility is to use (polymeric) colloids, typically spheres, and to tune the conductive percolative channels using self-assembly. Finally, one should emphasize the different ways of achieving continuous networks within polymeric matrices using block copolymers and self-assembling moieties [52, 123–127].

In relation to the optical properties, self-assembly can allow a variety of different properties, for example photonic bandgap materials (or dielectric mirrors), materials for photovoltaics, nonlinear optics, polymeric light-emitting devices, optical storage and sensors and photorefractive materials. Some of them are discussed here in the perspective of polymeric self-assembly.

A final remark is that due to the broad field, all facets of this rapidly developing field cannot be covered within the present limited space. In particular, we do not deal with layer-by-layer and self-assembled monolayer techniques, Langmuir–Blodgett techniques, electrochromic and photorefractive materials, and electroluminant polymers are mentioned only in passing.

3.2
Transport of Ions and Protons within Self-assembled Polymer Systems

3.2.1
Self-assembled Protonically Conducting Polymers

Proton conductivity or proton transport is of major interest in several fields, ranging from biological systems to materials science combining chemistry and physics, some of them having important technological potential, for example in fuel cells (for the extensive literature on various applications see for example [128–156] and more generally for ion-containing polymers [157]). Responsive smart materials will also be presented, illustrating possibilities offered by phase transitions to tune the properties.

There are several stringent requirements for the fuel cell membrane material between the anode and the cathode: the material should be highly protonically conducting but still not electronically conducting, should be mechanically strong to withstand robust use and should withstand harsh chemical environments and the economics should be acceptable. To add one further requirement: the feasible operating temperature can be in excess of $100\,°C$, where the possible detrimental effects of CO on the catalytic materials are reduced, depending on the fuel cell concept and materials used.

Widely studied protonically conducting materials are polymeric perfluorosulfonic acids which contain fluorocarbon chains and where sulfonic acid groups are grafted on the side-chains [158]. It becomes conducting as the sulfonic acids are highly hygroscopic and the polymer can be considered as a percolating medium for water-mediated proton conductivity. Water has been the most studied proton conducting medium [130, 159]. Protonated water species, such as H_3O^+

and $H_5O_2^+$, can diffuse, which is called the vehicular mechanism. On the other hand, water molecules form hydrogen-bonded networks as they are both hydrogen bonding acceptors and hydrogen-bonding donors. In such a medium, proton hopping from the protonated water species to nonprotonated H_2O species can take place, followed by their reorientation, i.e. the Grotthus mechanism [130, 159]. In order to exhibit substantial conductivity across the membrane, the water clusters have to form connected networks, i.e. they have to percolate. This requires that during the operational conditions, there has to be a sufficient level of moisture to allow water absorption and the connectivity of the protonically conducting channels should be controllable and stable. The latter aspect has been the subject of intense study in perfluorinated membranes already for an extended period [146, 158].

If the optimal operating temperature is high, water is not a preferred protonically conducting medium due to its low boiling-point. There has been an intense search for alternative amphoteric hydrogen-bonded compounds capable of rendering high proton conductivity. Phosphoric acid, pyrazole, imidazole and benzimidazole have been studied [143, 149, 160–162]. Phosphoric acid is relatively highly conducting [163]. It, and other acids, have been blended with poly(ethylene oxide), poly(vinyl alcohol), polyacrylamide and basic polymers rendering salts, such as polyethylenimine, polydiallyldimethylammonium salts, poly(4-vinylimidazole), poly(4-vinylpyridine) and polybenzimidazole [164–175]. The acid–base complexes have aroused interest even in the anhydrous case [176]. Another concept has been based on imidazoles and it has also been immobilized using polysiloxane backbones to which imidazoles are connected using flexible oligoethylene spacers [161]. To enhance transport, one may favor systems with low proton density and a large number of binding sites [131].

Methods to achieve conducting domains based on self-assembly are addressed next. Undecylimidazole protonated by a small amount of monododecylphosphoric acid leads to lamellar self-assembly and conductivity under anhydrous conditions [177]. Extending to polymers, there are several challenges: A narrow polydispersity is required to control the structures. This typically requires living polymerization, which may not be straightforward using polymers with the chemical stability required in fuel cells. Second, self-assembly renders only local structures and it is nontrivial to achieve continuous conducting channels across the membrane. An elegant approach could be to construct a gyroid structure (Fig. 3.2) but it may be nontrivial to achieve in large-scale applications even using more straightforward materials. Another scheme would be to use a cylindrical structure with the cylinders aligned across the membranes.

There have been extensive studies on block copolymers consisting of polyolefinic and polystyrene blocks where the latter blocks are partially sulfonated for enhanced water absorption and protonic conductivity (e.g. [178–201]). Typical materials are hydrogenated triblock copolymers, such as polystyrene-b-polybutadiene-b-polystyrene (SBS), polystyrene-b-polyisobutene-b-polystyrene (SIBS) and polystyrene-b-polyethylene/butene-b-polystyrene (SEBS), and also architectures other than linear have been used (Fig. 3.5). Taking SEBS with 28% polystyrene, where the polystyrene is partially sulfonated, a well-defined lamellar self-assembly is obtained on casting the membrane from THF [186, 198]. However, the conductivity across the membrane, i.e. across the lamellae, remains rather low, which is not surprising, as the lamellae tend to be aligned along the external surfaces [186, 198]. Methanol and water swelling change the morphology to a more disordered co-continuous-like structure and the conductivity increases across the membrane. However, the methanol transport also increases which is undesirable. The structures and conductivity can be tuned by changing the degree of sulfonation [185, 188] and the methanol permeation can be reduced [188] by chemical modifications using maleic anhydride groups within the ethylene/butene midblocks. Sulfonated SBS in combination with UV cross-linked butene groups has been used to immobilize the cylindrical protonically conducting sulfonated PS domains [189]. The conductivity is comparable to that of Nafion and the methanol permeability is strongly reduced [189]. A systematic study dealing the correlation between the structure and conductivity was recently reported using sulfonated SIBS [195, 201] using a range of degrees of sulfonations and different solvents to cast the membranes. Well-defined cylindrical and lamellar self-assembled ion channels could be obtained but typically they are parallel to membrane surfaces, leading to reduced conductivity across the mem-

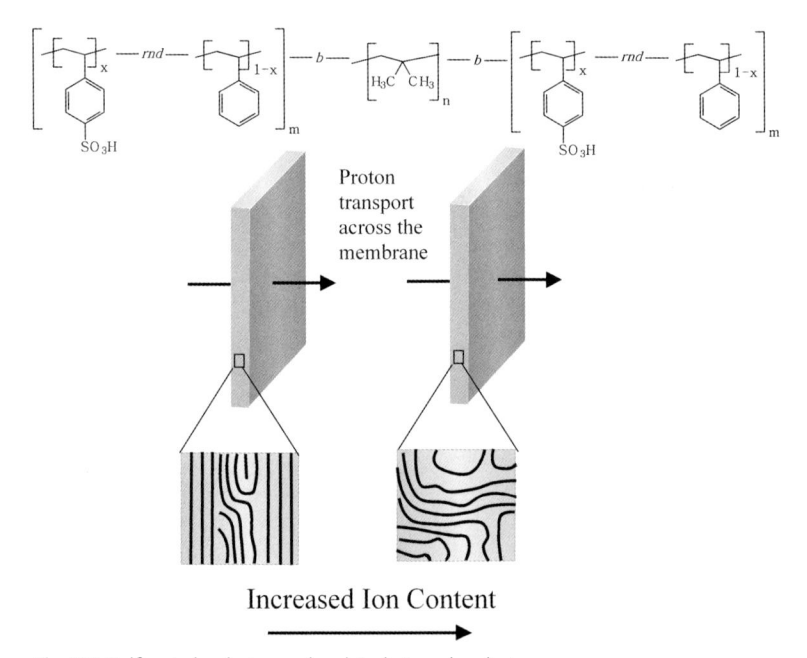

Fig. 3.5 Sulfonated polystyrene-*b*-polyisobutene-*b*-polystyrene leads to well defined self-assembled structures at low levels of sulfonation and the domains are aligned parallel to the membrane. Increasing the sulfonation level leads to less periodic structures but also to increased proton transport [201].

brane. At higher ion contents, the structures become less well defined, but the conductivity across the membranes increases substantially. Such sulfonated triblock copolymers can render conductivity values approaching those of Nafion, but the methanol transport and chemical stability may still limit the practical use in fuel cells.

Another effort to control the structure of the conducting sulfonated polystyrene domains is based on the polystyrene backbone, where well-defined polystyrene sulfonate grafts are connected using controlled spacings (Fig. 3.6) [202]. This leads to 5–10-nm wide sulfonated polystyrene channels through the polystyrene matrix which are connected as a continuous ionic network, thus providing conductivity 0.24 S cm^{-1} with a styrene sulfonic acid loading of 19.1 mol% and which contains only 37 vol.% water. This conductivity is 3–5 times larger than that of Nafion 117 for a similar water content.

The conducting channel orientation and connectivity are critical for high conductivity. It is nontrivial to obtain perpendicular self-assembled structures across block copolymer films and this requires careful consideration of solvents, surface energies, film thicknesses, nanoparticles and potentially using an external field [112–117, 122, 203–206]. That aligned protonically conducting self-assembled structures can lead to anisotropic transport has been separately investigated [87, 175, 195, 207]. A model material related to fuel cells is polystyrene-b-poly(4-vinylpyridine) mixed with phosphoric acid, where a lamellar self-assembly was obtained having alternating lamellar domains of polystyrene and an acid–base mixture of P4VP and phosphoric acid [175]. The pure P4VP(H$_3$PO$_4$)$_x$ is relatively well conducting, i.e. 10^{-2} S cm^{-1} at 100 °C for $x=2.5$. The lamellar self-assembled structures were flow aligned and the alternating conducting and PS lamellae are both parallel to the flow after the alignment. An anisotropy in the conductivity was observed, but the anisotropy was only one order of magnitude. This suggests that it may be challenging to have macroscopic nanoscale domains that do not have "dead-ends" even if the structures are aligned.

Another approach for well-defined alignment of self-assembled conducting domains is based on thin films where the interfacial energies of the external sur-

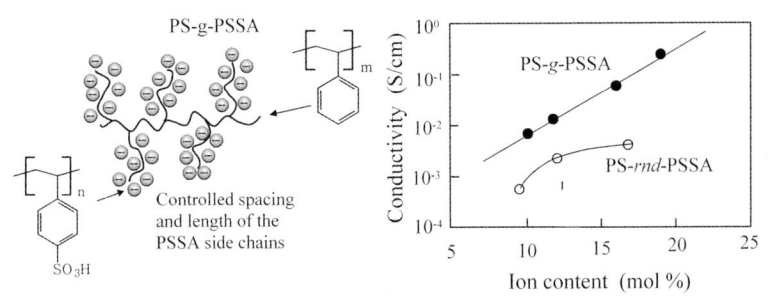

Fig. 3.6 A concept to achieve conducting hydrated ionic poly(styrenesulfonic acid) (PSSA) channels through a polystyrene (PS) matrix using grafting [202].

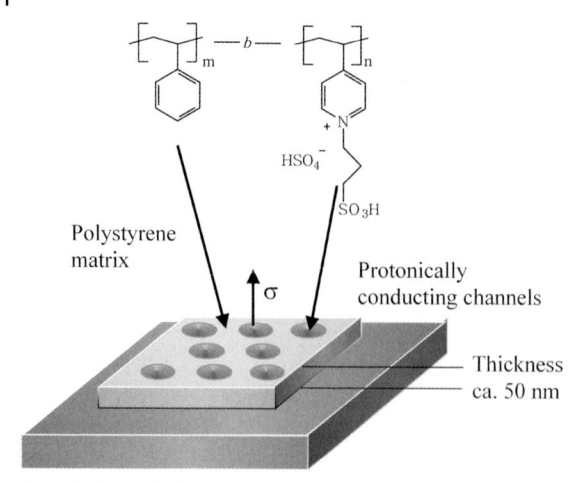

Fig. 3.7 Protonically conducting thin membrane, showing high conductivity anisotropy across the membrane [208].

faces can be used [204, 205]. To achieve self-assembled protonically conducting membranes, the pyridine groups of polystyrene-b-poly(4-vinylpyridine) were ionically functionalized, leading to cylindrically self-assembled ionic domains (Fig. 3.7) [208]. The resulting films show highly anisotropic conductivity, i.e. 0.015 S cm^{-1} across the membrane and very low conductivity $<10^{-15}$ S cm^{-1} to the lateral direction at 25 °C under 45% relative humidity. We emphasize that the concept is based on the very thin films (50 nm) and, in fact, the aspect ratio of the channels is small in this case.

Chemically more stable materials are preferred for fuel cell membranes, such as fluorinated materials, but even aromatic polymers may suffice for less stringent applications [137, 151, 154, 209].

Recently there have been efforts to construct block copolymers based on fluorinated blocks, aiming to combine self-assembly and chemical stability. The proton conductivity of sulfonated polysulfones possessing a relatively low degree of sulfonation can be enhanced by block copolymerization with poly(vinylidene difluoride) (PVDF) [210]. This may be due to the self-assembly between the sulfonated and hydrophobic blocks. Sulfonated poly(aryl ether ketone) copolymers based on 4,4′-(hexafluoroisopropylidene)diphenol, 1,3-bis(4-fluorobenzoyl)benzene and disulfonated difluorobenzophenone have good thermal and oxidative stability [211]. Amphiphilic block copolymers comprising sulfonated poly(vinylidene difluoride-co-hexafluoropropylene)-b-polystyrene were synthesized having moderately low polydispersity index [212, 213]. Transmission electron microscopy studies suggested interconnected networks of ion channels, each 8–15 nm wide and the conductivity was of the same order as that of Nafion. The proton conductivities reach $>3.3\times10^{-2}$ S cm^{-1} at 80 °C. The films show structure at two length scales based on X-ray scattering: the structure at a length scale of ap-

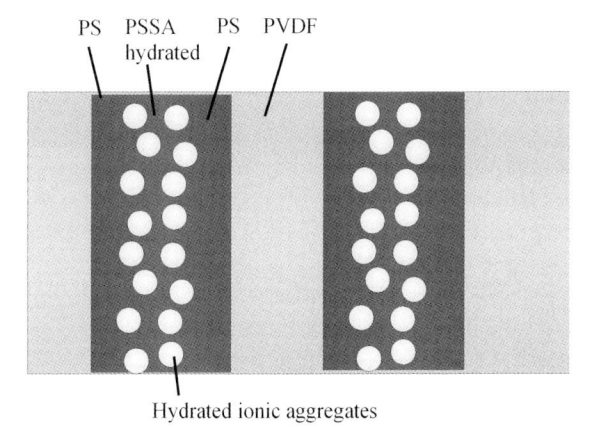

Fig. 3.8 Hierarchical self-assembly of protonically conducting poly(vinylidene difluoride-co-hexafluoropropylene)-b-polystyrene [213].

proximately 4 nm is due to the immiscibility of the polystyrene and the fluorinated blocks and a substructure within the sulfonated polystyrene domains exists due to segregation of the hydrated ionic groups and the hydrophobic polystyrene chains (Fig. 3.8). The longer length scale morphology shows relatively high ordering whereas within the sulfonated polystyrene blocks there is a more disordered structure.

A different concept for combining polymeric self-assembly and protonic conduction is based on colloidal self-assembly (Fig. 3.9) [214]. Lightly cross-linked ca. 500-nm polystyrene colloidal spheres were dispersed in ethanol with polystyrene-b-poly(2-vinylpyridine) block copolymer, which has a short polystyrene and a long poly(2-vinylpyridine) block. After solvent removal and under appropriate annealing, the colloidal system was deformed to polyhedra where the poly(2-vinylpyridine) surface layers percolate. Conductivity was obtained upon doping the latter using sulfonic acid [214]. The colloidal concept has been developed further to lead relatively high protonic conductivity [215].

Protonically conducting self-assembled polymers allow also smart and responsive materials. The pyridine groups of diblock copolymer polystyrene-b-poly(4-vinylpyridine), PS-b-P4VP were ionically complexed with methanesulfonic acid (MSA) and the resulting poly(4-vinylpyridinium methanesulfonate) was hydrogen bonded with an amphiphile, pentadecylphenol (PDP). The latter molecule consists of a nonpolar pentadecyl alkyl tail and a polar end-group, consisting of hydrogen-bonding phenol. This leads to hierarchical lamellar-*within*-lamellar self-assembly at two length scales: the first self-assembly between the PS and P4VP(MSA)(PDP)-block, leading to a periodicity of ca. 35 nm, and within the latter block another level of self-assembly with a periodicity of 4.8 nm takes place, where the PDP and P4VP(MSA) form alternating layers. A sequence of phase transitions takes place upon heating and cooling, which manifests in the

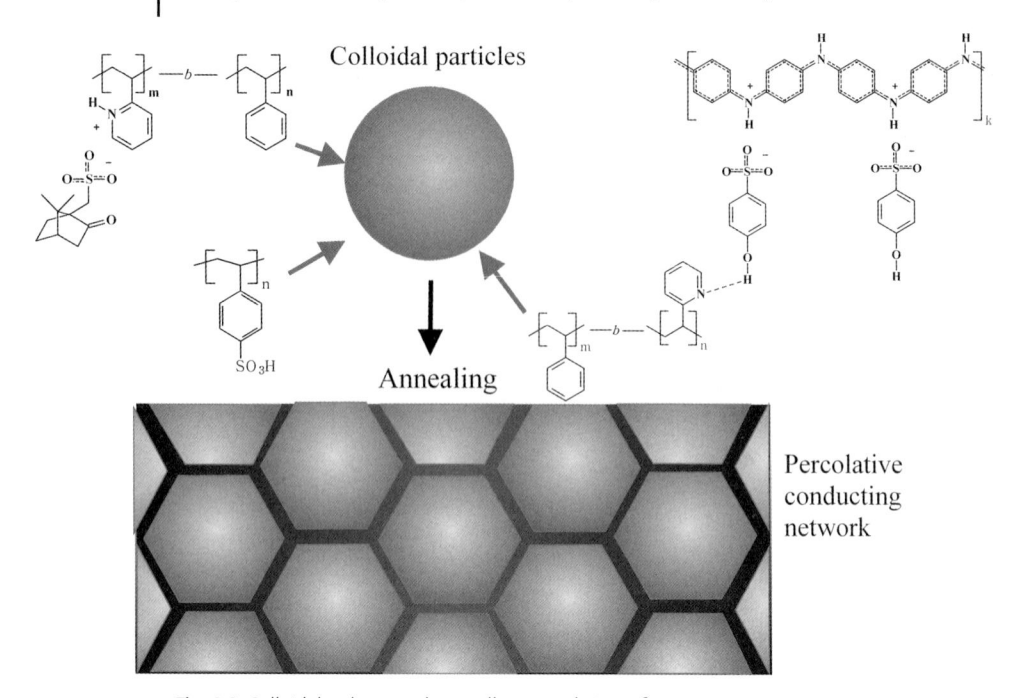

Fig. 3.9 Colloidal polymer spheres allow templating of protonically or conjugated polymers to allow percolated networks [214–216].

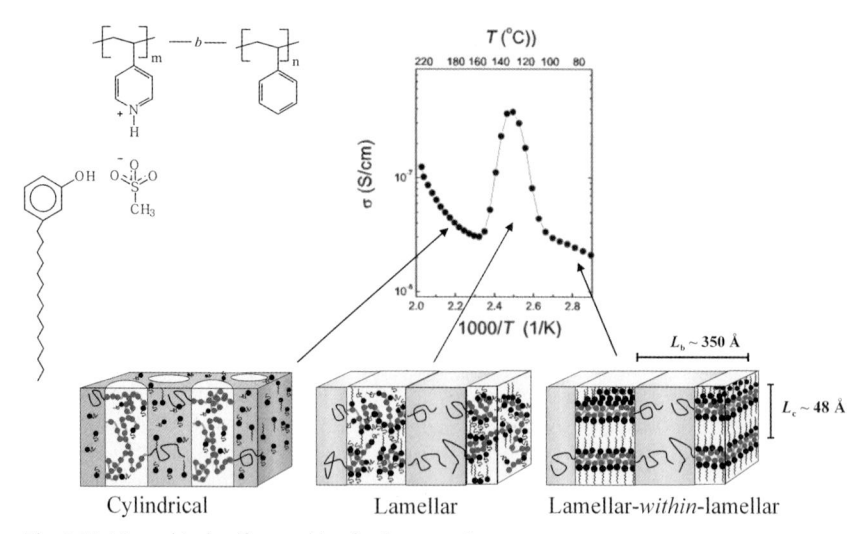

Fig. 3.10 Hierarchical self-assembly of polystyrene-b-poly(4-vinylpyridinium methanesulfonate) hydrogen bonded with pentadecylphenol amphiphile [34]. As a function of temperature different self-assembled phases are obtained, which consequently affects the conductivity.

conductivity (Fig. 3.10). More generally, related hierarchically self-assembled systems allow tuning of the properties by controlling the structures at different length scales [44, 87, 207, 217–222]. Polyelectrolyte–surfactant complexes self-assemble into various nanoscale structures and they can exhibit electrical conductivity [223, 224].

3.2.2
Self-assembled Ionically Conducting Polymers

Self-assembled ionic conductors have recently been reviewed [21, 22, 225]. Ionic conductors are required in various electrochemical applications based on polymeric electrolytes, especially for Li^+ batteries [98, 226–231]. A widely studied system is a salt-in-polymer system where a lithium salt, e.g. CF_3SO_3Li, has been solvated in the poly(ethylene oxide) (PEO) matrix to form solid polymer electrolytes [98, 231–236]. However, enhanced ionic conductivity requires high chain mobility and at room temperature the conductivity can be reduced due to crystallization of PEO. To allow amorphous materials with low glass transition temperatures has been tackled in many ways, e.g. using plasticizers or using short PEO chains that have increased mobility and suppressed crystallization where the graft and branched architectures have been useful [237–240]. Also, self-assembly is used where amorphous ethylene oxide-containing domains are incorporated within block copolymeric structures, leading to synergistic properties [225, 241–243].

There were early efforts to combine PEO within block copolymer structures, for example by grafting short PEO grafts to the middle block of polystyrene-*b*-polybutadiene-*b*-polystyrene triblock copolymer [244, 245]. Upon solvating CF_3SO_3Li to the PEO, conductivity in the order of 10^{-5} S cm^{-1} at ambient temperature was obtained and self-assembly and suppressed PEO crystallization were observed. Polystyrene-*b*-polyhydroxystyrene-*b*-polystyrene with PEO grafts within the middle block showed self-assembly and relatively high ionic conductivity upon solubilizing $LiClO_4$ in the PEO grafts [246]. Polyethylene-*b*-poly(ethylene oxide-*co*-propylene oxide)-*b*-polyethylene triblock copolymers with 10 wt.% of the end blocks leads to gelation based on physical networking [247]. Lithium bis(trifluoromethylsulfonyl)imide salt (LiTFSI) was solvated therein to allow conductivity approaching 10^{-5} S cm^{-1} at 20 °C [247, 248]. A comb-shaped architecture of gel electrolyte consists of a polystyrene backbone where PEO or poly(ethylene oxide-*co*-propylene oxide) chains are connected [249]. All the polyether side-chains were terminated by hydrocarbon (C_{16}) chain ends. LiTFSI salt solvation leads to ion conductivity that reached 10^{-2} S cm^{-1} at 20 °C. A related approach is based on amphiphilic methacrylate polymers with oligo(ethylene oxide) segments terminated by a linear alkyl chain [250]. The polymers self-assemble into nanometer-thick lamellar structures with alternating layers of conductive ethylene oxide and insulating alkyl layers.

In an effort to combine rubbery mechanical macroscopic properties and self-assembled ionically conducting domains, poly(lauryl methacrylate)-*b*-poly[oli-

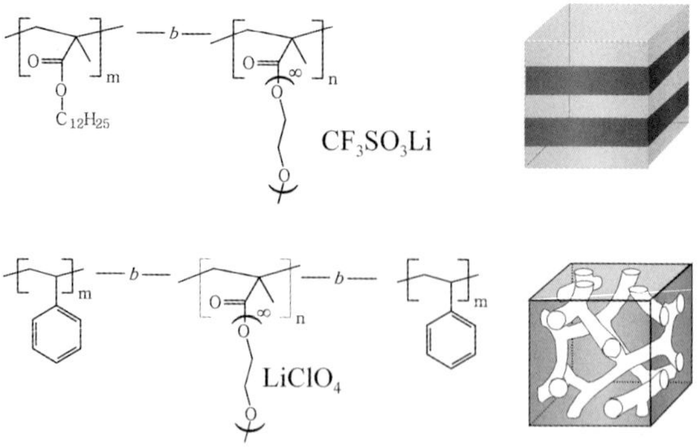

Fig. 3.11 Examples of self-assembled Li-conducting polymer electrolytes [242, 251].

go(oxyethylene) methacrylate] diblock copolymers containing CF_3SO_3Li were studied using different molecular weights and relatively narrow molecular weight distributions (Fig. 3.11) [241]. Reducing the length of the lauryl side-chain to be only methyl, i.e. using poly(methyl methacrylate)-b-poly[oligo(oxyethylene) methacrylate], reduces the repulsion between the blocks and self-assembly does not occur [242]. However, adding CF_3SO_3Li leads to self-assembly and conductivity. Therefore, both the structure and conductivity can be tuned by solvating the salt. Single-ion conducting block copolymer electrolytes were prepared using poly(lauryl methacrylate)-b-poly(lithium methacrylate)-b-poly[(oxyethylene)$_9$ methacrylate], in which case the counterions for lithium ions were bound to the polymer backbone [243]. This allows a lithium transfer number of unity and a conductivity of 10^{-5} S cm^{-1} at 70 °C. Various further modifications have recently been reviewed [225]. An effort to use the gyroid phase for Li$^+$ conductors was described using polystyrene-b-polyethylene oxide and LiClO$_4$ [251]. The conductivity level is relatively high, i.e. more than 10^{-4} S cm^{-1} at room temperature. The polystyrene domains promote mechanical strength and the PEO domains allow ion conduction in the polymer electrolyte without any plasticizer.

The phase transitions of the self-assembled structures can have a major effect on the ionic conductivity, as in the case of protonically conducting self-assemblies [34]. A diblock copolymer was synthesized having a linear PEO block and dendron-like block with a controlled number and length of alkyl chains (Fig. 3.12) [252]. CF_3SO_3Li was solvated in PEO. Such amphiphilic dendrons showed complex sequences of self-assembled phases ranging from crystalline lamellar, through cubic micellar (*Pm3n*), hexagonal columnar and continuous cubic (*Ia3d*) to fluid lamellar mesophases, until a disordered structure, each having their characteristic conductivity.

Extensive research has been carried out to achieve a helical PEO-like environment for the Li$^+$ ions and to combine it with self-assembly [236, 253–259]. The building blocks are poly[2,5,8,11,14-pentaoxapentadecamethylene(5-alkyloxy-1,3-phenylene)] chains, which typically have hexadecylmethylene (C$_{16}$) side-chains (Fig. 3.13). Due to the comb-shaped architecture consisting of the backbone and

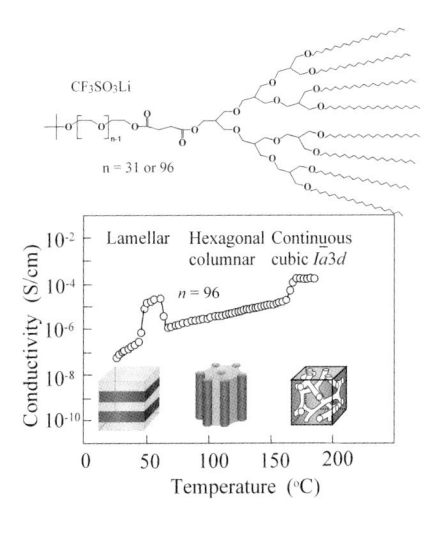

Fig. 3.12 An example of dendron-containing block copolymers and the observed ionic conductivity as a function of temperature upon mixing with CF_3SO_3Li [252].

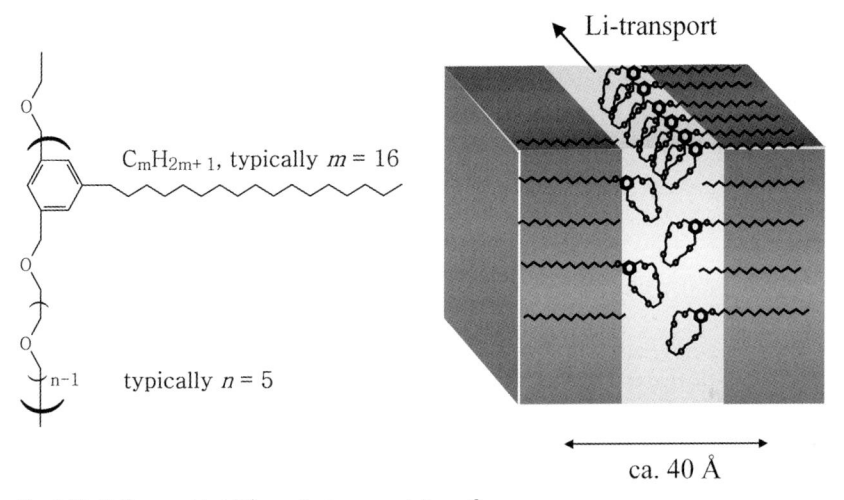

Fig. 3.13 Self-assembled Li^+ conductors consisting of poly[2,5,8,11,14-pentaoxapentadecamethylene(5-alkyloxy-1,3-phenylene)] [236, 253–259].

the repulsive dense set of side-chains, lamellar self-assembly is achieved where the alkyl chains interdigitate. The polymer backbone forms helices which incorporate the Li^+ ions. The compositions may also include additional polymer components or alkanes to tune the properties. The conductivity levels are in excess of 10^{-4} S cm^{-1} at 40 °C, where additional alkyl groups can be useful within the alkyl tail domains. A copolymer was also synthesized based on poly[2,5,8,11,14-pentaoxapentadecamethylene-(5-alkyloxy-1,3-phenylene)] and poly[2-oxatrimethylene(5-alkyloxy-1,3-phenylene)], where the alkyl side-chains are hexadecyl or mixed dodecyl–octadecyl (50:50) [260]. A conduction mechanism is proposed

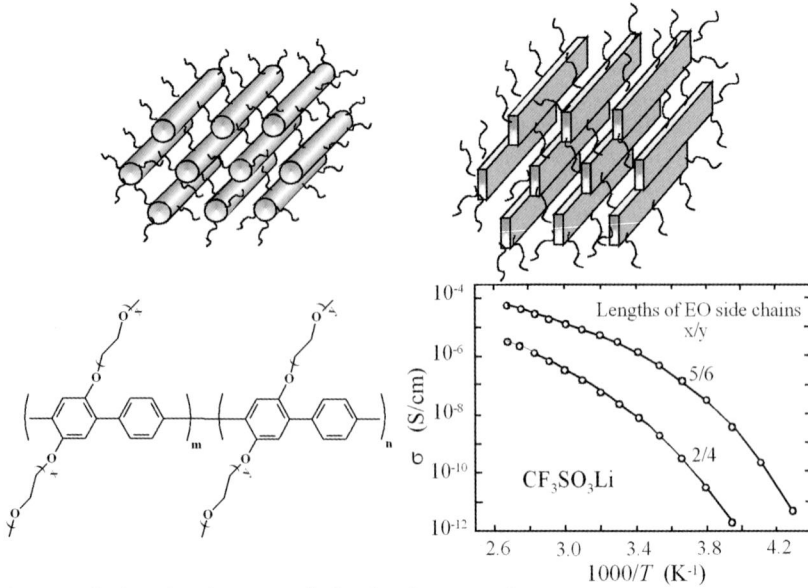

Fig. 3.14 If poly(p-phenylene) is grafted with a dense set of side-chains, lamellar or cylindrical self-assemblies can be obtained [264]. Poly(p-phenylene) containing two different lengths of ethylene oxide side-chains and the resulting ionic conductivity values obtained upon mixing CF_3SO_3Li for side-chains of lengths 5 and 6 ethylene oxide repeat units [266].

whereby Li⁺ hopping takes place along rows of decoupled aggregates (dimers/quadrupoles).

It can be challenging to ensure the alignment and connectivity of the ionically conducting self-assembled domains within a matrix polymer. This encourages the study of another route where the ionically conducting polyelectrolyte phase is molecularly reinforced using self-assembled rigid rod polymers. One concept is based on grafting short ethylene oxide chains to a rigid rod poly(p-phenylene) (PPP) polymeric backbone (Fig. 3.14). Unfunctionalized PPP is not soluble and does not melt. However, incorporating a dense set of side-chains leads to "hairy rods" [261–263], i.e. a comb-shaped architecture, where the mutual aggregation of the rods is reduced by the side-chain spacers [264, 265]. By using a statistical copolymer $PPP(EO)_{x/y}$ involving two different side-chain lengths (x and y) of oligo(ethylene oxide) side-chains, self-assembled materials have been introduced where the tendency for the side-chain crystallization is suppressed. Li⁺ conduction is achieved by adding LiTFSi [266]. For longer side-chains, consisting of e.g. $x=5$ and $y=6$ ethylene oxide repeat units, the materials exhibit an order–disorder transition (in the range 90–160 °C) to a disordered isotropic state upon heating. If the side-chains are shorter, e.g. $x=4$ and $y=2$, the lamellar structure prevails up to 240 °C. In another version, the anionic counterions have been incorporated within the polymeric backbone [267].

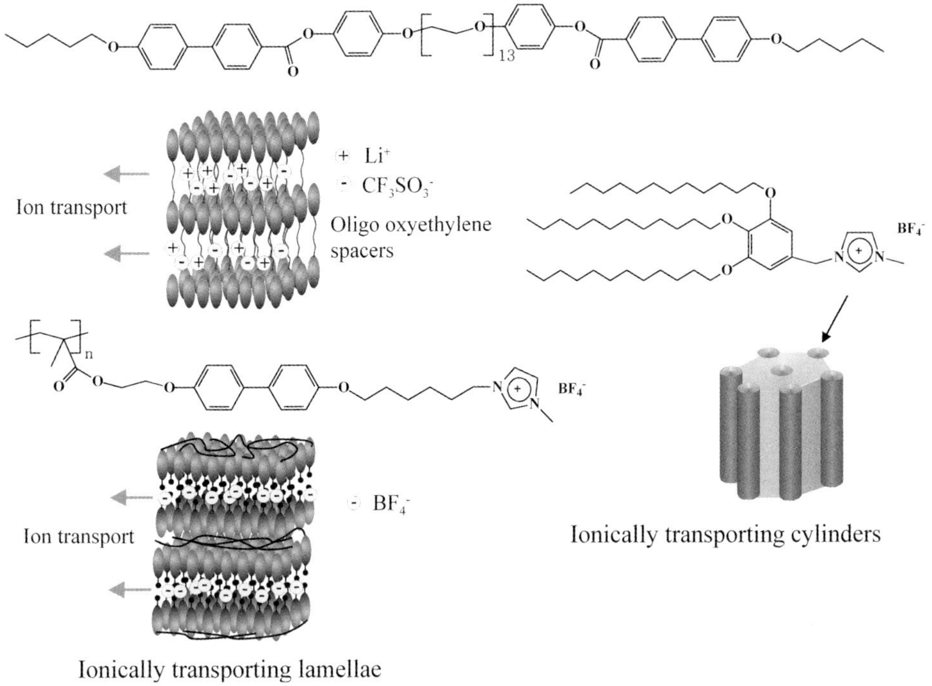

Fig. 3.15 Examples of concepts to control the dimensionality of ionic conductivity [21, 268–270].

Mesogenic groups are versatile for controlling self-assembly when various functionalities and anisotropic properties are pursued; for a recent review, see [22]. Mesomorphic dimeric compounds consisting of rigid mesogenic cores and flexible oxyethylene chains are able to complex CF_3SO_3Li to form a lamellar smectic self-assembly (Fig. 3.15) [268]. The ionic conductivity is anisotropic and reaches 5.5×10^{-4} S cm^{-1} parallel to the layers. By tailoring the chemical structure of the self-assembling groups, it is possible rationally to design the dimensionality of the conductivity [21]. One can also connect the structure-directing mesogens as side-chains of polymers to allow lamellar ion conducting layers (Fig. 3.15) [269].

3.3
Self-assembly of Conjugated Polymers

There is a recent and comprehensive review [47] on various aspects of self-assembled and supramolecular conjugated systems and in this discussion only a few of the concepts are dealt with. Conjugated polymers have either a totally rigid rod-like conformation or a semi-rigid coiled conformation and they typically include aromatic or heteroaromatic groups capable of π-stacking [100]. There-

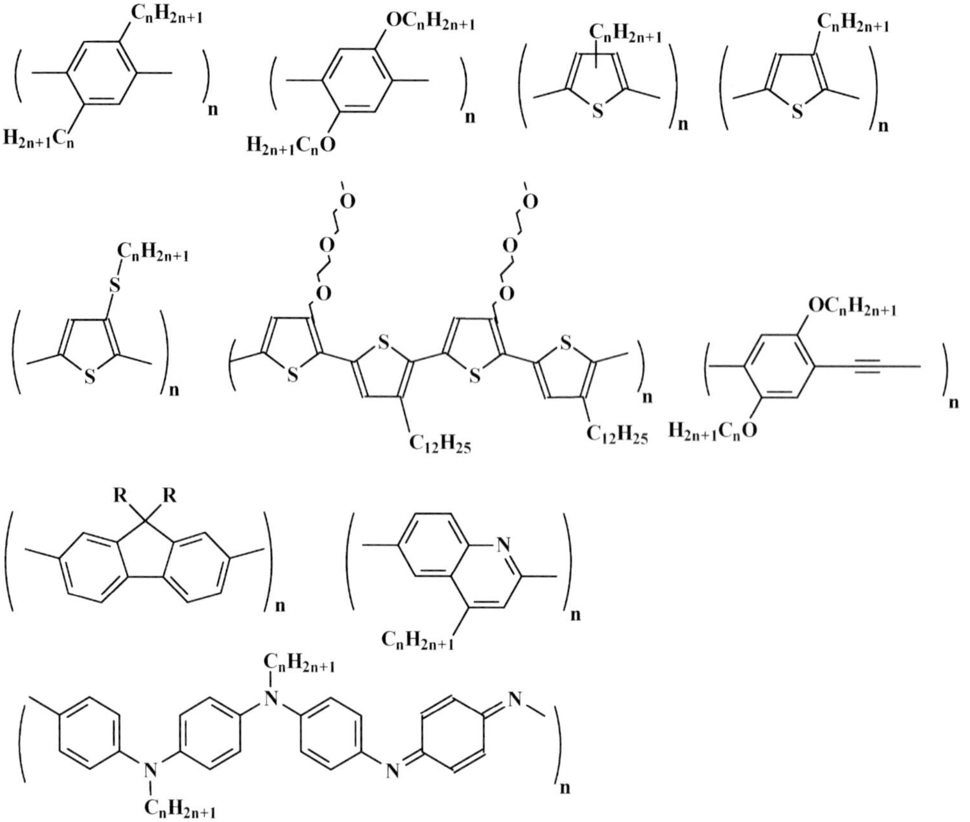

Fig. 3.16 Examples of hairy rod polymers consisting of conjugated polymers with covalently bonded flexible side-chains. The backbones consist of poly(p-phenylene)s, polythiophenes, poly(phenylene vinylene)s, polyfluorenes, polyquinolines and polyanilines.

fore, they are prone to aggregation and it is not surprising that in the general case they do not melt and they can be poorly soluble in common solvents. The aggregation can be controlled by incorporating proper side-chains to the conjugated backbone and such a comb-shaped architecture is also denoted "hairy rods" (Figs. 3.14 and 3.16) [262, 264–266, 271, 272]. The covalently bonded side-chains can also be regarded as "bound solvent molecules" and they also render surface activity towards another solvent phase. But additionally, the side-chains can lead to self-assembly as the flexible side-chains are repulsive [262, 273]. For example, poly(p-phenylene) (PPP) is an insoluble and infusible conjugated polymer. Several types of hairy rods have been prepared (Fig. 3.16) [262, 264–266, 274–277]. The polymers self-assemble to cylindrical or lamellar phases depending on the type and length of the alkyl chains [264, 266]. The nature and length

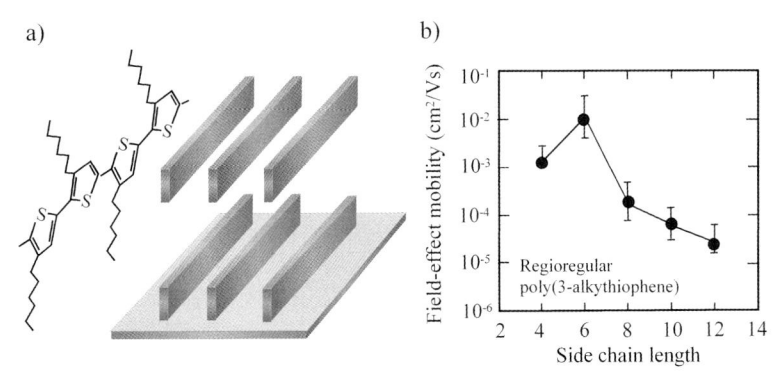

Fig. 3.17 (a) Lamellar self-assembly of regioregular poly(3-hexylthiophene) which leads to enhanced charge transport [301]. (b) Effect of side-chain length on hole mobility [299, 301].

of the covalently bonded side-chains have a large effect on the phase behavior [273].

Alkyl-substituted polythiophenes (Fig. 3.16) were first studied in the 1980s and the self-assembled lamellar structures in bulk were revealed using X-ray scattering [278–283]. For example, regiorandom poly(octylthiophene) shows a lamellar self-assembly at room temperature with a periodicity of approximately 2.2 nm and an order–disorder transition takes place at approximately 150 °C. The transition is reversible upon cooling/heating cycles. Implicitly, the materials are melt processable due to the fluid state in the disordered state.

Major recent efforts concern control of regioregularity, as it has profound effects on the structure and charge carrier mobility [284–301]. Self-assembly of poly(3-hexylthiophene)s results in a lamellar structure on SiO_2/Si substrates and relatively high charge carrier mobility of approximately 0.1 $cm^2 V^{-1} s^{-1}$ was achieved using highly regioregular polymers and processing conditions that lead to the parallel alignment of the lamellae relative to the substrates (Fig. 3.17) [301]. This suggests high mobilities based on two-dimensional transport within self-assembled conjugated lamellae, which could be important for applications for polymer transistors within logic circuits and active-matrix displays. The selection of solvents [290, 298] and alkyl side-chain lengths is important for enhanced transport [299]: The average hole mobility varies from $1.2 \times 10^{-3} cm^2 V^{-1} s^{-1}$ in poly(3-butylthiophene) and $1 \times 10^{-2} cm^2 V^{-1} s^{-1}$ in poly(3-hexylthiophene) to $2.4 \times 10^{-5} cm^2 V^{-1} s^{-1}$ in poly(3-dodecylthiophene). Therefore, the hexyl side-chain seems to be optimal for charge transport. As already pointed out, self-assembly generally leads only to local order. This was clearly manifested in STM studies using regioregular poly(3-hexylthiophenes) on graphite, which demonstrated "folds" of the chains and differently aligned self-assembled domains [291–293].

The polyalkylthiophenes can be doped for conductivity. Regioregular poly(3-alkylthiophene)s have been p-doped with I_2 vapor and they display high room

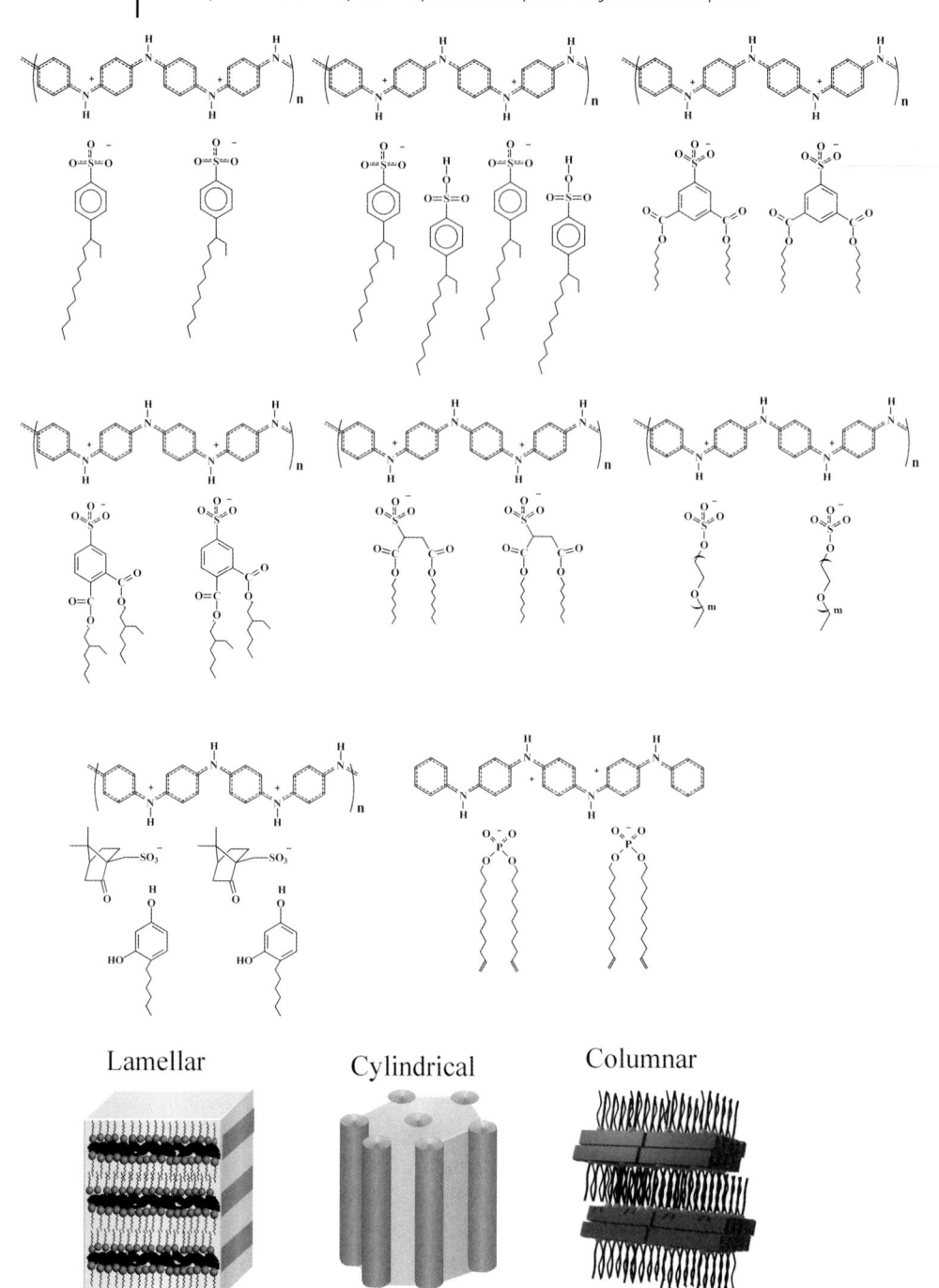

Fig. 3.18 Various acid doped complexes for polyaniline and oligomeric anilines [58, 82, 102, 175, 321, 324–330, 332, 333].

temperature conductivity (100–750 S cm^{-1}) [285, 286]. Langmuir–Blodgett techniques allow films with conductivity of 67–100 S cm^{-1} for the head-to-tail poly(3-hexylthiophene)–stearic acid compounds [287]. The side-chains allow tuning of the functionalities; one can also prepare "Janus-type" polythiophenes that have both hydrophobic alkyl side-chains and hydrophilic oligo(oxyethylene) side-chains [302]. They allow self-assembled monolayers on surfaces using Langmuir–Blodgett techniques. The side-chains can also be semifluorinated [303]. Side-chains have been connected to a wide variety of different polymers to allow hairy rods or comb-shaped polymers, self-assembly and tunable properties; for conjugated polymers see [100, 262, 271, 275, 304, 305] and for flexible polymers [17, 306]. Polyfluorenes with different alkyl chains are among the most feasible conjugated polymers for light-emitting devices and they also undergo self-assembly [307–312]. Comb-shaped regioregular poly(4-alkylquinoline-2,6-diyl)s show lamellar structures and π-stacking and emit yellow light at 542–557 nm [313]. Finally, it is pointed that the planarity can be controlled by supramolecules interactions [314, 315].

In the previous examples, the self-assembly is achieved based on the covalently connected repulsive side-chains. Self-assembly can also be achieved given that there exists a sufficiently strong physical attractive interaction to balance the repulsive interaction. Polyaniline is probably the most studied conjugated polymer in this context. The undoped state of polyaniline consists of benzenediamine and quinonediimine moieties [100, 316–318]. It can be doped by a redox reaction by electron transfer but, importantly, its salts with strong acids are electronically conducting due to protonation of the iminic nitrogen and a subsequent redox reaction along the chain [316, 319]. Note, however, that even earlier conducting polyaniline sulfate had been used in electrochemical applications [320]. An important finding was that the dopant can also have another functional group besides the protonating acid group [321]: It can contain, for example, alkyl spacers to reduce aggregation, it can contain surface-active groups for compatibilization, hydrogen-bonding sites, mesogens or dyes. A widely used functional counterion is dodecylbenzenesulfonic acid (DBSA) (Fig. 3.18). The bonding is due to ionic interaction and therefore the material can also be classified as a polyelectrolyte–surfactant complex. The stoichiometric complex where only the iminic nitrogens are protonated leads to infusible and solid crystalline material [322]. Adding more DBSA, i.e. using one DBSA molecule corresponding to each repeat unit of polyaniline, does not lead to phase separation, obviously as the additional molecules are bonded by hydrogen bonding to the sulfonates and amines. Such a composition leads to lamellar fluid-like self-assembly (Fig. 3.18) [58].

The counterion engineering allows one to tailor a balance between the conductivity and processability. The concept allows solid films using common solvents with conductivities as high as approximately 200–400 S cm^{-1} [321]. For example, using octanesulfonic acid [321] the solubility in low-polarity solvents may be reduced. However, the conductivity can simultaneously be increased for shorter side-chains, obviously due to a larger hopping conductivity between the chains [323]. Upon increasing the total number of methyl groups in the side-chains, such as in dinonylnaphthalenesulfonic acid, the complexes become highly soluble in several solvents, but the conductivity is strongly reduced [324]. The effect of adding two alkyl chains with different lengths and different branchings has been systematically studied (Fig. 3.18) [102, 325–328]. The self-assembly, transport and other properties can also be tuned by additional amphiphilic compounds due to attractive and repulsive interactions. In this case the problem is to identify compounds that have sufficiently strong attractive interaction to the protonated main chain [82, 329, 330]. The phase behavior of supramolecular hairy rods has been studied theoretically [331]. The functionalized counterions allow even more complex structural tuning: self-assembly within colloidal suspensions can be used to template networks of electrically conducting sulfonic acid-doped polyaniline (see Fig. 3.9) [216].

The previous examples concern a comb-shaped architecture where the side-groups are either covalently or physically connected and which leads to self-assembly, given that the side groups are sufficiently long and repulsive. Next we address the linear architecture where the polymeric blocks are connected end-to-end. In an effort to combine electronic conductivity, there have already been early efforts to combine charge-transfer salts selectively within block copolymeric self-assembled domains. Polybutadiene-b-polyvinylpyridine block copolymer was complexed using 7,7′,8,8′-tetracyanoquinodimethane [334, 335]. Self-assembled morphologies were obtained with conductivities of 10^{-3}–10^{-4} S cm^{-1}. Mixed ionic and electronic conductors were prepared [336–338], for example by using poly(2,5,8,11,14,17,20,23-octaoxapentacosyl methacrylate)-b-poly(4-vinylpyridine) where LiClO$_4$ was added to the oxyethylene phase to obtain ionic conductors with conductivities of approximately 5×10^{-6} S cm^{-1} and the 4-vinylpyridine phase was complexed with 7,7′,8,8′-tetracyanoquinodimethane to obtain electronic conductivities of approximately 10^{-6} S cm^{-1} at 25 C [336].

Examples have been shown here of how substitution of polythiophenes using short alkyl chains render solubility and self-assembly. Instead of the comb-shaped architecture, one could consider also the linear block copolymeric architecture where the solubilizing groups are at the ends. Well-defined triblock copolymers consisting of narrow molecular weight polystyrene end-groups of 30 repeat units and a central block of 11 thiophene groups form micelles of diameters of 12 nm in chloroform solution and in the solid state, consisting of a thiophene core (Fig. 3.19) [339]. Regioregular polyalkylthiophenes can also be used as blocks in block copolymers. Polystyrene-b-poly(3-hexylthiophene), polystyrene-b-poly(3-hexylthiophene)-b-polystyrene and poly(methyl acrylate)-b-poly(3-hexylthiophene)-b-polystyrene block copolymers have been studied, each having a narrow molecular weight distribution (Fig. 3.19) [340]. Nanoscale fibrillar morphology is obtained owing to aggregation of the conjugated domains due to stacking. I$_2$ doping renders conductivities that range from 4.6×10^{-5} to 110 S cm^{-1}. Based on atom-transfer radical polymerization (ATRP), poly(3-hexylthiophene)-b-polystyrene diblock copolymer with a polystyrene majority phase has been synthesized [341]. Nanoscale ribbons are observed where poly(3-hex-

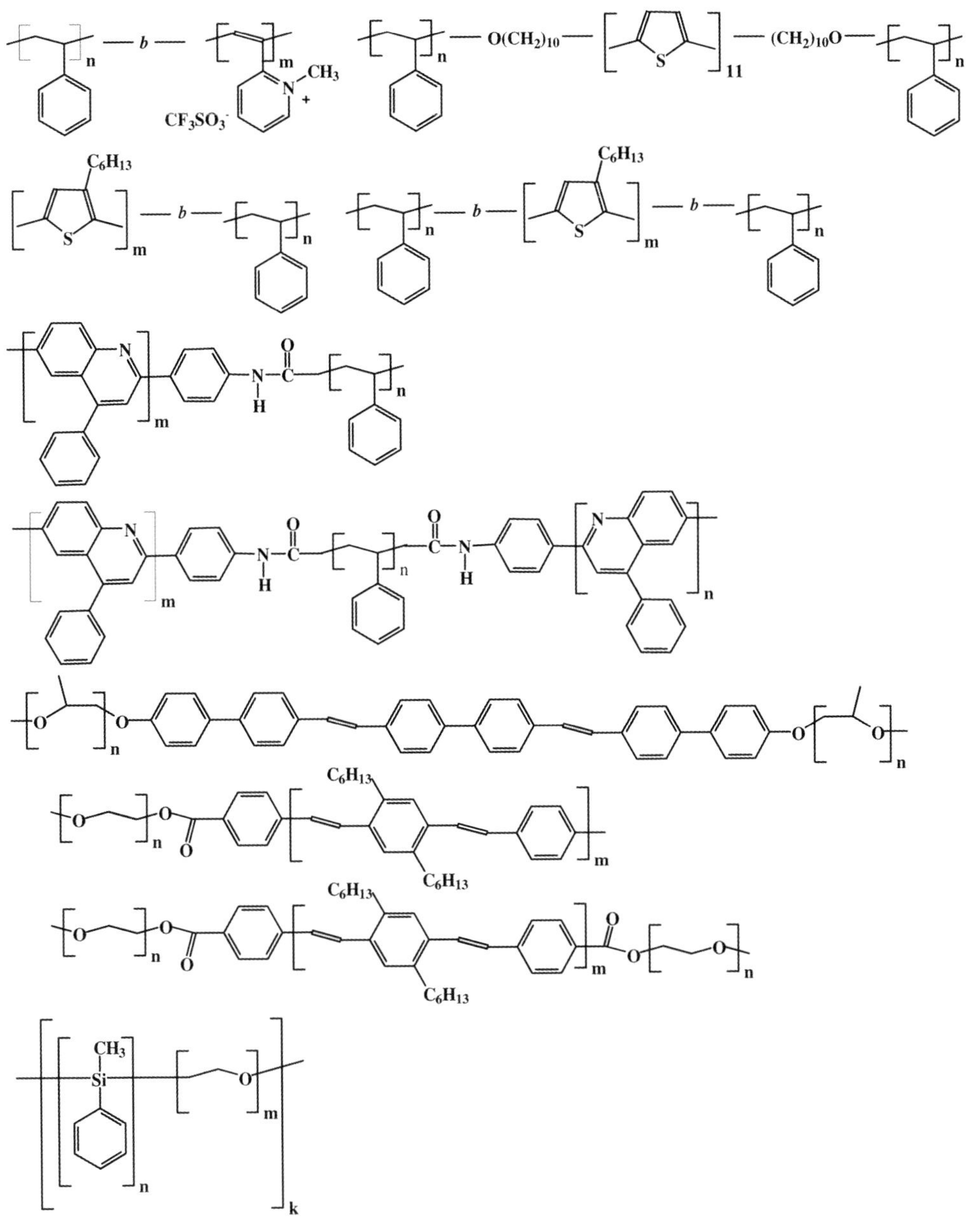

Fig. 3.19 Examples of block copolymers containing conjugated blocks [339–343, 345–349].

ylthiophenes) form the cores of the self-assembled ribbon-like entities. The structure is sensitive to selection of the solvents, which drastically reflects in the conductivity. I_2 doping leads to conductivity of the order 10^{-2} S cm^{-1} but it is reduced if the ribbon-like morphology is not obtained. Aggregates of amphiphilic diblock copolymers based on styrene and thiophene were prepared by injection of polymer solution into water and by electrochemically induced formation of vesicles [342].

A low molecular weight diblock copolymer consisting of flexible polystyrene block and pyridinium trifluoromethanesulfonate-substituted polyacetylene has been synthesized (Fig. 3.19) [343]. Conductivity suggests an alternating insulator–semiconductor layered structure of the deposited films. A comprehensive discussion on functionalization of polyacetylene by functional pendant groups has been presented recently [344]. They show various functional properties such as liquid crystallinity, photoconductivity, light emission, photoresistance, chromism, helical chirality, optical nonlinearity, self-assembly, cytocompatibility and bioactivity.

Amphiphilic rod–coil diblock and coil–rod–coil triblock copolymers of oligo-(phenylenevinylene)-b-poly(ethylene glycol) and poly(ethylene glycol)-b-oligo-(phenylenevinylene)-b-poly(ethylene glycol) self-assemble into long cylindrical micelles [350]. The micelles have diameters of approximately 8–10 nm and are composed of an conjugated block surrounded by a poly(ethylene glycol) corona. An even more rigid conjugated block was used in triblock copolymers where the central block consists of three biphenyls connected using vinylene groups and the end blocks are poly(propylene oxide) [349]. A brilliant blue emission was observed with a fluorescence maximum at 479 nm. The molecules self-assemble into successively one-dimensional lamellar, two-dimension hexagonal and three-dimension tetragonal structures with increasing coil-to-rod volume fraction (Fig. 3.20).

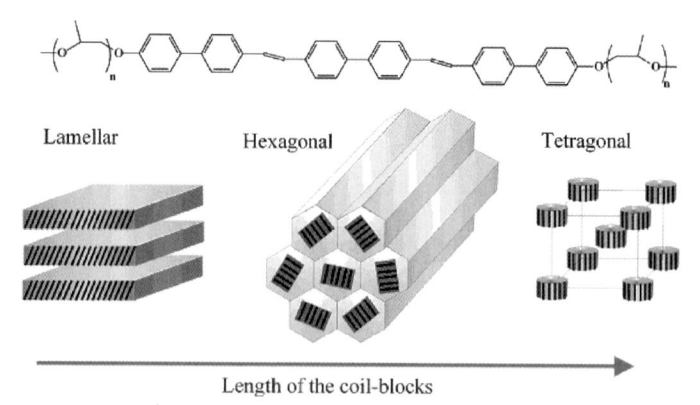

Fig. 3.20 An example of ABA-type coil–rod–coil triblock copolymer with a conjugated central block and the structures obtained [349].

Poly(phenylquinoline)-*b*-polystyrene rod–coil block copolymers have been extensively studied [346, 351]. They can self-assemble into hollow spherical micelles having diameters of a few micrometers. Including the third block, i.e. polyquinoline–NHCO–polystyrene–NHCO–polyquinoline rod–coil–rod triblock copolymer, spherical vesicles in solution are formed [347].

The structures of block copolymers consisting of various conjugated blocks and coiled blocks in thin films have recently been reported [352, 353]. The conjugated blocks are side-group modified polyphenylene, poly(phenyleneethynylene), polyfluorene or polyindenofluorene and the coiled blocks are polydimethylsiloxane, PEO or polystyrene [352, 353]. The copolymers organize as ribbons with constant width and height in the nanometer range (Fig. 3.21). The comparison of the experimental data with the theoretical modeling indicates that the ribbons are made of regular stacks of conjugated chains surrounded by coiled non-conjugated segments. In addition to π-conjugated polymers, δ-conjugated polymers allow self-assembly and interesting properties. An amphiphilic multi-block copolymer consisting of alternating PEO segments and polydisperse polymethylphenylsilane segments self-assemble as vesicles, micellar rods and helixes in water-based solvents, even if the polymethylphenylsilane blocks have large polydispersity [348].

Dendronic side-groups in polymers allow detailed control of self-assembly [354–356]. Organic donor or acceptor groups can be connected to the apexes of the dendrons (Fig. 3.22). By selecting a polymer with complementary electron acceptor or donor groups as side-groups, electron-donor stacks can be formed within self-assembled columns that lead to enhanced charge carrier mobilities [355]. Conjugated dendritic thiophene derivatives have also been studied [357].

Finally, polymeric self-assembly can be used as a template to grow inorganic and metallic structures based on selective complexation of metal salts within block copolymer domains and consequent reduction (e.g. [358–363]). These and other deposition concepts can be applied to self-assembled structures to achieve conductivity [364]. Diblock copolymers consisting of dendritic bulky and flexible

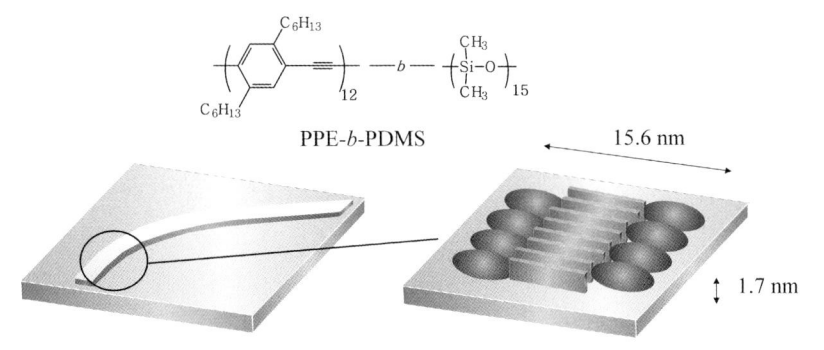

Fig. 3.21 Scheme of nanoscale ribbons based on hexyl-substituted polyphenyleneethynylene-*b*-polydimethylsiloxane block copolymer in thin films [352, 353].

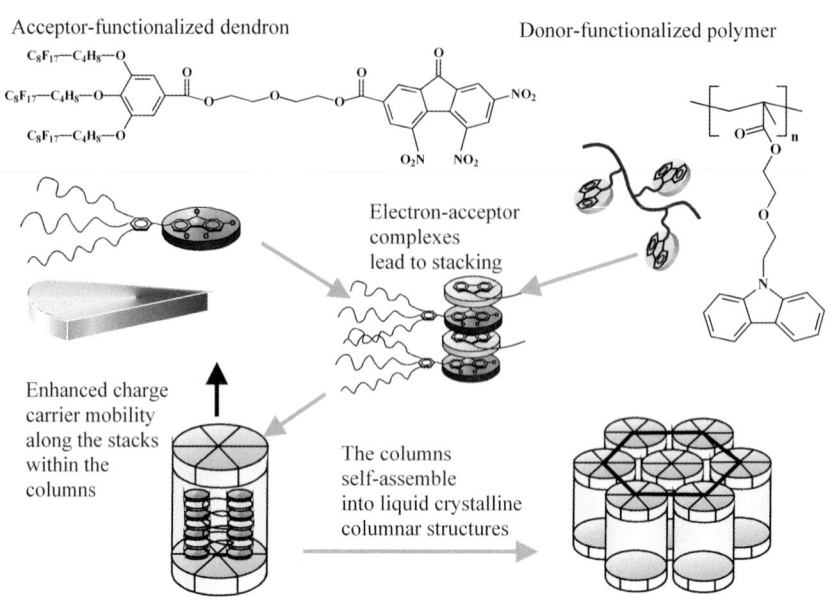

Fig. 3.22 An example of how electron acceptor and donors can be stacked within self-assembled domains for enhanced charge carrier mobility [355].

nonpolar carbosilane block and rigid helical polyisocyanide blocks form complexes with silver ions and lead to nanowires [365]. Distinct nanowires can be achieved by using polystyrene-*b*-poly(4-vinylpyridine) where pentadecylphenol is hydrogen bonded to the pyridines, which leads to hierarchical self-assembly [366]: Polystyrene cylinders within poly(4-vinylpyridine)–pentadecylphenol can be obtained and, after aligning and removing the pentadecylphenol, distinct polystyrene wires with a poly(4-vinylpyridine) corona are obtained [86, 109, 367], which can be metallized with CdSe, Pd, Ag and Au [368, 369]. The self-assembled domains of symmetrical polystyrene-*b*-poly(4-vinylpyridine) diblock copolymers have been used as templates to prepare Ag layers by electroless deposition and anisotropic conductivity was achieved along the lamellae [370]. Just in passing, we point out that self-assembling amyloid protein fibers, polyelectrolyte chains and even DNA can be used to construct nanowire elements [12, 371].

3.4
Optical Properties of Self-assembled Polymer Systems

As there exists a vast literature on the optical properties of polymers, such as on materials for light-emitting devices, electrochromism, liquid crystallinity, optical switching, photorefractive applications, photovoltaics and photoconductivity, we limit the present discussion to three topics which have an intimate connection with self-assembly, i.e. photonic crystals or bandgap materials, nonlinear optical materials and photovoltaic materials.

Well-defined dielectric structures at the optical length scale are able to manipulate flow of light [372–378]. This has attracted much attention over recent years in the context of optical communication, sensing, optical limiting and even coatings. Photonic crystals are long-range ordered structures whose refractive index varies in a spatially periodic manner. Taken that the periodicity matches the optical wavelengths, that the structure has proper symmetry and that sufficient dielectric contrast exists, the resulting photonic crystals may exhibit a forbidden bandgap where no photons are allowed to propagate. In combination with controlled defect structures, a wealth of applications in photonics are expected, e.g. capability to confine, guide and control light. If optical wavelengths are aimed for, the required periodicity is in the range 100–200 nm. Detailed structures and defects with sufficient dielectric contrast can be constructed using lithographic and etching techniques. For certain large area applications, simpler techniques could be more suitable. Spontaneous self-assembly of colloids [375], synthetic opals [379–384], inverted opals [379, 385–387] and block copolymers [218, 219, 221, 378, 388–402], on the other hand, allows the preparation of small enough structures. Although self-assembly leads to a well-defined local order and offers a potentially low-cost method for the production of photonic crystals, it is nontrivial to achieve perfectly ordered structures over the macroscopic length scale combining carefully engineered defects.

The simplest periodic dielectric structures capable of interacting with optical waves are dielectric mirrors, quarter-wave stacks or 1D photonic crystals which consist of alternating layers of high and low refractive index materials [218, 219, 221, 374, 388, 390, 392, 395, 397, 403]. It is a challenge to achieve high periodicity in the range 100–200 nm using purely block copolymers, as very high molecular weight polymers would be needed. High molecular weight usually leads to very slow achievement of the equilibrium structure during annealing and therefore it is challenging to obtain good enough overall order. In this respect, it can be helpful to construct the self-assembly using block copolymers with smaller molecular weights but where additionally homopolymers, oligomeric plasticizers or physically bonded amphiphiles or liquid crystals are added to the domains [218, 219, 221, 388, 390–392, 395, 397, 403] (Fig. 3.23). The blocks can also consist of rigid mesogenic moieties [395]. Roll casting [399] or various annealing procedures promote sufficient macroscopic order. Another critical aspect is the dielectric contrast. Most organic polymers have their refractive index in a narrow range around $n \approx 1.5$. Therefore, the dielectric contrast tends to remain small, but still allowing some bandgap effects. In order to increase the dielectric contrast, nanoparticles have been selectively sequestered to the structures [392, 397, 404]. For example, Au nanoparticles with polystyrene brushes are confined within the polystyrene layers of the self-assembled structures of polystyrene-b-poly(ethylene/propylene), $C_{18}H_{37}$-functionalized Au nanoparticles are confined in the interfaces and methyl-functionalized SiO_2 particles in the poly(ethylene/propylene) domains. In this way, the contrast can be increased. There have been

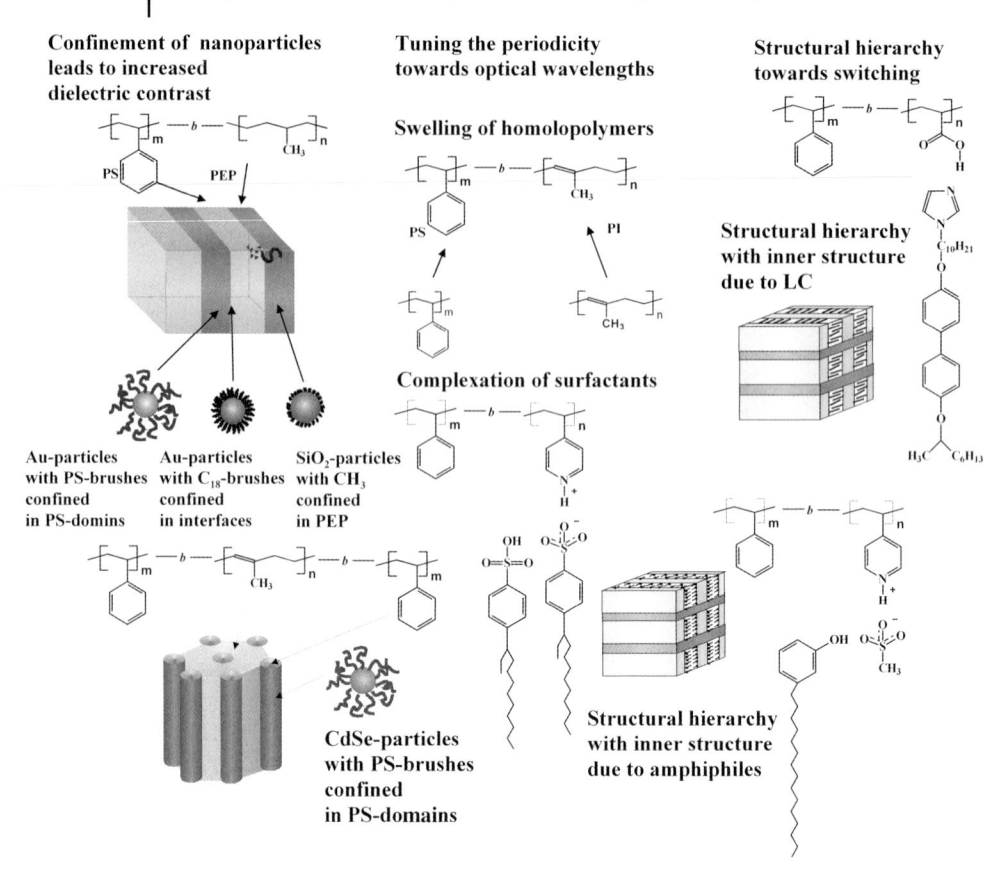

Fig. 3.23 Examples of block copolymeric systems aimed at photonic bandgap materials [218, 221, 378, 392, 395, 397].

recent efforts to achieve controllable and switchable photonic bandgap structures where abrupt color changes have been obtained upon heating and cooling due to phase transitions within the self-assembled structures [221, 395]. Such concepts might pave the way also towards smart coatings. In the purely layered geometry, it is not possible to achieve a complete photonic bandgap, which would stop the light propagation in all directions. Therefore, higher dimensional block copolymer structures have been pursued [374, 377, 393, 398, 400, 402]. Even 2D materials consisting of self-assembled cylinders do not allow a complete bandgap [399] and in this respect 3D network-like structures are currently being studied in detail [402]. An interesting concept, although not directly related to self-assembly, concerns the preparation of photonic crystals using holographic lithography [401].

As another example, optically nonlinear materials are discussed. Over the years, there have been considerable efforts to construct optically nonlinear polymer-based materials for novel optoelectronic devices (for reviews, see [405–409]).

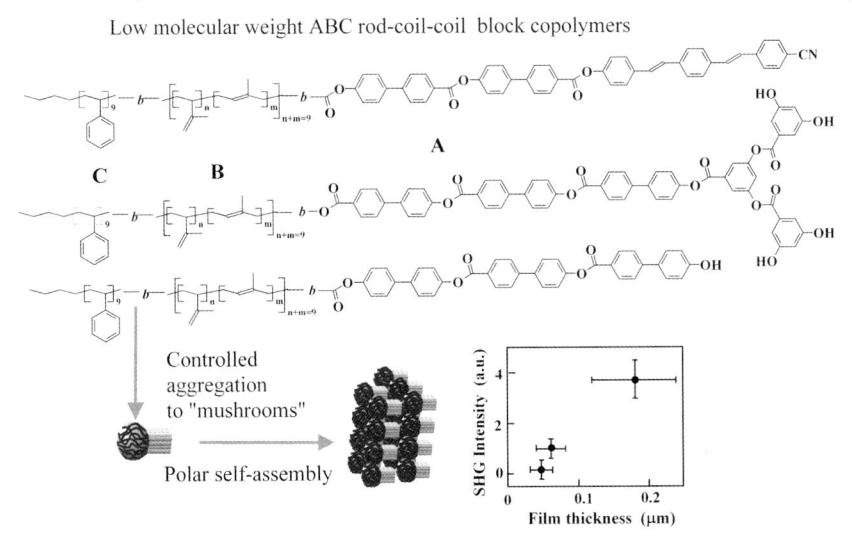

Fig. 3.24 Oligomeric rod–coil–coil triblock copolymers, which form "mushroom-like" structures and noncentrosymmetric self-assembly [419–428]. The materials show second-order harmonic generation (SHG) [419].

In particular, second-order nonlinear optical materials can double the frequency of the incident light and allow, for example, electro-optic modulators and laser frequency doublers. Such materials can be accomplished using chromophores which consist of electron acceptor and electron donor groups that are connected using a conjugated linker, given that the chromophores form noncentrosymmetric, i.e. polar, structures. Noncentrosymmetric structures are challenging to obtain and require specific materials and processes. Common techniques include alignment of the chromophores embedded in a polymeric matrix by an electric field, i.e. poling, and the subsequent "freezing" of their alignment by, for example, cooling the system to a glassy state, chemically immobilizing the chromophores by cross-linking or controlled deposition of the poled layers by Langmuir–Blodgett techniques [405–409]. Importantly, the poled chromophores are prone to relaxation towards the thermodynamic nonpolar equilibrium state, which can be a serious limitation considering the applications. In an effort to control the relaxation of the chromophore alignment, they have been covalently bonded to rigid rod polymers that undergo lamellar self-assembly [410–413]. Given that the system is poled before the self-assembly is frozen, the relaxation can be effectively suppressed to be only two-dimensional within the self-assembled lamellar layers.

There have been efforts to find concepts that would lead directly to a spontaneous equilibrium noncentrosymmetric assembly [33, 414–417]. Note that collagen possesses a relatively strong second-order nonlinear susceptibility [418]. In the present context, low molecular weight triblock copolymers consisting of a

A **B** **C**

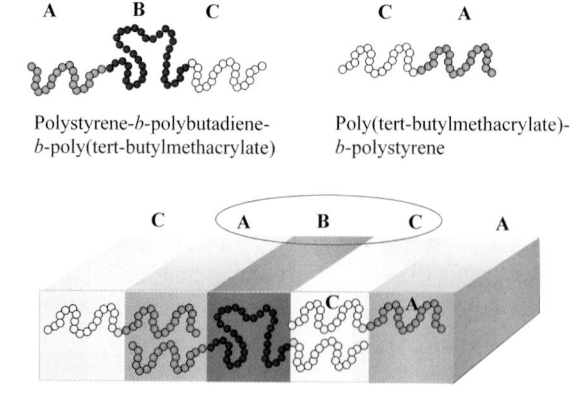

Polystyrene-*b*-polybutadiene-
b-poly(tert-butylmethacrylate)

C **A**

Poly(tert-butylmethacrylate)-
b-polystyrene

Fig. 3.25 Schematics for obtaining noncentrosymmetric self-assembly by mixing an ABC triblock copolymer and an AC diblock copolymer [430].

rod-like end block, flexible polyisoprene middle block and flexible but more bulky polystyrene end block are interesting (Fig. 3.24) [419–428]. The rods tend to aggregate mutually, but as the coils take up a larger lateral space, the packing becomes hindered. Therefore, a concurrent stretching of the coils takes place and the number of aggregated chains becomes controlled, e.g. approximately 100. This concept leads to mushroom-shaped supramolecular amphiphiles. They are reported to self-assemble curiously in monolayers in a noncentrosymmetric fashion [419–428]. As the rods can be functionalized, the concept offers spontaneous noncentrosymmetric self-assembly and second-order nonlinear optical properties (Fig. 3.24) [419, 426, 427], and also piezoelectricity [429]. Dendron-functionalized rod–coil–coil triblock oligomers can also bind nanocrystals and promote ultraviolet lasing [428].

Noncentrosymmetric self-assembly can also be obtained using mixtures of block copolymers. If one uses pure ABC triblock copolymers or pure AC diblock copolymers, the structures are symmetric. However, upon mixing ABC and AC block copolymers, macroscopic phase separation does not necessarily take place and the polymers can form noncentrosymmetric self-assembly within the mixtures, as shown schematically in Fig. 3.25 [430]. This concept can pave the way for various new functionalities.

The final example concerns polymeric photovoltaics (for recent reviews, see [431–435]) where block copolymeric self-assembly turns out to be interesting due to the enhanced amount of available interfaces [436–445].

In order to achieve a photovoltaic effect, charge generation to electrons and holes has to take place upon illumination. The electron–hole pair, or exciton, is strongly bound as the dielectric constant of organic matter is low. It is critical to collect the electrons and holes separately to the opposite current collectors, so avoiding recombination. This can be accomplished using domains of electron-accepting and electron-donating materials. If the exciton is created sufficiently

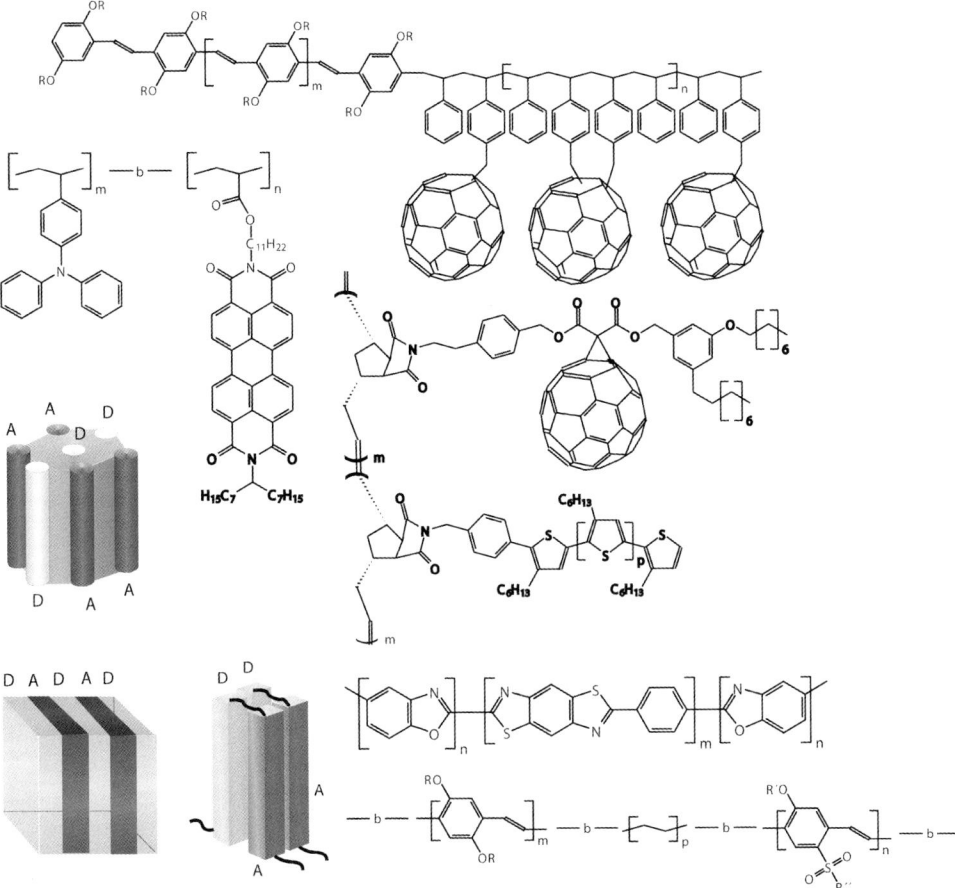

Fig. 3.26 Examples of various block copolymers containing electron-accepting (A) groups and electron-donating groups (D), and schemes for self-assembly that can lead to enhanced photovoltaic properties for charge collection to the external electrodes [436, 439, 444, 446].

close to such interfaces (less than approximately 10 nm), the exciton can be dissociated so that the electrons enter the electron-accepting material and the holes the electron-donating material. By applying an electric field and given that sufficient charge carrier diffusion exists within the two domain, the charges can be collected to the external collectors. There have been a variety of different routes to allow a high interfacial area between electron-acceptor and -donor materials, such as interpenetrating networks. Self-assembly could offer a feasible route, if also the alignment of the domains could be controlled. An ABA triblock conjugated copolymer, poly(2,5-benzoxazole)-*b*-poly(benzobisthiazole-2,6-diyl-1,4-phenylene)-*b*-poly(2,5-benzoxazole), shows self-assembly and efficient energy transfer [345]. A diblock copolymer consisting of alkoxy-substituted poly(*p*-phenylenevinylene) (PPV) and fullerene-functionalized polystyrene block has been described (Fig. 3.26) [436–438]. There the PPV blocks act as electron donors and the fullerenes are electron acceptors. Photophysical measurements show that there is efficient electron transfer from the excited PPV blocks to the fullerenes [436]. Multiblock copolymers consisting of electron-accepting alkoxy-substituted poly(p-phenylenevinylene) and electron-donating alkylsulfone-containing poly-(p-phenylenevinylene) with alkyl bridging groups have been studied (Fig. 3.26) [439, 440]. The authors also considered feasible self-assemblies to enhance the charge carrier collection to the external electrodes. Poly(vinyltriphenylamine)-b-poly(triphenyl-1,3,5-triazine) self-assembles to allow hole-conducting moieties in the first domains and electron-conducting moieties in the latter domains [443]. Another variant consists of similar hole-conducting blocks but the electron-conducting domains contain perylene-3,4:9,10-tetracarboxylic moieties [446]. Transmission electron microscopy shows self-assembly, although a number of defects can be observed. Finally, diblock copolymers were recently synthesized containing poly(3-hexylthiophene) and fullerenes within the side-groups [444, 445].

3.5
Conclusion

We have discussed a few examples where self-assembled polymers and supramolecules are useful for tailoring properties for electrical and photonics applications. They allow the required periodic structures for photonic bandgap materials; they allow synergistic behavior for electrically conducting materials, as the conductivity and other properties can in principle be tailored separately; they allow processability of conjugated polymers by controlling the separation and packing between them, which is important for optical properties and electrical transport; and they allow noncentrosymmetric self-assembly, which leads to enhanced second-order nonlinear optical properties and piezoelectric behavior. In order to provide wide applications, especially on a larger scale, the economics is an important issue and there the recent achievements with living polymerization techniques may be relevant. Finally, beyond the self-assembly which leads to the local order, processing techniques have to be developed to control the overall order and to reduce the defects. Due to progress in these topics, self-assembled polymers are finding their way to applications.

Acknowledgments

Financial support from the Academy of Finland (Centre of Excellence "Bio- and Nanopolymers Research Group", 77317), National Technology Agency (Finland), the Dutch Polymer Institute (DPI) and the Dutch Organization for Scientific Research (NWO) is gratefully acknowledged.

References

1 I. W. Hamley, *The Physics of Block Copolymers*, Oxford University Press, Oxford, **1998**.
2 N. Hadjichristidis, S. Pispas, G. Floudas, *Block Copolymers: Synthetic Strategies, Physical Properties and Applications*, Wiley, New York, **2002**.
3 G. Holden, H. R. Kricheldorf, R. P. Quirk (Eds.), *Thermoplastic Elastomers*, Hanser, Munich, **2004**.
4 J.-M. Lehn, *Supramolecular Chemistry*, VCH, Weinheim, **1995**.
5 F. Vögtle, *Supramolecular Chemistry*, Wiley, Chichester, **1993**.
6 R. P. Sijbesma, F. H. Beijer, L. Brunsveld, B. J. B. Folmer, J. H. K. K. Hirschberg, R. F. M. Lange, J. K. L. Lowe, E. W. Meijer, *Science* **1997**, *278*, 1601–1604.
7 M. Rehahn, *Acta Polym.* **1998**, *49*, 201–224.
8 M. Rehahn, *Mater. Sci. Technol.* **1999**, *20*, 319–374.
9 J. H. K. K. Hirschberg, F. H. Beijer, H. A. van Aert, P. C. M. M. Magusin, R. P. Sijbesma, E. W. Meijer, *Macromolecules* **1999**, *32*, 2696–2705.
10 L. Brunsveld, B. J. B. Folmer, E. W. Meijer, R. P. Sijbesma, *Chem. Rev.* **2001**, *101*, 4071–4097.
11 R. P. Sijbesma, E. W. Meijer, *Chem. Commun.* **2003**, 5–16.
12 C. M. Niemeyer, C. A. Mirkin (Eds.), *Nanobiotechnology: Concepts, Applications and Perspectives*, Wiley-VCH, Weinheim, **2004**.
13 H.-A. Klok, *J. Polym. Sci., Polym. Chem. Ed.* **2005**, *43*, 1–17.
14 T. J. Deming, *Adv. Drug Deliv. Rev.* **2002**, *54*, 1145–1155.
15 Y. Gnanou, S. Lecommandoux, *BIOforum Eur.* **2004**, *8*, 38–40.
16 A. Skoulios, in *Advances in Liquid Crystals*, Vol. 1 (Ed. G. H. Brown), Academic Press, New York, **1975**, 169–188.
17 V. P. Shibaev, L. Lam (Eds.), *Liquid Crystalline and Mesomorphic Polymers*, Springer, New York, **1994**.
18 K. Anderle, J. H. Wendorff, *Macromol. Chem. Macromol. Symp.* **1994**, *96*, 165–168.
19 V. V. Tsukruk, J. H. Wendorff, *Trends Polym. Sci.* **1995**, *3*, 82–89.
20 W. Brostow (Ed.), *Mechanical and Thermophysical Properties of Polymer Liquid Crystals*, Chapman and Hall, London, **1998**.
21 T. Kato, *Science* **2002**, *295*, 2414–2418.
22 T. Kato, N. Mizoshita, K. Kishimoto, *Angew. Chem. Int. Ed.* **2006**, *45*, 38–68.
23 A. G. Petrov, *The Lyotropic State of Matter*, Gordon and Breach, Amsterdam, **1999**.
24 I. W. Hamley, *Introduction to Soft Matter*, Wiley, Chichester, **2000**.
25 M. Muthukumar, C. K. Ober, E. L. Thomas, *Science* **1997**, *277*, 1225–1232.
26 A. Babloyantz, *Molecules, Dynamics and Life: An Introduction to Self-organization of Matter*, Wiley, New York, **1986**.
27 S. Camazine, J.-L. Deneubourg, N. R. Franks, J. Sneyd, G. Theraulaz, E. Bonabeau, *Self-organization in Biological Systems*, Princeton University Press, Princeton, NJ, **2001**.
28 G. M. Whitesides, B. Grzybowski, *Science* **2002**, *295*, 2418–2421.
29 G. M. Whitesides, J. P. Mathias, C. T. Seto, *Science* **1991**, *254*, 1312–1319.
30 L. N. Christophorov, *BioSystems* **1995**, *35*, 171–174.
31 V. V. Tsukruk, *Prog. Polym. Sci.* **1997**, *22*, 247–311.
32 V. V. Tsukruk, V. N. Bliznyuk, *Prog. Polym. Sci.* **1997**, *22*, 1089–1132.
33 S. I. Stupp, V. LeBonheur, K. Walker, L. S. Li, K. E. Huggins, M. Keser, A. Amstutz, *Science* **1997**, *276*, 384–389.
34 J. Ruokolainen, R. Mäkinen, M. Torkkeli, T. Mäkelä, R. Serimaa, G. ten Brinke, O. Ikkala, *Science* **1998**, *280*, 557–560.
35 M. W. Matsen, C. Barrett, *J. Chem. Phys.* **1998**, *109*, 4108–4118.
36 M. Lee, B.-K. Cho, Y.-S. Kang, W.-C. Zin, *Macromolecules* **1999**, *32*, 7688–7691.
37 O. Ikkala, G. ten Brinke, *Science* **2002**, *295*, 2407–2409.
38 A. F. Thünemann, *Prog. Polym. Sci.* **2002**, *27*, 1473–1572.
39 S. T. Selvan, in *Nanoscale Materials* (Eds. L. M. Liz-Marzan, P. V. Kamat, Kluwer, Norwell), **2003**, 247–272.

40 C. Park, J. Yoon, E. L. Thomas, *Polymer* **2003**, *44*, 6725–6760.
41 I. W. Hamley, *Angew. Chem. Int. Ed.* **2003**, *42*, 1692–1712.
42 C. F. J. Faul, M. Antonietti, *Adv. Mater.* **2003**, *15*, 673–683.
43 M. Antonietti, *Nat. Mater.* **2003**, *2*, 9–10.
44 O. Ikkala, G. ten Brinke, *Chem. Commun.* **2004**, 2131–2137.
45 R. M. Kramer, M. O. Stone, R. R. Naik, *Proc. SPIE* **2004**, *5331*, 106–111.
46 H. D. Abruna, *Anal. Chem.* **2004**, *76*, 310A–319A.
47 F. J. M. Hoeben, P. Jonkheijm, E. W. Meijer, A. P. H. J. Schenning, *Chem. Rev.* **2005**, *105*, 1491–1546.
48 F. S. Bates, G. H. Fredrickson, *Annu. Rev. Phys. Chem.* **1990**, *41*, 525–557.
49 F. S. Bates, G. H. Fredrickson, *Phys. Today* **1999**, *52*, 32–38.
50 M. Lee, B.-K. Cho, W.-C. Zin, *Chem. Rev.* **2001**, *101*, 3869–3892.
51 K. Matyjaszewski, J. Xia, *Chem. Rev.* **2001**, *101*, 2921–2990.
52 A.-V. Ruzette, L. Leibler, *Nat. Mater.* **2005**, *4*, 19–31.
53 G. Wegner, *Makromol. Chem. Macromol. Symp.* **1986**, *1*, 151–171.
54 C. G. Bazuin, F. A. Brandys, *Chem. Mater.* **1992**, *4*, 970–972.
55 K. Iwasaki, A. Hirao, S. Nakahama, *Macromolecules* **1993**, *26*, 2126–2131.
56 M. Antonietti, J. Conrad, A. Thünemann, *Macromolecules* **1994**, *27*, 6007–6011.
57 C. G. Bazuin, A. Tork, *Macromolecules* **1995**, *28*, 8877–8880.
58 W.-Y. Zheng, R.-H. Wang, K. Levon, Z. Y. Rong, T. Taka, W. Pan, *Makromol. Chem. Phys.* **1995**, *196*, 2443–2462.
59 M. Antonietti, C. Burger, J. Effing, *Adv. Mater.* **1995**, *7*, 751–753.
60 O. Ikkala, J. Ruokolainen, G. ten Brinke, M. Torkkeli, R. Serimaa, *Macromolecules* **1995**, *28*, 7088–7094.
61 M. C. M. van der Sanden, C. Y. Yang, P. Smith, A. J. Heeger, *Synth. Met.* **1996**, *78*, 47–50.
62 F. A. Brandys, C. G. Bazuin, *Chem. Mater.* **1996**, *8*, 83–92.
63 M. Antonietti, S. Henke, A. Thünemann, *Adv. Mater.* **1996**, *8*, 41–45.
64 A. Wenzel, M. Antonietti, *Adv. Mater.* **1997**, *9*, 487–490.
65 H.-L. Chen, M.-S. Hsiao, *Macromolecules* **1999**, *32*, 2967–2973.
66 M. Knaapila, J. Ruokolainen, M. Torkkeli, R. Serimaa, L. Horsburgh, A. P. Monkman, W. Bras, G. ten Brinke, O. Ikkala, *Synth. Met.* **2001**, *121*, 1257–1258.
67 T. Kato, J. M. J. Fréchet, *Macromolecules* **1989**, *22*, 3818–3819.
68 T. Kato, J. M. J. Fréchet, *Macromolecules* **1990**, *23*, 360.
69 T. Kato, H. Kihara, T. Uryu, A. Fujishima, J. M. J. Fréchet, *Macromolecules* **1992**, *25*, 6836–6841.
70 U. Kumar, T. Kato, J. M. J. Fréchet, *J. Am.Chem. Soc.* **1992**, *114*, 6630–6639.
71 R. V. Tal'roze, N. A. Platé, *Polym. Sci.* **1994**, *36*, 1479–1486.
72 R. V. Tal'roze, S. A. Kuptsov, T. I. Sycheva, V. S. Bezborodov, N. A. Platé, *Macromolecules* **1995**, *28*, 8689–8691.
73 M. Pfaadt, G. Moessner, D. Pressner, S. Valiyaveettil, C. Boeffel, K. Müllen, H. W. Spiess, *J. Mater. Chem.* **1995**, *5*, 2265–2274.
74 T. Kato, T. Nakano, T. Moteki, T. Uryu, S. Ujiie, *Macromolecules* **1995**, *28*, 8875–8876.
75 T. Kato, J. M. J. Fréchet, *Macromol. Symp.* **1995**, *98*, 311–326.
76 J. Ruokolainen, G. ten Brinke, O. Ikkala, M. Torkkeli, R. Serimaa, *Macromolecules* **1996**, *29*, 3409–3415.
77 R. V. Talroze, S. A. Kuptsov, T. L. Lebedeva, G. A. Shandryuk, N. D. Stepina, *Macromol. Symp.* **1997**, 219–228.
78 J. Ruokolainen, M. Torkkeli, R. Serimaa, E. B. Komanschek, G. ten Brinke, O. Ikkala, *Macromolecules* **1997**, *30*, 2002–2007.
79 J. Ruokolainen, J. Tanner, O. Ikkala, G. ten Brinke, E. L. Thomas, *Macromolecules* **1998**, *31*, 3532–3536.
80 O. Ikkala, J. Ruokolainen, M. Torkkeli, J. Tanner, R. Serimaa, G. ten Brinke, *Colloids Surf. A* **1999**, *147*, 241–248.
81 O. Ikkala, M. Knaapila, J. Ruokolainen, M. Torkkeli, R. Serimaa, K. Jokela, L. Horsburgh, A. P. Monkman, G. ten Brinke, *Adv. Mater.* **1999**, *11*, 1206–1210.
82 H. Kosonen, J. Ruokolainen, M. Knaapila, M. Torkkeli, K. Jokela, R. Serimaa, G. ten Brinke, W. Bras, A. P. Monkman,

O. Ikkala, *Macromolecules* **2000**, *33*, 8671–8675.
83 H. Kosonen, J. Ruokolainen, M. Knaapila, M. Torkkeli, R. Serimaa, W. Bras, A. P. Monkman, G. ten Brinke, O. Ikkala, *Synth. Met.* **2001**, *121*, 1277–1278.
84 H.-L. Chen, C.-C. Ko, T.-L. Lin, *Langmuir* **2002**, *18*, 5619–5623.
85 E. Polushkin, G. Alberda van Ekenstein, I. Dolbnya, W. Bras, O. Ikkala, G. ten Brinke, *Macromolecules* **2003**, *36*, 1421–1423.
86 G. Alberda van Ekenstein, E. Polushkin, H. Nijland, O. Ikkala, G. ten Brinke, *Macromolecules* **2003**, *36*, 3684–3688.
87 T. Ruotsalainen, M. Torkkeli, R. Serimaa, T. Mäkelä, R. Mäki-Ontto, J. Ruokolainen, G. ten Brinke, O. Ikkala, *Macromolecules* **2003**, *36*, 9437–9442.
88 C.-S. Tsao, H.-L. Chen, *Macromolecules* **2004**, *37*, 8984–8991.
89 J. Ruokolainen, J. Tanner, G. ten Brinke, O. Ikkala, M. Torkkeli, R. Serimaa, *Macromolecules* **1995**, *28*, 7779–7784.
90 D. G. Kurth, P. Lehmann, M. Schütte, *Proc. Natl. Acad. Sci. USA* **2000**, *97*, 5704–5707.
91 J.-F. Gohy, B. G. G. Lohmeijer, S. K. Varshney, B. Decamps, E. Leroy, S. Boileau, U. S. Schubert, *Macromolecules* **2002**, *35*, 9748–9755.
92 J.-F. Gohy, B. G. G. Lohmeijer, U. S. Schubert, *Makromol. Chem. Rapid Commun.* **2002**, *23*, 555–560.
93 S. Valkama, O. Lehtonen, K. Lappalainen, H. Kosonen, P. Castro, T. Repo, M. Torkkeli, R. Serimaa, G. ten Brinke, M. Leskelä, O. Ikkala, *Macromol. Rapid Commun.* **2003**, *24*, 556–560.
94 J. H. K. K. Hirschberg, L. Brunsveld, A. Ramzi, J. A. J. M. Vekemans, R. P. Sijbesma, E. W. Meijer, *Nature* **2000**, *407*, 167–170.
95 S. Schmatloch, M. F. Gonzalez, U. S. Schubert, *Makromol. Chem. Rapid Commun.* **2002**, *23*, 957–961.
96 M.-J. Brienne, J. Gabard, J.-M. Lehn, I. Stibor, *J. Chem. Soc., Chem. Commun.* **1989**, 1868–1870.
97 D. Jones (Ed.), *Topical Issue: "Polymer Membranes II"*, Fuel Cells **2005**, *5*.
98 F. M. Gray, *Solid Polymer Electrolytes*, VCH, Weinheim, **1991**.
99 H. S. Nalwa (Ed.), *Handbook of Advanced Electronic and Photonic Materials and Devices*, Vol. 10, Academic Press, San Diego, **2001**.
100 T. A. Skotheim, R. L. Elsenbaumer, J. R. Reynolds, *Handbook of Conducting Polymers*, Marcel Dekker, New York, **1998**.
101 G. G. Wallace, G. M. Spinks, P. R. Teasdale, *Conductive Electroactive Polymers; Intelligent Materials Systems*, Technomic Publishing, Lancaster, **1997**.
102 A. Pron, P. Rannou, *Prog. Polym. Sci.* **2002**, *27*, 135–190.
103 V. K. Gupta, R. Krishnamoorti, J. A. Kornfield, S. D. Smith, *Macromolecules* **1995**, *28*, 4464–4474.
104 Z.-R. Chen, J. A. Kornfield, S. D. Smith, J. T. Grothaus, M. M. Satkowski, *Science* **1997**, *277*, 1248–1253.
105 Y. Zhang, U. Wiesner, *J. Chem. Phys.* **1997**, *106*, 2961–2969.
106 J. Sänger, W. Gronski, H. Leist, U. Wiesner, *Macromolecules* **1997**, *30*, 7621–7623.
107 H. Leist, K. Geiger, U. Wiesner, *Macromolecules* **1999**, *32*, 1315–1317.
108 K. de Moel, R. Mäki-Ontto, M. Stamm, O. Ikkala, G. ten Brinke, *Macromolecules* **2001**, *34*, 2892–2900.
109 K. de Moel, G. O. R. Alberda van Ekenstein, H. Nijland, E. Polushkin, G. ten Brinke, R. Mäki-Ontto, O. Ikkala, *Chem. Mater.* **2001**, *13*, 4580–4583.
110 R. Mäkinen, J. Ruokolainen, O. Ikkala, K. de Moel, G. ten Brinke, W. De Odorico, M. Stamm, *Macromolecules* **2000**, *33*, 3441–3446.
111 A. Jain, L. M. Hall, C. B. W. Garcia, S. M. Gruner, U. Wiesner, *Macromolecules* **2005**, *38*, 10095–10100.
112 T. Thurn-Albrecht, J. Schotter, G. A. Kästle, N. Emley, T. Shibauchi, L. Krusin-Elbaum, K. Guarini, C. T. Black, M. T. Tuominen, T. P. Russell, *Science* **2000**, *290*, 2126–2129.
113 E. Schäffer, T. Thurn-Albrecht, T. P. Russell, U. Steiner, *Nature* **2000**, *403*, 874–877.
114 P. Mansky, J. DeRouchey, T. P. Russell, J. Mays, M. Pitsikalis, T. Morkved,

H. Jaeger, *Macromolecules* **1998**, *31*, 4399–4401.

115 A. Böker, A. Knoll, H. Elbs, V. Abetz, A. H. E. Müller, G. Krausch, *Macromolecules* **2002**, *35*, 1319–1325.

116 A. Böker, H. Elbs, H. Hänsel, A. Knoll, S. Ludwigs, H. Zettl, V. Urban, V. Abetz, A. H. E. Müller, G. Krausch, *Phys. Rev. Lett.* **2002**, *89*, 135502.

117 K. Schmidt, A. Boeker, H. Zettl, F. Schubert, H. Hänsel, F. Fischer, T. M. Weiss, V. Abetz, A. V. Zvelindovsky, G. J. A. Sevink, G. Krausch, *Langmuir* **2005**, *21*, 11974–11980.

118 R. A. Segalman, H. Yokoyama, E. J. Kramer, *Adv. Mater.* **2001**, *13*, 1152–1155.

119 S. Stangler, V. Abetz, *Rheol. Acta* **2003**, *42*, 569–577.

120 S. O. Kim, H. H. Solak, M. P. Stoykovich, N. J. Ferrier, J. J. de Pablo, P. F. Nealey, *Nature* **2003**, *424*, 411–414.

121 D. Sundrani, S. B. Darling, S. J. Sibener, *Langmuir* **2004**, *20*, 5091–5099.

122 C. Osuji, P. J. Ferreira, G. Mao, C. K. Ober, J. B. Vander Sande, E. L. Thomas, *Macromolecules* **2004**, *37*, 9903–9908.

123 C. Y. Yang, Y. Cao, P. Smith, A. J. Heeger, *Synth. Met.* **1993**, *53*, 293–301.

124 O. T. Ikkala, J. Laakso, K. Väkiparta, E. Virtanen, H. Ruohonen, H. Järvinen, T. Taka, P. Passiniemi, J.-E. Österholm, Y. Cao, A. Andreatta, P. Smith, A. J. Heeger, *Synth. Met.* **1995**, *69*, 97–100.

125 F. S. Bates, W. W. Maurer, P. M. Lipic, M. A. Hillmyer, K. Almdal, K. Mortensen, G. H. Fredrickson, T. Lodge, *Phys. Rev. Lett.* **1997**, *79*, 849–852.

126 J. Tanner, O. T. Ikkala, J. Laakso, P. Passiniemi, in *Electrical, Optical and Magnetic Properties of Organic Solid Materials III* (Eds. A. K.-Y. Jen, C. Y.-C. Lee, L. R. Dalton, M. F. Rubner, G. E. Wnek, L. Y. Chiang), Materials Research Society, Boston, **1996**, 565–570.

127 F. Bates, *MRS Bull.* **2005**, *30*, 525–532.

128 P. Colomban (Ed.), *Proton Conductors: Solids, Membranes and Gels – Materials and Devices*, Cambridge University Press, Cambridge, **1992**.

129 T. Kudo, in *CRC Handbook of Solid State Electrochemistry* (Eds. P. J. Gellings, H. J. M. Bouwmeester), CRC Press, Boca Raton, FL, **1997**, 195–221.

130 K.-D. Kreuer, *Chem. Mater.* **1996**, *8*, 610–641.

131 K. D. Kreuer, *Solid State Ionics* **1997**, *94*, 55–62.

132 K. D. Kreuer, *Solid State Ionics* **1997**, *97*, 1–15.

133 O. Savadogo, *J. Power Sources* **2004**, *127*, 135–161.

134 C. G. Granqvist, A. Azens, A. Hjelm, L. Kullman, G. A. Niklasson, D. Rönnow, M. Strømme Mattsson, M. Veszelei, G. Vaivars, *Solar Energy* **1998**, *63*, 199–216.

135 J. R. Stevens, W. Wieczorek, D. Raducha, K. R. Jeffrey, *ACS Symp. Ser.* **1999**, *726*, 51–70.

136 T. Norby, *Solid State Ionics* **1999**, *125*, 1–11.

137 M. Rikukawa, K. Sanui, *Prog. Polym. Sci.* **2000**, *25*, 1463–1502.

138 K. D. Kreuer, *J. Membr. Sci.* **2001**, *185*, 29–39.

139 J. A. Kerres, *J. Membr. Sci.* **2001**, *185*, 3–27.

140 J. C. Lassegues, J. Grondin, M. Hernandez, B. Maree, *Solid State Ionics* **2001**, *145*, 37–45.

141 W. Wieczorek, G. Zukowska, R. Borkowska, S. H. Chung, S. Greenbaum, *Electrochim. Acta* **2001**, *46*, 1427–1438.

142 F. Opekar, K. Stulik, *Crit. Rev. Anal. Chem.* **2002**, *32*, 253–259.

143 M. F. H. Schuster, W. H. Meyer, *Annu. Rev. Mater. Res.* **2003**, *33*, 233–261.

144 J. Roziere, D. J. Jones, *Annu. Rev. Mater. Res.* **2003**, *33*, 503–555.

145 F. Ciuffa, F. Croce, A. D'Epifanio, S. Panero, B. Scrosati, *J. Power Sources* **2004**, *127*, 53–57.

146 A. S. Ioselevich, A. A. Kornyshev, J. H. G. Steinke, *J. Phys. Chem. B.* **2004**, *108*, 11953–11963.

147 Q. Li, R. He, J. O. Jensen, N. Bjerrum, *Fuel Cells* **2004**, *4*, 147–159.

148 M. A. Hickner, H. Ghassemi, Y. S. Kim, B. R. Einsla, J. E. McGrath, *Chem. Rev.* **2004**, *104*, 4587–4612.

149 K.-D. Kreuer, S. J. Paddison, E. Spohr, M. Schuster, *Chem. Rev.* **2004**, *104*, 4637–4678.

150 K. Miyatake, M. Watanabe, *Electrochemistry* **2005**, *73*, 12–19.

151 W. L. Harrison, M. A. Hickner, Y. S. Kim, J. E. McGrath, *Fuel Cells* **2005**, *5*, 201–212.
152 P. Jannasch, *Fuel Cells* **2005**, *5*, 248–260.
153 Y. Yang, S. Holdcroft, *Fuel Cells* **2005**, *5*, 171–186.
154 B. Smitha, S. Sridhar, A. A. Khan, *J. Membr. Sci.* **2005**, *259*, 10–26.
155 J. A. Asensio, P. Gomez-Romero, *Fuel Cells* **2005**, *5*, 336–343.
156 A. L. Rusanov, D. Likhatchev, P. V. Kostoglodov, K. Mullen, M. Klapper, *Adv. Polym. Sci.* **2005**, *179*, 83–134.
157 A. Eisenberg, J.-S. Kim, *Introduction to Ionomers*, Wiley, New York, **1998**.
158 K. A. Mauritz, R. B. Moore, *Chem. Rev.* **2004**, *104*, 4535–4585.
159 N. Agmon, *Chem. Phys. Lett.* **1995**, *244*, 456–462.
160 K. D. Kreuer, A. Fuchs, M. Ise, M. Spaeth, J. Maier, *Electrochim. Acta* **1998**, *43*, 1281–1288.
161 H. G. Herz, K. D. Kreuer, J. Maier, G. Scharfenberger, M. F. H. Schuster, W. H. Meyer, *Electrochim. Acta* **2003**, *48*, 2165–2171.
162 M. Schuster, T. Rager, A. Noda, K. D. Kreuer, J. Maier, *Fuel Cells* **2005**, *5*, 355–365.
163 D. T. Chin, H. H. Chang, *J. Appl. Electrochem.* **1989**, *19*, 95–99.
164 P. Donoso, W. Gorecki, C. Berthier, F. Defendini, C. Poinsignon, M. Armand, *Solid State Ionics* **1988**, *28–30*, 969–974.
165 S. Petty-Weeks, J. J. Zupancic, J. R. Swedo, *Solid State Ionics* **1988**, *31*, 117–125.
166 J. S. Wainright, J.-T. Wang, R. F. Savinell, M. Litt, H. Moaddel, C. Rogers, *Proc. Electrochem. Soc.* **1994**, 255–264.
167 J. S. Wainright, J.-T. Wang, D. Weng, R. F. Savinell, M. Litt, *J. Electrochem. Soc.* **1995**, *142*, L121–L123.
168 J. R. Stevens, W. Wieczorek, D. Raducha, K. R. Jeffrey, *Solid State Ionics* **1997**, *97*, 347–358.
169 M. F. Daniel, B. Desbat, F. Cruege, O. Trinquet, J. C. Lassegues, *Solid State Ionics* **1988**, *28–30*, 637–641.
170 M. F. Daniel, B. Desbat, J. C. Lassegues, *Solid State Ionics* **1988**, *28–30*, 632–636.
171 A. Bozkurt, M. Ise, K. D. Kreuer, W. H. Meyer, G. Wegner, *Solid State Ionics* **1999**, *125*, 225–233.
172 A. Bozkurt, W. H. Meyer, *Solid State Ionics* **2001**, *138*, 259–265.
173 A. Bozkurt, Ö. Ekinci, W. H. Meyer, *J. Appl. Polym. Sci.* **2003**, *90*, 3347–3353.
174 H. Erdemi, A. Bozkurt, W. H. Meyer, *Synth. Met.* **2004**, *143*, 133–138.
175 M. Tiitu, M. Torkkeli, R. Serimaa, T. Mäkelä, O. T. Ikkala, *Solid State Ionics* **2005**, *176*, 1291–1299.
176 D. Rodriguez, C. Jegat, O. Trinquet, J. Grondin, J. C. Lassegues, *Solid State Ionics* **1993**, *61*, 195–202.
177 M. Yamada, I. Honma, *J. Phys. Chem. B* **2004**, *108*, 5522–5526.
178 R. A. Weiss, A. Sen, L. A. Pottick, C. L. Willis, *Polym. Commun.* **1990**, *31*, 220–223.
179 R. A. Weiss, A. Sen, L. A. Pottick, C. L. Willis, *Polymer* **1991**, *32*, 2785–2792.
180 A. Mokrini, J. L. Acosta, *Polymer* **2001**, *42*, 9–15.
181 A. Mokrini, J. L. Acosta, *Polymer* **2001**, *42*, 8817–8824.
182 C. A. Edmondson, J. J. Fontanella, S. H. Chung, S. G. Greenbaum, G. E. Wnek, *Electrochim. Acta* **2001**, *46*, 1623–1628.
183 C. A. Edmondson, J. J. Fontanella, *Solid State Ionics* **2002**, *152–153*, 355–361.
184 A. Mokrini, J. L. Acosta, *J. Appl. Polym. Sci.* **2002**, *83*, 367–377.
185 J. Kim, B. Kim, B. Jung, *J. Membr. Sci.* **2002**, *207*, 129–137.
186 J. Kim, B. Kim, B. Jung, Y. S. Kang, H. Y. Ha, I.-H. Oh, K. J. Ihn, *Macromol. Rapid Commun.* **2002**, *23*, 753–756.
187 J. M. Sloan, Y. Elabd, E. Napadensky, *Proc. 60th Annu Tech Conf Soc. Plast. Eng.* **2002**, *3*, 3939–3941.
188 J. Won, S. W. Choi, Y. S. Kang, H. Y. Ha, I.-H. Oh, H. S. Kim, K. T. Kim, W. H. Jo, *J. Membr. Sci.* **2003**, *214*, 245–257.
189 J. Won, H. H. Park, Y. J. Kim, S. W. Choi, H. Y. Ha, I.-H. Oh, H. S. Kim, Y. S. Kang, K. J. Ihn, *Macromolecules* **2003**, *36*, 3228–3234.
190 Y. A. Elabd, E. Napadensky, J. M. Sloan, D. M. Crawford, C. W. Walker, *J. Membr. Sci.* **2003**, *217*, 227–242.

191 A. Nacher, P. Escribano, C.D. Rio, A. Rodriguez, J.L. Acosta, *J. Polym. Sci., Part A: Polym. Chem.* **2003**, *41*, 2809–2815.

192 M. Pan, X.S. Yang, C.H. Shen, R.Z. Yuan, *Key Eng. Mater.* **2003**, *249*, 385–390.

193 J.-E. Yang, J.-S. Lee, *Electrochim. Acta* **2004**, *50*, 617–620.

194 T. Fujinami, D. Miyano, T. Okamoto, M. Ozawa, A. Konno, *Electrochim. Acta* **2004**, *50*, 627–631.

195 Y.A. Elabd, C.W. Walker, F.L. Beyer, *J. Membr. Sci.* **2004**, *231*, 181–188.

196 A. Mokrini, C. Del Rio, J.L. Acosta, *Solid State Ionics* **2004**, *166*, 375–381.

197 P.G. Escribano, A. Nacher, C. Del Rio, L. Gonzalez, J.L. Acosta, *J. Appl. Polym. Sci.* **2004**, *93*, 2394–2402.

198 B. Kim, J. Kim, B. Jung, *J. Membr. Sci.* **2005**, *250*, 175–182.

199 H. Bashir, J.L. Acosta, A. Linares, *J. Membr. Sci.* **2005**, *253*, 33–42.

200 D. Sangeetha, *Eur. Polym. J.* **2005**, *41*, 2644–2652.

201 Y.A. Elabd, E. Napadensky, C.W. Walker, K.I. Winey, *Macromolecules* **2006**, *39*, 399–407.

202 J. Ding, C. Chuy, S. Holdcroft, *Chem. Mater.* **2001**, *13*, 2231–2233.

203 T. Thurn-Albrecht, R. Steiner, J. DeRouchey, C.M. Stafford, E. Huang, M. Bal, M. Tuominen, C.J. Hawker, T.P. Russell, *Adv. Mater.* **2000**, *12*, 787–791.

204 M.J. Fasolka, A.M. Mayes, *Annu Rev. Mater. Res.* **2001**, *31*, 323–355.

205 A. Knoll, A. Horvat, K.S. Lyakhova, G. Krausch, G.J.A. Sevink, A.V. Zvelindovsky, R. Magerle, *Phys. Rev. Lett.* **2002**, *89*, 035501.

206 Y. Lin, A. Böker, J. He, K. Sill, H. Xiang, C. Abetz, X. Li, J. Wang, T. Emrick, S. Long, Q. Wang, A.C. Balazs, T.P. Russell, *Nature* **2005**, *434*, 55–59.

207 R. Mäki-Ontto, K. de Moel, E. Polushkin, G. Alberda van Ekenstein, G. ten Brinke, O. Ikkala, *Adv. Mater.* **2002**, *14*, 357–361.

208 G. Cho, K.-P. Park, J. Jang, S. Jung, J. Moon, T. Kim, *Electrochem. Commun.* **2002**, *4*, 336–339.

209 O. Yamada, Y. Yin, K. Tanaka, H. Kita, K.-I. Okamoto, *Electrochim. Acta* **2005**, *50*, 2655–2659.

210 Y. Yang, Z. Shi, S. Holdcroft, *Macromolecules* **2004**, *37*, 1678–1681.

211 P. Xing, G.P. Robertson, M.D. Guiver, S.D. Mikhailenko, S. Kaliaguine, *Polymer* **2005**, *46*, 3257–3263.

212 Z. Shi, S. Holdcroft, *Macromolecules* **2005**, *38*, 4193–4201.

213 L. Rubatat, Z. Shi, O. Diat, S. Holdcroft, B.J. Frisken, *Macromolecules* **2006**, *39*, 720–730.

214 R. Mezzenga, J. Ruokolainen, G.H. Fredrickson, E.J. Kramer, *Macromolecules* **2003**, *36*, 4466–4471.

215 J. Gao, D. Lee, Y. Yang, S. Holdcroft, B.J. Frisken, *Macromolecules* **2005**, *38*, 5854–5856.

216 R. Mezzenga, J. Ruokolainen, G.H. Fredrickson, E. Kramer, D. Moses, A.J. Heeger, O. Ikkala, *Science* **2003**, *299*, 1872–1874.

217 J. Ruokolainen, G. ten Brinke, O.T. Ikkala, *Adv. Mater.* **1999**, *11*, 777–780.

218 H. Kosonen, S. Valkama, J. Ruokolainen, M. Torkkeli, R. Serimaa, G. ten Brinke, O. Ikkala, *Eur. Phys. J.* **2003**, *10*, 69–75.

219 H. Kosonen, S. Valkama, J. Ruokolainen, G. ten Brinke, O. Ikkala, *Mater. Res. Soc. Symp. Proc.* **2003**, *775*, 147–152.

220 G. ten Brinke, O. Ikkala, *Macromol. Symp.* **2003**, *203*, 103–109.

221 S. Valkama, H. Kosonen, J. Ruokolainen, M. Torkkeli, R. Serimaa, G. ten Brinke, O. Ikkala, *Nat. Mater.* **2004**, *3*, 872–876.

222 G. ten Brinke, O. Ikkala, *Chem. Rec.* **2004**, *4*, 219–230.

223 M. Antonietti, M. Neese, G. Blum, F. Kremer, *Langmuir* **1996**, *12*, 4436–4441.

224 M. Antonietti, M. Maskos, F. Kremer, G. Blum, *Acta Polym.* **1996**, *47*, 460–465.

225 D.R. Sadoway, *J. Power Sources* **2004**, *129*, 1–3.

226 P.V. Wright, *Br. Polym. J.* **1975**, *7*, 319–327.

227 M. Ratner, D.F. Shriver, *Chem. Rev.* **1988**, *88*, 109–124.

228 B. Scrosati (Ed.), *Applications of Electroactive Polymers*, Chapman and Hall, London, **1993**.

229 C.A. Angell, K. Xu, S.-S. Zhang, M. Videa, *Solid State Ionics* **1996**, *86–88*, 17–28.

230 M. A. Ratner, P. Johansson, D. F. Shriver, *MRS Bull.* **2000**, *25*, 31–37.
231 P. V. Wright, *MRS Bull.* **2002**, *27*, 597–602.
232 J. R. MacCallum, C. A. Vincent (Eds.), *Polymer Electrolyte Reviews*, Vol. 1, Elsevier, Amsterdam, **1987**.
233 A. Ferry, M. M. Doeff, *Curr. Trends Polym. Sci.* **1998**, *3*, 117–130.
234 P. V. Wright, Y. Zheng, D. Bhatt, T. Richardson, G. Ungar, *Polym. Int.* **1998**, *47*, 34–42.
235 W. Xu, M. D. Williams, C. A. Angell, *Chem. Mater.* **2002**, *14*, 401–409.
236 Y. Zheng, J. Lui, G. Ungar, P. V. Wright, *Chem. Rec.* **2004**, *4*, 176–191.
237 M. Kono, E. Hayashi, M. Watanabe, *J. Electrochem. Soc.* **1998**, *145*, 1521–1527.
238 A. Nishimoto, M. Watanabe, Y. Ikeda, S. Kohjiya, *Electrochim. Acta* **1998**, *43*, 1177–1184.
239 A. Nishimoto, K. Agehara, N. Furuya, T. Watanabe, M. Watanabe, *Macromolecules* **1999**, *32*, 1541–1546.
240 Y. Ikeda, *J. Appl. Polym. Sci.* **2000**, *78*, 1530–1540.
241 P. P. Soo, B. Huang, Y.-I. Jang, Y.-M. Chiang, D. R. Sadoway, A. M. Mayes, *J. Electrochem. Soc.* **1999**, *146*, 32–37.
242 A.-V. G. Ruzette, P. P. Soo, D. R. Sadoway, A. M. Mayes, *J. Electrochem. Soc.* **2001**, *148*, 537–543.
243 S.-W. Ryu, P. E. Trapa, S. C. Olugebefola, J. A. Gonzalez-Leon, D. R. Sadoway, A. M. Mayes, *J. Electrochem. Soc.* **2005**, *152*, A158–A163.
244 J. R. M. Giles, F. M. Gray, J. R. MacCallum, C. A. Vincent, *Polymer* **1987**, *28*, 1977–1981.
245 F. M. Gray, J. R. MacCallum, C. A. Vincent, J. R. M. Giles, *Macromolecules* **1988**, *21*, 392–397.
246 K. Hirahara, A. Takano, M. Yamamoto, T. Kazama, Y. Isono, T. Fujimoto, O. Watanabe, *React. Funct. Polym.* **1988**, *37*, 169–182.
247 P. Jannasch, *Polymer* **2002**, *43*, 6449–6453.
248 P. Jannasch, *Chem. Mater.* **2002**, *14*, 2718–2724.
249 P. Jannasch, W. Loyens, *Solid State Ionics* **2004**, *166*, 417–424.
250 G. Liu, M. T. Reinhout, G. L. Baker, *Solid State Ionics* **2004**, *175*, 721–724.
251 T. Niitani, M. Shimada, K. Kawamura, K. Dokko, Y.-H. Rho, K. Kanamura, *Electrochem. Solid-State Lett.* **2005**, *8*, A385–A388.
252 B.-K. Cho, A. Jain, S. M. Gruner, U. Wiesner, *Science* **2004**, *305*, 1598–1601.
253 F. B. Dias, S. V. Batty, J. P. Voss, G. Ungar, P. V. Wright, *Solid State Ionics* **1996**, *85*, 43–49.
254 F. B. Dias, S. V. Batty, A. Gupta, G. Ungar, J. P. Voss, P. V. Wright, *Electrochim. Acta* **1998**, *43*, 1217–1224.
255 Y. Zheng, F. Chia, G. Ungar, P. V. Wright, **2000**, *16*, 1459–1460.
256 Y. Zheng, F. Chia, G. Ungar, P. V. Wright, *J. Power Sources* **2001**, *97–98*, 641–643.
257 F. S. Chia, Y. Zheng, J. Liu, G. Ungar, P. V. Wright, *Solid State Ionics* **2002**, *147*, 275–280.
258 Y. Zheng, J. Liu, Y.-P. Liao, G. Ungar, P. V. Wright, *Dalton Trans.* **2004**, 3053–3060.
259 Y. Zheng, J. Liu, Y.-P. Liao, G. Ungar, P. V. Wright, *J. Power Sources* **2005**, *146*, 418–422.
260 J. Liu, Y. Zheng, Y.-P. Liao, D. C. Apperley, G. Ungar, P. V. Wright, *Electrochim. Acta* **2005**, *50*, 3815–3826.
261 M. Ballauff, G. F. Schmidt, *Makromol. Chem. Rapid Commun.* **1987**, *8*, 93–97.
262 M. Ballauff, *Angew. Chem. Int. Ed.* **1989**, *28*, 253–267.
263 M. Wenzel, M. Ballauff, G. Wegner, *Makromol. Chem.* **1987**, *188*, 2865–2873.
264 T. Vahlenkamp, G. Wegner, *Makromol. Chem. Phys.* **1994**, *195*, 1933–1952.
265 T. F. McCarthy, H. Witteler, T. Pakula, G. Wegner, *Macromolecules* **1995**, *28*, 8350–8362.
266 U. Lauter, W. H. Meyer, G. Wegner, *Macromolecules* **1997**, *30*, 2092–2101.
267 P. Baum, W. H. Meyer, G. Wegner, *Polymer* **2000**, *41*, 965–973.
268 T. Ohtake, M. Ogasawara, K. Ito-Akita, N. Nishina, S. Ujiie, H. Ohno, T. Kato, *Chem. Mater.* **2000**, *12*, 782–789.
269 K. Hoshino, M. Yoshio, T. Mukai, K. Kishimoto, H. Ohno, T. Kato, *J. Polym. Sci., Polym. Chem. Ed.* **2003**, *41*, 3486–3492.

270 T. Ohtake, Y. Takamitsu, M. Ogasawara, K. Ito-Akita, K. Kanie, M. Yoshizawa, T. Mukai, H. Ohno, T. Kato, *Macromolecules* **2000**, *33*, 8109–8111.

271 G. Wegner, *Thin Solid Films* **1992**, *216*, 105–116.

272 S. Vanhee, R. Rulkens, U. Lehmann, C. Rosenauer, M. Schulze, W. Köhler, G. Wegner, *Macromolecules* **1996**, *29*, 5136–5142.

273 R. Stepanyan, A. Subbotin, M. Knaapila, O. Ikkala, G. ten Brinke, *Macromolecules* **2003**, *36*, 3758–3763.

274 P. Bäuerle, G. Götz, M. Hiller, S. Scheib, T. Fischer, U. Segelbacher, M. Bennati, A. Grupp, M. Mehring, M. Stoldt, C. Seidel, F. Geiger, H. Schweizer, E. Umbach, M. Schmelzer, S. Roth, H. F. Egelhaaf, D. Oelkrug, P. Emele, H. Port, *Synth. Met.* **1993**, *61*, 71–79.

275 G. Wegner, *Makromol. Chem. Macromol. Symp.* **1996**, *106*, 1415–1419.

276 H. Menzel, in *Polymer Materials Encyclopedia* (Ed. J. C. Salamone), CRC Press, Boca Raton, FL, **1996**, 2916–2927.

277 W. Hu, H. Nakashima, K. Furukawa, Y. Kashimura, K. Ajito, Y. Liu, D. Zhu, K. Torimitsu, *J. Am. Chem. Soc.* **2005**, *127*, 2804–2805.

278 K. Y. Jen, G. G. Miller, R. L. Elsenbaumer, *J. Chem. Soc., Chem. Commun.* **1986**, 1346–1347.

279 S. D. D. V. Rughooputh, M. Nowak, S. Hotta, A. J. Heeger, F. Wudl, *Synth. Met.* **1987**, *21*, 41–50.

280 S. Hotta, S. D. D. V. Rughooputh, A. J. Heeger, *Synth. Met.* **1987**, *22*, 79–87.

281 K. Yoshino, S. Nakajima, R.-I. Sugimoto, *Jpn. J. Appl. Phys.* **1986**, *26*, L1038–L1039.

282 T. J. Prosa, M. J. Winokur, J. Moulton, P. Smith, A. J. Heeger, *Macromolecules* **1992**, *25*, 4364–4372.

283 S.-A. Chen, J.-M. Ni, *Macromolecules* **1992**, *25*, 6081–6089.

284 R. D. McCullough, S. Tristram-Nagle, S. P. Williams, R. D. Lowe, M. Jayaraman, *J. Am. Chem. Soc.* **1993**, *115*, 4910–4911.

285 X. Wu, T.-A. Chen, R. D. Rieke, *Macromolecules* **1995**, *28*, 2101–2102.

286 X. Wu, T.-A. Chen, R. D. Rieke, *Macromolecules* **1996**, *29*, 7671–7677.

287 M. Rikukawa, M. Nakagawa, Y. Tabuchi, K. Sanui, N. Ogata, *Synth. Met.* **1997**, *84*, 233–234.

288 R. D. McCullough, *Adv. Mater.* **1998**, *10*, 93–116.

289 R. D. McCullough, in *Handbook of Oligo- and Polythiophenes* (Ed. D. Fichou), Wiley-VCH, Weinheim **1999**, 1–44.

290 H. Sirringhaus, R. J. Wilson, R. H. Friend, M. Inbasekaran, W. Wu, E. P. Woo, M. Grell, D. D. C. Bradley, *Appl. Phys. Lett.* **2000**, *77*, 406–408.

291 E. Mena-Osteritz, A. Meyer, B. M. W. Langeveld-Voss, R. A. J. Janssen, E. W. Meijer, P. Bäuerle, *Angew. Chem., Int. Ed.* **2000**, *39*, 2680–2684.

292 E. Mena-Osteritz, *Adv. Mater.* **2002**, *14*, 609–616.

293 B. Grevin, P. Rannou, R. Payerne, A. Pron, J.-P. Travers, *Adv. Mater.* **2003**, *15*, 881–884.

294 G. Barbarella, in *Electronic and Optical Properties of Conjugated Molecular Systems in Condensed Phases* (Ed, S. Hotta), Research Signpost, Trivandrum **2003**, 79–97.

295 F. C. Krebs, S. V. Hoffmann, M. Jorgensen, *Synth. Met.* **2003**, *138*, 471–474.

296 D. W. Breiby, E. J. Samuelsen, O. Konovalov, B. Struth, *Synth. Met.* **2003**, *135–136*, 363–364.

297 M. Mas-Torrent, D. Den Boer, M. Durkut, P. Hadley, A. P. H. J. Schenning, *Nanotechnology* **2004**, *15*, S265–S269.

298 J.-F. Chang, B. Sun, D. W. Breiby, M. M. Nielsen, T. I. Sölling, M. Giles, I. McCulloch, H. Sirringhaus, *Chem. Mater.* **2004**, *16*, 4772–4776.

299 A. Babel, S. A. Jenekhe, *Synth. Met.* **2005**, *148*, 169–173.

300 B. S. Ong, Y. Wu, P. Liu, *Proc. IEEE* **2005**, *93*, 1412–1419.

301 H. Sirringhaus, P. J. Brown, R. H. Friend, M. M. Nielsen, K. Bechgaard, B. M. W. Langeveld-Voss, A. J. H. Spiering, R. A. J. Janssen, E. W. Meijer, P. Herwig, D. M. de Leeuw, *Nature* **1999**, *401*, 685–688.

302 T. Bjoernholm, D. R. Greve, N. Reitzel, T. Hassenkam, K. Kjaer, P. B. Howes, N. B. Larsen, J. Boegelund, M. Jayaraman, P. C. Ewbank, R. D. McCullough, *J. Am. Chem. Soc.* **1998**, *120*, 7643–7644.

303 Hong, J.C. Tyson, J.S. Middlecoff, D.M. Collard, *Macromolecules* **1999**, *32*, 4232–4239.

304 W.-Y. Zheng, K. Levon, J. Laakso, J.-E. Österholm, *Macromolecules* **1994**, *27*, 7754–7768.

305 W.-Y. Zheng, K. Levon, T. Taka, J. Laakso, J.-E. Österholm, *J. Polym. Sci., Polym. Phys. Ed.* **1994**, *33*, 1289–1306.

306 N.A. Platé, V.P. Shibaev, *Comb-Shaped Polymers and Liquid Crystals*, Plenum Press, New York, **1987**.

307 M. Grell, D.D.C. Bradley, G. Ungar, J. Hill, K.S. Whitehead, *Macromolecules* **1999**, *32*, 5810–5817.

308 M. Grell, W. Knoll, D. Lupo, A. Meisel, T. Miteva, E. Neher, H.-G. Nothofer, U. Scherf, A. Yasuda, *Adv. Mater.* **1999**, *11*, 671–675.

309 M. Knaapila, B.P. Lyons, K. Kisko, J.P. Foreman, U. Vainio, M. Mihaylova, O.H. Seeck, L.-O. Pålsson, R. Serimaa, M. Torkkeli, A.P. Monkman, *J. Phys. Chem. B* **2003**, *107*, 12425–12430.

310 D. Neher, *Makromol. Chem. Rapid Commun.* **2001**, *22*, 1365–1385.

311 M.J. Winokur, J. Slinker, D.L. Huber, *Phys. Rev. B* **2003**, *67*, 184106.

312 M. Knaapila, K. Kisko, B.P. Lyons, R. Stepanyan, J.P. Foreman, O.H. Seeck, U. Vainio, L.-O. Pålsson, R. Serimaa, M. Torkkeli, A.P. Monkman, *J. Phys. Chem. B* **2004**, *108*, 10711–10720.

313 Y. Zhu, M.M. Alam, S.A. Jenekhe, *Macromolecules* **2003**, *36*, 8958–8968.

314 D.A.P. Delnoye, R.P. Sijbesma, J.A.J.M. Vekemans, E.W. Meijer, *J. Am. Chem. Soc.* **1996**, *118*, 8717–8718.

315 M. Moroni, J. Le Moigne, T.A. Pham, J.-Y. Bigot, *Macromolecules* **1997**, *30*, 1964–1972.

316 J.-C. Chiang, A.G. MacDiarmid, *Synth. Met.* **1986**, *13*, 193–205.

317 A.J. Epstein, A.G. MacDiarmid, *Synth. Met.* **1995**, *69*, 179–182.

318 A.G. MacDiarmid, *Rev. Mod. Phys.* **2001**, *73*, 701–712.

319 W.R. Salaneck, I. Lundström, W.-S. Huang, A.G. MacDiarmid, *Synth. Met.* **1986**, *13*, 291–297.

320 M. Doriomedoff, F. Hautiere-Cristofini, R.D. Surville, M. Jozefowicz, L.-T. Yu, R. Buvet, *J. Chim. Phys. Physicochim. Biol.* **1971**, *68*, 1055–1069.

321 Y. Cao, P. Smith, A.J. Heeger, *Synth. Met.* **1992**, *48*, 91–97.

322 C.Y. Yang, P. Smith, A.J. Heeger, Y. Cao, J.-E. Österholm, *Polymer* **1994**, *35*, 1142–1147.

323 A. Kobayashi, X. Xu, H. Ishikawa, M. Satoh, E. Hasegawa, *J. Appl. Phys.* **1992**, *72*, 5702–5705.

324 P.J. Kinlen, J. Liu, Y. Ding, C.R. Graham, E.E. Remsen, *Macromolecules* **1998**, *31*, 1735–1744.

325 I. Kulszewicz-Bajer, M. Zagórska, J. Niziol, A. Pron, W. Luzny, *Synth. Met.* **2000**, *114*, 125–131.

326 T.E. Olinga, J. Fraysse, J.P. Travers, A. Dufresne, A. Pron, *Macromolecules* **2000**, *33*, 2107–2113.

327 B. Dufour, P. Rannou, P. Fedorko, D. Djurado, J.P. Travers, A. Pron, *Chem. Mater.* **2001**, *13*, 4032–4440.

328 B. Dufour, P. Rannou, D. Djurado, H. Janeczek, M. Zagorska, A. de Geyer, J.-P. Travers, A. Pron, *Chem. Mater.* **2003**, *15*, 1587–1592.

329 J. Hartikainen, J. Ruokolainen, K. Rissanen, O. Ikkala, *Synth. Met.* **2001**, *121*, 1275–1276.

330 J. Hartikainen, M. Lahtinen, M. Torkkeli, R. Serimaa, J. Valkonen, K. Rissanen, O. Ikkala, *Macromolecules* **2001**, *34*, 7789–7795.

331 A. Subbotin, R. Stepanyan, M. Knaapila, O. Ikkala, G. ten Brinke, *Eur. Phys. J. E* **2003**, *12*, 333–345.

332 Z.-X. Wei, T. Laitinen, B. Smarsly, O. Ikkala, C.F.J. Faul, *Angew. Chem. Int. Ed.* **2005**, *44*, 751–756.

333 M. Vilkman, H. Kosonen, A. Nykänen, J. Ruokolainen, M. Torkkeli, R. Serimaa, O. Ikkala, *Macromolecules* **2005**, *38*, 7793–7797.

334 S. Kempf, W. Gronski, *Polym. Bull.* **1990**, *23*, 403–410.

335 S. Kempf, H.W. Rotter, S.N. Magonov, W. Gronski, H.J. Cantow, *Polym. Bull.* **1990**, *24*, 325–332.

336 J. Li, I.M. Khan, *Makromol. Chem.* **1991**, *192*, 3043–3050.

337 J. Li, S. Arnold, I.M. Khan, *Proc. SPIE* **1994**, *2189*, 126–133.

338 S. Arnold, L. M. Pratt, S. M. Khan, J. Li, I. M. Khan, *Proc. SPIE* **1995**, *2441*, 23–32.

339 M. A. Hempenius, B. M. W. Langeveld-Voss, J. A. E. H. van Haare, R. A. J. Janssen, S. S. Sheiko, J. P. Spatz, M. Möller, E. W. Meijer, *J. Am. Chem. Soc.* **1998**, *120*, 2798–2804.

340 J. Liu, E. Sheina, T. Kowalewski, R. D. McCullough, *Angew. Chem. Int. Ed.* **2002**, *41*, 329–332.

341 T. Kowalewski, R. D. McCullough, K. Matyjaszewski, *Eur. Phys. J. E* **2003**, *10*, 5–16.

342 D. M. Vriezema, A. Kros, J. Hoogboom, A. E. Rowan, R. J. M. Nolte, *Polym. Prepr.* **2004**, *45*, 749–750.

343 L. Balogh, L. Samuelson, K. S. Alva, A. Blumstein, *Macromolecules* **1996**, *29*, 4180–4186.

344 J. W. Y. Lam, B. Z. Tang, *Acc. Chem. Res.* **2005**, *38*, 745–754.

345 X. L. Chen, S. A. Jenekhe, *Macromolecules* **1996**, *29*, 6189–6192.

346 S. A. Jenekhe, X. L. Chen, *Science* **1998**, *279*, 1903–1907.

347 X. L. Chen, S. A. Jenekhe, *Macromolecules* **2000**, *33*, 4610–4612.

348 N. A. J. M. Sommerdijk, S. J. Holder, R. C. Hiorns, R. G. Jones, R. J. M. Nolte, *Macromolecules* **2000**, *33*, 8289–8294.

349 M. Lee, J.-W. Kim, I.-W. Hwang, Y.-R. Kim, N.-K. Oh, W.-C. Zin, *Adv. Mater.* **2001**, *13*, 1363–1368.

350 H. Wang, H. H. Wang, V. S. Urban, K. C. Littrell, P. Thiyagarajan, L. Yu, *J. Am. Chem. Soc.* **2000**, *122*, 6855–6861.

351 S. A. Jenekhe, X. L. Chen, *Science* **1999**, *283*, 372–375.

352 P. Leclere, A. Calderone, K. Müllen, J. L. Bredas, R. Lazzaroni, *Mater. Sci. Technol.* **2002**, *18*, 749–754.

353 P. Leclere, E. Hennebicq, A. Calderone, P. Brocorens, A. C. Grimsdale, K. Müllen, J. L. Bredas, R. Lazzaroni, *Prog. Polym. Sci.* **2003**, *28*, 55–81.

354 V. Percec, C.-H. Ahn, G. Ungar, D. J. P. Yeardley, M. Möller, S. S. Sheiko, *Nature* **1998**, *391*, 161–164.

355 V. Percec, M. Glodde, T. K. Bera, Y. Miura, I. Shiyanovskaya, K. D. Singer, V. S. K. Balagurusamy, P. A. Heiney, I. Schnell, A. Rapp, H.-W. Spiess, S. D. Hudson, H. Duan, *Nature* **2002**, *419*, 384–387.

356 V. Percec, A. E. Dulcey, V. S. K. Balagurusamy, Y. Miura, J. Smidrkal, M. Peterca, S. Nummelin, U. Edlund, S. D. Hudson, P. A. Heiney, H. Duan, S. N. Maganov, S. A. Vinogradov, *Nature* **2004**, *430*, 764–768.

357 C. Xia, X. Fan, J. Locklin, R. C. Advincula, A. Gies, W. Nonidez, *J. Am. Chem. Soc.* **2004**, *126*, 8735–8743.

358 M. Moller, J. P. Spatz, A. Roescher, S. Mossmer, S. T. Selvan, H. A. Klok, *Macromol. Symp.* **1997**, *117*, 207–218.

359 L. M. Bronstein, S. N. Sidorov, P. M. Valetsky, J. Hartmann, H. Cölfen, M. Antonietti, *Langmuir* **1999**, *15*, 6256–6262.

360 J. P. Spatz, *Angew. Chem. Int. Ed.* **2002**, *41*, 3359–3362.

361 R. Glass, M. Möller, J. P. Spatz, *Nanotechnology* **2003**, *14*, 1153–1160.

362 R. Glass, M. Arnold, E. A. Cavalcanti-Adam, J. Blümmel, C. Haferkemper, C. Dodd, J. P. Spatz, *New J. Phys.* **2004**, *6*, 101.

363 L. M. Bronstein, S. N. Sidorov, V. Zhirov, D. Zhirov, Y. A. Kabachii, S. Y. Kochev, P. M. Valetsky, B. Stein, O. I. Kiseleva, S. V. Polyakov, E. V. Shtykova, E. V. Nikulina, D. I. Svergun, A. R. Khokhlov, *J. Phys. Chem. B* **2005**, *109*, 18786–18798.

364 T. Hashimoto, H. Hasegawa, *Trans. Mater. Res. Soc. Jpn.* **2004**, *29*, 77–82.

365 J. J. L. M. Cornelissen, R. Van Heerbeek, P. C. J. Kamer, J. N. H. Reek, N. A. J. M. Sommerdijk, R. J. M. Nolte, *Adv. Mater.* **2002**, *14*, 489–492.

366 J. Ruokolainen, M. Saariaho, O. Ikkala, G. ten Brinke, E. L. Thomas, M. Torkkeli, R. Serimaa, *Macromolecules* **1999**, *32*, 1152–1158.

367 W. van Zoelen, G. A. van Ekenstein, E. Polushkin, O. Ikkala, G. ten Brinke, *Soft Matter* **2005**, *1*, 280–283.

368 A. W. Fahmi, H.-G. Braun, M. Stamm, *Adv. Mater.* **2003**, *15*, 1201–1204.

369 A. W. Fahmi, M. Stamm, *Langmuir* **2005**, *21*, 1062–1066.

370 S.-H. Yun, S. M. Yoo, B.-H. Sohn, J. C. Jung, W.-C. Zin, S.-Y. Kwak,

T. S. Lee, *Langmuir* **2005**, *21*, 3625–3628.

371 T. Scheibel, R. Parthasarathy, G. Sawicki, X.-M. Lir, H. Jaeger, S. L. Lindquist, *Proc. Natl. Acad. Sci. USA* **2003**, *100*, 4527–4532.

372 E. Yablonovitch, *Phys. Rev. Lett.* **1987**, *58*, 2059–2062.

373 S. John, *Phys. Rev. Lett.* **1987**, *58*, 2486–2489.

374 J. D. Joannopoulos, R. D. Meade, J. N. Winn, *Photonic Crystals*, Princeton University Press, Princeton, NJ, **1995**.

375 *MRS Bull.* **1998**, *23*. Topical issue.

376 J. D. Joannopoulos, *Nature* **2001**, *404*, 257–258.

377 E. Yablonovitch, *Nat. Mater.* **2003**, *2*, 648–649.

378 J. Yoon, W. Lee, E. L. Thomas, *MRS Bull.* **2005**, *30*, 721–726.

379 S. John, K. Busch, *J. Lightwave Technol.* **1999**, *17*, 1931–1943.

380 K. Yoshino, Y. Kawagishi, S. Tatsuhara, H. Kajii, S. Lee, M. Ozaki, Z. V. Vardeny, A. A. Zakhidov, *Superlatt. Microstruct.* **1999**, *25*, 325–341.

381 B. Gates, S. H. Park, Y. Xia, *Adv. Mater.* **2000**, *12*, 653–656.

382 B. Gates, Y. Xia, *Appl. Phys. Lett.* **2001**, *78*, 3178–3180.

383 M. Müller, R. Zentel, T. Maka, S. G. Romanov, C. M. Sotomayor Torres, *Chem. Mater.* **1999**, *12*, 2508–2512.

384 Y. Xia, Y. Yin, Y. Lu, J. McLellan, *Adv. Funct. Mater.* **2003**, *13*, 907–918.

385 A. A. Zakhidov, R. H. Baughman, Z. Iqbal, C. Cui, I. Khayrullin, S. O. Dantas, J. Marti, V. G. Ralchenko, *Science* **1998**, *282*, 897–901.

386 Y. A. Vlasov, X.-Z. Bo, J. C. Sturm, D. J. Norris, *Nature* **2001**, *414*, 289–293.

387 H. Miguez, F. Meseguer, C. López, F. López-Tejeira, J. Sánchez-Dehesa, *Adv. Mater.* **2001**, *13*, 393–396.

388 Y. Fink, A. M. Urbas, M. G. Bawendi, J. D. Joannopoulos, E. L. Thomas, *J. Lightwave Technol.* **1999**, *17*, 1963–1969.

389 A. Urbas, Y. Fink, E. L. Thomas, *Macromolecules* **1999**, *32*, 4748–4750.

390 A. Urbas, R. Sharp, Y. Fink, E. L. Thomas, M. Xenidou, L. J. Fetters, *Adv. Mater.* **2000**, *12*, 812–814.

391 A. C. Edrington, A. M. Urbas, P. DeRege, C. X. Chen, T. M. Swager, N. Hadjichristidis, M. Xenidou, L. J. Fetters, J. D. Joannopoulos, Y. Fink, E. L. Thomas, *Adv. Mater.* **2001**, *13*, 421–425.

392 M. Bockstaller, R. Kolb, E. L. Thomas, *Adv. Mater.* **2001**, *13*, 1783–1786.

393 M. Maldovan, A. M. Urbas, N. Yufa, W. C. Carter, E. L. Thomas, *Phys. Rev. B* **2002**, *54*, 165123.

394 A. M. Urbas, M. Maldovan, P. DeRege, E. L. Thomas, *Adv. Mater.* **2002**, *14*, 1850–1853.

395 C. Osuji, C.-Y. Chao, I. Bita, C. K. Ober, E. L. Thomas, *Adv. Funct. Mater.* **2002**, *12*, 753–758.

396 A. M. Urbas, E. L. Thomas, H. Kriegs, G. Fytas, R. S. Penciu, L. N. Economou, *Phys. Rev. Lett.* **2003**, *90*, 108302.

397 M. R. Bockstaller, E. L. Thomas, *Proc. SPIE* **2003**, *5222*, 94–107.

398 M. Maldovan, C. K. Ullal, W. C. Carter, E. L. Thomas, *Nat. Mater.* **2003**, *2*, 664–667.

399 T. Deng, C. Chen, C. Honeker, E. L. Thomas, *Polymer* **2003**, *44*, 6549–6553.

400 M. Maldovan, E. L. Thomas, C. W. Carter, *Appl. Phys. Lett.* **2004**, *84*, 362–364.

401 C. K. Ullal, M. Maldovan, E. L. Thomas, G. Chen, Y.-J. Han, S. Yang, *Appl. Phys. Lett.* **2004**, *84*, 5434–5436.

402 M. Maldovan, E. L. Thomas, *J. Opt. Soc. Am. B* **2005**, *22*, 466–473.

403 Y. Fink, J. N. Winn, S. Fan, C. Chen, J. Michel, J. D. Joannopoulos, E. L. Thomas, *Science* **1998**, *282*, 1679–1682.

404 M. R. Bockstaller, E. L. Thomas, *Phys. Rev. Lett.* **2004**, *93*, 166106.

405 H. F. Mark, N. M. Bikales, C. G. Overberger, G. Menges, J. I. Kroschwitz, *Encyclopaedia of Polymer Science and Engineering*, Wiley, New York **1987**.

406 C. Bosshard, K. Sutter, P. Prêtre, J. Hulliger, M. Flörsheimer, P. Kaatz, P. Günter (Eds.), *Organic Nonlinear Optical Materials (Advances in Nonlinear Optics)*, Gordon and Breach, Basel, **1995**.

407 L. R. Dalton, A. W. Harper, R. Ghosn, W. H. Steier, M. Ziari, H. Fetterman, Y. Shi, R. V. Mustacich, A. K.-Y. Jen, K. J. Shea, *Chem. Mater.* **1995**, *7*, 1060–1081.

408 J. A. Delaire, K. Nakatani, *Chem. Rev.* **2000**, *100*, 1817–1845.
409 Topical issue K.-S. Lee (Ed.), *Adv. Polym. Sci.* **2003**, *161*.
410 C.-S. Kang, C. Heldmann, H.-J. Winkelhahn, M. Schulze, D. Neher, G. Wegner, R. Wortmann, C. Glania, P. Kraemer, *Macromolecules* **1994**, *27*, 6156–6162.
411 C.-S. Kang, H.-J. Winkelhahn, M. Schulze, D. Neher, G. Wegner, *Chem. Mater.* **1994**, *6*, 2159–2166.
412 C. Heldmann, M. Schulze, G. Wegner, *Macromolecules* **1996**, *29*, 4686–4696.
413 C. Heldmann, D. Neher, H.-J. Winkelhahn, G. Wegner, *Macromolecules* **1996**, *29*, 4697–4705.
414 R. G. Petschek, K. M. Wiefling, *Phys. Rev. Lett.* **1987**, *59*, 343–346.
415 A. Halperin, *Macromolecules* **1990**, *23*, 2724–2731.
416 F. Tournilhac, L. M. Blinov, J. Simon, S. V. Yablonskii, *Nature* **1992**, *359*, 621–623.
417 J. Prost, R. Bruinsma, F. Tournilhac, *J. Phys. II (Paris)* **1994**, *4*, 169–187.
418 K. M. Reiser, P. Stoller, P. Celliers, A. Rubenchik, C. Bratton, D. Yankelevich, *Proc. SPIE* **2003**, *5212*, 149–156.
419 S. I. Stupp, K. E. Huggins, L. S. Li, L. H. Radzilowski, M. Keser, V. Lebonheur, S. Son, *NATO ASI Ser. C* **1997**, *499*, 219–240.
420 S. I. Stupp, M. Keser, G. N. Tew, *Polymer* **1998**, *39*, 4505–4508.
421 G. N. Tew, S. I. Stupp, *ACS Symp. Ser.* **1998**, *704*, 218–226.
422 G. N. Tew, L. Li, S. I. Stupp, *J. Am. Chem. Soc.* **1998**, *120*, 5601–5602.
423 G. N. Tew, M. U. Pralle, S. I. Stupp, *Angew. Chem. Int. Ed.* **2000**, *39*, 517–521.
424 E. R. Zubarev, M. U. Pralle, L. Li, S. I. Stupp, *Science* **1999**, *283*, 523–526.
425 L. S. Li, S. I. Stupp, *Macromolecules* **1997**, *30*, 5313–5320.
426 L. Li, S. I. Stupp, *Appl. Phys. Lett.* **2001**, *78*, 4127–4129.
427 L. Li, E. R. Zubarev, B. A. Acker, S. I. Stupp, *Macromolecules* **2002**, *35*, 2560–2565.
428 L. Li, E. Beniash, E. R. Zubarev, W. Xiang, B. M. Rabatic, G. Zhang, S. I. Stupp, *Nat. Mater.* **2003**, *2*, 689–694.
429 M. U. Pralle, K. Urayama, G. N. Tew, D. Neher, G. Wegner, S. I. Stupp, *Angew. Chem. Int. Ed.* **2000**, *39*, 1486–1489.
430 T. Goldacker, V. Abetz, R. Stadler, I. Erukhimovich, L. Leibler, *Nature* **1999**, *398*, 137–139.
431 C. J. Brabec, N. S. Sariciftci, in *Semiconducting Polymers* (Eds. G. Hadziioannou, P. F. Van Hutten), Wiley-VCH, Weinheim **2000**, 515–560.
432 K. M. Coakley, M. D. McGehee, *Chem. Mater.* **2004**, *16*, 4533–4542.
433 R. A. J. Janssen, J. C. Hummelen, N. S. Sariciftci, *MRS Bull.* **2005**, *30*, 33–36.
434 S. E. Shaheen, D. S. Ginley, G. E. Jabbour, *MRS Bull.* **2005**, *30*, 10–19.
435 J. Roncali, *Chem. Soc. Rev.* **2005**, *34*, 483–495.
436 U. Stalmach, B. de Boer, C. Videlot, P. F. van Hutten, G. Hadziioannou, *J. Am. Chem. Soc.* **2000**, *122*, 5464–5472.
437 B. de Boer, U. Stalmach, P. F. van Hutten, C. Melzer, V. V. Krasnikov, G. Hadziioannou, *Polymer* **2001**, *42*, 9097–9109.
438 G. Hadziioannou, *MRS Bull.* **2002**, *27*, 456–460.
439 S.-S. Sun, *Solar Energy Mater. Solar Cells* **2003**, *79*, 257–264.
440 S. Sun, Z. Fan, Y. Wang, J. Haliburton, *J. Mater. Sci.* **2005**, *40*, 1429–1443.
441 R. A. Segalman, C. Brochon, G. Hadziioannou, *Opt. Sci. Eng.* **2005**, *99*, 403–420.
442 M. Behl, E. Hattemer, M. Brehmer, R. Zentel, *Macromol. Chem. Phys.* **2002**, *203*, 503–510.
443 M. Behl, R. Zentel, *Makromol. Chem. Phys.* **2004**, *205*, 1633–1643.
444 K. Sivula, Z. T. Ball, N. Watanabe, J. M. J. Fréchet, *Adv. Mater.* **2006**, *18*, 206–210.
445 Z. T. Ball, K. Sivula, J. M. J. Fréchet, *Macromolecules* **2006**, *39*, 70–72.
446 S. M. Lindner, M. Thelakkat, *Macromolecules* **2004**, *37*, 8832–8835.

4
Atomic Force Microscopy of Polymers: Imaging, Probing and Lithography

Sergei S. Sheiko and Martin Möller

4.1
Introduction

Modern technologies rely on our ability to tailor materials structure and properties on the nanometer scale. This in turn has set greater demands on polymer chemistry and also on characterization methods such as microscopy. In addition to controlling architecture and self-organization of macromolecules, there is general aspiration to measure properties, monitor processes and modify structures on the smallest possible length scales. In this regard, atomic force microscopy (AFM) represents a unique experimental technique which combines in one instrument three areas of applications, i.e. the visualization of surface morphology, probing physical properties and maskless lithography (Scheme 4.1). Since its development in 1986 [1], AFM has become an indispensable tool both in academia and industry. The recent proliferation of nano- and biotechnologies is partly due to the invention of AFM, which made it possible to study nanometer-sized objects under various environmental conditions.

AFM's most significant areas of applications are those involving the visualization of structures and processes. This is due to its high spatial resolution, which

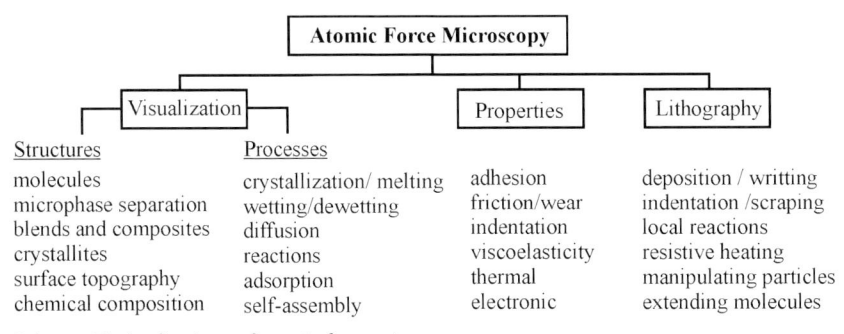

Atomic Force Microscopy			
Visualization		Properties	Lithography
Structures	Processes	adhesion	deposition / writting
molecules	crystallization/ melting	friction/wear	indentation /scraping
microphase separation	wetting/dewetting	indentation	local reactions
blends and composites	diffusion	viscoelasticity	resistive heating
crystallites	reactions	thermal	manipulating particles
surface topography	adsorption	electronic	extending molecules
chemical composition	self-assembly		

Scheme 4.1 Applications of atomic force microscopy.

for soft polymers varies within 1–10 nm. With this, AFM opens access to the least explored range of structures those that are smaller than 100 nm (see Section 4.3). In this range of length scales, materials are full of intriguing electronic, magnetic and optical properties promising new advancements in microlithography, data storage, photonics and molecular electronics. Furthermore, the imaging conditions do not require any sample modification or special environment such as those needed for electron microscopy. As such, native structures can be visualized in their natural medium, which is especially vital in biology. Not only is AFM effective at small length scales, it is also able to visualize larger structures such as crystallites, micelles and blends, making it possible to establish direct correlations between the molecular structure and macroscopic properties. Along with visualization of static structures, AFM is able to monitor various processes such as crystallization and conformational transitions allowing studies of dynamic properties of polymeric materials. The visualization abilities can be further enhanced by adding chemical contrast through modified AFM probes [2] and spectroscopic imaging [3, 4].

The ability to measure physical properties is based on a great variety of contrast mechanisms offered by AFM. This extends the microscope's utilization beyond morphological studies, making it possible to distinguish between soft and hard domains, between hydrophilic and hydrophobic surface patterns, and between viscous and elastic behaviors (see Section 4.4). In addition, AFM can be used to stretch polymer chains and compress cells [5–8], measure adhesive and frictional forces [9–11] and run indentation and wear test on sub-100 nm length scales [12–14]. It is important to emphasize that the measurements of physical properties can be performed concurrently with visualization of the surface morphology, which is vital for accurate structure–property relations, for example, between the roughness and chemical composition of the surface and its adhesion [15].

Utilization of AFM as a maskless lithographic tool is based on the combination of nanometer-sized contact along with precision scanning and positioning of a sharp AFM tip. The latter can be used for either writing various patterns with small organic molecules akin to a fountain pen or excavating wells and trenches, and also for inducing chemical reactions through a catalyst bound to the probe apex (see Section 4.5). One can also apply an electric field to induce locally oxidation [16, 17] and resistive heating [18]. By applying a controlled force to a pressure-sensitive material, one can generate a well-defined pattern of aligned nanoparticles [19]. The ability of AFM to draw various surface patterns on the nanometer scale has already found practical applications such as dip-pen nanolithography [20], fountain pen nanopatterning [21], and millipede data storage [22]. Furthermore, monitoring and manipulation of single macromolecules and nanoparticles is now possible [23–25].

Within all three application areas, AFM techniques frequently parallel other methods such as light and X-ray scattering, optical and electron microscopy, quartz microbalance and ellipsometry. In many cases, AFM allows measurements which are difficult or impossible through other techniques, e.g. conformational analysis and probing of individual macromolecules at the sub-10 nm scale [26–30]. In some cases, AFM provides information similar to other techniques; however, the data can be obtained in a much easier way by avoiding special sample treatments and special surrounding conditions. In every case, one should keep in mind that AFM is a surface technique, which only partially reveals the bulk properties. In order to avoid misinterpretation of AFM images, the findings should be correlated with other methods. For example, a thickness of surface-exposed crystallites obtained with AFM needs to be compared with the lamellar spacing in the bulk material determined by small-angle X-ray scattering (see Section 4.3.2) or the dimensions of surface-confined micelles and molecules should be compared with the corresponding dimensions in solution measured by light scattering (see Section 4.3.4.1).

This chapter reviews the recent progress made in the AFM of polymeric materials. Here, we distinguish three areas of applications: visualization of polymer morphologies, measuring interfacial properties and lithography. Even though in this chapter we deliberately focus only on AFM techniques for synthetic polymers, one should not disregard laser scanning confocal microscopy and near-field optical microscopy, which in recent years have expanded their application range to nanometer dimensions [31]. Along with the advances in synthetic materials and physical sciences made possible by AFM, it has made seminal contributions to the life sciences, providing high-resolution insights into self-organization principles and functions of biological macromolecules and colloids [32–35].

4.2
General Principles of AFM

AFM is a scanning probe technique, wherein a sharp needle mounted on a flexible cantilever scans across the substrate and probes interactions between the needle and the substrate (Fig. 4.1a). The interaction forces, regardless their nature, are measured through deflection of the cantilever. Due to the universal character of the tip–sample interactions, AFM became more useful in materials science than scanning tunneling microscopy (STM), which is limited to studies of conducting and semiconducting samples. However, AFM pays a heavy price for this universality. As mentioned above, the cantilever senses all kinds of interactions that are able to cause deflection of the cantilever. These include van der Waals, Coulomb, solvation, capillary and indentation forces. Even thermal noise of a free cantilever can be detected. In many systems, AFM measures a superposition of several forces, although some of the forces dominate in a particular range of distances between the tip and surface. For example, in polyelectrolytes electrostatic double-layer interactions dominate at larger distances of ~ 10 nm, where the van der Waals forces are almost non-existent. However, when the tip is brought into contact with a substrate, the attraction due to van der Waals and capillary forces and repulsion due to indentation become significant (Fig. 4.1b).

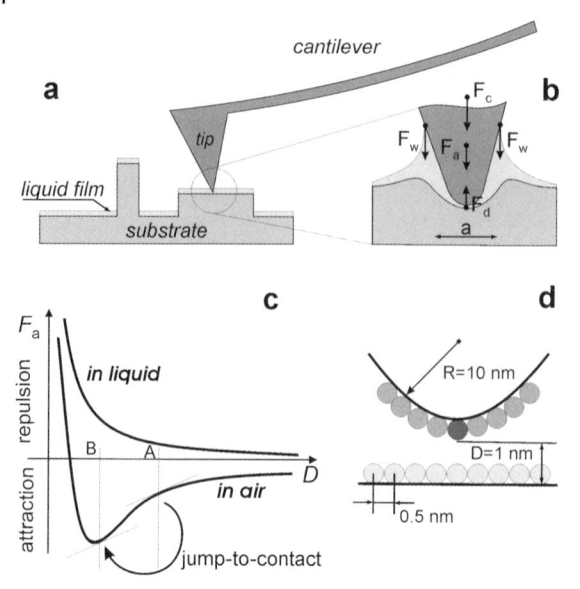

Fig. 4.1 (a) AFM probe (cantilever with a sharp tip at the end) scans across the substrate surface to measure its topography. (b) In the contact area, one observes deformation of the sample due to the tip indentation and formation of the meniscus due to condensation of the surrounding vapors (usually water). As such, the AFM tip experience a combination of repulsive forces F_d due to surface deformation indentation and attractive van der Waals forces F_a and surface tension wetting forces F_w. These forces cause deflection of the cantilever, $F_c = F_w + F_a - F_d$, which is measured by the position-sensitive detector.

4.2.1
Resolution

The resolution in AFM is a complex issue which includes many phenomena, although it is ultimately determined by the force–distance dependence between the tip and the substrate, i.e. the greater the slope of the force–distance curve the higher is the resolution. This situation is similar to STM, where the strong decay of the tunneling current with distance results in superior resolution. Of the many phenomena involved in the resolution of AFM, one should mention four. First, in addition to the outmost atom at the probe apex, there are many other atoms at the tip that interact with the substrate (Fig. 4.1d). Due to the long-range nature of the interaction forces, the contribution of the background atoms is significant and leads to an increase in the effective contact area. For a spherical tip, the contact area is given by

$$A_{\text{eff}} = 2\pi RD/(n-5)$$

where R is the tip radius (typically $R \approx 10$ nm), D is the surface separation ($D \approx 0.5$ nm at contact) and n is the power coefficient in the intermolecular po-

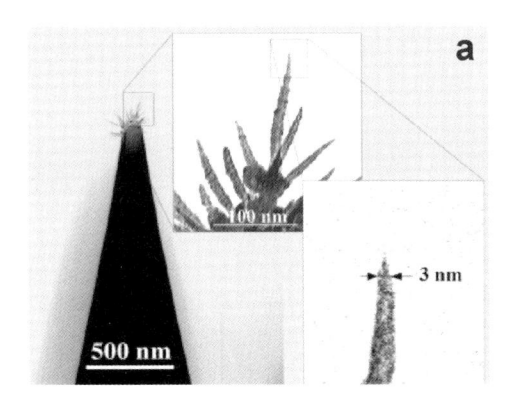

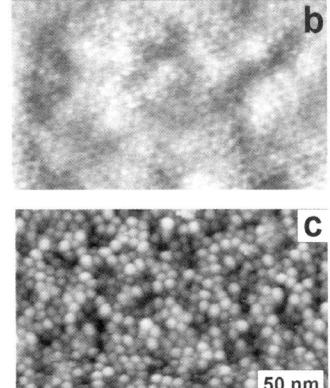

Fig. 4.2 (a) TEM image shows ultra-sharp needles with a radius $R < 3$ nm which were plasma-grown on top of a regular Si probe. The longest needle is used for scanning the sample surface. Height images of carbosilane dendrimers measured by (b) a regular Si tip and (c) a grown ultra-sharp tip.

tential $W(D) \sim 1/D^n$ ($n=6$ for van der Waals forces) [36]. Therefore, even without sample deformation, the radius of the effective contact area is about 3 nm. Second, the force–distance dependence is not monotonic. As can be seen in Fig. 1c, the slope of the $F(D)$ curve steadily changes both its magnitude and sign as the tip approaches the substrate, which makes it difficult to control the tip–substrate distance. Third, there is a range of distances wherein the stiffness, i.e. spring constant, of the cantilever is smaller that the stiffness of the tip–substrate interaction potential. In this situation, the soft cantilever spontaneously jumps to contact with the sample as the tip approaches the sample surface. The uncontrolled jump-to-contact phenomenon results in deformation of the substrate and formation of meniscus (Fig. 4.1b). Both effects lead to the increase in the contact area and decrease in resolution. Even at a zero force applied by the cantilever (zero deflection), the contact area can be significant due to indentation of the tip by the van der Waals and capillary forces (Fig. 4.1b). Fourth, there is always instrumental noise in force measurements due to thermal motion and electronics.

There are various ways to enhance resolution in AFM. In contact AFM, one should use softer cantilevers and sharper tips. Softer cantilevers reduce noise and cause smaller sample deformation. Typically, one uses cantilevers with a spring constant of ~ 0.1 N m^{-1}, which is much softer than the effective spring constant of a typical sample (~ 10 N m^{-1}). Sharp tips allow operation at a smaller contact area and lower adhesion. Commercial Si and Si$_3$N$_4$ tips have a radius of about 10 nm. Recently, high-resolution AFM probes have been developed by growing ultra-sharp needles on top of a regular Si tip in a plasma environment (Fig. 4.2a) [37]. These probes feature multiple needles; however, one of these needles is typically longer than the others and actually probes the surface structure with molecular resolution (Fig. 4.2b and c). The adhesion can be further reduced by scanning samples under liquids (Fig. 4.1c). The liquid environment eliminates capillary forces, reduces the van der Waals attraction and prevents the jump-to-contact phenomenon.

An alternative path to a higher resolution emerged with the introduction of frequency modulation AFM (FM-AFM) [38–40]. In general, dynamic AFM modes are less subject to noise than quasi-static contact AFM. Furthermore, oscillating the cantilever at a large enough amplitude (typically ~ 10 nm) prevents jump-to-contact. The amplitude should be larger than the A–B distance in Fig. 4.1c. However, more recently, it has been shown that operation of FM-AFM at smaller amplitudes would have additional advantages as it allows both the signal-to-noise ratio and the relative contribution of the short-range forces to be increased [41]. In order to reduce the amplitude value and still avoid jump-to-contact, one should increase the stiffness of the cantilever and thus reduce the jump-to-contact range. By using cantilevers with a spring constant of 1 kN m^{-1} one can operate at amplitudes below 100 pm and achieve true atomic resolution (see Section 4.3.1.1). If stiff cantilevers are not available, one can operate at higher harmonic oscillations, which also increase the contribution of short-range forces from the outmost atoms at the probe apex.

4.2.2
Probe Calibration

For quantitative measurements of both the interaction forces and surface topography, the tip shape and the spring constant need to be calibrated. Various approaches to characterize the tip shape are described elsewhere [42]. Concurrently, a series of methodologies, including the added mass methods [43, 44], unloaded resonance [45] and reference spring [46, 47] techniques, in addition to thermal noise analysis [48], were developed to quantify the spring constant. More recently, new methods have been proposed for the calibration of both torsional and normal spring constants with an uncertainty below 10% [49–53]. Furthermore, the torsional response of an externally driven cantilever in a viscous medium was analyzed [54, 55].

4.2.3
Imaging and Operation Modes

Operational principles, various imaging modes and theoretical models of AFM are thoroughly described in many reviews and textbooks [42, 56–60]. Here, we will briefly describe the most typical scanning and imaging modes that are currently used in polymer research and also present the latest developments, such as ultra-fast scanning and spectroscopic imaging. The most common scanning modes include contact mode and intermittent contact modes. Experiments under non-contact conditions remain very challenging due to technical difficulties in maintaining a stable tip–sample distance. The scanning modes can be further broken into different imaging modes, such as height mode to image surface topography, force/amplitude modes to measure changes in the interaction forces and phase mode to sense variations in adhesion and mechanical properties of the sample.

4.2.3.1 Contact Mode

In the contact mode of operation, a sharp tip on the end of a flexible cantilever is brought into permanent contact with the substrate. During scanning with the tip across the substrate, the cantilever flexes as the tip goes over higher and lower areas and thus maps out the height profile of the sample. One may encounter difficulties in the interpretation of topographic images when scanning chemically heterogeneous samples, e.g. polymer blends. Domains that exhibit repulsive interaction or weaker attraction with the tip will be imaged as protrusions above the sample surface. While scanning in contact mode, one can apply low-frequency force modulations (~ 8 kHz) to probe the mechanical properties of the underlying substrate. Changes in the deflection amplitude will be larger when the probe comes into contact with a harder surface region than with a softer one. In contact mode, one can also measure lateral tip–sample forces to study the friction on different surfaces. Although friction measurements are a valuable asset of AFM, the permanent contact between the tip and surface generates shear deformation at the sample surface, which may lead to modification of and even damage to soft polymers and thus represents a serious limitation of contact AFM. Nevertheless, gentle scanning and force modulation are still very useful for the characterization of relatively hard and large morphologies, such as lamellar structures of semicrystalline polymers, polymer blends and composite materials [61, 62].

4.2.3.2 Intermittent Contact Mode

In the second mode of operation, known as the tapping mode (Fig. 4.3), instead of bringing the tip into contact with the surface, the cantilever is oscillated above the surface at its resonant frequency (typically within the 10–400 kHz range) by a piezoceramic element near the cantilever base. As the oscillating tip approaches the substrate, its interaction with the tip changes the resonance frequency and thus reduces the oscillation amplitude at the scanning frequency. The variations in the amplitude are then transformed into the height profile of the sample. In addition, one can measure variations in the phase of the oscillating probe (Fig. 4.3 b), which is sensitive to both the chemical composition and physical properties of the sample surface. The relationship between the phase shift and the nature of the tip–specimen interaction is complex; however, phase images provide qualitative information about the composition of heterogeneous polymer systems [63, 64], as the phase shift is proportional to the energy dissipated into a sample.

Since in the tapping mode the AFM tip touches the surface only intermittently, it almost entirely eliminates lateral forces, allowing the characterization

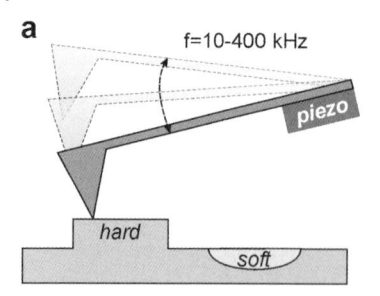

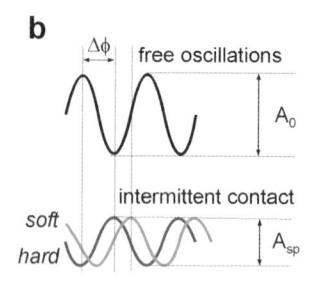

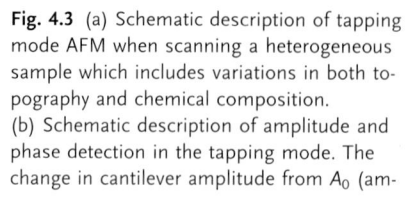

Fig. 4.3 (a) Schematic description of tapping mode AFM when scanning a heterogeneous sample which includes variations in both topography and chemical composition. (b) Schematic description of amplitude and phase detection in the tapping mode. The change in cantilever amplitude from A_0 (amplitude of free-oscillating probe) to A_{sp} (setpoint amplitude selected by the operator) is used for the feedback mechanism which tracks surface topography. The phase shift, which is larger on viscous, i.e. energy-dissipative, materials provides material contrast.

of soft and weakly adhered samples. Even operation at lower set points, i.e. with stronger tip–sample interactions, was shown not to be harmful [65]. However, the normal contact force during operation at lower set points can exceed 200 nN [66] and dissipated energy can be of the order of 100 kT per tap [67, 68]. At such high forces, the AFM tip can distort or permanently destroy the sample structure. The force exerted on the substrate can be lowered by increasing the effective quality factor of an oscillating cantilever [32, 69]. Quality factor enhancement is particularly important when working in liquids wherein additional damping of the cantilever motion reduces the quality factor from typically a few hundred in air to around 1 in liquid. This results in broadening of the resonant amplitude peak and thus weaker phase contrast.

4.2.3.3 Chemical Force Microscopy

Through the use of chemically modified probes, the chemical composition of the surface can be mapped (Fig. 4.4a). Functional groups chemically attached to the probe apex exhibit stronger interactions with definite chemical groups at the surface via system-specific forces such as hydrogen bonding, acid–base interactions and antibody–antigen interactions. Usually these forces dominate nonspecific van der Waals interactions and thus enhance chemical contrast in AFM micrographs, hence the name chemical force microscopy (CFM). Recently, energy dissipation imaging has been used to differentiate the subtle differences between hydroxyl- and carboxyl-terminated regions in a patterned self-assembled monolayer [70] (Fig. 4.4b and c). CFM allows imaging of the lateral distribution of chemical groups at the sample surface with a resolution down to 10–20 nm; however, the chemical contrast may interfere with topographic and mechanical contrasts. In this case, one should separately evaluate the surface topography (using unmodified tips) and use low forces to reduce sample deformation. In

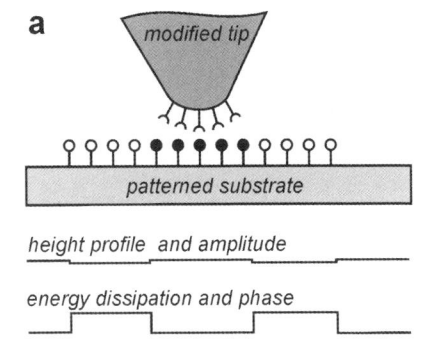

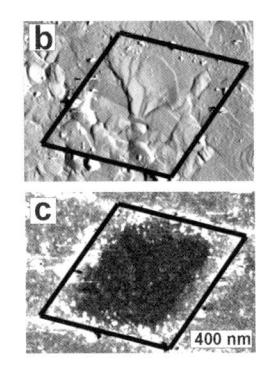

Fig. 4.4 (a) Schematic description of chemical force microscopy. The functionalized tip exhibits specific interactions with particular molecular groups at the substrate surface. The resulting variations in the interaction forces (usually attraction) result in corresponding changes in the cantilever deflection (amplitude) and phase. As such, chemically patterned surfaces may demonstrate height contrast, even though the surface is geometrically flat. However, much stronger contrast is observed in phase and energy dissipation images. AFM micrographs demonstrate (b) amplitude and (c) energy dissipation images of a patterned SAM surface of hydroxyl surrounding a carboxyl square, which was measured with an 11-mercapto-1-undecanol-modified AFM tip. The black square highlights the edges of the pattern [70].

addition, it is advisable to apply complementary techniques for surface studies. These include IR and Raman spectroscopy, SIMS, MALDI mass spectrometry, X-ray photoelectron spectroscopy, and near-field optical techniques.

4.2.3.4 Electric Force Microscopy

Compositional mapping of heterogeneous polymer materials in AFM is usually based on differences in the mechanical properties of individual components. Such mapping is also possible using local differences in sample dielectric properties [71, 72]. When a conductive probe is biased with respect to the conducting sample, its resonant frequency is shifted to lower frequency due to long-range attractive forces. This shift in frequency and the related shift of the phase are caused by the gradient of the electric field near the surface of the sample using a sharp conductive probe. The electrostatic force $F(z)$ between the tip and surface at a relative separation z is related to the local capacitance C between the tip and the sample by

$$F(z) = -\frac{1}{2}\frac{\partial C}{\partial z}U^2$$

where U is the potential difference between the tip and the sample. These forces are weak and might be screened by the tip–sample force interactions occurring during tapping. Therefore, a two-pass lift mode technique is used to separate the effects of the electric and mechanical forces (Fig. 4.5a). EFM is

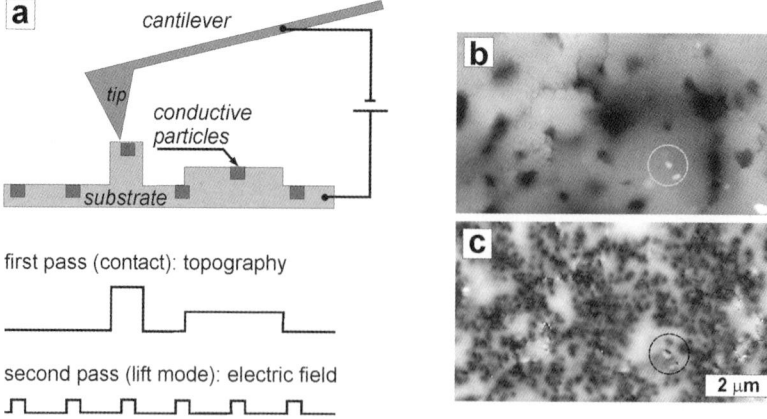

Fig. 4.5 (a) With electric force microscopy one obtains two types of images: height and phase images. The height image is taken in the first pass in contact mode and then the phase image is taken in the second pass in lift mode (typically 5–50 nm above the surface) giving local field variation. EFM was used to study bulk morphology of a commercial thermoplastic vulcanizate filled with carbon black particles (10% by volume). (b) The height image reveals corrugated topography of the sample surface in the 400-nm range. (c) Dark spots in the phase image obtained at a lift height of 50 nm and sample voltage of 10 V indicate the locations of conductive particles, which may contribute to a percolation network responsible for the conductivity of this sample. Dashed circles highlight three particles as reference points [76].

usually applied to conductive materials such as semiconductors or composite materials containing conductive particles [73–76]. Figure 4.5b and c demonstrate application of EFM to a conductive rubber filled with carbon black particles. However, it is also possible to use EFM for to study dielectric systems deposited on a conductive substrate [77].

4.2.3.5 Spectroscopic Imaging

The range of practical applications expands significantly when AFM is used in combination with IR absorption and Raman scattering spectroscopy. Here it is possible to measure vibrational spectra and thus perform quantitative analysis of the chemical composition of surfaces with AFM resolution. Even though in recent years optical microscopy has made significant progress in expanding their application range to nanometer dimensions [31], the spatial resolution is fundamentally limited by the diffraction of light, which does not allow the resolution of features smaller than 200 nm [78]. This limitation was alleviated with the introduction of scanning near-field optical microscopy (SNOM). Here, one distinguishes between apertured (Fig. 4.6a) and apertureless (Fig. 4.6b) SNOM techniques. Traditional SNOM uses a small sub-wavelength aperture at a close proximity to the surface, to obtain optical response from the nanometer-sized area of the sample. In this case, spatial resolution is limited by the aperture

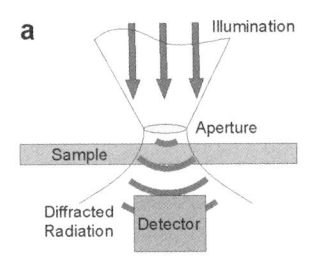

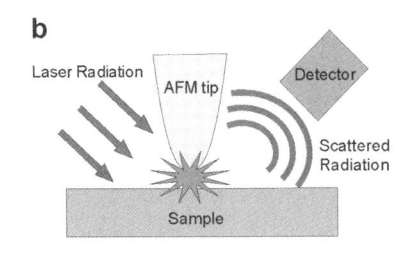

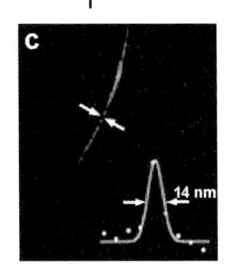

Fig. 4.6 Schematic drawings of (a) apertured and (b) apertureless detection of light in scanning near-field microscopes. (c) Near-field Raman image of a single SWNT deposited on a glass coverslip. The contrast in the images reflects the local intensity of the Raman G' band (2640 cm^{-1}). The inset in (c) shows a Gaussian fit to the line section shown [80].

size, which at best is on the scale of hundreds of nanometers. Concurrently, it was proposed to use apertureless SNOM to increase resolution by focusing laser light on to a metal-coated AFM tip [79–81]. As shown in Fig. 4.6b, the AFM probe is used as a miniature antenna for local enhancement of the incident light. The radiation strikes the AFM tip, causing a local evanescent field around the tip, which is then scattered or absorbed by the sample, which is then detected. Here, the spatial resolution is limited by the tip size, which, with new advances in the AFM tips, can approach just a few nanometers. The ability to perform high-resolution spectroscopic imaging is demonstrated in Fig. 4.6c. A near-field Raman image of a single single-wall nanotube (SWNT) was recorded with a spatial resolution of ∼14 nm.

4.2.3.6 Rapid Scanning AFM

In many AFM experiments, the ultimate goal is to follow processes *in situ* and in real time. This is particularly vital for processes, such as polymer crystallization and chemical reactions, which demonstrate multiple pathways through a number of transient and metastable states. However, the time resolution of conventional AFM is hindered by the serial nature of the data collection – the very factor that allows AFM and STM to circumvent the diffraction limit of optical techniques [82]. Two factors limit the imaging rate: the response time of the cantilever and the resonant frequency of the scanning stage. In order to reduce the response time, one can either use higher frequency cantilevers [83] or electronically enhance the quality of the cantilever [84]. Regarding the second limitation, the mechanics of the microscope can be redesigned and optimized for rapid imaging with the use of smaller cantilevers and high resonant frequency scan stages [85]. This pushed the image rate of conventional AFM to 25 frames s^{-1}, allowing temporal resolution of the motion of motor proteins [86]. Recently, the possibility of obtaining SPM images with millisecond resolution has been realized [87, 88], giving access to macromolecular relaxation time scales. However, one may argue that complete details of molecular-scale motions by fast-scanning AFM are still out of reach, since the diffusive hopping of an adsorbed atom can be as fast as 10^{-13} s. However, video-rate imaging is very appropriate for high molecular weight polymers, which exhibit significantly slower molecular motion. For example, a polymer chain with a molar mass of about 10^5 g mol^{-1} requires about 0.1 s to move a distance on the order its own size (~ 30 nm) [89]. This so-called characteristic reptation time is close to the currently possible 10 frames s^{-1} imaging rate.

4.3
Visualization of Polymer Morphologies

Imaging of surface morphologies is currently the most important area of application of AFM. Indeed, the principle "seeing is believing" represents the most traditional and effective approach in experimental sciences. As mentioned above, imaging can be performed on scales from hundreds of micrometers down to nanometers, permitting observations of surface structures as large as spherolites and phase-separated domains in polymer blends and as small as block copolymer micelles and single macromolecules. Another attractive feature of this technique is that it has relatively simple sample preparation and does not require selective staining, which is usually required for transmission electron microscopy and fluorescence optical microscopy studies. Along with topographic imaging, one can perform compositional mapping of heterogeneous materials such as semicrystalline polymers, polymer blends, composites and block copolymers. In addition to plain visualization, AFM micrographs allow quantitative characterization of polymer structural properties including absolute molecular weight [90], degree of branching [91, 92], crystallization rates [93], friction coefficient from flow and molecular diffusion measurements [94], and covalent bond strength [95]. It is important to note that the visualized morphologies may correspond to an equilibrium state or, more likely, represent a transient structure frozen in a metastable state due to slow diffusion of large polymer molecules. Since, in many cases, it is difficult to distinguish between these two situations, additional experimental tests for equilibrium are required for accurate data interpretation.

4.3.1
Molecular Visualization

In recent years, AFM has been successfully used for the visualization of both natural [96–106] and synthetic [107–146] molecules. The unique advantage of AFM is that it allows the visualization of molecules in great detail, including the contour length and the local curvature. Fluorescence microscopy is another viable technique to study various molecular-scale phenomena [147–157]. However, although the analysis of molecular conformation and dynamics is reliable for long and stiff molecules, optical techniques are inaccurate in the case of short and flexible chains, whose overall size and local curvature are below the optical resolution limit. The role of molecular visualization has grown to be especially profound with the synthesis of complex molecules [158–167] whose structures are difficult to confirm using conventional characterization techniques such as NMR and light scattering. This is especially true for molecules that are branched, heterogeneous and polydisperse. Here, AFM images provide unambiguous proof of the molecular architecture along with accurate analysis of size, conformation and ordering of molecules on surfaces.

4.3.1.1 Molecular Resolution

In molecular visualization experiments, one should distinguish the resolution of individual molecules within a dense sample and the visualization of single, i.e. isolated, molecules adsorbed on a substrate. These two different experiments face different challenges and use different approaches in data interpretation.

True molecular resolution Interpretation of molecular-scale images of dense samples is often ambiguous because one may observe an ordered array, even though the studied material is supposed to be amorphous or observations that the visualized lattice is not consistent with crystallographic data. One of the best examples is graphite, where every second atom in the hexagonal surface unit cell is not visible in STM images, but can be resolved with low-temperature

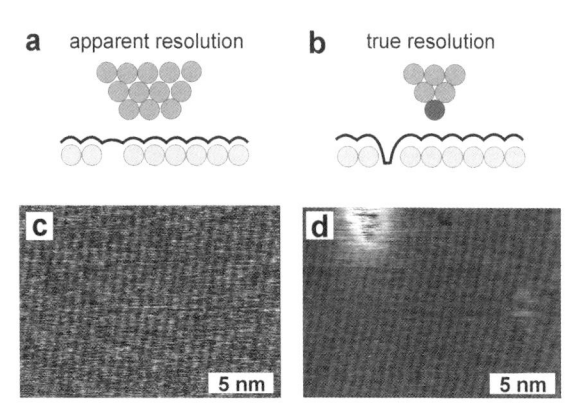

Fig. 4.7 Two schematics of tips and a crystalline sample illustrate the origin of (a) apparent and (b) true resolution in AFM. When scanning with a sharp tip (b), one clearly visualizes the vacancy on the sample surface. When scanning with a dull tip (a), AFM correctly reproduces the periodic structure of the lattice, while the vacancy is hardly visible. (c, d) Height AFM images of the bc plane of a polydiacetylene crystal were recorded in tapping mode with (c) an etched Si probe and (d) a high-resolution ultra-sharp probe. The contrast in the images in (c) and (d) reflects the height corrugations in the 0–0.3 nm and 0–0.6 nm scale, respectively. Two white arrows in (d) indicate the molecular-size defects.

AFM with pN force sensitivity [168]. In molecular scale imaging of crystalline samples, one should distinguish between the true molecular resolution and the apparent resolution (Fig. 4.7a and b). One of the criteria for the true resolution is reproducible visualization of lattice defects such as vacancies, terrace edges and dislocations. Although many, especially early, results correctly reproduce the crystallographic lattice of the studied samples, reproducible observations of lattice defects are rare. Another viable criterion for true molecular resolution is reproducible visualization of molecules within an amorphous sample. Obviously, probes with multiple contacts will fail to reproduce the molecular packing, whereas sharp probes can. As shown in Fig. 4.2c, specially sharpen probes are able to resolve 6-nm spherical molecules within an amorphous melt of carbosilane dendrimer. The visualized molecules are relatively large; however, the 6-nm resolution is not a limit for AFM and molecular resolution of amorphous materials on significantly smaller scales can be expected in the near future.

As discussed in Section 4.2, there are two general paths to true molecular resolution. The first path is to use quasi-static contact mode AFM to minimize the forces exerted on the sample. To reduce the forces, one should use sharp tips and soft cantilevers and also reduce adhesion by either modifying cantilevers or doing measurements in a liquid medium. It is also highly desirable that the sample surface is clean, inert and hard. Using sharp tips allows minimization of the contact area and also adhesion forces. Recently, true molecular resolution has been demonstrated in a tapping-mode AFM study of polydiacetylene crystal using ultra-sharp probes with an extremity of approximately 1 nm [37]. In Fig. 4.7c and d, the image features correspond to the edges of the individual side-groups of the polymer chains that form the crystal surface. It is evident that the molecular rows within the unit cell are resolved only in the images obtained with the ultra-sharp probe (Fig. 4.7d). In addition, the molecular size defects were observed on this surface as a proof of the true molecular resolution. When the Si probe (radius $\sim$10 nm) is applied, a continuous strip is seen instead of two rows (Fig. 4.7c).

The second path to molecular resolution is to use frequency modulation AFM. This technique also requires sharp tips and clean, weakly adhering samples, while it sets opposite requirements for cantilever stiffness. As discussed in Section 4.2.1, FM-AFM techniques employ very stiff cantilevers to prevent jump-to-contact and allow stable operation at small amplitudes. Figure 4.8 shows a clear picture of an Si surface with a defect and large corrugation [170]. The resolution achieved is similar to or even greater than those achieved by STM. Three reasons were identified for this advantage [41]. First, AFM is sensitive to all electrons whereas STM senses the loosely bounded electrons at the Fermi level. Second, FM-AFM allows operation at much smaller tip–sample distances without destroying the tip than in STM. Third, tip–sample interaction forces include repulsive forces with a very short decay length comparable to those for tunneling current in STM.

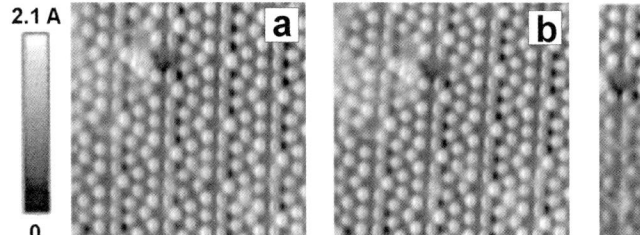

Fig. 4.8 Series of topographic images of Si(111) (7×7) observed by FM-AFM under different scanning conditions. (a) Frequency shift $\Delta f=-160$ Hz, a defect is seen in the left upper section of the image and there are dual maxima for each adatom. (b) $\Delta f=-160$ Hz, image similar to (a). (c) Before the start of this scan, the tip was displaced to the right by a distance of 2 nm; consequently, the defect appears shifted to the left by a distance of 2 nm.

Visualization of single macromolecules One of the main challenges in the visualization of single molecules is physical contact between a molecule and the AFM tip, which may induce significant perturbations in the molecular conformation and even displace the molecule. One should realize that every time the tip touches the substrate in tapping mode, there is energy of order 100 kT dissipated in the contact area (Section 4.2.3). Energies of such magnitude exceed by far both the thermal energy used for molecular diffusion and adhesion energy required for molecular desorption. However, if molecules are robust, strongly adhering and possess a well-defined conformation, accurate interpretation of molecular images is possible.

The first images of single molecules were obtained for DNA molecules due to their well-defined and easily recognizable conformation. During the last few years, significant progress has also been achieved in the visualization and monitoring of synthetic molecules, although these experiments remain very challenging, for several reasons. First, synthetic polymers usually do not form well-defined tertiary structures, except densely branched macromolecules such as dendrimers, monodendron-jacketed linear chains and molecular bottle-brushes (see below). Second, the persistence length and hence the radius of curvature of the polymer backbone is usually smaller than the average AFM resolution of about 5 nm. Third, flexible chain-like molecules usually adopt a coiled conformation. Usually, samples should be prepared in a special way to allow visualization of a polymer chain in its extended conformation. One can simply rely on strong adhesion, which may quench conformation when adsorbing molecules on a solid substrate from solution. One can also use external fields to extend molecules. Another viable alternative is to decorate the linear chain with bulky side-groups through either covalent [114] or non-covalent [171] bonding. In this case, the steric repulsion between the side-groups stretches the backbone enhancing its apparent persistence length. Moreover, adsorbed side-groups separate the molecule's backbones and thus facilitate their lateral resolution. Different types of covalently bonded macromolecules are discussed in the next section. Here,

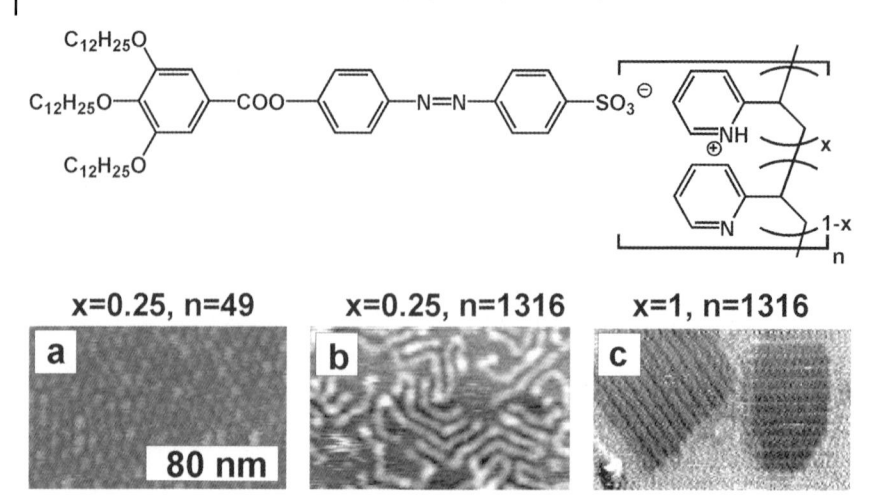

Fig. 4.9 Phase tapping mode images of individual supramolecular structures formed upon complexation of wedge-shaped mesogenes with a linear flexible chain. The shape of the structures depends on the degree of polymerization of the chain and degree of complexation x: (a) spheres; (b) semi-flexible cylinders; (c) rod-like particles.

Fig. 4.9 shows differently shaped supramolecular structures formed by self-assembly and concurrent complexation of wedge-shaped amphiphilic sulfonic acid molecules with poly(2-vinylpyridine) (P2VP) homopolymer via proton transfer [172].

4.3.1.2 Using Molecular Visualization to Prove Synthetic Strategies

Modern technologies and materials rely on our ability to design macromolecules which adopt a particular conformation and execute predetermined functions. However, designer macromolecules often demonstrate complex structures, which may include multiple branching, heterogeneous chemical composition and specific functional groups. This complicates the verification of such structures with conventional techniques such as GPC, LS, NMR and IR adsorption. Here, molecular visualization represents a very effective tool as it not only provides pictorial evidence for the molecular architecture, but also enables its characterization. Figure 4.10 shows a gallery of images of various designer and natural molecules visualized by AFM. Carbosilane dendrimers in Fig. 4.10a represent one of the most challenging samples because the molecules are small and form a disordered fluid at room temperature [173]. Only recently was it possible to achieve clear resolution of individual dendrimers using ultra-sharp tips. Figure 4.10b shows monodendron-jacketed linear chains wherein AFM confirmed self-organization of dendritic side-groups into cylindrical morphology [113, 114]. In Fig. 4.10c, AFM reveals the detailed structure of energy-converting ATPase

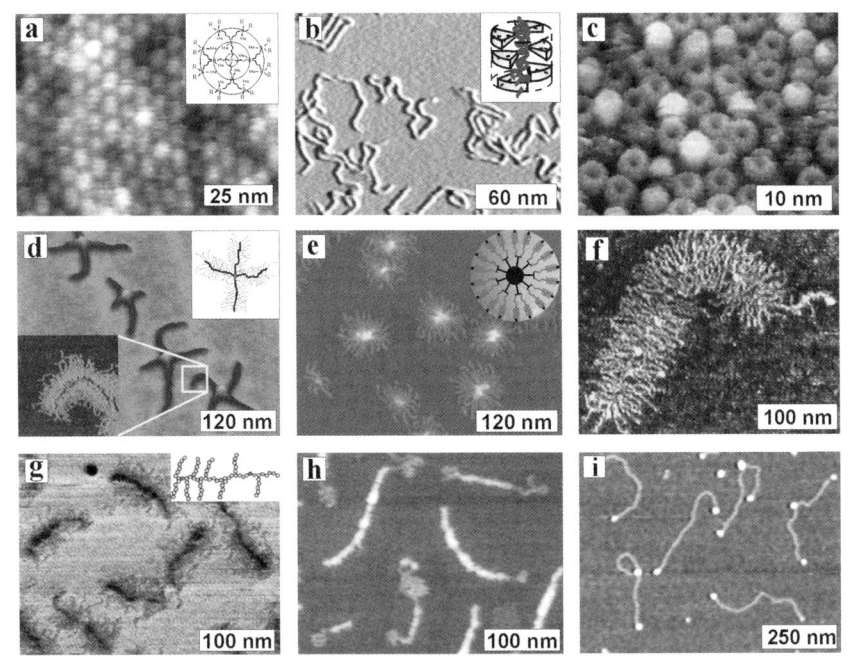

Fig. 4.10 AFM images of individual molecular architectures: (a) height image of carbosilane dendrimers of sixth generation; (b) amplitude image of monodendron jacketed-linear chains; (c) AFM confirms the rotor stoichiometry of *Ilyobacter tartaricus* ATP synthase and shows that the cylindrical Na^+-driven rotor comprises 11 c-subunits; (d) four-arm star-like molecular brushes; (e) multi-arm block copolymer with p(BMA-b-PEGMA) arms; (f) an individual fetal bovine aggrecan monomer adsorbed on mica showing ~ 100 chondroitin sulfate glycosaminoglycan (CS-GAG) chains that are covalently bound at extremely high densities (2–4 nm molecular separation distance) to a 250-kDa protein core; (g) phase image of molecular brushes with gradient grafting density clearly demonstrate bulky head and fuzzy tail; (h) height image of pBA brushes end-capped with linear chains of poly(octadecyl methacrylate); (i) λ-DNA molecules end-capped with streptavidin molecules.

rotor rings [174]. In Fig. 4.10d and e, one can see two examples of star-like macromolecules: four-arm molecular brushes [91] and multi arm block copolymer stars [92]. Most of the visualized molecules possessed the targeted number of arms; however, molecules with fewer arms were also observed, indicating possible termination reactions. Figure 4.10f presents the biological counterpart of the synthetic brush molecules – proteoglycans, complex hybrids of protein backbone and polysaccharide side-chains [102]. Similarly to synthetic brushes (inset in Fig. 4.10d), one can clearly see the hairy morphology which is thought to be responsible for the particular viscoelastic properties of mucus layers in lung airways and for the exceptional stiffness, toughness, resilience and lubrication properties of the articular cartilage. Figure 4.10g demonstrates the most spectacular structures – brushes with a gradient of grafting density along the backbone [175]. In addition to the clearly visualized "hairy" structure, AFM discerned the bulky head and fuzzy tail, providing unambiguous proof of its gradient composition, which also results in a tadpole conformation in compressed monolayers [176]. Figure 4.10h shows cylindrical brushes with poly(octadecyl methacrylate) linear chains on both ends [177]. These molecules spontaneously associate to form linear and branched multimers due to crystallization of p(ODMA) chain ends. Finally, in Fig. 4.10i, one can see DNA molecules wherein both biotinylated ends anchor streptavidin–horseradish peroxidase fusion protein [26]. The image allows one to distinguish between the DNA chain and the molecular recognition groups at the chain ends. Other examples of successfully visualized molecules and molecular assemblies include arborescent graft polymers [178], block copolymer micelles [179] and brushes with block copolymer side-chains [180].

4.3.1.3 Molecular Characterization

In addition to verifying synthetic strategies, molecular visualization permits accurate measurements of molecular weight, size and conformation. The unique advantage of AFM is that one obtains molecular dimensions in direct space, affording more opportunities for statistical analysis. The pictorial resolution allows fractionation of the visualized molecules by size, branching topology and chemical composition in addition to sorting out the irrelevant species.

Measuring molecular weight by AFM The molecular weight distribution of brush-like molecules was determined by using a combination of the Langmuir–Blodgett (LB) technique and AFM [90]. The LB technique provided mass density information (m/S), whereas visualization of monolayer by AFM enabled accurate measurements of the number of molecules per unit area (N/S). From the ratio of the mass density to the molecular density, the number-average molecular weight was determined as $M_n = m/S$. As shown in Fig. 4.11, the results obtained were in good agreement with those of gel permeation chromatography

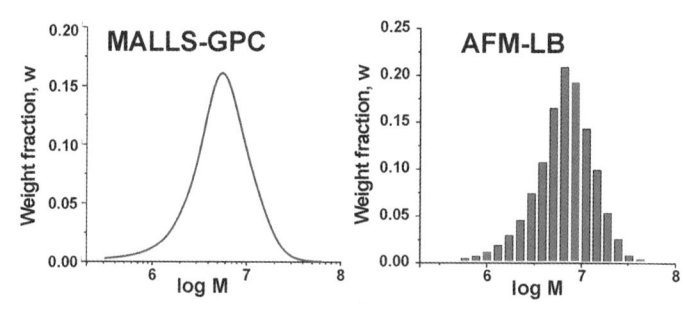

Fig. 4.11 Molecular weight distribution measured by two independent techniques, MALLS-GPC and AFM-LB, which demonstrated an excellent agreement.

(GPC) using a multi-angle laser light scattering (MALLS) detector. This approach can be applied to a large variety of molecular and colloidal species. In particular, the visualization-based approach is useful for relatively large species ($>10^7$ Da) that are difficult to measure by light scattering and GPC. The AFM–LB method also requires smaller sample amounts than conventional techniques.

Length distribution of individual arms Matyjaszewski et al. prepared a series of star-like brushes (Fig. 4.10 d) wherein the number of arms varied from two to four, while the lengths of the side-chains were kept approximately constant [181]. Characterization of individual arms (Fig. 4.12), separately from the entire molecule, can only be done by means of molecular visualization. In agreement with the Schulz–Flory theory of condensation polymerization, the polydispersity of the total length was significantly lower compared with that of the arm length, confirming

$$PDI_{star} = \frac{PDI_{arm}}{f} + \frac{f-1}{f}$$

as derived for random coupling of f arms. It was also shown that the variation in polydispersity had an effect on molecular ordering [182].

Fractal dimension and persistence length Polymer chains are fractal objects whose dimensionality is determined by interactions between the monomeric units and solvent molecules [183]. Figure 13a and b show typical AFM images of brush molecules making direct analysis of their conformation in two dimensions possible [28, 95]. Through molecular visualization, one can measure both the persistence length and fractal dimension of single molecules in various environments (Fig. 4.13c and d). For molecular brushes, one can tune the persistence length from 10 to 5000 nm by increasing the side-chain length and grafting density. This is 0.2–100 times larger than the persistence length of DNA

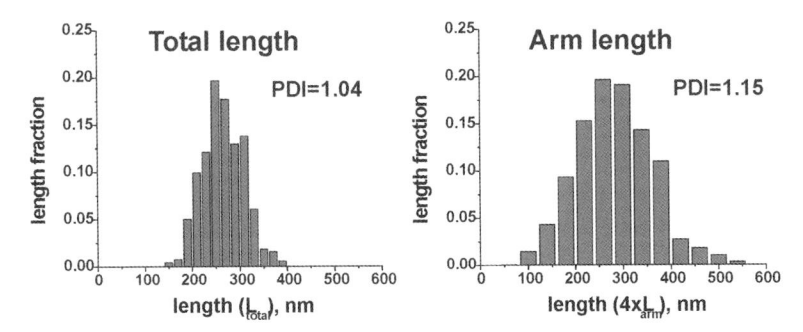

Fig. 4.12 Visualization of four-arm star-like molecules allowed independent measurements of the total length distribution along with the length distribution of individual arms.

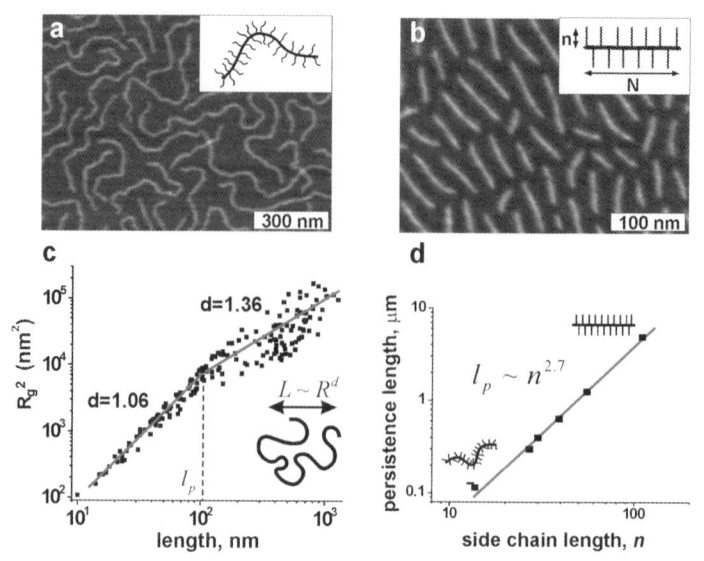

Fig. 4.13 Brush molecules change from (a) flexible to (b) rod-like conformation with increasing degree of polymerization of side-chains n. (c) Persistence length l_p is determined as the distance along the chain at which the fractal dimension changes from $d \approx 1$ to $d \approx 1.33$. (d) The persistence length increase with the side-chain length as $l_p \sim n^{2.7}$ [95].

(~ 50 nm) and approaches that of F-actin (~ 10 μm). Furthermore, the micrographs provide a unique opportunity to measure local orientation of the molecular contour and verify the recently developed theory of chain conformation in dilute and semi-dilute solutions [184].

4.3.1.4 Visualization Molecular Processes

The dynamic behavior of polymer chains at interfaces is poorly understood [185]. Most significant experiments on the molecular scale were performed on DNA molecules using fluorescence microscopy [147–150, 153, 154, 186–191]. However, the applicability of optical techniques is limited due to low spatial resolution. In this respect, AFM is an appropriate technique which potentially can provide real-space and real-time information about translational, rotational and bending motions of molecules at the nanometer length scales. Currently, AFM is making the first steps in the visualization of molecular processes, such as diffusion, conformational transitions, chemical reactions, and self-assembly.

Conformational transitions Conformational transitions represent one of the classical topics in polymer science with significant implications in biology, wherein secondary and tertiary structures determine packaging and functions of biomolecules. The character of the transition depends on the molecular archi-

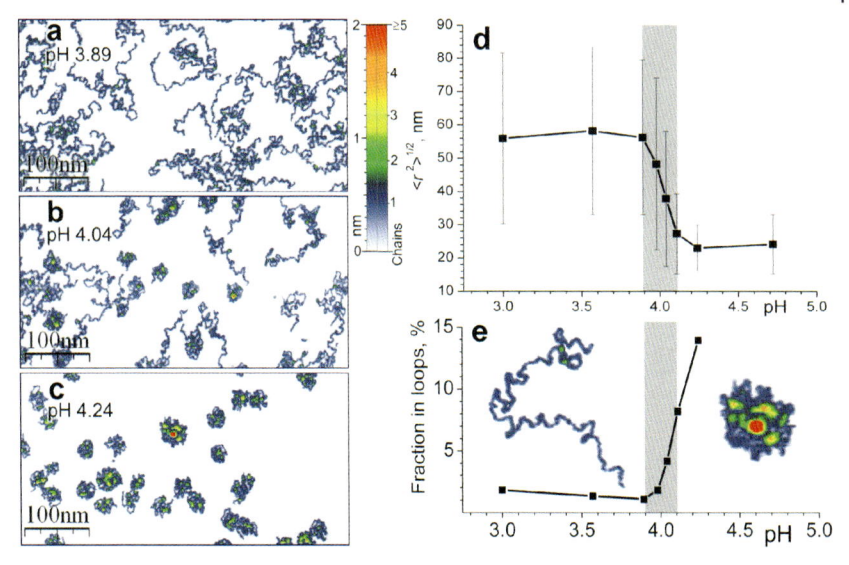

Fig. 4.14 AFM-visualized conformations of adsorbed P2VP molecules: (a) pH 3.89, extended coils; (b) pH 4.04, intermediate state; (c) pH 4.24, compact coils. Z-scale bar shows a number of superposed chains assuming the height increment of 0.4 nm. (d) Directly measured values of r.m.s. end-to-end distance of P2VP single molecules adsorbed on mica surface versus pH. (e) Fraction of monomer units in loops versus pH. The gray zone in (d) and (e) is the pH range of the conformation transition.

tecture, chain stiffness [192] and charge density [193]. It also depends on the surrounding conditions (pH, ionic strength and solvent quality) and dimensionality of the system [194]. The most interesting and relevant data are obtained if one controls the surrounding condition within an AFM measuring cell. Recently, Roiter and Minko reported the observation of a coil–globule transition for protonated poly(2-vinylpyridine) macromolecules which were adsorbed from aqueous solutions with different pH [195]. As shown in Fig. 4.14, the highly protonated poly(2-vinylpyridine) chains possess the conformation of a two-dimensional coil, whereas at a low degree of protonation, molecules are in the conformation of a strongly compressed 3D globule. These data confirm the earlier observations performed in bulk solution using light scattering, IR adsorption and Raman spectroscopy [196, 197].

AFM also allows *in situ* monitoring of conformational transitions provided that the molecules are sufficiently slow and adhere strongly to the substrate to prevent tip-induced perturbations. Recently, environment-controlled AFM has been used to study conformational changes of single poly(methacrylate)-*graft*-poly(*n*-butyl acrylate) brush molecules on mica. As shown in Fig. 4.15, the molecules transform reversibly from an extended worm-like conformation to a compact globular conformation on changing the relative vapor pressure of water and ethanol in the surrounding atmosphere. Adsorption of water and ethanol changes the surface properties of the mica substrate and thus attraction of the

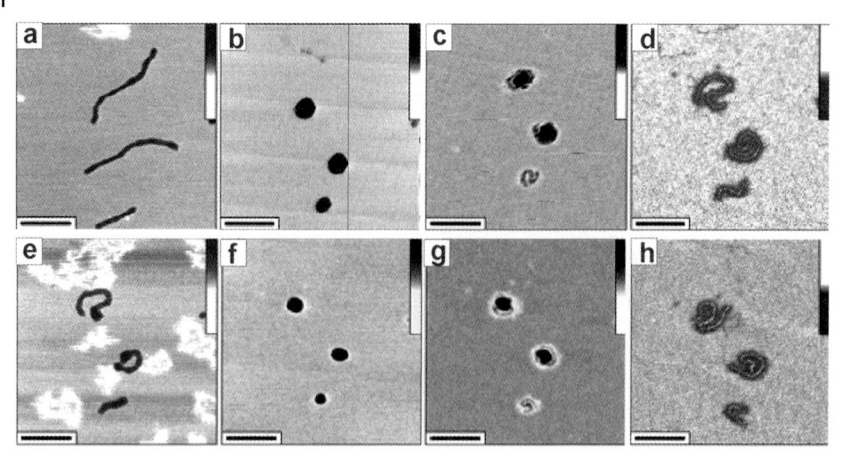

Fig. 4.15 AFM visualizes reversible conformational transition caused by variation of the ethanol vapor pressure in the surrounding atmosphere. When the extended, tightly adsorbed poly(n-butyl acrylate) brush molecules are exposed to ethanol vapor, the macromolecules swell and contract to form compact globules. Exchanging the ethanol vapor to a humid atmosphere caused the molecules to extend again to a worm-like two-dimensional conformation. The sequence of phase images demonstrates two cycles (first, **a–d**; second, **e–h**) of collapse and subsequent extension for three individual pBA brush molecules on mica in vapor saturated with humidified ethanol (80 vol%) and water.

side-chains to the substrate. The coexistence of collapsed and extended strands within the same molecule indicates a first-order transition, in agreement with previous studies of Langmuir films [141].

Monitoring association and dissociation reactions Next-generation chemical and biochemical assays on chips will involve chemical reactions one molecule at a time. Scientists are working on developing a new methodology to study surface-mediated chemical reactions on the molecular level directly and *in situ*. Until now, AFM was successfully used to follow the progress of a reaction in *ex situ* manner. Figure 4.16 demonstrates physical association of end-functionalized brushes on a solid substrate due to crystallization of octadecyl tails [198]. The molecules spontaneously associate to form multimers such as chains and branches.

In a different series of experiments, molecular visualization was successfully used to monitor the reverse process, i.e. scission of covalent C–C bonds of the polymer backbone brush-like macromolecules upon adsorption on a substrate [95]. Figure 4.17a shows a series of AFM images obtained for different incubation times of a brush molecule with particularly long side on a water–propanol substrate. As the time spent on the substrate increases, one sees that the molecules become progressively shorter while the number of species per unit area correspondingly increases, indicating to scission of the backbone (Fig. 4.17b).

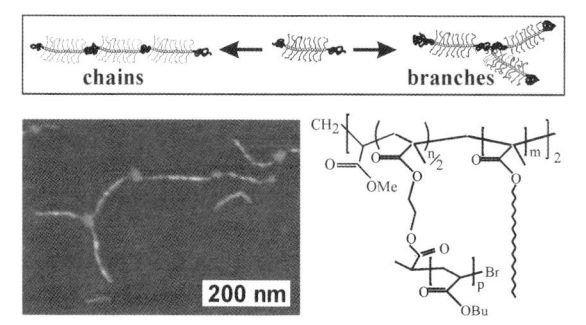

Fig. 4.16 Association of poly(n-butyl acrylate) brushes due to crystallization of linear poly(octadecyl methacrylate) chains on both ends of the brush backbone [198].

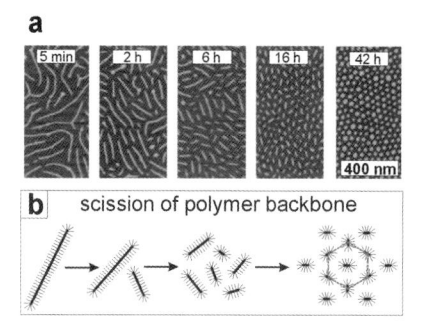

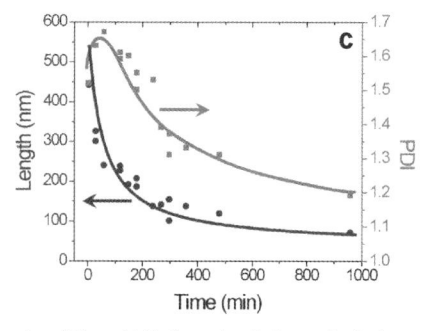

Fig. 4.17 (a) Height AFM images of the brush-like macromolecules with long side-chains ($n=140$) were measured at different exposure times to the water–propanol (99.8:0.2, w/w) substrate. (b) The molecules become shorter with longer exposure time to the substrate, which is attributed to spontaneous scission of the polymer backbone. From the molecular images one obtains (c) the number-average contour length, $L_n = (\Sigma L_i \times n_i)/(\Sigma n_i)$, and polydispersity index of the system, $PDI = L_w/L_n$. The experimentally measured time dependence of the contour length (●) is fitted with $A = A_0 e^{-kt}$ (lower line) at $k = 2.3 \times 10^{-5}$ s^{-1}. The measured polydispersity index $PDI = L_w/L_n$ (■) shows good agreement with those obtained by a computer model designed to simulate random scission (upper line).

By measuring the length decrease as a function of time, the scission process was shown to be a first-order reaction with a rate constant of $k = 2.3 \times 10^{-5}$ s^{-1} (Fig. 4.17c). The macromolecule's destruction occurs because its side-chains stretch the polymer backbone to maximize the number of contacts with the substrate. This phenomenon opens intriguing opportunities for molecular-scale studies of dissociation kinetics and for the design of macromolecules that induce tension and fracture on surfaces in a specific way [199].

To summarize, through molecular visualization one can directly confirm synthetic strategies, characterize molecular conformation and also monitor chemical reactions on surfaces. For many systems, the impact of molecular visualization is especially unique and cannot be reproduced by other techniques. This was demonstrated for brush-like molecules whose design greatly benefits from the synthetic control of the chain stiffness, branching functionality and the chemical structure of the block-ends and also from our ability to visualize them.

4.3.2
Crystallization of Polymer Chains

Semicrystalline polymers constitute the largest fraction of industrial plastics used for the fabrication of fibers, films, blends and composites. As such, a lot of studies in the last 50 years have been focused on the fundamental understanding of structural organization, crystallization, melting and processing of semicrystalline polymers. Their behavior is essentially different from those of small molecules because long chains are much larger than the crystal thickness and different portions of the chain can participate in different crystals. The fundamentals of the structure and crystallization of polymers were established in the 1950s and 1970s, respectively [200]; however, this area of science still attracts much attention, not only due to its technological significance, but also due to many unresolved questions. In recent years, research activities is this field have been intensified due to the development of new techniques such as high-speed calorimetry, synchrotron radiation microfocus beams and AFM. The last two techniques allow the resolution of crystalline superstructures down to the micrometer and nanometer range, respectively.

Morphology The morphological hierarchy of semicrystalline films, molds and fibers is very complex, resulting in many unresolved issues, including the thickness of polymer crystals, growth sectors of folded chain lamellae, lamellar branching and bending, spherolite organization and the morphology of amorphous phases. Understanding of the morphological hierarchy and connectivity of structural components on all levels is a key to fabricating materials with superior mechanical properties, e.g. silk fibers [201–204].

It is known that kinetic trapping during the crystallization of semicrystalline polymers leads to crystalline lamellae of finite thicknesses (typically 5–50 nm), with a significant portion of chain segments folding back into the crystals [205, 206]. Other emerging segments accumulate in the amorphous layers as loose loops, dangling segments or tie molecules [207, 208]. From both kinetics and equilibrium points of view, there are arguments that suggest variations of crystal thickness for crystallizing both homopolymers and copolymers [209]. However, systematic SAXS [210, 211], TEM [212] and AFM [213, 214] studies show that isothermal crystallization leads to crystals with uniform thickness. The only exception is polymers, e.g. polyethylene (PE), with active sliding motion of chains within crystals, which leads to crystal thickening and eventually to a thickness distribution. As shown in Fig. 4.18, the thickness distribution of poly(ethylene terephthalate) lamellae is narrow and the mean thickness is constant throughout the whole crystallization process [213]. Real-time AFM studies clearly show equal thicknesses for

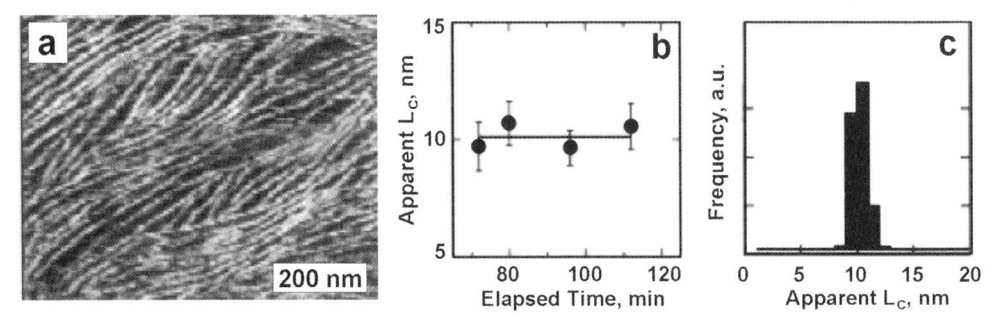

Fig. 4.18 Crystallization of poly(ethylene terephthalate) was monitored by tapping mode AFM. (a) Phase images were recorded *in situ* at 23 °C. Image analysis gives (b) time dependence of the mean crystal thickness and (c) thickness distribution in the final state.

both dominant and subsidiary crystallites, which refute the assumption that the first grown lamellae should be thicker than the secondary crystallites.

Prior to AFM, transmission electron microscopy in combination with X-ray diffraction had been applied to the in-depth examination of single PE crystals. The combination of techniques made it possible to observe the sectorization and determine the polymer chain orientation inside individual crystal sectors. However, many questions related to the organization of single crystals remain open. One such question concerns the structure of the lamellar surface, which is presumably formed by chain folding according to the adjacent re-entry model. The other question is the chain packing and microstructure of the lamellar bulk. The sensitivity of AFM to height measurements made it possible to monitor chain unfolding *in situ* with exceptional precision. Real-time imaging of single PE crystals at elevated temperatures revealed lamellar thickening caused by unfolding of individual chains from a kinetically formed folded state to an energetically favorable extended-chain conformation [215]. Figure 4.19a and b demonstrate that holes appear simultaneously with thickening of the adjacent locations, which is consistent with the earlier TEM data [216]. Thickening proceeds gradually after a stepwise change at 115 °C, as reflected in the height histograms in Fig. 4.19c. Studies of the thickening mechanisms in various polymers can be useful in understanding the role played in these processes by the crystal/amorphous interphase [217] and the polymer nature of the reorganizing species.

An effective approach in the elucidation of fine structural features of polymer crystals and the mechanisms of polymer crystallization is to use model molecules. For example, ultra-long alkanes (C_nH_{2n+2}, n > 150) are considered to be an appropriate model for PE. Recent AFM studies have shown that the structures of single crystals of $C_{390}H_{782}$ and PE are similar, although the details of their thermal behavior are completely different [218]. Upon annealing, alkane crystals undergo a complete series of transformations corresponding to stepwise unfolding from the folded-in-five conformation toward the fully extended-chain crystal,

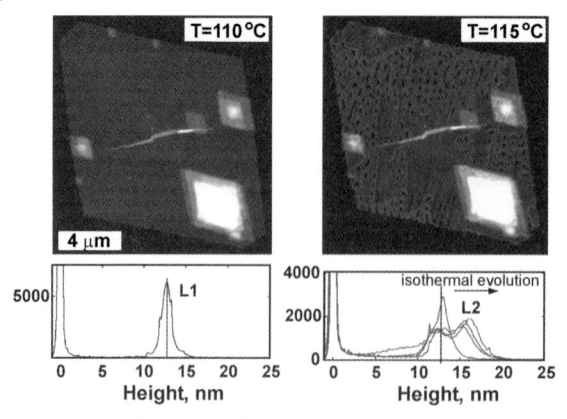

Fig. 4.19 Height images of dry single crystals of polyethylene measured at 110 °C and after 1.5 h of annealing at 115 °C. Height histograms corresponding to AFM show the evolution of lamellar thickness after annealing at different temperatures.

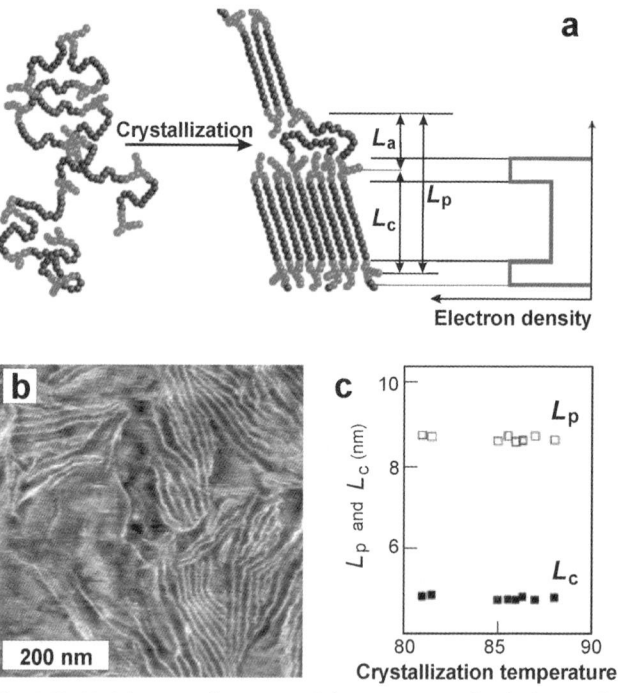

Fig. 4.20 (a) Schematic illustration of the trajectory of a chain within P44/5-Prop crystals before and after crystallization. Superimposed on the right is a pictorial representation of the electron-density profile corresponding to the crystalline lamellae, where L_a is the thickness of the amorphous layer. (b) L_p and L_c as a function of crystallization temperature (measurements at room temperature). L_c values were obtained from the fit of the SAXS.

while the chain unfolding in PE crystals is a continuous and slower process. Another series of model compounds was proposed to control selectively the chain folding of polymers [219]. Structural instructions were encoded in a linear backbone which includes alternating crystallizable, long alkyl sequences of monodisperse sizes separated by short spacers containing side-chains and acting as stops and fold-controlling units (Fig. 4.20a). This code translates during a crystallization process to generate a semicrystalline morphology with a structure-controlled crystal thickness of ~ 5 nm that remains constant over a wide temperature range (Fig. 4.20b). This approach allows control of the lamella thickness by steric interactions only, in contrast to previous attempts aimed at engineering polymer crystallization through hydrogen bonding [220, 221].

Crystallization There is great interest in the factors controlling the lamella thickness during crystallization [222], and also the role of the precursor mesophase in the formation of crystals [223, 224]. Even though the Lauritzen–Hoffman theory remains as the only theory that correctly interprets the kinetics of polymer crystal growth [225], the mechanism of polymer crystallization remains largely unknown [226]. Along with nanometer resolution in real space, AFM allows *in situ* measurements over a wide range of temperatures and thus provides both temperature and temporal information about the crystallization kinetics. One is able to resolve minute steps in crystallization, which are usually obscured or overlooked in the global morphological parameters obtained by small-angle X-ray scattering (SAXS). Figure 4.21 shows initial steps in the development of spherulites of poly(bis-phenol octane ether) starting from nucleation and growth of the founding lamella followed by its branching and splaying [227]. It was also shown that not every nucleus, like those in Fig. 4.21a, develops into a lamella. Some of the nuclei disintegrate in this stage of crystallization, which is consistent with homogeneous nucleation. Finally, the growing object becomes a spherulite whose morphology depends strongly on whether it emerges through homogeneous or heterogeneous nucleation [228]. The processes of lamellar perfection, e.g. lamellar thickening and merging of lamellar fragments, are revealed during the secondary crystallization stage [229].

Additional information was obtained by monitoring the spherulite growth using rapid scanning AFM through active damping of the cantilever quality factor (Section 4.2.3.6). Rapid imaging allowed the full temperature range of growth to be explored [230], revealing the sequential buildup of spherulites including rapidly growing dominant lamellae along with slower growing secondary lamellae (Fig. 4.22). The relative contribution of the dominant and secondary lamellae was shown to depend on the crystallization temperature. At higher temperatures, one expects a lower branching rate and thus preferential growth of the dominant lamellae.

AFM provides evidence that crystallization is preceded by formation of a precursor mesophase. High-resolution AFM images revealed a granular structure that was attributed to a columnar mesophase [231, 232]. Recently, the role of a precursor mesophase has been further investigated by measuring selected-area

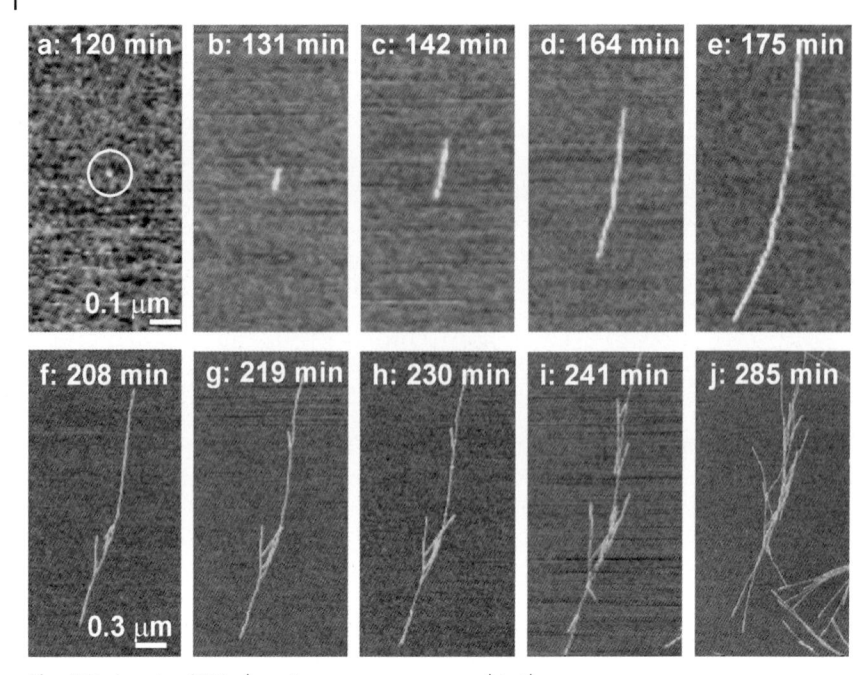

Fig. 4.21 A series AFM phase images were measured in the same area of poly(bis-phenol octane ether) at 22 °C, i.e. at a high degree of supercooling, $\Delta T = 61$ °C. (a) An embryo develops in (b) a short lamella of 60 nm in length. (c–e) This so-called founding lamella grows in length to ~ 1 μm. (f–j) The lamella grows further into a lamellar sheaf through branching and splaying.

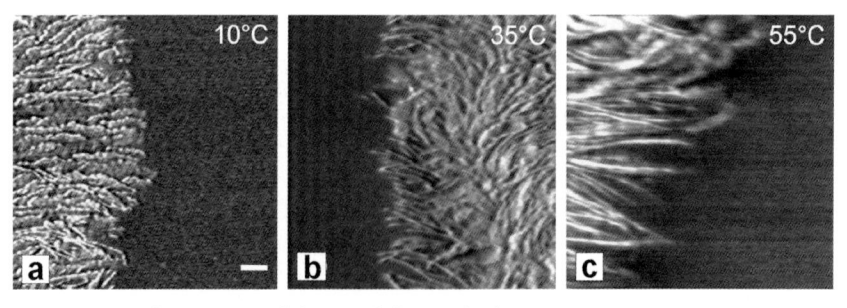

Fig. 4.22 AFM phase images of the growth fronts of spherulites of poly(hydroxybutyrate-co-valerate) developing at different temperatures. Scale bar = 100 nm.

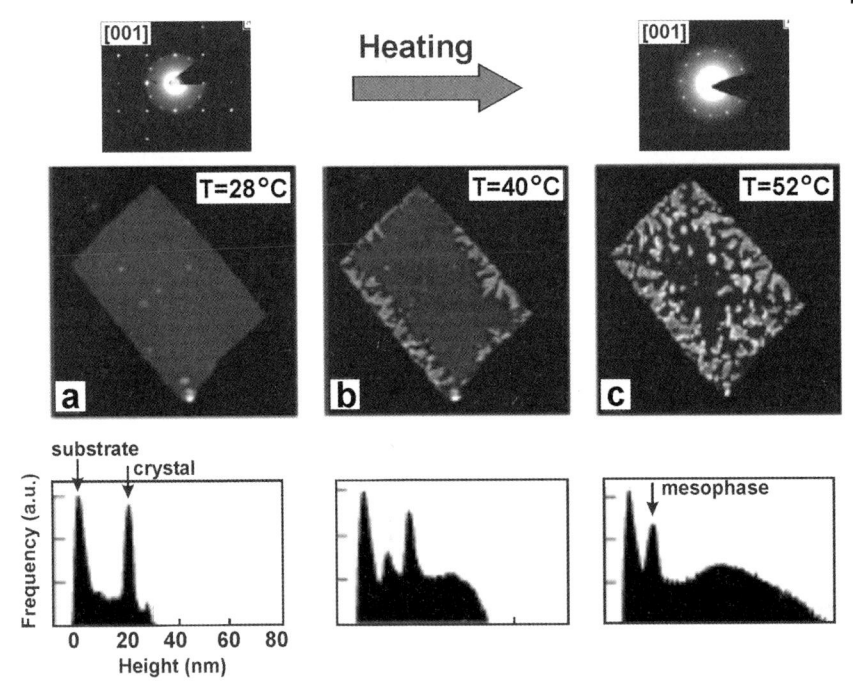

Fig. 4.23 Topographic tapping mode AFM images (13×13 µm²) show the morphological evolution of a single crystal of PDPS as a function of temperature. Concurrently measured electron-diffraction patterns demonstrate the transformation of the pseudotetragonal R phase of PDPS into the hexagonal columnar mesophase. Height histograms obtained from the AFM micrographs after preliminary background correction. The positions of the substrate, the initial value of the crystal thickness and the mesomorphic regions are indicated by arrows.

electron diffraction and variable-temperature AFM in single crystals of polydipropylsiloxane (PDPS) [233]. The crystal-to-hexagonal columnar mesophase transition was observed upon heating, whereas crystal thickening occurred via a short-term dwelling in the mesophase (Fig. 4.23). The crystallization of PDPS from the hexagonal mesophase resulted in very thick (100–150 nm thick) crystalline lamellae, wherein polymer chains are fully extended [234]. This feature, which is found in a few polymers, makes PDPS similar to systems such as HDPE at high pressure or 1,4-*trans*-polybutadiene. The high chain mobility in the mesophase can account for the exceptional range of accessible crystal thicknesses of PDPS, which is much larger than that observed for PE single crystals.

Recently, a method has been presented for controlling the initiation and kinetics of polymer crystal growth using dip-pen nanolithography and an AFM tip coated with poly-(DL-lysine hydrobromide) (PLH) [235]. As shown in Fig. 4.24, triangular prisms of the polymer grow epitaxially on freshly cleaved mica substrates, while their in-plane and out-of-plane growth rates are controlled by raster scanning the coated tip across the substrate.

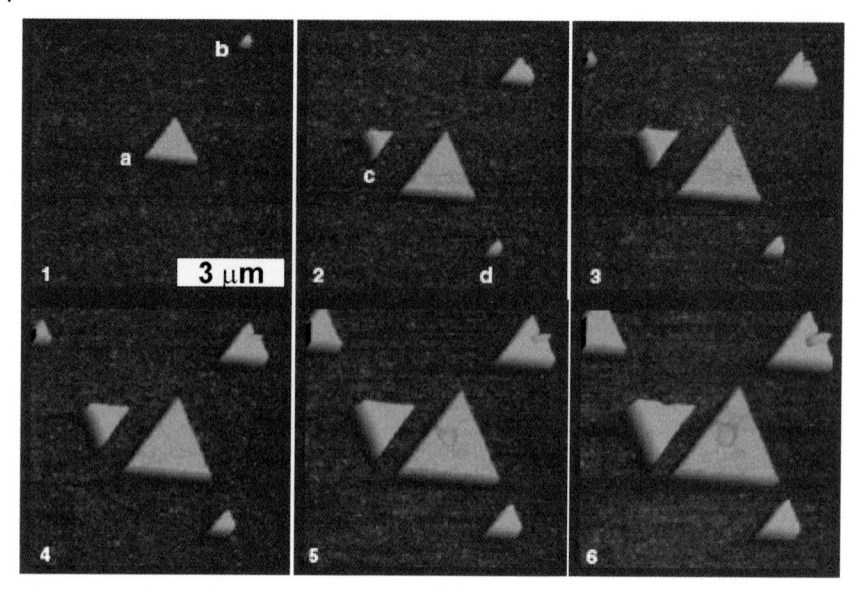

Fig. 4.24 Panels 1–6 show a series of 8×8-mm 3D topographic AFM images of mica taken at 256-s intervals (relative humidity, 30%; temperature, 20 °C), obtained by continuously scanning an AFM tip coated with PLH molecules in tapping mode. Crystals (labeled a, b, c and d) were chosen for kinetic studies. PLH triangles with edge lengths ranging from 100 nm to 10 mm and with heights from 5 to 50 nm were generated. Lower humidity favors the formation of smaller triangles.

4.3.3
Wetting and Dewetting Phenomena

Wetting phenomena at polymer surfaces are and is an established research area which has great practical relevance [236]. The formation and stability of thin polymer coatings are vital for coatings, lithography, application of pesticides and cosmetics. Various experimental techniques have been developed to explore both equilibrium wetting structures and kinetics of spreading and dewetting processes [237]. Equilibrium thermodynamics describes film thickness, drop profile, surface segregation and wetting transitions. Here, AFM became a mandatory technique as it allows accurate observation of very thin films (< 10 nm) and microscopic drops (< 100 nm) that are hardly visible with optical techniques. Kinetic studies deal with characteristic rates of spreading and dewetting, energy dissipation and molecular dynamics. When studying wetting phenomena, one should distinguish the behavior of thick films, thin films, and monolayers. As films become thinner, the effect of the substrate becomes significant, causing phase separation and alignment of molecules with the films. In monolayers, the substrate effect is particularly strong in controlling both the conformation and dynamics of polymer molecules.

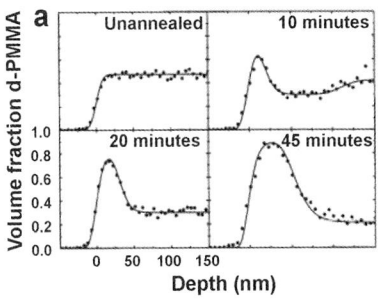

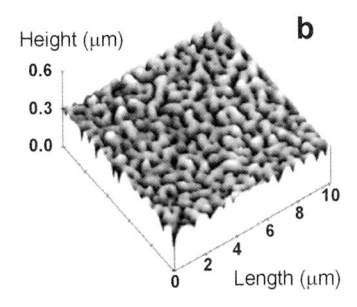

Fig. 4.25 (a) Volume fraction profiles for d-PMMA-SAN blend films annealed at 463 K as measured by LE-FReS. The rapidly increasing surface excess is indicative of hydrodynamic flow. (b) AFM height image of a d-PMMA-SAN blend film with the d-PMMA removed after being first annealed for 1 h at 458 K. The acetic acid treatment reveals the bicontinuous morphology of the SAN underneath the PMMA wetting layer.

4.3.3.1 Polymer Blends

In the last 20 years, significant research efforts have been focused on understanding the film structure of polymer blends [236]. Various issues resulting from concurrent effects of phase separation, surface segregation and wetting transitions were studied in detail both theoretically and experimentally. A combination of forward recoil spectrometry (FReS) and AFM represents one of the most powerful approaches to monitor the morphology of the blend film. Figure 4.25a shows low-energy FReS data describing the growth of a wetting layer of deuterated PMMA from a blend with styrene-acrylonitrile (SAN) copolymer [238]. Complementarily, AFM was used to image the samples, along with the selective dissolution of the PMMA in acetic acid. As can be seen in Fig. 4.25b, the SAN surface is much rougher than the surface of the PMMA wetting layer and exhibits a bicontinuous morphology.

More recently, interest in thin films of polymer blends has been revived by directed phase separation in the presence of electric fields. Electrohydrodynamic instability of the interface between two dielectric fluids generates patterns of oriented microdomains with a wavelength determined by the interfacial energy and the electric field gradient at the interface [239]. When combined with a topographically patterned electrode, this strategy can be used to replicate patterns of ∼ 100 nm [240]. As shown in Fig. 4.26, the pattern formation is a hierarchical process which leads to lateral structures that with two independent characteristic dimensions determined by the electrode spacing, the electrode surface profile and the initial film thickness of the PMMA layer.

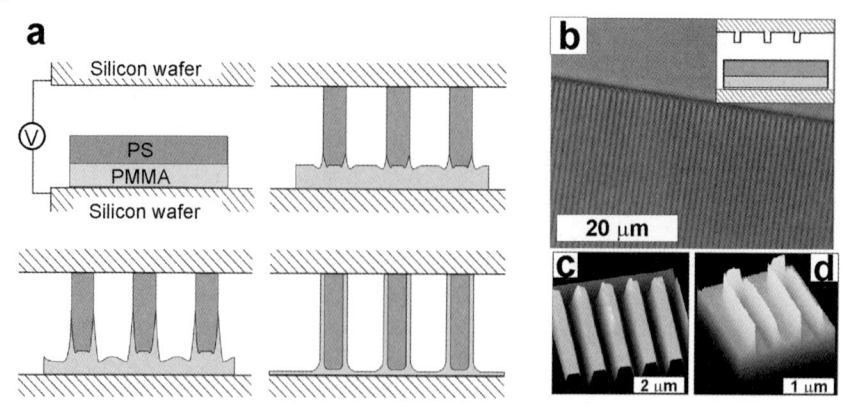

Fig. 4.26 (a) Starting from a polymer bilayer, the polymer–air surface is first destabilized by an electric field. The initial instability results in the formation of columns spanning from the surface of the lower layer to the top electrode. In a secondary instability, the deformation of the lower layer is enhanced by the electric field, driving the polymer upward on the outside of the columns. (b) A topographically structured electrode (inset) provides additional means to control the pattern formation process. The optical micrograph shows a line pattern replicated by the electrohydrodynamic instability. (c) The AFM image shows five replicated lines. (d) After removing the PS phase by washing the sample in cyclohexane, the secondary PMMA structure is revealed in the scanning electron micrograph in (c) (height 160 nm, width 100 nm).

4.3.3.2 Spreading

The wetting transition, discussed previously for polymer blends, is also relevant to homopolymers. The spreading behavior of thick films is well known and can be rigorously described by continuum hydrodynamic theories [241–248]. These theories work well for thick films (with a thickness of the order of 10 nm); however, their use for thin films, particularly monolayers, is ambiguous [249–259]. In some systems, the spreading rate was found to be independent of the chain length, which was ascribed to plug flow on homogeneous substrates [260]. However, in other systems, a strong increase with the chain length is observed and attributed to reptation through a network of substrate-pinned molecular segments [261, 262]. This and many other controversies necessitate experimental data for motion of fluids on the molecular scale.

Typically, optical ellipsometry [263, 264] and, more recently, phase-modulated microscopy [265] are used to monitor motion of thin films. However, optical techniques fail to answer questions on the molecular mechanism of flow. Here, AFM demonstrates its unique advantage as it allows visualization of molecules in great detail, including the contour length and the local curvature. Although this advantage was immediately recognized, the first AFM studies of flows did not attain molecular resolution [266, 267]. Unlike single molecules, visualization of molecules within dense samples is challenging due to lower contrast and overlapping of polymer chains. It is even more challenging to visualize molecules in flowing monolayers. A breakthrough in molecular visualization of

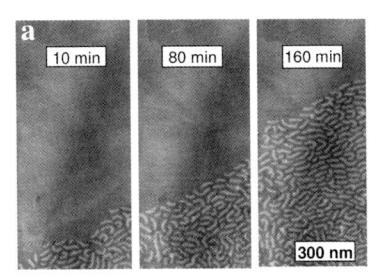

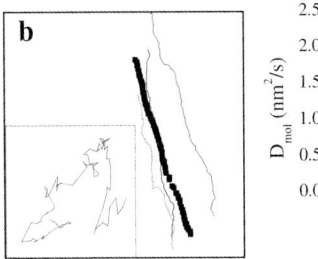

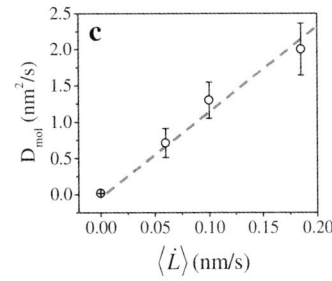

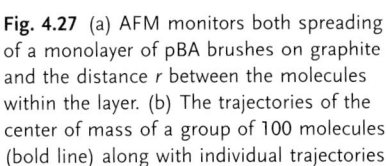

Fig. 4.27 (a) AFM monitors both spreading of a monolayer of pBA brushes on graphite and the distance r between the molecules within the layer. (b) The trajectories of the center of mass of a group of 100 molecules (bold line) along with individual trajectories of three molecules (thin lines). Inset: molecular path within the flowing monolayer. (c) The molecular diffusion coefficient, which was measured in the frame of the flowing film, increases with the flow velocity.

spreading was achieved by using brush-like macromolecules that are thicker, stiffer and move more slowly than regular polymer chains [94]. Through molecular visualization, both the displacement of the contact line and the movement of individual molecules within the film were simultaneously measured (Fig. 4.27a). Whereas the film length was shown to follow the classical dependence $L=(D_{spread}t)^{1/2}$, the individual molecules moved in a random-walk fashion in the frame of the flowing film (Fig. 4.27b). The time dependence of the mean square intermolecular distance $\langle r^2(t)\rangle = 4D_{mol}t$ gives the diffusion coefficient $D_{mol}=1.3$ nm^2 s^{-1}. The origin of the molecular diffusion within the flowing monolayer is still a subject of debate, although the diffusion is clearly enhanced by the flow, as evidenced by the linear increase of the diffusion coefficient with the flow velocity (Fig. 4.27c).

Spreading instability in polymer films is another poorly understood issue, which impacts coating, lubrication and microfluidics. In some applications, instabilities lead to uneven surface coverage, whereas in others instabilities act as new tools for surface patterning [268–270] and mixing enhancement [271]. The instability principles are known for thick films where variations in viscosity, surface tension and mass density, and also fluid inertial and viscoelastic effects [272, 273], can destabilize the flow [274–279]. In thin films there are still many unanswered questions due to the intimate coupling of the driving and frictional forces to surface-confined molecular conformation and dynamics. In this respect, a particularly interesting system is that of compressible monolayers wherein molecules change their conformation in response to variations of the film pressure. Recently, a new type of flow fingering instability in polymer monolayers as they spread on a solid substrate has been observed [280]. Through the use of AFM, the development of the instability was monitored over a broad range of length scales from the millimeter-long precursor film all the way down to the movements of individual molecules within the film (Fig. 4.28). The tracing of the evolution of the instability pattern on the molecular scale per-

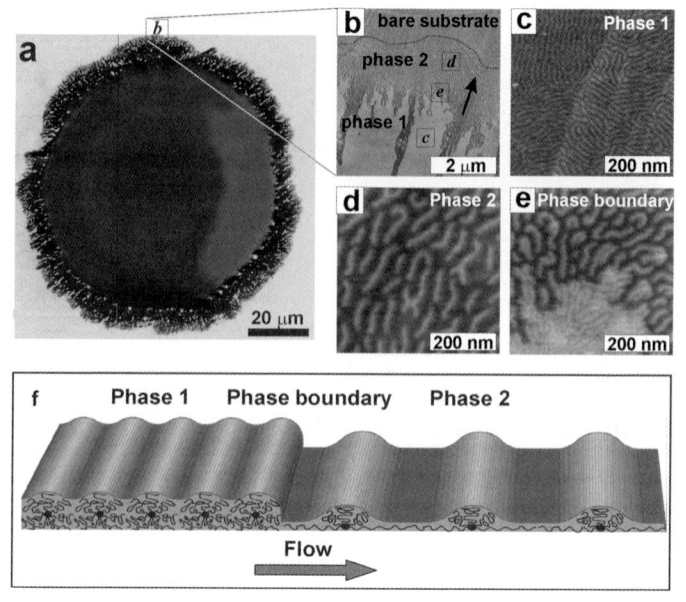

Fig. 4.28 (a) AFM images a spreading drop of brush molecules on graphite. After 7 h the drop reveals a precursor film with a dendritic structure, characteristic of the flow instability. (b–e) Higher magnification AFM images of the precursor film show two conformationally different phases (c, d) separated by a sharp phase boundary (e). (f) Schematics of the conformational phase transition.

mitted an understanding of the underlying physical mechanism. The instability was shown to be triggered by conformational changes of flowing macromolecules, which lead to an instantaneous jump both in the monolayer thickness and in the flow velocity across the phase boundary.

4.3.3.3 Dewetting

In addition to the surface wetting for many applications it is also necessary to understand what causes a polymer films to dewet surfaces. Even though dewetting is also a very mature topic, there are still many unsettled issues such as dynamics of hole growth and fingering instability of the rim [281]. Typically, optical microscopy is used to observe the break-up of polystyrene films on silicon substrates as a function of film thickness [282]. In these experiments, the break-up of the films was attributed to a spinodal process [283]. Here, the contribution of AFM is very important as it is able to quantify the characteristic wavelength of instability on much smaller length scales and to monitor *in situ* the dewetting process at much earlier stages [284]. AFM provided solid evidence for spinodal dewetting as a mechanism for the break-up of polymer films by determining the fastest growing length as a function of film thickness [285]. AFM revealed uniformly distributed surface undulations, in contrast to the random distribution of holes observed by optical microscopy. There have been several other studies of the dewetting of polystyrene from silicon substrates, although not all of these are attributable to spinodal dewetting [286–289].

4.3.4
Block Copolymers

Block copolymers represent one the most interesting classes of engineered polymeric materials. The first copolymers were synthesized in an attempt to mix chemically different species on the molecular scale and thus obtain novel materials with intermediate properties. In addition to new properties, nanostructured materials with microphase separation both in solution and in a solvent-free state were obtained. In solution, block copolymers form micelles provided that the solvent is selective, i.e. a good solvent for one block and a poor solvent for the other. Depending on the composition and solvent quality, micelles can adopt different shapes such as spheres, cylinders and vesicles (Fig. 4.29a). These are the most typical shapes, although other geometries such as toroids and discs have recently been observed and even more unusual morphologies are sure to be discovered in the near future. In addition to solution micelles, there are also so-called surface micelles that are formed upon adsorption from a non-selective solvent on to a selective substrate. Depending on surface coverage, one observes either a star-like (Fig. 4.29b) or stripe-like (Fig. 4.29c) morphology with a globular core of the weakly adhering block and a corona of the strongly adhering block. In solvent-free bulk, block copolymers undergo microphase separation

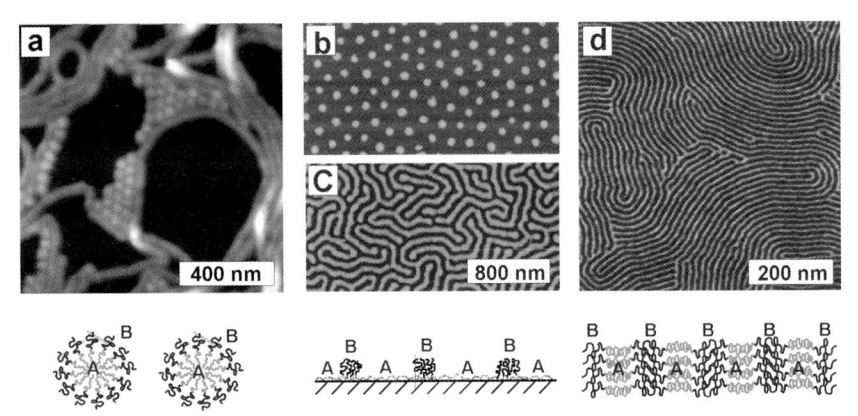

Fig. 4.29 Various block copolymer morphologies and their structural interpretation (cartoons). (a) Coexistence of spherical and cylindrical micelles of PS-*b*-PI in solution. (b, c) Spherical and cylindrical surface micelles of PS-*b*-P2VP in ultra-thin films and (c) lamellar morphology of a thick SBS film.

wherein the individual blocks segregate into nanometer-sized domains (Fig. 4.29 d). Usually, these domains are uniform in size and organize into a mesoscopic lattice with a spacing in the range 10–100 nm. The range of practical applications of block copolymers is very broad and is continuously extending to new areas. These include engineering plastics, dispergent agents, emulsifiers, drug delivery, data storage and lithographic templates.

AFM is a very appropriate and informative technique for block copolymer studies as it provides reliable resolution of ∼10-nm sized microdomains and is able to distinguish between the chemically different domains. AFM images also provide three-dimensional information about the shape and size of block copolymer morphologies.

4.3.4.1 Micelles

Although block copolymer micelles hold great potential, they are difficult to study. This is caused by several factors, one of which is the wide variety of the micelles size from 10-nm spheres to cylinders and vesicles larger than 10 µm. This requires the use of multiple techniques to cover the full range of length scales. Micelles are also nanoheterogeneous systems that consist of chemically different core and shell components, which in turn can be swollen to different extents with solvent. Another important issue is that the micelles are dynamic systems, which can change their weight, size and shape in response to changes in the surrounding environment, i.e. temperature and solvent quality. This dynamic nature of micelles means that they must be studied under carefully controlled conditions using experimental techniques that can preserve native structures. Usually, a combination of microscopic and scattering techniques is employed to study the full range of polymer micelle sizes and morphologies both *ex situ* and in real time. Microscopy, such as cryogenic TEM, is a technique that is usually used to visualize micelles [290–294]. AFM has the added benefit of being a 3-D technique, allowing for not only visualization of an entire micelle, but also the determination of the volume that the micelles occupy [295].

One of the most intriguing aspects of micelles is whether or not the morphology can be changed in an easy manner. Being able to switch reversibly between the various morphologies and micelle size would open up a wide avenue of possible applications. Possible stimuli to induce morphological changes include temperature and pH in aqueous systems. Many studies have shown that the morphology can be changed through the addition of a co-solvent or homopolymer [296–299]. While these studies show that it is possible to change from one morphology to another, the only way to recover the original morphology is through the addition of more solvent or diblock. Recently, it has been demonstrated that similar morphological transitions can be induced by changing temperature while keeping the solvent and polymer concentration constant [300]. The original morphologies can then be recovered by reversing the temperature change (Fig. 4.30). Using a similar strategy, it is also possible to study the transition between cylinders and vesicles demonstrating unusual pathways and transient structures.

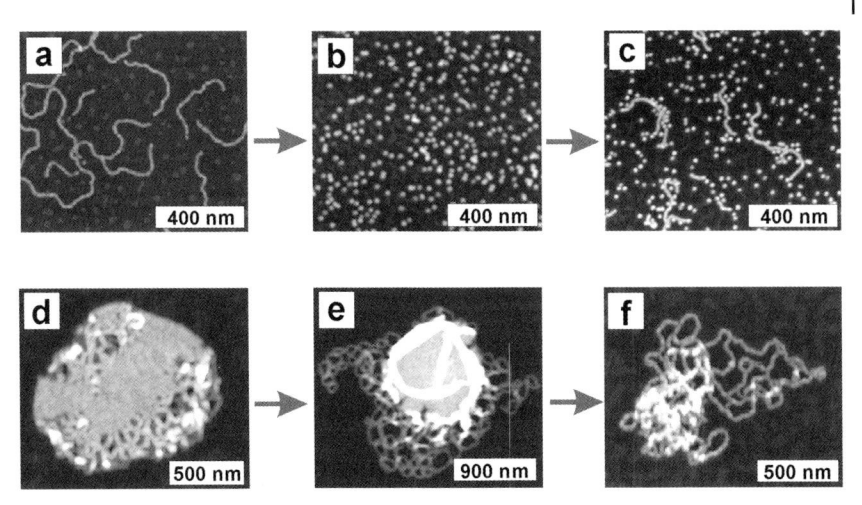

Fig. 4.30 AFM images show a morphological transition from cylindrical micelles to spherical micelles upon (a, b) heating and a slow return to cylindrical micelles upon (c) cooling. AFM images in (d, f) show vesicles at various stages of the morphological transition to cylindrical micelles. After crossing the phase boundary, the capillary instability leads to perforated vesicles (d) followed by a network (e) and then individual cylindrical micelles (f).

4.3.4.2 Alignment of Microdomains

Thin films of block copolymers are widely investigated for their potential uses in nanolithography and as high-density magnetic storage media, where the polymer film acts as a mask for etching nanometer-sized features [301]. These systems are convenient because both the size and periodicity of self-assembled microdomains can be tuned by changing the copolymer composition at length scales below 100 nm. A challenge that remains is controlling the long-range order of the microdomains. Alignment of block copolymer domains over micrometer length scales has been achieved using graphoepitaxy, shear alignment and chemically patterned substrates. Graphoepitaxy is a technique that uses topographically patterned substrates to confine a single layer of the block copolymer domains laterally over micrometer length scales, which are much larger than the periodicity of the individual microdomains. A topographically patterned silicon substrate was used to template a large, single-crystal grain of styrene–2-vinylpyridine block copolymer spherical domains [302]. The presence of a hard edge was shown to impart translational order to both the hexatic and the liquid phases, which faded as the distance from the edge increased. Large-scale alignment of bilayer block copolymer films was also achieved by applying in-plane shear (Fig. 4.31a) [303]. This method was shown to eliminate all disclinations and most dislocations from arbitrarily large samples, as evidenced by AFM observation of the microdomain lattice (Fig. 4.31b). The resulting films can act as ordered templates for nanolithography.

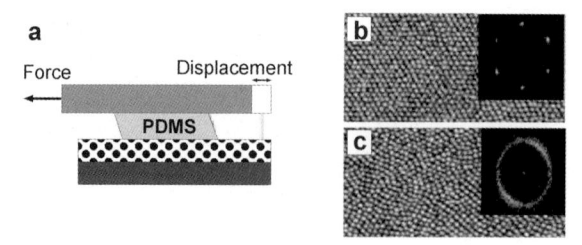

Fig. 4.31 (a) The shear alignment setup. (b) Tapping mode AFM image shows shear-induced ordering of a hexagonal bilayer of polystyrene-*block*-poly(ethylene-*alt*-propylene) diblock copolymer (PS spheres are white). (c) Slightly thinner or thicker films show a completely disordered lattice.

Unlike spherical microdomains, it is more difficult to orient cylindrical microdomains, particularly in the direction perpendicular to the substrate. Thermal annealing [304] and electric fields [305] were applied to enhance the degree of order and induce specific alignment of the cylinders. In thin films, interfacial interactions strongly affect the alignment of microdomains. Recently, a simple approach has been developed for modifying solid surfaces, based on an ultrathin cross-linkable film of a random copolymer, which does not rely on specific surface chemistries. Thin films of styrene (S) and methyl methacrylate (MMA) with 2% reactive benzocyclobutene (BCB) functionality randomly incorporated along the backbone were spin coated or transferred, then thermally cross-linked on a wide variety of metal, metal oxide, semiconductor and polymeric surfaces [306]. In this way, one obtains a coating with a controlled thickness and well-defined surface energy. In the situation when both blocks are equally attracted by the substrate, cylinders may spontaneously align perpendicular to the substrate film. The coatings were then used with PMMA-*b*-PS, which in the bulk form hexagonally packed cylindrical microdomains of PMMA in a PS matrix with a lattice spacing $L_0 = 34.1$ nm. As shown in Fig. 4.32b, the surfaces coated with P(S-*r*-BCB-*r*-MMA), on which the interfacial interactions are balanced, produce films in which the cylindrical microdomains are oriented normal to the surface, regardless of the underlying substrate. When these films were exposed to ultraviolet radiation, the PS was cross-linked and the PMMA was degraded; after rinsing with acetic acid, pores were produced that extend from the film surface to the underlying substrate. Recently, as was widely announced in the press, IBM has implemented a full development program for flash memory chips based on self-assembling block copolymer templates prepared by living free-radical chemistry.

An elegant alternative for obtaining hierarchically structured and aligned materials is to combine microphase separation of block copolymers and self-organization oligomeric amphiphiles through hydrogen bonding, ionic and coordination interactions [167, 171]. Recently, Möller et al. have demonstrated that this approach can be used to control the alignment of microdomains in a block co-

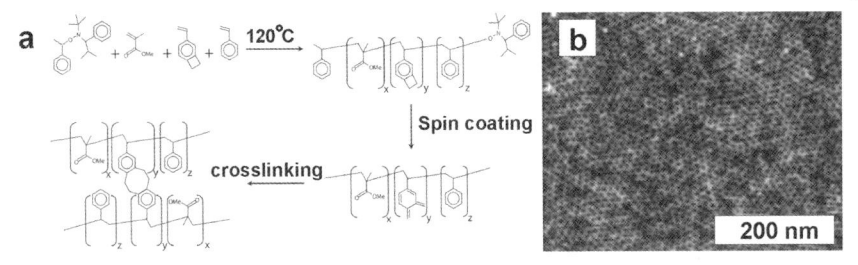

Fig. 4.32 (a) Synthesis and primary cross-linking reaction of the P(S-r-BCB-r-MMA), which has a composition of PS:BCB:PMMA 56:2:42. (b) AFM height image of a 33-nm thick film of P(S-b-MMA) copolymer prepared on substrates coated with a film of cross-linked random copolymer mat regardless of the underlying substrate. PMMA was selectively removed by exposing the film to UV radiation and rinsing with acetic acid as a good solvent for PMMA. This nanoporous PS film constitutes a template or scaffold for the fabrication of nanostructured materials [306].

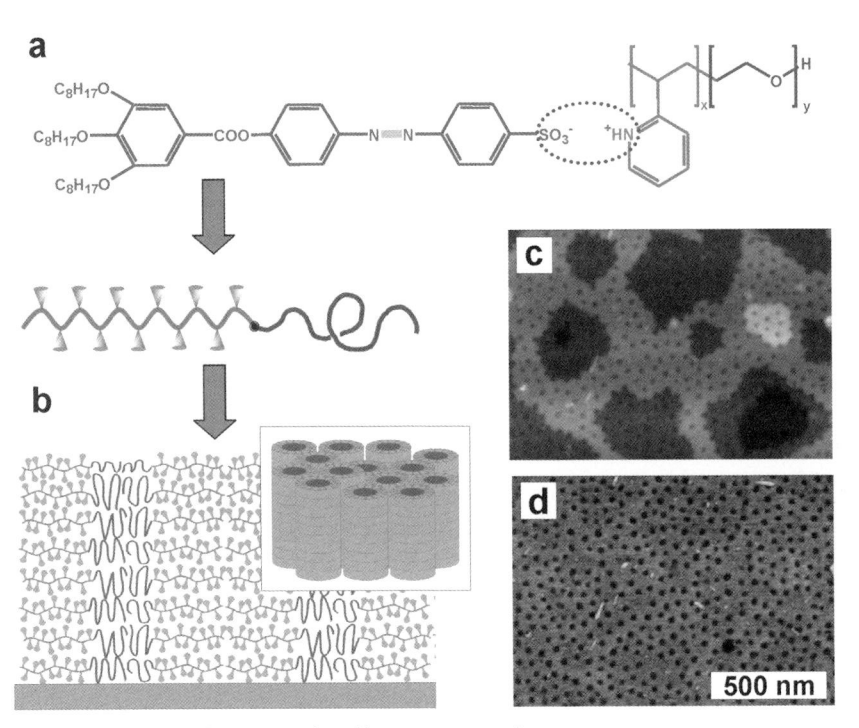

Fig. 4.33 (a) Chemical structure, (b) self-organization in thin films and (c) AFM height images of P2VP$_{180}$-b-PEO$_{560}$/(1-H)$_{0.5}$ complex after 12 h of benzene vapor annealing.

polymer thin film by incorporation of wedge shaped molecules having a sulfonic group at the tip into a poly(2-vinylpyridine)-*b*-poly(ethylene oxide) (P2VP-*b*-PEO) diblock copolymer via proton transfer at different degrees of neutralization. The formed complexes exhibited strong phase-segregated patterns comprising the liquid crystalline layers of complexed P2VP fragments and cylindrical domains of PEO. For a complex with a degree of neutralization 0.50, the PEO cylinders were oriented perpendicular to the substrate (Fig. 4.33). By decreasing the degree of neutralization to 0.25, the PEO cylinders partially merged and a mixed lamellar–cylindrical morphology was formed.

4.4
Measuring Properties

Unlike the widely recognized ability of AFM to visualize surface morphologies, probing physical properties is not yet a routinely used procedure. The major technical challenges that hinder quantitative studies include definition of the true contact area, monitoring the probe–substrate distance, preventing thermal drift, controlling the loading and unloading rates, and calibration of the force sensor. In addition, there are many problems in interpretation of the data due to the simultaneous contribution of various interaction forces. For example, a simple indentation test involves a complex tensor of elastic and inelastic strains, friction at the probe–sample interface, specific adhesion, and capillary forces. Due to the intricate nature and instrumental complexity of the highlighted issues, AFM studies typically provide qualitative information which allows a distinction between the hard and soft materials or between an attractive and a repellant surface. Quantitative measurements are far more challenging and represent one of the active topics of current AFM studies. In addition to direct probing of physical properties, it is possible to measure some properties indirectly through visualization of processes and structures. For example, measurements of molecular diffusion coefficient D_m or spreading rate D_s of fluid monolayers give the friction coefficient as $\zeta \cong k_B T/AD_m$ or $\zeta \cong S/D_s$, where S is the spreading coefficient and A is the molecular area [94].

4.4.1
Adhesion and Colloidal Forces

AFM is an ideal tool for the characterization of adhesion on the nanometer scale, which is vital to the stiction problems in MEMS devices [307]. One of the key parameters in these measurements is the true contact area between the probe and substrate. If one deals with a flat substrate and smooth spherical probe, then an appropriate model from contact mechanics can be applied to analyze the pull-off force data [308]. Based on this method, the so-called "colloidal probe technique" was developed by using a micrometer-sized spherical particle glued to the end of an AFM cantilever as the force sensor [309]. However, both the probe and substrate usually demonstrate irregular shape and ill-defined roughness, which reduces the contact area and results in lower adhesion forces than are expected from ideal geometries. Recently, systematic studies have been conducted to elucidate the effect of surface roughness on the distribution of pull-off forces acquired by AFM [310–312]. Furthermore, elastic contact has to be extended to take viscoelastic effects into account and use it for the evaluation of work of adhesion along with the modulus of elastomeric films [313].

Along with the work of adhesion, the AFM-based colloidal probe technique allows measurements of various colloidal forces, which become significant when the probe interacts with a substrate through a fluid medium. If three phases meet at the contact area, the probe–substrate interaction is also affected by capillary forces. In order to understand the contribution of capillary forces, the influence of the relative humidity on amplitude and phase of the cantilever oscillation in tapping mode AFM was investigated [314]. For hydrophilic AFM tip and sample, one observed a transition from a regime with a net attractive force between tip and sample to a net repulsive regime with decreasing relative humidity. Numerical simulations show that this behavior can be explained by assuming the intermittent formation and rupture of a capillary neck in each oscillation cycle of the AFM cantilever.

In this case, simple pull-off force measurements are not sufficient. Atomic force spectroscopy (AFS) represents a tool which measures the dependence of the interaction force on the probe–sample distance [315]. Another variation of AFM is chemical force microscopy (CFM), which uses the probe tip of a force microscope covalently modified with specific functional groups, providing a flexible approach for studying interactions between specific chemical functionalities. The progress in CFM in recent years as it applies to adhesion of soft materials, including both experimental and theoretical approaches, has recently been surveyed elsewhere [2, 316]. Along with high-throughput adhesion measurements and mapping of full intermolecular potentials from van der Waals forces and hydrogen bonding, this technique allows studies of the kinetics of the unbinding processes and contributions from the medium that are usually considered secondary. Recently, a strong temperature effect on the probe–substrate interaction has been observed and attributed to an entropic contribution from the ordered solvent layers that form on the probe and sample surfaces upon detachment [317].

Polymers are difficult samples for adhesion measurements due to intrinsic roughness and chemical heterogeneity on the nanoscale along with the complex viscoelastic nature of the mechanical contact. Unlike polymers, self-assembled monolayers (SAMs) are ideal samples for AFM measurements of adhesion and colloidal forces. The monolayers offer more possibilities to control the chemical composition of the surface and to induce well-defined chemical patterning. SAM also allows quantitative studies of intermolecular interactions between specific chemical functionalities that are involved in molecular recognition and protein folding.

4.4.2
Friction

A fundamental understanding of friction on the atomic [318] and nanometer scales [319] is critical for designing micro- and nano-electromechanical systems. There are two powerful instruments in nanotribology – the surface force apparatus and the atomic force microscope. AFM experiments allow high-resolution mapping of friction properties between solids on the nanometer scale by friction force microscopy (FFM) and adhesion by CFM. In FFM, as a sharp tip scans across a surface in the direction perpendicular to the long axis of the cantilever, the latter twists due to the frictional resistance of the tip. Friction data are then obtained by subtracting the lateral force reverse trace from the forward trace. These measurements depend on both the externally applied load and intrinsic load due to adhesion. The latter often causes nonlinear friction behaviors for negative loads up to the pull-off force [320]. When analyzing the friction data, only the portion of the friction traces that corresponds to steady sliding regime is usually considered, thereby excluding the static friction forces. As with adhesion measurements, the tip shape and the cantilever force constant must be calibrated to ensure quantitative analysis of lateral forces. Frictional properties are also very sensitive to environmental conditions and sample structure. Therefore, self-assembled monolayers with controlled functionality represent the most popular model systems to study the way in which molecular organization influences nanoscale friction [321–325]. These studies present many experimental and theoretical challenges, including the characterization of the true contact area and limited applicability of the continuum contact mechanics models.

The sensitivity of AFM sliding friction to a solvent-induced (water) glass–rubber transition was explored (Fig. 4.34) [326]. The sliding friction force on a semicrystalline poly(vinyl alcohol) (PVA) surface was measured as a function of relative humidity (RH) up to 75% (above which scan-induced wear was problematic). The results are shown in Fig. 4.34c, a comparison of sliding friction force on highly crystalline regions (2) with that on more amorphous regions (1) in which a glass–rubber transition is expected. Little RH dependence is observed on the crystalline regions, whereas a dramatic increase in friction is observed on the amorphous regions above 60% RH. At this humidity a moisture content of $\sim 8\%$ would shift the glass transition from around 80 °C in anhydrous PVA to 20–22 °C. Thus, above 60% RH, segmental dissipative motions are activated by shear stress. In Fig. 4.34d, the kinetics of the glass–rubber transition are examined by comparing results at four scan velocities. The onset of rising friction shifts to lower RH as the scan velocity slows by three decades.

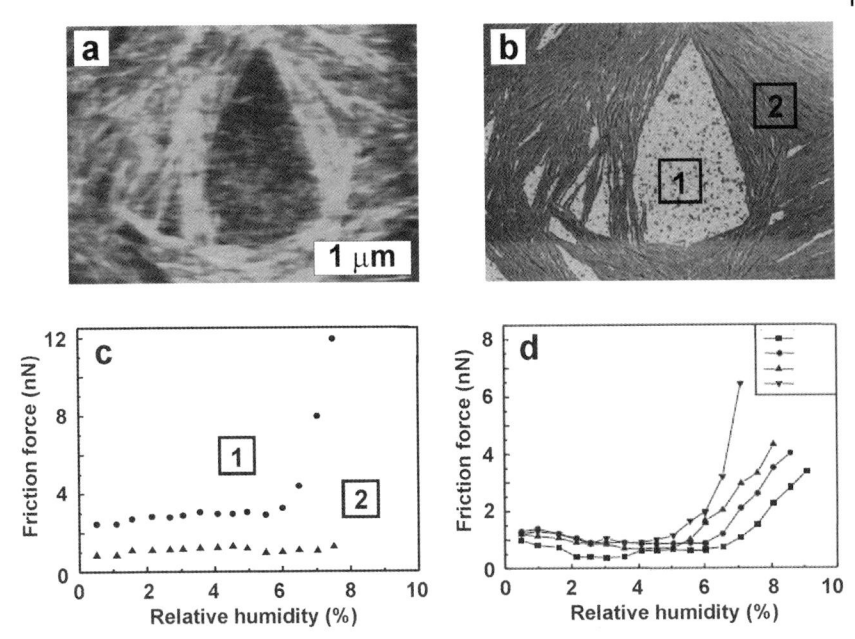

Fig. 4.34 AFM height (a) and friction (b) images on a semicrystalline PVA surface. (c) Comparison of RH dependence of friction on predominantly amorphous (1) versus highly crystalline (2) surface regions in (b). (d) Scan velocity dependence of friction versus RH on (1).

4.4.3
Mechanical Properties

The ability to probe surface mechanical properties with nanometer-scale lateral and vertical resolutions is critical for many emerging applications. For nanomechanical probing experiments, one usually exploits either AFM or microindentation techniques [327, 328]. AFM permits the characterization of the mechanical properties of the sample through probe indentation and shear. Mechanical tests include indentation, wear and force–distance curves measured upon normal and shear loading [13].

Quantitative characterization of polymeric materials is difficult due to numerous technical and material issues outlined above. Nevertheless, a number of successful applications have recently been demonstrated, including nanomechanical probing of spin-coated and cast polymer films, organic lubricants, self-assembled monolayers, polymer brushes, biological tissues and individual tethered macromolecules. Absolute values of the elastic modulus have been measured for polymer surfaces in the range from 0.01 to 30 000 MPa [329]. In addition, force modulation techniques have been applied to enhance instrument sensitivity and allow measurements of time-dependent materials properties, such as

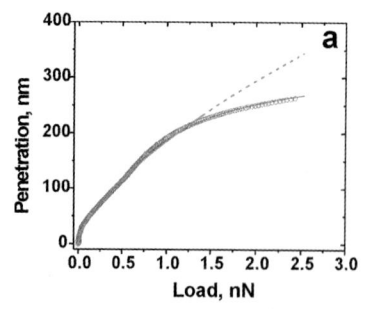

Fig. 4.35 A binary polymer brush layer on a silicon wafer was prepared from rubbery poly(methyl acrylate) (PMA) and glassy poly(styrene-co-2,3,4,5,6-pentafluorostyrene) (PSF) using the "grafting from" approach. A series of force–distance curves were collected before and after the experiments to confirm the deformation was elastic. (a) The experimental loading curve (circles), fitting with the tri-layered model (solid line, almost completely buried by experimental data points) and Hertzian model (dashed line). (b) Experimental depth distribution of the elastic modulus for the polymer brush layer (circles) and the best fitting with the tri-layered model (solid line) showing a slight increase in the elastic modulus near the surface and a sharp increase in proximity to a stiff substrate.

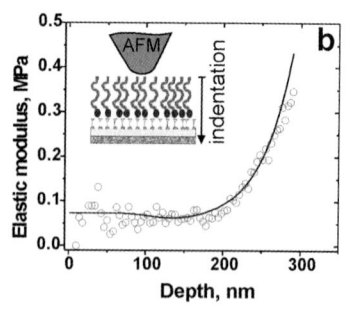

loss modulus and damping coefficients, not readily obtained with quasi-static indentation techniques [330]. Shear modulation AFM allows measurements of surface glass transition temperatures, and relaxation times.

Recently, a model for analyzing microindentation of layered polymer films has been developed and tested for various dry and wet systems with elastic moduli ranging from 0.05 to 3000 MPa and with layer thickness down to 20–50 nm. An example of the force–distance curve for the same binary polymer brush in a good solvent for one of the grafts is presented in Fig. 4.35. The best fitting of the loading curve and the depth profile was obtained by using a very low elastic modulus value of 0.07 MPa for initial deformations not exceeding 200 nm. The compliant region is followed by rising elastic resistance for larger indentation depth caused by the presence of the underlying solid substrate. Slightly increased elastic modulus near the surface related to minor brush layering was attributed to segregation of the tougher polymer near the surface.

4.4.4
Molecular Force Spectroscopy

Over the past 20 years, various single-molecule techniques have been developed to measure directly the mechanical properties of individual molecules and even individual bonds. These include optical and magnetic tweezers, glass microneedles and micropipettes, the biomembrane force probe, hydrodynamic methods and the AFM-based techniques. Basic principles of the techniques along with the accessible force, frequency and length ranges have been reviewed elsewhere [331, 332]. AFM is one of the most sensitive techniques as it allows measurements of nanonewton forces with piconewton precision in a dynamic range above 10 µs [333]. Single-molecule force spectroscopy has found many specific applications in biophysics, providing new insights into molecular motors [334], DNA base pairing [335–339], protein folding [340–342] and molecular recognition [343, 344]. More recently, this method has been successfully used to study general properties of macromolecules such as strength of molecular bonds [345–348], elasticity of polymer chains [349–351] and adsorption of polymer molecules on a solid substrate [352].

Pulling an adsorbed polymer chain involves two processes, i.e. partial desorption of the chain and extension of the desorbed section of the chain (Fig. 4.36a). The interplay of various forces, which is mediated by the surrounding environment, is particularly intricate in the case of charged polymers such as polyelectrolytes [353, 354] and polyampholytes [355]. For polyelectrolytes on oppositely charged substrates, it was found that the magnitude of desorption forces increases with the Debye screening length and with the linear charge density of the polyelectrolyte chains [354]. In agreement with theory, the force demonstrated a linear dependence for pulling of an overstretched chain in a strong field and a square-root dependence for stretching a coiled chain in a weaker field.

The chain-stretching experiments provide information about the contour length of adsorbed macromolecules (Fig. 4.36b) [356, 357]. However, these measurements depend strongly on the solvent quality and on the interaction with the substrate [352]. For poly(acrylic acid) molecules, it was shown that under both poor and good solvent conditions the contour length can be underestimated (Fig. 4.36c). In poor solvents, the monomer–monomer interactions are stronger than the monomer–substrate interactions. This leads to a collapsed polymer globule which desorbs prior to complete extension. By making experiments under good solvent conditions, e.g. at $pH > pK_a$ for polyacids, one reduces monomer–monomer attraction; however, one also reduces attraction to a similarly charged substrate. In very good solvents, chains may desorb prior complete extension. Under intermediate conditions, e.g. moderately good solvents and strong adsorption forces, one can fully extend the adsorbed chain and thus measure its contour length. One should note that the chain-extension forces may depend on the pulling rate akin to the unbinding forces of ligand–receptor bonds [358–360]. However, the force showed a weak dependence since desorption of an individual monomeric unit occurs much faster than the pulling rates accessible in the AFM experiment. Unlike the force, the measured chain length revealed a stronger dependence on the pulling rate, which was ascribed to the fact that the rate of the final unbinding becomes comparable to the experimental pulling speed [352].

In addition to studies of chain elasticity and adsorption forces, today there are increasing numbers of experimental and theoretical studies devoted to the investigation of covalent, metallic and coordinative chemical bonds [361–366]. Two different routes have been explored to measure covalent binding forces at the single-bond level. The binding forces between surface atoms of solid-state mate-

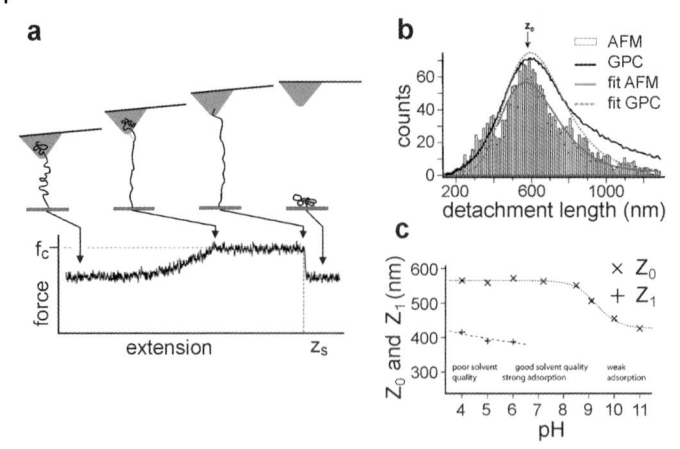

Fig. 4.36 (a) A force–distance curve is obtained in a single molecule desorption experiment with a surface-grafted polymer monolayer and a nonfunctionalized AFM tip. Upon contacting the tip with the substrate, surface-grafted polyelectrolyte chains are allowed to adsorb on the tip. Upon retracting the tip, one observes a nonlinear increase in the stretching force followed by a plateau of constant force due to successive desorption of adsorbed monomers, whereas the complete desorption of the entire polymer chain results in a sudden drop of the adhesive force. (b) Molecular weight–length distributions were obtained by AFM at pH 6 (histogram) and by GPC (black curve). The data fit resulted in maxima positions z_0 of 572 nm (AFM) and 592 nm (GPC). (c) Plot of the maximum position z_0 taken from the experimentally obtained length histograms against the pH of the buffer solution. z_1 corresponds to the second peak observed at lower pH.

rials have been determined by bringing the front atom of an AFM tip into close proximity of atoms of a planar surface and mapping the interatomic forces. In order to probe the short-range chemical-binding potentials, one usually uses stiff cantilevers with spring constants of the order of 50–100 N m^{-1} [367] and also applies force feedback mechanisms to reduce the noise level and increase force resolution [368, 369]. In the other type of experiment, covalent forces within molecules have been determined by stretching individual polymers, until either the polymer backbone or the covalent surface anchor ruptures as a consequence of the applied tensile force. In these experiments, the AFM tip and substrate surface are separated by tens of nanometers, where the long-range van der Waals and electrostatic forces no longer interfere with the short-range chemical forces investigated. In order to quantify the covalent binding forces within molecules, Bensimon et al. used a receding water meniscus to stretch and rupture individual DNA molecules, which were attached at both ends to a functionalized glass surface [370]. They estimated the forces required to rupture the backbone of double-stranded DNA by analyzing the deformations of the ruptured molecules and comparing them with the bending energy of DNA. Their value for the maximum tensile force was 476 pN, which is lower than the 800 pN applied in stretching experiments without rupturing [371]. Grandbois et al. studied two types of covalent anchors between the substrate and carboxyamylose molecules, i.e. aminosilane coupling to a glass surface and aminothiol coupling to a gold surface [372]. At the applied force-loading rate of 10 nN s^{-1}, the mean bond-rupture force was measured as 2.0±0.3 and 1.4±0.3 nN, respectively, indicating that the surface attachment was the weakest link in the chain.

4.4.5
Thermal Properties

In addition, to mechanical and adhesion properties, AFM can also be used to study the thermal properties of polymers on nanometer and even molecular scales. Various modes of AFM, such as friction force microscopy and shear modulation force microscopy [373–375], were used to measure local glass transitions and melting temperatures of polymers at the surface. It was observed that the surface T_g is significantly lower than the bulk value at a film thickness below 17 nm. Scanning local acceleration microscopy [376] and micro-thermal analysis [377] can measure the thermal properties of polymers with high lateral resolution but the measurements are affected by physical contact between the tip and sample. In one of the recent developments, AFM was operated in the non-contact mode with the cantilever oscillating above the sample at its resonance frequency [378]. By measuring the frequency as a function of temperature, one observes a plateau (glass transition) or a sharp kink (melting point) in the otherwise linear decreasing function. The frequency variation is attributed to a change in the viscoelastic properties. This method is especially useful for the characterization of heterogeneous polymers such as blends, block copolymers and composites. Figure 4.37 shows the resonance frequency curves for core–shell latex particles with a particle size of about 80 nm. They have two plateaus, which are consistent with the predicted glass transition temperatures of the core (T_g=49 °C) and shell (T_g=3 °C).

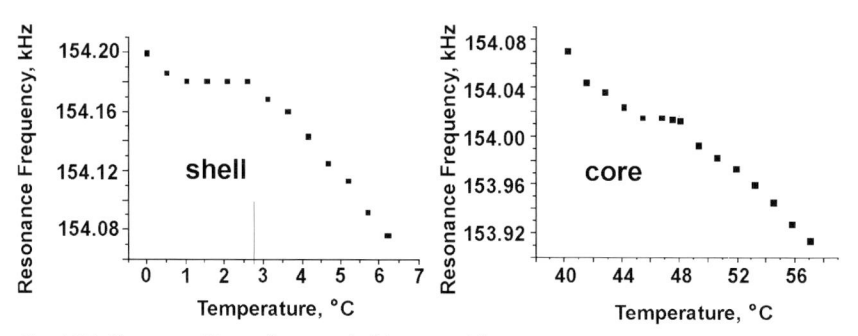

Fig. 4.37 Glass transitions of a core-shell latex particle composed of styrene, butyl acrylate and methacrylic acid: (a) the shell shows a transition between 1 and 3 °C, while (b) the core undergoes a transition between 45 and 48 °C.

4.5
Scanning Probe Lithography

The ability of AFM to create various surface patterns either through mechanical scratching-and-indentation or physical-and-chemical adsorption was promptly recognized in the first development of the technique [379–381]. In recent years, this area of AFM applications has been industrialized, offering devices for microelectronics, data storage and biotechnology. Of the industrial applications of scanning probe lithography, one should especially mention "dip-pen" nanolithography (DPN) [20] and "millipede" data-storage technology [22].

4.5.1
Writing with Molecular Inks

In DPN, molecules (typically alkanethiols) are delivered from the AFM tip to a solid substrate via capillary transport in a manner analogous to that of a dip pen (Fig. 4.38a). The technique combines ambient operation and resolution matching that of e-beam lithography and allows one to generate nanoscale arrays of a variety of materials ranging from oligonucleotides through proteins to conjugated polymers (Fig. 4.38b–e). Originally, the developed technology allowed lithography in a positive printing mode with 30-nm linewidth resolution [20]. The recent developments in DPN have pushed the limits to 10-nm linewidth and 5-nm lateral resolution. Furthermore, using an array of many thousands of AFM tips (currently up to 50 000) allows patterning of large areas in a parallel fashion within a reasonable time span [382]. Recently, parallel DPN has been used to generate protein arrays in a high-throughput manner at 14 000 dots in 10 min [383]. To achieve the ultimate goal of being able to deliver a different ink to each probe in a multipen array, new microfluidic arrays of addressable ink wells were developed.

Tip–substrate molecular transport is a complicated process influenced by many parameters. Fundamental studies in this area are focused on a better understanding of ink transport processes to optimize new ink–substrate combinations and thus improve the resolution. The greatest technological challenges include performing parallel-probe lithography, developing reliable tip–surface alignment protocols, maintaining tip feedback, patterning a wider variety of molecules and characterizing large arrays of nanostructures. Independent control of each probe tip is the next step and can be accomplished using piezoelectric, capacitive or thermoelectric actuation, whereby resistive heating of a multilayer cantilever causes bending of the probe. Using this approach, a range of complex patterns can be generated at high speed because the contact between each tip and the writing surface is independently controlled. DPN was also performed in the tapping mode, which is very suitable for deposition on soft surfaces [384]. A synthetic peptide was deposited through control of the tapping amplitude, yielding lines with a width of 70 nm. Tapping mode DPN permits not only deposition but also subsequent imaging and manipulation of biomolecules using a gentler mode of AFM more conducive to biological imaging.

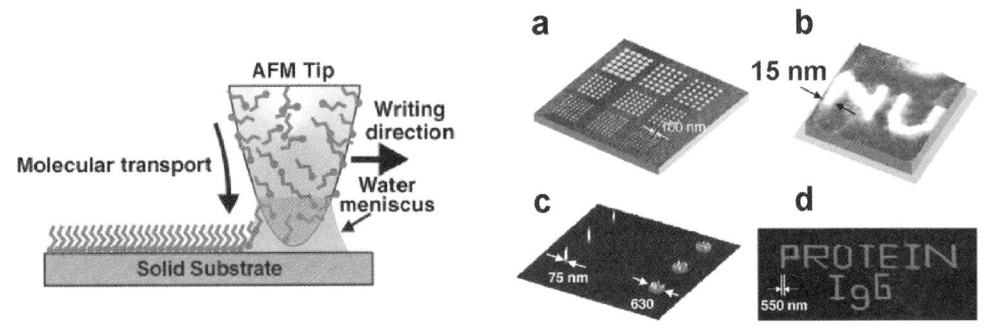

Fig. 4.38 (a) Schematic representation of the DPN process. A water meniscus forms between the AFM tip which is coated with "ink" molecules and the solid substrate. (b, c) Nanoscale dot arrays and letters written on a polycrystalline Au surface with mercaptohexadecanoic acid. (d) TM-AFM image of 25- and 13-nm gold nanoparticles hybridized to surface DNA templates generated with direct-write DPN. (e) Fluorescence image of direct-write DPN patterns of fluorescently labeled immunoglobulin G (IgG) on SiO_x.

In parallel, other approaches to the controlled delivery of chemicals in localized regions were developed. These include the fountain pen technique wherein liquid or gaseous materials are continuously delivered from a reservoir to a substrate through a cantilevered micropipette [21, 385]. This approach allows control of the ink supply and the mass of the probe, which is important for dynamic applications. Recently, the activity of an enzyme has been combined with the accuracy in positioning a tip in AFM to develop a novel technique referred to as enzyme-assisted nanolithography [386]. By use of other enzymes, this method will open the possibility of chemically modifying surfaces on a nanometer scale.

4.5.2
Thermomechanical Surface Patterning

There is an unquenchable demand for increasing storage capacity, which roughly doubles every 3–4 months. In traditional magnetic recording technologies, physical limitations of the bit size are rapidly being reached. At IBM, a novel AFM-based technology for ultra-high-density storage devices is being developed by using a thermal mechanical read–write strategy [387]. The so-called Millipede Project uses a heated cantilever to produce a nanoscopic indentation in a thin polymer film which represents a bit of information (Fig. 4.39). To record data, a tip is heated to over 400 °C, while the reading process is accomplished by operating the tips at slightly lower temperatures. The recording speed is greatly increased by employing an array of many cantilevers with each cantilever individually addressable in a parallel fashion – hence the name Millipede. This technology permits data storage densities of up to 1 Tbit in^{-2}, high recording rates and the possibility to read back and erase data [388].

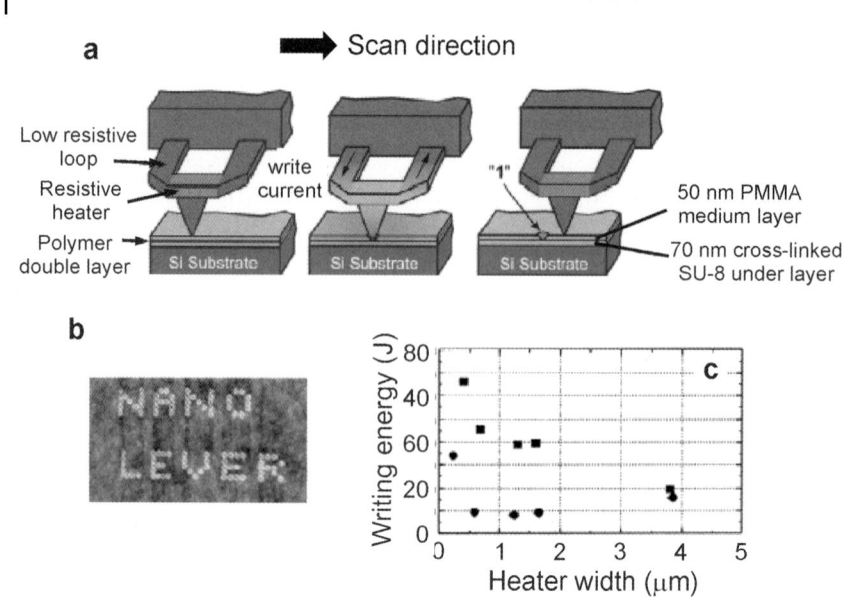

Fig. 4.39 Principle of thermomechanical writing. A thin, writable PMMA layer is deposited on top of an Si substrate separated by a cross-linked film of epoxy photoresist. A combination of tip heating to soften the polymer and the pressure exerted by the tip on the polymer causes the tip to sink into the polymer and write a bit. (a) Thermomechanical bit-writing experiment in a thin polymer. The pitch between bits is 70 nm, corresponding to a 2-bit density of 150 Gbit in^{-1}. (b) Energy needed to write a bit for cantilevers having different heater widths, with (circles) and without (squares) thermal constriction. A writing energy of less than 10 nJ was achieved.

Fundamental studies in this area are focused on understanding the heat transport mechanisms during writing and reading, resulting in improvements in cantilever design for increased speed, sensitivity and reduced power consumption in both writing and reading operation [389, 390]. The present cantilever designs operate with an array data rate 35 Mbit s^{-1} at a power of 330 mW and can operate at < 100 mW. New cantilevers have also been developed that include a heater for writing and a piezoelectric sensor for reading [391], along with carbon nanotube tips that allow indentation sizes of 22 nm in diameter [392]. However, problems still remain with the polymeric recording layer. Currently, little is known about the behavior, stability of the nanometer-sized holes in thin films and the underlying principles of the writing and erasing process. All of these fundamental materials questions are critically important since a useful operating device would require the ability to read and write repeatedly. Using living radical polymerization chemistry, Hawker and coworkers developed a novel polymeric recording layer which contains thermally reversible cross-linkages [393].

The other mechanical AFM-based lithography techniques involve scratching of a sample with a sharp AFM tip. These include wearing, shaving and grafting

[394]. In wearing, sample molecules are removed by the AFM tip through repeated scanning along the same area at low force. In shaving, adsorbate molecules are displaced by the tip during one scan at higher forces exceeding the displacement threshold. In grafting, AFM tips are also used to shave molecules that concurrently become replaced with other molecules available in a surrounding medium. In these studies, a self-assembled monolayer is an excellent model system which allows systematic studies of displacement thresholds, self-healing of shaved areas and heterogeneous surface patterning.

References

1 Binnig, G., Quate, C. F., Gerber, Ch. *Phys. Rev. Lett.* **1986**, *56*, 930.
2 Vancso, G. J., Hillborg, H., Schönherr, H. *Adv. Polym. Sci.* **2005**, *182*, 55.
3 Mehtani, D., Lee, N., Hartschuh, R. D., Kisliuk, A., Foster, M. D., Sokolov, A. P., Maguire, J. Scanning nano-Raman spectroscopy of silicon and other semiconducting materials, Chapter 6 in *Tip Enhancements*, Eds. Kawata, S., Shalaev, V. M., Elsevier, Amsterdam, **2006**.
4 Anderson, N., Hartschuh, A., Novotny, L. *Mater. Today* **2005**, *8*, 50.
5 Rief, M., Gautel, M., Oesterhelt, F., Fernandez, J. M., Gaub, H. E. *Science* **1997**, *276*, 1109.
6 Winfree, E., Liu, F., Wenzler, L. A., Seeman, N. C. *Nature* **1998**, *394*, 539.
7 Meadows, P. Y., Bemis, J. E., Al-Maawali, S., Walker, G. C. *ACS Symp. Ser.* **2005**, *897*, 148.
8 Mahaffy, R. E., Park, S., Gerde, E., Käs, J., Shih, C. K. *Biophys J.* **2004**, *86*, 1777.
9 Kappl, M., Butt, H.-J. *Part. Part. Syst. Charact.* **2002**, *19*, 129.
10 Segeren, L. H. G. J., Siebum, B., Karssenberg, F. G., Van den Berg, J. W. A., Vancso, G. J. *J. Adhesion Sci. Technol.* **2002**, *16*, 793.
11 Grierson, D. S., Flater, E. E., Carpick, R. W. *J. Adhesion Sci. Technol.* **2005**, *19*, 291.
12 Bhushan, B., Israelachvili, J. N., Landman, U. *Nature* **2002**, *374*, 607.
13 Bhushan, B. *Wear* **2005**, *259*, 1507.
14 Ebenstein, D. M., Wahl, K. J. *J. Colloid Interface Sci.* **2006**, *298*, 652.
15 Sato, F., Okui, H., Akiba, U., Suga, K., Fujihira, M. *Ultramicroscopy* **2003**, *97*, 303.
16 Bo, X.-Z., Rokhinson, L. P., Yin, H., Tsui, D. C., Sturm, J. C. *Appl. Phys. Lett.* **2002**, *81*, 3263.
17 Legrand, B., Deresmes, D., Stievenard, D. *J. Vac. Sci. Technol. B* **2002**, *20*, 862.
18 Lyuksyutov, S. F., Vaia, R. A., Paramonov, P. B., Juhl, S., Waterhouse, L., Ralich, R. M., Sigalov, G., Sancaktar, E. *Nat. Mater.* **2003**, *2*, 468.
19 Cavallini, M., Biscarini, F., Leon, S., Zerbetto, F., Bottari, G., Leigh, D. A. *Science* **2003**, *299*, 531.
20 Piner, R. D., Zhu, J., Xu, F., Hong, S., Mirkin, C. A. *Science* **1999**, *283*, 661.
21 Deladi, A., Tas, N. R., Berenschot, J. W., Krijnen, G. J. M., de Boer, M. J., de Boer, J. H., Peter, M., Elsenpoek, M. C. *Appl. Phys. Lett.* **2004**, *85*, 5361.
22 Vettiger, P., Cross, G., Despont, M., Drechsler, U., Dürig, U., Gotsmann, B., Häberle, W., Lantz, M. A., Rothuizen, H. E., Stutz, R., Binnig, G. K. *IEEE Trans. Nanotechnol.* **2002**, *1*, 39.
23 Samori, P., Rabe, J. P. *J. Phys. Condens. Matter* **2002**, *14*, 9955.
24 Barner, J., Mallwitz, F., Shu, L., Schlüter, A. D., Rabe, J. P. *Angew. Chem. Int. Ed.* **2003**, *42*, 1932.
25 Luscher, S., Fuhrer, A., Held, R., Heinzel, T., Ensslin, K., Bichler, M., Wegscheider, W. *Microelectron. J.* **2002**, *33*, 319.
26 Rivetti, C., Guthold, M., Bustamante, C. *J. Mol. Biol.* **1996**, *264*, 919.
27 Scheuring, S., Fotiadis, D., Möller, C., Muller, S. A., Engel, A., Müller, D. J. *Single Mol.* **2001**, *2*, 59.
28 Sheiko, S. S., Möller M. *Chem. Rev.* **2001**, *101*, 4099.

29 Kumaki, J., Hashimoto, T. *J. Am. Chem. Soc.* **2003**, *125*, 4907.
30 Minko, S., Roiter, Y. *Curr. Opin. Colloid Interface Sci.* **2005**, *10*, 9.
31 Ito, S., Aoki, H. *Adv. Polym. Sci.* **2005**, *182*, 131.
32 Hörber, J. K. H., Miles, M. J. *Science* **2003**, *302*, 1002.
33 Frederix, P. L. T. M., Akiyama, T., Staufer, U., Gerber, Ch., Fotiadis, D., Müller, D. J., Engel, A. *Curr. Opin. Chem. Biol.* **2003**, *7*, 641.
34 Hogan, J. *Nature* **2006**, *440*, 14.
35 Janovjak, H., Kedrov, A., Cisneros, D. A., Sapra, K. T., Struckmeier, J., Müller, D. J. *Neurobiol. Aging* **2006**, *27*, 546.
36 Israelachvili, J. N. *Intermolecular and Surface Forces*, 2nd edn, Academic Press, New York, **1992**.
37 Klinov, D., Magonov, S. N. *Appl. Phys. Lett.* **2004**, *84*, 2697.
38 Martin, Y., Williams, C. C., Wickramasinghe, H. K. *J. Appl. Phys.* **1987**, *61*, 4723.
39 Albrecht, T. R. *J. Appl. Phys.* **1991**, *69*, 668.
40 Durig, U., *J. Appl. Phys.* **1992**, *72*, 1778.
41 Giessibl, F. J. *Materials Today* **2005**, *8*, 32.
42 Sheiko S. S. *Adv. Polym. Sci.* **1999**, *151*, 61.
43 Cleveland, J. P., Manne, S., Bocek, D., Hansma, P. K. *Rev. Sci. Instrum.* **1993**, *64*, 403.
44 Senden, T. J., Ducker, W. A. *Langmuir* **1994**, *10*, 1003.
45 Sader, J. E., Chon, J. W. M., Mulvaney, P. *Rev. Sci. Instrum.* **1999**, *70*, 3967.
46 Gibson, C. T., Watson, G. S., Myhra, S. *Nanotechnology* **1996**, *7*, 259.
47 Torii, A., Sasaki, M., Hane, K., Okuma, S. *Meas. Sci. Technol.* **1996**, *7*, 179.
48 Hutter, J. L., Bechhoefer, J. *Rev. Sci. Instrum.* **1993**, *64*, 1868.
49 Gotszalk, T., Grabiec, P., Rangelow Ivo, W. *Ultramicroscopy* **2003**, *97*, 385.
50 Cumpson, P. J., Hedley, J., Clifford, C. A. *J. Vac. Sci. Technol. B* **2005**, *23*, 1992.
51 Jericho, S. K., Jericho, M. H. *Rev. Sci. Instrum.* **2002**, *73*, 2483.
52 Burnham, N. A., Chen, X., Hodges, C. S., Matei, G. A., Thoreson, E. J., Roberts, C. J., Davies, M. C., Tendler, S. J. B. *Nanotechnology* **2003**, *14*, 1.
53 Asay, D. A., Kim, S. H. *Rev. Sci. Instrum.* **2006**, *77*, 043903.
54 Ballato, A. *Ceram. Trans.* **2003**, *136*, 533.
55 Green, C. P., Sader, J. E. *J. Appl. Phys.* **2002**, *92*, 6262.
56 Sarid, D. *Scanning Force Microscopy: with Applications to Electric, Magnetic and Atomic Forces*, Oxford University Press, New York, **1991**.
57 Magonov, S. N. *Surface Analysis with STM and AFM: Experimental and Theoretical Aspects of Image Analysis*, VCH, Weinheim, **1996**.
58 Meyer, E., Hug, H. J., Bennewitz, R. *Scanning Probe Microscopy: the Lab on a Tip*, Springer, Berlin, **2003**.
59 Giessibl, F. J. *Rev. Mod. Phys.* **2003**, *75*, 949.
60 Hofer, W. A., Foster, A. S., Shluger, A. L. *Rev. Mod. Phys.* **2003**, *75*, 1287.
61 Paige, M. F. *Polymer* **2003**, *44*, 6345.
62 Opdahl, A., Koffas, T. S., Amitay-Sadovsky, E., Kim, J., Somorjai, G. A. *J. Phys.: Condens. Matter* **2004**, *16*, R659.
63 Li, J., Liang, W., Meyers, G. F., Heeschen, W. A. *Polym. News* **2004**, *29*, 335.
64 Achalla, P., McCormick, J., Hodge, T., Moreland, C., Esnault, P., Karim, A., Raghavan, D. *J. Polym. Sci., Part B* **2006**, *44*, 492.
65 Su, C., Huang, L., Kjoller, K., Babcock, K. *Ultramicroscopy* **2003**, *97*, 135.
66 Stark, M., Stark, R. W., Heckl, W. M., Guckenberger, R. *Proc. Natl. Acad. Sci. USA* **2002**, *99*, 8473.
67 Anczykowski, B., Gotsmann, B., Fuchs, H., Cleveland, J. P., Elings, V. B. *Appl. Surf. Sci.* **1999**, *140*, 376.
68 Balantekin, M., Atalar, A. *Phys. Rev. B* **2003**, *67*, 193404.
69 Rodriguez, T. R., Garcia, R. *Appl. Phys. Lett.* **2003**, *82*, 4821.
70 Ashby, P. D., Lieber, C. M. *J. Am. Chem. Soc.* **2005**, *127*, 6814.
71 Martin, Y., Abraham, D. W., Wickramasinghe, H. K. *Appl. Phys. Lett.* **1988**, *52*, 1103.
72 Colchero, J., Gil, A., Baró, A. M. *Phys. Rev. B* **2001**, *64*, 245403.
73 Viswanatha, R., Heaney, M. B. *Phys. Rev. Lett.* **1995**, *75*, 4433.

74 Bachtold, A., Fuhrer, M. S., Plyasunov, S., Forero, M., Anderson, E. H., Zettl, A., McEuen, P. L. *Phys. Rev. Lett.* **2000**, *84*, 6082.
75 Lei, C. H., Das, A., Elliott, M., Macdonald, J. E. *Appl. Phys. Lett.* **2003**, *83*, 482.
76 Yerina, N., Magonov, S. N. *Rubb. Chem. Technol.* **2003**, *76*, 846.
77 Krayev, A. V., Talroze, R. V. *Polymer* **2004**, *45*, 8195.
78 Brakenhoff, G. J., Blom, P., Barends, P. *J. Microsc.* **1979**, *117*, 219.
79 Ichimura, T., Hayazawa, N., Hashimoto, M., Inouye, Y., Kawata, S. *Phys. Rev. Lett.* **2004**, *92*, 220801.
80 Anderson, N., Hartschuh, A., Cronin, S., Novotny, L. *J. Am. Chem. Soc.* **2005**, *127*, 2533.
81 Mehtani, D., Lee, N., Hartschuh, R. D., Kisliuk, A., Foster, M. D., Sokolov, A. P., Maguire, J. F. *J. Raman Spectrosc.* **2005**, *36*, 1068.
82 Besenbacher, F., Laegsgaard, E., Stensgaard, E. *Mater. Today* **2005**, *8*, 26.
83 Yang, J. L., Despont, M., Drechsler, U., Hoogenboom, B. W., Frederix, P. L. T. M., Martin, S., Engel, A., Vettiger, P., Hug, H. J. *Appl. Phys. Lett.* **2005**, *86*, 134101.
84 Sulchek, T. *Appl. Phys. Lett.* **2000**, *76*, 1473.
85 Walters, D. A. *Rev. Sci. Instrum.* **1996**, *67*, 3583.
86 Ando, T. *Proc. Natl. Acad. Sci. USA* **2001**, *98*, 12468.
87 Humphris, A. D. L., Hobbs, J. K., Miles, M. J. *Appl. Phys. Lett.* **2003**, *83*, 6.
88 Humphris, A. D. L., Miles, M. J., Hobbs, J. K. *Appl. Phys. Lett.* **2005**, *86*, 034106.
89 Lodge, T. P. *Phys. Rev. Lett.* **1999**, *83*, 3218.
90 Sheiko, S. S., daSilva, M., Shirvaniants, D. G., LaRue, I., Prokhorova, S. A., Beers, K., Matyjaszewski, K. *J. Am. Chem. Soc.* **2003**, *125*, 6725.
91 Matyjaszewski, K., Qin, S. H., Boyce, J. R., Shirvanyants, D., Sheiko, S. S. *Macromolecules* **2003**, *36*, 1843.
92 Kreutzer, G., Ternat, C., Nguyen, T. Q., Plummer, C. J. G., Månson, J.-A. E., Castelletto, V., Hamley, I. W., Sun, F., Sheiko, S. S., Klok, H.-A. *Macromolecules* **2006**, *39*, 4507.
93 Ivanov, D., Magonov, S. Atomic force microscopy studies of semicrystalline polymers at variable temperature, in *Polymer Crystallization, Lecture Notes in Physics*, Eds. Reiter, G., Sommer, J. U., Springer, Berlin, **2003**, pp. 98–129.
94 Xu, H., Shirvaniants, D., Beers, K., Matyjaszewski, K., Rubinstein, M., Sheiko, S. S. *Phys. Rev. Lett.* **2004**, *93*, 206103.
95 Sheiko, S. S., Sun, F., Randal, A., Shirvaniants, D., Matyjaszewski, K., Rubinstein, M. *Nature* **2006**, *440*, 191.
96 Rivetti, C., Guthold, M., Bustamante, C. *J. Mol. Biol.* **1996**, *264*, 919.
97 Fang, Y., Yang, J. *J. Phys. Chem. B* **1997**, *101*, 441.
98 Erie, D. A., Bustamante, C. *Science* **1995**, *269*, 989.
99 Müller, D. J., Schabert, F. A., Muldt, G., Engel, A. *Biophys. J.* **1995**, *68*, 1681.
100 Yoshimura, H. *Adv. Biophys.* **1996**, *34*, 93.
101 Maeda, H. *Langmuir* **1997**, *13*, 4150.
102 Ng, L., Grodzinsky, A. J., Patwari, P., Sandy, J., Plaas, A., Ortiz, C. *J. Struct. Biol.* **2003**, *143*, 242.
103 Camesano, T. A., Wilkinson, K. J. *Biomacromolecules* **2001**, *2*, 1184.
104 Wadu-Mesthrige, K., Pati, B., McClain, M., Liu, G.-Y. *Langmuir* **1996**, *12*, 3511.
105 Brown, H. G., Hoh, J. H. *Biochemistry* **1997**, *36*, 15035.
106 Heller, E. J., Cromie, M. F., Lutz, C. P., Eigler, D. M. *Nature* **1994**, *369*, 464.
107 Bopp, M. A., Meixner, A. J., Tarrach, G., Zschokke-Graenacher, I., Novotny, L. *Chem. Phys. Lett.* **1996**, *263*, 721.
108 Gimzewski, J. K., Jung, T. A., Cuberes, M. T., Schlittler, R. R. *Surf. Sci.* **1997**, *386*, 101.
109 Swartzentruber, B. S. *Phys. Rev. Lett.* **1996**, *76*, 459.
110 Ala-Nissila, T., Ferrando, R., Ying, S. C. *Adv. Phys.* **2002**, *51*, 949.
111 Hashimoto, T., Okumura, A., Tanabe, D. *Macromolecules* **2003**, *36*, 7324.
112 Minko, S., Kiriy, A., Gorodyska, G., Stamm, M. *J. Am. Chem. Soc.* **2002**, *124*, 3218.
113 Prokhorova, S. A., Sheiko, S. S., Möller, M., Ahn, C.-H., Percec, V. *Makromol. Rapid Commun.* **1998**, *19*, 359.

114 Percec, V., Ahn, C.-H., Ungar, G., Yeardley, D. J. P., Möller, M., Sheiko, S. S. *Nature* **1998**, *391*, 161.
115 Percec, V., Dulcey, A. E., Balagurusamy, V. S. K., Miura, Y., Smidrkal, J., Peterca, M., Nummelin, S., Edlund, U., Hudson, S. D., Heiney, P. A., Duan, H., Magonov, S. N., Vinogradov, S. A. *Nature* **2004**, *430*, 764.
116 Zhuang, W., Ecker, C., Metselaar, G. A., Rowan, A. E., Nolte, R. J. M., Samori, P., Rabe, J. P. *Macromolecules* **2005**, *38*, 473.
117 Zhang, A., Barner, J., Goessl, I., Rabe, J. P., Schlüter, A. D. *Angew. Chem. Int. Ed.* **2004**, *43*, 5185.
118 Jaeckel, F., Ai, M., Wu, J., Müllen, K., Rabe, J. P. *J. Am. Chem. Soc.* **2005**, *127*, 14580.
119 Ornatska, M., Bergman, K. N., Rybak, B., Peleshanko, S., Tsukruk, V. V. *Angew. Chem. Int. Ed.* **2004**, *43*, 5246.
120 Genson, K. L., Holzmüller, J., Leshchiner, I., Agina, E., Boiko, N., Shibaev, V. P., Tsukruk, V. V. *Macromolecules* **2005**, *38*, 8028.
121 Ko, H., Jiang, C., Tsukruk, V. V. *Chem. Mater.* **2005**, *17*, 5489.
122 Koch, N., Heimel, G., Wu, J., Zojer, E., Johnson, R. L., Bredas, J.-L., Müllen, K., Rabe, J. P. *Chem. Phys. Lett.* **2005**, *413*, 390.
123 Drager, A. S., Zangmeister, R. A. P., Armstrong, Neal R., O'Brien, D. F. *J. Am. Chem. Soc.* **2001**, *123*, 3595.
124 Li, J., Piehler, L. T., Qin, D., Baker, J. R. Jr., Tomalia, D. A., Meier, D. J. *Langmuir* **2000**, *16*, 5613.
125 Viville, P., Leclere, P., Deffieux, A., Schappacher, M., Bernard, J., Borsali, R., Bredas, J.-L., Lazzaroni, R. *Polymer* **2004**, *45*, 1833.
126 Ornatska, M., Peleshanko, S., Rybak, B., Holzmüller, J., Tsukruk, V. V. *Adv. Mater.* **2004**, *16*, 2206.
127 Mourran, A., Tartsch, B., Gallyamov, M., Magonov, S., Lambreva, D., Ostrovskii, B. I., Dolbnya, I. P., de Jeu, W. H., Möller, M. *Langmuir* **2005**, *21*, 2308.
128 Yoshida, M., Fresco, Z. M., Ohnishi, S., Fréchet, J. M. J. *Macromolecules* **2005**, *38*, 334.
129 Watanabo, S., Regen, S. L. *J. Am. Chem. Soc.* **1994**, *116*, 8855.
130 Tsukruk, V. V. *Adv. Mater.* **1998**, *10*, 253.
131 Hierlemann, A., Campbell, J. K., Baker, L. A., Crooks, R. M., Ricco, A. J. *J. Am. Chem. Soc.* **1998**, *120*, 5323.
132 Saville, P. M., White, J. W., Hawker, C. J., Wooley, K. L., Fréchet, J. M. J. *J. Phys. Chem.* **1993**, *97*, 293.
133 Iyer, J., Hamond, P. T. *Langmuir* **1999**, *15*, 1299.
134 Schenning, A. P. H. J., Elissen-Román, C., Weener, J.-W., Baars, M. W. P. L., van der Gaast, S. J., Meijer, E. W. *J. Am. Chem. Soc.* **1998**, *120*, 8199.
135 Sheiko, S. S., Buzin, A. I., Muzafarov, A. M., Rebrov, E. A., Getmanova, E. V. *Langmuir* **1998**, *14*, 7468.
136 Wang, P. W., Liu, Y. J., Devadoss, C., Bharathi, P., Moore, J. S. *Adv. Mater.* **1996**, *8*, 237.
137 Ponomarenko, S. A., Boiko, N. I., Shibaev, V. P., Magonov, S. *Langmuir* **2000**, *16*, 5487.
138 Sheiko, S. S., Gauthier, M., Möller, M. *Macromolecules* **1997**, *30*, 2343.
139 Sidorenko, A., Zhai, X. W., Peleshanko, S., Greco, A., Shevchenko, V. V., Tsukruk, V. V. *Langmuir* **2001**, *17*, 5924.
140 Dziezok, P., Sheiko, S. S., Fischer, K., Schmidt, M., Möller, M. *Angew. Chem. Int. Ed. Engl.* **1997**, *109*, 2812.
141 Sheiko, S. S., Prokhorova, S. A., Beers, K. L., Matyjaszewski, K., Potemkin, I. I., Khokhlov, A. R., Möller, M. *Macromolecules* **2001**, *34*, 8354.
142 Percec, V., Ahn, C.-H., Cho, W.-D., Jamieson, A. M., Kim, J., Leman, T., Schmidt, M., Gerle, M., Möller, M., Prokhorova, S. A., Sheiko, S. S., Cheng, S. Z. D., Zhang, A., Ungar, G., Yeardley, D. J. P. *J. Am. Chem. Soc.* **1998**, *120*, 8619.
143 Bo, Z., Zhang, C., Severin, N., Rabe, J. P., Schlüter, A. D. *Macromolecules* **2000**, *33*, 2688.
144 Yin, R., Zhu, Y., Tomalia, D. A. *J. Am. Chem. Soc.* **1998**, *120*, 2678–2679.
145 Bao, Z., Amundson, K. R., Lovinger, A. J. *Macromolecules* **1998**, *31*, 8647.
146 Loos, J. *Adv. Mater.* **2005**, *17*, 1821.
147 Perkins, T. T., Smith, D. E., Chu, S. *Science* **1994**, *264*, 819.
148 Perkins, T. T., Quake, S. R., Smith, D. E., Chu, S. *Science* **1994**, *264*, 822.

149 Kas, J., Strey, H., Sackmann, E. *Nature* **1994**, *368*, 226.
150 Kas, J., Strey, H., Tang, J.X., Finger, D., Ezzell, R., Sackmann, E., Janmey, P.A. *Biophys. J.* **1996**, *70*, 609.
151 Stancik, E.J., Widenbrant, M.J.O., Laschitsch, A.T., Vermant, J., Fuller, G.G. *Langmuir* **2002**, *18*, 4372.
152 Michalet, X. *Science* **1997**, *277*, 1518.
153 Maier, B., Rädler, J.O. *Phys. Rev. Lett.* **1999**, *82*, 1911.
154 Maier, B., Rädler, J.O. *Macromolecules* **2000**, *33*, 7185.
155 Humphrey, D., Duggan, C., Saha, D., Smith, D., Käs, J. *Nature* **2002**, *416*, 413.
156 Jeon, S., Bae, S.C., Granick, S. *Macromolecules* **2002**, *35*, 8401.
157 Nykypanchuk, D.N., Strey, H.H., Hoagland, D.A. *Science* **2002**, *297*, 987.
158 Percec, V., Glodde, M., Johansson, G., Balagurusamy, V.S.K., Heiney, P.A. *Angew. Chem. Int. Ed.* **2003**, *42*, 4338.
159 Gillies, E.R., Jonsson, T.B., Fréchet, J.M.J. *J. Am. Chem Soc.* **2004**, *126*, 11936.
160 Stupp, S.I., LeBonheur, V., Walker, K., Li, L.S., Huggins, K.E., Keser, M., Amstutz, A. *Science* **1997**, *276*, 384.
161 Park, S., Lim, J.-H., Chung, S.-W., Mirkin, C.A. *Science* **2004**, *303*, 348.
162 Zimmerman, S.C., Wendland, M.S., Rakow, N.A., Zharov, I., Suslick, K.S. *Nature* **2002**, *418*, 399.
163 Pochan, D.J., Chen, Z., Cui, H., Hales, K., Qi, K., Wooley, K.L. *Science* **2004**, *306*, 94.
164 Iatrou, H., Mays, J.W., Hadjichristidis, N. *Macromolecules* **1998**, *31*, 6697.
165 Zhang A.F., Barner J., Goessl, I., Rabe, J.P., Schlüter, A.D. *Angew. Chem. Int. Ed.* **2004**, *43*, 5185.
166 Newkome, G.R., Mishra, A., Moorefield, C.N. *Polym. Mater. Sci. Eng.* **2001**, *84*, 1.
167 Ikkala, O., ten Brinke, G. *Science* **2002**, *295*, 2407.
168 Yan, D., Zhou, Y., Hou, J. *Science* **2004**, *303*, 65.
169 Hembacher, S., Giessibl, F.J., Mannhart, J., Quate, C.F. *Proc. Natl. Acad. Sci. USA* **2003**, *100*, 12539.
170 Giessibl, F.J., Hembacher, S., Bielefeldt, H., Mannhart, J. *Science* **2000**, *289*, 422.
171 Ruokolainen, J., Mäkinen, R., Torkkeli, M., Mäkelä, T., Serimaa, R., ten Brinke, G., Ikkala, O. *Science* **1998**, *280*, 557.
172 Zhu, X., Tartsch, B., Beginn, U., Möller, M. *Chem. Eur. J.* **2004**, *10*, 3871.
173 Sheiko, S.S., Eckert, G., Muzafarov, A.M., Räder, H.J., Möller, M. *Macromol. Chem. Rapid Commun.* **1996**, *17*, 283.
174 Stahlberg, H., Müller, D.J., Suda, K., Fotiadis, D., Engel, A., Meier, T., Matthey, U., Dimroth, P. *EMBO Rep.* **2001**, *2*, 229.
175 Börner, H.G., Duran, D., Matyjaszewski, K., DaSilva, M., Sheiko, S.S. *Macromolecules* **2002**, *35*, 3387.
176 Lord, S.J., Sheiko, S.S., LaRue, I., Qin, S., Lee, H.-I., Matyjaszewski, K. *Macromolecules* **2004**, *37*, 4235.
177 Qin, S., Matyjaszewski, K., Xu, H., Sheiko, S.S. *Macromolecules* **2003**, *36*, 605.
178 Li, J., Gauthier, M., Teertstra, S.J., Hu, X., Sheiko, S.S. *Macromolecules* **2004**, *37*, 795.
179 Klok, H.-A., Vandermeulen, G.W.M., Nuhn, H., Rösler, A., Hamley, I.W., Castelletto, V., Xu, H., Sheiko, S.S. *Faraday Discuss.* **2005**, *128*, 29.
180 Börner, H.G., Beers, K.L., Matyjaszewski, K., Sheiko, S.S., Möller, M. *Macromolecules* **2001**, *34*, 4375.
181 Matyjaszewski, K., Qin, S., Boyce, J.R., Shirvanyants, D., Sheiko, S.S. *Macromolecules* **2003**, *36*, 1843.
182 Boyce, J.R., Shirvanyants, D., Sheiko, S.S., Ivanov, D.A., Qin, S.H., Börner, H.G., Matyjaszewski, K. *Langmuir* **2004**, *20*, 6005.
183 Rubinstein, M., Colby, R. *Polymer Physics*, Oxford University Press, New York, **2003**.
184 Wittmer, J.P., Meyer, H., Baschnagel, J., Johner, A., Obukhov, S., Mattioni, L., Müller, M., Semenov, A.N. *Phys. Rev. Lett.* **2004**, *93*, 147801.
185 Granick, S. *J. Polym. Sci. B* **2003**, *41*, 2755.
186 Stancik, E.J., Widenbrant, M.J.O., Laschitsch, A.T., Vermant, J., Fuller, G.G. *Langmuir* **2002**, *18*, 4372.
187 Michalet, X. *Science* **1997**, *277*, 1518.

188 Humphrey, D., Duggan, C., Saha, D., Smith, D., Käs, J. *Nature* **2002**, *416*, 413.
189 Jeon, S., Bae, S. C., Granick, S. *Macromolecules* **2002**, *35*, 8401.
190 Sukhishvili, S. A., Chen, Y., Müller, J. D., Gratton, E., Schweizer, K. S., Granick, S. *Nature* **2000**, *406*, 146.
191 Granick, S., Zhao, J., Xie, A. F., Bae, S. C. *Macromol. Symp.* **2003**, *201*, 89.
192 Ueda, M., Yoshikawa, K. *Phys. Rev. Lett.* **1996**, *77*, 2133.
193 Dobrynin, A. V., Rubinstein, M., Obukhov, S. P. *Macromolecules* **1996**, *29*, 2974.
194 Yoshikawa, K., Takahashi, M., Vasilevskaya, V. V., Khokhlov, A. R. *Phys. Rev. Lett.* **1996**, *76*, 3029.
195 Roiter, Y., Minko, S. *J. Am. Chem. Soc.* **2005**, *127*, 15688.
196 Puterman, M., König, J. L., Lando, J. B. *J. Macromol. Sci., Phys.*, **1979**, *B16*, 89.
197 Puterman, M., Garcia, E., Lando, J. B. *J. Macromol. Sci., Phys.*, **1979**, *B16*, 117.
198 Qin, S., Matyjaszewski, K., Xu, H., Sheiko S. S. *Macromolecules* **2003**, *36*, 605.
199 Granick, S., Bae, S. C. *Nature* **2006**, *440*, 160.
200 Hoffman, J., Davis, G., Lauritzen, J. The rate of crystallization of linear polymers with chain folding, in *Treatise on Solid State Chemistry*, Vol. 3, Ed. Hannay, N. B., Plenum Press, New York, **1976**, pp. 497–614.
201 Simmons, A. H., Michal, C. A., Jelinski, L. W. *Science* **1996**, *271*, 84.
202 Vollrath, F., Knight, D. P. *Nature* **2001**, *410*, 541.
203 Jin, H.-J., Kaplan, D. L. *Nature* **2003**, *424*, 1057.
204 Termonia, Y. *Biomacromolecules* **2004**, *5*, 2404.
205 Wunderlich, B. *Macromolecular Physics. Vol. 1: Crystal Structure, Morphology, Defects*, Academic Press, New York, **1973**.
206 Bassett, D. C. *CRC Crit. Rev. Solid State Mater. Sci.* **1984**, *12*, 97.
207 Mandelkern, L. *Faraday Discuss.* **1979**, *68*, 310.
208 Gautam, S., Balijepalli, S., Rutledge, G. C. *Macromolecules* **2000**, *33*, 9136.

209 Strobl, G. *Prog. Polym. Sci.* **2006**, *31*, 398.
210 Heck, B., Strobl, G., Grasruck, M. *Eur. Phys. J. E* **2003**, *11*, 117.
211 Ivanov, D. A., Hocquet, S., Dosière, M., Koch, M. *Eur. Phys. J. E* **2004**, *13*, 363.
212 Patel, D., Bassett, D. C. *Proc. R. Soc. London, Ser. A* **1994**, *445*, 577.
213 Ivanov, D., Amalou, Z., Magonov, S. N. *Macromolecules* **2001**, *34*, 8944.
214 Mareau, V. H., Prud'homme, R. E. *Macromolecules* **2005**, *38*, 398.
215 Tian, M., Loos, J. *J. Polym. Sci. Phys.* **2001**, *39*, 763.
216 Geil, P. H. *Polymer Single Crystals*, Wiley, New York, **1963**.
217 Ivanov, D. A., Pop, T., Yoon, D., Jonas, A. *Macromolecules* **2002**, *35*, 9813.
218 Magonov, S. N., Yerina, N. A., Ungar, G., Reneker, D. H., Ivanov, D. A. *Macromolecules* **2003**, *36*, 5637.
219 Le Fevere De Ten Hove, C., Penelle, J., Ivanov, D. A., Jonas, A. M. *Nat. Mater.* **2004**, *3*, 33.
220 Krejchi, M. T. *Science* **1994**, *265*, 1427.
221 Rathore, O., Sogah, D. Y. *J. Am. Chem. Soc.* **2001**, *123*, 5231.
222 Hauser, G., Schmidtke, J., Strobl, G. *Macromolecules* **1998**, *31*, 6250.
223 Keller, A., Hikosaka, M., Rastogi, S., Toda, A., Barham, P., Goldbeck-Wood, G. *J. Mater. Sci.* **1994**, *29*, 2579.
224 Imai, M., Kaji, K., Kanaya, T., Sakai, Y. *Phys. Rev. B* **1995**, *52*, 12696.
225 Lauritzen, J. I., Hoffman, J. D. *J. Appl. Phys.* **1973**, *44*, 4340.
226 Muthukumar, M. *Eur. Phys. J. E* **2000**, *3*, 199.
227 Lei, Y., Chan, C., Li, J., Ng, K., Wang, Y., Jiang, Y. *Macromolecules* **2002**, *35*, 6751.
228 Lei, Y., Chan, C., Wang, Y., Ng, K., Jiang, Y., Li, L. *Polymer* **2003**, *44*, 4673.
229 Basire, C., Ivanov, D. A. *Phys. Rev. Lett.* **2000**, *85*, 5587.
230 Hobbs, J. Following processes in synthetic polymers with scanning probe microscopy, in *Applications of Scanned Probe Microscopy to Polymers*, Ed. Batteas, J. D., Oxford University Press, Oxford, **2005**, pp. 194–206.
231 Hugel, T., Strobl, G., Thomann, R. *Acta Polym.* **1999**, *50*, 214.

232 Magonov, S. N., Godovsky, Y. K. *Am. Lab.* **1999**, *31*, 52.
233 Gearba, R. I., Dubreuil, N., Anokhin, D. V., Godovsky, Y. K., Ruan, J.-J., Thierry, A., Lotz, B., Ivanov, D. A. *Macromolecules* **2006**, *39*, 978.
234 Gearba, R. I., Anokhin, D. V., Bondar, A. I., Godovsky, Y. K., Papkov, V. S., Makarova, N. N., Magonov, S. N., Bras, W., Koch, M. H. J., Masin, F., Goderis, B., Ivanov, D. A. *Macromolecules* **2006**, *39*, 988.
235 Liu, X., Zhang, Y., Goswami, D. K., Okasinski, J. S., Salaita, K., Sun, P., Bedzyk, M. J., Mirkin, C. A. *Science* **2005**, *307*, 1763.
236 Geoghegan, M., Krausch, G. *Prog. Polym. Sci.* **2003**, *28*, 261.
237 Stamm, M. *Adv. Polym. Sci.* **1992**, *100*, 357.
238 Wang, H., Composto, R. J. *Phys. Rev. E* **2000**, *61*, 1659.
239 Dickey, M. D., Gupta, S., Leach, K. A., Collister, E., Willson, C. G., Russell, T. P. *Langmuir* **2006**, *22*, 4315.
240 Morariu, M. D., Voicu, N. E., Schaeffer, E., Lin, Z., Russell, T. P., Steiner, U. *Nat. Mater.* **2003**, *2*, 48.
241 Blacke, T. D., de Coninck, J. *Adv. Colloid Interface Sci.* **2002**, *96*, 21.
242 Ruckenstein, E. *J. Colloid Interface Sci.* **1996**, *179*, 136.
243 de Gennes, P. G. *Rev. Mod. Phys.* **1985**, *57*, 827.
244 Joanny, J. F., de Gennes, P. G. *J. Phys. (Paris)* **1986**, *47*, 121.
245 Leger, L., Joanny, J. F. *Rep. Prog. Phys.* **1992**, *55*, 431.
246 Yerushalmi-Rozen, R., Kerle, T., Klein, J. *Science* **1999**, *285*, 1254.
247 Shull, K. R., Karis, T. E. *Langmuir* **1994**, *10*, 334.
248 Brochard, F., de Gennes, P. G. *J. Phys. Lett.* **1984**, *45*, L597.
249 de Gennes, P. G., Cazabat, A. M. *C. R. Acad. Sci., Ser. 2* **1990**, *310*, 1601.
250 Heslot, F., Fraysse, N., Cazabat, A. M. *Nature* **1989**, *338*, 640.
251 Belyi, V. A., Witten, T. A. *J. Chem. Phys.* **2004**, *120*, 5476.
252 Voue, M., de Coninck, J. *Acta Mater.* **2000**, *48*, 4405.
253 Cazabat, A. M., Valignat, M. P., Villette, S., de Coninck, J., Louche, F. *Langmuir* **1997**, *13*, 4754.
254 Voué, M., Valignat, M. P., Oshanin, G., Cazabat, A. M. *Langmuir* **1999**, *15*, 1522.
255 Burlatsky, S. F., Oshanin, G., Cazabat, A. M., Moreau, M., Reinhardt, W. P. *Phys. Rev. E* **1996**, *54*, 3832.
256 Cazabat, A. M., Fraysse, N., Heslot, F., Carles, P. *J. Phys. Chem.* **1990**, *94*, 7581.
257 Drelich, J. *Pol. J. Chem.* **1997**, *71*, 525.
258 Cazabat, A. M., de Coninck, J., Hoorelbeke, S., Valignat, M. P., Villette, S. *Phys. Rev. E* **1994**, *49*, 4149.
259 Voue, M., Semal, S., de Coninck, J. *Langmuir* **1999**, *15*, 7855.
260 Valignat, M. P., Oshanin, G., Villette, S., Cazabat, A. M., Moreau, M. *Phys. Rev. Lett.* **1998**, *80*, 5377.
261 Dao, T. T., Archer, L. A. *Langmuir* **2001**, *17*, 4042.
262 Bruinsma, R. *Macromolecules* **1990**, *23*, 276.
263 Heslot, F., Fraysse, N., Cazabat, A. M. *Nature* **1989**, *338*, 640.
264 Ausserre, D., Picard, A. M., Leger, L. *Phys. Rev. Lett.* **1986**, *57*, 2671.
265 Kavehpour, H. P., Ovryn, B., McKinley, G. H. *Phys. Rev. Lett.* **2003**, *91*, 196104.
266 Villette, S., Valignat, M. P., Cazabat, A. M. *Physica A* **1997**, *236*, 123.
267 Glick, D., Thiansathaporn, P., Superfine, R., *Appl. Phys. Lett.* **1997**, *71*, 3513.
268 Kataoka, D. E., Troian, S. M. *Nature* **1999**, *402*, 794.
269 Karthaus, O. *Chaos* **1999**, *9*, 308.
270 Mitov, Z., Kumacheva, E. *Phys. Rev. Lett.* **1998**, *81*, 3427.
271 Shinbrot, T., Alexander, A., Muzzio, F. I. *Nature* **1999**, *397*, 675.
272 Pakdel, P., McKinley, G. H. *Phys. Rev. Lett.* **1996**, *77*, 2459.
273 Shull, K. R., Flanigan, C. M., Crosby, A. J. *Phys. Rev. Lett.* **2000**, *84*, 3057.
274 Saffman, P. G., Taylor, G. *Proc. R. Soc. London, Ser. A* **1958**, *245*, 312.
275 Homsy, G. M. *Ann. Rev. Fluid Mech.* **1987**, *19*, 271.
276 Cazabat, A. M. *Nature* **1990**, *346*, 824.
277 Troian, S. M., Herbolzheimer, E., Safran, S. A. *Phys. Rev. Lett.* **1990**, *65*, 333.
278 Huppert, H. *Nature* **1982**, *300*, 427.

279 Diez, J. A., Kondic, L. *Phys. Rev. Lett.* **2001**, *86*, 632.
280 Xu, H., Shirvanyants, D., Beers, K. L., Matyjaszewski, K., Dobrynin, A. V., Rubinstein, M., Sheiko, S. S. *Phys. Rev. Lett.* **2005**, *94*, 237801.
281 Green, P. F., Ganesan, V. *Eur. Phys. J. E* **2003**, *12*, 449.
282 Reiter, G. *Phys. Rev. Lett.* **1992**, *68*, 75.
283 Brochard-Wyart, F., Daillant, J. *Can. J. Phys.* **1990**, *68*, 1084.
284 Gu, X., Raghavan, D., Douglas, J. F., Karim, A. *J. Polym. Sci., Polym. Phys.* **2002**, *40*, 2825.
285 Xie, R., Karim, A., Douglas, J. F., Han, C. C., Weiss, R. A. *Phys. Rev. Lett.* **1998**, *81*, 1251.
286 Seemann, R., Herminghaus, S., Jacobs, K. *Phys. Rev. Lett.* **2001**, *86*, 5534.
287 Karapanagiotis, I., Evans, D. F., Gerberich, W. W. *Langmuir* **2001**, *17*, 3266.
288 Müller, M., MacDowell, L. G., Müller-Buschbaum, P., Wunnike, O., Stamm, M. *J. Chem. Phys.* **2001**, *115*, 9960.
289 Masson, J.-L., Green, P. F. *Phys. Rev. Lett.* **2002**, *88*, 205504.
290 Putaux, J.-L., Borsali, R., Schappacher, M., Deffieux, A. *Faraday Discuss.* **2005**, *128*, 163.
291 Price, C. *Pure Appl. Chem.* **1983**, *55*, 1563.
292 Oostergetel, G. T., Esselink, F. J., Hadziioannou, G. *Langmuir* **1995**, *11*, 3721.
293 Danino, D., Gupta, R., Talmon, Y. *J. Colloid Interface Sci.* **2002**, *249*, 180.
294 Kesselman, E., Talmon, Y., Bang, J., Abbas, S., Li, Z. B., Lodge, T. P. *Macromolecules* **2005**, *38*, 6779.
295 LaRue, I., Adam, M., Sheiko, S. S., Rubinstein, M. *Macromolecules* **2004**, *37*, 5002.
296 Zhang, L. F., Eisenberg, A. *Macromolecules* **1999**, *32*, 2239.
297 Lodge, T. P., Bang, J., Li, Z., Hillmyer, M. A., Talmon, Y. *Faraday Discuss.* **2005**, *128*, 1.
298 Ding, J., Liu, G., Yang, M. *Polymer* **1997**, *38*, 5497.
299 Hu, Z., Jonas, A. M., Varshney, S. K., Gohy, J. F. *J. Am. Chem. Soc.* **2005**, *127*, 6526.
300 LaRue, I., Adam, M., Pitsikalis, M., Hadjichristidis, N., Rubinstein, M., Sheiko, S. S. *Macromolecules* **2006**, *39*, 309.
301 Hawker, C. J., Russell, T. P. *MRS Bull.* **2005**, *30*, 952.
302 Segalman, R. A., Hexemer, A., Kramer, E. J. *Macromolecules* **2003**, *36*, 6831.
303 Angelescu, D. E., Waller, J. H., Register, R. A., Chaikin, P. M.. *Adv. Mater.* **2005**, *17*, 1878.
304 Harrison, C., Adamson, D. H., Cheng, Z., Sebastian, J. M., Sethuraman, S., Huse, D. A., Register, R. A., Chaikin, P. M. *Science* **2000**, *290*, 1558.
305 Xiang, H., Lin, Y., Russell, T. P. *Macromolecules* **2004**, *37*, 5358.
306 Ryu, D. Y., Shin, K., Drockenmuller, E., Hawker, C. J., Russell, T. P. *Science* **2005**, *308*, 236.
307 Drelich, J., Mittal Brill, K. L. (Eds.) *Atomic Force Microscopy in Adhesion Studies*, Academic Press, New York, **2005**.
308 Shi, X., Zhao, Y.-P. *J. Adhesion Sci. Technol.* **2004**, *18*, 55.
309 Kappl, M., Butt, H.-J. *Part. Part. Syst. Charact.* **2002**, *19*, 129.
310 Segeren, L. H. G. J., Siebum, B., Karssenberg, F. G., Van den Berg, J. W. A., Vancso, G. J. *J. Adhesion Sci. Technol.* **2002**, *16*, 793.
311 Sato, F., Okui, H., Akiba, U., Suga, K., Fujihira, M. *Ultramicroscopy* **2003**, *97*, 303.
312 Tormoen, G. W., Drelich, J., Nalaskowski, J. *J. Adhesion Sci. Technol.* **2005**, *19*, 215.
313 Ebenstein, D. M., Wahl, K. J. *J. Colloid Interface Sci.* **2006**, *298*, 652.
314 Zitzler, L., Herminghaus, S., Mugele, F. *Phys. Rev. B* **2002**, *66*, 155436.
315 Leite, F. L., Herrmann, P. S. P. *J. Adhesion Sci. Technol.* **2005**, *19*, 365.
316 Vezenov, D. V., Noy, A., Ashby, P. *J. Adhesion Sci. Technol.* **2005**, *19*, 313.
317 Noy, A., Zepeda, S., Orme, C. A., Yeh, Y., De Yoreo, J. J. *J. Am. Chem. Soc.* **2003**, *125*, 1356.
318 Bennewitz, R. *Mater. Today* **2005**, *8*, 42.
319 Grierson, D. S., Flater, E. E., Carpick, R. W. *J. Adhesion Sci. Technol.* **2005**, *19*, 291.

320 Schwarz, U. D., Zwörner O., Köster, P., Wiesendanger, R. *Phys. Rev. B* **1997**, *56*, 6987.
321 McDermott, M. T., Green, J.-B. D., Porter, M. D. *Langmuir* **1997**, *13*, 2504.
322 Shon, Y. S., Lee, S., Colorado, S. S., Perry, S. S., Lee, T. R. *J. Am. Chem. Soc.* **2000**, *122*, 7556.
323 Van der Vegte, E. W., Subbotin, A., Hadziioannou, G., Ashton, P. R., Preece, J. A. *Langmuir* **2000**, *16*, 3249.
324 Leggett, G. J. *Anal. Chim. Acta* **2003**, *479*, 17.
325 Brukman, M. J., Marco, G. O., Dunbar, T. D., Boardman, L. D., Carpick, R. W. *Langmuir* **2006**, *22*, 3988.
326 Haugstad, G., Hammerschmidt, J. A., Gladfelter, W. L. *ACS Symp. Ser.* **2001**, *781*, 230.
327 Van Landingham, M. R., Villarrubia, J. S., Guthrie, W. F., Meyers, G. F. *Macromol. Symp.* **2001**, *167*, 15.
328 Chen, Xi, Vlassak, J. J. *J. Mater. Res.* **2001**, *16*, 2974.
329 Du, B., Tsui, O. K. C., Zhang, Q., He, T. *Langmuir* **2001**, *17*, 3286.
330 Syed Asif, S. A., Wahl, K. J., Colton, R. J., Warren, O. L. *J. Appl. Phys.* **2001**, *90*, 1192.
331 Merkel, R. *Phys. Rep.* **2001**, *346*, 344.
332 Clausen-Schaumann, H., Seitz, M., Krautbauer, R., Gaub, H. E. *Curr. Opin. Chem. Biol.* **2000**, *4*, 524.
333 Janshoff, A., Neitzert, M., Oberdorfer, Y., Fuchs, H. *Angew. Chem. Int. Ed.* **2000**, *39*, 3213.
334 Mehta, A. D., Rief, M., Spudich, J. A., Smith, D. A., Simmons, R. M. *Science* **1999**, *283*, 1689.
335 Bockelmann, U., Essevaz-Roulet, B., Heslot, F. *Phys. Rev. Lett.* **1997**, *79*, 4489.
336 Williams, M. C., Rouzina, I. *Curr. Opin. Struct. Biol.* **2002**, *12*, 330.
337 Hansma, H. G., Kasuya, K., Oroudjev, E. *Curr. Opin. Struct. Biol.* **2004**, *14*, 380.
338 Harris, S. A. *Contemp. Phys.* **2004**, *45*, 11.
339 Sattin, B. D., Pelling, A. E., Goh, M. C. *Nucleic Acids Res.* **2004**, *32*, 4876.
340 Zhuang, X. W., Rief, M. *Curr. Opin. Struct. Biol.* **2003**, *13*, 88.
341 Carrion-Vazquez, M., Oberhauser, A. F., Fisher, T. E., Marszalek, P. E., Li, H., Fernandez, J. M. *Prog. Biophys. Mol. Biol.* **2000**, *74*, 63.
342 Kellermayer, M. S. Z., Bustamante, C., Granzier, H. L. *Biochim. Biophys. Acta, Bioenergy* **2003**, *1604*, 105.
343 Moy, V. T., Florin, E. L., Gaub, H. E. *Science* **1994**, *266*, 257.
344 Florin, E. L., Moy, V. T., Gaub, H. E. *Science* **1994**, *264*, 415.
345 Janshoff, A., Neitzert, M., Oberdörfer, Y., Fuchs, H. *Angew. Chem. Int. Ed.* **2000**, *39*, 3213.
346 Beyer, M. K., Clausen-Schaumann, H. *Chem. Rev.* **2005**, *105*, 2921.
347 Ray, C., Brown, J. R., Akhremitchev, B. B. *J. Phys. Chem. B* **2006**, *110*, 17578.
348 Cui, S., Albrecht, C., Kuehner, F., Gaub, H. E. *J. Am. Chem. Soc.* **2006**, *128*, 6636.
349 Brant, D. A. *Curr. Opin. Struct. Biol.* **1999**, *9*, 556.
350 Rief, M., Oesterhelt, F., Heymann, B., Gaub, H. E. *Science* **1997**, *275*, 1295.
351 Abu-Lail, N. I., Camesano, T. A. *J. Microsc.* **2003**, *212*, 217.
352 Sonnenberg, L., Parvole, J., Borisov, O., Billon, L., Gaub, H. E., Seitz, M. *Macromolecules* **2006**, *39*, 281.
353 Minko, S., Roiter, Y. *Curr. Opin. Colloid Interface Sci.* **2005**, *10*, 9.
354 Chatellier, X., Senden, T. J., Joanny, J.-F., Meglio, J.-M. *Europhys. Lett.* **1998**, *41*, 303.
355 Ozon, F., di Meglio, J.-M., Joanny, J.-F. *Eur. Phys. J. E* **2002**, *8*, 321.
356 Li, H., Liu, B., Zhang, X., Gao, C., Shen, J., Zou, G. *Langmuir* **1999**, *15*, 2120.
357 Al-Maawali, S., Bemis, J. E., Akhremitchev, B. B., Leecharoen, R., Janesko, B. G., Walker, G. C. *J. Phys. Chem. B* **2001**, *105*, 3965.
358 Evans, E. *Annu. Rev. Biophys. Biomol. Struct.* **2001**, *30*, 105.
359 Janovjak, H., Struckmeier, J., Müller, D. J. *Eur. Biophys. J.* **2005**, *34*, 91.
360 Zou, S., Schönherr, H., Vancso, G. J. *J. Am. Chem. Soc.* **2005**, *127*, 11230.
361 Bustamante, C., Bryant, Z., Smith, S. B. *Nature* **2003**, *421*, 423.
362 Zhang, W., Zhang, X. *Prog. Polym. Sci.* **2003**, *28*, 1271.

363 Allemand, J.F., Bensimon, D., Croquette, V. *Curr. Opin. Struct. Biol.* **2003**, *13*, 266.
364 Dufrene, Y.F. *Curr. Opin. Microbiol.* **2003**, *6*, 317.
365 Seitz, M. Force spectroscopy, in *Nanobiotechnology*, Eds. Niemeyer, C.M., Mirkin, C., Wiley-VCH, Weinheim, **2004**, 404–428.
366 Fotiadis, D., Scheuring, S., Müller, S.A., Engel, A., Müller, D.J. *Micron* **2002**, *33*, 385.
367 Lantz, M.A., Hug, H.J., Hoffmann, R., van Schendel, P.J.A., Kappenberger, P., Martin, S., Baratoff, A., Güntherodt, H.J. *Science* **2001**, *291*, 2580.
368 Jarvis, S.P., Yamada, H., Yamamoto, S.L., Tokumoto, H., Pethica, J.B. *Nature* **1996**, *384*, 247.
369 Erlandsson, R., Yakimov, V. *Phys. Rev. B* **2000**, *62*, 13680.
370 Bensimon, D., Simon, A.J., Croquette, V., Bensimon, A. *Phys. Rev. Lett.* **1995**, *74*, 4754.
371 Clausen-Schaumann, H., Rief, M., Tolksdorf, C., Gaub, H.E. *Biophys. J.* **2000**, *78*.
372 Grandbois, M., Beyer, M., Rief, M., Clausen-Schaumann, H., Gaub, H.E. *Science* **1999**, *283*, 1727.
373 Hammerschmidt, J., Gladfelter, W., Haugstadt, G. *Macromolecules* **1999**, *32*, 3360.
374 Overney, R.M., Buenviaje, C., Luginbuhl, R., Dinelli, F. *J. Thermal Anal. Calorim.* **2000**, *59*, 205.
375 Ge, S., Pu, Y., Zhang, W., Rafailovich, M., Sokolov, J., Buenviaje, C., Buckmaster, R., Overney, R.M. *Phys. Rev. Lett.* **2000**, *85*, 2340.
376 Oulevey, F., Burnham, N., Gremaud, G., Kulik, A., Pollock, H., Hammiche, A., Reading, M., Song, M., Hourston, D. *Polymer* **2000**, *41*, 3087.
377 Pollock, H., Hammiche, A. *J. Phys. D* **2001**, *34*, 23.
378 Meincken, M., Balk, L.J., Sanderson, R.D. *Surf. Interface Anal.* **2003**, *35*, 1034.
379 Ho, W. *Acc. Chem. Res.* **1998**, *31*, 567.
380 Nyffenegger, R.M., Penner, R.M. *Chem. Rev.* **1997**, *97*, 1195.
381 Jaschke, M., Butt, H.-J. *Langmuir* **1995**, *11*, 1061.
382 Salaita, K., Lee, S.W., Wang, X., Huang, L., Dellinger, T.M., Liu, C., Mirkin, C.A. *Small* **2005**, *1*, 940.
383 Lee, S.W., Oh, B.-K., Sanedrin, R.G., Salaita, K., Fujigaya, T., Mirkin, C.A. *Adv. Mater.* **2006**, *18*, 1133.
384 Agarwal, G., Sowards, L.A., Naik, R.R., Stone, M.O. *J. Am. Chem. Soc.* **2003**, *125*, 580.
385 Lewis, A., Kneifetz, Y., Shambrodt, E., Radko, A., Khatchatryan E., Sukenik, C. *Appl. Phys. Lett.* **1999**, *75*, 2689.
386 Riemenschneider, L., Blank, S., Radmacher, M. *Nano Lett.* **2005**, *5*, 1643.
387 Vettiger, P., Despont, M., Drechsler, U., Durig, U., Haberle, W., Lutwyche, M.I., Rothuizen, H.E., Stutz, R., Widmer, R., Binnig, G.K. *IBM J. Res. Dev.* **2000**, *44*, 323.
388 Gotsmann, B. *Nova Acta Leopoldina* **2005**, *92*, 69.
389 King, W.P., Kenny, T.W., Goodson, K.E., Cross, G.L.W., Despont, M., Dürig, U.T., Rothuizen, H., Binnig, G., Vettiger, P. *J. Microelectromech. Syst.* **2002**, *11*, 765.
390 Drechsler, U., Burer, N., Despont, M., Dürig, U., Binnig, G., Robin, F., Vettiger, P. *J. Microelectron. Eng.* **2003**, *67*, 397.
391 Lee, C.S., Nam, H.-J., Kim, Y.-S., Jin, W.-H., Cho, S.-M., Bu, J.-U. *Appl. Phys. Lett.* **2003**, *83*, 4839.
392 Lantz, M., Gotsmann, B., Dürig, U., Vettiger, P., Nakayama, Y., Shimizu, T., Tokumoto, H. *Appl. Phys. Lett.* **2003**, *83*, 1266.
393 Frommer, J., Miller, R.D., Hawker, C., Dürig, U.T., Gotsmann, B., Vettiger, P., Lantz, M.A. *US Pat. Appl. 2005047307*, **2005**.
394 Liu, G.-Y., Xu, S., Qian, Y. *Acc. Chem. Res.* **2000**, *33*, 457.

5
Scattering from Polymer Systems

Megan L. Ruegg and Nitash P. Balsara

5.1
Introduction

Scattering is a ubiquitous process resulting from the interaction between waves and matter. The derivation of scattering laws is covered in many standard text books [1–5] and is not repeated here. Our main objective is to present a comprehensive set of equations that can be used to interpret the results of scattering experiments performed on polymer-containing samples.

Collimated beams of light, X-rays and neutrons travel through homogeneous media in a straight line. We will assume that the incident radiation is directed along the x-direction and the wave is represented by

$$A(x) = \text{Re}[A_0 \exp(-i2\pi x/\lambda_0)] \qquad (1)$$

where λ_0 is the wavelength of the incident radiation in air, A_0 is the magnitude of the incident wave, $i = \sqrt{-1}$ and Re takes the real part of the imaginary number. The typical ranges of available wavelengths are 400–650 nm for light, 0.07–0.15 nm for X-rays and 0.5–1.4 nm for neutrons. This covers the range of radiation that is typically used to study polymers. When the incident beam encounters heterogeneities, scattered waves traveling in directions that are different from x are created. The objective of scattering experiments is to determine the structure and composition of the heterogeneities by studying the intensities of the scattered beams.

Figure 5.1 shows schematic views of two typical scattering experiments. The angle between the incident and scattered beams, θ, is called the scattering angle. We assume that both beams are perfectly collimated by the introduction of appropriate optical units. The scattering volume, V, is the portion of the sample illuminated by the incident beam that intersects the detector collimation. In Fig. 5.1a, we show a wide-angle scattering experiment wherein θ values between 0 and 180° can, in principle, be accessed. In this geometry, V depends on θ. In Fig. 5.1b we show a forward scattering (small-angle) experiment where θ is restricted to values below 90 °C. In this geometry, V does not depend on θ.

Macromolecular Engineering. Precise Synthesis, Materials Properties, Applications.
Edited by K. Matyjaszewski, Y. Gnanou, and L. Leibler
Copyright © 2007 WILEY-VCH Verlag GmbH & Co. KGaA, Weinheim
ISBN: 978-3-527-31446-1

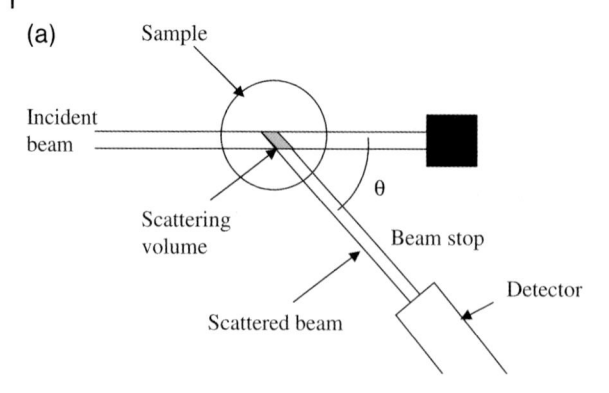

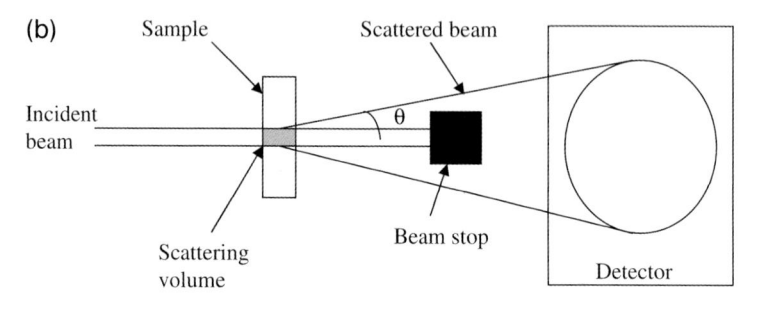

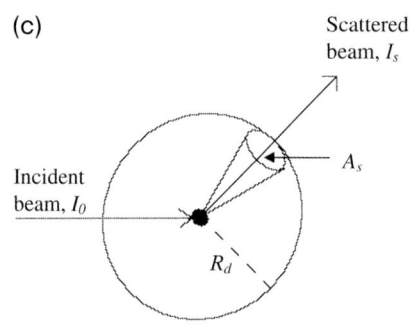

Fig. 5.1 (a) Wide-angle scattering experimental set-up and (b) small-angle scattering experimental set-up, where θ is the scattering angle. (c) Scattering geometry, where I_0 is the intensity of the incident beam, I_s is the intensity of the scattered beam, A_s is the detector area and R_d is the distance from the detector to the sample.

We will restrict our analysis to elastic scattering where the wavelengths of the scattered and incident beams are identical (or nearly so). When the beam encounters a point heterogeneity, radiation is scattered uniformly in all directions. We focus our attention on coherent scattering where the incident and scattered beams are in-phase [6]. The detector selects scattered waves that are oriented along a unit vector $\mathbf{a_s}$. The scattering vector, $\mathbf{q}$, is defined as

$$\mathbf{q} = \frac{2\pi}{\lambda}(\mathbf{a_s} - \mathbf{a_x}) \tag{2}$$

where $\mathbf{a_x}$ is a unit vector along the incident (x) direction and λ is the wavelength of the beam in the sample; $\lambda = \lambda_0/\eta$ (η is the average refractive index of the sample). It is straightforward to show that the magnitude of $\mathbf{q}$ is given by

$$q = \frac{4\pi}{\lambda}\sin(\theta/2) \tag{3}$$

Consider a sphere of radius R_d centered at the sample as shown in Fig. 5.1c. A detector with area A_s is located on this sphere. The incident intensity, I_0, is equal to the number of particles (photons or neutrons), N_0, arriving on the exposed area of the sample, A, in a given amount of time, t. The scattered rays reaching the detector are contained in a cone as shown in Fig. 5.1c. $\Omega = A_s/R_d^2$ is the solid angle subtended by the detector at the sample location. The scattered intensity is normalized by Ω rather than A_s, as the number of particles arriving at area A_s depends on R_d. The detected scattered intensity, I_s, along a given direction, is defined in terms of the number of scattered particles along that direction, N_s, in time t, per unit solid angle:

$$I_0 = \frac{N_0}{At}$$
$$I_s = \frac{N_s}{\Omega t} = \frac{N_s R_d^2}{A_s t} \tag{4}$$

I_s/I_0 has units of (length)^2 (and is usually expressed in cm^2). We assume that the detector is in the far field so that R_d is much greater than the characteristic length scale of the structures within the sample. The normalized scattering intensity, I, which has units of (length)$^{-1}$ (and is usually expressed in cm^{-1}), is given by

$$I = \frac{I_s}{I_0 V} \tag{5}$$

where V is the scattering volume.

5.2
Scattering from Binary Mixtures

Let us assume for simplicity that the scattering sample is composed of two chemically distinct, perfectly monodisperse components labeled 1 and 0. For example, 1 could be a flexible polymer such as polystyrene or a colloidal sphere and 0, which we will label the background component, could be either a solvent such as toluene or another polymer such as polyisoprene. For purposes of

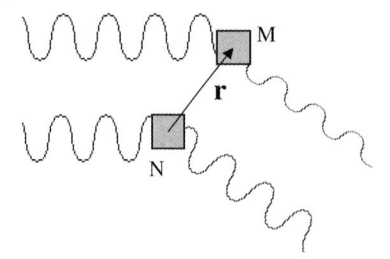

Fig. 5.2 Interference calculation with two particles, N at the origin and M at a distance **r** from the origin. The amplitudes of the scattered waves depend on the scattering length densities of the particles N and M.

modeling the scattering process, we will assume a coarse-grained view of the system. We divide the scattering volume into a large number of reference volume units with volume v_{ref}. We assume that v_{ref} is small enough that each reference volume unit can be regarded as a point heterogeneity. Consider two such units, one located at the origin (N) and the other located at vector **r**(M), as shown in Fig. 5.2. The amplitudes of the coherently scattered waves emanating from points N and M are proportional to the scattering lengths b_N and b_M that are characteristics of the units located there. The intensity of the scattered wave at the detector is given by

$$I_s(\mathbf{q}) = I_0 b_N b_M \exp(-i\mathbf{q} \cdot \mathbf{r}) \tag{6}$$

If we have n reference volume units in our scattering volume each located at vector $\mathbf{r}_i$ relative to a laboratory coordinate frame ($i=1-n$), then the scattered intensity is given by summing Eq. 6 over all pairs of units:

$$\langle I_s(\mathbf{q}_0) \rangle = I_0 \sum_{k=1}^{n}\sum_{l=1}^{n} \langle b_k b_l \exp[-i\mathbf{q}\cdot(\mathbf{r}_k - \mathbf{r}_l)]\rangle = I_0 \sum_{k=1}^{n}\sum_{l=1}^{n} \langle b_k b_l \exp(-i\mathbf{q}\cdot\mathbf{r}_{kl})\rangle \tag{7}$$

where $\mathbf{r}_{kl}$ is the vector connecting reference volume unit l to unit k and the angle brackets correspond to time/ensemble averages, because the scattering units are, in general, mobile. We will omit the angle brackets in subsequent equations for simplicity of notation. Throughout this paper we will use indices k and l to refer to reference volume units.

Our scattering volume will typically contain two or more components. At any given time, each reference volume unit can be occupied by only one of the components. The two-point spatial correlation function between reference volume units of components i and j separated by a vector **r** is $\rho_{ij}(\mathbf{r})$. Throughout this paper we will use indices i and j to refer to the components in the mixture. Given a reference volume unit of component 1 at the origin, $\rho_{11}(\mathbf{r})$ is the ensemble-averaged probability of finding another reference volume unit of component 1 at **r**. Similarly, given a reference volume unit of component 1 at the origin, $\rho_{10}(\mathbf{r})$ is the ensemble averaged probability of finding a reference volume unit of component 0 at **r**. The normalized structure factor $S_{ij}(\mathbf{q})$ is the three-dimensional Fourier transform of the corresponding correlation function:

$$S_{ij}(q) \propto \int \rho_{ij}(\mathbf{r}) \exp(-i\mathbf{q} \cdot \mathbf{r}) d\mathbf{r} \qquad (i = 0,1 \text{ and } j \, 0,1) \qquad (8a)$$

$$S_{ij}(q) \propto \int_0^\infty \rho_{ij}(r) \frac{\sin(qr)}{qr} (4\pi r^2) dr \qquad (i = 0,1 \text{ and } j \, 0,1) \qquad (8b)$$

where Eq. 8b is applied to the case of an isotropic system when ρ is a function of the magnitude of $\mathbf{r}$ ($r=|\mathbf{r}|$).

In the case of a binary mixture where each reference volume unit contains either component 1 or 0, it is adequate to locate the units that are of one type, say component 1. This is because all other units are, by definition, occupied by component 0. This is often referred to as the "incompressibility" constraint. Our coarse-grained approach therefore does not account for microscopic voids that are present in all liquids and solids. Note that the term "incompressible" does not imply unresponsive to pressure. In fact, the present approach can be used to interpret scattering results at any pressure. In incompressible binary mixtures $S_{11}(\mathbf{q}) = -S_{10}(\mathbf{q})$ because given reference volume unit 1 at the origin, the probability of obtaining a component 1 reference volume unit at $\mathbf{r}=1-$ (probability of obtaining a type 0 reference volume unit at **r**). Similar considerations lead to the conclusion that $S_{11}(\mathbf{q}) = S_{00}(\mathbf{q})$. There is therefore only one independent structure factor in incompressible binary mixtures and Eq. 7 reduces to

$$I_s(\mathbf{q}) = I_0(b_1 - b_0)^2 S_{u,11}(\mathbf{q}) = I_0(b_1 - b_0)^2 \sum_{k=1}^{n_1} \sum_{l=1}^{n_1} \langle \exp(-i\mathbf{q} \cdot \mathbf{r}_{kl}) \rangle \qquad (9)$$

where the summation is now restricted to reference volume units of type 1 only and n_1 is the number of reference volume units of type 1. $S_{u,11}(\mathbf{q})$ is the un-normalized structure factor, as defined by Eq. 9.

We have divided the total scattering volume into n reference volume units with volume v_{ref}. Any component, whether it is a polymer chain, a hard colloidal sphere or another type of geometry, can be defined as a collection of reference volume units. We will use the term "molecule" of type i to refer to a contiguous collection of reference volume units of type i. Each molecule of type i is composed of N_i reference volume units. In the case of polymer chains, the chain can be split up into reference volume units with a volume v_{ref} and the number of reference volume units per molecule of component i is equal to $N_i = N_{\text{mon},i} v_{\text{mon},i}/v_{\text{ref}}$, where $N_{\text{mon},i}$ and $v_{\text{mon},i}$ are the number of chemical repeat units (monomers) per molecule and monomer volume of i, respectively. We will use this reference volume-based definition of N_i in this chapter. Analogously, a spherical particle with volume $= (4/3)\pi R_i^3$ (R_i is the radius of the sphere) can be divided into reference volume units with $N_i = 4\pi R_i^3/3v_{ref}$. For either a polymer chain or a colloidal sphere, the number of molecules of type i in the system is defined as $n_{m,i} = n_i/N_i$.

The summation in Eq. 9 can be split into two types of terms, one where the indices k and l lie within a given molecule and the other where the indices lie on different molecules:

$$I_s(\mathbf{q}) = I_0(b_1 - b_0)^2 \left[n_{m,1} \sum_{k=1}^{N_1} \sum_{l=1}^{N_1} \langle \exp(-i\mathbf{q} \cdot \mathbf{r}_{kl}) \rangle_{\text{same\_molecule}} \right.$$

$$\left. + (n_{m,1}^2 - n_{m,1}) \sum_{k=1}^{N_1} \sum_{l=1}^{N_1} \langle \exp(-i\mathbf{q} \cdot \mathbf{r}_{kl}) \rangle_{\text{different\_molecule}} \right] \quad (10\,a)$$

Since $n_{m,1}^2 \gg n_{m,1}$, we can neglect $n_{m,1}$ in front of the second term in Eq. 10a:

$$I_s(\mathbf{q}) = I_0(b_1 - b_0)^2 \left[n_{m,1} \sum_{k=1}^{N_1} \sum_{l=1}^{N_1} \langle \exp(-i\mathbf{q} \cdot \mathbf{r}_{kl}) \rangle_{\text{same\_molecule}} \right.$$

$$\left. + n_{m,1}^2 \sum_{k=1}^{N_1} \sum_{l=1}^{N_1} \langle \exp(-i\mathbf{q} \cdot \mathbf{r}_{kl}) \rangle_{\text{different\_molecule}} \right] \quad (10\,b)$$

If we assume that component 1 is dilute and the individual molecules are uncorrelated, then the second term in Eq. 10b is negligible because $\langle \mathbf{r}_{kl} \rangle_{\text{different\_molecule}}$ is much greater than $\langle \mathbf{r}_{kl} \rangle_{\text{same\_molecule}}$ and the normalized scattering intensity, $I(q)$, which depends only on the magnitude of $\mathbf{q}$, is given by

$$I(q) = \frac{I_s(q)}{I_0 n v_{ref}} = (B_1 - B_0)^2 \frac{\phi_1 v_{ref}}{N_1} \left[\sum_{k=1}^{N_1} \sum_{l=1}^{N_1} \langle \exp(-i\mathbf{q} \cdot \mathbf{r}_{kl}) \rangle_{\text{same\_molecule}} \right]$$

$$= (B_1 - B_0)^2 v_{ref} \phi_1 N_1 P_1(q) \quad (11)$$

where nv_{ref} is the total sample volume, $\phi_1 = n_{m,1} N_1/n$ is the volume fraction of component 1, $B_i = b_i/v_{ref}$ is the scattering length density of component i and $P_1(q)$ is the form factor of a molecule of type 1 that accounts for intramolecular correlations only:

$$P_1(q) = \frac{1}{N_1^2} \sum_{k=1}^{N_1} \sum_{l=1}^{N_1} \langle \exp(-i\mathbf{q} \cdot \mathbf{r}_{kl}) \rangle_{\text{same\_molecule}} \quad (12)$$

Note, however, that Eq. 11 is only valid in the limit of extreme dilution because the number of intermolecular correlation terms in Eq. 10b ($n_{m,1}^2$) is much larger than the number of intramolecular terms ($n_{m,1}$).

The scattering length density, B_i, is determined by both the chemical nature of the components and the nature of the radiation. For light, X-rays and neutrons, B_i is given by

$$B_i(\text{light}) = \frac{2\pi \eta_i \eta}{\lambda_0^2} \tag{13a}$$

$$B_i(\text{X-rays}) = \frac{r_e \sum_j Z_j}{v_{\text{mon}}} \tag{13b}$$

and

$$B_i(\text{neutrons}) = \frac{\sum_j b_j}{v_{\text{mon}}} \tag{13b}$$

The scattering length density of materials for X-rays and neutrons can be predicted from atomic composition whereas the optical scattering length density cannot. The optical scattering length density in Eq. 13a is thus expressed in terms of refractive indices of the mixture, η, and the refractive index of component i, η_i, which can be measured independently. Equation 13a is only applicable when the difference between η and η_i is small compared with η. It is more convenient to compute X-ray and neutron scattering length densities on the basis of a chemical repeat unit rather than our reference volume unit. The ratios in the right-hand sides of Eqs. 13b and c will not depend on the definition of v used in the denominator provided that the numerator is also based on the same volume. Hence, in Eqs. 13b and c, we will use the monomer unit as the repeat unit with volume v_{mon}. Z_j is the number of electrons in each of the atoms in volume v_{mon}, r_e is the classical radius of the electron, which is 2.81×10^{-13} cm for $\lambda = 0.154$ nm radiation (r_e is a weak function of wavelength), and b_i is the neutron scattering length of each of the atoms in component i in volume v_{mon}. For completeness, we give the refractive indices of typical polymers and solvents in Table 5.1 and the neutron scattering lengths of elements typically found in polymers in Table 5.2. The summations in Eqs. 13b and c are over all of the

Table 5.1 Refractive indices of polymers and solvents.

Polymer or solvent	η_i ($\lambda = 590$ nm, $T = 20\text{--}25\,^\circ$C)
Polystyrene	1.59
Polyisoprene	1.52
Poly(1,3-butadiene)	1.52
Poly(methyl methacrylate)	1.49
Polypropylene, atactic	1.47
Toluene	1.49
Benzene	1.50
Hexane	1.37
Tetrahydrofuran	1.40
Cyclohexane	1.42

Table 5.2 Neutron scattering lengths.

Element	β_i (10^{-12} cm)
C	0.665
^{1}H	−0.374
^{2}D	0.667
O	0.580
N	0.936

atoms in volume v_{mon}. For a solution of 10% polystyrene in toluene at room temperature, B_{PS}(light) at $\lambda_0 = 500$ nm is 6.0×10^9 cm^{-2}. For polystyrene at 140 °C where the monomer volume $v_{mon} = 0.179$ nm^3, B_{PS}(X-rays) $= 8.8 \times 10^{10}$ cm^{-2} and B_{PS}(neutrons) $= 1.3 \times 10^{10}$ cm^{-2}.

It is convenient to express the normalized scattering intensity as a product of three terms: a contrast factor that depends on the scattering length densities of the components, the reference volume v_{ref} and normalized structure factors, $S_{ij}(\mathbf{q}) = S_{u,ij}(\mathbf{q})/n$, that include all other contributions. For binary mixtures

$$I(\mathbf{q}) = (B_1 - B_0)^2 v_{ref} S_{11}(\mathbf{q}) \tag{14}$$

and in the dilute and uncorrelated limit

$$S_{11}(q) = N_1 \phi_1 P_1(q) \tag{15}$$

Form factors for objects/molecules that are typically encountered in polymeric materials are as follows [7–11]:

$$P_{1\_chain}(x) = \frac{2(e^{-x} + x - 1)}{x^2} \quad \text{(Gaussian polymers)} \tag{16a}$$

$$P_{1\_sphere}(x) = \frac{9(\sin x - x \cos x)^2}{x^6} \quad \text{(spheres)} \tag{16b}$$

$$P_{1\_disk}(x) = \frac{2}{x^2}\left(1 - \frac{J_1(2x)}{x}\right) \quad \text{(thin disks)} \tag{16c}$$

$$P_{1\_shell}(x) = \left(\frac{\sin x}{x}\right)^2 \quad \text{(thin shells)} \tag{16d}$$

$$P_{1\_ellipsoid}(x) = \int_0^{n/2} P_{1\_sphere}(x) \cos \beta \, d\beta \quad \text{(ellipsoids of revolution)} \tag{16e}$$

$$P_{1\_rod}(q) = \frac{2}{x}\left(\int_0^x \frac{\sin \beta}{\beta} d\beta - \frac{1 - \cos x}{x}\right) \quad \text{(thin rods)} \tag{16f}$$

In Eq. 16a, $x = q^2 N_1 l_1^2/6$, where l_1 is the statistical segment length (l used as an index, e.g. in Eqs. 7 and 9–12, should not be confused with the statistical segment length) and $R_{g,1}$ is the radius of gyration of the chain based on a reference volume v_{ref}. In Eqs. 16b, c and d, $x = qR$, where R is the radius of the sphere, disk and shell, respectively. In Eq. 16c, J_1 is a first-order Bessel function:

$$J_1(2x) = \sum_{l=0}^{\infty} \frac{(-1)^l}{2^{2l+1} l! (1+l)!} (2x)^{2l+1}$$

In Eq. 16e, $x = qa[\cos^2 \beta + (b^2/a^2) \sin^2 \beta]^{1/2}$, where the ellipsoid of revolution is obtained by revolving an ellipse with semi-axes a and b about the a axis. In Eq. 16f, $x = qL$, where L is the length of the cylinder. In Eqs. 16e and f, β is a variable of integration.

All of the $P(q)$ expressions (Eqs. 16a–f) approach unity as q approaches zero. Note that if η, the refractive index, of 0–1 mixtures is a linear function of ϕ_1, then $(d\eta/d\phi_1)^2 = (\eta_1 - \eta_0)^2$ and the familiar Zimm equation for light scattering from dilute polymer chains in terms of refractive index increments is recovered from Eqs. 13a, 14, 15 and 16a [12].

For a non-dilute, homogeneous blend of two Gaussian homopolymers [13]:

$$\frac{1}{S_{11}(q)} = \frac{1}{N_1 \phi_1 P_1} + \frac{1}{N_0 \phi_0 P_0} - 2\chi_{10} \tag{17}$$

where the subscript 0 represents quantities that describe component 0 and χ_{10} is the Flory–Huggins interaction parameter between the two components. Equation 17 was originally derived by de Gennes using the random phase approximation (RPA) [13]. For mixtures of a homopolymer and a solvent, it is convenient to set $v_{ref} = v_{sol}$, the volume occupied by a solvent molecule, with $N_1 = N_{mon} v_{mon}/v_{sol}$ and $N_0 = 1$. Excluded volume interactions obtained when polymers are dissolved in good solvents are not accounted for in Eq. 17 [2, 3, 14]. For mixtures of two homopolymers, it is convenient to choose a particular value

Table 5.3 Common polymer characteristics (at 140 °C).

Polymer	v_{mon} (nm^3)	l (nm) [a]
Polystyrene	0.179	0.50
Polyisoprene (5% 1,2-addition)	0.136	0.56
Polybutadiene (10% 1,2-addition)	0.111	0.66
Polypropylene	0.176	0.57
Polyethylene	0.119	0.77
Poly(ethylene oxide)	0.069	0.72
Poly(methyl methacrylate)	0.149	0.54

a) The l values are based on $v_{ref} = 0.1$ nm^3.

Table 5.4 Flory–Huggins interaction parameters at 140°C (based on $v_{ref}=0.1$ nm^3).

Polymer 1	Polymer 2	χ_{12}
Polystyrene	Polyisoprene (with 7% 1,2-addition)	0.0505
Polystyrene	Polybutadiene (with 95% 1,2-addition)	0.0296
Polyisobutene	Head-to-head polypropylene	−0.0007
Poly(ethylene oxide)	Polystyrene	0.0401

of v_{ref}. In [15] values of χ_{ij} and l_i, based on $v_{ref}=0.1$ nm^3, are presented. Typical values of v_{mon} and l_i are given in Table 5.3 and χ_{ij} values in Table 5.4. We propose adopting the same reference volume (0.1 nm^3) for calculating the scattering intensity from polymer mixtures.

Regardless of the topology of the object, in the low-q limit, the scattering intensity of a dilute suspension of objects is given by [2, 16]

$$I(q) = (B_1 - B_0)^2 v_{ref} N_1 \phi_1 \left(1 - \frac{q^2 R_g^2}{3}\right) \tag{18}$$

where R_g is the radius of gyration of the objects. For convenience, we list R_g^2 of the objects listed in Eq. 16: $Nl^2/6$ for Gaussian polymer chains, $3R^2/5$ for spheres, $R^2/2$ for thin disks, R^2 for thin shells, $(a^2+2b^2)/5$ for ellipsoids of revolution and $L^2/12$ for thin rods. R_g of flexible objects such as polymer chains can be strongly affected by the nature of the solvent. The Gaussian state, where $R_g \propto N^{1/2}$ is obtained in solvents at a temperature where $\chi=\frac{1}{2}$ [17]. Flory called this the θ temperature and we therefore refer to the $R_{g\theta}$ obtained under these conditions as $R_{g\theta}$. In good solvents where $\chi<\frac{1}{2}$, R_g of polymer chains is significantly larger than $R_{g\theta}$ and is given by [17]

$$\left(\frac{R_g}{R_{g\theta}}\right)^5 - \left(\frac{R_g}{R_{g\theta}}\right)^3 = \frac{9\sqrt{6}}{16} \frac{v_{ref}}{l^3} (1 - 2\chi) N^{1/2} \tag{19}$$

where the subscript 1 has been dropped for convenience and χ, N and l are based on v_{ref}, which in Eq. 19 is set equal to the molecular volume of the solvent.

5.3
Homogeneous Multicomponent Polymer Mixtures

Expressions for the scattering profile of homogeneous multicomponent mixtures were developed by Benoit et al. [14] and Akcasu and Tombakoglu [18]. A comprehensive review of this approach is contained in the work of Hammouda. [19] We define a component to be a contiguous collection of chemically identical reference volume units. An A–B diblock copolymer thus has two components. An A–B–A triblock copolymer has three components even if the two A blocks are the same length and chemically identical. The structure factor of the mixture is represented by a matrix due to the existence of many components in the mixture. One of the components is chosen to be the background component and is labeled component 0. The incompressibility constraint enables elimination of correlations with the background component. The expressions that we provide below apply specifically to the case where the background component is not connected to any other component (e.g. a homopolymer). The structure factor matrix $[\underline{S}(q)]$, which describes correlations between all of the components in the mixture except the background component, is written as a function of the structure factor matrix in the absence of interactions $[\underline{S}^0(q)]$ and an interaction matrix describing all of the interactions between the components in the mixture $[\underline{V}(q)]$:

$$\underline{S}(q) = [\underline{S}^0(q)^{-1} + \underline{V}(q)]^{-1} \tag{20}$$

In the absence of interactions, there will only be correlations between reference volume units if they are part of the same connected chain. For Gaussian coils, S_{ii}^0 is given by

$$S_{ii}^0(q) = N_i \phi_i P_i(q) \tag{21}$$

as previously described in Eq. 15, where $P_i(q)$ was given in Eq. 16a. Equation 21 is used to describe correlations between reference volume units of the same component i (this could describe correlations between reference volume units comprising a homopolymer chain or, in the case of a block copolymer, it could describe the correlations between reference volume units of the same component within each block).

Correlations in the absence of interactions also exist between reference volume units of different types when they comprise a connected chain. Consider the situation of a triblock copolymer in which each block is chemically distinct. Figure 5.3 illustrates this situation in which the blocks are labeled i, p and j. Equation 22 describes the correlations in the absence of interactions between the outer components (i and j) of the triblock copolymer:

$$S_{ij}^0(q) = (N_i \phi_i N_j \phi_j)^{1/2} F_i(q) F_j(q) E_p(q) \qquad (i \neq j \neq p) \tag{22}$$

where $F_i(q) = [1 - \exp(-x_i)]/x_i$, $E_p(q) = \exp(-x_p)$, $x_i = q^2 N_i l_i^2/6$ and $x_p = q^2 N_p l_p^2/6$. Note that Eq. 22 can also be used to describe a diblock copolymer

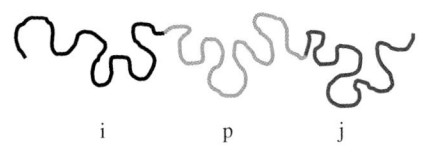

Fig. 5.3 A triblock copolymer with blocks i, p and j.

of blocks i and j in which $E_p(q) = 1$. More exotic chain architectures (such as a star copolymer) can be described by Eq. 22 if each correlation can be broken down into correlations between two connected chemically distinct chains ($E=1$) or correlations between chains which are separated by a chemically distinct chain ($E \neq 1$). When considering reference volume units of components i and j that are not on the same connected chain, $S_{ij}^0(q) = 0$.

The interaction matrix is constructed from the following equations, which can be used for any chain architecture provided that the reference component (0) is a Gaussian homopolymer chain:

$$V_{ii}(q) = \left[\frac{1}{N_0 \phi_0 P_0(q)} - 2\chi_{i0} \right] \tag{23}$$

$$V_{ik}(q) = \left[\frac{1}{N_0 \phi_0 P_0(q)} - \chi_{i0} - \chi_{k0} + \chi_{ik} \right] \quad (i \neq k) \tag{24}$$

A column vector $\underline{\Delta B}$ is used to describe the scattering length densities of the components. The elements of $\underline{\Delta B}$ are the scattering length density differences of each component i as compared to the background component 0:

$$\Delta B_i = B_i - B_0 \tag{25}$$

where B_i is the scattering length density of component i and was defined in Eqs. 13a–c.

The coherent scattering profile from a multicomponent mixture can now be written:

$$I(q) = v_{\text{ref}} \underline{\Delta B}^T \underline{\underline{S}}(q) \underline{\Delta B} \tag{26}$$

The background component (0) enters into the above equations only where written explicitly (i.e. $i,j,p \neq 0$ for Eqs. 21–25).

The RPA equations can be written for many different scenarios. We will now illustrate a few different examples of how the above equations can be used.

The first example involves the simplest situation of a blend of two homopolymers. In this case the homopolymers are labeled 0 (the background component) and 1. The following equations describe the components of $\underline{\underline{S}}^0(q)$, $\underline{\underline{V}}(q)$ and $\underline{\Delta B}$ (as there are only two components in this mixture, these matrices and vector become scalar quantities):

$$S_{11}^0(q) = N_1 \phi_1 P_1(q) \tag{27}$$

$$V_{11}(q) = \left[\frac{1}{N_0 \phi_0 P_0(q)} - 2\chi_{10} \right] \tag{28}$$

$$\Delta B_1 = B_1 - B_0 \tag{29}$$

Therefore,

$$I(q) = v_{ref}\underline{\Delta B}^T \underline{\underline{S}}(q)\underline{\Delta B} = (B_1 - B_0)v_{ref}\left[\frac{1}{N_1\phi_1 P_1(q)} + \left(\frac{1}{N_0\phi_0 P_0(q)} - 2\chi_{10}\right)\right]^{-1}(B_1 - B_0) \tag{30}$$

Simplifying:

$$I(q) = (B_1 - B_0)^2 v_{ref}\left[\frac{1}{N_1\phi_1 P_1(q)} + \frac{1}{N_0\phi_0 P_0(q)} - 2\chi_{10}\right]^{-1} \tag{31}$$

which is identical with the result of substituting Eq. 17 into Eq. 14 [13].

We will now consider the situation in which a homopolymer is blended with a diblock copolymer. The homopolymer will be the reference component, 0, and the blocks of the diblock copolymer will be labeled 1 and 2. In the interest of brevity, we will not write out all of the equations; however, Eqs. 20–26 are used directly to describe $I(q)$. In this case $\underline{\underline{S}}^0(q)$ and $\underline{\underline{V}}(q)$ are now 2×2 matrices and $\underline{\Delta B}$ is a two-component column vector. The scattering intensity obtained from a pure diblock copolymer is then described by taking the limit $\phi_0 \to 0$ of $I(q)$ obtained from the homopolymer–diblock copolymer blend. In this case, the background homopolymer can be chosen to have simple properties such as $N_0 = 1$ and $\chi_{i0} = 0$. The result of taking this limit is given by Eq. 32, the scattering intensity from a pure diblock copolymer:

$$I(q) = (B_1 - B_2)^2 v_{ref}\left[\frac{S_{11}^0 + S_{22}^0 + 2S_{12}^0}{S_{11}^0 S_{22}^0 - (S_{12}^0)^2} - 2\chi_{12}\right] \tag{32}$$

which is the well-known result for scattering from disordered diblock copolymers derived by Leibler [20].

Due to divergence problems (ϕ_0 appears in the denominator of many terms), this limit is difficult to take analytically for any situations more complicated than a homopolymer–diblock copolymer blend. For instance, one could imagine calculating the structure factor of a mixture of two diblock copolymers. In this case, a background homopolymer component would be included initially and again the limit $\phi_0 \to 0$ of $I(q)$ would be taken to obtain the pure diblock copolymer mixture result. For this relatively simple case, it is difficult to obtain an analytical result even after using software such as Mathematica. In such cases, $I(q)$ can be computed numerically using Eq. 26 by simply setting ϕ_0 equal to a very small number (e.g. 10^{-4}; the number chosen for the computation will depend on the number of significant figures maintained in the calculations) and ensuring that the computed $I(q)$ does not depend on the value of ϕ_0. Other approaches for computing scattering from complex mixtures are outlined in [21].

The above procedures can be used to describe scattering from a variety of mixtures. Although we have not explicitly discussed molecules that are not

Gaussian chains, the framework developed is perfectly general and can be used to study multicomponent mixtures containing these kinds of molecules, as described elsewhere [19]. The effect of polydispersity on the RPA equations is described in [22].

5.4
Mixtures Containing Non-overlapping Hard Objects

In the case where the objects of interest are not deformable and cannot overlap (e.g. suspensions of hard spheres, hard rods, etc.), the intra- and intermolecular contributions to scattering in Eqs. 10 and 14 can be factored into a product of two terms:

$$I(\mathbf{q}) = (B_1 - B_2)^2 v_{\text{ref}} \phi_1 N_1 P_1(q) S_{s,11}(\mathbf{q}) \tag{33}$$

where $S_{s,11}(\mathbf{q})$ is the portion of the structure factor that captures the relative arrangement of the center of masses of the non-overlapping objects. Note that Eq. 33 does not apply in the case of overlapping objects such as polymer blends where molecules freely invade the "space" occupied by neighboring molecules. It is straightforward to see that Eq. 31 is not consistent with Eq. 33.

Equation 33 can be used to quantify the organization of hard objects. Measurements of $I(q)$ can be used to determine $S_{s,11}(q)$ provided that the contrast factor and $P(q)$ are known, as is often the case (see Eqs 16a–f, for example). In a few cases, analytical expressions for $S_{s,11}(q)$ are available. For a homogenous mixture of non-interacting hard spheres, $S_{s,11}(q)$ is a function only of the magnitude of $\mathbf{q}$ and is approximated by the Percus–Yevick equation [23]:

$$S_{s,11}(q) = \frac{1}{1 + 24\phi G/2qR} \tag{34}$$

where R is the radius of the spheres and

$$\begin{aligned}G = &\{a(\phi)/[4(qR)^2]\}[\sin(2qR) - 2qR\cos(2qR)] \\&+ \{\beta(\phi)/[8(qR)^3]\}\{4qR\sin(2qR) + [2 - 4(qR)^2]\cos(2qR) - 2\} \\&+ \gamma(\phi)/[32(qR)^5](-16(qR)^4\cos(2qR) \\&+ 4\{[12(qR)^2 - 6]\cos(2qR) + [8(qR)^3 - 12qR]\sin(2qR) + 6\})\end{aligned} \tag{35}$$

$$a = (1 + 2\phi)^2/(1 - \phi)^4 \tag{36}$$

$$\beta = -6\phi(1 + \phi/2)^2/(1 - \phi)^4 \tag{37}$$

$$\gamma = (\phi/2)(1 + 2\phi)^2/(1 - \phi)^4 \tag{38}$$

5.5
Disordered Isotropic Systems

In some systems, assuming that the two-point correlation function has a particular functional form, Eq. 8b can be used to calculate the structure factor based on the two-point correlation function. The structure factor can then be inserted directly into Eq 14 to find $I(q)$. A typical example where this approach is valid is in the vicinity of critical points of binary mixtures where the fluctuations are much larger in size than that of individual molecules. The Ornstein–Zernike scattering equation, which is applicable to binary mixtures near the critical point, is based on the following correlation function [24]:

$$p(r) = \frac{1}{r}\exp(-r/\xi) \tag{39}$$

where $r = |\mathbf{r}|$ and ξ is the correlation length. This results in the following scattering intensity profile:

$$I(q) = \frac{C}{(1 + q^2\xi^2)} \tag{40}$$

If

$$p(r) = \exp(-r/\xi) \tag{41}$$

an expression that is often used to describe phase-separated mixtures, then $I(q)$ is given by the Debye–Beuche equation [25]:

$$I(q) = \frac{C}{(1 + q^2\xi^2)^2} \tag{42}$$

If

$$p(r) = \exp[(-r/\xi)^2] \tag{43}$$

then $I(q)$ is given by the Guinier equation [4]:

$$I(q) = C\exp(-q^2\xi^2/4) \tag{44}$$

Finally, for a periodic structure with limited long-range order, a correlation function that combines the Ornstein-Zernike correlation function with a sine wave is used:

$$p(r) = \frac{d}{2\pi r}\exp(-r/\xi)\sin(2\pi r/d) \tag{45}$$

where d is the periodicity. This results in the Teubner–Strey intensity profile used to describe microemulsions [26]:

$$I(q) = \frac{C}{a_2 + c_1 q^2 + c_2 q^4} \qquad (46)$$

where a_2, c_1 and c_2 are constants.

In Eqs. 40, 42, 44 and 46, C is a constant related to the contrast.

5.6
Ordered Arrays

Equation 33 also applies to ordered arrays of objects with either crystalline or liquid crystalline symmetry. For the case of crystals, $S_{s,11}(\mathbf{q})$ contains delta functions at values of $\mathbf{q}$ that satisfy

$$\begin{pmatrix} q_1 \\ q_2 \\ q_3 \end{pmatrix} = 2\pi \begin{pmatrix} h/a_1 \\ k/a_2 \\ l/a_3 \end{pmatrix} \qquad (47)$$

where $\mathbf{a_1}$, $\mathbf{a_2}$ and $\mathbf{a_3}$ are vectors that define the unit cell, q_1, q_2 and q_3 are components of vector $\mathbf{q}$ resolved along the 1, 2 and 3 directions of the unit cell, respectively, and h, k and l are integers also known as Miller indices. In liquid crystalline systems, periodic order is restricted to either one or two dimensions. In these systems, Eq. 47 applies along the directions of order only.

The scattering from crystalline and liquid crystalline systems is analyzed using standard crystallographic techniques [4, 27]. Ordered arrays in polymer systems are typically found in crystalline polymers, block copolymer melts and multicomponent mixtures containing block copolymers. Scattering from unoriented samples is analyzed based on powder diffraction methods, whereas scattering from oriented samples is analyzed on the basis of fiber and single-crystal diffraction, depending on the nature of the sample. Peaks in the $S_{s,11}(q)$ are modulated by the functional form of $P(q)$. In most cases, the symmetry of the ordered array is determined from values of $\mathbf{q}$ (both magnitude and direction in the case of aligned phases) where scattering peaks are observed. If the ordered objects and matrix are composed of pure components then Eq. 33 is applicable. However, in many cases the ordered objects are themselves mixtures of polymer chains. For example, there is considerable mixing of chains in ordered phases formed by block copolymers in the vicinity of the order–disorder transition. Strictly, analysis of such systems should be begin with Eq. 10, as neither Eq. 33 nor equations for homogeneous mixtures such as Eqs. 31 and 32 are applicable.

5.7
Examples

We will now discuss various applications of the scattering equations presented above. All of the examples are derived from previous neutron scattering work by our group.

5.7.1
Scattering from Homogeneous Binary Homopolymer Blends

The simplest application involves determining Flory–Huggins interaction parameters from the scattering profiles obtained from binary homogeneous polymer blends [28, 29]. We study three binary homogeneous blends composed of the following polymers: A is saturated polybutadiene with 89% 1,2-addition (labeled hPB89 or dPB89 depending on whether it is hydrogenated or deuterated), B is polyisobutene (labeled PIB) and C is saturated polybutadiene with 63% 1,2-addition (labeled hPB63 or dPB63). The characteristics of polymers A, B and C used here are given in Table 5.5 and the compositions of the binary blends are given in Table 5.6. One of the components of the blends is deuterated in order to provide neutron scattering contrast (the other component is completely hydrogenous). The scattering profiles obtained from the three blends as a function of temperature are shown in Fig. 5.4. The solid curves are the result of fitting Eq. 31 to the data at each temperature for each blend. The values of χ_{ij} and l_i were obtained by fitting the experimental data to Eq. 31. The temperature de-

Table 5.5 Characteristics of polymers used in examples [a].

Name	M_w (kg mol^{-1})	PDI	ρ (g mL^{-1})	1,2-Addition (%)	n_D
hPB89(10)	10.1	1.01	0.8625	89.1	NA
dPB89(10)	10.6	1.01	0.9020	89.1	2.54
dPB89(24)	25.3	1.01	0.9070	90.4	2.79
hPB89(35)	34.9	1.02	0.8639	90.1	NA
dPB89(35)	36.5	1.02	0.9037	90.1	2.56
dPB63(10)	10.5	1.02	0.9125	61.6	3.44
PIB(13)	12.5	1.04	0.9134	NA	NA
PIB(24)	24.0	1.05	0.9131	NA	NA
PIB(45)	44.6	1.04	0.9140	NA	NA
hPBPB(79–66)	78.5–65.4	1.01	0.8639	89.7–63.9	NA
hPBPB(41–38)	41.3–37.6	1.01	0.8633	91.9–62.7	NA
dPBPB(41–38)	43.5–39.6	1.01	0.9098	91.9–62.7	2.99

a) Polymer names are based on the hydrogenated polymer molecular weight; M_w is the weight-average molecular weight; PDI is the polydispersity index, PDI = M_w/M_n, where M_n is the number-average molecular weight; ρ is the average density; n_D is the number of deuterium atoms per C_4 repeat unit.

Table 5.6 Compositions of binary blends.

Blend	Component A	Component B	Component C	ϕ_A	ϕ_B	ϕ_C
B1	dPB89(10)	PIB(45)	–	0.673	0.327	–
B2	hPB89(10)	–	dPB63(10)	0.493	–	0.507
B3	–	PIB(13)	dPB63(10)	–	0.477	0.523

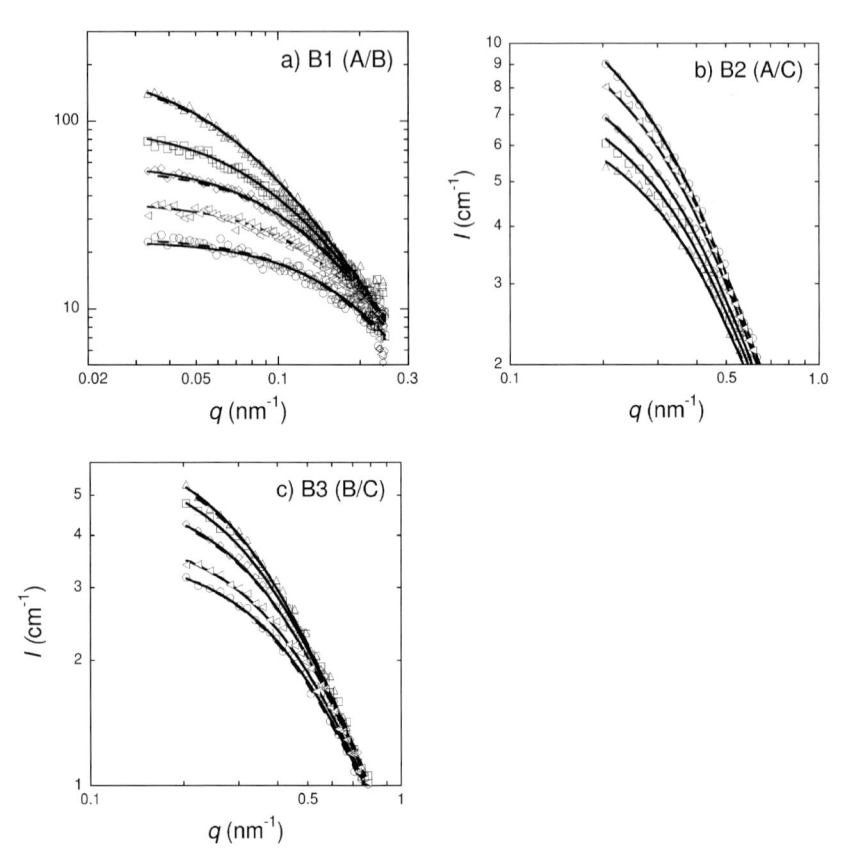

Fig. 5.4 SANS from binary blends of (a) A–B homopolymers, (b) A–C homopolymers and (c) B–C homopolymers at selected temperatures. The curves are the random phase approximation fit to the data with χ_{mn} and l_m as adjustable parameters. Temperatures: (a) 30 (○), 60 (△), 90 (◇), 119 (□), 169 °C (△); (b) and (c) 29 (○), 50 (△), 92 (◇), 134 (□), 175 °C (△).

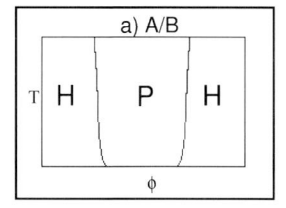

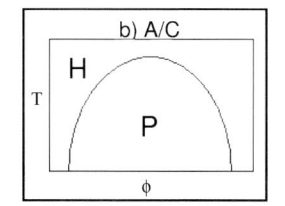

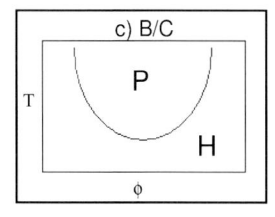

Fig. 5.5 Binary phase diagrams resulting from the experimentally determined Flory–Huggins interaction parameters for the (a) A–B blend, (b) A–C blend and (c) B–C blend.

pendences of the three Flory–Huggins interaction parameters thus obtained were as follows [29]:

$$\chi_{AB} = 0.00034 + 3.94\frac{1}{T} - 817\frac{1}{T^2} \tag{48}$$

$$\chi_{AC} = 0.00209 - 1.18\frac{1}{T} + 747\frac{1}{T^2} \tag{49}$$

$$\chi_{BC} = -0.00085 + 6.87\frac{1}{T} - 2480\frac{1}{T^2} \tag{50}$$

Typical uncertainties in measurement of χ range between 5 and 10% [30]. Furthermore, the values of $l_{mon,i}$ (based upon a monomer repeat unit) were found to be temperature independent: $l_{mon,A}=0.55$ nm, $l_{mon,B}=0.58$ nm and $l_{mon,C}=0.75$ nm.

Figure 5.5 is a qualitative drawing of the binary phase diagrams resulting from the above temperature dependences of χ_{ij}. We see three very different phase diagrams due to the qualitative differences in the temperature dependence of χ_{ij}. χ_{AB} is almost independent of temperature and therefore the A–B phase diagram is nearly independent of temperature. χ_{AC} is positive at low temperatures, indicating repulsion between A and C chains and then upon heating the degree of repulsion weakens as χ_{AC} decreases. Therefore, we would expect that an A–C blend would be macrophase separated at low temperatures and then upon heating would homogenize (Fig. 5.5 b). Finally, χ_{BC} is negative at low temperature, indicating attraction between B and C chains. This attraction lessens upon heating as χ_{BC} increases. Figure 5.5c indicates that the B–C blend would be homogeneous at low temperatures and macrophase separated at high temperatures.

These three blends indicate the wide range of phase diagrams that one can obtain in binary homopolymer blends. The RPA-based equations can be used to quantify the degree of attraction or repulsion between the components in binary blends. We will now examine the manifestation of these interaction parameters in multicomponent blends.

5.7.2
Homogeneous Multicomponent Blends

We consider a homogeneous multicomponent blend (labeled blend M1) consisting of an A homopolymer, a B homopolymer and an A–C diblock copolymer [29]. A, B and C have the same chemical structures as in the binary blends utilized in the χ characterization (Section 5.7.1). However, higher molecular weight polymers were utilized in the multicomponent blend. The composition of the multicomponent blend is given in Table 5.7. We restrict our attention to mixtures where $\phi_A/\phi_B=1$, which is equal to the critical composition for a blend with no block copolymer. The main source of scattering contrast in this blend is due to the fact that homopolymer A is deuterated. However, there are non-negligible contributions to the scattering intensity due to the scattering length density differences between B and C chains in spite of the fact that they are not labeled with deuterium. The scattering profiles obtained from blend M1 at selected temperatures are shown in Fig. 5.6. The resulting profiles exhibit a broad peak that is consistent with either a homogeneous blend with periodic concentration fluctuations or a microphase separated structure with limited or no long-range order. Similar confusion reigned in the early measurements of scattering from diblock copolymer melts near the order–disorder transition [31]. Multicomponent RPA can be used to predict the scattering profile of homogeneous A/B/A–C mixtures as we have pre-determined all of the parameters that are needed. Equations 20–26 are used to calculate $I(q)$. In this case $\underline{S}^0(q)$ and $\underline{V}(q)$ are now 3×3 matrices $\underline{\Delta B}$ and is a three-component column vector. The parameters needed for the calculations are polymer characteristics that can be determined independently (N_i and ϕ_i) and parameters that we have already determined from binary blends (χ_{ij} and l_i). This approach makes the assumption that the molecular weight and compositional dependence of χ_{ij} is negligible. In addition, it assumes that χ_{AC} obtained from a binary blend can be used to describe the interactions of the A and C blocks of the diblock copolymer. Finally, the effect of deuteration on χ_{ij} has been ignored.

Comparisons between theory (curves) and experiment (symbols), obtained at 150, 169 and 189 °C, are shown in Fig. 5.6. The solid curves are theory with no adjustable parameters. The RPA theory accurately captures the low-q plateau of $I=10$ cm^{-1}, the location of the scattering peak at $q \approx 0.12$ nm^{-1} and the $I \approx q^{-2}$ tail at high q for all three temperatures. The theory also captures the peak intensity at 169 and 189 °C, but over-predicts the peak intensity at 150 °C. The pre-

Table 5.7 Compositions of multicomponent blends.

Blend	Component A	Component B	Component A–C	ϕ_A	ϕ_B	ϕ_{A-C}
M1	dPB89(24)	PIB(24)	hPBPB(79–66)	0.246	0.254	0.500
M2	dPB89(35)	PIB(45)	hPBPB(41–38)	0.316	0.284	0.400
M3	hPB89(35)	PIB(45)	dPBPB(41–38)	0.316	0.284	0.400

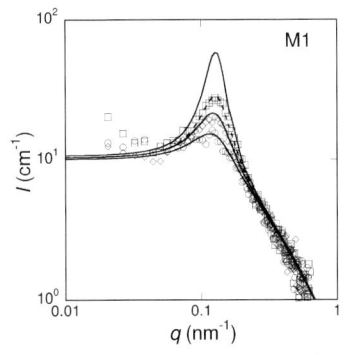

Fig. 5.6 RPA predictions with no adjustable parameters (solid curves) and SANS profiles (open circles) for blend M1 at 150 (□), 169 (◇) and 189 °C (○). The dotted curve corresponds to adjusting χ_{AB} such that the theory matches the data.

sence of the scattering peak indicates the presence of periodic concentration fluctuations with a characteristic length scale, $d = 2\pi/0.12 \text{ nm}^{-1} = 52$ nm. The increasing peak intensity with decreasing temperature is an indication of microphase separation. The agreement between theory and experiment in Fig. 5.6 is remarkable considering that all of the parameters needed for the RPA calculations were obtained from independent experiments. There are slight differences between theory and experiment at 150 °C. If we hold two of the three χ parameters constant and change the third, the theory exactly overlaps the small-angle neutron scattering (SANS) data. If we hold χ_{AC} and χ_{BC} constant, the value of χ_{AB} required to fit the data is 0.0049. The result of this fit is shown as the dotted line in Fig. 5.6. The value of χ_{AB} determined from the binary homogeneous blend at 150°°C was 0.0051. Hence the differences in the χ parameters obtained from fitting the multicomponent data at 150 °C are within the error of measuring χ from binary homogeneous blends. We take the quantitative agreement between RPA and the measured SANS profiles as an indication that blend M1 is homogeneous at temperatures between 150 and 189 °C. These data also demonstrate that χ parameters obtained from binary homopolymer blends can be used to predict quantitatively the scattering from homogeneous multicomponent mixtures.

5.7.3
Microphase Separated Multicomponent Phases

Here we discuss neutron scattering from microphase separated phases observed from A/B/A–C blends. [28] The compositions of the blends discussed here are also given in Table 5.7. A, B and C have the same chemical structures as in the binary blends utilized in the χ characterization (Section 5.7.1). However, higher molecular weight polymers were utilized in the multicomponent blend. Two analogous multicomponent blends were created, one in which homopolymer A was deuterated (blend M2) and the second with the diblock copolymer deuterated (blend M3). In all other respects, the two blends are identical (Table 5.7). The scattering profiles obtained from blends M2 and M3 are shown in Fig. 5.7. We will assume that blends M2 and M3 have the same structure. By labeling different components we are only emphasizing different aspects of the same structure.

We conducted two independent runs on sample M2. SANS profiles from the two separated runs on sample M2 at selected temperatures are shown in Fig. 5.7a. At temperatures between 30 and 90 °C, the SANS data for both runs show a peak at q values between 0.10 and 0.12 nm^{-1}, indicating the presence of periodic structures. The intensities of the peaks at 30 and 50 °C for the two runs differ substantially. In addition, the two-dimensional scattering profile from one run was azimuthally inhomogeneous whereas that from the other was homogeneous. These are characteristics of periodic phases with long-range order (e.g. lamellae). In contrast, at 70 and 90 °C the scattering profiles for the two runs are identical and azimuthally homogeneous. These are characteristics of disordered phases (e.g. a microemulsion).

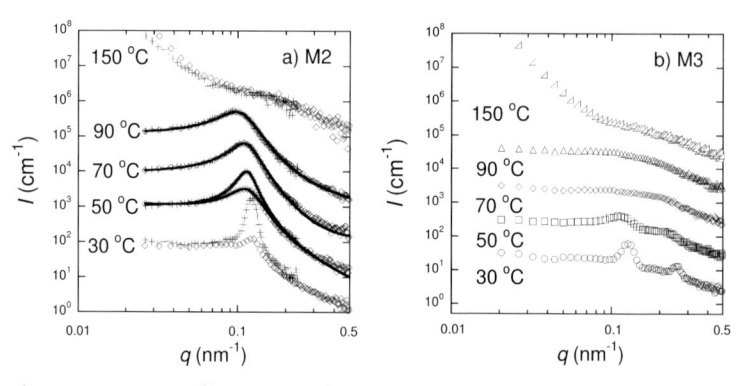

Fig. 5.7 (a) SANS profiles obtained from two independent runs of blend M2 at selected temperatures (30, 50, 70, 90 and 150 °C). The open diamonds are the data taken from run 1 and the plus signs are the data taken from run 2. The solid lines are the Teubner–Strey scattering profile fitted to the data. In order to delineate the data, each data set has been multiplied by the following factors: 1 (30 °C), 10 (50 °C), 10^2 (70 °C), 10^3 (90 °C), 10^5 (150 °C). (b) SANS profiles obtained from blend M3 at selected temperatures: 30 °C (○), 50 °C (data set multiplied by 10) (□), 70 °C (data set multiplied by 10^2) (◇), 90 °C (data set multiplied by 10^3) (△) and 150 °C (data set multiplied by 10^4) (⊿).

The scattering from microemulsions in the vicinity of the primary maximum is often described by the Teubner–Strey (TS) equation (Eq. 46). The curves in Fig. 5.7a are the least-squares fits of Eq. 46 through the data with a, b and c as adjustable constants. We take agreement between the TS equation and our data to be indicative of the presence of a microemulsion. Based on this, the agreement between the scattering profiles obtained from M2 and their azimuthal homogeneity, we conclude that blend M2 is a microemulsion at 70 and 90 °C. It is perhaps appropriate to add here that our attempts to fit the 30 °C data of blend M2 to the TS equation were unsuccessful.

In order to understand the nature of the phase at 30 °C in blend M2, we also examined blend M3 with neutron scattering. The scattering data from blend M3 are shown in Fig. 5.7b. At 30 °C the presence of two peaks, at $q_1 = 0.127$ nm^{-1} and $q_2 = 0.254$ nm^{-1}, is observed. The fact that $q_2 = 2q_1$ indicates the presence of a lamellar phase. This is consistent with our analysis of the data from blend M2 at 30 °C. We therefore conclude that blend M2 is lamellar at 30 °C.

The state of M2 at 50 °C is not clear. The two scattering peaks at $q_1 = 0.113$ nm^{-1} and $q_2 = 0.218$ nm^{-1} ($q_2 = 2q_1$ within experimental error) observed from blend M3 (Fig. 5.7b) and our observations of the scattering profiles from blend M2 (Fig. 5.7a) suggest that we have a lamellar phase. However, the blend M2 data are consistent with the TS equation, which suggests that the blend is a microemulsion. Based on the criteria that we have established above, we therefore cannot uniquely determine the structure of blend M2 at 50 °C. The Gibbs phase rule requires a region of coexistence between two multicomponent one phase systems. Perhaps our difficulties in determining the structure of blend M2 at 50 °C are due to the coexistence of lamellae and microemulsions. The data obtained from blend M3 at 70 and 90 °C shown in Fig. 5.7b do not show pronounced features and hence do not affect our conclusion based on blend M2 that the sample is a microemulsion at these temperatures.

There are many aspects of the scattering profiles of blends M2 and M3 that cannot be understood on the basis of standard crystallography. It is clear, for example, that the Bragg peaks from M2 and M3 lie on top of a monotonically decaying scattering background. This is made clear in Fig. 5.8, where we plot $I(q)$ obtained from M3 for two selected temperatures (30 and 70 °C). The background is undoubtedly due to the connectivity of the chains and the fact that the lamellar phases contain mixtures of A, B and C chains.

In order to focus on the Bragg peaks, we use Eq. 16a to approximate the background. Lacking a full theory of scattering from multicomponent lamellar phases, we assume that the background is given by

$$I_{\text{background}}(q) = KP_{1\_\text{chain}}(x) = K\frac{2(e^{-x} + x - 1)}{x^2} \tag{51}$$

where $x = q^2 R_{g,1}^2$ and K is a fitting parameter that is related to the scattering length densities of the components in the mixture. The curves in Fig. 5.8 are fits of Eq. 51 with K and $R_{g,1}$ as adjustable parameters. The difference between

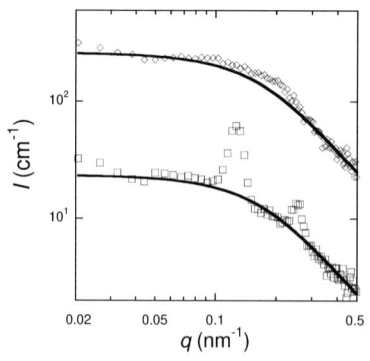

Fig. 5.8 SANS data obtained from M3 at 30 (□) and 70 °C (◇). The data at 70 °C were multiplied by a factor of 10. The Debye function, $I = K[\exp(-R_g^2 q^2) + R_g^2 q - 1]/(R_g^2 q^2)^2$, was fitted to the data with the radius of gyration (R_g) and contrast (K) as adjustable parameters (solid lines).

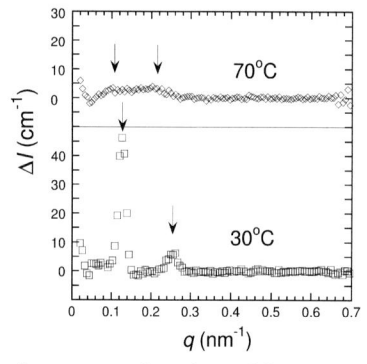

Fig. 5.9 SANS data obtained from M3 at 30 (□) and 70 °C (◇) with the Debye function subtracted from the data. For clarity, each data set has a separate ΔI axis. The arrows point to the locations of the primary and secondary peaks determined through methods described in the text.

the SANS intensity (I) and the Debye function ($I_{background}$) allows us to focus on changes in the structure of blend M3 with temperature. The quantity $\Delta I = I - I_{background}$ allows us to focus on the lamellar phase. In Fig. 5.9 we plot ΔI versus q of M3. As expected the 30 °C data in Fig. 5.9 show a primary peak at $q = 0.127$ nm^{-1} and a second order peak at $q = 0.254$ nm^{-1}. In Fig. 5.9 we also show a ΔI versus q plot obtained at 70 °C from M3. Not surprisingly, both primary and second-order peaks in ΔI diminish in intensity as temperature is increased. The relative changes in the scattering intensity of the two peaks are, however, surprising. One generally expects the second-order peak to decrease in intensity faster than the primary peak. In sample M3 we see that ΔI of the primary peak at $q = 0.127$ nm^{-1}, ΔI_1, changes from 47 to 3 cm^{-1} when the sample temperature is increased from 30 to 70 °C. In contrast, ΔI of the secondary peak at $q = 0.245$ nm^{-1}, ΔI_2, changes from 6 to 3 cm^{-1} in the same temperature window.

The scattering profiles from M2 and M3 therefore contain two puzzling observations:

1. why is the higher order peak seen in M3 and not M2?
2. why is the intensity of the second-order peak in M3 less sensitive to temperature than that of the primary peak?

M2 and M3 are microphase separated at temperatures between 30 and 90 °C [28]. Self-consistent field theory (SCFT) [28, 29, 32–34] provides a convenient framework for determining the equilibrium composition of the microphases. For simplicity, we conducted the SCFT calculations in one dimension (see [28, 29] for details). We use a Cartesian coordinate frame with the z-axis oriented along the lamellar normal (in our previous publications, z was a dimensionless variable; in this chapter, z is dimensional). The SCFT calculations determined the partition function of the A/B/A–C system by modeling Gaussian polymer chains in an external field. The external field accounted for all of the interactions in the system, which have been previously characterized through the Flory–Huggins χ parameters. There were therefore no adjustable parameters. From the partition function of the system, the SCFT calculations allowed the determination of the z dependence of the volume fractions of all of the four components in our A/B/A–C mixture. Figure 5.10 shows a typical theoretical composition profile calculated using SCFT. The volume fraction profiles, $\phi_i(z)$, are plotted for blends M2 and M3 at 30 °C with a lamellar spacing of 50 nm ($\phi_i(z)$ is assumed to be the same for both blends). We only show $\phi_i(z)$ of half a lamella because the other half is a mirror image of the results presented in Fig. 5.10. Knowledge of the scattering length density of each component allows the determination of the scattering length density profiles within the microphases. The SCFT-based scattering length density profiles, $B(z)$, which are the average scattering length densities of the sample at position z, are obtained by summing over the scattering length density of each component multiplied by the volume fraction of that component at position z. $B(z)$ is shown for M2 and M3 in Figs. 5.11 and 5.12, respectively. We only show $B(z)$ of half a lamella because the other half is a mirror image of the results presented in Figs. 5.11 and 5.12. The calculated lamellar sizes (twice the z_{max} for each profile) change with temperature, as can be seen in Figs. 5.11, 5.12. These changes are in quantitative agreement with experimental measurements ($d = 2\pi/q_1$). The qualitative differences in $B(z)$ of M2 and M3 are clear. In M2, the deuterated component is concentrated in the A-rich domain and hence the $B(z)$ profile is similar to a $\cos(2\pi z/d)$ wave. In M3, the deuterated components are concentrated at the A–B interface and there are two such interfaces in each period. Hence the $B(z)$ profile is similar to a $\cos(4\pi z/d)$ wave.

Quantitative comparisons of the SCFT-based calculations and the scattering profiles are possible by Fourier analysis. $B(z)$ profiles were fitted to a four-term Fourier series:

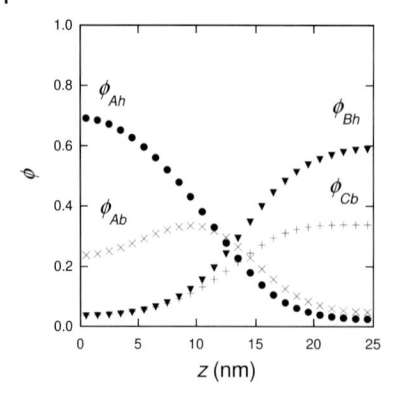

Fig. 5.10 Volume fraction profiles of the components, calculated using SCFT for blends M2 and M3 at 30 °C with a box size of 25 nm. The components are homopolymer A (Ah) (•), homopolymer B (Bh) (▼), the A block of the diblock copolymer (Ab) (×) and the C block of the diblock copolymer (Cb) (+).

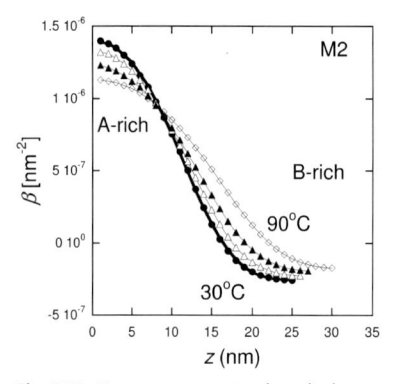

Fig. 5.11 Neutron scattering length density, versus distance, z, for half a lamellar spacing for blend M2 at 30 (•), 50 (△), 70 (▲) and 90 °C (◇). The solid line through the 30 °C data is a fit using Eq. 52.

$$B(z) = B_0 + \sum_{n=1}^{3} \Delta B_n \cos\left(\frac{2\pi z}{d/n}\right) \tag{52}$$

where d is the lamellar spacing, B_0 is a constant and ΔB_n are the fitted amplitudes of the cosine terms. We found that all of the features of the SCFT-based $B(z)$ profiles were captured by this truncated series. The curves in Figs. 5.11 and 5.12 are fits of Eq. 52 through the SCFT calculations. These curves permit the determination of ΔB_n.

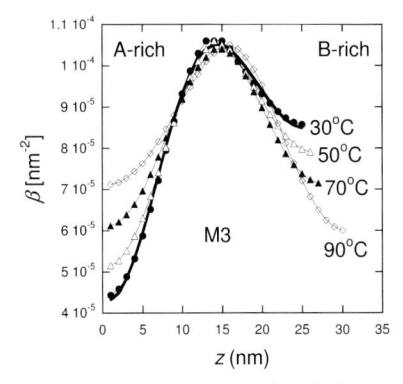

Fig. 5.12 Neutron scattering length density, B versus distance, z, for half a lamellar spacing for blend M3 at 30 (●), 50 (△), 70 (▲) and 90 °C (◇). The solid line through the 30 °C data is a fit using Eq. 52.

The amplitudes ΔB_1, ΔB_2 and ΔB_3 for blend M2 at 30 °C were 8.7×10^{-7}, 1.0×10^{-7} and -4.2×10^{-8}, respectively. Since the scattering intensity is proportional to the square of the difference in scattering length densities, I at $q = 2\pi n/d$ is proportional to $(\Delta B_n)^2$ [1]. In M2 we therefore expect that the intensities for the second- and third-order peaks will be almost two orders of magnitude smaller than that for the primary peak. The relative magnitudes of ΔB_1, ΔB_2 and ΔB_3 did not change significantly between 30 and 90 °C. It is therefore no surprise that higher order peaks were not observed in blend M2 (Fig. 5.7 a). In contrast, the Fourier amplitudes ΔB_1, ΔB_2 and ΔB_3 for blend M3 at 30 °C are 1.9×10^{-7}, 2.0×10^{-7} and 1.8×10^{-8}, respectively. Note that ΔB_1 and ΔB_2 are now comparable in magnitude, whereas ΔB_3 is negligible. We therefore expect to observe both primary and secondary peaks in blend M3 (Fig. 5.7 b).

Careful examination of the SCFT results at 30 °C for M3 show that the scattering length density profile is clearly asymmetric (Fig. 5.12). The scattering length density on the A-rich side of the lamellae is 4.4×10^{-5} nm^{-2}, whereas that on the B-rich side is 8.7×10^{-5} nm^{-2}, because there is much more diblock copolymer in the B-rich than in the A-rich region. The scattering intensity at the primary peak location in M3, $q = 2\pi/d$, is due to this asymmetry. The asymmetry in $B(z)$ decreases with increasing temperature. At 70 °C, the difference in scattering length densities on the A- and B-rich sides is less than 1.0×10^{-5} nm^{-2}. The disappearance of the primary peak in blend M3 at 70 and 90 °C is due to the fact that the scattering contrast between the coexisting microphases, which is responsible for the primary scattering peak, diminishes. In contrast, there is little change in the peak in $B(z)$ at $z = 15$ nm (Fig. 5.12). This indicates that the scattering intensity at $q = 4\pi/d$ should be unaffected by temperature.

In Fig. 5.13, we compare the experimentally determined changes in ΔI_1 and ΔI_2 of M3 with SCFT-based calculation of ΔB_1^2 and ΔB_2^2. Both quantities ΔI_i and ΔB_i^2 are normalized so that they are set to unity at 30 °C. The experimental

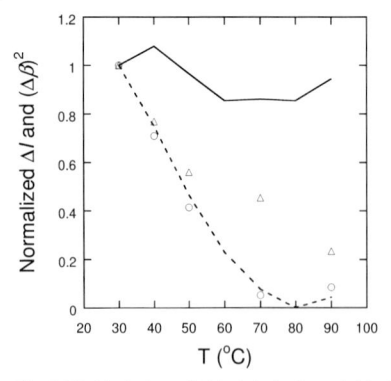

Fig. 5.13 Variation of ΔI_1 (circles) and ΔI_2 (triangles) with temperature obtained by SANS and $(\Delta B_1)^2$ (dotted curve) and $(\Delta B_2)^2$ (solid curve) as predicted from the SCFT neutron scattering length density profile. Both ΔI_i and $(\Delta B_i)^2$ were normalized with the value at 30 °C.

results in Fig. 5.13 (symbols) re-emphasize the fact that ΔI_2 (triangles) decreases more slowly with increasing temperature than ΔI_1 (circles). This qualitative trend is also seen in the SCFT calculations (curves). ΔB_1^2 decreases rapidly with increasing temperature (dotted line) whereas ΔB_2^2 does not (solid line). The agreement between the temperature dependence of ΔI_1 and ΔB_1^2 is remarkably good. The experimentally determined temperature dependence of ΔI_2 is considerably more drastic than that of ΔB_2^2. We therefore see that our approach alone does not capture quantitatively all of the experimental observations.

The Fourier analysis thus provides quantitative explanations for the puzzling nature of scattering profiles from M2 and M3:
1. The second-order peak is seen in M3 and not M2 because $\Delta B_2^2/\Delta B_1^2$ of M3 is significantly larger than that of M2.
2. The intensity of the second-order peak in M2 is less sensitive to temperature than that of the primary peak because ΔB_1^2 is a sensitive function of temperature whereas ΔB_2^2 is not.

Our analysis of scattering from microphase-separated systems is by no means as rigorous as that from homogeneous systems. The effect of chain connectivity is lumped into $I_{background}(q)$ in a manner that is not rigorous. Better agreement between theory and experiment may be obtained by conducting experiments on aligned single crystals. Nevertheless, we show here that careful analysis of the peak intensity of ordered phases gives unique insight into the nature of the microphase separated states. This is in contrast to most of the literature on scattering experiments from ordered polymer samples, where it is customary to analyze the values of **q** where peaks appear and ignore the intensity at the peak.

5.8
Conclusion

We have described a comprehensive framework for analyzing light, X-ray and neutron scattering from a variety of polymer systems. This includes homogeneous single-phase systems, microphase-separated systems and ordered and disordered suspensions of hard objects. We have made no attempt to summarize the vast literature on scattering from polymer systems. We have restricted our attention to coherent elastic scattering wherein the scattering data can be used to obtain structural information. Many important branches of scattering, such as inelastic scattering, resonant scattering and scattering from grazing incident beams, have not been covered.

References

1 Roe, R. J. *Methods of X-Ray and Neutron Scattering in Polymer Science*, Oxford University Press, New York, **2000**.
2 Higgins, J. S., Benoit, H. C. *Polymers and Neutron Scattering*, Clarendon Press, Oxford, **1994**.
3 Graessley, W. W. *Polymer Liquids and Networks, Structure and Properties*, Garland Science, New York, **2003**.
4 Guinier, A. *X-Ray Diffraction in Crystals, Imperfect Crystals and Amorphous Bodies*, Dover, New York, **1994**.
5 Kerker, M. *The Scattering of Light*, Academic Press, New York, **1969**.
6 X-rays have a $\pi/2$ phase shift.
7 Debye, P. *Journal of Physical and Colloid Chemistry* **1947**, *51*, 18–32.
8 Rayleigh. *Proceedings of the Royal Society of London Series A* **1914**, *90*, 219–225.
9 Kratky, O., Porod, G. *Journal of Colloid Science* **1949**, *4*, 35–70.
10 Neugebauer, T. *Annalen der Physik* **1942**, *42*, 509–533.
11 Mazur, J., McIntyre, D., Wims, A. M. *Journal of Chemical Physics* **1968**, *49*, 2896.
12 Meier, G., Momper, B., Fischer, E. W. *Journal of Chemical Physics* **1992**, *97*, 5884–5897.
13 de Gennes, P. G. *Scaling Concepts in Polymer Physics*, Cornell University Press, Ithaca, NY, **1979**.
14 Benoit, H., Benmouna, M., Wu, W. L. *Macromolecules* **1990**, *23*, 1511–1517.
15 Eitounim, H. B., Balsaram, N. P., Thermodynamics of polymer blends, in *Physical Properties of Polymers Handbook*, Mark, J. E. (Ed.), Springer, New York, **2006**, Chapter 19; based on Balsara, N. P., Thermodynamics of polymer blends, in *Physical Properties of Polymers Handbook*, Mark, J. E. (Ed.), AIP Press, New York, **1996**, Chapter 19, pp. 257–268.
16 Doi, M. *Introduction to Polymer Physics*, Clarendon Press, Oxford, **1995**.
17 Flory, P. J. *Principles of Polymer Chemistry*, Cornell University Press, Ithaca, NY, **1953**.
18 Akcasu, A. Z., Tombakoglu, M. *Macromolecules* **1990**, *23*, 607–612.
19 Hammouda, B. *Advances in Polymer Science* **1993**, *106*, 87–133.
20 Leibler, L. *Macromolecules* **1980**, *13*, 1602–1617.
21 Mori, K., Tanaka, H., Hashimoto, T. *Macromolecules* **1987**, *20*, 381–393.
22 Burger, C., Ruland, W., Semenov, A. N. *Macromolecules* **1990**, *23*, 3339–3346.
23 Percus, J. K., Yevick, G. J. *Physical Review* **1958**, *110*, 1–13.
24 Ornstein, L. S., Zernike, F. *Proceedings of the Koninklijke Akademie van Wetenschappen te Amsterdam* **1914**, *17*, 793–806.
25 Debye, P., Bueche, A. M. *Journal of Applied Physics* **1949**, *20*, 518–525.
26 Teubner, M., Strey, R. *Journal of Chemical Physics* **1987**, *87*, 3195–3200.

27 Cullity, B. D. *Elements of X-Ray Diffraction*, Prentice Hall, Upper Saddle River, NJ, **2001**.
28 Reynolds, B. J., Ruegg, M. L., Balsara, N. P., Radke, C. J., Shaffer, T. D., Lin, M. Y., Shull, K. R., Lohse, D. J. *Macromolecules* **2004**, *37*, 7401–7417.
29 Ruegg, M. L., Reynolds, B. J., Lin, M. Y., Lohse, D. J., Balsara, N. P. *Macromolecules* **2006**, *39*, 1125–1134.
30 Balsara, N. P., Fetters, L. J., Hadjichristidis, N., Lohse, D. J., Han, C. C., Graessley, W. W., Krishnamoorti, R. *Macromolecules* **1992**, *25*, 6137–6147.
31 Roe, R. J., Fishkis, M., Chang, J. C. *Macromolecules* **1981**, *14*, 1091–1103.
32 Helfand, E. *Journal of Chemical Physics* **1975**, *62*, 999–1005.
33 Evers, O. A., Scheutjens, J., Fleer, G. J. *Macromolecules* **1990**, *23*, 5221–5233.
34 Matsen, M. W. *Journal of Physics – Condensed Matter* **2002**, *14*, R21–R47.

6
From Linear to (Hyper) Branched Polymers: Dynamics and Rheology

Thomas C. B. McLeish

6.1
Introduction

The fascinating rheology of fluids containing flexible polymers flows from both necessity and beauty. Born of the rapid growth in synthetic polymer materials in the post-War years, the need to understand and control the processing of such highly viscoelastic liquids as polymer melts led rapidlyto the fundamental investigations of Flory [1], Stockmayer [2] and Edwards [3]. These in turn were building on work of Kuhn [4] (how large would macromolecules, linear or branched, be?) and Zimm [5] and Rouse [6] (how would such giant molecules move?). These pioneers were already using a beautiful notion that was to take hold of condensed-matter physics in the mid-20th century – that of *universality*, or the independence of physical phenomena from local, small-scale details. The emergence of universal properties is usually associated with "critical phenomena" [7], since near phase transitions, the spatial scale of correlated fluctuations may hugely exceed molecular dimensions. Properties that depend on these fluctuations, such as compressibility of a fluid near its critical point, are then insensitive to molecular detail. Although there is at first glance no apparent neighboring critical point in the case of polymeric fluids, both universality in exponents and renormalized quantities appear in abundance. Moreover, there is a natural large number associated with mesoscopic, rather than microscopic, length scales. The defining feature of a polymer is, after all, its large "degree of polymerization", N, or *molecular weight, M,* of the chains.

Such an approach has been particularly effective in the realm of *topological* effects. The polymer melts of industrial polymer processing are very highly overlapped on the molecular level, where it becomes immediately apparent that molecular relaxation processes controlling elastic stress are prolonged to very long times indeed. The classical "relaxation modulus" $G(t)$, measuring stress linear-response to a step strain, records a "plateau" value before a terminal relaxation time that increases rapidly with molecular weight. Experiments restricted to the time-scales of the plateau are hardly able to distinguish between the polymer

Macromolecular Engineering. Precise Synthesis, Materials Properties, Applications.
Edited by K. Matyjaszewski, Y. Gnanou, and L. Leibler
Copyright © 2007 WILEY-VCH Verlag GmbH & Co. KGaA, Weinheim
ISBN: 978-3-527-31446-1

melt and a rubber, in which the chains are permanently cross-linked to each other at very rare points, sufficiently for each chain to be permanently immobilized from large-scale diffusion. Conceptually, the absent "cross-links" were replaced in the minds of engineers and physicists alike by "entanglements" [8]. These loosely defined objects were assumed to represent the topological constraint that covalently bonded molecular chains may not pass through each other. The effective distance between these objects could be calculated, employing rubber elasticity theory, as

$$G_N^{(0)} = \frac{RT\rho}{M_e}$$

to deduce the degree of polymerization between entanglements, N_e, or the equivalent "entanglement molecular weight", M_e. The number N_e consistently turned out to be of the order of 10^2, a significant number, because it shows that such chains are already near-universal polymers. A remarkable universality also emerged in measurements of the scaling of melt viscosity η on the molecular weight M of many different polymer chemistries [8]:

$$\eta \sim M^1 \quad M < M_c$$
$$\eta \sim M^{3.4} \quad M > M_c \qquad (1)$$

For each material, a critical molecular weight, M_c, emerged, above which the viscosity rises very steeply with molecular weight. Moreover, within experimental error, this explicitly dynamic observation was linked phenomenologically to the essentially static measurements of the plateau modulus by the correlation $M_c \simeq 2M_e$. This connection between essentially dynamic (M_c) and static (M_e) experiments, observed over a wide range of chemistries, is strong evidence that topological interactions dominate both the molecular dynamics and the viscoelasticity at the 10-nm scale in polymer melts (and at correspondingly larger scales for concentrated solutions).

These results illustrate the importance of rheology as a window into the dynamics of entangled polymers. For as the topological constraints are relaxed by the relative diffusive motion of the polymers after a bulk strain to a sample, so the deformed subchains are able to re-equilibrate their orientational distribution. As this happens, so the measured elastic stress will fall from the rubber elastic plateau to zero in a way that matches exactly the ensemble average over the molecular dynamics. This time-dependent function, the "relaxation modulus":

$$G(t) = \sum_{i=1}^{N} g_i e^{-t/\tau_i} \qquad (2)$$

contains, by inversion of Eq. (2) the time-scale distribution for entanglement loss.

Without going beyond rheological measurements on bulk samples, there has long been other very strong evidence that molecular topology is the dominant physics in melt dynamics. This emerges from the phenomenology of "long-chain branched" (LCB) melts. These materials, commonly used in industry, possess identical molecular structure to their linear cousins on the local scale, but contain rare molecular branches. The density of branching varies from one branched carbon in every 10 000 to one in 1000. This level is chemically all but undetectable, yet the melt rheology is changed out of all recognition if the molecular weight is high enough [9]. Provided that $M \gg M_e$, the limiting low-shear viscosity may be much higher for the same molecular weight. Moreover, in strong extensional flows, the melt responds with a much higher apparent viscosity than in linear response. This phenomenon, vital for the stable processing properties of branched melts, is called "extension hardening". The effect is all the more remarkable because in shear flows, branched, in addition to linear, melts exhibit a lower stress than would be predicted by a continuation of their linear response [10] (they are "shear-thinning"). A fascinating example of the difference between linear and branched entangled melts is well known from flow visualization experiments. The velocity field in a strong "contraction flow" of a linear polymer melt resembles that of a Newtonian fluid, where that of a branched polymer sets up large vortices situated in the corners of the flow field. Slight changes to the topology of the molecules themselves give rise to qualitatively different features in the macroscopic fluid response. Driven by these results, careful anionic polymerization of star-shaped polymers began in the 1950s to create the simplest possible branched architecture [11–15]. The enormous increase in viscosity was at first thought to be due to uncontrolled gelation, but comparison between different chemistries indicates that it is the number of entanglements along the star polymer *arms* that dominates universally:

$$\eta \sim e^{v' M_a / M_e} \tag{3}$$

where M_a is the molecular weight of a single arm of the star and v' an order-one coefficient. More remarkably, the number of arms affects neither the viscosity nor the relaxation spectrum [16] (provided that this is not more than ~ 30, when new low-frequency effects and structures do arise [17, 18]). When the viscosity is plotted against the *span* molecular weight, defined as twice the arm molecular weight (Fig. 6.1), the viscosity is always greater than that of the analogous linear melt. The tracer diffusion constant of entangled stars is also exponentially dependent on the arm molecular weight, in the same manner as in Eq. (3), but the values of D can be many orders of magnitude smaller in an immobile matrix than in a homopolymer star melt.

The most successful of theoretical accounts of these observations has been the *tube model* of Edwards [19] and de Gennes [20]. A confining tube around each chain is invoked to represent the sum of all topological non-crossing constraints active with neighboring chains, and the tube radius, a, is of the order of the end-to-end length of a chain of molecular weight M_e. In this way, only

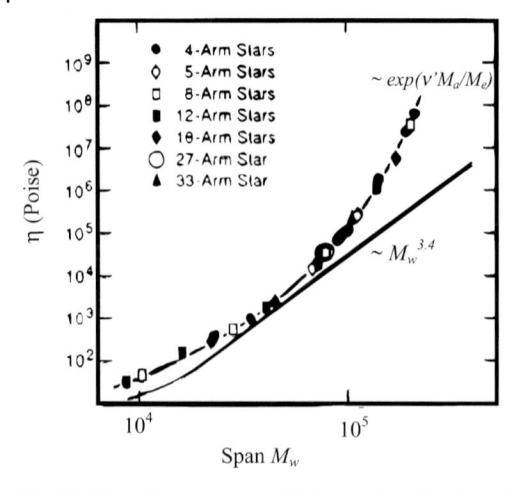

Fig. 6.1 Viscosity versus span (twice arm) molecular weight for a series of PI star melts of arm-number from 4 to 33. Linear polymer result shown by the light line. From [16].

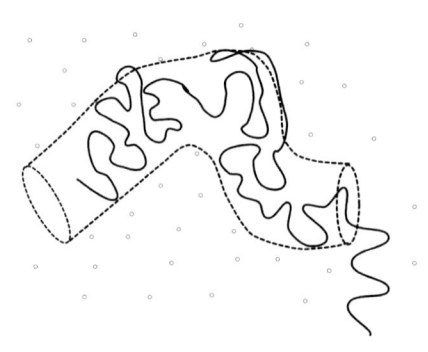

The plot shows η (Poise) vs Span M_w, with markers for 4-Arm Stars, 5-Arm Stars, 8-Arm Stars, 12-Arm Stars, 18-Arm Stars, 27-Arm Star, and 33-Arm Star. Scaling indicated as $\sim \exp(\nu' M_a/M_e)$ for stars and $\sim M_w^{3.4}$ for linear.

Fig. 6.2 A tube-like region of constraint arises around any selected polymer chain in a melt due to the topological constraints of other chains (small circles) in its neighborhood. Courtesy of R. Blackwell.

chains of higher molecular weight than M_e are strongly affected by the topological constraints (Fig. 6.2). The tube would suppress any motion perpendicular to the tube's local axis beyond a distance of a, but permit both local curvilinear chain motions and center-of-mass diffusion along the tube. de Gennes coined the term "reptation" for this snake-like wriggling of the chain under Brownian motion. The theory gives immediately a characteristic time-scale for disengagement from the tube τ_d (and so a viscosity $\eta \simeq G_N^{(0)} \tau_d$) proportional to the cube of the molecular weight of the trapped chain (compare Eq. 1).

In the late 1970s, Doi and Edwards developed the tube concept into a theory of entangled melt dynamics and rheology for monodisperse, linear chains [21].

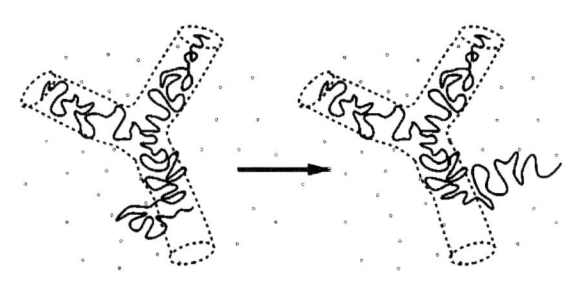

Fig. 6.3 The process of arm retraction predicted by the tube model for the case of dangling entangled arms, as from the branch point of a star polymer. Unlike in reptation, reconfiguration of the outer parts of the arm occurs many times for one relaxation of deeper segments.

Very significantly, de Gennes also realized that a tube-like confining field would endow a dangling arm, fixed to the network at one end, or belonging to a star-shaped polymer in a network, with exponentially slow relaxations. In this topology, reptation would be suppressed by the immobile branch point [22], and only exponentially-rare retractions of the dangling arm would disengage it from its original tube (Fig. 6.3). This immediate and powerful result of the simple tube idea explained at a blow why linear and branched chains exhibit such different dynamics – in many ways this is more convincing evidence for its validity than agreement over reptation dynamics prediction for $G(t)$ of linear chains. It immediately explains the great universality over different chemistries exhibited by entangled rheology: the key notion is that the dynamical modes responsible for viscoelasticity are *coarse-grained* at the level of the chain. The relevant dynamical degrees of freedom comprise the set of chain configurations $\mathbf{R}_a(n,t)$ that map the arclength position of the nth monomer on the ath chain on to its spatial position $\mathbf{R}_a$ at time t, but averaged over fluctuations within tube segments of end-to-end distance a and subchain degree of polymerization $\tilde{N} = N_e$. Then we may replace subchain stretches $\Delta \mathbf{R}/\tilde{N}$ with $\tilde{N}\partial \mathbf{R}/\partial n$, bearing in mind that this notation implies a choice of subchain $\tilde{N}$ below which we do *not* take the usual limit of calculus (the Gaussian chain is formally non-differentiable in the limit $b \to 0$; $N^2 b$ fixed). Macroscopic measurements relate directly to selected ensemble averages of $\mathbf{R}_a(n,t)$.

The tube model also points to the reason why careful polymerization chemistry has been essential to the progress of this field. To develop a simple theory for star polymers into an understanding of the rich complexity of a multiply branched industrial melt requires experiments on a hierarchy of topologies together with rheological and other experiments on them that provide close tests of the quantitative implementation of the tube model in each case [23]. In the following we review some of the important chemistry involved [but see the chapters by Fontanille (Vol. I, Chapter 2) and Hadjichristidis (Vol. II, Chapter 6) for more details], then review the current state of the models in the light of experiment. We close with some current challenges.

6.2
Synthesis and Rheology of Controlled Topology Polymers

Nearly monodisperse (single molecular weight) polymers are a natural experimental requirement for molecular theories of polymer dynamics. As soon as substantial polydispersity is introduced, the number of variables also rises (think of the set of moments of the molecular weight distribution). Accurate values of the moments higher than the third are notoriously hard to acquire, even if a theory is sophisticated enough to account for them in a predictive way. It is far better to work with materials in which the polydispersity is reduced to a perturbative quantity. The breadth of the molecular weight distribution is usually summarized by the "polydispersity index", PI, defined as the ratio of number-average and weight-average molecular weights. For a perfectly monodisperse material, $PI=1$. In the common "free radical" polymerization route, every chain first begins growth, then chemically adds monomers until a random termination event fixes its the molecular weight. Like all linear Markov processes (it is isomorphic to radioactive decay in time), the resulting distribution of chain lengths is exponential, giving $PI=2$. Other polymerization processes such as scission and two-chain interactions always broaden the distribution from this value. Industrial polyethylenes are common with PIs as high as 30. Even values of 2 would be ineffective tests for dynamic theories, since the lower molecular weights may act as unentangled solvent, and the occasional high molecular weight chains affect the elasticity of the melt out of proportion to their volume fraction. Fortunately, another family of polymerization methods, termed "living polymerizations" is able to deliver much sharper distributions. In this process, every chain begins growth on a single initiator molecule, then adds monomer on to a single "living" end at a uniform mean rate until supplies are exhausted, or until the polymerization is quenched by addition of a terminator (which is sometimes a polar solvent such as methanol). Now the distribution is Poissonian, with very small normalized variance if the degree of polymerization is large. The most common version used to make model materials for experiment is anionic polymerization [24], indicating the charge on the living chain end. Polymers commonly prepared anionically are polystyrene (PS), polyisoprene (PI) and polybutadiene (PB), a useful series since they span a wide range of entanglement molecular weights ($M_e=17\,100$, 4500 and 1800, respectively). Values of the PI as low as 1.01 are routinely quoted, although in very clean polymerizations the true values may be even smaller, since this resolution is set by the separation columns used in characterization, rather then the intrinsic spread of molecular weights. Although it is not possible to prepare kilogram or tonne quantities of monodisperse material in this way, 10 g or even 100 g are possible,

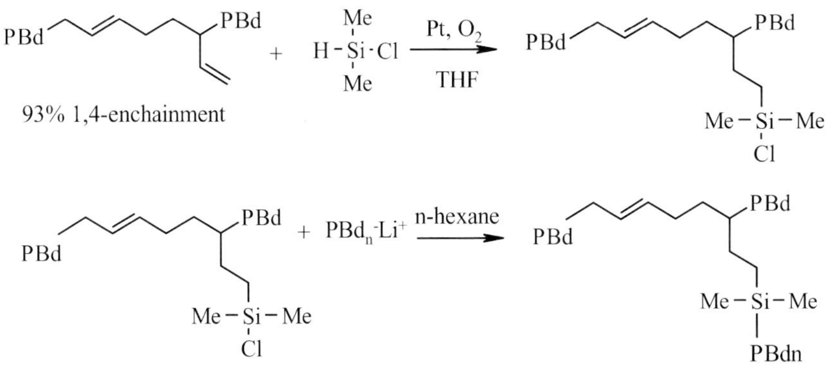

6.2 Synthesis and Rheology of Controlled Topology Polymers

Fig. 6.4 A reaction scheme for combs. Occasional pendant double bonds (vinyls) from anionically polymerized PB are bound to chlorosilanes. These in turn act as coupling sites for separately synthesised arm material. Courtesy of C.M. Ferneyhough [26].

permitting even the relatively material-hungry measurements of nonlinear extensional flows to be attempted on model materials (see below).

The second great advantage of anionic methods is their ability to construct well-defined branched structures by controlled coupling reactions at the living chain ends. Chlorosilanes are typical coupling agents, with each living end (which hosts a metal ion from the initiator, reacting with a chlorine from the coupling molecule. Since it is possible to synthesize very complex chlorosilanes, many monodisperse chains may be joined to the same coupler, forming "star" polymers of controlled functionality as high as 256 [13, 16]. More complex architectures may also be built from anionically polymerized components. Starting with a difunctional initiator will grow two simultaneous chains from the same point. The two living ends may then be chlorosilane-coupled to separately synthesized "arms", to give the H-shaped structure that has been very significant in identifying the linear and nonlinear physics of branching [25]. As an illustration of what may be achieved, Fig. 6.4 shows a reaction scheme for the synthesis of comb topologies from PB [26]. The small fraction of pendant vinyl groups on a PB linear chain, to become the comb "backbone", are chlorinated so that they become attachment points for the living ends of separately synthesized linear material destined to become the comb "arms". This route for comb synthesis provides a good yield of material with monodisperse backbone and arm segments, but has the drawback that the number of arms grafted on to each backbone is stochastic. An alternative approach here is to assemble the whole structure from pre-synthesized macromonomers, which has also been used to create more complex architectures [27]. Other desirable structures include "pom-poms" [28] and arborescent polymers [29] (Fig. 6.5).

More recently, serious attention has been paid to complex architectures of entangled polymers that are also polydisperse, but in a controlled and calculable

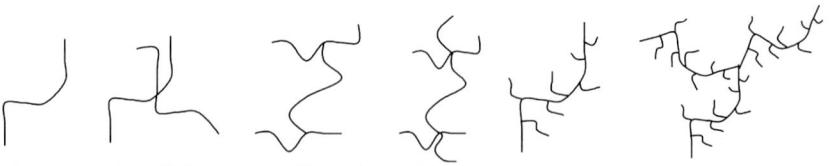

Fig. 6.5 Topological illustrations of branched polymer types synthesized to date by anionic polymerization: from left to right, linears, stars, Hs, pom-poms, combs and star-combs.

way. If well-entangled, monodisperse polymers are lightly cross-linked so that the links are truly uncorrelated, the resulting ensemble follows the statistical distribution of "mean-field gelation" [30, 31]. Initially unentangled chains when cross-linked follow instead different "percolation statistics" [31]. Both of these ensembles have power-law polydispersity of the form

$$f(M) \sim M^{-\tau} f_c\left(\frac{M}{M_{\max}}\right) \tag{4}$$

where $f_c(x)$ is a cut-off function at the maximum molecular weight of the ensemble. Recent rheological studies have followed both linear [32] and nonlinear [33] response of controlled randomly branched melts. All of these materials are in the "A_3" class of cross-linking of Flory (there is no preferred direction at any junction point in the clusters). Recently, another well-controlled polymerization of the "AB_2" class has become available to experiment (in this case the three strands at each junction point may be divided into two types, A and B, that only react with each other). This is the single-site metallocene family of materials. The reaction scheme bears some similarity with that of the comb materials described above in that growing chains on catalytic sites may be thrown off with pendant double bonds (like the living ends of the comb side-branches). These may be re-incorporated in an equivalent way to ordinary monomers at other chain growth sites, leading to tree-like families of molecules with self-similar structure [34, 35]. Remarkably, the entire family of structures is parameterized by only two numbers, which may be taken as a probability of meeting a branch point moving against the polymerization direction, b^U, and the mean degree of polymerization between branch points, N_x [36]. The self-similarity implicit in the polymerization of the branched polymers generates recursion relations for their statistics in a natural way. Many are analytically soluble. As an example, we quote here the bivariate distribution for the number-density of polymers containing N monomers and β branch points:

$$P(N,\beta) = \frac{N^{2\beta}}{N_x^{2\beta+1}\beta!(\beta+1)!}\left(b^U\right)^\beta\left(1-b^U\right)^{\beta+1}\exp(-N/N_x) \tag{5}$$

More complex reaction schemes that include chain scission at present defy analytic enumeration of statistics, but may be amenable at least to stochastic simulation. This approach has been applied to low-density polyethylene [37].

6.2.1
Linear Rheology

The majority of experimental data on entangled polymer dynamics are accounted for by rheological measurements of linear response of stress to an imposed strain, or equivalents (details of the technique are available in a number of texts, such as [38], and a comprehensive survey of data in [8]). In a typical experiment, a small ($\sim$1 g) sample of material is compressed between parallel circular plates. One plate is driven around its axis by small angles; the other feeds a torque transducer. Although the magnitude of the strain imposed on the material locally increases towards the perimeter of the plates, the displacement is controlled so that even the most highly strained material remains in linear response. An alternative, "stress-controlled" arrangement imposes a fixed stress and measures the response in strain of the sample. In either case, the parallel-plate geometry imposes a simple shear deformation at all points within the sample.

At $t = 0$ a small step-strain γ (usually shear – but in linear deformation the geometry is of no consequence up to a prefactor) is imposed and sustained and the resulting decaying stress $\sigma(t)$ measured. If the material is in true linear response, the limit of zero strain may be taken so that $\sigma(t) = G(t)\gamma$. The function $G(t)$ is the *relaxation modulus*, and decreases monotonically with time. In the case of entangled flexible polymers, the rubber-elastic expression for the stress from the statistical mechanics of the melt chains yields a direct interpretation of the normalized stress relaxation function in terms of the coarse-grained chain variables discussed above:

$$G(t) = \frac{G_N^{(0)}}{S_{xy}(0)} S_{xy}(t) \quad \text{with} \quad S_{xy}(t) \equiv \left\langle \frac{\partial R_x(n,t)}{\partial n} \frac{\partial R_y(n,t)}{\partial n} \right\rangle \tag{6}$$

where the average is taken over all chains and monomers in the chains. The most common strain history used to extract information equivalent to $G(t)$ is the harmonic oscillation of strain $\gamma(t) = Re(\gamma_0 e^{i\omega t})$, which gives a more natural experimental response function of the "complex modulus" $G^*(\omega) = i\omega \int_0^\infty G(t) e^{-i\omega t} dt$. If we write $G^*(\omega) = G'(\omega) + iG''(\omega)$, then we can identify the real part, G', as the in-phase (elastic) part of the modulus and the imaginary part G'' as the out-of-phase (dissipative) part. In general, both will be frequency dependent, crossing over from viscous (dissipative) behavior at low frequencies (where $G'' > G'$) to elastic behavior at high frequencies (where $G'' < G'$). The key pedagogical example is the fluid of a single relaxation time $G(t) = G_0 e^{-t/\tau}$. The Fourier transform of $G(t)$ is readily done to yield

$$G'(\omega) = G_0 \frac{\omega^2 \tau^2}{1 + \omega^2 \tau^2}; \quad G''(\omega) = G_0 \frac{\omega \tau}{1 + \omega^2 \tau^2} \tag{7}$$

Note that the correct elastic and viscous behavior are recovered at high and low frequency, respectively. The characteristic time emerges from this plot as the in-

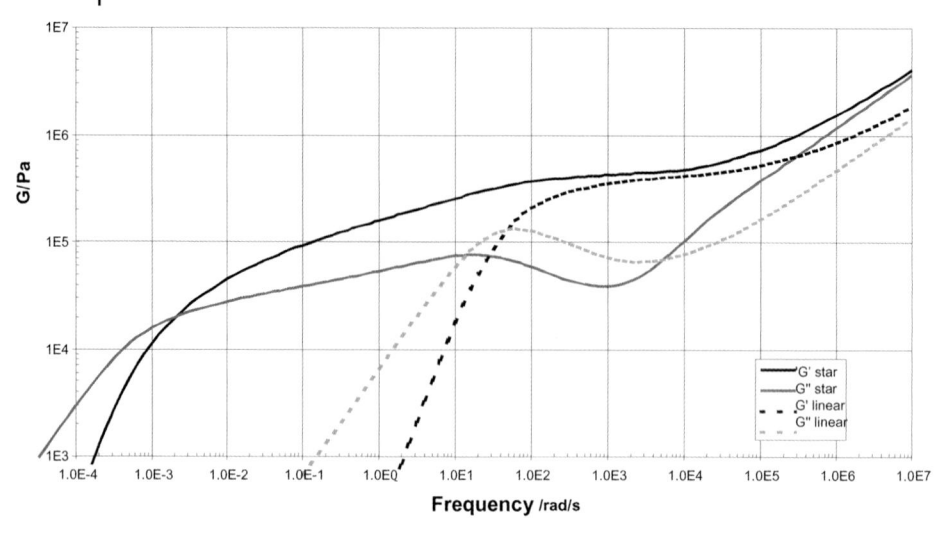

Fig. 6.6 Comparison of G' and G'' for monodisperse linear (broken lines) and star (continuous lines) polyisoprene melts. The linear molecule and the *span* of the star molecule both comprise about 40 entanglements. Note the much broader range of relaxation times for the star polymer.

verse of the frequency at which the curves cross (or the maximum in G'' in this case). The result for the terminal viscosity $\eta = G\tau$ is in fact general: it is always true that $\eta \approx G\tau$, where G is an effective modulus and τ a characteristic relaxation time of the fluid.

More realistic examples are furnished by the elastic and loss modulus for a range of polymer-like materials. It is possible in many cases to extract effective information on relaxations covering many decades of frequency in polymers because of *time–temperature superposition*. For most polymers above both their melting-point and glass transition temperature, T_g, the time-scales of *all* viscoelastic relaxations shift with temperature by the same factor $a_T = Ae^{T-T_0}$, for material-dependent values of A and T_0 (using the Vogel–Fulcher or WLF form of the shift function [8]). By this method, up to 12 effective decades in frequency are accessible for polymers with very low T_g, such as PI and PB, even though the experimental frequency range of mechanical oscillation available in the laboratory may not exceed four decades.

Figure 6.6 shows results from the linear rheology of a linear and a star-branched polyisoprene (PI) melt where the molecular weight of the linear is closely matched with the span of the star (twice the arm). The linear polymer shows an elastic plateau (and at higher frequencies the characteristic $\sqrt{\omega}$ behavior of the free chain dynamics of local segments). At around $\omega = 10^2$ s^{-1}, the peak in $G''(\omega)$ betrays the near-single-exponential signature of the reptation process. In the star architecture, only one carbon atom out of the $\sim 10^4$ present in the molecule carries a long-chain branch, yet the response function is clearly qualitatively different from that of the linear polymer. The maximum in $G''(\omega)$ is no longer anywhere near the cross-over, indicating a much broader superposition of relaxation modes over three orders of magnitude. This is also suggested from the form of $G''(\omega)$ itself – to reconstruct the broad, sloping shoulder of this function would require the superposition of Maxwell-like responses over three orders of magnitude in frequency. The terminal time is also much longer in the case of the star polymer, and no clear plateau emerges in $G'(\omega)$, even for high degrees of entanglement. In addition, comparison of star polymers of different molecular weight and arm number gives an astonishing result: the viscosity and terminal time are *not* dependent on the overall molecular weight of the stars, but only on the molecular weight of the *arms*, M_a [12]. Moreover, the dependence is roughly exponential, in contrast to the power law of linear chains, following $\eta \sim \exp(0.6 M_a/M_e)$. All these radically different rheological signatures in star polymers are now very well attested [14, 16, 39]. It is possible to turn the insights from the tube concept that explain the radically different observations into a quantitative theory (see below).

More complex monodisperse branched polymers such as H-shaped melts [25, 40], "pom-poms" [28] and combs [26, 41, 42] have been manufactured anionically and measured by linear rheology. In these cases, clusters of relaxation process appear, in some cases as if the melts were composed of bimodal blends of linear chains. Like star polymers, the terminal times and viscosities are roughly exponentially dependent on the length of the dangling arms.

6.2.2
Non-linear Rheology

The coarse-grained molecular expression for the stress, Eq. (6), is applicable whenever the chain segments on the scale of entanglements, or tube diameters, are well approximated by Gaussian chains, a criterion that may hold under even fairly large deformations. This is because the chain configuration is only very weakly perturbed from equilibrium at the length scale of links. Such a local linearity condition applies for segments containing $\tilde{N}$ links in *macroscopic* extensions of up to a local strain of $\sqrt{\tilde{N}}$, which may be as large as 10 or more in entangled melts. This is well into a highly nonlinear range of response for the entanglement structure, even though the local subchains are still in linear response, and bearing a simple coarse-grained molecular interpretation! So rheology in highly nonlinear response is a promising tool for the investigation of entanglement structure under high strains.

Three limiting cases of flow geometry are important to our study: (i) shear, (ii) uniaxial extension and (iii) planar extension. Perhaps they are best visualized as the local deformations in the situations of sliding parallel plates, fiber-spinning and film-drawing, respectively. Rheometers are designed to impose either shear (relatively easy) or extensional flows (more challenging) on material. The latter pose a more difficult experimental problem, because of the need to re-

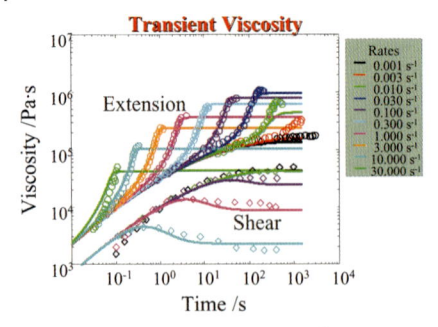

Fig. 6.7 Startup stress growth coefficients in extension (upper curves) and shear (lower curves) for a commercial low-density polyethylene.

spect the deforming free surfaces of the sample. Yet properties of entangled polymers can be radically different in extension and shear and especially for branched polymers it is important to measure both. In the case of both entangled linear and branched polymers, the key information yielded by nonlinear shear rheology in a "cone-and-plate" rheometer [38] is contained in the transient behavior of the shear-stress $\sigma_{xy}(t)$ and normal stress $N_1(t) \equiv \sigma_{xx}(t) - \sigma_{yy}(t)$ on startup of steady shear flow and the following steady-state values (all in the frame where the shear flow is given by $v_x = \dot{\gamma} y$ in terms of the shear rate $\dot{\gamma}$).

Extensional rheometers have been much harder to develop to a point at which good, reproducible data are attainable. This is due, as we have seen, to the necessity for free surfaces over most of the sample in an extensional flow. It is also exacerbated by the requirement that material points separate exponentially with time, in contrast to shear flow, from the simple integration of the velocity field $v_x = \dot{\varepsilon} x$. However, extensional rheology is an important independent measure of the nonlinear rheology of many materials, and good devices are now available [43–48]. For example, branched entangled polymers (see below) may be strain-hardening in extension (the effective "viscosity" defined as stress/strain-rate increases with strain), but strain-softening in shear. This is true of the canonical commercial branched material low-density polyethylene (LDPE) (upper curves in Fig. 6.7) together with the transient shear response (lower curves). Here the extensional stress difference $\sigma_{xx}(t) - \sigma_{yy}(t)$ divided by the extension rate $\dot{\varepsilon}$ (the "stress-growth coefficient"), is plotted against time. For the transient shear experiment we similarly plot $\sigma_{xy}(t)/\dot{\gamma}$. This way of representing data ensures that the curves superimpose at early times when the material is in purely linear response. The extensional data show strong "hardening" at the higher of two extension rates, but none at the lower. In distinction, the shear responses at high rates are all softer than in linear response.

We find in this nonlinear response yet another strong qualitative effect of branching. However, limited measurements of the nonlinear extensional response of star polymer melts indicate that extension hardening requires more than one branch point per molecule.

Harder still is the challenge of making quantitative measurements in strong planar extension. However, this, too, has proved very important, at least in a transitionary period in which phenomenological integral constitutive equations suggested that if the uniaxial and shear responses of material differed greatly, then the planar response should lie alongside the shear response, rather than the extensional. This is because the two-dimensionality of both shear and planar extensional flows induced the same invariant structure. In fact, experiments on LDPE in both tubular [49] and sheet [50] geometries have shown that the (hardening) stress-growth coefficients in planar flows are almost identical with those in uniaxial. This has had considerable impact on the development of molecular-based constitutive equations.

An attempt to reduce the complication of experiments in which both strains and strain-rates are in nonlinear response (the double nonlinearity arises because the dimensionless strain rates $\dot{\gamma}\tau_{max}$ and also the strains $\gamma \gg 1$) has classically been made by taking the experimental limit of very fast "step" strains, and looking at the response in stress as a function of strain only, comparing at fixed time following the step strain. Some materials display "time–strain separation" after some time τ_k [51], so that the dependence on time following the step strain is just the same as in linear response:

$$\sigma(\gamma, t) = \gamma h(\gamma) G(t) \tag{8}$$

where $h(\gamma)$ is the "damping function". In the case of monodisperse materials, the damping function was observed to possess a universal, highly softening, form (termed "A-type" damping in a classic review by Osaki) [52, 53] before any theoretical suggestions that this might be expected on molecular grounds. Moreover, the molecular weight dependence of the time τ_k was found, in these cases, to grow with molecular weight as $\tau_k \sim M^2$. Suggestively, this is the same scaling as that of the time-scale for stretch relaxation of a chain of molecular weight M. Polydispersity tends to reduce the severity of the strain dependence, as do branching and disentangling by lowering molecular weight or polymer concentration. The exception to this is the case of entangled melts of *star* polymers. These possess damping functions very similar to those of monodisperse linear polymers [54], and are both well accounted for by the theoretical "Doi–Edwards" damping function [55]. Damping function measurements of monodisperse complex architecture polymers have as yet yielded only small amounts of data, but there are indications that architectures such as the "H" or "pom-pom" can yield time–strain factorability separately in more than one region of time following a step strain [25]. Figure 6.8 shows an "early time" and a "late time" damping function extracted from a PI H-polymer melt, together with the "Doi–Edwards" function seen in monodisperse linear polymers and stars and a stiffer "early time" damping function from a detailed tube model for this architecture.

DAMPING FUNCTION FOR PIH 111B20A AT 25C
(1deg CP)

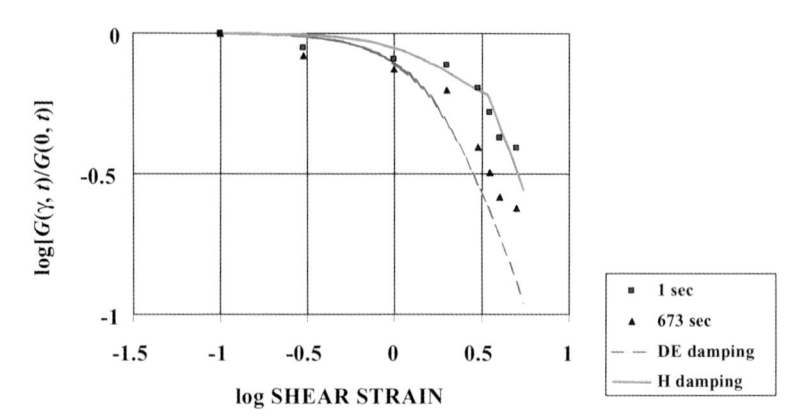

Fig. 6.8 Damping functions observed in a PI H-polymer melt [25] at early and late times, and comparison with the highly shear-thinning Doi–Edwards damping function.

6.3
Entangled Branched Chains in Linear Response

We have seen how the tube model can account qualitatively for some of the qualitative observations made on star polymer melts. The key observation is that the branch point of a q-armed star within the confining potential of the tube is *localized* [22] to a region of order a in size for time-scales faster than all disengagement processes. This is a consequence of the equilibrium tension f_{eq} induced by the free ends of the tube: the entropic gain per tube segment from explorations of chain ends that increase the tube length. The reptation of the branch point along any one of the arms' confining tubes by a "primitive path" (or distance along the coarse-grained tube) s would require a entropic fluctuation of free energy of magnitude $f_{eq}(q-1)s = (3k_BT/a)(q-1)s$. We will see that this is typically far greater than the free energy barrier posed to independent retractions of individual arms. However, the mechanism for configurational (and stress) relaxation is unchanged – tube segments must be visited by chain ends. In entangled star polymers this can only happen by fluctuations in the primitive path length – the chain forms unentangled loops along and emerging from the tube so that the free end retraces the tube contour before re-emerging again into new tube (see Fig. 6.3). Clearly, shallow retractions will happen much more frequently than deep retractions – this is the origin of the huge spread in relaxation times, and the deepest retractions themselves will become extremely rare as the molecular weight of the arm increases. More quantitative treatments require a calculation of the dynamics of Brownian motion in a deep potential well, which we consider in the next section.

6.3.1
A Tube Model for Star Polymers and Critical Tests

To develop the theory we will need the rate for the escape time of a single degree of freedom (e.g. diffusing particle) over a barrier. Calculations have often written $\tau_{\rm esc} \simeq \tau_0 e^{U/k_{\rm B}T}$ [22, 56], where U is the barrier height and τ_0 some local microscopic hopping rate. This is often good enough, but there are times when the pre-factors to the expression will be important. We need the full result for the mean first passage time for a diffusing particle (of self-diffusion constant D), released at $s=0$ to attain a point s within a potential $U(s)$ [57, 58]:

$$\tau(s) = \frac{1}{D} \int_0^s dx' e^{U(x')} \int_{-\infty}^{x'} dx e^{-U(x)} \tag{9}$$

which can be approximated when the gradient of $U(s)$ is finite as

$$\tau(s) \simeq \frac{e^{U(s)}}{D U'(s)} \sqrt{\frac{2\pi}{U''(0)}}$$

The effective potential $U(s)$ for the length of the primitive path s inwards from the equilibrium value arises from including both the (quadratic) curvilinear rubber-elastic term and the (linear) end-tension term [59]. Defining a fractional coordinate $0 < x < 1$ along the arm primitive path (so that complete retractions correspond to $x=1$), the potential for a star arm becomes

$$U(x) = k_{\rm B} T \nu_Q Z_{\rm arm} x^2 \tag{10}$$

where ν_Q is a geometric constant of value 3/2 in a Gaussian model and $Z_{\rm arm} = M_{\rm arm}/M_{\rm e}$. We bear in mind that there may be corrections to the quadratic potential of order 0.1 that will make a difference to the rate of deep retractions [60]. The other major assumption in this approach to star-arm dynamics is that only the slowest Rouse mode acts as an effective dynamic variable for deep retractions. This assumption seems appropriate, since Rouse modes of index p have effective elastic potentials that vary as p^2 [55]. However, the higher modes may renormalize the barrier-hopping time-scales [58], and may be required in future, more accurate, theories of branched polymers.

Once such approximations have been made, the observations above can be rapidly turned into a semiquantitative theory for star-polymer stress relaxation [22] which is amenable to more quantitative refinement [61]. Apart from small displacements of the end, the diffusion to any position x along the arm will now need to be activated and so is exponentially suppressed, possessing its own characteristic stress relaxation time $\tau(s)$ given by the average first passage time of the diffusing free end to x. The effective diffusion constant, D, of the Brownian particle we can identify as the Rouse diffusion constant of the arm, modified by a factor 2 since the mean displacements of monomers in deep retractions are proportional to their distance from the branch point. Simple substitution of the arc-length potential into the general result gives a dominant exponential term to the longest relaxation time $\tau(1)$ of

$$\tau_{\text{late}}(1) = \frac{\pi^{5/2}}{\sqrt{6}} \tau_e Z_{\text{arm}}^{3/2} \exp\left(\nu_Q Z_{\text{arm}}\right) \tag{11}$$

(note: the prefactor is not insignificant!), where τ_e is the Rouse time of an entanglement segment. We have chosen the notation $\tau_{\text{late}}(x)$ to distinguish this expression from the result for non-activated rapid retractions $\tau_{\text{early}}(x)$. The relaxation modulus in the star melt can then be written

$$G(t) = G_N^{(0)} \int_0^1 p(x,t) dx \tag{12}$$

where $p(x,t)$ is the survival probability of the tube segment at the fractional coordinate x (the probability that it has not been visited by the free end before time t). In the limit of high potential barriers $U(x)$ the form of $p(x,t)$ becomes proportional to $e^{-t/\tau(x)}$ in approximation that is asymptotically exact [58]. This form of $p(x,t)$ is in turn well approximated for highly entangled arms by a step function in x: all segments exterior to the segment $x(t)$ given by the inverse of the function $\tau(x)$ are almost certain to have relaxed, because their relaxation time-scales are exponentially shorter, whereas segments nearer to the core are conversely almost certainly unrelaxed.

Unfortunately, although qualitatively promising, this version of the theory fails disastrously at the quantitative level. A glance at the polyisoprene ($M_e = 5000$) star data above (Fig. 6.6) will suffice: a star with this arm molecular weight (105 000) is predicted to carry an exponential term for the terminal time of approximately 10^{18}. Yet this must describe roughly the width of the "relaxation shoulder" in $G''(\omega)$ in the figure, which is only six decades broad. Pearson and Helfand [61] accommodated this anomaly by allowing the pre-exponential factor ν in Eq. (10) to take on experimentally adjusted values, finding $\nu = 0.5$ to fit the data well (using our definition of M_e). However, the real problem lies with the need to account for "constraint release" – the effect of the transience of the tube constraints themselves on a single star arm. Fortunately, it is sometimes much simpler to treat in the case of star polymers.

The contribution of constraint release to the dynamics of entangled star polymers is very much greater than that to linear polymers. This arises from the very broad distribution of relaxation time-scales we have discussed above and the barrier-hopping character of branched arm relaxation. Fortunately, the same breadth of time-scales provides a simple way of calculating the effect [62, 63]. As a consequence of the exponential separation of relaxation time-scales along a star arm, by the time that a given tube segment x in the population is relaxing, all segments of tube x' such that $x' < x$ (nearer a chain end) have renewed their configurations typically many times. So chain segments at x and nearer the star cores *do not entangle* with these fast segments at the time-scale $\tau(x)$ and beyond. Alternatively, we can say that the tube is dilated due to this effective dilution of the entanglement network: fast-relaxing segments act as solvent for the slower relaxing ones.

The new information necessary to make this approach quantitative is the dependence of the effective entanglement molecular weight on the concentration, Φ, of unrelaxed segments. This is known from experiments on dilution of polymer melts by theta-solvents to be approximately $M_e(\Phi) = M_{e0}/\Phi^a$ (with $a = 1$ and $a = 4/3$ two popular choices [59]), which corresponds to the approximately quadratic concentration dependence of $G_0 \sim \Phi^{a+1}$ when a melt is diluted with a theta-solvent. At any stage in the relaxation dynamics of a melt of identical star polymers, therefore, when a segment x is currently relaxing for the first time, the effective entanglement molecular weight is $M_e(x) = M_{e0}/(1-x)^a$. This means that the effective number of entanglements along the arm is reduced at this time-scale and is $Z_{arm}(x) = Z_{arm}(1-x)^a$. To recompute the relaxation times $\tau(x)$ with the dynamic dilution assumption, we consider the activated diffusion in a *hierarchical* way. To retract from x to $x + dx$, the attempt frequency is $\tau(x)^{-1}$ and the activated probability for this diffusive step is $\exp\{(-1/kT)[U(x+dx; M_e(x)) - U(x; M_e(x))]\}$, where the notation for U indicates that the running value of the entanglement molecular weight is kept. Taking the limit of dx small gives the differential equation

$$\frac{dU_{eff}}{dx} = \frac{\partial U}{\partial x}[x; M_e(x)] \qquad (13)$$

where the dependence on an x-dependent value for M_e arises from the "dynamic dilution" of the tube increasing the effective value of M_e. Taking the dilution exponent $a = 1$, integration leads to an effective renormalization of the potential $U(x)$, which is now a cubic in x:

$$U_{eff}(x) = \nu Z_{arm}\left(x^2 - \frac{2}{3}x^3\right) \qquad (14)$$

The terminal time and viscosity are dominated, as we saw above, by the potential at complete retraction, $U_{eff}(1)$. This is now given by $1/2(M_a/M_e)$, in much closer agreement with experiments. The equation for the relaxation modulus also needs modifying since each element of chain ds contributing to the stress relaxation now does so in an environment diluted by $(1-x)$, and so picks up this factor within the integrand as a coefficient of $p(s,t)$. The general result is

$$G(t) = G_N^{(0)} \int_0^1 \frac{\partial G}{\partial \Phi} \frac{\partial \Phi(x)}{\partial x} \exp\left(-\frac{t}{\tau(x)}\right) dx \qquad (15)$$

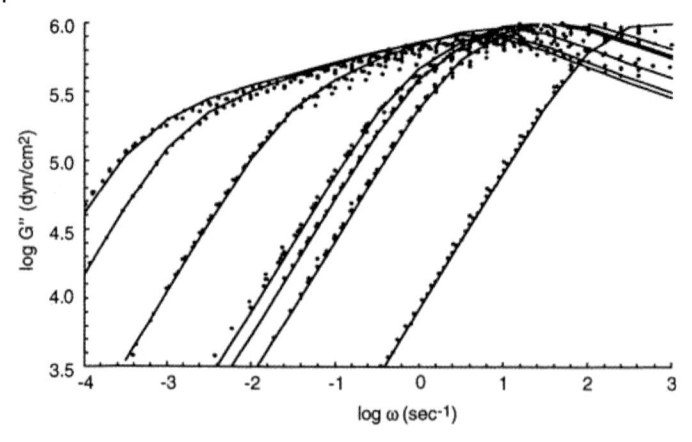

Fig. 6.9 Data on series of PI star polymers from [16] and corresponding theoretical predictions using the theory of [16]. In order of decreasing frequency of the terminal times, the arm molecular weights $10^{-3}M_a = 11, 17, 37, 44, 48, 95$ and 105.

with $\tau(x)$ calculated from the full expression for barrier-hopping, Eq. (9), but using the dilated effective potential $U_{\text{eff}}(x)$. The shape of the relaxation spectrum predicted by this procedure does indeed fit rheological data on pure star melts better than the quadratic expression calculated for stars in permanent networks [17, 63, 64], especially when corrected for at high frequencies by a crossover to unactivated tube loss by Rouse-like motion, $\tau_{\text{early}}(x) \sim x^4$, near the free end.

The curves through the experimental points in the data for $G''(\omega)$ on the PI stars (Fig. 6.9) were calculated via this scheme [64], and fit the experiments using literature values of the two fitting parameters required, τ_e (a horizontal shift on the figure) and $G_N^{(0)}$ (a vertical shift). Other chemistries of star melts are equally well accounted for by this approach [17, 65, 66]. The great strength of this remarkably powerful theoretical framework is that, in principle, only these two parameters are required for all molecular weights and topologies of a single chemistry, so that the theory is very constrained.

Rather stringent tests are supplied by bimodal blends involving a monodisperse star-polymer component. Rheological results are available on the cases of star–star [67, 68] and on star–linear [67, 69] blends. The highly cooperative nature of tube-dilation CR in star arm retraction leads to a remarkable prediction in the case of polydisperse stars (of which bimodal blends are the simplest case). A strong *motional narrowing* occurs, with the slower component relaxing exponentially faster in the presence of the faster than it does in a homopolymer melt, and the converse for the smaller. As an immediate result [62], the terminal time of the blend carries a form of exponential dependence on arm molecular weight that is identical with that of the homopolymer, but with Z_{arm} replaced by its weight average so that

$$\tau(1) \sim \exp\left[\frac{\nu}{3}\langle Z_{\text{arm}}\rangle_w\right]$$

The detailed relaxation of each component in the blend is calculated by a transformation from the fractional path length coordinate of species i, x_i, to the coordinate $z_i = Z_{a,i}^{1/2} x_i$ in which the calculation of the renormalized potential becomes universal.

However, not all the possible experimental probes support the predictions of a theory invoking maximal tube dilation. When the self-diffusion of monodisperse stars is compared with their rheology, it is possible to examine the dilation of the tube at the terminal time rather carefully [70]. The product ηD_{self} cancels the exponential dependence of the two dynamic quantities, leaving only the pre-exponential dependences on Z_{arm}. In the case of diffusion, the additional physical assumption in a calculation of D_{self} is the size of the hop made by the branch point. The natural assumption to make for the jump length if tube dilation holds is the value of the tube diameter at the disentanglement transition we identified above at $x < 1 - (2/3Z_{\text{arm}})^{1/2}$. Such an assumption gives $\eta D_{\text{self}} \sim Z_{\text{arm}}^{-1/2}$, but if the hops take place in the bare undiluted tube, the prediction is the much steeper $\eta D_{\text{self}} \sim Z_{\text{arm}}^{-3/2}$. The data in [71] are very clear: the scaling with Z_{arm} is much closer to the undiluted result than the dilated case, although the prefactor indicates that the typical jumps are larger than the bare value of a by a factor of about 3.

A purely rheological experiment that is nonetheless sensitive to the diffusive properties of the branch point is the relaxation of a melt of three-arm stars with one arm shorter than the other two. If considerably shorter, the retraction of the third arm permits further orientational relaxation by reptation of the remaining linear object, albeit slowed by the larger effective drag of the branch point. This invokes a new process: rheological relaxation times emerging from the effective friction of an entangled side-arm (this will also arise in the context of H-polymers and combs below). This introduces a further dimensionless constant into the model, by writing the effective curvilinear diffusion constant of the branch point as

$$D_{\text{bp}} = \frac{p^2 a^{*2}}{2q\tau_a(1)} \tag{16}$$

The new constant p is the fraction of the current tube diameter a^* jumped on average by a branch point at each jump time $q\tau_a(1)$. The dependence of the effective drag of the branch point on q in a melt is taken as simply proportional in this approximation. In fact, we expect p to be rather smaller than unity from the nature of the projection of spatial diffusion on to the curvilinear tube, and a value of $p^2 = 1/12$ was originally suggested as optimal [25]. Recent studies of just such series of PEP [72] and PI [73] asymmetric stars confirmed the accuracy of the dynamic dilution theory for nearly symmetric stars, but found that short arms of just a few entanglements appeared to contribute to the drag to a much greater extent than predicted by this assumption. The reptation rate of the asymmetric star after relaxation of the short branch also depends, of course, on whether the hopping process occurs in a dilated or undilated tube, as in the discussion of star diffusion above. Very recent results indicate that a surprisingly low value of the hopping constant of $p^2 = 1/40$, but in the physically motivated dilated tube, can account for these data and also those on more complex architectures [101].

Perhaps the most powerful test of effective tube dilation is dielectric relaxation of chains with dipoles parallel to the chain axis. This is because, to a good approximation, the dielectric relaxation function measures the surviving tube fraction directly (at least at time-scales shorter than the CR Rouse time of the tube) so that $\varepsilon(t) \sim \mu(t)$, while the rheological relaxation function picks up any relaxation function (e.g. from constraint release) of the tube itself $G(t) \sim \mu(t)R(t)$. Since, for the case of full tube dilation we have $R(t) = [\mu(t)]^a$, where a is the modulus dilution exponent ($a \simeq 1$), we predict

$$G(t) \sim [\varepsilon(t)]^{a+1} \qquad (17)$$

when tube dilation holds. Recent results on PI stars indicate a departure from the tube dilation prediction for $x \gtrsim 2/3$ [74] for $Z_{arm} = 8$. Increasing Z_{arm} to 16 makes the picture even clearer (Fig. 6.10). Although the rheological response indicates the apparent continuation of tube dilation beyond $x = 2/3$, the dielectric relaxation for this final one-third of the arm is almost monomodal at the terminal time, showing a clear peak in $\varepsilon''(\omega)$!

Even without the experimental difficulties, as soon as one goes beyond rheology, the original treatment of dynamic dilution in stars reviewed above [62, 63] leaves some important questions unanswered, and is based on a physically intuitive, but non-formal, treatment of the Kramers problem. In particular, it leaves the physical interpretation of the effective potential unclear, and does not appear to visit the problem of the balance between effective primitive path length and effective drag. Furthermore, the existence of a wide range of relaxation times for constraint release really implies that the relaxation is not properly of a single degree of freedom, but that the Kramers space is of higher dimension. A recent reappraisal of the original calculation in the light of both experimental and theoretical concerns does indeed predict that the dynamic tube dilation approximation should hold for shorter time-scales, but that it should break down for star polymers well before the terminal time [75]. We are left with a very persuasive, and quantitatively accurate, approximation that has been applied to much more complex architectures than star polymers (see below), but that has yet to be controlled in a satisfactory way!

6.3.2
H-Polymers and Combs

The attraction of the H-polymer melt is that the idea of hierarchical relaxation by arm retraction, and the consequent renormalization of the effective degree of

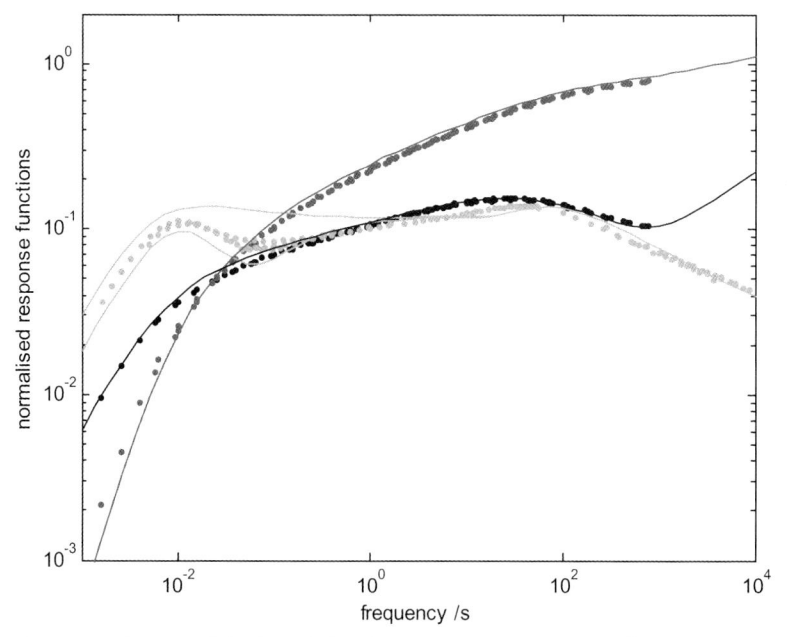

Fig. 6.10 Dielectric and rheological data for a six-arm PI star melt with $M_a/M_e = 16$. Curves associated with the lower two functions (G' and G'') are predictions of the rheology by dynamic dilution [63], the upper (light) curve is its prediction for the upper data set for $\varepsilon''(\omega)$. The peaked (light) curve is the prediction for $\varepsilon''(\omega)$ of a theory in which dynamic dilution (see Section 6.3.1) is arrested for the final one-third of the star arm, and all remaining response relaxed with the last entanglement. Data courtesy of H. Watanabe.

entanglement of the melt, generalizes in an appealing way to this architecture. It also provides the simplest branched polymer containing the essential new feature of a segment "trapped" between two branch points. Dynamic tube dilution suggests that H-shaped polymer at long time-scales becomes just an ensemble of the cross-bar sections moving against drag concentrated at the attachment points of the arms (Fig. 6.11). Before making any lengthy calculations, therefore, we can see that the early-time dynamics will be similar to that of star melts (apart from the presence of slow cross-bar material in the environment of the arms), and that the late-time dynamics may even be reptation of the effectively linear cross-bars in supertubes defined only by their mutual entanglements. A very similar picture will present itself in the case of entangled combs, with the difference that the effective late-time drag will be located at the branch points along the comb backbone.

An early set of data on H-polystyrenes by Roovers [40] was obtained on a series of molecular weights with equal amounts of material in each arm M_a and cross-bar M_b. The materials did indeed exhibit additional relaxations at frequencies greater than star arm terminal times, and the viscosity of the series de-

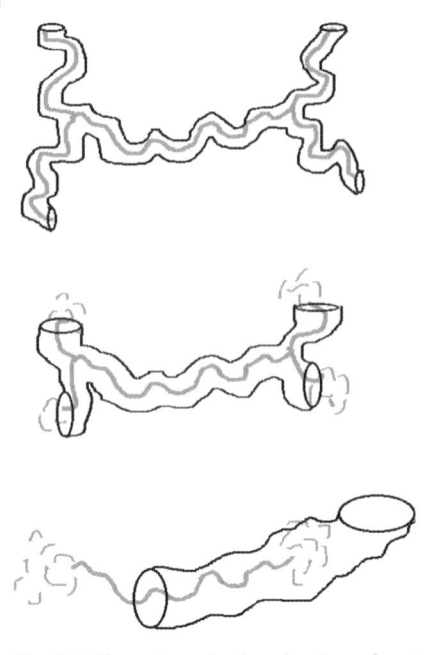

Fig. 6.11 Three stages in the relaxation of an H-polymer melt. Arm-length fluctuation precedes cross-bar reptation in a progressively widening tube.

pended on arm molecular weight in a stronger way than in a melt of pure star $\eta \sim \exp(\nu Z_{\mathrm{arm}})$ with $\nu > 0.5$ (rather than 0.3). Later, other groups synthesized H-architectures from more highly entangled PI [25, 76] and comb and pom-pom architectures from PB [26, 28, 41, 42]. These exhibited clearer features in the rheology, including a $G''(\omega)$ peak arising from cross-bar (or backbone) reptation.

To outline the theoretical development of the tube model for these architectures [25, 77], we take the dynamic processes that progressively liberate tube segments in the H-polymer in order of time-scale. They are represented pictorially in Fig. 6.11.

Stage (i): At early times (following a small step strain) stress will relax by path-length fluctuation in the dangling arms.

Stage (ii): The rapid path-length fluctuation crosses over to an exponentially slow "activated diffusion" for the deeper arm fluctuations. The effective tube diameter grows continuously and self-consistently throughout this regime, and the cross-bars remain immobile.

Stage (iii): After the star-like arms have completely retracted, the physical picture for the molecules changes to consider the mobility of the cross-bars in widened tubes defined only by their mutual entanglements. All of the effective friction is concentrated at the branch points.

Stage (iv): The free curvilinear diffusion of the chain ends is suppressed when path-length fluctuations become slowed by the effective elastic potential (so thinking of the cross-bar as a "two-arm star" [78]). Central portions of the cross-bar are relaxed by reptation.

The strategy for turning this physics into a quantitative theory within the DTD approximation requires two steps: (1) we must calculate a hierarchy of time-scales $\tau(x)$ in terms of an arc coordinate or coordinates, x, that trace through the molecular segments from the extremities to the center; (2) we then write a form for the effective modulus of an entanglement network diluted to a concentration $\Phi(x)$, where $G(\Phi) = G_N^{(0)}[\Phi(x)]^{a+1}$. For H-polymers it is natural to divide the arc coordinate into two sections: x_a runs along the arms from the ends to the branch points ($0 < x_a < 1$) and x_b runs from the branch point to the middle of the cross-bar ($0 < x_b < 1$). Now Eq. (15) becomes

$$G(t) = G_N^{(0)}(a+1)\left[\int_0^1 \phi_b^{a+1}(1-x_b)^a e^{-t/\tau_b(x_b)}dx_b + \int_0^1 (1-\phi_a x_a)^a e^{-t/\tau_a(x_a)}dx_a\right] \quad (18)$$

where ϕ_a and ϕ_b are the volume fractions of arms and cross-bar, respectively. The first term comes from the contribution of cross-bar material and the second from the relaxation of the dangling arms. In each integral, the term in parentheses represents the effective concentration of unrelaxed entangling network surrounding a segment relaxing on a time-scale $\tau_{a/b}(x)$. Details of the calculations can be found in [25] and with a more refined treatment of polydispersity in [73]. Here we draw attention to a few key features of the physics:

Stage (ii): It is easy to see, for example, that the potential well for the arm fluctuation is deeper and steeper than for pure star polymers, due to the fraction of cross-bar material that behaves as permanent network for the relaxing arms. In this way, the fraction of arm material enters into the exponent for the longest relaxation time among arm retractions. The form for the arm retraction time with a general dilution exponent a gives

$$\tau_a(1) \sim \exp\left[3Z_{\text{arm}}\frac{1-\phi_b^\beta(1+\beta\phi_a)}{\beta(1+\beta)\phi_a^2}\right] \quad (19)$$

where $\beta = 1 + a$. The mean first-passage time for diffusers over such a potential barrier as $U_{\text{eff}}(x_a)$ can be calculated analytically for all x_a in terms of integrals over the potential [25].

Stage (iii): When the path length of the dangling arms eventually fluctuates to zero, corresponding to the free end retracing a path through the melt to the branch point itself, the branch point may make a diffusive hop through the melt. At this time-scale of $\tau_a(1)$, the tube diameter is set only by other cross-bars (if maximal tube dilation is valid, which is currently predicted to be the case provided that $\phi_a < 2/3$, as we saw above) so has a value $a^* = a_0\phi_b^{-a/2}$.

There is a slight uncertainty in the O(1) number that relates the time-scale of the hop to the mean hopping distance (just as in the case of diffusion of star polymers). A ratio of this hopping distance to the tube diameter, usually noted as p, of 1/12 was found for the PI materials of [25] (although this is misprinted as 1/6 in the original paper).

The results of this procedure, plus a careful treatment of fluctuations in cross-bar path length, which introduces no new parameters into the theory, apart from the O(1) number p, may be compared with the linear rheological spectra for real entangled H-polymers.

Figure 6.12 shows the frequency-dependent moduli calculated from the scheme outlined above for two values of arm and cross-bar molecular weights. The prominent "shoulder" feature at higher frequencies in $G''(\omega)$ is a clear signature of the arm relaxations; its logarithmic width increases with the arm molecular weight. The low-frequency peak comes from the cross-bar: lengthening the cross-bar takes the peak to lower frequencies, whereas reducing it both speeds up the cross-bar relaxation and weakens the magnitude of the peak as the volume fraction ϕ_b reduces, and with it its contribution to the modulus. So the H-polymer spectrum we expect contains features reminiscent of the spectra of both star [a broad shoulder in $G''(\omega)$] and linear polymers [a well-defined peak in $G''(\omega)$] at the frequencies corresponding to the inverse relaxation time of the structural components that resemble them (arms and cross-bar, respectively). Plotted alongside the data are predictions from the tube model outlined above.

For H-polymers, accounting for polydispersity is very important indeed. The dashed lines in the figures give the model predictions in the absence of polydispersity and the solid lines the predictions when even small polydispersity ratios of order 0.01 are assumed in the calculation. The reason for the strong effect is the exponential dependence of the branch-point diffusion constant D_{bp} on the arm molecular weight. Even small increases in Z_{arm} can greatly influence D_{bp}, and consequently the reptation time of the cross-bar. Even when the polydispersity is small, however, its effect is tricky to calculate [73]. There are two opposite tendencies: the motional narrowing of the self-entanglements of the arms works to reduce the effect of polydispersity, whhereas their entanglements with the slow cross-bar material increases it. When the volume fraction of cross-bar material is significant, these two effects can be cast into an approximation for the terminal time of the arms, which, up to exponential factors, becomes

$$\langle \tau_a(1) \rangle \approx \exp\left[\frac{\nu \phi_a \bar{Z}_{arm}}{3}\right] \exp\left[\nu \phi_b \bar{Z}_{arm} + \frac{\nu^2 \phi_b^2 \bar{Z}_{arm}^2 \varepsilon_a}{2}\right]$$

The rheological results are certainly consistent with measured values of the polydispersity and this calculation. It should also be noted that the anomalously slow diffusion noted in the case of asymmetric stars does not arise in the H-polymer experiments.

Very similar observations were made in the case of "pom-pom" architectures [28]. The theoretical development reviewed above applies without change to this

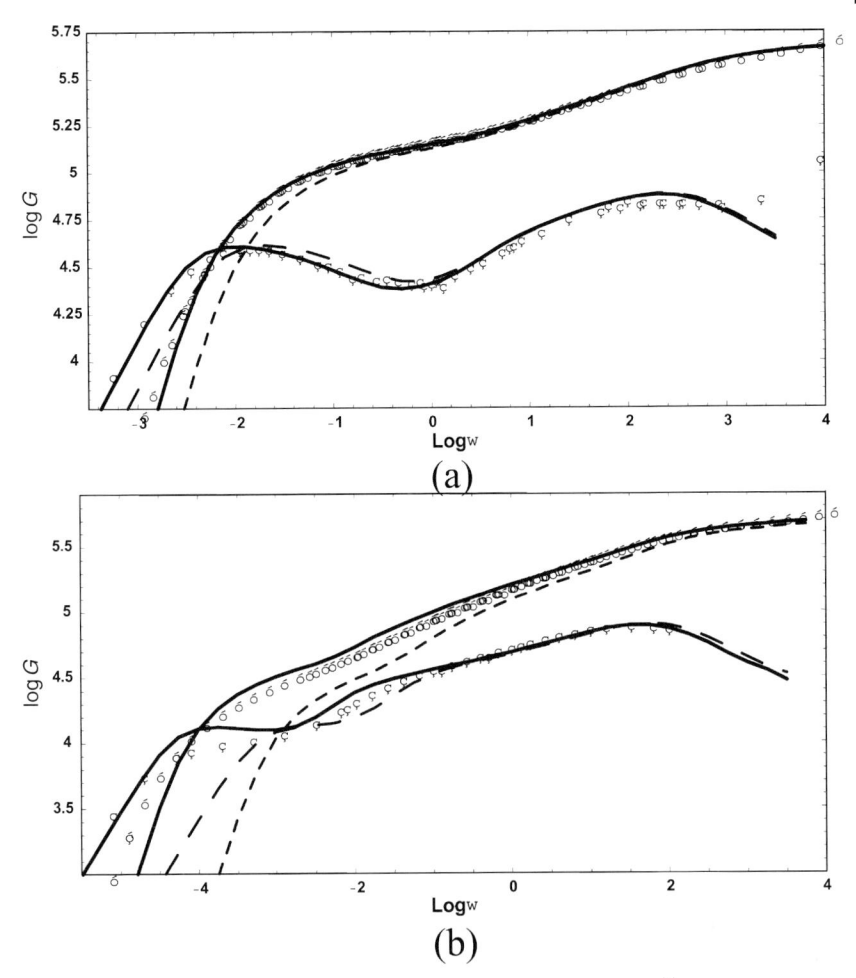

Fig. 6.12 Experimental and theoretical complex moduli for the two H-polymers of the table. Both are PIs with cross-bar molecular weight of 110 000, with arm molecular weights of 20 000 (a) and 52 000 (b). Fitting parameters of $G_N^{(0)}$ and τ_e were consistent with literature values and $p^2 = 1/12$. Polydispersity corrections were omitted to produce the dashed curves.

architecture, since the number of arms attached at each end of the cross-bar, q, has been kept variable.

The case of comb melts is very similar to that of high-q pom-poms, with the exception that the drag on the comb backbone section is distributed at all branching points with dangling arms, rather than concentrated at the end. The evidence of several studies supports the exponential separation of relaxation modes of the entangled arms via deep CLF, combined with either renormalized Rouse or reptation dynamics for the backbones, depending on whether they are self-entangled or not ($Z_b\phi_b^a \lesssim 1$) [26, 41, 42, 79]. The theoretical application of the tube model to the case of combs has supported this at the quantitative level [26, 81–83]. Both comb and pom-pom melts have an additional degree of freedom over the H-polymer in the number of arms, which allows independent tuning of the backbone fraction ϕ_b from the arm molecular weight. Hence the terminal modulus ($\sim \phi_b^{a+1}$) and relaxation time ($\sim e^{\nu Z_a}$) can be independently set by the molecular architecture. The same careful treatment of polydispersity is necessary, with the additional feature of most anionic samples made to date that the number of arms, q, is also polydisperse. One other subtle feature is that the random attachment of arms to the backbone results in a more severe polydispersity of the final dangling ends of the backbone, which themselves constitute topologically effective arms. Even carefully synthesized comb samples therefore point the way to algorithms capable of implementing the tube model in cases of highly random branching.

6.3.3
Complex Topologies – the Seniority Distribution

Beyond comb and pom-pom architectures lie an infinite family of more complex topologies that are also, in principle, amenable to the rules of the tube model. The picture of hierarchical retraction dynamics with dynamic dilution can be generalized to arbitrary topologies of branched polymers, although the irregularities of real polydisperse branched melts present tough challenges for implementation of the theory. For structures with many branch points, a simplification is to treat the relaxation in discrete stages, calculating the time-scales at which arm retraction has penetrated to each layer of segments. This, of course, only applies when the molecular chains between branch points are monodisperse – for real architectures (where these segments are typically exponentially distributed) this approach is at best a severe approximation. At each stage the effective topology of the molecule simplifies: all faster relaxing (outer) segments relax much faster than the current time-scale, and so are not part of the entangled network for longer time-scales, but instead dilute the current value of M_e and so the effective tube diameter, $a^*(t)$. In an arbitrary branched molecule, entangled in a melt, any segment will be relaxed by deep, renormalized retraction from whichever of the two trees it is connected to relaxes first. This special statistic (equivalent to the longest path to the exterior of the molecule within the relaxing tree, measured in topological steps between branch points) has been termed the *seniority* of the segment [84]. An example of assignment of seniorities on a randomly branched molecule is given in Fig. 6.13.

The distribution of seniorities in a melt uniquely determines the entangled dynamics in linear response if the chain sections between branch points are monodisperse. For example, the Cayley tree of n layers of branching, and of branch points of functionality f [85], contains f^n segments in its outermost layer and $(f^{n+1} - f)/(f - 1)$ segments altogether. The effective concentration of unrelaxed segments after m levels have relaxed is $C(m) = (f^{n-m+1} - f)/$

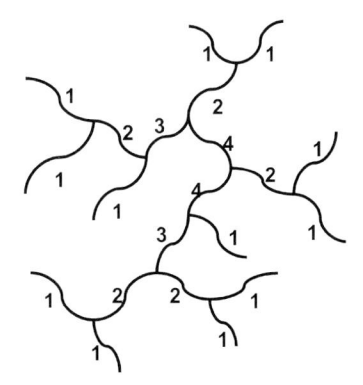

Fig. 6.13 Illustration of the assignation of seniority values to topologically distinct subchains in a branched molecule. After [30].

$(f^{n+1} - f) \simeq f^{-m}$ when n is large. Solving the star-arm like retraction from level m to level $m+1$ with the approximation that the effective concentration at level m is valid throughout that stage of the hierarchy gives the set of induction relations

$$\tau_{m+1} = \tau_m e^{(\nu/2)(Z_x)f^{-m}} \tag{20}$$

which have the solution

$$\tau_m = \tau_0 \exp\left[\nu Z_x \left(\frac{1-f^{-m}}{1-f}\right)\right]$$

where ν is the principal path parameter (usually taken as 3/2) and Z_x the (bare) number of tube segments between branch points on the tree. At this level of approximation we may also take $G(t) \simeq G_0[C(m(t))]^{a+1}$. This leads to the logarithmic form of stress relaxation $G(t) = G_0 \ln(\tau_{\max}/t)^\theta$, where $\tau_{\max}$ is the (finite) relaxation time for τ_m as $m \to \infty$ and θ is a "topological exponent" with a value of $1+a$ in this case. A more careful treatment of the Cayley tree case, including the fast fluctuation contributions, has recently been given [86]. Synthesis of dendritic polymers has been successful in producing rheologically relevant quantities in PS [87] and in hyperbranched (randomly linked) form from AB_2 condensation [88], but in both cases the threshold for entanglement was not reached, and a modified Rouse behavior dominated the rheology. It turns out that other, less regular, tree-like structures also have the logarithmic form for $G(t)$ for their entanglement-dominated relaxations, but with different values for θ. For example, the ensemble of randomly branched trees just at the classical mean-field gelation point has $\theta = 3 + a$ [84]. The finite value for the maximum entangled time-scale at the critical point for the gelation ensemble is significant industrially (the viscosity divergence at gel point is weak, and due only to slow, and very dilute, unentangled dynamics). It arises from the steep tail in the seniority distribution: $C(m) \sim m^{-2}$ for gelation, whose integral converges.

If the subchains between branch points are polydisperse, the time-scale hierarchy and seniority hierarchy are not equally ordered. However, an important consequence was noted very early in application of tube models to branched polymers [89]: if the rheology is dominated by single, long, dangling arms in the presence of a network (so that CR is moderated) and if the polydispersity is exponential in form [i.e. $P(Z_a)e^{-Z_a/\bar{Z}_a}$], then the exponentially rare presence of very long arms is countered by their exponentially long relaxation times to give an effective power-law relaxation

$$G(t) \sim t^{-u} \quad \text{with} \quad u = u(v, Z_a) \simeq \frac{1}{vZ_a} \tag{21}$$

This is approximately true in the case of the mean-field gelation ensemble [84], where, near the percolation threshold (where an incipient gel forms [90], the prediction for the apparent dynamic exponent that is valid for several decades of the logarithmic form of $G(t)$ is $u = \gamma/(vZ_x)$ (note that this is actually *independent* of the topological exponent θ, and so is of general application to highly branched entangled polymers). The prediction that a strongly entangled gelation ensemble would exhibit effectively power-law relaxation is amusing, because this is also the Rouse result for unentangled percolation clusters [31, 91, 92]. In this case the result is independent of Z_x:

$$u = \frac{d_f(\tau - 1)}{d_f + 2} \simeq 0.67$$

where d_f is the fractal dimension of the clusters (expected value 2 in 3-D percolation ensembles) and τ the exponent of the power-law molecular weight distribution function (expected value 2.2 in 3-D percolation). This prediction is in accord with experimental results on cross-linked systems with $Z_x < 1$ [93]. Cross-linking linear chains that are already entangled, however, does indeed lead to a much more striking enhancement of viscosity, consistent with the exponentially slow retraction mode of dangling arms [94, 95]. Experiments on reacting monodisperse precursor polymers have shown that careful cross-linking of linear segments using a urethane condensation technique does indeed create a much more slowly relaxing melt when $Z_x > 1$, and that, for the systems so far explored up to $Z_x = 13.6$, the dynamic exponent $u \simeq 1/Z_x$ with a prefactor close to unity according to [96, 97]. The comparable work using a polycondensing polyester system found equally good agreement, but with a prefactor closer to 0.5 [32].

A more readily accessible system is now available in the form of the long-chain-branched metallocene catalysed ensemble [34, 35, 98]. In the ideal circumstances of a continually stirred steady-state reaction with vinyl incorporation of branches, the topological distribution of molecules a well-defined one-parameter family [36], in the same class as AB_2 condensation. The resulting seniority distribution is different from that of random gelation. This is due to the special, directional, way in which molecules are synthesized. The relaxing retraction that frees any given segment is far more likely to come from the direction opposite to that of the molecule's prior synthesis at the catalyst site! This is due to the greater number of routes through the molecule that end with a chain end in that direction (in the synthesis direction there is only one – the final terminating monomer). This apparently innocent difference from the gelation ensemble leads to a major divergence in the predicted, and measured, rheology. For the seniority distribution in a highly branched ensemble of this type follows $C(m) \sim m^{-1}$ so that the continuous limit of the induction equation for relaxation time-scales (Eq. 20) takes the form

$$\frac{d \ln \tau}{dm} = vZ_x[C(m)]^a \qquad (22)$$

and produces a formally divergent relaxation time if $a = 1$. This, apparently formal, piece of statistical physics turns out to have important industrial consequences: the danger of unmanageably high viscosities is much greater in this catalytic branching process than in a cross-linking one, as managers of pilot plants dedicated to their development know to their cost!

One should bear in mind, however, that all the results above are calculated within the approximation of fixed molecular weight between branch points, and so are at best qualitative given the exponential sensitivity to this quantity in all processes dominated by activated retraction. The challenge for theory is the same in metallocene ensembles as from the randomly cross-linked gelation experiments, in that the molecular weight of the segments between cross-links in the most ideal real materials is exponentially distributed. The call for a quantitative evaluation of the tube dilation theory from these complex, but well-defined, materials had been the motivation for the development of algorithms that address truly polydisperse branched structures. The first challenge arises from how to handle the renormalization of the retraction dynamics of a dangling branch when a side branch from it relaxes, turning from a topological fixed point into a source of drag, as pictured in Fig. 6.14. One way of accounting for this is to insert a discontinuity, or "waiting time" in the assignation of local relaxation times in terms of path length from the molecule's extremities $\tau(x)$. This is the basis of a method by Larson and colleagues, in which to make approximate (numerical) calculations of linear rheology for arbitrary admixtures of architectures [99, 100]. However, this approach does not generalize easily to the more complex "branch-on-branch" structures that arise in densely branched ensembles.

There are other physical processes that must be accounted for in a tube dilation approach to arbitrary ensembles, which we may summarize as follows: (i) the effect of the geometric placement of a side branch on the relaxation of a main branch, (ii) the renormalization at long times of branched structures into

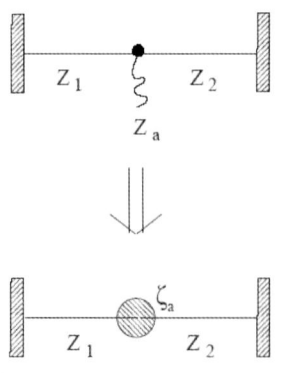

Fig. 6.14 An arm Z_a relaxes on two other arms Z_1 and Z_2, both of which might be connected to other relaxing networks. After Z_a has relaxed completely, it is replaced by a drag point with a friction ζ_a, represented as a shaded sphere. Shaded rectangles represent arbitrary networks of connected arms.

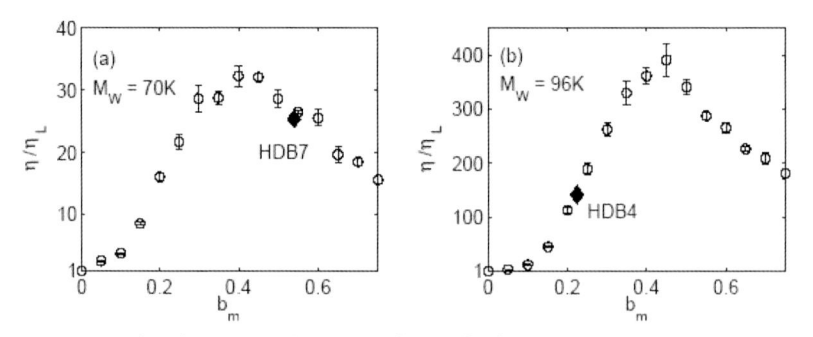

Fig. 6.15 Predicted viscosity enhancement for two fixed average molecular weights against the average number of branch points per molecule b_m in a metallocene PE ensemble. Calculations from [101] and data from [98].

effectively linear chains that relax by reptation, (iii) high-frequency chain-end and fluctuation relaxations and (iv) the appearance of "disentanglement transitions" when tube dilation occurs faster than spatial exploration by the renormalized chains that the tube contain. This last process generally occurs in the terminal time regime, but may also arise at intermediate time-scales, depending on the distribution of structures in the melt. A recent version of the hierarchical algorithm addresses all of these points by the device of maintaining two, rather than one, "relaxation fronts" passing from the extremities to the interiors of the molecules [101]. In addition to the front separating relaxed from oriented segments, a deeper front marks the current effective "root" of possible contour-length fluctuations. When applied to a model system of metallocene branched ensembles, the linear rheology is quantitatively predicted. Each sample may be plotted on a one-parameter graph of the viscosity enhancement (at fixed overall molecular weight) in terms of the average number of branch points per molecule, b_m, that rapidly approaches a maximum at $b_m \simeq 0.5$ before reducing gently once more. The magnitude of the peak itself is an exponentially strong function of the molecular weight (Fig. 6.15). Application of the model and algorithm to the gelation ensemble experiments of [32] also provides a quantitative account of existing data [102], providing corrections to Eq. (21) and failing only (and interestingly) in the "Ginzburg region" close to the gel point where mean-field statistics no longer apply.

6.4
Long-chain Branching in Nonlinear Response

A current challenge for molecular rheology is the nonlinear flow of highly branched polymers [10, 104]. The central issue, as we have seen, is that in both uniaxial and planar extensional flows, LDPE and other branched polymer melts are strain-*hardening* [49], while retaining a *softening* characteristic in shear. In this respect, the insights of the special molecular features of long-chain-branched polymers under high strain make a molecular approach doubly appealing. In particular, we will find that chain stretch has a special role to play in entangled branched polymers, and introduces a further distinct molecular mode of dynamics, that we term "branch point withdrawal". The specially synthesized model polymers are beginning to play as important a role here as they have in linear response.

6.4.1
Stretch and Branch-point Withdrawal (BPW): the Priority Distribution

The process of chain retraction can be applied to more complex topologies of entangled polymers under the same assumptions as made for linear polymers [77, 80]. Assuming that the tube deforms affinely with the bulk polymer implies that under a large step strain the mean primitive path length always increases. A linear chain is free to retract under the fast (non-diffusive) Rouse dynamics to restore its equilibrium path length. Doi and Edwards calculated that the result of this for the stress was a molecular weight- and chemistry-independent functional form for the damping function (see Eq. 8) [55]. In the case of star polymers, retraction may proceed just as for linear chains, and the damping function is expected to be of the universal "Doi–Edwards" type. This has been confirmed experimentally [54]. However, for polymers with higher levels of branching the situation is different. In spite of being stretched along its contour by a large bulk strain, a segment one level further into a branched molecule than the outermost arms cannot in general retract, because this would mean drawing those outer segments into its tube. It may only do this when its tension exceeds the sum of the (equilibrium) tensions in the impeding arms. This criterion in turn is only met beyond a critical strain. This strain will be propor-

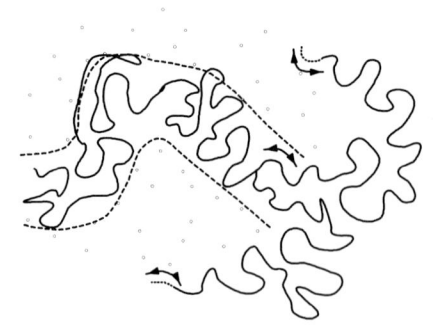

Fig. 6.16 The process of branch point withdrawal: a segment with greater than equilibrium tension pulls attached dangling arms some distance into its own tube, thus shortening their effective entangled path length.

tional to the number of arms attached at the outermost branch point, because of the Hookean elastic response of the chain's entropic tension. This process is illustrated in Fig. 6.16 for a local entanglement structure composed of two dangling arms joined to a deeper segment of seniority 2. When a bulk strain is applied, the primitive path length of the tube containing the seniority-2 segment increases in length by a geometrically calculable function of strain $\lambda(\gamma)$. Since, for small deformations, the branch points are trapped at the confluences of their tubes, the seniority-2 segment is stretched by the same amount and its entropic tension increased in proportion. When the tension $f = \lambda(\gamma)f_{eq}$ equals the sum of the equilibrium tensions of the dangling arms, and not before, the arm configuration may be partially collapsed and the branch points withdrawn into the tube originally occupied by the seniority-2 segment.

In a manner akin to the hierarchical relaxations in a highly branched polymer, this new form of retraction via "branch point withdrawal" also happens hierarchically, although as a function of strain rather than time. A segment sited two levels into the molecule may only retract when its tension exceeds that of all the first-level segments attached to it, and so on. As a consequence of this balance of entropic tensions, the strain at which any segment withdraws its outermost branch point (and so the functional form it contributes to the strain dependence of the stress) just depends on the number of free ends at the edge of the tree to which it is connected. Just as in the seniority distribution, there are in general two such trees: in this case it is the one that first permits BPW that determines the critical strain (we recall that any segments in a branched polymer is connected to two trees). This statistic (of the lesser number of free ends of the two trees) has been termed the "priority" distribution [80], in analogy with the " seniority" distribution which controls the relaxation times (Fig. 6.17). In general, a branched molecule has different seniority and priority distributions – a knowledge of the former is required to predict the linear stress-relaxation function, of the latter to predict the nonlinear strain response.

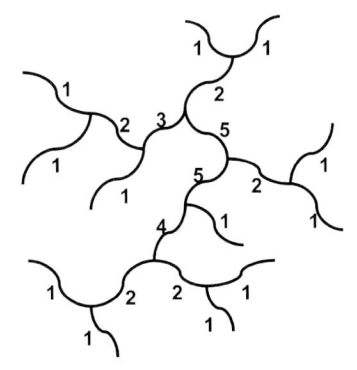

Fig. 6.17 The same topology of molecule as in Fig. 6.13, labeled this time with the segment *priorities*.

A general equation giving the damping function in terms of the priority distribution of an arbitrary branched melt is given in [80], where results for a range of topologies from Cayley trees to combs is given. In the case of the percolation ensemble, for example, the priority distribution is a universal power-law with exponent $-3/2$. In the case of metallocene branched melts it is possible to find analytic forms of the joint seniority and priority distributions [36]. It is generally true that $h(\gamma)$ for all classes of branched polymer lie above the Doi–Edwards result for linear chains.

An important, and startling, prediction is that for most monodisperse branched structures, time–strain factorability will in general be lost, although there may be regimes of time-scales between the stretch relaxation of one seniority and the orientational relaxation of the next, during which a local separability can define a "time-local" damping function. As we already saw for the important example of the H-polymer structure, a melt of the model polymer in Fig. 6.11 will behave as a diluted and slowed down system of entangled linear chains at times much longer than the longest relaxation time of the arms. Hence the nonlinear response in step strain at these time-scales must be described by the (much more thinning) Doi–Edwards damping function! The strain response exhibits a higher effective modulus at short times than at long times in nonlinear step strain, with a discontinuity in the gradient of the "early time" damping function, corresponding to the critical strain for onset of BPW. The difference in the two stress responses is connected by the time-scales of stretch relaxation of the cross-bar section, inhibited by the exponentially large drag at the branch points arising from the star-like arm retractions.

Experiments on H-polymer melts have confirmed this expectation [25] (for the damping function results, see Fig. 6.8), together with a subtle and interesting feature: the stretch relaxation time depends itself rather strongly on strain. The higher the strain, the sooner is the transition from a rubbery response to a Doi–Edwards-like strain thinning. For high strains beyond the level at which branch point withdrawal occurs, this is not difficult to understand, as the dangling segments which control the effective drag on the branch points are smaller. However, even at smaller strains the branch point will tend to withdraw the dangling arms by up to one entanglement length. This is not a minor perturbation to the effective drag on the branch point because of its exponential dependence on the dangling primitive path length (we recall the underlying dynamics are those of an entangled star arm). This subtle partial retraction is, however, the sort of conjecture that requires more than rheological measurement to confirm satisfactorily. Given the clear observation of non-factorable $G(t, \gamma)$ in model monodisperse branched polymers, the puzzle which then arises is to explain why the highly branched LDPE *does* exhibit time–strain factorability [103]. Experiments and calculations on branched melts with controlled polydispersity will be the next stage, since it is known that summing the response of intrinsically nonfactorable rheological models with a sufficiently broad range of relaxation times can lead to an overall response that is indistinguishable from a factorable one [104].

In this case of specific molecular configurations under a nonlinear strain, it has been essential to employ direct structural experimental probes on the model materials in addition to the indirect rheological ones. Such anisotropic and partially collapsed structure on the length scale of whole molecules is picked up by small-angle neutron scattering on quenched strained samples. Early experiments were able to contribute direct structural evidence that small translational rearrangements of branch points do occur for all strains [25], but indicated that the experimental anisotropy was actually about twice as large as that which was predicted to arise from branch point withdrawal. More careful experiments on both deuterium-labeled PI and PB H-polymers (where the cross-bar portions alone carried the deuterium labels) [105, 106] also showed that the anisotropy continued to increase well after the Rouse relaxation times of the chain, while the stress itself monotonically decreased (Fig. 6.18). Such experiments demand

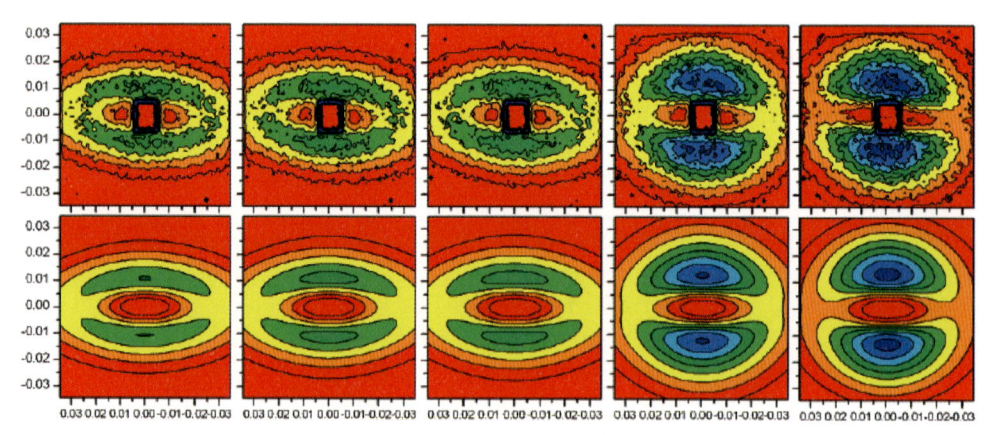

Fig. 6.18 2D SANS intensity patterns from a uniaxially-stretched melt of deutero-labelled H-polymers with contours spaced at intensities of 10, 12.5, 15, ..., 27.5 cm^{-1}. The stretch direction is vertical. Relaxation times vary from immediately after the stretch to a full relaxation of a dangling arm [106].

very careful temperature control of the sample in the neutron beam itself, close to its glass transition temperature, so that rapid processes may be slowed and quenched to relevant exposure times for neutrons. Accounting for these remarkable neutron scattering results quantitatively requires a quenched-variable extension of the "RPA" approximation, together with careful accounting of all the quenched and deformed constraints on the H-polymer as its structural relaxation continues [107] (bottom row in Fig. 6.18). It also shows how scattering is sensitive to aspects of melt structure to which rheology is not, since the amplification of the anisotropy results from elastic inhomogeneities in the entanglement network at the length scale of the tube diameter [106].

6.4.2
Branched Polymers – a Minimal Model

It is possible to call on the linear and nonlinear physics investigated by the model chemistries discussed above to develop full constitutive equations for arbitrary flow histories in simple cases of branched architecture. For a first case we need to remove the complications of polydispersity in molecular weight and topology, but must have sufficient complexity to capture the new phenomena of uniform segment stretch and branch point withdrawal suggested by the tube model. This has recently proved a very effective line of attack on the general problem of LCB entangled viscoelasticity, since the physics may be checked against model architectures [25, 108, 109], while supplying theoretical structures that may be generalized and applied directly to commercial polymers [110–113]. From the considerations above on step-strain response, it will be clear that monodisperse star polymers do not satisfy our requirements since they possess no segment without a free end. So a more fruitful choice is a family of architectures based on the H-polymer, but with a variable number of arms, q, attached symmetrically to each end of the cross-bar portion. Such "pom-pom" polymers permit us to explore the notion of "degree of branching", contained in q, while keeping simplicity of structure [28, 108]. The other structural parameters will be the number of entanglements of the cross-bar $Z_b = M_b/M_e$ and arm $Z_a = M_a/M_e$, as illustrated in Fig. 6.19. Further simplifications may be made if

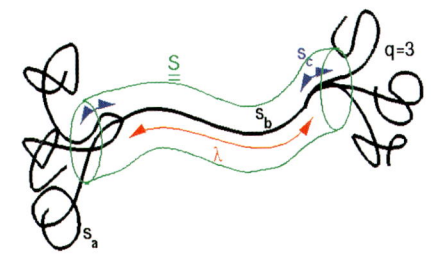

Fig. 6.19 Schematic diagram of the pom-pom molecule illustrating structural parameters and dynamic variables.

restriction is made to the *first* nonlinearities that appear in the viscoelastic response.

From our review of the H-polymer melt in linear response above, a direct generalization of the outlines of a theory to the pom-pom architecture is trivial. The arms relax by retraction, in the presence of the effectively permanent cross-bar segments, and the branch points themselves behave as slow diffusers whose fundamental step time is given by the first-passage time for arm retraction. The only modifications required from the H-polymer results are (i) an account of the number of arms in the cross-bar volume fraction and (ii) an increase in the effective drag coefficient of the branch points so that $\zeta_b \sim 1/(q-1)$. Hence the slowest relaxing segments in the melt (by typically exponentially large ratios) are the cross-bars themselves. The nonlinear response of the melt will be dominated by the cross-bars at deformation rates at which the arms are hardly perturbed from equilibrium for most of their length. So for a first theory of nonlinear response we ignore the contribution to stress from the tube occupied by the arms (apart from a trivial addition of the linear viscoelasticity, calculated similarly to Eq. (18) and concentrate solely on the cross-bars. The only exception to this is required when deformation rates are so high that branch-point withdrawal occurs, when the strain-induced renormalization of the arm drag will require some new physics to quantify BPW. This theory will be valid for timescales $\tau > \tau_a$, or alternatively deformation rates $\dot{\varepsilon} < \tau_a^{-1}$.

In this regime (see Fig. 6.11), the molecules become topologically *linear* polymers, entangled in the diluted tubes arising from cross-bar entanglement alone, but with special dissipation: all the effective drag is located at the extremities of the chains. In nonlinear response this drag will depend on the state of deformation itself via BPW. The polymers will therefore reptate, with a near single-exponential relaxation modulus, and stretch curvilinearly within their tubes. The stretch may be represented as the simple scalar ratio of current path length L to equilibrium length L_0, $\lambda = L/L_0$. In this regard they resemble entangled versions of the simple "dumbbell" models of dilute polymer solutions, except for the variability of drag, which we shall see overcomes many of the shortcomings of those earlier phenomenological models [114]. Curvilinear stretching will affect the stress contributed per molecule in the two ways we saw in the case of linear polymers with chain stretch. First, since the segments are Gaussian chains, stretch increases the effective chain tension, so increasing each component of the stress tensor in proportion. Second, a stretched cross-bar will have an increased primitive path, contributing additionally as the number of tube segments of length a^* that it spans. The (unit trace) geometry of the stress tensor $\mathbf{S}$ will, as in all tube models, be given by the second moment average of the tube-segment orientations $\mathbf{u}$, $\mathbf{S} = \langle \mathbf{uu} \rangle$. A final contribution arises from arm material dragged into tube once occupied by cross-bar segments. As we have seen, this occurs once the cross-bar has attained its maximum stretch. A convenient measure is the time-dependent number of entanglement segments so withdrawn, Z_c. These segments are aligned with the cross-bar tube, but *not* stretched, and so carry only one factor of the stretch. Hence, as the orientation moment (controlled by reptation), the stretch (a measure of curvilinear retraction) and the withdrawn arm length will vary with time, the stress has the structure

$$\sigma(t) = G[\lambda(t)]^2 \mathbf{S}(t) \left[1 + 2\frac{Z_c(t)}{L_0 \lambda(t)}\right] \quad (23)$$

where G is an effective modulus. In a monodisperse melt of real pom-pom architecture polymers, the effective modulus G depends on the true plateau modulus of the polymer $G_N^{(0)}$ and the volume fraction of cross-bar material, since at the relevant time-scales arm material is acting as a diluent. If $\beta = 1 + \alpha$ is the dilution exponent (taken variously as 2 or 7/3), then we have $G = (15/4) G_N^{(0)} \phi_b^\beta$. No further reduction in variables is possible than that of Eq. (23) apart from the good approximation in which the last term, representing stress from aligned arm material, is discarded. Doing this does not change the special factorized structure of the stress. In particular, it will not be possible to write a closed equation for the stress tensor itself; instead, we need one dynamic equation for the orientation and one for the (scalar) stretch. We now review the derivation of dynamic equations for the three structural variables $\mathbf{S}(t)$, $\lambda(t)$ and $X_c(t)$.

A differential approximation for the evolution of the cross-bar orientation (useful for intensive computations such as in finite-element solvers for complex flow geometries [109, 115]) functions by calculation of an auxiliary tensor $\mathbf{A}(t)$, then projecting on to a unit-trace tensor (this is an approximation to the molecular process of retraction) to find the approximation for $\mathbf{S}(t)$:

$$\frac{D}{Dt}\mathbf{A} - \mathbf{K}.\mathbf{A} - \mathbf{A}.\mathbf{K}^T = -\frac{1}{\tau_b}(\mathbf{A} - \mathbf{I}); \quad \mathbf{S}(t) = \frac{\mathbf{A}(t)}{\text{trace}[\mathbf{A}(t)]} \quad (24)$$

The essential feature captured by this approximation is the asymptotic form in shear that $S_{xy} \sim \dot{\gamma}^{-1}$. This is important in controlling the behavior of stretch in shear flows. In cases where the differential approximation is compared with real data, it is important to note that the expression for G changes to $G = 3G_0 \phi_b^\beta$. Other differential approximations have recently been applied, and other choices are necessary if the second normal stress difference $N_2 = \sigma_{yy} - \sigma_{zz}$ is to be non-zero [113]. The relaxation time for orientation arises from reptation of the cross-bar in a diluted tube, under a drag dominated by arm retraction on time-scales τ_a:

$$\tau_b = \frac{4q\tau_a \phi_b^{2\alpha} Z_b^2}{\pi^2 p^2} \quad (25)$$

Note that although this is a *reptation* time, it scales as Z_b^2, not Z_b^3, because the effective drag arises from the branch points only, and is independent of Z_b.

To derive the dynamic equation for the stretch ratio $\lambda(t)$, we employ the same notion as above, that each branch point is a source of effective drag ζ_b, via an

Einstein relation $\zeta_b = k_B T/D_b$, modified by BPW. The branch point diffusers are also subjected to drag from stretching tube around them, and from the entropic elasticity of the chain connecting them. The final result is

$$\dot{\lambda} = \langle \mathbf{K} : \mathbf{S} \rangle \lambda - \frac{1}{\tau_s}(\lambda - 1) \tag{26}$$

with the stretch relaxation time

$$\tau_s = \frac{1}{3} \frac{q\phi_b{}^a \tau_a}{p^2} e^{-k^*(\lambda-1)/(q-1)} \tag{27}$$

(note that in earlier publications a different prefactor arises due to the cleaner definition of M_e adopted here). Just as in the calculation of the damping function, we must override the dynamic equation for the stretch by the non-analytic BPW as soon as the tension in the cross-bar equals the sum of equilibrium tensions of the dangling arms. This is just when $\lambda = q$, so Eq. (26) operates only until it would violate $\lambda < q$, in which case the maximal stretch is maintained until the driving flow would cause stretch relaxation below q. The non-analyticity arises, as usual in statistical physics, when the dynamics result from integrating out many microscopic variables.

The set of equations (Eqs. 23–27) determine the constitutive behavior of the pom-pom model. Remarkably, for the physically relevant ordering of $\tau_s < \tau_d$, it exhibits strong shear softening (yet with overshoots in shear and normal stress), and equally strong extension hardening, up to a maximum set by the degree of branching, q. These are all the special qualitative features observed in the extensive studies of the much more topologically polydisperse LDPE [10]. Significantly, the hardening is shared equally in uniaxial and planar extension, also the case in branched melts generally [49].

Tests on monodisperse pom-pom-like architectures have to date been limited to studies on H-polymers, but the nonlinear transients in extension and shear together with their overshoots and maxima were quantitatively accounted for using a value of $k^* \simeq 0.36$ [25, 105]. Together with the SANS data discussed above, they represent strong support of the underlying model.

However, perhaps the most important legacy of this model study, that led to the pom-pom equations, is their use as elements in mathematical models for LDPE and other polydisperse LCB polymers. The insights gained in studies of the model architectures above may be applied to commercial, random branched polymers, even without detailed knowledge of their architectures. Of course, an eventual goal of this programme of work is the ability to predict the rheological response of any melt of topologically complex polymers if their distribution of structures is known. However, the insights we have gained into the H (and in general "pom-pom") family of structures are informative straight away in that they point to a generic feature of the rheology of branched polymers. This is that:

An entangled segment of a branched polymer in the melt will contribute to the bulk stress via both its orientation (tensor property) and stretch (scalar property). These quantities will relax with different characteristic time-scales. The segment will stretch in flow, but only up to a maximum ratio, given by the effective number of free ends attached to its delimiting branch points.

The "pom-pom" constitutive equation was based on these assumptions and showed how such a structure could lead to a polymer melt that was simultaneously extension hardening and shear thinning. To model existing random branched polymers such as LDPE this way one can take the linear relaxation spectrum $G(t) = \sum_i g_i \exp(-t/\tau_i)$ as an indication of the distribution of segment orientation times, and "decorate" each mode i with the nonlinear parameters of a "buried" branched segment, i.e. a stretch relaxation time τ_{si} and a maximum stretch ratio q_i. This has been done recently by several groups [110, 111, 113]. With the same set of parameters, the weaker response in shear and the second normal stress in planar extension are modeled, in addition to the stiffer response in the first normal stress of the extensional flows. Similar fits are achieved at a range of deformation rates. Of particular note is the consequence of the q_i distribution: the extensional curves typically show a rapid change of gradient when the dominant segments for that extension rate reach their maximum extension. As we have seen, this would cause a free extending film to break, and end the experiment. Hence a possible consequence is that the break points in an extensional data set are direct measures of the molecular distributions of the "priority" distribution q_i. Such "spectra" of nonlinear parameters with time-scales are proving to be a very useful way of "fingerprinting" a branched polymer. An example of such application of a LCB "modeling toolkit" is shown against nonlinear data for the LDPE of Fig. 6.7.

6.5
Application to Other Topologies and Challenges

In many ways, the success of the multi-mode version of the pom-pom model in accounting quantitatively for the nonlinear rheology of commercial LCB polymers is a great surprise. After all, we have seem time and again that the exponential hierarchy jointly imposed on the melt by the topological constraints and the presence of branching mean that the effective topology of the molecules depends on the time-scale with which they are observed. An H-polymer melt, for example, may behave as a linear polymer at low frequency. Such "topological renormalization" would appear to militate against any constitutive formulation that used a fixed set of priority parameters q at all flow-rates. A second concern is that, in practice, the elements of the pom-pom ensemble with different relaxation times occupy typically positions of seniority within the same molecules, and so will be directly coupled in a flow. A study aimed at investigating how these effects would emerge from a more rigorous application of the tube model considered the case of a melt of monodisperse Cayley tree molecules of three levels [86]. The coupling of segments of the three different seniorities that emerges is of two kinds. First, *orientation* is advected by the flow from inner to outer segments. This is responsible for a much longer time to achieve steady state in a shear flow than in a decoupled model, since very high strains are required to deform tube segments originally around high-seniority chain to the outside of the molecule. But second, information on chain *stretch* is convected in the reverse sense. It is the time-dependent stretch of the outer segments, rather than their nominal priorities, that determines the maximum stretch of deeper segments. In consequence, although it is possible to approximate the behavior of such a melt by a decoupled ensemble of pom-pom modes, the elements used are not typically identifiable with particular seniorities in the molecules, and much of the agreement is possible due to the necking instability in extension hiding later behavior that would otherwise distinguish the models.

Such calculations are a useful caution when pursuing more advanced goals, such as the prediction of nonlinear rheology from polydisperse LCB melts. From the LCB metallocene ensemble we reviewed in Section 6.3.3, it is possible to derive not only the seniority but also the priority distributions of the ensemble at any time-scale of observation [36]. Using the two parameters of plateau modulus $G_N^{(0)}$ and entanglement segment Rouse time τ_e, an estimation of the rheology may be made by mapping the bivariate distribution of seniority and priority on to a pom-pom ensemble with the identical bivariate distribution. Only the physics of seniority and priority coupling is omitted. Initial results are promising, correctly predicting, for example, that the regime of extension-hardening switches from low extension rates to high at a value of the branching parameter b_U of 0.1. An example of the quality of prediction by this technique is given in Fig. 6.20, where the lighter, non-hardening set of curves result from topological renormalization of the ensemble by one level of branching. The surprising apparent lack of the topological renormalization at lower flow-rates, when one would have naïvely thought it necessary (see Fig. 6.20), is a current challenge. The effective amplification of local (tube scale) deformation rates greatly above the bulk rate identified in [86] may be a clue. However, a fully quantitative account of nonlinear flow in complex admixtures of branched polymers will certainly require a proper treatment of convective constraint release, whereby the retraction of surrounding chains from the flow accelerates the dynamics of the tube constraints themselves. Well developed in the case of monodisperse linear polymers [116], there are severe conceptual challenges in applying the idea to branched polymers.

Even in the case of linear response, there is a need to sharpen our understanding of the physics of effective drag from entangled side arms. Even the language of the debate about "thin tube" and "dilated tube" hopping is tied to the framework of DTD, which we have seen may not be adequate beyond our current level of experiments on simple architectures [75]. A recent suggestion that the dimensionless hopping parameter p (see Eq. 16) is dependent on the size of the backbone of the polymer that the branch point sits on [117] clearly gives the wrong scaling for the total drag on such a backbone in the limit of

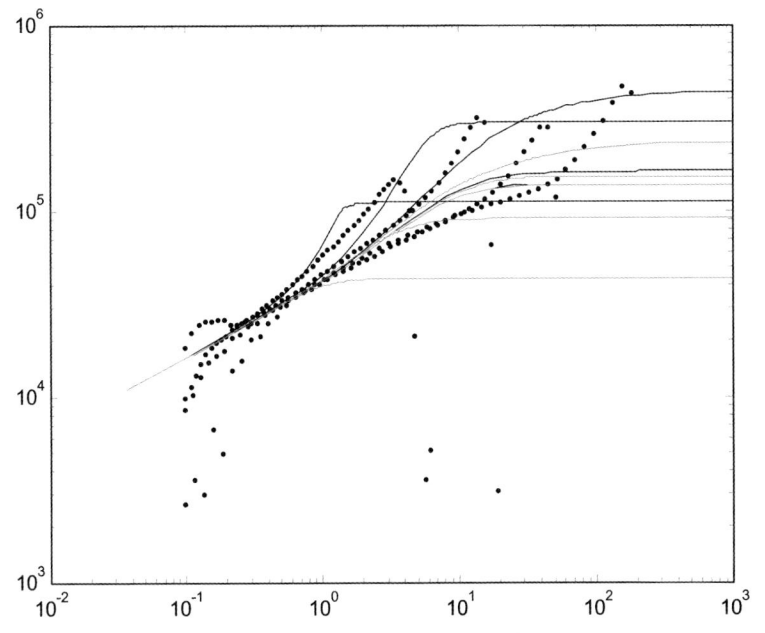

Fig. 6.20 Data on extensional transient flow (stress growth coefficient in Pa with time in s) for an LCB metallocene melt with $b_U \approx 0.1$. Predictions with the full priority distribution are the dark lines, with the outer branched relaxed in light. Extension rates are 0.03, 0.1, 0.3 and 1.0 s^{-1}.

large polymers, but is indicative of the degree of uncertainty still resident in the available data sets (but see [101] for a consistent treatment of as wide a range as currently available). Extending the physics of branch point motion to the nonlinear regime will rely on further careful SANS rheology experiments on key architectures, in addition to using the increasing power of simulation [118].

There is a need for nonlinear experiments on well-defined architectures intermediate in complexity between the simple H and pom-pom materials and the fully polydisperse commercial melts currently available. Nonlinear measurements on well-defined combs will probably be the next step, providing a degree more structure than the H-melts, but model systems with more than two levels of seniority and priority, such as Cayley trees, but yet are well entangled, remain a material dream still to be realized.

References

1 Flory, P.J., **1953**, *Principles of Polymer Chemistry*, Cornell University Press, Ithaca, NY.
2 Zimm, B.H., Stockmayer, W.H., **1949**, *J. Chem. Phys., 17*, 1301.
3 Edwards, S.F., **1976**, The configuration and dynamics of polymer chains, in *Molecular Fluids*, Balian, R., Weill, G. (Eds.), Gordon and Breach, London, pp. 151–208.
4 Kuhn, W., **1934**, *Kolloid Z., 68*, 2.
5 Zimm, B.H., **1956**, *J. Chem. Phys., 24*, 269.
6 Rouse, P.E., **1953**, *J. Chem. Phys., 21*, 1272.
7 Zinn-Justin, J., **1993**, *Field Theory and Critical Phenomena*, Clarendon Press, Oxford.
8 Ferry, J.D., **1986**, *Viscoelastic Properties of Polymers*, Wiley, New York.
9 Small, P.A., **1975**, *Adv. Polym. Sci., 18*, 1.
10 Meissner, J., **1975**, *Pure Appl. Chem., 42*, 551.
11 Hadjichristidis, N., Roovers, J.E.L., **1974**, *J. Polym. Sci., Polym. Phys. Ed. 12*, 2521.
12 Kraus, G., Gruver, J.T., **1965**, *J. Polym. Sci. A3*, 105.
13 Roovers, J.E.L., Bywater, S., **1972**, *Macromolecules, 4*, 385.
14 Graessley, W.W., Masuda, T., Roovers, J.E.L., Hadjichristidis, N., **1976**, *Macromolecules, 9*, 127.
15 Roovers, J.E.L., **1981**, *Polymer, 22*, 1603.
16 Fetters, L.J., Kiss, A.D., Pearson, D.S., Quack, G.F., Vitus, F.J., **1993**, *Macromolecules, 26*, 647.
17 Vlassopoulos, D., Pakula, T., Fytas, G., Roovers, J., Karatasos, K., Hadjichristidis, N., **1997**, *Europhys. Lett., 39*, 617.
18 Pakula, T., Vlassopoulos, D., Fytas, G., Roovers, J., **1998**, *Macromolecules, 31*, 8931.
19 Edwards, S.F., **1967**, *Proc. R. Soc. London, 92*, 9.
20 de Gennes, P.G., **1971**, *J. Chem. Phys., 55*, 572.
21 Doi, M., Edwards, S.F., **1978**, *J. Chem. Soc., Faraday Trans. 2, 74*, 1789.; Doi, M., Edwards, S.F., **1978**, *J. Chem. Soc., Faraday Trans. 2, 74*, 1802; Doi, M., Edwards, S.F., **1978**, *J. Chem. Soc., Faraday Trans. 2, 74*, 1818; Doi, M and Edwards, S.F., **1979**, *J. Chem. Soc., Faraday Trans. 2, 75*, 38.
22 de Gennes, P.G., **1975**, *J. Phys. (Paris), 36*, 1199.
23 McLeish, T.C.B., **2002**, *Adv. Phys., 51*, 1379.
24 Quack, G., Hadjichristidis, N., Fetters, L.J., Young, R.N., **1980**, *Ind. Eng. Chem. Prod. Res. Dev., 19*, 587.
25 McLeish, T.C.B., Allgaier, J., Bick, D.K., Bishk, G., Biswas, P., Blackwell, R., Blottière, B., Clarke, N., Gibbs, B., Groves, D.J., Hakiki, A., Heenan, R., Johnson, J.M., Kant, R., Read, D.J., Young, R.N., **1999**, *Macromolecules, 32*, 6734.
26 Ferneyhough, C.M., Young, R.N., Poche, D., Degroot, A.W., Bosscher, F., **2001**, *Macromolecules, 34*, 7034.
27 Koutalas, G., Lohse, D.J., Hadjichristidis, N., **2005**, *J. Polym. Sci., Polym. Chem. Ed., 43*, 4040.
28 Archer, L.A., Varshney, S.K., **1998**, *Macromolecules, 31*, 6348.
29 Hempenius, M.A., Michelberger, W., Möller, M., **1997**, *Macromolecules, 30*, 5602.
30 Stockmayer, W.H., **1943**, *J. Chem. Phys., 11*, 45; Stockmayer, W.H., **1944**, *J. Chem. Phys., 12*, 125.
31 Rubinstein, M., Colby, R.H., Gillmor, J., **1989**, in *Space–Time Organisation in Macromolecular Fluids*, pp. 66–74, Tanaka, F., Ohta, T., Doi, M. (Eds.), Springer, Berlin.
32 Luisignan, C.P., Mourey, T.H., Wilson, J.C., Colby, R.H., **1999**, *Phys. Rev. E, 60*, 5657.
33 Kasehagen, L.J., Macosko, C.W., **1998**, *J. Rheol., 42*, 1303.
34 Soares, J.B.P., Hamilec, A.C., **1995**, *Polymer, 36*, 2257.
35 Soares, J.B.P., Hamilec, A.C., **1996**, *Macromol. Theory Simul., 5*, 547.
36 Read, D.J., McLeish, T.C.B., **2001**, *Macromolecules, 34*, 1928.
37 Iedema, P.D., Wulkow, M., Hoefsloot, H., **2000**, *Macromolecules, 33*, 7173.
38 Macosko, C.W., **1994**, *Rheology Principles, Measurements and Applications*, Wiley, New York.

39. Graessley, W. W., Roovers, J. E. L., **1979**, *Macromolecules, 12*, 959.
40. Roovers, J., **1984**, *Macromolecules, 17*, 1196.
41. Roovers, J., Graessley, W.W., **1981**, *Macromolecules, 14*, 766.
42. Islam, M., Juliani, T., Archer, L. A., Varshney, S. K, **2001**, *Macromolecules, 34*, 6438.
43. Münstedt, H., **1979**, *J. Rheol., 23*, 421.
44. Sridhar, T., Tirtaatmadja, V., Nguyen, D. A., Gupta, R. K., **1991**, *J. Non-Newtonian Fluid Mech., 40*, 271.
45. Spiegelberg, S. H., McKinley, G. H., **1996**, *J. Non-Newtonian Fluid Mech., 67*, 49.
46. Anna, S. L., McKinley, G. H., Nguyen, D. A., Sridhar, T., Muller, S. J., Huang, J., James, D. F., **2001**, *J. Rheol., 45*, 83.
47. Meissner J., **1997**, *Rheology, 5*, 120.
48. Schulze, J. S., Lodge, T. P., Macosko, C. W., Hepperle, J., Münstedt, H., Bastian, H., Ferri, D., Groves, D. J., Kim, Y. H., Lyon, M., Schweizer, T., Virkler, T., Wassner, E., Zoetelief, W., **2001**, *Rheol. Acta, 40*, 457.
49. Laun, H. M., Schuch, H., **1989**, *J. Rheol., 33*, 119.
50. Wagner, M. H., Ehrecke, P., Hachmann, P., Meissner, J., **1998**, *J. Rheol., 42*, 621.
51. Osaki, K., **1993**, *Rheol. Acta, 32*, 429.
52. Osaki, K., Nishizawa, K., Kurata, M., **1982**, *Macromolecules, 15*, 1068.
53. Fukuda, M., Osaki, K., Kurata, M., **1975**, *J. Polym. Sci., Polym. Phys. Ed., 13*, 1563.
54. Osaki, K., Takatori, E., Kurata, M., Watanabe, H., Yoshida, H., Kotaka, T., **1990**, *Macromolecules, 23*, 4392.
55. Doi, M., Edwards, S. F., **1986**, *The Theory of Polymer Dynamics*, Oxford University Press, Oxford.
56. Doi, M., Kuzuu, N. Y., **1980**, *J. Polym. Sci., Polym. Lett. Ed., 18*, 775.
57. Kramers, H. A., **1940**, *Physica (Amsterdam), 7*, 284.
58. Hanggi, P., Talkner, P., Borkovec, M., **1990**, *Rev. Mod. Phys., 62*, 251.
59. McLeish, T. C. B., Milner, S. T., **1998**, *Adv. Polym. Sci., 143*, 195.
60. Rubinstein, M., Helfand, E., **1985**, *J. Chem. Phys., 82*, 2477.
61. Pearson, D. S., Helfand, E., **1984**, *Macromolecules, 19*, 888.
62. Ball, R. C., McLeish, T. C. B., **1989**, *Macromolecules, 22*, 1911.
63. Milner, S. T., McLeish, T. C. B., **1997**, *Macromolecules, 30*, 2159.
64. Milner, S. T., McLeish, T. C. B., **1998**, *Macromolecules, 31*, 7479.
65. Adams, C. H., Hutchings, L. R., Klein, P. G., McLeish, T. C. B., Richards, R. W., **1996**, *Macromolecules, 29*, 5717.
66. Vega, D. A., Sebastian, J. M., Russel, W. B., Register, R. A., **2002**, *Macromolecules, 35*, 169.
67. Struglinski, M. J., Graessley, W. W., Fetters, L. J., **1988**, *Macromolecules, 21*, 783.
68. Blottière, B., McLeish, T. C. B., Hakiki, A., Young, R. N., Milner, S. T., **1998**, *Macromolecules, 31*, 9295.
69. Milner, S. T., McLeish, T. C. B., Johnson, J., Hakiki, A., Young, R. N., **1998**, *Macromolecules, 31*, 9345.
70. Frischknecht, A., Milner, S. T., **2001**, *Macromolecules, 33*, 9764.
71. Bartels, C. B., Crist, B., Fetters, L. J., Graessley, W. W., **1986**, *Macromolecules, 19*, 785.
72. Gell, C. B., Graessley, W. W., Efstratiadis, V., Pitsikalis, M., Hadjichristidis, N., **1997**, *J. Polym. Sci., Part B, 35*, 1943.
73. Frischknecht, A., Milner, S. T., Pryke, A., Young, R. N., Hawkins, R., McLeish, T. C. B., **2002**, *Macromolecules, 35*, 4801.
74. Watanabe, H., Matsumiya, Y., Osaki, K., **2000**, *J. Polym. Sci. B, Polym. Phys. Ed., 38*, 1024.
75. McLeish, T. C. B., **2002**, *J. Rheol., 47*, 177.
76. Hakiki, A., Young, R. N., McLeish, T. C. B., **1996**, *Macromolecules, 29*, 6348.
77. McLeish, T. C. B., **1988**, *Macromolecules, 21*, 3639.
78. Milner, S. T., McLeish, T. C. B., **1998**, *Phys. Rev. Lett., 81*, 725.
79. Roovers, J., Toporowski, P. M., **1987**, *Macromolecules, 20*, 2300.
80. Bick, D. K., McLeish, T. C. B., **1996**, *Phys. Rev. Lett., 76*, 2587.
81. Yurasova, T. A., McLeish, T. C. B., Semenov, A. N., **1994**, *Macromolecules, 27*, 7205.
82. Kapnistos, M., Vlassopoulos, D., Roovers, J., Leal, L. G., **2005**, *Macromolecules, 38*, 7852.

83 Inkson, N. J., Graham, R. S., McLeish, T. C. B., Groves, D. J., Fernyhough, C. M., **2006**, *Macromolecules*, 39, 4217.
84 Rubinstein, M., Zurek, S., McLeish, T. C. B., Ball R. C., **1990**, *J. Phys. (Paris)*, 51, 757.
85 McLeish, T. C. B., **1988**, *Europhys Lett.*, 6, 511.
86 Blackwell, R. J., Harlen, O. G., McLeish, T. C. B., **2001**, *Macromolecules*, 34, 2579.
87 Dorgan, J. R., Knauss, D. M., Al-Muallem, H. A., Huang, T., Vlassopoulos, D., **2003**, *Macromolecules*, 36, 380.
88 Suneel, Buzza, D. M. A., Groves, D. J., McLeish, T. C. B., Parker, D., Keeney, A. J., Feast, W. J., **2002**, *Macromolecules*, 35, 9605.
89 Curro, J. G., Pincus, P., **1983**, *Macromolecules*, 16, 559.
90 Stauffer, D., **1985**, *Introduction to Percolation Theory*, Taylor and Francis, Philadelphia, PA.
91 Cates, M. E., **1985**, *J. Phys. (Paris)*, 46, 1059.
92 Martin, J. E., Adolf, D., Wilcoxon, J. P., **1988**, *Phys. Rev. Lett.*, 76, 2587.
93 Durand, D., Delsanti, M., Adam, M., Luck, J. M., **1987**, *Europhys Lett.*, 3, 297.
94 Valles, E. M., Macosko, C. W., **1979**, *Macromolecules*, 12, 521.
95 Mours, M., Winter, H. H., **1996**, *Macromolecules*, 29, 7221.
96 Nicol, E., Nicolai, T., Durand, D., **2001**, *Macromolecules*, 34, 5205.
97 Gasilova, E., Benyahia, L., Durand, D., Nicolai, T., **2002**, *Macromolecules*, 35, 141.
98 Costeux, S., Wood-Adams, P., Beigzadeh, D., **2002**, *Macromolecules*, 35, 2514.
99 Park, S. J., Shanbhag, S., Larson, R. G., **2005**, *Rheol. Acta*, 44, 319.
100 Larson, R. G., **2001**, *Macromolecules*, 34, 4556.
101 Das, C., Inkson, N. J., Read, D. J., Kelmanson, M. A., McLeish, T. C. B., **2006**, *J. Rheol.*, 50, 207.
102 Das, C., Read, D. J., Kelmanson, M. A., McLeish, T. C. B., **2006**, *Phys. Rev. E*, 74, 011404.
103 Rubio, P., Wagner, M. H., **2000**, *J. Non-Newtonian Fluid Mech.*, 92, 245.
104 Larson, R. G., **1987**, *J. Non-Newtonian Fluid Mech.*, 23, 249.
105 Heinrich, M., Pyckhout-hinzen, W., Richter, D., Straube, E., Read, D. J., McLeish, T. C. B., Groves, D. J., Blackwell, R. J., Wiedenmann, A., **2002**, *Macromolecules*, 35, 6650.
106 Heinrich, M., Pyckhout-hinzen, W., Allgaier, J., Richter, D., Straube, E., McLeish, T. C. B., Wiedenmann, A., Blackwell, R. J., Read, D. J., **2004**, *Macromolecules*, 37, 6650.
107 Read, D. J., **1999**, *Eur. J. Phys. B*, 12, 431.
108 McLeish, T. C. B., Larson, R. G., **1998**, *J. Rheol.*, 42, 81.
109 Bishko, G., McLeish, T. C. B., Harlen, O. G., Larson, R. G., **1997**, *Phys. Rev. Lett.*, 79, 2352.
110 Inkson, N. J., McLeish, T. C. B., Groves, D. J., Harlen, O. G., **1999**, *J. Rheol.*, 43, 873.
111 Blackwell, R. J., McLeish, T. C. B., Harlen, O. G., **2000**, *J. Rheol.*, 44, 121.
112 Graham, R. S., McLeish, T. C. B., Harlen, O. G., **2001**, *J. Rheol.*, 45, 275.
113 Verbeeten, W. M. H., Peters, G. W. M., Baaijens, F. P. T., **2001**, *J. Rheol.*, 45, 823.
114 Bird, R. B., Curtiss, C., Armstrong, R., Hassager, O., **1977**, *Dynamics of Polymeric Liquids*, Vols. I and II, Wiley, New York.
115 Lee, K., Mackley, M. R., McLeish, T. C. B., Nicholson, T. M., Harlen, O. G., **2001**, *J. Rheol.*, 45, 1261.
116 Graham, R. S., Likhtman, A. E., McLeish, T. C. B., Milner, S. T., **2003**, *J. Rheol.*, 47, 1171.
117 Lee, J. H., Fetters, L. J., Archer, L. A., **2005**, *Macromolecules*, 38, 10763.
118 Kremer, K., **2000**, in *Soft and Fragile Matter*, Cates, M. E., Evans, M. R. (Eds.), IOP, Bristol, p. 145.

7
Determination of Bulk and Solution Morphologies by Transmission Electron Microscopy

Volker Abetz, Richard J. Spontak, and Yeshayahu Talmon

7.1
Introduction

The properties of materials are related to their structure. The structures themselves depend on the chemical composition of the materials and their (self)organization. Single-component systems in the condensed state may differ in their state of aggregation (liquid, amorphous or crystalline solid). In addition, multicomponent materials may be phase segregated and display different superstructures (morphologies), which depend to a large extent on their relative compositions and the degree of incompatibility between the different constituent species. In addition, the processing history often strongly influences morphology. This latter aspect is especially true for polymer materials, which are processed in solution or in the molten state. The length scale of domains in these superstructures is typically in the range from a few to a few hundred nanometers. These length scales are appropriate for investigation by electron microscopy. Many techniques based on electrons as a source of radiation have been developed since the first introduction of transmission electron microscopy (TEM) by Ernst Ruska in the 1940s [1]. Besides conventional techniques that probe a sample with elastically scattered electrons to generate images and scattering (diffraction) patterns, energy-dispersive images can be acquired, where the specific characteristic absorption (ionization) behavior of chemically different atoms can be used to discern their spatial distribution in the sample. This technique is based on the detection of incident electrons that have lost a characteristic energy due to inelastic interactions with electrons from different types of atoms. It is especially useful for atoms with low atomic numbers. The ionization of atoms by the incoming electron beam also gives rise to the emission of characteristic X-rays, which can be measured and mapped by energy-dispersive X-ray detectors. This method is especially well suited for elements with high atomic numbers. The spatial resolution of such elemental maps depends on, among other parameters, the electron beam (probe) size in the scanning mode of the microscope.

Macromolecular Engineering. Precise Synthesis, Materials Properties, Applications.
Edited by K. Matyjaszewski, Y. Gnanou, and L. Leibler
Copyright © 2007 WILEY-VCH Verlag GmbH & Co. KGaA, Weinheim
ISBN: 978-3-527-31446-1

Two-dimensional projections of a sample do not always allow unambiguous conclusions about its structure. In such cases, tomography can be performed by recording tilt series of the sample in the electron microscope. Here, very high precision of the sample holder is required and the sample must "survive" the cumulative electron exposure of such a tilt series. Mathematical procedures derive a three-dimensional reconstruction of the sample from the tilt series. Organic and biological samples are particularly sensitive to the electron beam and suffer extensively from beam-induced damage. Low-dose techniques taking advantage of very sensitive CCD cameras in combination with cryogenic conditions in the sample-holder are strategies to reduce damage and even permit *in situ* examination of solvated specimens. In this chapter, we discuss the following techniques in more detail:
- background of electron microscopy;
- conventional TEM of bulk materials;
- cryo-TEM and freeze-fracture TEM of solutions;
- transmission electron microtomography;
- analytical electron microscopy.

7.2
Background of Electron Microscopy

The spatial resolution g of a transmission microscope is given by Abbe's equation

$$g = 0.61 \frac{\lambda}{n \sin \alpha}$$

where λ denotes the wavelength of the radiation and $n \sin \alpha$ is the numerical aperture.

In the diffraction-limited far-field, it is impossible to use visible light ($\lambda = 400$–800 nm) to investigate structures on the nanometer scale. Technically, it is still challenging to achieve better resolution by using radiation with a shorter wavelength, such as X-rays, because of problems associated with the design of appropriate lenses. Due to the wave–particle duality, electrons also possess a wave nature, as expressed by the de Broglie relationship $\lambda = h/mv$, where h is Planck's constant and m and v are the rest-mass and velocity of an electron, respectively. This wavelength is tunable by the electron velocity, which can, in turn, be controlled by an accelerating voltage. Ernst Ruska was the first to develop an electromagnetic lens and thus laid the basis for building the first microscope using electrons instead of light [2].

The relationship between the kinetic energy of an electron and its energy in an electric field is $1/2 mv^2 = eU$, where e is the electric charge of the electron and U is the accelerating voltage. It follows, as an example, that for an acceleration voltage of 80 kV, the wavelength is 0.0043 nm. However, resolution is limited in practice to about 0.1 nm because of lens imperfections (aberrations). Hence TEM is a suitable analytical technique to study multiphase polymeric materials. The use of very high acceleration voltages results in very high spatial resolutions. Modern microscopes can ideally resolve distances even down to sub-ångstrom levels. A disadvantage of very high voltages is the loss of phase contrast. Although high (atomic) resolution is important in studies of ceramics and metals, the atomic length scale is of limited importance when studying the morphological features of multicomponent polymeric systems. Lower acceleration voltages (between ca. 80 and 200 kV) are often sufficient for routine analyses. Instruments with an even lower acceleration voltage are commercially available. At a voltage of 5 kV, a spatial resolution of a few nanometers has been reported, which is sufficient for many polymer systems [3]. Lower voltages also yield improved contrast between different materials that do not exhibit sufficient natural contrast at higher acceleration voltages. However, a drawback with using low accelerating voltages is the radiation damage to the samples, which is significantly greater than at higher voltages.

An integrated TEM system consists basically of three units: the electron beam source, illumination system and imaging system [4]. A scheme is shown in Fig. 7.1.

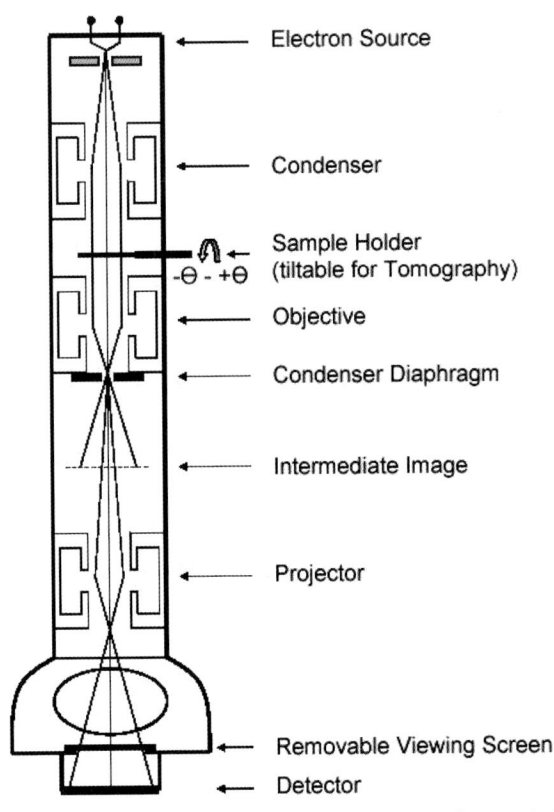

Fig. 7.1 Scheme of the beam path in a conventional transmission electron microscope.

High vacuum is required throughout the optical path. While the first electron microscopes used tungsten filaments as electron sources, brighter LaB_6 crystals and so-called field emission guns (FEGs) have become increasingly more routine, ultimately leading to better defined electron wavelengths. Well-defined wavelengths improve the energy resolution necessary for electron-based spectroscopic techniques such as energy-filtered transmission electron microscopy (EFTEM) and electron energy-loss spectroscopy (EELS). The illumination system consists of a set of electromagnetic lenses and apertures to focus the beam onto the sample. The imaging system finally magnifies the real-space representation of the sample. In addition to the advantages of high energy resolution for EFTEM, a very narrow energy distribution of the electron beam is essential for minimizing the chromatic aberration of the lenses. The spherical aberration cannot be corrected completely, but it can be reduced to tolerable levels by proper aperture settings, focal length and acceleration voltage. New high-resolution TEM instruments are equipped with special spherical aberration correctors that can improve resolution to the atomic level. This type of aberration limits the spatial resolution to a higher value than the theoretical value given by Abbé. The astigmatism of the electromagnetic lenses can be corrected by so-called stigmator coils around the TEM column. In particular, the astigmatism of the objective lens must be corrected carefully to avoid introducing errors in acquired images. While images can still be collected on photographic film, modern instruments are typically equipped with high-quality CCD cameras for image acquisition.

TEM can also be used to perform electron diffraction, which is a useful complement to high-resolution (HR) TEM. This is an interesting option for studies of crystalline order in, for example, individual single-walled carbon nanotubes [5]. It allows correlation of supramolecular features with the atomic or molecular arrangement in crystals. A reoccurring problem with many polymers is their sensitivity to the electron beam and for this reason polymeric crystals tend to become amorphous or decompose in the beam before a pristine scattering pattern can be recorded. Examples of electron diffraction on crystalline polymers are studies of large polycyclic aromatic hydrocarbons [6]. If crystallographic information is not restricted to a localized sample volume of a few nm^3, X-ray scattering is often preferable (cf. Chapter 5, Volume 3).

In the TEM imaging mode, different types of real-space representations of specimens can be recorded, depending on the way in which incident electrons are scattered by the sample and the choice of electrons used to form an image. *Elastic scattering* occurs if electrons from the beam pass through atoms with no loss of energy. In most cases, however, incident electrons pass very close to the nucleus, where the high charge density leads to a change in electron momentum. Depending on the angular change in direction, imaging can be conducted in bright-field or dark-field mode. In the bright-field mode, only electrons scattered over a small angle are detected. The angular change is small if electrons are scattered by a nucleus with a small charge density (small number of protons). In the dark-field mode, electrons scattered to larger angles are selectively used for image formation. Here, the electrons are scattered by nuclei with large

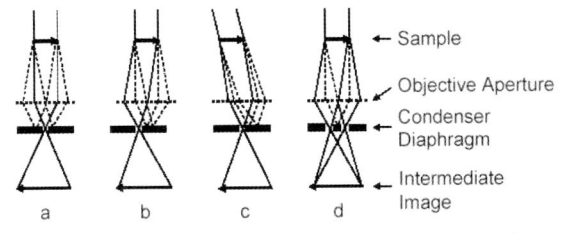

Fig. 7.2 Different modes for detecting elastically scattered electrons. Bright-field mode (a) and dark-field modes (b–d), with shifted condenser diaphragm (b), tilted beam (c) and beam stop (d).

atomic numbers (and therefore high charge densities). The type of scattered electrons used for image acquisition is selected by an aperture placed after the objective lens (Fig. 7.2).

In the elastic dark-field mode, the beam is tilted so that a diffraction condition is aligned with the objective aperture and resultant images reveal the real-space crystalline features responsible for electron diffraction. In this chapter, elastic dark-field imaging is not considered further. *Inelastic scattering* occurs if the electrons from the incoming beam interact with the electrons of the atoms and undergo substantial energy losses during these interactions. Using beam sources with a very narrow energy distribution, specific energy windows can be selected so that specific absorptions can be used to generate chemical images that show the spatial distribution of elements according to the fundamental principle of EELS. This imaging mode is particularly useful for light elements and is discussed in more detail below.

In addition to TEM, there is another way to obtain images of polymeric specimens with an electron beam. In scanning electron microscopy (SEM), which was developed later than TEM, an electron beam is rastered across the sample and various response signals are detected. There are different ways to detect and discriminate various types of signals. Reverse-trajectory electrons from the original beam (backscattered electrons) with limited chemical specificity can be detected or secondary electrons with a much lower energy than the incident electrons can be collected to provide topographical information. Each scanned point of the sample is related to the intensity of scattered electrons detected and together these point intensities generate an image. The spatial resolution of the image depends on the probe size (i.e. the diameter of the incident beam), accelerating voltage and the material under investigation. Depending on the electron density of the material, the scattering volume of the incident beam can differ greatly. Controlled positioning of the electron beam on the sample is achieved with scanning coils, which are illustrated schematically in Fig. 7.3.

In conventional SEM instruments the scattered electrons are detected above the sample. Newer instruments are also capable of detecting transmitted electrons generated upon rastering an electron beam [referred to as scanning transmission electron microscopy (STEM)]. In fact, STEM options are currently available in both

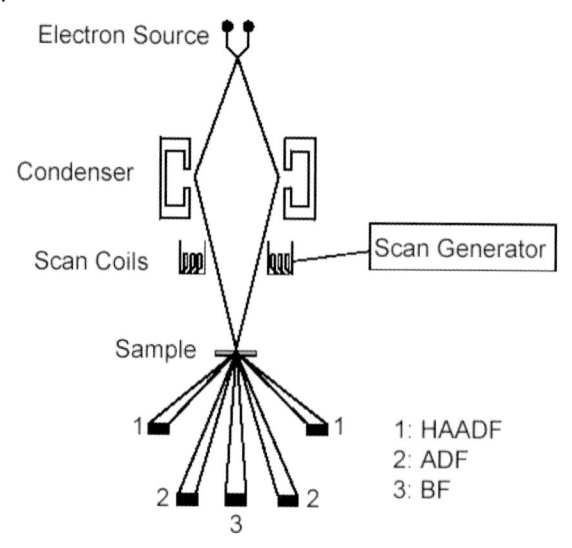

Fig. 7.3 Scheme of the scanning mode or an electron microscope with different detectors: high-angle annular dark field (HAADF), annular dark field (ADF) and bright field (BF).

TEM and SEM instruments. With the introduction of FEGs, very intense electron beams and narrow probes can be obtained, thereby leading to improved spatial resolution on a nanometer scale. In comparison with atomic force (or surface force/probe) microscopy (cf. AFM in Chapter 4, Volume 3), S(T)EM provides access to global views of specimens in relatively short times, whereas AFM affords a variety of complementary imaging modes to probe a specimen surface over a small area. Therefore, the combination of S(T)EM and AFM is often very useful.

7.3
Conventional TEM of Bulk Materials

Samples for TEM should generally be very thin (in the range of several tens of nanometers) to avoid multiple interactions of electrons and sample atoms as the electrons traverse the sample. Multiple interactions necessarily lead to smeared information and a loss of spatial resolution and compositional information. Thin films of bulk materials can be obtained by ultrathin sectioning of a sample with a glass or diamond knife at ambient or cryogenic temperatures. An alternative to sectioning is focused ion beam (FIB) sectioning, which employs a high-energy beam rather than a knife. Dip-coating a solution on a transparent amorphous carbon film is one way to obtain very thin samples of a polymer film or polymer particles. The morphology of particles dried in this fashion will, however, differ from their shape in solution. Electron-transparent films can also be produced by spin-coating or shearing onto a solid substrate (e.g. KCl) that can subsequently be dissolved, thereby allowing the film to be picked up on a TEM support grid. Cryogenic TEM or TEM imaging of metal replicas obtained from freeze-fractured samples are also possibilities to visualize solvated morphologies in a state very close to its unperturbed state. These techniques are discussed later in this chapter.

7.3.1
Sectioning of Samples

Ultramicrotomes or cryoultramicrotomes are typically the instruments of choice to prepare ultrathin sections of bulk polymeric materials. These instruments are used in conjunction with glass or, preferably, diamond knives. Glass knives are often freshly prepared from larger glass stock and must be free of structural inhomogeneities especially along the fracture edge. The main advantage of these knives is their low price. Their disadvantages are low stability and inability to have a controlled cutting angle at the edge of the knife. If a diamond knife is used, well-defined angles are available (e.g. 35° and 45°). These angles are smaller than those achieved with glass knives and thus lead to reduced sample compression during cutting. Diamond knives are more stable than their glass analogues and can be used for significantly longer times (especially if they are not subjected to hard inorganics that dull the knife edge). A practical disadvantage of diamond knives is the much higher price.

The preparation of polymeric specimens for TEM depends strongly on sample shape and mechanical properties. If the samples are too soft, they can be hardened by methods such as γ-irradiation or chemical treatment. In some cases, hardening may be induced by staining the samples for improved phase contrast, if chemical cross-links are created. Extremely small or very thin samples must be embedded before they can be sectioned in an ultramicrotome. The most common way is to insert the sample in a secondary polymer medium such as an epoxy resin, poly(methyl methacrylate), methylcellulose or a latex. An important advantage of epoxies is that their final hardness can be adjusted by judicious selection of the accelerator, hardener and polymerization temperature. It is critically important that the specimen must be incompatible with embedding medium to avoid sample swelling and accompanying changes in structure. In addition, the hardness of the embedding medium must be selected to be as close as possible to the hardness of the embedded sample.

Before mounting the sample into an ultramicrotome, it should be shaped into a frustum of a pyramid-like piece with an area in the range of 0.5×0.5 mm. The sample is mounted on a stage that is moved vertically, as well as towards the knife, in a controlled, continuous fashion. Both the knife and the sample can be cooled separately in a cryoultramicrotome, which is cooled by liquid nitrogen. Since thin, unaltered sections are required for TEM, cooling is essential to harden soft specimens such as elastomers and other low-T_g polymers, and also biologically inspired materials and improve sectioning [7]. If the samples are cut under dry conditions, collection of ultrathin sections and their transfer on to TEM support grids can be hindered by electrostatic charging. Collection of such sections is facilitated if the sections are floated on a liquid during cutting. In this case, only the top of the knife is above the level of the surrounding liquid. The liquid (DMSO–water or other mixtures that freeze at low temperatures) is contained in a trough just behind the edge of the knife, from where floating sections can be easily "fished" with a TEM grid. Alternatively, sections can be loaded directly from the knife edge onto a grid maintained at low temperature. Rapid and careful removal of the grid from the cryoultramicrotome, followed by its placement on a metal block (which serves as a heat sink), allows the sections to relax on the grid without section alteration due to condensation or exposure to external flotation agents.

Depending on the orientation of the bulk sample, different projections of a morphology may be observed in a TEM image. As an example, different views of a lamellar morphology are evident in Fig. 7.4 [8].

Co-continuous morphologies constitute another example where, in addition to an existing orientation, the finite specimen thickness influences the projection. In block copolymer systems, sample thickness is often less than the size of the periodic unit cell of the morphology, thus yielding different images as a function of position within the unit cell. For this reason, characterization of periodic co-continuous morphologies has benefited tremendously from the simulation of TEM images, which has proven to be an important and powerful analytical tool. Figure 7.5 shows TEM images and corresponding simulations of the gyroid morphology in a microphase-ordered block copolymer blend [9].

Sectioning may lead to distortions of the native specimen morphology. By freezing the sample or hardening it in other (e.g. chemical) ways, these distortions can be kept tolerably small. However, sometimes cutting traces of the knife are visible, especially if the knife edge contains defects. Fourier filtering of images that exhibit this artefact can be used to erase such parallel lines [7].

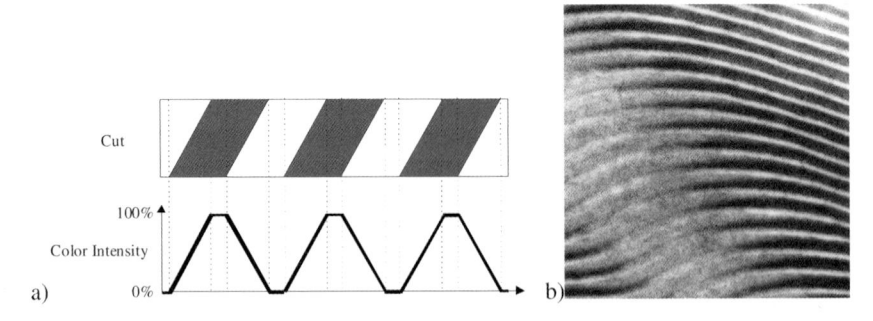

Fig. 7.4 (a) Scheme of a lamellar morphology, which is cut at an oblique angle and the resulting contrast; (b) TEM image of a lamellar polybutadiene-b-poly(methyl methacrylate) diblock copolymer showing different projections. The smeared left part corresponds to the situation illustrated in (a), whereas the right portion corresponds to a situation where the specimen is sectioned normal to the lamellae, thus leading to maximum contrast [8].

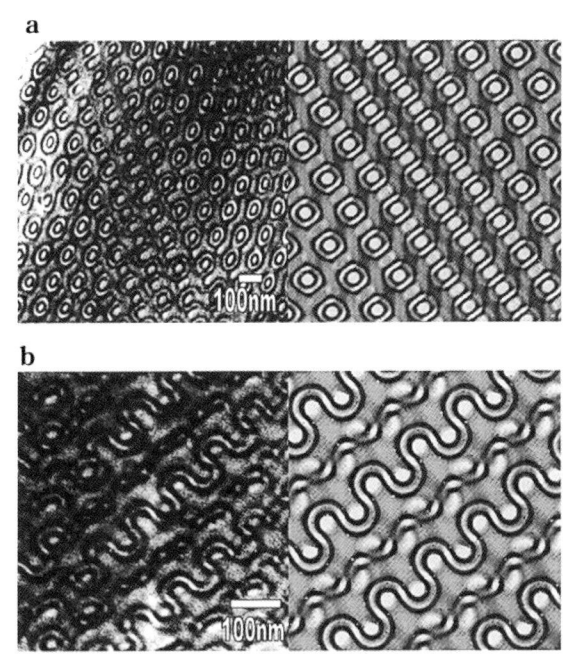

Fig. 7.5 Core-shell double-gyroid morphology of a blend composed of a polystyrene-*b*-polybutadiene-*b*-poly(methyl methacrylate) triblock terpolymer and a polybutadiene-*b*-poly(methyl methacrylate) diblock copolymer. (a) Left, TEM image along the [110] projection; right, TEMsim simulation of the [1.02 0.99 0.05] projection, 50% translation of the unit cell and 67% thickness of the unit cell. (b) Left, TEM image along the [112] projection; right, TEMsim simulation of the [112] projection, 50% translation of the unit cell and 72% thickness of the unit cell. Reproduced with permission from [9], Copyright 1999 American Chemical Society.

7.3.2
Staining of Samples

In conventional TEM, elastically scattered electrons are primarily used for generating a magnified image of the sample. A commonly encountered problem in TEM studies of organic materials is the intrinsically low contrast between constituent species, which is due to small differences in the electron density of different organics or polymers. The situation changes markedly if, for example, a two-component system consists of one component with a high atomic number. Atoms possessing a high charge density in the nucleus promote scattering, in which case features containing these atoms scatter to a greater extent than features having no (or fewer) of these atoms. Natural phase contrast may be sufficient to distinguish between different features of a morphology, as is shown for example in Fig. 7.6a for a polystyrene-*b*-polyisoprene-*b*-poly(phenylmethylsiloxane)-*b*-poly(2-vinylpyridine) tetrablock quaterpolymer [10].

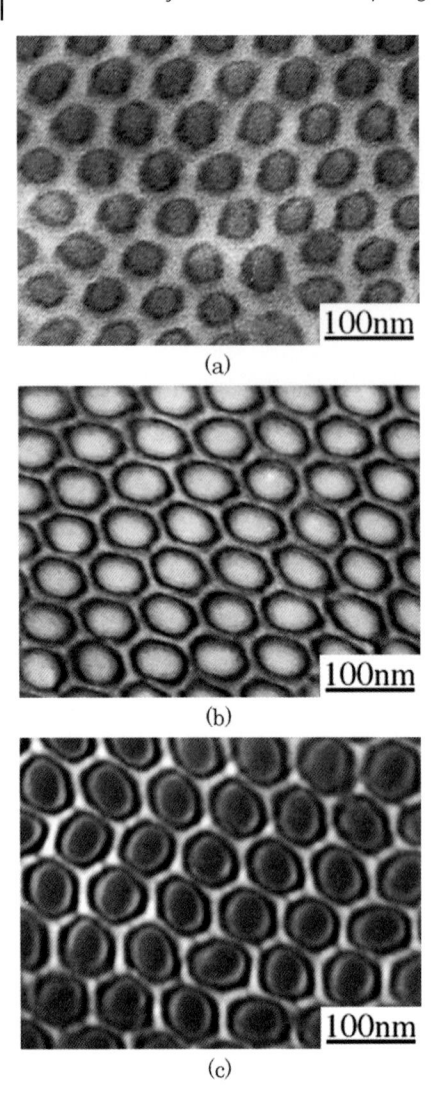

Fig. 7.6 Bright-field TEM images of a tetrablock quaterpolymer consisting of styrene, isoprene, dimethylsiloxane and 2-vinylpyridine (a) without staining, (b) stained with OsO_4 and (c) stained first with OsO_4 and subsequently with CH_3I. Reproduced with permission from [10], Copyright 2002 American Chemical Society.

Polymers containing other heavy atoms such as chlorine, bromine and sulfur provide additional examples where intrinsic phase contrast relative to conventional polymers composed purely of carbon and hydrogen may emerge. Another source of "natural" contrast between different components in a multicomponent polymer system arises from differences in electron beam damage. For instance, poly(methyl methacrylate) is very sensitive to high-energy electrons and decomposes quickly during routine TEM investigation. The result is sample thinning at locations of beam-sensitive material, which provides phase contrast on the basis of variations in specimen thickness.

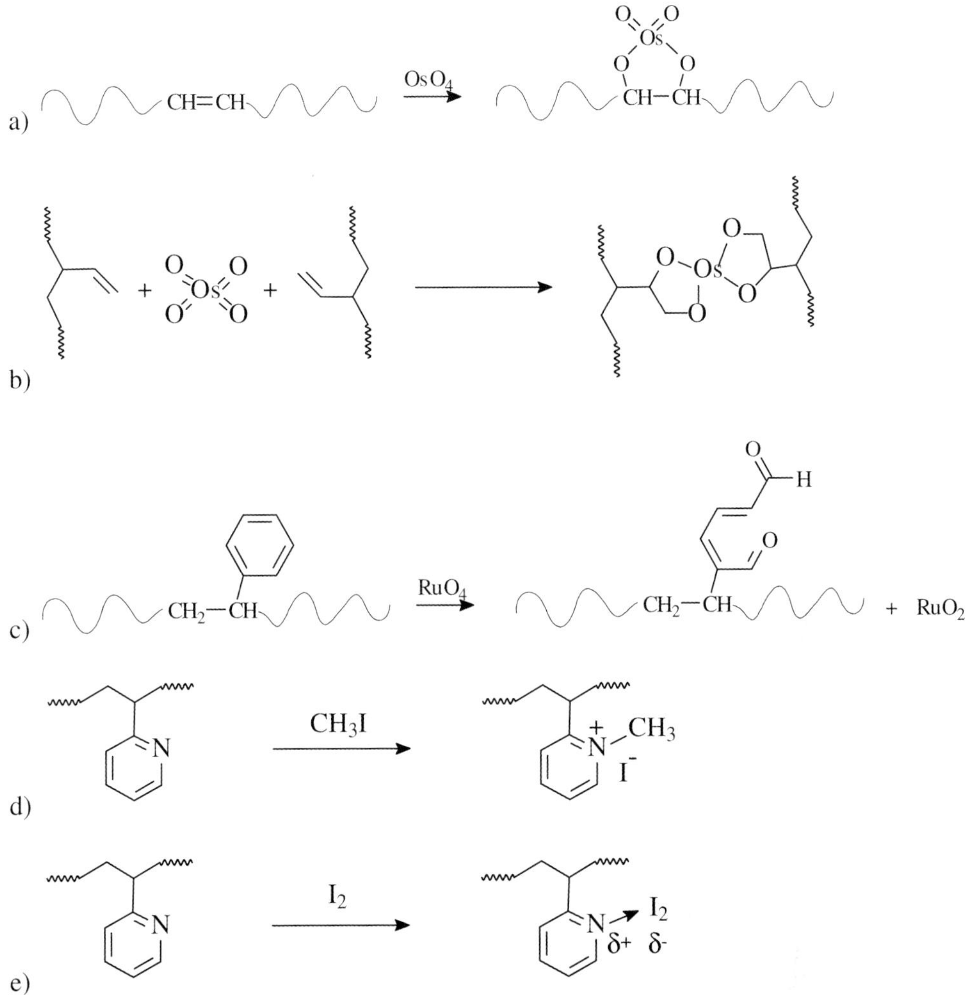

Fig. 7.7 Staining of polymers by various heavy-atom compounds: (a) staining and (b) cross-linking of olefinic groups in a polymer by OsO_4, (c) staining of phenyl ring with RuO_4, (d) staining of N atoms with CH_3I and (e) staining of N atoms with I_2.

In most polymer blends or block copolymers, the intrinsic contrast is insufficient. In these cases, selective staining of one or several components with reactive species containing heavy atoms is necessary to visualize the morphology. Several staining reactions are included for illustrative purposes in Fig. 7.7.

A frequently used selective staining agent is OsO_4 [11]. This compound is a subliming solid that can be brought in contact with a sample by enclosing them together under vacuum. Under ambient conditions, the OsO_4 is often dissolved

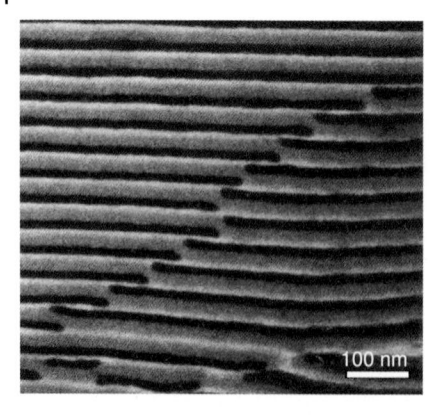

Fig. 7.8 Bright-field TEM image of a polystyrene-b-polybutadiene-b-poly(tert-butyl methacrylate) triblock terpolymer blended with a polystyrene-b-poly(tert-butyl methacrylate) diblock copolymer and stained with OsO$_4$.

in an aqueous solution. This highly reactive staining agent has a high vapor pressure (which is why it must be used with caution) and selectively combines with olefinic groups in the sample (typically within seconds to minutes, cf. Fig. 7.7a). An OsO$_4$ molecule can react twice and thus generate a chemical cross-link between neighboring chains (Fig. 7.7b). Examples demonstrating the utility of this staining agent in morphological studies are displayed in Fig. 7.4b, left images in Figs. 7.5a and b, 7.6b and 7.8, where in the last case the bright-field image shows dark, OsO$_4$-stained polybutadiene domains and bright, unstained polystyrene and poly(tert-butyl methacrylate) domains in a blend of a polystyrene-b-polybutadiene-b-poly(tert-butyl methacrylate) triblock terpolymer and a polystyrene-b-poly(tert-butyl methacrylate) diblock copolymer forming a non-centrosymmetric, periodic lamellar superstructure [12].

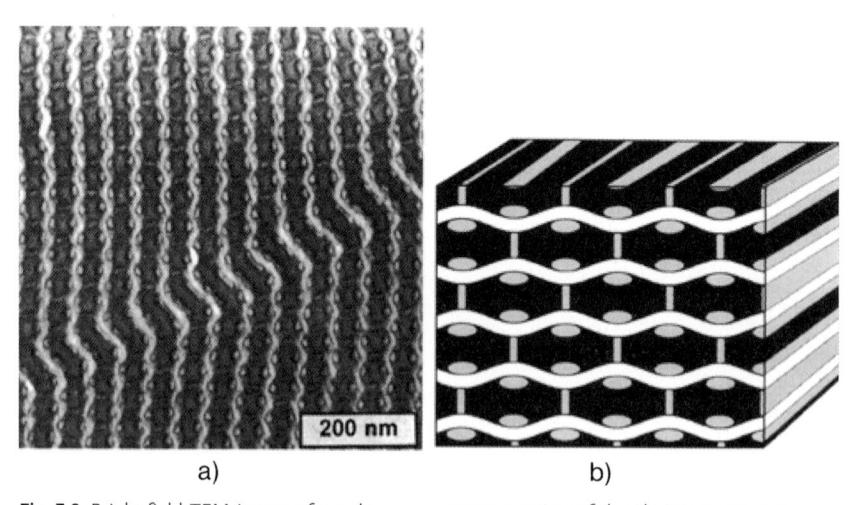

Fig. 7.9 Bright-field TEM image of a polystyrene-b-poly(ethylene-co-butene)-b-poly-(methyl methacrylate) triblock terpolymer stained with RuO$_4$ (a) and a schematic representation of the "knitting pattern" morphology (b). Reproduced with permission from [18], Copyright 2001, American Chemical Society.

Another important staining agent is RuO_4, which is prepared *in situ* by reacting $RuCl_3$ with aqueous NaOCl [13, 14]:

$$2\,RuCl_3 \cdot 3\,H_2O + 8\,NaClO \rightarrow 2\,RuO_4 + 8\,NaCl + 3\,Cl_2 + 3\,H_2O$$

In this case, samples are either put into or above an aqueous RuO_4 solution for a predetermined time (minutes to hours). The utility of this staining agent is that it reacts with the phenyl ring of aromatic compounds (Fig. 7.7c). This staining technique is routinely applied to samples that do not contain olefinic groups, such as hydrogenated Kraton copolymers, e.g. polystyrene-*b*-poly(ethylene-*co*-butylene)-*b*-polystyrene (SEBS) triblock copolymers. The morphology of a RuO_4-stained polystyrene-*b*-poly(ethylene-*co*-butylene)-*b*-poly(methyl methacrylate) triblock terpolymer (SEBM) is presented in Fig. 7.9. This material exhibits the so-called knitting pattern morphology [15–19].

Note that the staining reaction occurs predominantly at the interface between polystyrene and the neighboring block. This is probably due to the density dif-

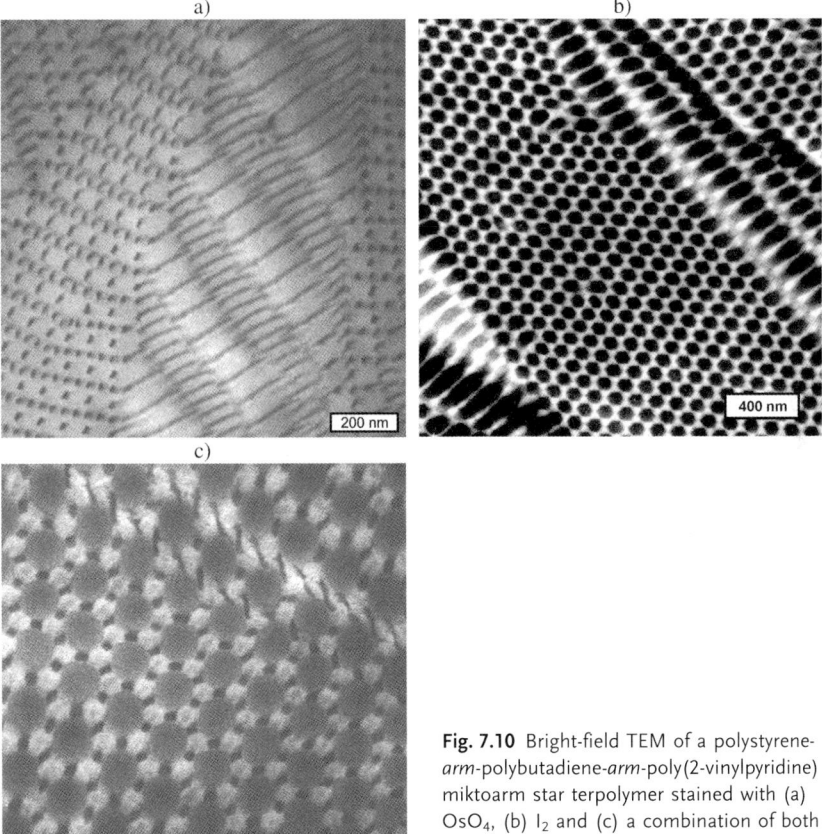

Fig. 7.10 Bright-field TEM of a polystyrene-*arm*-polybutadiene-*arm*-poly(2-vinylpyridine) miktoarm star terpolymer stained with (a) OsO_4, (b) I_2 and (c) a combination of both OsO_4 and CH_3I. Reproduced with permission from [21], Copyright 2000, Wiley-VCH.

ference, which enables the staining agent to access the interface more easily than the bulk material of the polystyrene domains. One drawback of RuO_4 that is not shared by OsO_4 is the propensity for RuO_4 to nucleate and grow into crystals on sample surfaces upon extended exposure to RuO_4 vapor.

Other staining agents include Br_2, which effectively reacts with and stains double bonds via a simple addition reaction. The presence of nucleophilic sites in a polymer can be used to induce staining by electrophilic agents such as CH_3I. The reaction produces a salt, as depicted in the example of poly(2-vinylpyridine) (Fig. 7.7 d) [20]. This staining strategy can be used in combination with OsO_4, as shown in Fig. 7.6 c. Another example of dual staining is provided with the polystyrene-*arm*-polybutadiene-*arm*-poly(2-vinylpyridine) miktoarm star terpolymer shown in Fig. 7.10 [21]. The same polymer reacts with I_2 vapor (Fig. 7.10) to produce a charge-transfer complex (Fig. 7.7 e) [22].

Although other staining agents have been proposed, the last one included here is phosphotungstic acid, which can selectively stain polyacrylonitrile (PAN), which has been demonstrated in blends of PAN with PMMA [23].

7.4
Cryo-TEM and Freeze-fracture TEM of Solutions

To characterize fully a liquid or a semi-liquid system by TEM, the sample must be made compatible with the operational requirements of the microscope, as described above. Cryogenic temperature (cryo) TEM is an excellent tool by which to obtain direct images of structured liquid or semi-liquid specimens, *thermally fixed* into a vitreous or quasi-solid state. Sample preparation for cryo-TEM (see below) makes it possible to preserve the nanostructure of the system in its native "hydrated" state, while making it compatible with the stringent requirements of the microscope. It avoids the pitfalls of "chemical fixation," namely adding chemicals not originally present in the system, and fairly often drying the specimen. Specimen preparation by chemical fixation often alters completely or distorts the nanostructures in the original system. Because cryo-TEM provides high-resolution direct images of the nano/microstructures in the system, it can elucidate the nature of the basic building blocks that comprise the systems, covering a wide range of length scales from a few nanometers to several micrometers.

The term "cryo-TEM" covers the two main classes of techniques that involve fast cooling of the specimen, as part of its preparation to be examined by TEM. The more widely used technique is direct imaging cryo-TEM, which involves rapid cooling of a thin specimen and its transfer at cryogenic temperature into the microscope, where it is maintained and observed at cryogenic temperatures. Alternatively, a larger sample of the system is quickly cooled, the frozen sample is fractured and a thin metal–carbon replica is prepared of the fracture surface. The sample is subsequently melted or dissolved away so that the replica is cleaned and dried prior to TEM observation at ambient temperature. That is called "freeze-fracture-replication cryo-TEM" or FFR. The next section describes in some detail the basics of direct imaging cryo-TEM. That is followed by a section describing FFR.

7.4.1
Direct Imaging Cryo-TEM

By fast cooling, one lowers the vapor pressure of liquid and semi-liquid systems to make them compatible with the high vacuum maintained in the microscope column, which is typically $<10^{-6}$ Pa. That cooling also arrests supramolecular motion that may cause blurring of the recorded image. As with all TEM specimens, cryo-TEM specimens must be thin, not thicker than about 300 nm for the usual accelerating voltage of 120 kV. In thicker specimens inelastic electron scattering deteriorates image quality. These inelastically scattered electrons may be filtered out by an in-column or post-column energy filter, as described later in this chapter.

Because nanostructured liquids such as polymer solutions and liquid crystalline phases are very sensitive to changes in composition, the specimen preparation method of choice is thermal fixation, i.e. ultra-fast cooling of the liquid specimens into a vitrified or quasi-solid state. This is achieved by rapidly plunging the specimen into a suitable cryogen. Another advantage of thermal fixation is that it is much more rapid than chemical fixation, because thermal diffusivities are higher than mass diffusivities. Of course, it likewise eliminates the addition of an alien compound to the system.

The cooling rate needed for vitrification of water is on the order of 100 000 K s^{-1}, as estimated theoretically [24]. The cooling rate during vitrification measured experimentally in an actual specimen preparation set-up [25] has been found to be on the order of 10^5 K s^{-1}, as needed for successful vitrification. A slow cooling rate leads to the formation of crystalline ice, hexagonal or cubic, in aqueous systems. In non-aqueous systems, various other crystalline matrices may form in the cooled specimen. Crystal formation constitutes a major change of the original system that causes optical artefacts, mechanical damage to the microstructure and redistribution of solutes. Those solutes are often expelled from the growing crystalline lattice and are deposited either in the crystal grains or often at grain boundaries.

The limited penetration power of even high-energy electrons requires thin films (up to 300 nm thick, as stated above). High-resolution imaging requires thinner samples. The high cooling rates needed for vitrification also require a very large surface area-to-volume ratio, as obtained in thin films. In practice, most direct-imaging vitrified specimens display a wide thickness range. While microscopes operating at 200, 300 and 400 kV are capable of imaging specimens thicker than specified above, image interpretation becomes increasingly more difficult with specimen thickness. The high depth of field of the TEM leads to superposition of information from many layers of thick specimens, all projected on the plane of the detector. In FFR, samples are usually much thicker, which is useful when thin films cannot be formed because of the high viscosity of the liquid phase or because large aggregates are suspended in the liquid phase. The greater thickness leads to slower cooling rates, hence vitrification is rarely achieved in FFR.

To vitrify the specimen successfully the cryogen must be at a very low temperature and well below its boiling point to avoid formation of an insulating gas film around the specimen during quenching (the "Leidenfrost effect"). Good cryogens have high thermal conductivity. Liquid nitrogen, which is safe and inexpensive, is, unfortunately, a poor cryogen for most applications (see exceptions below) because of the narrow temperature range between its freezing and boiling points. For aqueous systems the cryogen of choice is liquid ethane, cooled to its freezing point (–183 °C) by liquid nitrogen. Its normal boiling point is about –90 °C, hence no gas can form around the specimen during vitrification. It does, however, dissolve most apolar organic solvents.

To preserve the nanostructure of the liquid specimen at the precise given conditions, namely temperature and concentration, cryo-TEM specimens are prepared in a controlled environment of prescribed temperature and atmosphere to prevent temperature change and loss of volatiles, e.g. water vapor. Not all laboratories use these precautions, however, as this requires special expertise and equipment, referred to as a controlled-environment vitrification system (CEVS). Several types are available, especially the relatively simple but very reliable systems based on that developed by Bellare et al. [26], modified by Talmon and coworkers over the years [25, 27], and the automatic "Vitrobot" of FEI Co. that was developed by Frederik and coworkers [28]. The CEVS can be operated from –10 to +70 °C and with various saturated or unsaturated atmospheres. It can also be used for the preparation of FFR specimens. Specimen preparation in the CEVS is performed as follows: 3–5 µL of a pre-equilibrated system is pipetted onto a perforated carbon film supported on a TEM copper grid, held by tweezers and mounted on a spring-loaded plunger. In some cases, a bare grid (i.e. a microscope grid not covered by a perforated film) is used. Most of the liquid is blotted away by a piece of filter-paper wrapped on a metal strip, leaving behind thin liquid films supported on the hole edges of the perforated carbon film. After blotting, the plunging mechanism is activated, a trap-door or a shutter opens simultaneously and the specimen is quickly driven into the cryogen and vitrified. For best results the minimum entry velocity of the specimen into the cryogen is about 2 m s^{-1}. One may opt to wait for various reasons before blotting and plunging (see below), while the specimen is kept in the controlled atmosphere of the CEVS or Vitrobot. Finally, the vitrified sample is transferred under liquid nitrogen to the workstation of a cold-stage where the TEM grid is loaded into a special cryo-holder and subsequently transferred into the microscope. More technical details can be found elsewhere [26, 29].

The blotting step is very important for successful sample preparation. It may be performed in a number of ways. The simplest way is wicking most of the liquid by simply touching the filter-paper on the edge of the grid supporting the sample drop. Viscoelastic fluids require blotting with a shearing or "smearing" action that temporarily reduces the viscosity of a shear-thinning liquid, thereby allowing the formation of a thin enough liquid film on the support. Another option is to press two pieces of blotting paper on both sides of the specimen, which can be performed either manually or, as in the case of the Vitrobot, automatically. This method usually produces more uniform films. The blotting process and the confinement of the liquid as a thin specimen may introduce artefacts of which one should be aware (discussed below). In addition to changes in the nature of the nanostructure, distortions of large objects, and also alignment of slender "one-dimensional" (rods or threads) or large "two-dimensional" (sheets) objects, may be introduced upon blotting.

After blotting, while the thin specimen is still liquid, it may be kept in the controlled environment of the CEVS for some time insofar as sample evaporation is avoided. In the case of many self-aggregating systems such as threadlike micelles, this is most useful to allow the specimen to relax following the shear and elongation that it may undergo during blotting [30, 31]. The specimen may also undergo different processes directly on the grid, such as chemical or physical reactions induced by different triggers, such as rapid heating [25, 27] or cooling [32], pH jumps [33] or gelation. Those processes may be stopped at an intermediate stage by plunging the specimen into the cryogen. The experiment may be repeated a number of times, each time allowing the process to proceed further towards completion. Thus, a sequence of vitrified specimens is produced, providing "time sectioning" of the process of interest. This sequence of experiments is also referred to as time-resolved cryo-TEM. Several variations of the CEVS have been built to expedite this type of cryo-TEM analysis [25, 27, 34]. High-viscosity, self-aggregating phases that cannot be directly prepared as a thin liquid film may also be formed by on-the-grid cooling (or, in rare cases, heating), starting with a thin film of a low-viscosity precursor.

There is increasing interest in molecular self-aggregation, including that of block copolymers, in non-aqueous systems. While direct imaging cryo-TEM can be extended to non-aqueous systems, liquid ethane, often the vitrification cryogen of choice, cannot be used, because it is a good solvent for many non-polar liquids. However, many systems, such as those composed of branched hydrocarbons, aromatics or glycerides, do not crystallize readily upon cooling and therefore can be vitrified even in liquid nitrogen [35]. That is also true for aqueous systems containing sufficiently high concentrations of glycols (>20%) [32].

7.4.2
Freeze-fracture Replication

As described above, FFR is an indirect route to cryo-TEM. In the same fashion as direct-imaging cryo-TEM, FFR involves fast cooling of a solvated specimen in an appropriate cryogen. In FFR, the specimen is larger than that of the former technique, in which case complete sample vitrification, especially in the case of aqueous systems, is not achieved and freezing artefacts need to be taken into consideration. The frozen specimen is fractured after fast cooling and a metal–carbon replica of the fracture surface is prepared by vapor deposition. Fracturing and replication are both performed in a vacuum chamber, while the specimen is maintained at cryogenic temperatures, typically below $-150\,°C$. First, a heavy metal (in most cases platinum or a mixture of platinum and carbon) is deposited, usually at an angle of about 45° to the horizon. This enhances contrast by a "shadowing" effect. A smaller angle must be used to bring out very fine details. A carbon layer is then deposited perpendicularly to the specimen to enhance the mechanical stability of the metal replica. In modern equipment, the sources required for the deposition of the metal and carbon are electron guns bombarding a suitable target. Sometimes an additional preparation step called etching, i.e. controlled removal of high vapor-pressure components by sublimation, is used. This requires warming the specimen for a few minutes to about $-100\,°C$ in the case of aqueous systems. At this temperature, the vapor pressure of ice becomes sufficiently high to yield an appreciable sublimation rate. When the replica is ready, the sample is melted or dissolved in a suitable solvent and the replica is cleaned, collected on a TEM support grid, dried and imaged in the microscope at ambient temperature. The process of fracturing the specimen and replication is conducted using dedicated commercial systems. Fast cooling may be performed in the CEVS to allow quenching from well-controlled conditions [33]. Recall that FFR is most useful when relatively high-resolution images are needed, but direct imaging cryo-TEM is not practical, e.g. in the case of high-viscosity systems or systems containing large features that cannot be accommodated in the thin specimens required for direct-imaging cryo-TEM. While the technique is an excellent complement for direct-imaging cryo-TEM [36], it has lost popularity over the last two decades and, regrettably, is used in only a few research laboratories. One of the primary reasons for this loss of interest is that the technique is labor intensive. It relies more than many other techniques on the technical skills of the user and it requires complex and expensive equipment. In recent years, commercial FFR units have furthermore required modifications before they could be successfully used. The gradual disappearance of the technique is unfortunate, because FFR–TEM is a valuable imaging technique.

7.4.3
Limitations, Precautions, Artefacts and Extensions of the Technique

Cryo-TEM is probably the most reliable technique to obtain direct, nanometric-resolution images of liquid systems. However, it cannot always be applied and, despite all the precautions taken, imaging artefacts are possible. Users must therefore be able to (1) tell when the technique can be applied and (2) distinguish between real nanostructures and artefacts. The distinction may be difficult and data from other techniques should always be used in parallel. It has already been mentioned that high-viscosity systems cannot be made into thin direct-imaging cryo-TEM specimens. Remember, however, that shear-thinning viscoelastic liquids can be processed into thin liquid films, even if their zero-shear viscosity is high. Care must be taken to allow sufficient relaxation time on the grid, prior to vitrification, to avoid flow-induced structures, unless one is indeed interested in such structures.

When the system conditions in the CEVS are close to a phase boundary, chemical-potential control of the gas phase in the CEVS becomes critically important. The best approach in this case is to saturate the air in the CEVS with the same solution as that under investigation. In many cases, when only a surfactant or polymer at low concentration in water is used, may the chamber be saturated with pure water. However, if highly volatile components such as an organic solvent with a high vapor pressure or a low molecular weight cosurfactant are also present in the system, failure to use the actual solution for saturation may lead to drying artefacts.

Phase contrast is an important factor in any TEM technique. In cryo-TEM this is especially crucial, as in most cases inherent contrast in the systems studied is low. The contrast becomes even lower when the continuous phase is not pure water, but a mixture of water and a relatively high fraction of organic materials such as ethylene glycol or glycerol. At high concentrations of glycerol (~ 30 wt% and higher) aggregates of surfactants or polymers become practically invisible, as contrast between the structural elements and the surrounding matrix approaches zero. Although contrast is also fairly poor in organic solvents, reasonable images can be recorded in many cases, as illustrated by the examples provided below. In some organic solvents, contrast is reversed, i.e. the continuous phase appears more optically dense than the nanoscale aggregate. An example of this is given in Fig. 7.11, which shows micelles and vesicles of a polystyrene–polyisobutene diblock copolymer (molecular weight ratio 13:79) in a mixture of phthalates [37]. With increasing acceleration voltage, contrast in the TEM image is systematically reduced. For imaging soft materials, we prefer to work at a moderately low accelerating voltage (120 kV).

Because of the inherently poor contrast in vitrified cryo-TEM specimens, we almost always rely on phase contrast, which is equivalent in the corresponding principle in light microscopy that enhances contrast by converting phase differences to amplitude differences. Phase contrast is achieved in TEM by defocusing the microscope objective lens. This must be performed with care to avoid loss of resolu-

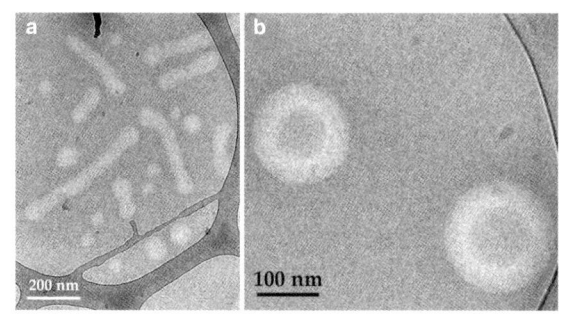

Fig. 7.11 Change in aggregation of an asymmetric poly(styrene-b-isoprene) diblock copolymer dissolved at 1 vol.% in solvents with varying styrene selectivity: dibutyl phthalate (DBP), diethyl phthalate (DEP) and dimethyl phthalate (DMP). In (a), micelles are observed in 1:1 DBP–DEP. In (b), vesicles form in 1:1 DEP–DMP. Note the contrast reversal between the aggregates (lighter) and the solvent (darker). After [37], with permission.

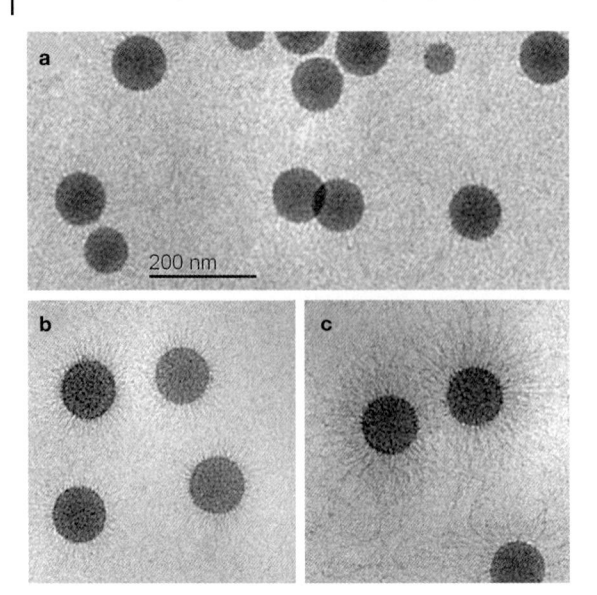

Fig. 7.12 Series of cryo-TEM images of vitrified 1 wt% particle suspensions of spherical polyelectrolyte brushes grafted on to polystyrene cores. The contrast is greatly enhanced compared with the original particles (a) by replacing the Na counterions of the polyelectrolyte chains by Cs cations (b) and, additionally, by incorporating BSA molecules (537 mg per gram of particles) that decorate the polyelectrolyte chains (c). After [38], with permission.

tion and introduction of imaging artefacts. One should avoid enhancing the contrast in liquid systems by staining (i.e. through the addition of a high electron density material to the system) in the hope that such external chemicals selectively attach to certain features in the specimen. Because most staining agents are either salts or acids of heavy elements, their addition to the system may severely alter the native nanostructure. However, in some cases the system can be altered slightly, with little effect on the nanostructure, to image desired structures. In the case of ionic polymers such as polyelectrolytes, it is safe to replace, for example, light sodium ions by much heavier cesium ions or chloride ions by iodide ions. In Fig. 7.12 we present an example where the contrast of polyelectrolyte brushes attached to polystyrene latex particles was enhanced by replacing Na^+ by Cs^+. The contrast was further enhanced by attaching bovine serum albumin to the brushes, as described by Wittemann et al. [38].

Another practical limitation encountered with cryo-TEM imaging, as well as a possible source of artefacts but occasionally a means by which to enhance contrast, is electron-beam damage. Organic compounds in vitrified aqueous or nonaqueous matrices are very susceptible to chemical and structural changes induced by the electron beam [39]. Hence all images must be recorded under low-dose conditions, namely at an exposure of just a few electrons per $Å^2$. Higher exposures to the electron beam may result in (1) the destruction of finer details,

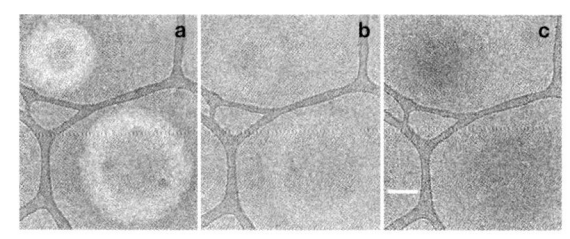

Fig. 7.13 Change in contrast of vesicles relative to the solvent in a vitrified specimen of 1% solution of an asymmetric poly(styrene-*b*-isoprene) diblock copolymer in diethyl phthalate after systematically increasing the electron exposure: 8 (a), 120 (b) and 320 electrons Å^{-2} (c). Scale bar is 50 nm. Reproduced with permission from [40], Copyright 2005, American Chemical Society.

(2) formation of new artefactual structures or, as been shown in some solvents, (3) contrast loss or reversal [40]. Contrast reversal is demonstrated in Fig. 7.13.

However, because electron-induced radiolysis occurs preferentially at the interface between organic materials and vitrified water, careful exposure of the specimen to the electron beam can promote contrast in areas where it is minimal [41].

The CEVS also allows us to quench liquid nanostructures from a given temperature of interest, which permits investigation into the effect of temperature changes on polymer self-aggregation in solution. This effect is evident in Fig. 7.14, which displays polymer nanoscale aggregates composed of a polystyrene solid core to which a network of cross-linked poly(isopropylacrylamide) (PNIPAM) was covalently attached to the surface [42]. The shell of these particles swells in water when the temperature is low. At elevated temperatures the corona is markedly collapsed, as seen in Fig. 7.14. To confirm this temperature-regulated effect, compare the corona at 5 and 45 °C.

Generally, TEM is not strictly a quantitative technique (with electron diffraction and lattice imaging as notable exceptions). Image magnification depends on the position of the imaged area on the optical axis of the microscope. Slight

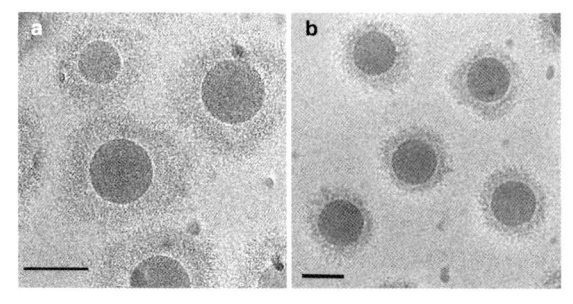

Fig. 7.14 Cryo-TEM images of a 0.2 wt% aqueous suspension of polystyrene–PNIPAM core–shell particles. The sample was vitrified at (a) 5 and (b) 45 °C. The core consists of polystyrene, whereas the corona is cross-linked PNIPAM with BIS. Note the change in corona size with temperature. Scale bars correspond to 100 nm.

deviations from this ideal position (the eucentric plane) result in changes in magnification that can be as high as ±10%. Additional errors may stem from imperfect calibration and electromagnetic lens hysteresis. Thus, quantitative data extracted from TEM images must be corroborated by data collected from more quantitative techniques, e.g. X-ray or neutron scattering. Reliable physical models required to interpret data from scattering techniques are based on real-space microscopy.

The apparent concentrations visible in cryo-TEM images may be misleading. One reason for this is the redistribution of suspended aggregates in the liquid, since particles tend to migrate from thinner to thicker areas. Larger particles tend to move faster to thicker areas, in which case size segregation is often encountered in vitrified specimens, with large particles being more numerous in thick areas, while smaller ones remain in thinner areas. The second reason is that the high depth of field causes all focused images of objects at different depths in the specimen to be projected on the detector. Such overlap often gives the impression of an elevated concentration relative to the actual concentration.

7.5
Transmission Electron Microtomography

7.5.1
Background

Conventional TEM imaging of periodic and aperiodic nanostructures formed by microphase-separated block copolymer systems, including their blends and solutions, relies heavily on phase contrast to yield 2D projections of copolymer morphologies for qualitative inspection and quantitative analysis. In many cases, the morphologies are selectively stained with heavy metals to enhance contrast, as discussed in the previous section. Since real-space visualization of these morphologies is critical to a fundamental understanding of how molecular and environmental factors govern copolymer self-organization and chain confinement, independent efforts have been made to elucidate the 3D characteristics of copolymer morphologies. One approach routinely employed in the study of nanostructured copolymers combines TEM with scattering profiles acquired by small-angle X-ray or neutron scattering (SAXS or SANS, respectively, described in Chapter 5, Volume 3) [43–46]. Real-space imaging confirms the appropriate selection of scattering models that are ultimately necessary to interpret reciprocal-space scattering patterns and extract relevant 3D structural information. These analytical methods have been shown to constitute a powerful combination capable of rendering both high-resolution and statistically meaningful information regarding block copolymer nanostructures. Caution must always be exercised, however, when using models to interpret scattering patterns, especially in the absence of TEM data. Another strategy employed to acquire 3D information from 2D TEM images uses computer models of various geometric surfaces [47, 48]. One of the most successful and detailed projects in this vein is TEMsim, which permits the user to compare slices of three-dimensionally complex surfaces along any projection axis with experimental TEM images for identification purposes. Since the geometric surfaces must be known, TEMsim is limited to ordered morphologies, as shown by the example of the gyroid morphology in Fig. 7.5.

Another approach to meet the ongoing challenge of 3D nanostructural characterization derives from the widespread success of 3D imaging in the medical and petroleum engineering fields to study non-invasively the human body and porous media, respectively. Positron emission tomography (PET) [49], magnetic resonance imaging (MRI) [50] and computer-assisted [X-ray] tomography (CAT) [51] refer to routinely encountered imaging methods that can, through collection of multiple images over a large angular range, generate 3D structural images of a subject irrespective of its complexity or periodicity. While these methods lack sufficient spatial resolution to be of any practical value in the study of nanostructured block copolymers, the principles on which they are founded can be readily extended to TEM. The resulting technique, referred to as 3D TEM imaging, transmission electron microtomography (TEMT) or simply electron tomography [52, 53], has its origins in morphological studies of biological assemblies [54, 55]. Although the inaugural 3D TEMT images [56] of a microphase-separated block copolymer morphology suffered from extensive beam damage and lacked high definition, more recent developments have yielded 3D images that are suitable for quantitative analysis. The objective of this section is to provide a description of how TEMT is performed and what structural information can be gleaned from 3D images of nanostructured block copolymers.

7.5.2
Methodology

The underlying premise of TEMT is that 3D reconstructions can be generated from a series of sequential 2D phase-contrast TEM images acquired at a fixed angular increment ($\Delta\theta$) over a large angular interval from $+\theta$ to $-\theta$ (depicted schematically in Fig. 7.1). Resolution considerations set both $\Delta\theta$ and θ. The number of projections collected during the course of a TEMT series (N), given by $2\theta/\Delta\theta+1$, is inversely related to the spatial resolution (d) of the ultimate 3D reconstruction by $N=\pi D/d$, where D is the characteristic size of the structural feature of interest. Moreover, θ dictates the noise level in the resultant reconstruction due to information lost from incomplete tilting. If a 3D reconstruction is performed with images collected along a single tilt axis, θ should ideally extend completely from $+90$ to $90°$. Since this is not practical, the volume of revolution generated along the single tilt axis is incomplete, thereby resulting in a wedge of missing information. Thus, although θ should be as close to $90°$ as possible, operational limitations reduce it. As a general rule of thumb, θ should not be less than $60°$. In this case, consider for illustrative purposes a 3D TEMT reconstruction of a nanostructured copolymer with structural elements measur-

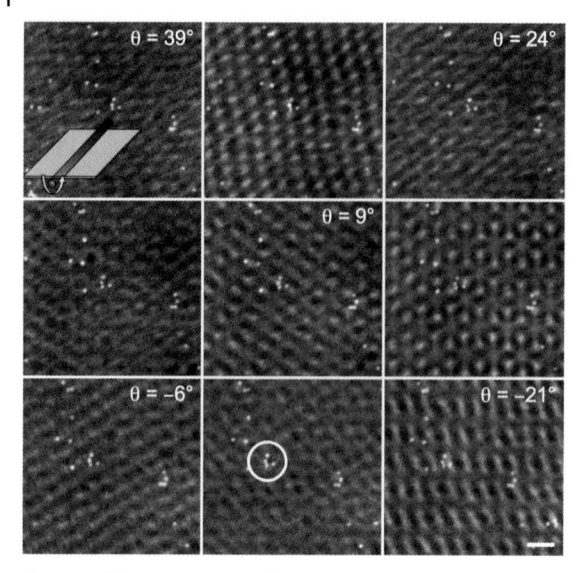

Fig. 7.15 Illustrative series of contrast-inverted tilt images (every fifth image with $\Delta\theta = 1.5°$) obtained from an ordered bicontinuous morphology observed in a lamellar SI diblock copolymer blended with polystyrene homopolymer. The schematic inset indicates that the images were acquired along a single tilt axis. The circle highlights the Au nanoparticles used as fiducial markers to align the images in the series. The scale marker corresponds to 100 nm.

ing 30 nm in size. If the features are to be resolved to 0.8 nm, ~ 118 sequential tilt images would be required. If $\theta = 60°$, then $\Delta\theta \approx 1°$, which identifies a very important requirement for performing TEMT: a high-precision goniometer. An example of a partial image tilt series obtained from a bicontinuous block copolymer morphology is displayed in Fig. 7.15 and demonstrates how the morphology changes due to variations in the projection axis upon tilting.

Accomplished TEM practitioners immediately recognize the danger of extensive beam damage associated with acquiring a large number of images from one area of a single organic specimen [57]. Developments in low-dose imaging and high-efficiency/area CCD cameras, coupled with automated focusing and specimen-translation functions and cryogenic specimen stages, can, however, greatly reduce the deleterious effects inflicted upon organic specimens by a high-energy electron beam. Recent advances in automated electron tomography utilize a predictive image motion strategy in conjunction with geometric modeling of the goniometer to minimize specimen exposure to the beam prior to or during acquisition of the full tilt series [58]. The extent to which a specimen is dimensionally altered during image acquisition can be ascertained during alignment of the image set. Careful alignment of the images within a dataset is necessary to account for slight shifts in specimen translation and electron optics during acquisition. While image cross-correlation can greatly facilitate such alignment, early TEMT methods have relied heavily upon alignment of fiducial

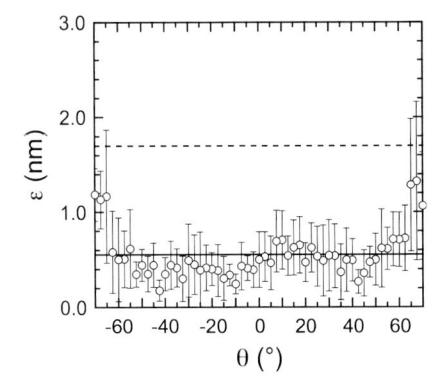

Fig. 7.16 Mean image alignment error (ε) presented as a function of tilt angle for a full tilt series ($\Delta\theta=2.5°$). Acquisition of the first set of images began at 0° and proceeded to +70°. The specimen was returned to 0° and the second set was collected to −70° and subsequently merged with the first image set. Alignment was performed with at least 10 Au nanoparticles. The solid line identifies the overall angle-averaged alignment error and the dashed line denotes the size of a single pixel in the CCD used to collect the digital images.

markers, typically Au nanoparticles decorated on the specimen surface prior to imaging [59, 60]. The positions of fiducial markers clearly identified at 0° tilt can be tracked as a function of tilt and thus used to align the images with considerable precision. The mean r.m.s. alignment error averaged over ~10 markers (ε) provides an immediate measure of specimen sensitivity under the electron beam. An example of $\varepsilon(\theta)$ is provided in Fig. 7.16 and reveals that the images at high tilt angles exhibit the most error due to an effective increase in specimen thickness along the electron beam (z) direction. The series-averaged error and the spatial resolution of a single pixel in the digital images (collected by a slow-scan CCD camera) are both included in the figure. It is important to note that the alignment error can be small (sub-pixel), indicating that the specimen did not change dimensionally to a discernible extent during acquisition of the entire image dataset. This is a critically important consideration for quantitative analysis purposes.

7.5.3
Reconstruction Fidelity

Once the images have been accurately aligned through the use of fiducial markers or cross-correlation algorithms, they are then reconstructed into a 3D volumetric element. One of the most successful approaches for achieving high-precision reconstructions from datasets collected along a single tilt axis employs the r-weighted (filtered) back-projection algorithm [61], although other reconstruction strategies have been proposed [62]. One current method employs a hybrid approach wherein the r-weighted back-projection algorithm is used to generate the first reconstruction, which is subsequently refined through the iterative use of other (real-space) algorithms. While detailed descriptions of these procedures are available, it is worth mentioning that the r-weighting (where r signifies the radial distance from the tilt axis) in back-projection reconstruction is important for resolution retention since it corrects for the finite number of images acquired. The resultant volumetric element must then be separated into structural features of interest by establishing a voxel threshold to generate an isosurface. In the case of selectively stained nanostructures, this threshold can often be identified by visual inspection or rigorous analysis (through the use of mathematical procedures such as the marching cubes algorithm) for systematic quantitation. Isosurfaces provide an independent and facile means by which to validate the selected threshold since they yield the volume fraction of structural details, which can then be compared with known material compositions. Comparison of a 3D TEMT isosurface image of the gyroid (G) morphology in a microphase-ordered SIS triblock copolymer (Fig. 7.17a) with a corresponding composition-corrected Schoen G surface (Fig. 7.17b), for instance, confirms the

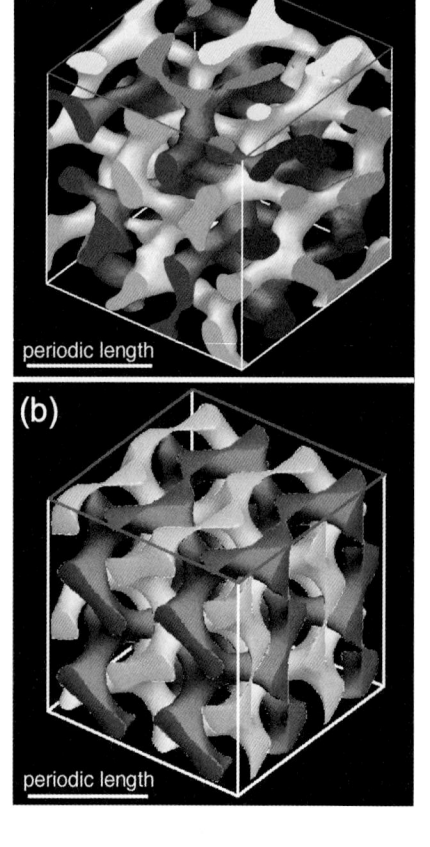

Fig. 7.17 Three-dimensional images of the bicontinuous G morphology obtained from (a) TEMT of an ordered SIS triblock copolymer and (b) a composition-matched Schoen G surface [21]. The minor S channels are pseudo-colored white and gray to illustrate that they interpenetrate but do not intersect. The I matrix is transparent to facilitate viewing. The periodic length is ~74 nm.

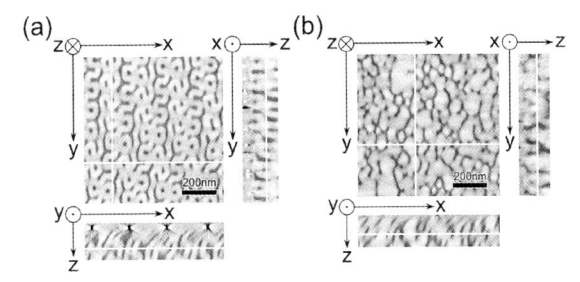

Fig. 7.18 Orthogonal slices of TEMT reconstructions generated from (a) an ordered bicontinuous morphology and (b) a disordered bicontinuous morphology, both observed to coexist in a SI diblock copolymer–polystyrene blend. The slice locations are identified as the solid white lines.

fidelity of the image reconstruction and threshold choice on the basis of composition (33 and 32 vol% S from the image and copolymer, respectively) [63]. In these images, the S channels are pseudo-colored white and gray, whereas the I matrix is transparent to demonstrate that the bicolored channels are interpenetrating, but not intersecting.

In addition to 3D isosurface representations of block copolymer morphologies, reconstructed volume elements can be examined along their orthogonal (x, y and z) axes, where the lateral x and y axes are normal to the electron beam and parallel to the specimen surface. Examples of x–y, y–z and x–z slices of a reconstruction are provided in Fig. 7.18a for the G morphology in a microphase-ordered SI diblock copolymer blended with polystyrene homopolymer. Periodic features expected for an ordered block copolymer nanostructure are evident in all three views, which further confirms the fidelity of the reconstruction. Included in Fig. 7.18b are TEMT results obtained from a disordered bicontinuous morphology that is found to coexist with the G morphology (Fig. 7.18a) in the same blend. Examination of the two image sets in Fig. 7.18 not only reveals structural differences between ordered and disordered morphologies in the same material, but also demonstrates a powerful feature of TEMT: because it is not model dependent, it is not limited by the shape or regularity of nanoscale structures of interest. This aspect of TEMT makes it particularly attractive in studies seeking to elucidate the 3D nature of block copolymer defects, and also aperiodic networks [64] and discrete dispersions (e.g. nanotubes [65]/nanofibers [66]) lacking long-range order. Three-dimensional images of such copolymer nanostructures derived from TEMT afford valuable qualitative insight into nanostructural connectivity and relevant aspect ratios, and also the real-space structural information required for accurate interpretation of small-angle scattering patterns. Careful analysis of 3D TEMT images can also directly yield crystallographic and topological information of periodic and aperiodic block copolymer nanostructures.

7.5.4
Quantitative Analysis

The 3D TEMT images of complex surfaces displayed in Figs. 7.17 and 7.18 are amenable to both local and global structural analyses [67]. Local topological analyses of such surfaces focus on structural characteristics that involve individual points or small patches of points on the surfaces. Two examples of topological metrics that can provide fundamental insight into the nature of a given surface include the mean (H) and Gaussian (K) curvatures. Both metrics are defined in terms of the radii of curvature r_1 and r_2 as follows: $H=(r_1^{-1}+r_2^{-1})/2$ and $K=(r_1 r_2)^{-1}$. Values of H and K at any point on a curved surface in a 3D image can be directly measured with procedures such as the sectioning and fitting method originally developed [68] for 3D laser-scanning confocal images of polymer blends undergoing spinodal decomposition. Repetition of this process yields probability density distributions such as the joint $P(H,K)$ distribution presented in Fig. 7.19 for the G morphology. This surface contour distribution can furthermore yield the independent $P_H(H)$ and $P_K(K)$ distributions portrayed in the figure to permit further evaluation of signature traits (e.g. the shape and symmetry of the distributions) as well as area-averaged quantities [63]. Whereas the area-averaged mean curvature of 0.034 nm^{-1} determined from the results in Fig. 7.19 could, in principle, be measured by small-angle scattering, the width

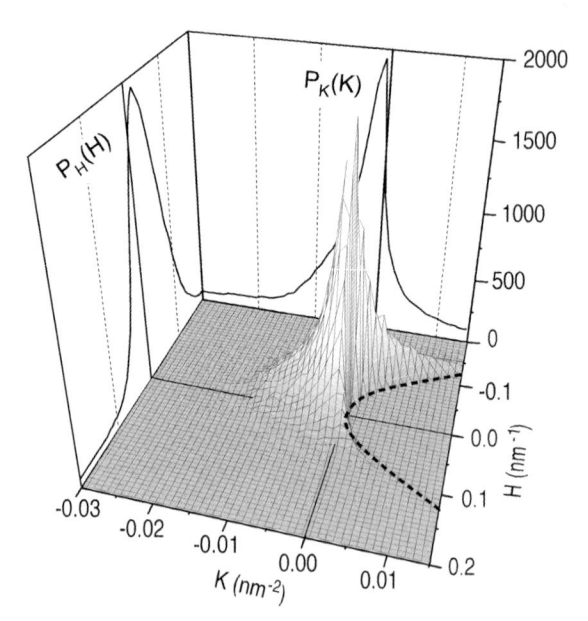

Fig. 7.19 Surface contour representation of the joint probability density, $P(H,K)$, measured from TEMT images of the G morphology observed in an ordered SIS triblock copolymer [21]. The individual probability densities, $P_H(H)$ and $P_K(K)$, are labeled. The dashed curve identifies the condition where $K=H^2$.

of $P_H(H)$, signified by its standard deviation σ_H (=0.042 nm^{-1}), could not. According to self-consistent field treatments [69], the value of σ_H relates directly to the stability of complex bicontinuous morphologies in block copolymer melts and is therefore of fundamental morphological and thermodynamic interest. Similar measurements can be performed on aperiodic morphologies for comparative purposes [64].

Global structural analyses include measurement of volume-fraction composition (discussed earlier), preferred orientation, channel coordination, genus (or the Euler–Poincaré characteristic), interfacial area density and crystallographic features (e.g. symmetry and periodicity or correlation length). Fourier analysis of the ordered morphology pictured in Fig. 7.18a yields the crystallographic identity and characteristic size of the unit cell. Peak positions located at $1 : \sqrt{6} : \sqrt{8}$ in the 3D Fourier transform of the isosurface used to generate Fig. 7.18a corroborate that the morphology is G. Moreover, the mean unit cell size, given by $2\pi(h^2+k^2+l^2)^{1/2}/q_{hkl}$ where h, k and l denote the Miller indices and q_{hkl} is the wavevector corresponding to the hkl spot, is 122 nm. Similar results are obtained [70] from mesoscale crystallography of the G morphology presented in Fig. 7.17a. Conversely, Fourier analysis of disordered sponge or network morphologies such as that in Fig. 7.18b yields a diffuse ring representative of the correlation length (~ 67 nm from Fig. 7.18b). Skeletonization of the 3D reconstructions in Figs. 7.17a and 7.18 further reveals that the channel coordination is primarily (89%) 3 in the three morphologies, suggesting that this is a naturally preferred structural arrangement. A channel coordination of 3 in ordered nanostructures is consistent with the G morphology. Another global topological metric is the Euler number (Eu), which is defined as 2–2g, where g corresponds to the genus. Measured values of Eu are -14 ± 0.4 for both G morphologies in Figs. 7.17a and 7.18a and -3.4 ± 0.2 for the disordered bicontinuous morphology shown in Fig. 7.18b. For reference, the value of Eu for the composition-adjusted Schoen G surface [21] in Fig. 7.17b is -16 ± 0.2. The examples discussed in this section demonstrate that high-fidelity TEMT images provide not only a 3D qualitative perspective of block copolymer (and other) nanostructures but also an opportunity to analyze, in rigorous quantitative fashion, relevant 3D features of these nanostructures.

7.5.5
Emerging Opportunities

A recent advance in TEMT is the use of dual-axis image acquisition [60]. All the previous examples discussed here employ a single tilt axis and the resulting 3D images suffer from resolution loss due to the wedge of missing information. Dual-axis image collection partially alleviates this shortcoming by collecting two tilt series along complementary orthogonal axes and combining the data to reduce the size of the wedge. Use of dual-axis TEMT has been shown to yield improved 3D images of block copolymer cylinders relative to images reconstructed along a single tilt axis. An issue that must be recognized in performing dual-

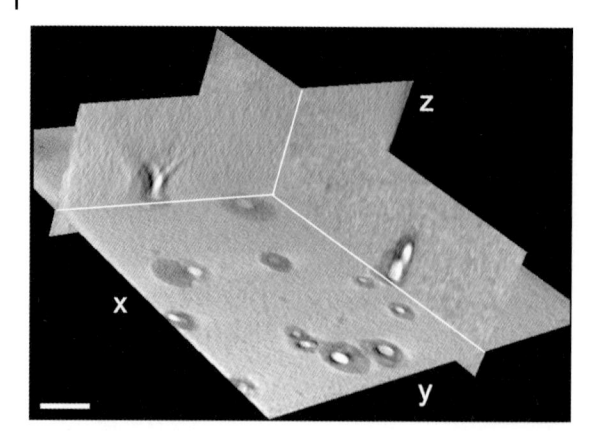

Fig. 7.20 Convergent orthogonal views of a TEMT reconstruction showing SM micelles formed in a polystyrene thin film as the film de-wets from a poly(methyl methacrylate) support. Micelles and copolymer patches along the polymer/polymer interface are visible, as are aggregated micelles. The scale marker corresponds to 500 nm.

axis TEMT is, however, the specimen damage incurred upon even longer exposure to the electron beam. Another emerging aspect of TEMT pertains to the specimens to which it is particularly well suited. Most studies of block copolymer nanostructures, for instance, endeavor to provide structural details of thermodynamically equilibrated morphologies for inclusion in phase diagrams or comparison with theoretical frameworks. A common understanding in macromolecular science is, however, that long, chain-like molecules may never attain true equilibrium within an experimentally accessible time frame. Unlike their (near-)equilibrated analogues, dynamically evolving specimens can provide valuable insight into the kinetics and mechanisms of molecular self-organization under various conditions [71]. It is in this latter spirit that TEMT can again prove highly beneficial, since it is not model restricted. Figure 7.20, for instance, shows three convergent axial views of SM block copolymer micelles that are frozen in the process of forming within a thin polystyrene film as the film begins to de-wet from a poly(methyl methacrylate) substrate in the melt. Although conventional TEM can be used to image these micelles as they form and evolve, TEMT alone permits assessment of the 3D size, shape and position of the micelles (relative to the polymer/air and polymer/polymer interfaces) as functions of parameters such as time, temperature, copolymer concentration and film thickness. One other unique benefit of TEMT is that it can be coupled with analytical methods capable of distinguishing species on the basis of their elemental composition to generate 3D energy-filtered (EF-TEMT) images. This aspect is discussed in the following section.

7.6
Analytical Electron Microscopy

7.6.1
Energy-dispersive X-Ray Mapping

Energy-dispersive X-ray spectroscopy, commonly abbreviated as EDS or EDX, relies on the collection of characteristic (i.e. element-specific) X-rays that are generated when a high-energy electron beam contacts a specimen under investigation by TEM, SEM or STEM [72]. The X-rays are collected with a detector capable of distinguishing X-rays on the basis of their energy, which is quantized due to core electron excitations and thus permits unambiguous identification of the element(s) responsible for X-ray production. As a chemical probe, an EDS analysis can be performed to yield elemental spectra from targeted point sources. If calibration standards are included in the analysis, these spectra are amenable to high-precision composition analysis. Since the focus of this section is visualization, however, only the second method by which EDS can be performed is considered further. In this case, the probe is raster scanned across a specimen and characteristic X-rays are collected as a function of lateral position. This process of generating an EDS X-ray map is repeated until sufficient statistics are attained. In the case of neat block copolymers or their blends with other polymers, this analytical microscopy technique has not been extensively used since most copolymer systems inherently consist of light elements, which are not always easily delineated by EDS. Moreover, specimen damage or drift during repeated scanning is problematic. The advent of bright field-emission guns and the growing need to identify inorganic additives (e.g. nanoparticles [73]) spatially modulated within nanostructured block copolymer matrices have, however, resulted in a renaissance of EDS mapping. An example of a lamellar diblock copolymer modified with surface-functionalized Au and Co nanoparticles is displayed in Fig. 7.21 to illustrate the utility of this analytical technique.

7.6.2
Energy-filtered Transmission Electron Microscopy

Another analytical method encountered only in TEM is electron energy-loss spectroscopy (EELS), which employs inelastically scattered electrons for composition analysis on the basis of elemental discrimination [74]. Transmitted electrons in TEM are unscattered (with no energy loss so that $\Delta E = 0$ eV), elastically scattered (also with $\Delta E = 0$ eV) or inelastically scattered (with $\Delta E > 0$ eV). In EELS, electrons transmitted through ultrathin specimens (to reduce the propensity for multiple scattering events) are separated in a spectrometer according to energy loss and the resulting spectra provide the same elemental information as EDS spectra. High-resolution EELS of fine spectral features can, however, also provide information on the local spatial environment (i.e. valence state) of elements of interest [75]. In contrast to EDS mapping, EELS can be spatially re-

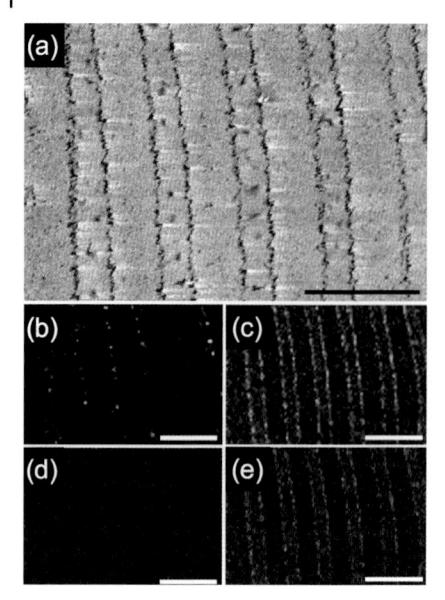

Fig. 7.21 Energy-dispersive X-ray analysis of an ordered poly(styrene-*b*-ethylene-*co*-propylene)–AuSPS–CoTOPO ternary nanocomposite. In (a), a bright-field STEM image confirms the lamellar morphology of the copolymer, and also interfacial and central segregation of the embedded nanoparticles. Corresponding EDS X-ray maps are provided for several different elements: (b) Co (K-edge), (c) Au (M-edge), (d) Si (K-edge) and (e) S (K-edge). The latter corresponds as expected to the Au map in (c). The scale marker corresponds to 250 nm. Courtesy of M. R. Bockstaller, Carnegie Mellon University.

solved through the use of an in-column or post-column energy filter to provide unique imaging opportunities collectively referred to as energy-filtered (EF) TEM [76]. The three most useful EFTEM imaging modes in block copolymer studies are individually considered below.

7.6.2.1 Zero-loss Imaging

Features of interest in most block copolymer systems are identified by phase contrast (with or without selective staining), in which case image quality (dictated by the signal-to-noise level) is dependent on the ratio of (un)scattered electrons with $\Delta E = 0$ eV to scattered electrons with $\Delta E > 0$ eV. In zero-loss (or elastic) imaging, only electrons comprising EELS spectra in the vicinity of $\Delta E = 0$ eV are used for image formation. Electron specificity is achieved through the use of an in-column electrostatic (Castaing–Henry–Ottensmeyer) or magnetic (Omega) energy filter or a post-column filter, wherein a polychromatic electron beam is dispersed according to energy loss. The location of these energy filters, and their operation, are depicted in Fig. 7.22. The resultant dispersion consequently permits removal of inelastically scattered electrons during image formation so that image quality is significantly improved in zero-loss imaging.

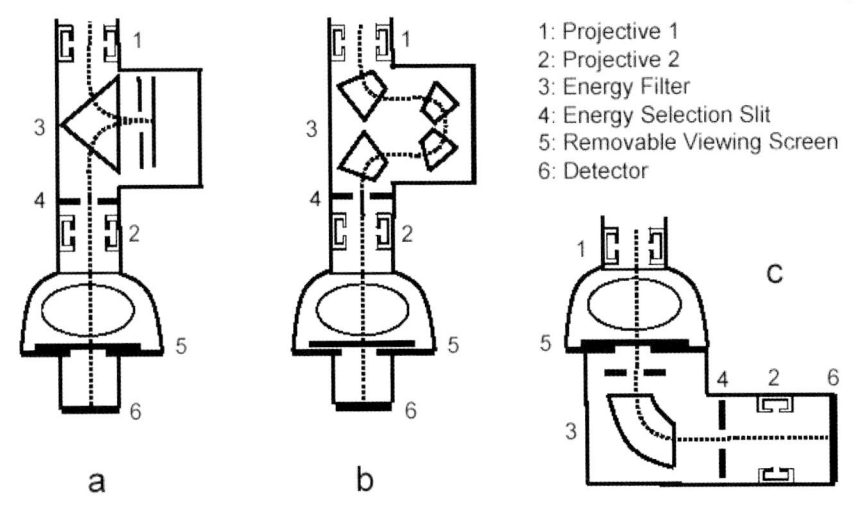

Fig. 7.22 Schematic illustration of different energy filters: (a) Castaing–Henry–Ottensmeyer (in-column), (b) Omega (in-column) and (c) post-column filter.

1: Projective 1
2: Projective 2
3: Energy Filter
4: Energy Selection Slit
5: Removable Viewing Screen
6: Detector

7.6.2.2 Structure-sensitive Imaging

In most systems of interest in soft-matter research, carbon is ubiquitous. While features composed of heavier elements (by atomic number) can often be identified within a carbon matrix, structure-sensitive imaging provides a valuable alternative. The primary carbon ionization edge is located at 284 eV. Imaging just below this edge removes contributions arising from electrons that are inelastically scattered from carbon [77], thereby highlighting the distribution of non-carbonaceous electrons in the same manner as high-angle annular dark-field (HAADF) imaging. Figure 7.23 illustrates the benefit of structure-sensitive imaging in a morphological investigation [78] of metallated block copolymer nanotubes cast from a selective solvent.

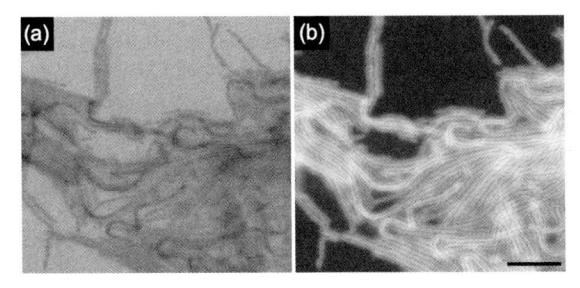

Fig. 7.23 Matched (a) zero-loss and (b) structure-sensitive EFTEM images [78] of highly asymmetric poly(ferrocenylsilane-b-dimethylsiloxane) nanotubes formed in a selective solvent. The scale marker corresponds to 500 nm.

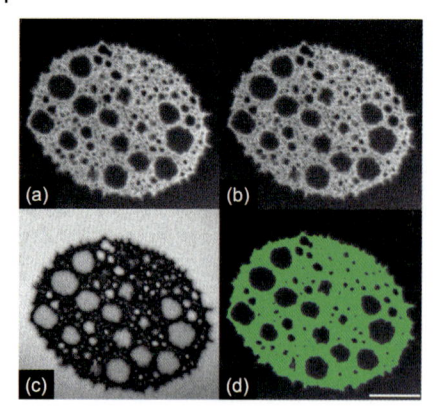

Fig. 7.24 Element-specific EFTEM image series of poly(methyl methacrylate) containing a semi-interpenetrating silicone network [79]. The images were acquired at different ΔE settings: (a) 93, (b) 105 and (c) 0 eV. The image displayed in (d) reveals the spatial distribution of Si in the blend. The scale marker corresponds to 1 μm.

7.6.2.3 Element-specific Imaging

As with EDS mapping, the spatial distribution of elements within a specimen can likewise be elucidated by EFTEM via element-specific imaging. In this imaging mode, 3–5 images are acquired at different ΔE intervals to (1) parameterize (fit) the EELS background, (2) interpolate a background image at a ΔE corresponding to the ionization edge of an element of interest (ΔE^*) and (3) identify the distribution of that element by subtracting the interpolated image from the actual image collected at ΔE^*. This image sequence is portrayed in Fig. 7.24.

The structure-sensitive images displayed in Fig. 7.24a and b are used to establish the characteristic EELS background. A zero-loss image of the specimen is shown in Fig. 7.24c and the spatial distribution of Si [79] in Fig. 7.24d. Practical advantages of EFTEM relative to EDS mapping include higher spatial resolution and substantially reduced acquisition times. For this reason, element-specific

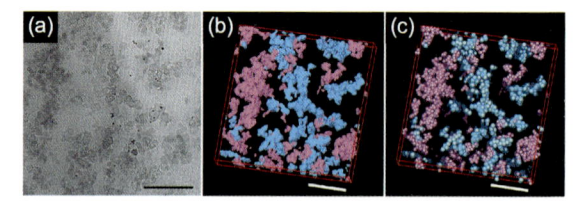

Fig. 7.25 EF-TEMT analysis of a nanocomposite composed of carbon black (CB) and silica nanoparticles embedded in a mixed polydiene matrix. A conventional TEM image of the system is shown in (a). The 3D EF-TEMT image in (b) reveals the spatial positions of the CB (blue) and Si (pink) nanoparticles. A Monte Carlo simulation (c) used in conjunction with the EF-TEMT permits more accurate estimation of the nanoparticle positions. The scale markers correspond to 200 nm Courtesy of H. Jinnai, Kyoto Institute of Technology.

imaging is preferable for studies of beam-sensitive specimens. An exciting synergy has been introduced recently through the combination of element-specific EFTEM and TEMT. By performing EFTEM at a fixed angular increment over a large tilt range, it is possible to generate 3D element-specific images [65, 80, 81], such as that presented for demonstrative purposes in Fig. 7.25. In this case, the 3D spatial distributions of two different nanoparticle species are clearly delineated. This approach has also been successfully extended [82] to microphase-ordered ABC triblock copolymers.

Acknowledgments

This work was supported by the Deutsche Forschungsgemeinschaft, the US National Science Foundation, the US Department of Energy and the Kenan Institute for Engineering, Technology and Science at North Carolina State University. The authors thank D. A. Agard and M. B. Braunfeld (University of California at San Francisco) and H. Jinnai (Kyoto Institute of Technology) for their participation in some of these studies and their valued contributions.

References

1 E. Ruska, *Angewandte Chemie International Edition in English* **1987**, *26*, 595.
2 E. Ruska, *Zeitschrift für Physik* **1934**, *87*, 580.
3 L. F. Drummy, J. Y. Yang, D. C. Martin, *Ultramicroscopy* **2004**, *99*, 247.
4 E. L. Thomas, in *Encyclopedia of Polymer Science and Engineering*, Vol. 5, 2nd edn. (Ed. Kroschwitz, J. I.), Wiley, New York, **1986**, 644–687.
5 J. C. Meyer, M. Paillet, G. S. Duesberg, S. Roth, *Ultramicroscopy* **2006**, *106*, 176.
6 C. Kuebel, K. Eckhardt, V. Enkelmann, G. Wegner, K. Müllen, *Journal of Materials Chemistry* **2000**, *10*, 879.
7 G. H. Michler, W. Lebek, *Ultramikrotomie in der Materialforschung*, Carl Hanser, Munich, **2004**.
8 T. Goldacker, PhD Thesis, University of Bayreuth, **1999**.
9 T. Goldacker, V. Abetz, *Macromolecules* **1999**, *32*, 5165.
10 K. Takahashi, H. Hasegawa, T. Hashimoto, V. Bellas, H. Iatrou, N. Hadjichristidis, *Macromolecules* **2002**, *35*, 4859.
11 K. Kato, *Journal of Polymer Science, Part B: Polymer Letters* **1966**, *4*, 35.
12 T. Goldacker, V. Abetz, R. Stadler, I. Erukhimovich, L. Leibler, *Nature* **1999**, *398*, 137.
13 J. S. Trent, J. I. Scheinbeim, P. R. Couchman, *Macromolecules* **1983**, *16*, 589.
14 C. Auschra, R. Stadler, *Macromolecules* **1993**, *26*, 2171.
15 U. Breiner, U. Krappe, R. Stadler, *Macromolecular Rapid Communications* **1996**, *17*, 567.
16 U. Breiner, U. Krappe, E. L. Thomas, R. Stadler, *Macromolecules* **1998**, *31*, 135.
17 L. F. Drummy, I. Voigt-Martin, D. C. Martin, *Macromolecules* **2001**, *34*, 7416.
18 H. Ott, V. Abetz, V. Altstädt, *Macromolecules* **2001**, *34*, 2121.
19 H. Ott, V. Abetz, V. Altstädt, Y. Thomann, A. Pfau, *Journal of Microscopy* **2002**, *205*, 106.
20 M. Kunz, M. Möller, H. J. Cantow, *Makromolekulare Chemie Rapid Communications* **1987**, *8*, 401.
21 H. Hückstädt, A. Göpfert, V. Abetz, *Macromolecular Chemistry and Physics* **2000**, *201*, 296.
22 M. Möller, R. W. Lenz, *Makromolekulare Chemie* **1989**, *190*, 1153.

23 W. P. Chen, M. F. Zhu, S. Song, B. Sun, Y. M. Chen, H. J. P. Adler, *Macromolecular Materials and Engineering* **2005**, *290*, 669.
24 D. R. Uhlmann, *Journal of Non-Crystalline Solids* **1972**, *7*, 337.
25 D. P. Siegel, W. J. Green, Y. Talmon, *Biophysical Journal* **1994**, *66*, 402.
26 J. R. Bellare, H. T. Davis, L. E. Scriven, Y. Talmon, *Journal of Electron Microscopy Technique* **1988**, *10*, 87.
27 M. H. Chestnut, D. P. Siegel, J. L. Burns, Y. Talmon, *Microscopy Research and Technique* **1992**, *20*, 95.
28 http://www.vitrobot.com/.
29 Y. Talmon, in *Modern Characterization Methods of Surfactant Systems* (Ed. B. P. Binks), Marcel Dekker, New York, **1999**, p. 147.
30 D. Danino, Y. Talmon, R. Zana, *Colloids and Surfaces A – Physicochemical and Engineering Aspects* **2000**, *169*, 67.
31 Y. Zheng, Z. Lin, J. L. Zakin, Y. Talmon, H. T. Davis, L. E. Scriven, *Journal of Physical Chemistry B* **2000**, *104*, 5263.
32 Y. Zhang, J. Schmidt, Y. Talmon, J. L. Zakin, *Journal of Colloid and Interface Science* **2005**, *286*, 696.
33 J. L. Burns, Y. Talmon, *Journal of Electron Microscopy Technique* **1988**, *10*, 113.
34 Y. Fink, Y. Talmon, in *Proceedings of the 13th International Congress on Electron Microscopy*, Vol. 1, **1994**, p. 37.
35 D. Danino, R. Gupta, J. Satyavolu, Y. Talmon, *Journal of Colloid and Interface Science* **2002**, *249*, 180.
36 S. Ruthstein, J. Schmidt, E. Kesselman, Y. Talmon, D. Goldfarb, *Journal of the American Chemical Society* **2006**, *128*, 3366.
37 J. Bang, S. M. Jain, Z. B. Li, T. P. Lodge, J. S. Pedersen, E. Kesselman, Y. Talmon, *Macromolecules* **2006**, *39*, 1199.
38 A. Wittemann, M. Drechsler, Y. Talmon, M. Ballauff, *Journal of the American Chemical Society* **2005**, *127*, 9688.
39 Y. Talmon, M. Adrian, J. Dubochet, *Journal of Microscopy* **1986**, *141*, 375.
40 E. Kesselman, Y. Talmon, J. Bang, S. Abbas, Z. B. Li, T. P. Lodge, *Macromolecules* **2005**, *38*, 6779.
41 K. Mortensen, Y. Talmon, *Macromolecules* **1995**, *28*, 8829.
42 J. J. Crassous, M. Ballauff, M. Drechsler, J. Schmidt, Y. Talmon, *Langmuir* **2006**, *22*, 2403.
43 S. Sakurai, H. Kawada, T. Hashimoto, L. J. Fetters, *Macromolecules* **1993**, *26*, 5796.
44 D. A. Hajduk, P. E. Harper, S. M. Gruner, C. C. Honeker, G. Kim, E. L. Thomas, L. J. Fetters, *Macromolecules* **1994**, *27*, 4063.
45 A. K. Khandpur, S. Forster, F. S. Bates, I. W. Hamley, A. J. Ryan, W. Bras, K. Almdal, K. Mortensen, *Macromolecules* **1995**, *28*, 8796.
46 J. H. Laurer, D. A. Hajduk, J. C. Fung, J. W. Sedat, S. D. Smith, S. M. Gruner, D. A. Agard, R. J. Spontak, *Macromolecules* **1997**, *30*, 3938.
47 E. L. Thomas, D. M. Anderson, C. S. Henkee, D. Hoffman, *Nature* **1988**, *334*, 598.
48 S. Ludwigs, A. Boker, V. Abetz, A. H. E. Muller, G. Krausch, *Polymer* **2003**, *44*, 6815.
49 S. S. Gambhir, *Nature Reviews Cancer* **2002**, *2*, 683.
50 J. C. Soares, J. J. Mann, *Biological Psychiatry* **1997**, *41*, 86.
51 R. A. Brooks, G. Dichiro, *Physics in Medicine and Biology* **1976**, *21*, 689.
52 J. Frank, *Electron Tomography: Three-dimensional Imaging with the Transmission Electron Microscope*, Plenum Press, New York, **1992**.
53 J. Frank, *Three-dimensional Electron Microscopy of Macromolecular Assemblies*, Oxford University Press, New York, **2006**.
54 M. Moritz, M. B. Braunfeld, J. W. Sedat, B. Alberts, D. A. Agard, *Nature* **1995**, *378*, 638.
55 B. F. McEwen, M. Marko, *Journal of Histochemistry and Cytochemistry* **2001**, *49*, 553.
56 R. J. Spontak, M. C. Williams, D. A. Agard, *Polymer* **1988**, *29*, 387.
57 L. C. Sawyer, D. T. Grubb, *Polymer Microscopy*, 2nd edn, Chapman and Hall, London, **1996**.
58 Q. X. S. Zheng, M. B. Braunfeld, J. W. Sedat, D. A. Agard, *Journal of Structural Biology* **2004**, *147*, 91.
59 R. J. Spontak, J. C. Fung, M. B. Braunfeld, J. W. Sedat, D. A. Agard, L. Kane, S. D. Smith, M. M. Satkowski, A. Ashraf,

D. A. Hajduk, S. M. Gruner, *Macromolecules* **1996**, *29*, 4494.

60 H. Sugimori, T. Nishi, H. Jinnai, *Macromolecules* **2005**, *38*, 10226.

61 H. H. Barrett, D. W. Wilson, B. M. W. Tsui, *Physics in Medicine and Biology* **1994**, *39*, 833.

62 P. A. Midgley, M. Weyland, J. M. Thomas, B. F. G. Johnson, *Chemical Communications* **2001**, 907.

63 H. Jinnai, Y. Nishikawa, R. J. Spontak, S. D. Smith, D. A. Agard, T. Hashimoto, *Physical Review Letters* **2000**, *84*, 518.

64 H. Jinnai, Y. Nishikawa, M. Ito, S. D. Smith, D. A. Agard, R. J. Spontak, *Advanced Materials* **2002**, *14*, 1615.

65 M. H. Gass, K. K. K. Koziol, A. H. Windle, P. A. Midgley, *Nano Letters* **2006**, *6*, 376.

66 E. A. Wilder, M. B. Braunfeld, H. Jinnai, C. K. Hall, D. A. Agard, R. J. Spontak, *Journal of Physical Chemistry B* **2003**, *107*, 11633.

67 S. Hyde, S. Andersson, K. Larsson, Z. Blum, T. Landh, S. Lidin, B. W. Ninham, *The Language of Shape*, Elsevier Science, Amsterdam, **1997**.

68 Y. Nishikawa, T. Koga, T. Hashimoto, H. Jinnai, *Langmuir* **2001**, *17*, 3254.

69 M. W. Matsen, F. S. Bates, *Journal of Chemical Physics* **1997**, *106*, 2436.

70 H. Jinnai, T. Kajihara, H. Watashiba, Y. Nishikawa, R. J. Spontak, *Physical Review E* **2001**, *64*, 10803.

71 T. Xu, A. V. Zvelindovsky, G. J. A. Sevink, K. S. Lyakhova, H. Jinnai, T. P. Russell, *Macromolecules* **2005**, *38*, 10788.

72 C. E. Lyman, D. E. Newbury, J. I. Goldstein, D. B. Williams, J. Romig, A. D., J. T. Armstrong, P. Echlin, C. E. Fiori, D. C. Joy, E. Lifshin, K.-R. Peters, *Scanning Electron Microscopy, X-ray Microanalysis and Analytical Electron Microscopy: a Laboratory Workbook*, Plenum Press, New York, **1990**.

73 M. R. Bockstaller, E. L. Thomas, *Physical Review Letters* **2004**, *93*.

74 R. F. Egerton, *Electron Energy-loss Spectroscopy in the Electron Microscope*, Plenum Press, New York, **1996**.

75 S. Wild, L. L. Kesmodel, G. Apai, *Journal of Physical Chemistry B* **2000**, *104*, 3179.

76 L. E. Reimer, *Energy-filtering Transmission Electron Microscopy*, Springer, Berlin, **1995**.

77 A. Du Chesne, *Macromolecular Chemistry and Physics* **1999**, *200*, 1813.

78 D. J. Frankowski, J. Raez, I. Manners, M. A. Winnik, S. A. Khan, R. J. Spontak, *Langmuir* **2004**, *20*, 9304.

79 R. Thomann, R. J. Spontak, in *Science, Technology and Education of Microscopy: an Overview* (Ed. A. Mendez-Vilas), Formatex, Badajoz, **2003**, p. 249.

80 M. Weyland, P. A. Midgley, *Microscopy and Microanalysis* **2003**, *9*, 542.

81 H. Jinnai, Y. Nishikawa, T. Ikehara, T. Nishi, *Advances in Polymer Science* **2004**, *170*, 115.

82 K. Yamauchi, S. Akasaka, H. Hasegawa, H. Iatrou, N. Hadjichristidis, *Macromolecules* **2005**, *38*, 8022.

8
Polymer Networks

Karel Dušek and Miroslava Dušková-Smrčková

8.1
Introduction

Covalent polymer networks are the largest molecules known because the whole macroscopic object is one molecule. Usually, a polymer network is composed of *branch points* of a certain *chemical functionality* ($f=3$, 4), which are connected by f chains to other branch points. The example of an ideal network having tetrafunctional ($f=4$) crosslinks is shown in Fig. 8.1. There, the differences in chemical composition are not displayed only the *connectivity* of the system of branch

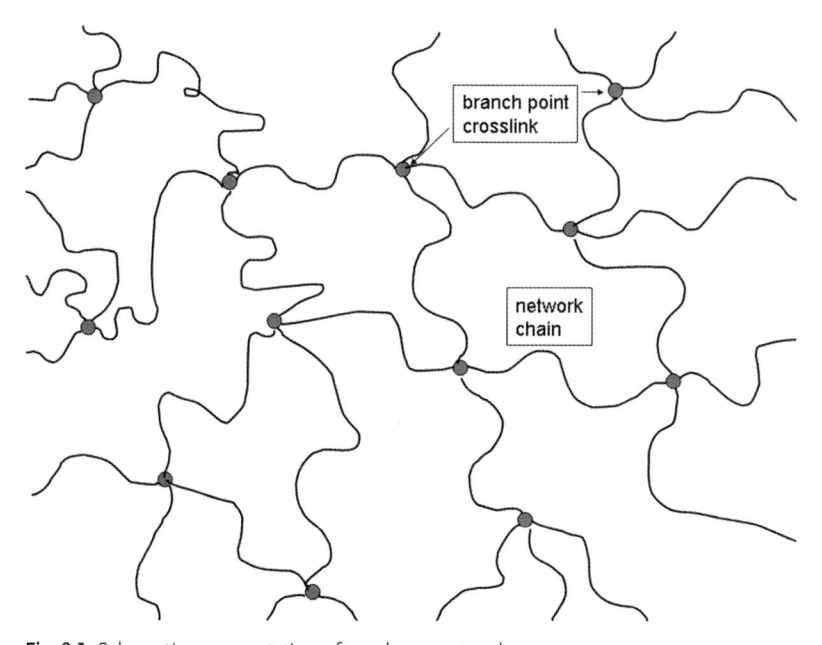

Fig. 8.1 Schematic representation of a polymer network.

Macromolecular Engineering. Precise Synthesis, Materials Properties, Applications.
Edited by K. Matyjaszewski, Y. Gnanou, and L. Leibler
Copyright © 2007 WILEY-VCH Verlag GmbH & Co. KGaA, Weinheim
ISBN: 978-3-527-31446-1

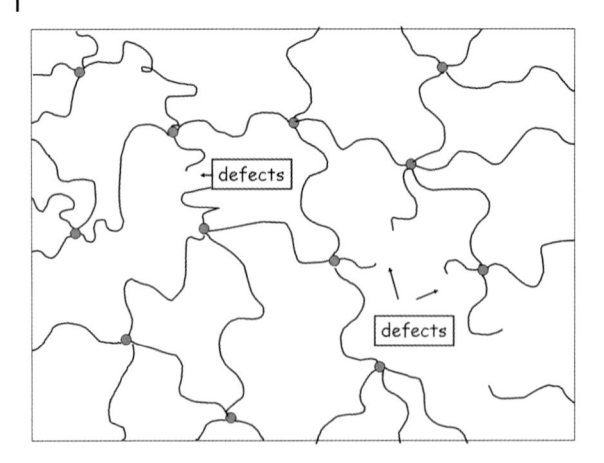

Fig. 8.2 A polymer network with defects.

points (junctions, crosslinks) and network chains by which the polymer network is characterized. The wavy lines indicate that the chains connecting crosslinks or junctions are flexible. With respect to connectivity, the flexibility of chains connecting crosslinks is irrelevant and a network can be visualized as a *graph* with *crosslinks, branch points* or *junctions* called *vertices* or *nodes*; the synonyms for *network chains* are *edges or lines*.

In reality the number of chains issuing from a certain junction (vertex) is not equal to the chemical functionality of the crosslink but can be lower (unreacted functional groups) and also the status of the chains may be different – through some of them we can continue via a sequence of chemical bonds to "infinity", through some others not. "Infinity" means the surface of the (macroscopic) sample. This brings two new elements to graph representation of networks – *coloring of edges and vertices* – and use of oriented graphs. Considering an edge associated with a vertex, we can look *out of* the vertex or *into* the vertex. "Coloring" is a graph-theoretical expression for assigning specific status or property to the vertex or edge.

A network with defects (with some *dangling chains*, different status of junctions) is illustrated by Fig. 8.2.

8.1.1
Polymer Networks as Materials

After a short excursion into the abstract world of graphs that describe polymer networks with respect to their connectivity, let us consider crosslinked polymers as materials.

Polymer networks range from soft gels and jellies, via rubbers to hard and tough materials; from highly swollen to dry systems. Physical gels are ones in which strong physical bonds replace covalent bonds and behave in a certain experimental window as covalent gels. The application areas of polymer networks are diverse, for example

- vulcanized rubbers (elastomers)
- thermosets and high-performance composites
- protective organic coatings
- sorbents and separation media
- biopolymer gels, drug delivery systems
- soft implants, hydrogel contact lenses
- solution concentrators
- electronics: printed circuits, information carriers.

The family of biogels is very broad and includes crosslinked protein and hybrid biopolymer gels where the performance is controlled by built-in protein motifs.

The main special features of polymer networks distinguishing the crosslinked polymers from the uncrosslinked ones are:
- memory of form (the crosslinked system remembers its shape and state at its birth)
- large reversible deformations in the rubbery state
- limited swelling and often higher chemical and physical durability.

There are analogies between a branching process resulting in a polymer network and networks assembled from other elements, like a resistor network, neural network, network of individuals with some common interest or target, or network of events developing in time.

In this chapter, we will be dealing with covalent polymer networks, their formation, structure and some structure-sensitive properties. Because of limited space, the explanation will be limited and references will be given to more detailed explanation. Recent reviews of polymer networks are given in Refs. [1–6].

8.2
Network Formation

For the formation of a polymer network, certain requirements concerning the network functionality and group reactivity must be fulfilled to make gelation possible. The presence of at least one precursor molecule with functionality higher than 2 is a necessary but not sufficient condition. Some off-stoichiometric systems or systems with a monofunctional component, even with full conversion of functional groups, do not gel at all. Some of the precursors are preformed by special synthesis; some others are synthesized by a subgel branching process. Examples are hyperbranched or off-stoichiometric highly-branched functional polymers.

8.2.1
Chemical Reactions Most Frequently Used for Preparation of Polymer Networks

Polymer networks can be formed by any chemical reaction by which covalent bonds are formed. Organic chemistry gives enough inspiration, more can be found in macromolecular chemistry sources, such as Ref. [7]. However, there exist differences between bond formation producing linear and crosslinked polymers, such as

1. Some crosslinking reactions are accompanied by formation of a low-molecular-mass by-product and its removal may be difficult.
2. Cyclization reactions can be much stronger and important.
3. Polydispersity is very large.
4. Transition from control by chemical reactivity of groups to diffusion control occurs earlier.
5. The topological limit of conversion of functional groups makes attainment of full conversion difficult.

These factors may alter the outcome of the reaction, promote some side reactions and make the characterization of the product difficult. Factor 1 disqualifies many reactions for use in the preparation of polymer networks. Reactions used for polymer network preparation can be divided into three categories:
1. step growth reactions
2. chain polymerization
3. crosslinking of preformed polymer chains (frequently step growth).

Here, we will only list the most important reactions used in practice, and will discuss in a little more detail the reactions of epoxide and isocyanate groups. For more details see Refs. [2, 7–9] and the respective chapters in Ref. [5]:
- reactions of epoxide groups [10, 11]
- reactions of isocyanate groups [12, 13]
- reactions of cyanate groups with phenols [14]
- esterification, and transesterification
- reactions of carboxy groups with aziridine
- chemistry of silicon-containing compounds – hydrolysis and condensation of alkoxysilanes and acyloxysilanes, hydrosilylation [15]
- reactions of phenols with formaldehyde, their intermediates [16]
- reactions of urea, thiourea, guanidine and melamine with formaldehyde
- addition of amino and mercapto groups on C=C bonds, tiol-ene combination of addition and polymerization mechanisms [17]
- vulcanization reactions of polydienes and other polymers containing unsaturation with sulfur, sulfur-containing compounds, peroxides
- crosslinking of saturated chains by ionizing radiation, peroxides
- free-radical polymerization and copolymerization of polyunsaturated monomers
- free-radical polymerization (photopolymerization) of bisunsaturated telechelics.

8.2.1.1 Epoxide Group

In the comprehensive review by May [10], many resins and curing systems are listed. The most frequent reaction for the curing of di- and triepoxides is the reaction with primary aromatic and aliphatic polyamines. Various di-, tri-, and tetraglycidyl ethers and glycidyl esters of different molecular mass, derivatives of diglycidylaniline (4,4,4′,4′-tetraglycidyldiaminodiphenyl methane, TGDDM), aromatic and aliphatic glycidyl ethers as well as various epoxidized vegetable oils are used as epoxide precursors. The epoxide group can also be introduced into polyfunctional precursors by copolymerization with glycidyl methacrylate or acrylate or by allyl glycidyl ether.

The addition of primary amine to an epoxy group occurs in two steps: in the first step, primary amine is transformed into an aminoalcohol having a secondary amine group and the remaining amine hydrogen reacts with another glycidyl group (Scheme 8.1).

The glycidyl group is more reactive than an epoxidized double bond; aromatic glycidyl ethers are more reactive than aliphatic glycidyl ethers. Polyamines are used as curing agents, most frequently diamines like 4,4′-diaminodiphenylmethane (DDM), 4,4′-diaminodiphenyl sulfone (DDS), aliphatic diamines, like 1,6-diaminohexane, diethylenetriamine $H_2NCH_2CH_2NHCH_2CH_2NH_2$ or technical grade diethylenetriamine with admixtures of other "polyethylene polyamines". The aliphatic amino group is generally more reactive than the aromatic one. For curing, the relative reactivities of the hydrogens on primary and formed secondary amino groups are important [18]. For aliphatic amines, the reactivity ratio is close to 1 or somewhat lower, whereas for aromatic amines it is lower. Expressing the molar concentrations of the primary amine as $[NH_2]$ and the hydrogens of the primary amine as $2[NH_2]$, the consumption of epoxy groups by a second order reaction can be written as

$$-\frac{d[ep]}{dt} = k_p[ep][NH_2] + k_s[ep][NH] \tag{1}$$

or per "amine hydrogen" as

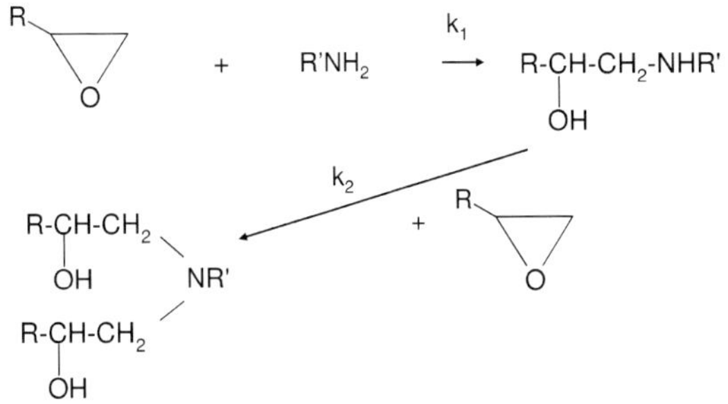

Scheme 8.1 Reaction of primary amino group with epoxide group.

$$-\frac{d[\text{ep}]}{dt} = k_p^*[\text{ep}]2[\text{NH}_2] + k_s[\text{ep}][\text{NH}] \tag{2}$$

Taking the "reactivity per amine hydrogen" as the reference state, then for an ideal state

$$k_p^*/k_s = \rho_H = 1 \quad \text{but} \quad k_p/k_s = \rho_A = 1/2$$

For reactions of aliphatic amines $\rho_H = 0.5-1$ and for aromatic amines it is about 0.3–0.5 [19], [20]. The reaction is catalyzed by proton donors including the hydroxy group. Because the hydroxy group is formed during the curing reaction, in the absence of catalyst (or in the case of its low content), the reaction is autocatalytic

$$-\frac{d[\text{ep}]}{dt} = (k_{p,\text{non-cat}} + k_{p,\text{cat}}([\text{ep}]_0 - [\text{ep}]))[\text{ep}][\text{NH}_2]$$
$$+ (k_{s,\text{non-cat}} + k_{s,\text{cat}}([\text{ep}]_0 - [\text{ep}]))[\text{ep}][\text{NH}] \tag{3}$$

the difference $[\text{ep}]_0-[\text{ep}]$ is equal to the concentration of reacted epoxide groups, i.e., to the concentration of [OH] groups that catalyze the epoxide ring opening. In the presence of proton donor catalyst, the reaction kinetics approaches second order.

For a stoichiometric system, or when amine groups are in excess, epoxide-amine addition is practically the only reaction. However, when the epoxide groups are in excess, the OH group generated by amine–epoxide addition initiates polyetherification – addition of OH group to the epoxide to give an ether bond [18, 21]. Because the OH group is regenerated, the addition reaction continues. Although the OH–epoxide reaction is considerably slower than epoxide–amine addition, it is important after all or almost all amine hydrogens have been consumed. Etherification is catalyzed, for example, by *tert*-amines or tetraalkylammonium salts. It adds crosslinks into the epoxide–amine network. The N–CH– or N–CH$_2$–bonds are durable against hydrolysis and oxidation and are used in high-performance composites and other engineering applications (adhesives, coatings, etc.).

Other important crosslinking reactions are reactions of the epoxide group with phenols, carboxyl groups and cyclic anhydrides. Reaction with a phenolic group is an etherification reaction. Due to its relative acidity, the reaction is much faster. In practice, polyepoxides are reacted with Novolacs (condensation products of phenols with formaldehyde under acid conditions).

A carboxyl group reacts with an epoxy group to give a hydroxy ester (Scheme 8.2). Aliphatic carboxyls are more reactive than aromatic carboxyls and the reaction of an epoxide group with a carboxyl group is slower than with an amine. The

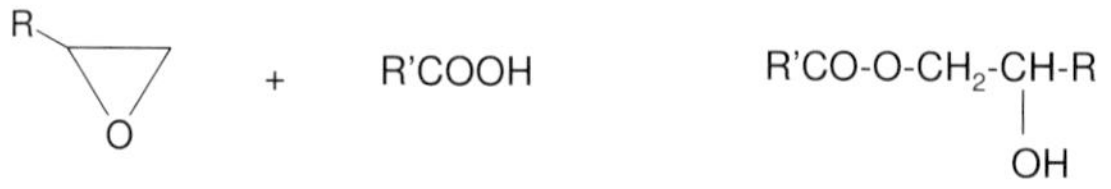

Scheme 8.2 Reaction of carboxyl group with epoxide.

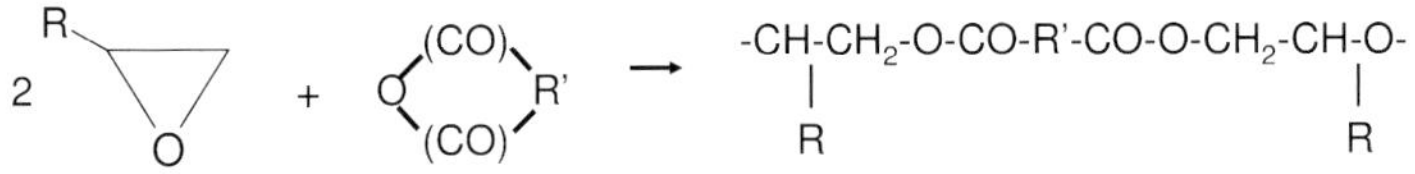

Scheme 8.3 Reaction of epoxide group with cyclic anhydride.

carboxyl–epoxide reaction is catalyzed by *tert*-amines, tetraalkylammonium salts, metal complexes of Cr, Co, Fe, Zn, etc. Etherification can be an important side reaction, especially for non-catalyzed systems. Transesterification continuing after esterification is completed is important. It results in an equilibrium between hydroxyester and diol and diester. From the point of view of branching and crosslinking, this is an important transformation [22]. When diacid and diepoxide are used, the diepoxide unit is transformed into tetraester, triester, monoester and tetrol. Although the average number of bonds per fragment remains the same, the redistribution of the number of bonds makes gelation and network formation possible, because the second-moment functionality average of diepoxide units $\langle f \rangle_2$ exceeds 2, whereas the first moment remains equal to 2 (for definition of functionality averages, see Eq. (9) in Section 8.5.2). The carboxy–epoxide reaction is important as a crosslinking reaction for carboxy-telechelic polymers as well as for branched polyester-based powder coatings. The crosslinker is usually triglycidyl isocyanurate.

The reaction of polyepoxides with cyclic anhydrides, like terahydrophthalic and hexahydrophthalic anhydrides and their derivatives is the most frequently used reaction for the preparation of durable polyester based thermosets and composites (Scheme 8.3).

Usually, traces of proton donors serve as initiators and the epoxide ring opening reaction is similar to the relatively slow carboxy–epoxide reaction. However, the reaction of the formed hydroxy group with anhydride is fast. The reaction is catalyzed by *tert*-amines, ammonium salts, phosphines and phosphonium salts. The mechanism of the non-catalyzed reaction, where the *tert*-amine catalyst adds to the epoxide ring to form first a zwitter ion which then reacts with anhydride [23], is interesting. The addition reaction then proceeds by an anionic mechanism.

There are a great number of other reactions involving the epoxide group which have been used in crosslinking such as the mercapto group (mercaptans react much faster than aliphatic alcohols) and thiiranes are used instead of oxiranes [24, 25]. Fast cationic polymerization with BF_3–amine or BF_3–alcohol catalysts, curing with dicyan diamide, ketimines (curing is released by water) [10], should also be mentioned.

8.2.1.2 Isocyanate Group

The isocyanate group is very reactive and its reactions are widely used in polymer network formation [8, 12, 13, 26, 27]. The most typical reaction paths are summarized in Scheme 8.4.

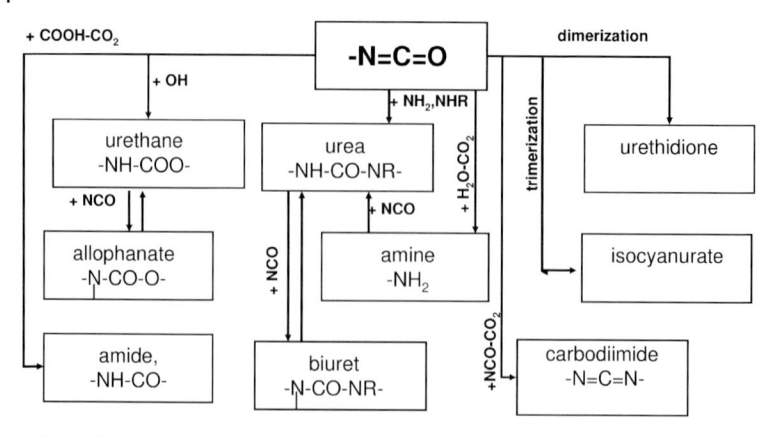

Scheme 8.4 Various reaction paths of the isocyanate groups in network formation reactions.

The most frequently used reaction is that of di- and polyisocyanates with aliphatic polyols to form urethane groups. Urethane bonds are stable against photo-oxidation and have moderate stability against hydrolysis. The isocyanate group reacts with amines giving stable urea groups; however, the reaction with aliphatic amines is very fast and a good mixing of reactants is not possible. Therefore, isocyanate groups are first blocked by a compound which forms a labile bond, such as oximes, lactams, or phenols. Blocked isocyanate is deblocked at elevated temperature and/or more active compounds. Cyclotrimerization by which three isocyanate groups form a ring is also sometimes used for curing. Knowledge of the intensity and the time sequence of various reaction paths is important for network buildup [28].

Among the diisocyanates there are some which have intrinsically different reactivities due to their nature, like the aliphatic–cycloaliphatic in isophorone diisocyanate (IPDI) or the position of the NCO group on the aromatic ring (2,4- vs. 2,6-toluene diisocyanate (TDI)) [8]. There can also exist a substitution effect in TDI – the reactivity of the remaining NCO is different when one of the NCO groups has reacted.

A difference in reactivity of functional groups is important for control of the gelation time and gel point conversion as well as the network structure. In simple cases, the gel point conversion is higher when one more and one less reactive group are combined in diisocyanate [29]. The commercial triisocyanates are not pure compounds and contain traces of by-product of higher and lower functionality. Small amounts of reaction products with moisture can also be present. For modeling of crosslinking the real functionality distribution or the respective averages must be determined.

Formation of urethanes from polyols is catalyzed by a wide range of catalysts of which the tin-based catalysts are most frequently used. The reactivity of the isocyanate group decreases in the series primary, secondary, tertiary and the same series holds for aliphatic alcohols; for them the reactivity ratio decreases in the proportion 100:30:1.5. Urethanes involving phenolic OH are less stable than the aliphatic ones; they are used rather as higher-temperature blocking agents. An important, and sometimes neglected, reaction is transurethanization occurring at relatively low temperature ($\geq 80\,^\circ$C) [30]. By this reaction the groups are interchanged (Scheme 8.5).

As was mentioned above, polyamines are too reactive towards polyisocyanates and have to be blocked. Blocking of the amine group is the other alternative. Blocking with aldehydes or ketones to form aldimines or ketimines is interesting. Ketimine is formed under elimination of water and moisture initiates the release of amine groups which react rapidly with isocyanate. The urea groups formed are stable and react with excess isocyanate groups to form biuret (Scheme 8.6).

The other curing reaction for polyisocyanates is cyclotrimerization used mainly for crosslinking of isocyanate endcapped low-molecular-mass polymers. It is catalyzed by quaternary ammonium salts, carboxylates, *tert*-amines, hindered *tert*-amines, Lewis acids, etc.

Of the other reactions shown in Scheme 8.4, the reaction with water is of the utmost importance: since the molar mass of water, $18\,\mathrm{g\,cm}^{-3}$, is two orders of magnitude less than the molar mass of most polyols. Therefore, 0.1% water on a weight basis gives 10% or more on a molar basis. Such an "excess of OH groups" may disturb the stoichiometric balance and eventually leave some alcoholic OH groups unreacted. Water reacts with NCO groups in two steps (Scheme 8.7).

The intermediate product – carbamic acid – decomposes rapidly to amine and carbon dioxide and the amine formed reacts rapidly with isocyanate, if it is still present; urea or biuret are the end products. Since formation and dissociation of allophanate and biuret are reversible reactions and biuret is more stable than allophanate, in many urethane systems, after a prolonged contact with moisture, urea groups are found.

R-NH-(C=O)-O-R' + R''-NH-(C=O)-O-R'''
  ↓↑
R-NH-(C=O)-O-R''+ R'-NH-(C=O)-O-R'''
              etc.

Scheme 8.5 Transurethanization reaction.

R-NH-(C=O)-O-R' + R-N=C=O ↔ R-N-(C=O)-O-R'
                                                  |
                                          (C=O)-NH-R

Scheme 8.6 Formation of biuret from urea.

R-N=C=O + H-O-H → [R-NH-(C=O)-OH] → R-NH$_2$ + CO$_2$

R-N=C=O + R'NH$_2$ → R-NH-(C=O)-NH-R'

Scheme 8.7 Reaction of isocyanate group with water.

8.2.2
What Do We Need to Know about Chemistry of Network-forming Reactions and Why?

For design and optimization of a crosslinking system, studies of monofunctional systems provide the basic information like feasibility of bond formation in dependence on temperature, catalysis, effect of solvent, possible side-reactions, which may be more important in crosslinking systems due to longer reaction time, and reactivity information. Systems involving low-functionality reactants and low-molecular-mass products are much simpler and easier to analyse. Examples of network formation studies based on this information are described in Ref. [31].

8.3
Polymer Networks Precursors

Polymer networks are prepared from precursors. A precursor is a chemical compound bearing functional groups. It is sometimes polydisperse in molecular mass and composition. Both the backbone (the part without functional groups) of the precursor and the functional groups are important for the polymer network to perform the desired function. The nature of the backbone can be a hydrocarbon, polyether, polyester, polysiloxane, etc. It brings in basic thermomechanical properties and durability. The architecture of a backbone of the same composition can be different. For instance, a polyester bearing OH group can be used in the form of a telechelic, a linear acrylate or methacrylate copolymer, a block copolymer, a star, a microgel, a hyperbranched polymer, etc. The importance of precursor architecture was recognized some time ago [32]. The major categories of current precursor architectures are shown in Figs. 8.3–8.5 with examples of real chemical structures.

The oligomeric or polymeric precursors, the telechelics, functional copolymer, stars and other similar precursors bring in the network a certain chemical environment. However, the telechelics may be polydisperse in molecular mass. The functional copolymers are polydisperse both in molecular mass and number of functional groups. Functional stars, microgels, dendrimers, hyperbranched polymers, off-stoichiometric functional polymers are internally branched. Upon crosslinking the branch points become "activated" and gradually contribute to the concentration of elastically active network chains (EANC) [33]. The branched precursors are usually distinguished by lower viscosity compared to linear polymers of the same molecular mass. Dendrimers are not used as conventional precursors because of their high price. Hyperbranched polymers formed from Ab_f monomers are somewhat similar in their high branching but polydisperse both in molecular mass and isomeric structures. Functional microgels are crosslinked particles of micron or submicron size. The last scheme shows polymer (hydrogel) networks with built-in protein motifs, a very fast developing group

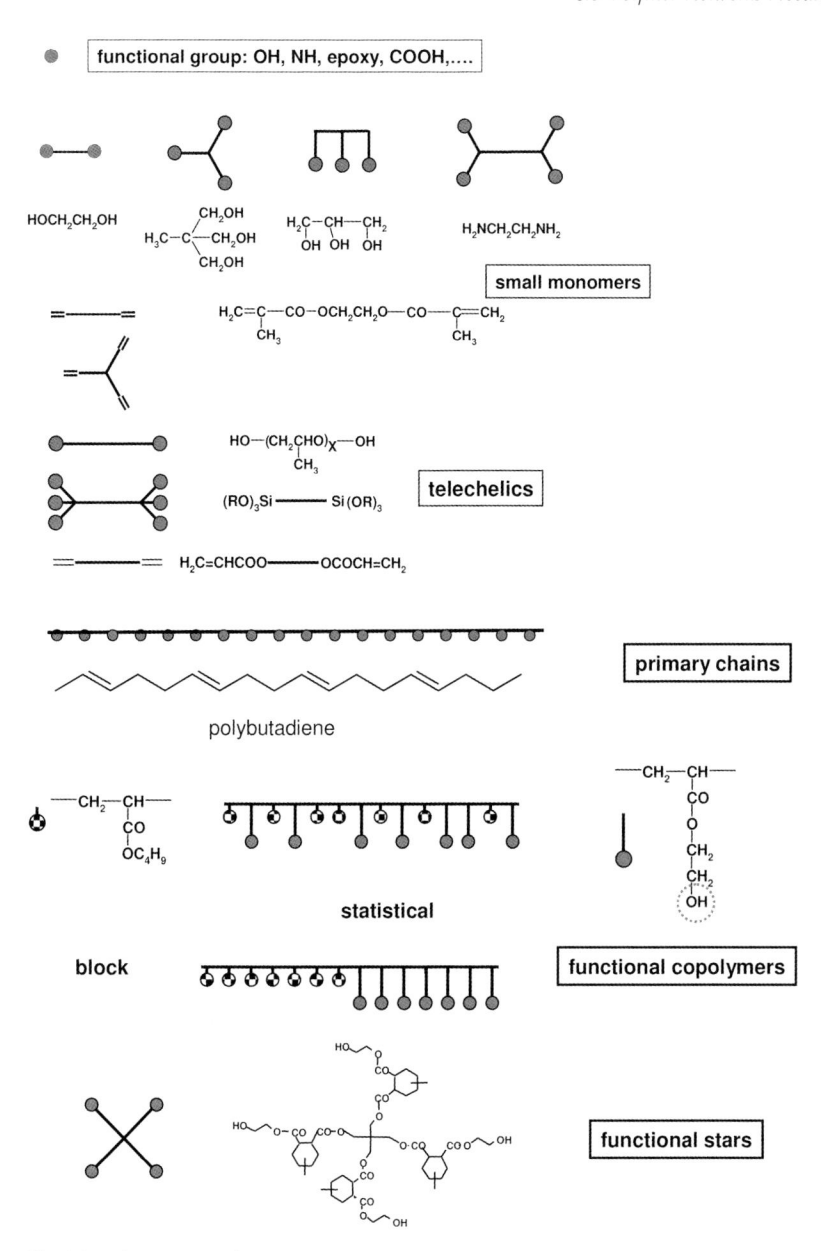

Fig. 8.3 Polymer networks precursors.

for various medical applications [34]. The coiled-coil associate is in fact a physical crosslink. It is prepared by building-in the motif into a polymerizable "monomer" such that the network is being formed by one of the processes discussed above. Not shown are functionalized nanoparticles, similar to microgels

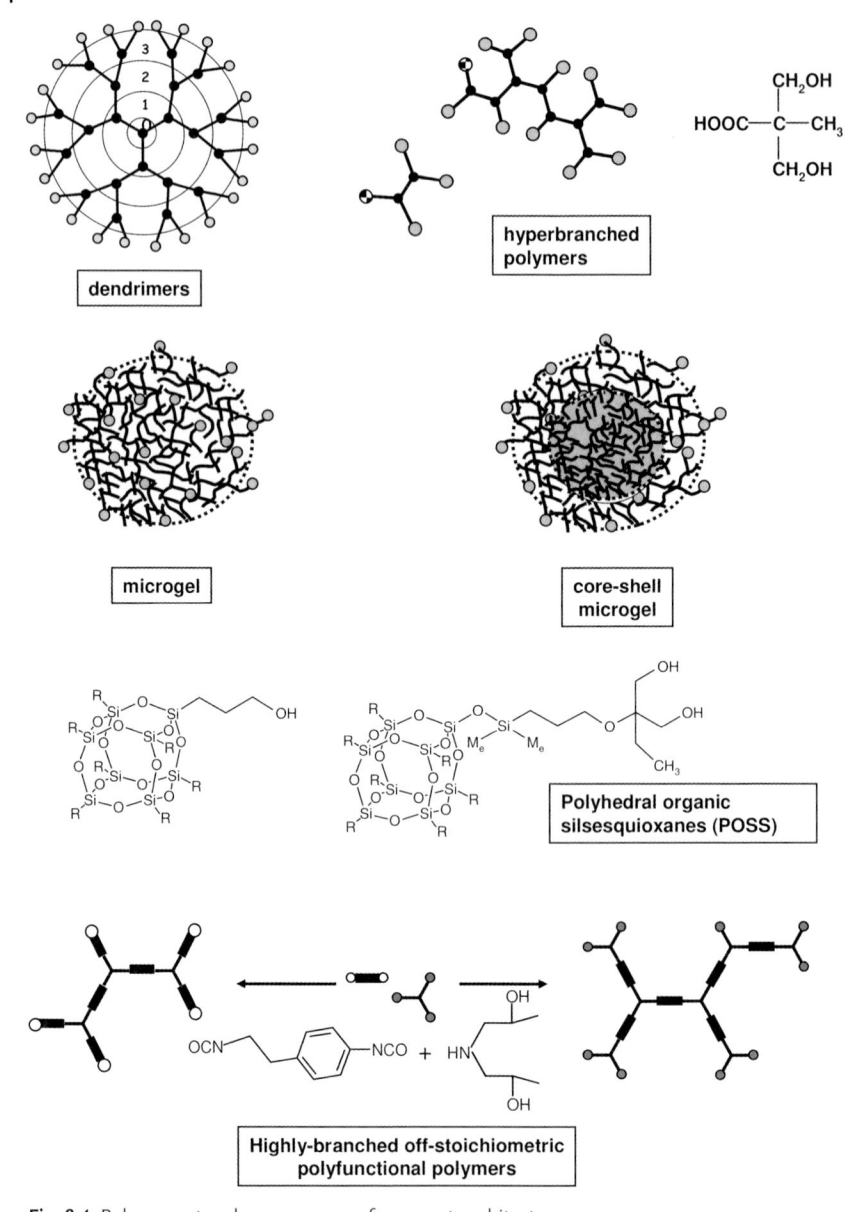

Fig. 8.4 Polymer networks precursors of compact architecture.

but with a stiff (inorganic) core. When discussing ways the networks can be made one should not forget the new concept of cyclic crosslinks of polyrotaxanes like cyclodextrin able to slide along the network chains [35, 36]. Such networks should be distinguished in the rubbery state (swollen gel) by good ultimate properties because of the absence of local stresses.

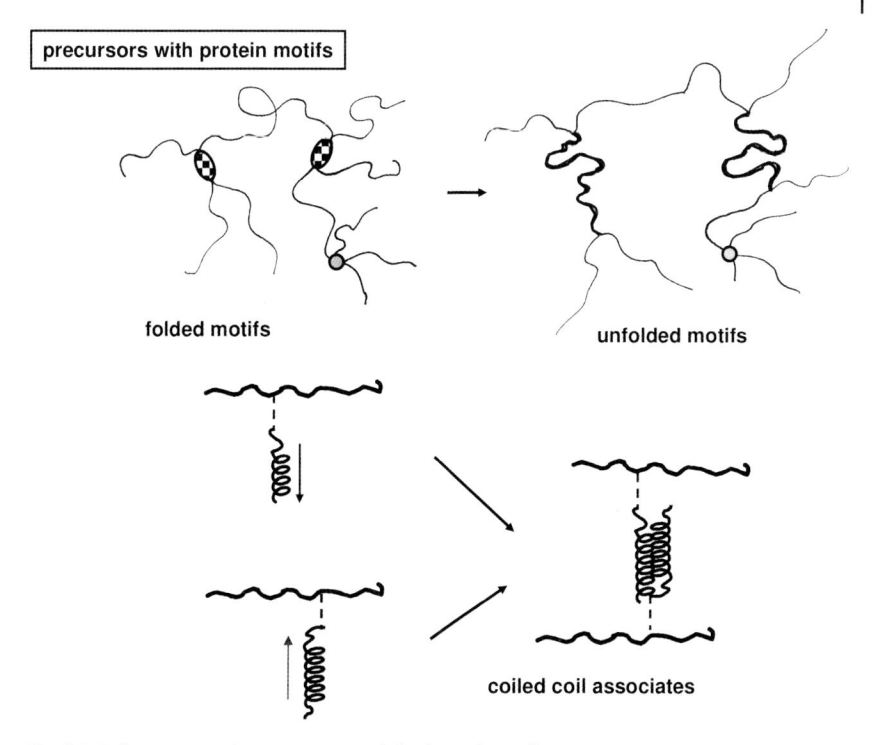

Fig. 8.5 Polymer networks precursors with biological motifs.

8.4
Crosslinking Kinetics

In chemistry, reactions are controlled by reactivity or diffusion (mobility). In macromolecular chemistry, it is similar but diffusion control is more frequent. When the transition from reactivity control to diffusion control sets in is very well described by the Eyring theory of transition (activated) complex

$$-A + B \underset{k_{AB,-D}}{\overset{k_{AB,D}}{\rightleftarrows}} \langle -AB- \rangle^* \xrightarrow{k_{AB,C}} -AB- \tag{4}$$

The group A attached to a chemical structure reacts with group B and an activated complex is formed. The reaction rate is controlled by the rate constant $k_{AB,D}$ (D means diffusion control). The formed activated complex $\langle -AB- \rangle^*$ is unstable and dissociates again back to the initial reactants with the rate controlled by the rate constant $k_{AB,-D}$. The complex has also another choice – to be transformed into a stable product, the bond $-AB-$. This happens with a rate controlled by the rate constant $k_{AB,C}$. (C means chemical control). When $k_{AB,D}, k_{AB,-D} \gg k_{AB,C}$, i.e. the reaction partners form the complex many times and dissociate again, the reaction outcome is controlled by $k_{AB,D}$, i.e., by chemical reactivity. However, when the diffusion is slow and $k_{AB,D^{\bullet}}$, $k_{AB,D^{\bullet}} < k_{AB,C^{\bullet}}$, the stable product is formed almost every time whenever the activated complex is formed by a rate controlled by $k_{AB,D}$ (encounter control). After some manipulation with the Eyring equation and using some simplifying assumptions, one arrives at the Rabinowitch equation [37]

$$\frac{1}{(k_{AB})_{app}} = \frac{1}{k'_{AB,C}} + \frac{1}{k_{AB,D}}, \quad \text{where} \quad \frac{1}{k'_{AB,C}} = \frac{1}{k_{AB,C}} \frac{k_{AB,-D}}{k_{AB,D}} \quad \text{and}$$

$$k_{AB,C}(T) = A\exp(-\Delta F_{act}/RT); \quad k_{AB,D}(T) = \phi_{mob}(T) \tag{5}$$

For the chemical reactivity control, the rate constant shows the Arrhenius dependence on temperature and is independent of conversion. For the mobility (diffusion) control, the Arrhenius dependence is no longer valid. For mobility control, the reaction rate depends on the temperature dependence of segmental mobility.

Thus, for a bimolecular reaction of groups bonded to polymer chains

$$\frac{d[-AB-]}{dt} = (k_{AB})_{app}[-A][B-] \tag{6}$$

The apparent rate constant is independent of conversion for the condition of the chemical reactivity control. Indeed, for many macromolecular systems where bonds are formed by step reactions ("slow" kinetics, the groups enter into the activated state many times before a stable bond is formed), the kinetic laws valid for low-molecular-mass systems are obeyed, although the values of the rate constants need not be the same. No change in functional group kinetics is observed in the vicinity of the gel point. Only at high conversions of functional groups does the difference between low-molecular-mass and high-molecular/mass kinetics become noticeable and a topological limit exists [4].

The mobility of molecules decreases with increasing size and increasing viscosity of the medium. The Trommdorff effect is a demonstration of diffusion control in linear chain polymerization. It is manifested by acceleration of polymerization due to the onset of diffusion controlled termination by free-radical recombination. It is observed above a certain molecular mass and monomer concentration and starts at a certain conversion when the viscosity is high enough. In crosslinking polymerization, the Trommsdorff effect starts earlier [38], close to the gel point, because of the increase in viscosity. Unlike in step growth, the reaction between two macroradicals is fast ($k_{AB,D} \ll k_{AB,C}$) and is controlled by macroradical diffusion (encounter).

According to this reasoning, the gelation phenomenon should always be connected with diffusion control because it involves the reactions between very large molecules embedded in a medium of large viscosity. Indeed, the study of critical exponents for the change in various quantities near the gel point shows [39] that they do not adhere to the mean field model on which the mass action law is based.

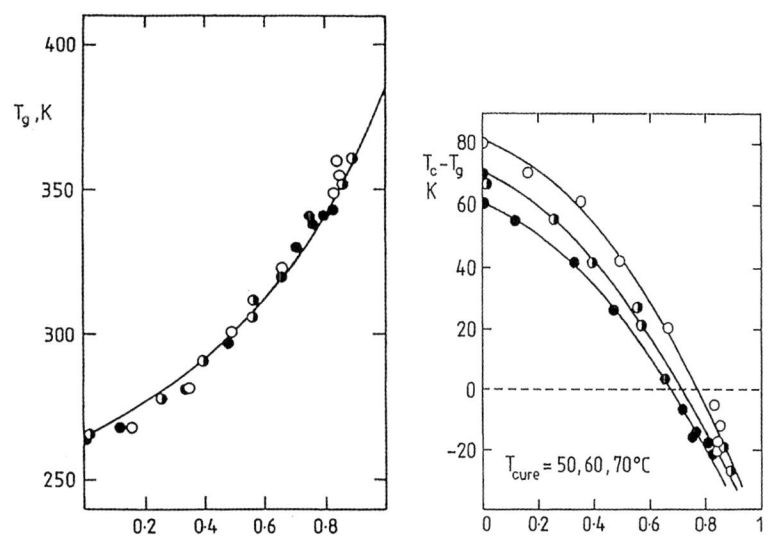

Fig. 8.6 Dependence of T_g of the reacting system bisphenol A diglycidyl ether–1,3-propanediamine on conversion of epoxide groups. Curing at temperatures T_c indicated. Reconstructed from data of Ref. [41].

The fact that we do not see any break on the kinetics consumption of functional groups near the gel point is due to the fact that the overwhelming majority of groups are bonded to "small" molecules which move fast enough [40].

Another case of diffusion control due to increase in viscosity of the reaction medium is the *transition of the reaction system into the glassy state* [41, 42]. Because the glass transition temperature, T_g, increases with increasing conversion, at a certain conversion T_g can become larger than T_c – the curing temperature (Fig. 8.6).

If we consider the segmental mobility from the point of view of the free-volume theory, the diffusion controlled rate constant is given by the following expression

$$k_D = k_{D0} \exp\left(\frac{-b}{f_g + a_f(T_c - T_g)}\right) \qquad (7)$$

where f_g is the fraction of free volume at T_g and a_f is the expansion coefficient of the free volume. Close to or soon after $T_g = T_c$ at isothermal cure, k_D starts to dominate the kinetics and k_{app} decreases because T_g continues increasing. Hypothetically, the reaction should stop when the denominator of Eq. (7) becomes zero (which is about 50 K below T_g), but the free-volume theory based on the WLF concept no longer holds and the volume relaxation of glass (physical aging of glass) rather follows another dependence on temperature and is also time dependent.

Summarizing the present knowledge about crosslinking kinetics:
- Many crosslinking reactions proceed as corresponding low-molecular-mass reactions and the passage through the gel point is not reflected on the kinetic curve.
- The passage through the gel point is sensed in the case of fast reactions, like termination in free-radical polymerization by recombination of two macroradicals, or for the change of properties dependent on mobility of the largest clusters resulting in gelation.
- Considerable slowing down of the crosslinking reaction is always seen when the system enters the glass transition region due to increasing group conversion or loss of solvent.

8.5
Buildup of Polymer Networks

Buildup of polymer networks is characterized by some common features:
- increase in molecular mass averages and broadening of the molecular-mass distribution
- reaching the gel point at which M_n is still finite but M_w and higher averages diverge
- beyond the gel point – existence of the infinite structure (gel) and finite molecules (sol) and gradual transition of sol to gel; decrease of molecular mass averages of the sol with increasing conversion
- increasing connectivity in the gel, formation of closed circuits, and decreasing concentration of dangling chains.

Figure 8.7 describes the network buildup from a trifunctional precursor. The brown path is a sequence of bonds with infinite continuation (continuation to the surface of the macroscopic sample). Figure 8.8 characterizes typical changes of various parameters as a function of conversion of functional groups. These changes are similar, irrespective of the composition of the crosslinking system, but the gel point on the conversion scale may change from very low values (10^{-4}, 10^{-3}) for crosslinking of long primary chains (any monomer unit can get crosslinked) up to values close to 1 when weak networks are formed in off-stoichiometric A + B systems, or when monofunctional components are present.

In a crosslinking system, the substructures can be classified with respect to the *connectivity of bonds* in which the units are engaged. We will first consider the bonds we examine when *looking out* of the unit through the bonds. If *none* of the bonds the unit is engaged in has *infinite continuation*, the unit is part of a sol, if *one and only one* bond the unit is engaged in has infinite continuation, the units is part of a *dangling chain*. If a unit is engaged *in two and only two* such bonds, the unit is part of an *elastically active network chain (EANC)*, and if the unit is engaged in 3 or more such bonds, the unit belongs to the category of *elastically active crosslinks (junctions)*. This is shown graphically in Fig. 8.9.

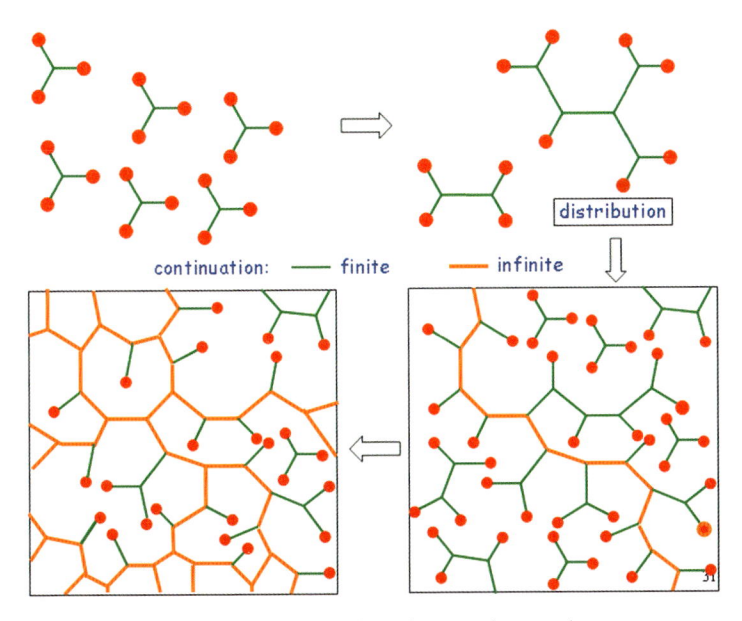

Fig. 8.7 Illustration of the network buildup from a trifunctional precursor.

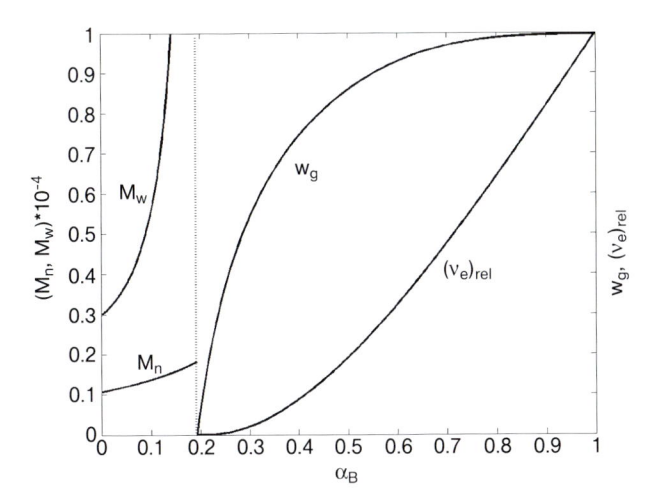

Fig. 8.8 Development of molecular mass averages, gel fraction, and relative concentration of EANCs for a statistical copolymer of units with primary and secondary OH groups (2:1) and a non-functional monomer, $M_{An}=2300$, $(f_A)_n=7.3$ crosslinked with a triisocyanate $M_{Bn}=567$; reactivity ratio of *sec.* to *prim.* OH equal to 0.1; $(v_e)_{rel}=v_e/(v_e)_{max}$, $(v_e)_{max}=3.83\times10^{-3}$ mol cm^{-3}.

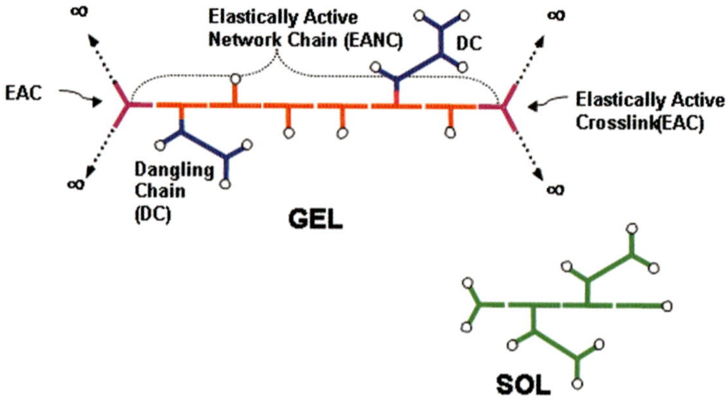

Fig. 8.9 Structure of a crosslinking system beyond the gel point.

8.5.1
Network Formation Theories and their Application

8.5.1.1 Classification of Network Formation Theories

Network formation theories describe quantitatively the evolution of the distribution of branched polymers, gelation, sol to gel transition, and the evolution of the gel structure. The evolution variable is the degree of conversion of functional groups or the reaction time. For equilibrium controlled reactions, conversion of functional groups uniquely describes the network structure, irrespective of previous history, but the structure formed by kinetically controlled reactions may depend on the way the given state (conversion) was reached. Yet, for many chemical systems the bond formation laws are simple and a mean-field approach based on chemical kinetics for units' transformation describes well the structure evolution, except some evolution steps near the critical point. The existing theories differ in the range of correlations caused by stochastic correlations and correlations in space [43].

1. *Statistical theories*: Branched and crosslinked structures are *generated from units in different reaction states* by proper combination of reacted functional groups (half-bonds) into bonds [44–51]. The *reaction states* of a unit are distinguished by the number of reacted groups and types of bonds they are engaged in. The *unit* (building unit) is a precursor unit, or its fragment or preformed substructure (superspecies used in combined method (cf., e.g., Refs. [56, 57]). The structures are generated at every degree of conversion from the distribution of the reaction states. This distribution is generated by equilibrium or kinetically (cf., upper part of Fig. 8.10). The assemblage of the network can be performed also beyond the gel point and a set of extinction probabilities can be calculated by recursive equations. This determines the probability that a given bond has finite or infinite continuation.
2. *Kinetic theories.* Branching and network formation is visualized as reactions between any possible pair of molecules (lower part of Fig. 8.10). Generation

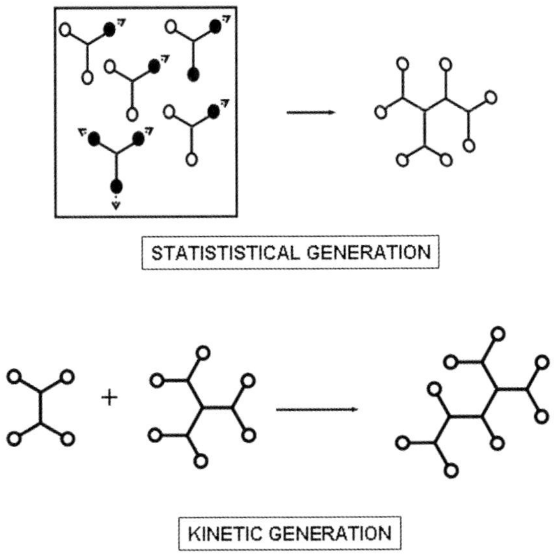

Fig. 8.10 Statistical generation from units in different reaction states and kinetic generation.

of branched and crosslinked structures is described by *infinite sets of differential equations* where *time* is the primary independent variable [52–58]. The sets of differential equations describe the distributions of molecules according to the parameters selected (e.g., degree of polymerization). In the differential equations, reactions between functional groups of every possible pair of species (molecules, gel) are considered and the "rate constants" are dependent not only on the number and type of functional groups, but can also be dependent on other properties of reacting species, such as their diffusivity, size and symmetry, or group accessibility. The complexity of the differential equations ranges from equations of chemical kinetics based on mass action law to generalized Smoluchowski equations [58, 59] in which the kernel contains information on the reaction ability of the molecules engaged in bond formation. The solution can be obtained analytically in a few cases, or by the moment method, or Monte Carlo simulation [54, 59].

3. *Computer simulation in space.* The structures are generated in space (usually 3-D) by performing "reactions" governed by rules similar to those used for Smoluchowski processes. Typical for this kind of simulation is percolation on lattice or off-lattice (cf. Fig. 8.11) [60–63]. The disadvantage of the presently used simulation in space is the static nature of structure growth. The weightings arising from various reaction paths, stochastic and space determined correlations, are possible in principle but difficult. In a new approach using molecular dynamics, this obstacle – absence of conformational rearrangements in bond formation – has been removed [64].

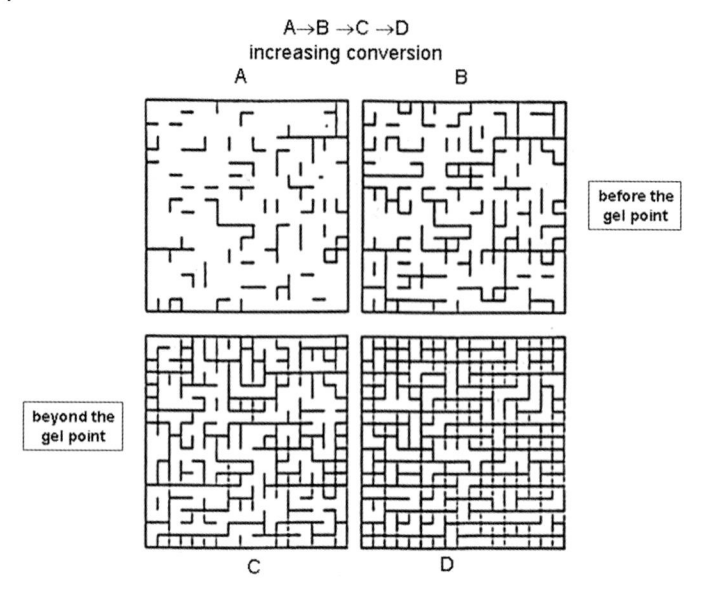

Fig. 8.11 Network formation as random bond percolation on lattice (in 2D).

The reference (ideal) system is one where: (i) The reactivities of all groups are the same and independent of the extent of crosslinking reaction. (ii) All reactions (before the gel point) are intermolecular; beyond the gel point the circuit closing is uncorrelated with the positions of reacting groups. (iii) There is no excluded-volume effect, the positions of any unit and any group are not restricted. (iv) The mobility of structures and substructures is much larger than the speed of bond formation, i.e., there is no diffusion control of bond formation.

The deviations from ideal behavior of the reference system are due to the short-range and long-range correlations. Stochastic correlations are independent of dimensionality of space, spatial correlations include cyclization, diffusion control, and excluded volume. The *statistical theories*, by their nature, can take into account only short-range correlations. The assemblage of units into sol molecules and a gel is a first-order Markov process. Perturbation methods exist within the statistical theories to account for some of the correlation. Kinetic methods treat stochastic correlations exactly and can simulate many spatial correlations. The advantage of the statistical theories is that they can describe (as an approximation) various substructures in the gel, such as dangling chains or elastically active network chains. The versions of this theory, most widely used in application, are the Theory of Branching Processes or Cascade Theory [4, 18, 31, 44–48] and Recursive Theory [50, 51]. These approaches are fully equivalent and the results are identical for the same starting assumptions. Only the mathematical formalism is somewhat different. In the TBP formalism, the distributions are described and manipulated by generating functions and RT formalism uses gfs only in more complicated cases when the precursor enters the system as a distribution [51].

8.5 Buildup of Polymer Networks

The difference between the theoretical approaches to gelation is seen in the critical exponents, ζ, of various quantities, X, such as molecular mass, gel fraction, number of EANCs, of the scaling relations valid in the vicinity of the critical (gel) point

$$X = \varepsilon^\zeta \tag{8}$$

where $\varepsilon = |a-a_c|$, a is conversion. The critical exponents for the mean field theories are different from percolation type exponents and critical exponents of the Smoluchowski aggregation processes vary in dependence on the form of the kernel extending from the mean-field. There is experimental evidence that critical exponents are different from the mean-field and closer to the percolation ones. This can be explained [40] by the immobility of the largest molecules – a typical feature for a percolation type process.

Cyclization. So far we have not discussed cyclization during network formation. Cyclization (intramolecular reaction) is an important phenomenon always accompanying intermolecular reaction [64–70, 31]. By intermolecular reactions the degree of polymerization and molecular mass increase, while ring closing within the same molecule does not affect the molecular mass. Figure 8.12 illustrates the competition between intermolecular crosslinking and cyclisation.

The relative intensity of cyclization, is given by the probability, P_c, that the two groups belong to the same molecule, relative to the case that they belong to different molecules. Considering the simplest case of a freely jointed chain, $P_c(x) \propto x^{-3/2}/c_0$ (Jacobson–Stockmayer dependence), where x is the number of

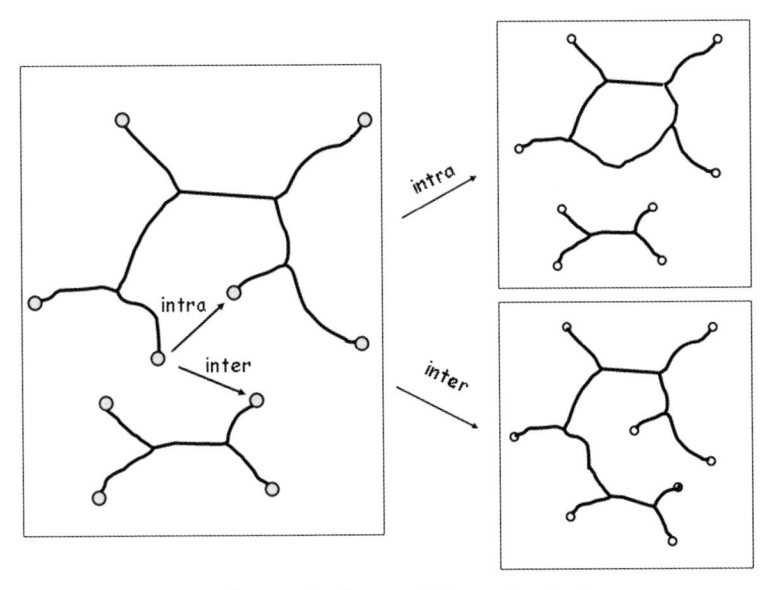

Fig. 8.12 Competition of intermolecular crosslinking and cyclization.

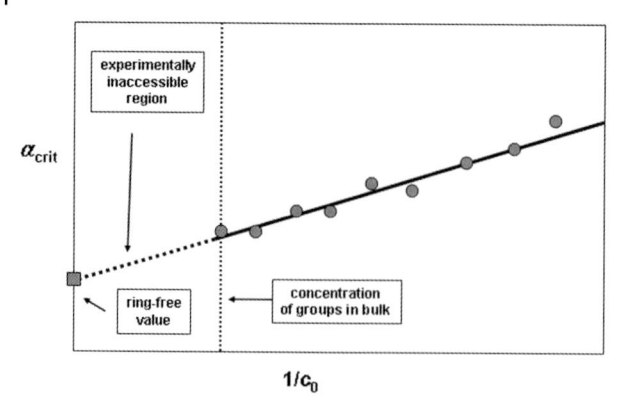

Fig. 8.13 Example of shift of the gel conversion to higher values as a function of the reciprocal concentration of functional groups and extrapolation to $1/c_0 \to 0$.

statistical segments in the sequence connecting the two groups and c_0 is the initial concentration of functional groups. In a branching system, the probability for formation of all possible ring sizes must be considered. Before the gel point, the problem is topologically clear and the number of cycles can be calculated from the number-average molecular mass and degree of conversion of functional groups into bonds, as well as from the shift of gel point conversion to higher values. Dilution of the reacting system (lower c_0) enhances cyclization. Therefore, extrapolation of the dependence of the gel point conversions, a_c, on the reciprocal concentration, $1/c_0$ to the hypothetical limit $1/c_0=0$ gives the ring-free values of the critical conversion (Fig. 8.13)

The extents of cyclization found from the experimental shift of the gel point range, for bulk systems, from practically zero (some epoxide–amine networks), small (e.g., polyether triol-diisocyanate, a few per cent bonds wasted in cycles), moderate (various polyfunctional stepwise systems), to some free-radical crosslinking copolymerizations with overwhelming cyclization at the beginning (see below). Cyclization is strong also in polysiloxane systems (see POSS polymers, for example). The slope of the dependence in Fig. 8.13 is also related to cyclization. There exist several theoretical treatments of cyclization discussed in Refs. [64–70]. So far, deviations from the ring-free values (e.g., caused by presence of diluent) have been ascribed to cyclization although the intermolecular reaction rate is affected by existing cycles. Also, cyclization information is an output of spatial simulation, but whenever lattice simulation is used, the outcome is affected by the lattice.

Much more complex is the situation beyond the gel point where, in the ideal situation, the concentration of EANCs is proportional to the cycle rank of the network, i.e. to the number of closed circuits. Experimentally, one can observe the effect of cycles beyond the gel point by (i) an increase in the sol fraction (sol molecules of the same molecular mass with cycles have a lower chance to

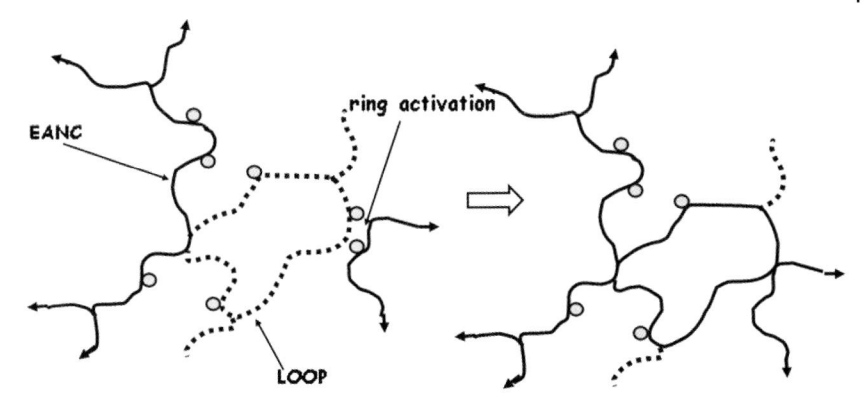

Fig. 8.14 Activation of cycles by crosslinking. Arrows: infinite continuation.

react with groups on an "infinite" gel, and (ii) lowering the concentration of EANCs due to the existence of elastically inactive loops. The cycles which were formed before or the loops formed after the gel point may be activated by continuing bond formation (Fig. 8.14).

8.5.2
Application of Network Formation Theories

When a crosslinking system is designed for a certain function, its examination by applying network formation theories is very useful, because the application of experimental characterization is limited compared to linear polymers due to wide distributions and the insolubility of the gel part. The information, which can be obtained theoretically, includes molecular mass averages, averages of radii of gyration, gel point conversion, sol and gel fractions, dangling chains, elastically active network chains and crosslinks, fraction of trapped entanglements, fractions of activated pre-existing branch points, and some other. There is good evidence that the statistical approaches based on first-order Markovian statistics give good predictions over a wide range of conversions for many experimental systems – but not for all and apparently not for critical exponents. Deviations exist for systems where some steps are diffusion controlled, like free-radical crosslinking (co)polymerization, systems where cyclization is dominating (cf., Section 8.7), systems with rigid chains, or systems with large fluctuations in composition. A simple and reliable method for treating cyclization, especially in the postgel state, is still needed. The combined method (combination of statistical and kinetic methods (cf., e.g., Refs. [56, 57]) helps to remove problems of complex kinetics. Application of the kinetic method is not difficult mathematically and it will grow in the future. Application of computer simulation is not easy if algorithms are to be developed, but easy for the user of commercial software. The disadvantage of the commercial software is that one can neither understand well the underlying algorithms nor modify or amend them.

The branching theories have been applied to many systems based on the precursors shown above. The limited space of this chapter does not allow for discussion of special features associated with individual precursors. Some common features are listed below:

- Some precursors are monodisperse and some others are *polydisperse* with respect to the number of *functional groups* and *molecular mass*. Often, such polydisperse precursors are formed by a polymerization process. In linear functional copolymers; the functionality distribution is often a linear function of DP distribution. Hyperbranched polymers, off-stoichiometric highly-branched functional polymers are formed by non-linear polymerization. In all these cases, the input information for crosslinking with respect to the polydispersities is preferably based on theoretical treatment of the precursor buildup [33].
- Effect of *functionality* (number of functional groups per precursor molecule): the gel point conversion is shifted to lower values when the functionality increases.

$$\langle f \rangle_1 = \sum_f f n_f, \quad \langle f \rangle_2 = \frac{\sum_f f^2 n_f}{\sum_f f n_f} \tag{9}$$

For functionality distributions, it is the second moment functionality average $\langle f \rangle_2$ which determines the increase in P_w and gel point conversion, whereas the w_g and the concentration of EANCs, v_e, have a mixed dependence on $\langle f \rangle_1$ and $\langle f \rangle_2$. M_w and w_g are also a function of molecular mass distribution. With increasing functionality the fraction of bonds wasted in cycles increases.

- *Polydispersity* in functional copolymers results in the existence of copolymer molecules having no or one functional group. During crosslinking such species contribute to the soluble fraction.
- *Off-stoichiometry* in alternating types of reactions makes the network looser; a critical molar ratio exists above which the network is not formed at all [71].
- *Reactivity differences*: the main practical purpose of the introduction of groups of the same type but of different reactivity is to control *time dependences* of various structural parameters. However, the gel point conversion also changes [48, 71].
- *Internal branching* in some precursors (e.g., functional copolymers, hyperbranched polymers, off-stoichiometric highly-branched polymers) gets gradually activated beyond the gel point and contributes to the concentration of EANCs [33]. The existence of internal branch points has no effect on gelation which is only a function of precursor functionality.

8.6
Properties of Polymer Networks

8.6.1
Structural Features

In the preceding section we discussed the structure of a polymer network from the point of view of the connectivity of precursor units through covalent bonds (network topology), as it develops during network formation. We saw that the structure can be influenced not only by chemical but also by physical factors (interactions). Even when these interactions are weak and the network has near to ideal structure, certain variations can exist. The junctions that are topological neighbors are also spatial neighbors, or the topological neighbors are far away from the spatial neighbors. Which of these situations arises is controlled (for instance, in the case of endlinking) by the length of telechelics and the functionality of the junctions; dilution is another variable. If the spatial neighbors are too far away, the bias is solved by cyclization or, possibly, by phase separation. This often happens in gels prepared in the presence of diluents; the c^*-theorem of DeGennes is valid, i.e. the EANCs are coils not interpenetrated by other chains (blobs). Gels – swollen weak networks – have been extensively studied by radiation scattering methods (static and dynamic light, small-angle X-ray, neutron scattering). It has been found that, in addition to transient thermal fluctuations, the gels show up static (quenched, frozen) fluctuations [1, 72–75] which are interpreted by constraints imposed on chain fluctuations by network connectivity. However, the static fluctuations depend on the type of network and the method of gel preparation and the correlation length exceeds the distance between crosslinks by an order of magnitude. Static fluctuations are due to any inhomogeneities differing in density. They have been explained by statistical fluctuations in the structure or by thermodynamic interactions. We are not going to discuss these experiments and theories in more depth and the reader is referred to the original literature. In this chapter, only the most typical properties of a polymer network – rubber elasticity and swelling will be discussed.

8.6.2
Equilibrium Rubber Elasticity

Reversibility of deformation is the most typical property of rubbery polymer networks. When a constant force is applied to a piece of crosslinked rubbery polymer, its deformation increases more or less rapidly and after some time constant deformation is reached. The statistical-mechanical theory of rubber elasticity is based on a freely-jointed chain or an equivalent chain which results in the distribution of chain end-to-end distances having a form of Gaussian function [6, 76]. The change in entropy of an EANC, ΔS_{def}, is related to the number of possible ways to realize a random walk between chain ends in the deformed and undeformed states ($\Delta S_{def} = k \Delta \ln W_{eted}$, where W_{eted} is the end-to-end distance distribution function). The treatments of Flory and James and Guth were generalized in the *junction-fluctuation theory* [6] (also *Flory–Erman theory*). The change in Helmholtz energy resulting from deformation of a network is given solely by the change in entropy, $\Delta F_{net} = -T\Delta S_{net}$

$$\Delta F_{net} = RTA\nu_e(\lambda_x^2 + \lambda_y^2 + \lambda_z^2 - 3) + B\nu_e \ln \lambda_x\lambda_y\lambda_z$$
$$\lambda_k = L_k/L_{k0}, \; k = x, y, z. \; \lambda_x\lambda_y\lambda_z = V/V_0 \qquad (10)$$

where ν_e is the concentration of elastically active network chains, L_k is the dimension of the sample along the k-axis, and $L_{k0} = L_0$ is the dimension in the isotropic reference state. The reference state is that at which the network chains are in a state of rest; it is usually assumed that the chains in homogeneous networks are in that state at network formation [76]. What we actually measure is the deformation relative to the isotropic (dry or swollen) state, Λ_k,

$$\Lambda_k = L_k/L_i; \; \lambda_k = (L_k/L_i)(L_i/L_d)(L_d/L_0) = \Lambda_k \varphi_2^{-1/3} \varphi_0^{1/3} \qquad (11)$$

where φ_2 is the volume fraction of polymer in a swollen sample and φ_0 is the volume fraction of a network polymer at network formation ($1-\varphi_0$ is the dilution). A and B are front factors which have the following values for phantom and affine networks

$$A_{ph} = (f_e - 2)/f_e, \; B_{ph} = 0 \quad A_{aff} = 1, \; B_{aff} = 2/f_e$$

where f_e is the average number of EANCs issuing from an elastically active crosslink [77]. It is equal to 3 just beyond the gel point and can reach, but need not, the value of chemical functionality at high conversions. Concerning the A and B factors, a *phantom network* is one in which there are no interactions between network chains, they can freely interpenetrate each other, and the fluctuations of junctions (crosslink) are determined only by network connectivity. In deformation, only the junctions mean positions are displaced affinely with the macroscopic deformation. An *affine network* is one in which the fluctuations of junctions are fully suppressed by transient interactions between network chains. The interactions are weakened by the presence of solvents and the affine behavior approaches the phantom behavior.

The stress–strain relations are obtained from ΔF_{net} accounting for the constraints between deformations in the x, y, and z directions by differentiation. Thus, the equilibrium stress $\sigma_{x,sw}$ (force per cross-section C_{sw}) for uniaxial extension or compression in the swollen state is given by

$$\frac{\partial \Delta F_{net}}{\partial L_x} = f_x; \; \sigma_{x,sw} = \frac{f_x}{C_{sw}} = RTA\nu_e\varphi_2^{1/3}\varphi_0^{2/3}(\Lambda_x - \Lambda_x^{-2}) \qquad (12)$$

According to Eq. (12), A varies between A_{ph} and A_{aff} and it is not easy to decide what value to take if ν_e is to be calculated from the stress–strain dependence. A

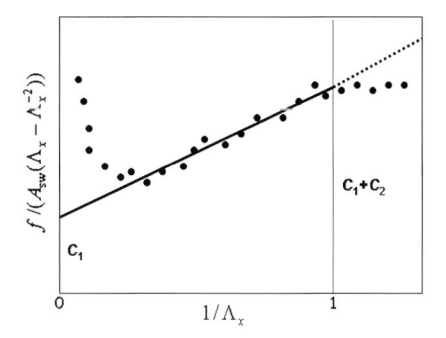

Fig. 8.15 Illustration of a Mooney–Rivlin plot of real crosslinked rubber.

is a function of a parameter κ which characterizes the strength of constraints on junction fluctuation, but it is difficult to obtain its value experimentally. The guideline is that the constraints diminish with increasing swelling and deformation in extension [6]. The dependence on tensile strain is similar to Money–Rivlin dependence (Fig. 8.15)

$$f/A_{sw} = (C_1 + C_2/\Lambda_x)(\Lambda_x - \Lambda_x^{-2}) \tag{13}$$

and thus, by linear extrapolation of the plot $\sigma_x/(\Lambda_x - \Lambda_x^{-2})$ against Λ_x^{-1} to $\Lambda_x \rightarrow A_{ph}$. value should be obtained and $C_1 = RTA_{ph}\nu_e\varphi_2^{1/3}\varphi_0^{2/3}(\Lambda_x - \Lambda_x^{-2})^\bullet$.

While the phantom network behavior is unambiguous, there are many instances when the junction fluctuations model does not reflect the effect of interchain interactions in full. The effect of transient interactions (entanglements) may also extend along network chains. This feature has been taken into account in the Erman and Monnerie theory and especially in the tube models [78]. Some entanglements may become entrapped between crosslinks; these are considered additive to the contribution by "chemical" (covalent) EANCs ($\nu_e = \nu_{e,chem} + \nu_{e,ent}T_{ent}$) [79–81]; $\nu_{e,ent}$ is calculated from the plateau modulus of uncrosslinked high-molecular-mass polymer and T_{ent} is the trapping factor derived from branching theory. However, trapped entanglements can slide along the chain during deformation (slip link model). In conclusion, much progress has been made in understanding and modeling various structural features of polymer networks, but a simple unambiguous way to determine crosslink density is questionable. For swollen networks prepared in solution, the problem of constraints by other chains is vanishingly small but other problems arise, such as the formation of cycles and inhomogeneities.

8.6.3
Dynamic Mechanical Properties

In the pregel region, the viscosity of the reacting system increases and the steady shear viscosity diverges at the gel point. Viscosity is an important processing property and some technologies are limited by too high or too small

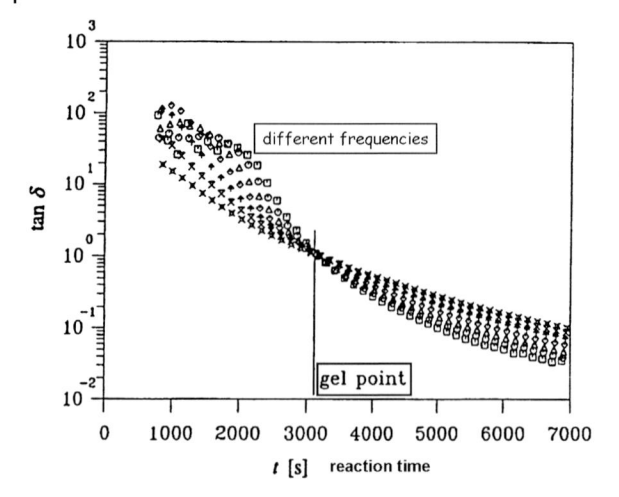

Fig. 8.16 Change in tanδ at different frequencies in dependence on reaction time for crosslinking of polybutadiene. Data by de Rosa [84], reproduced from Ref. [83] with permission.

(monomers in the initial stages) viscosity. The viscosity is a function of M_w and the dependence on temperature and M_w can be factored out

$$\eta = \eta_0(T)(M_w/M_{w0})^a \quad (14)$$

with the exponent $a=2.4–3.5$ [82]. The dynamic mechanical properties, expressed by the storage modulus, G', and the loss modulus, G'', or the tangent of the loss angle tanδ, also change. G', and G'' depend on the conversion of functional groups and on frequency. It is interesting that at the gel point tanδ becomes frequency independent (see Fig. 8.16 which shows tanδ dependence on reaction time). The gel time corresponds to the point where all curves cross, and tanδ becomes frequency independent (Winter–Chambon phenomenon) [83]

This method is very convenient for determination of gel point time or conversion, in contrast to the time-consuming determination of gel fraction by extraction, and extrapolation to zero gel fraction. The glass transition region widens with increasing network imperfectness (off-stoichiometric networks). The structure of dangling chains (their number and length can be controlled by manipulating with the functionality distribution [47] and, possibly, by off-stoichiometry. For more details on viscoelasticity see Ref. [85]).

8.6.4
Equilibrium Swelling

Uncrosslinked polymers dissolve in solvents without limits or take up a limited amount of solvent. The magnitude of solvent take-up is determined by the tendency of polymer molecules to mix with the solvent molecules. The Flory–Huggins mean-field model [86] is most widely used. The Gibbs energy of mixing for a polymer of infinite degree of polymerization is

$$\Delta G_{mix}/RT = \varphi_1 \ln \varphi_1 + g(\varphi_2)\varphi_1\varphi_2 \tag{15}$$

where φ_1 and φ_2 are volume fractions of solvent and polymer, respectively, and g is the interaction function [87], more frequently denoted as χ but introduced into the expression for chemical potential. If g and χ are independent of concentration, g=χ. From ΔG_{mix}, the chemical potentials of the solvent are obtained by differentiation of ΔG_{mix}; m is the ratio of the molar volumes of polymer and solvent molecules and we will assume here that $m \to \infty$

$$\Delta\mu_1/RT = \ln(1-\varphi_2) + \varphi_2 + \left(g - \varphi_1\frac{\partial g}{\partial \varphi_2}\right)\varphi_2^2 = \ln(1-\varphi_2) + \varphi_2 + \chi(\varphi_2)\varphi_2^2$$

$$\Delta\mu_2/mRT = -\varphi_2 + \left(g - \varphi_2\frac{\partial g}{\partial \varphi_1}\right)\varphi_1^2 = -\varphi_2 + \chi(\varphi_2)\varphi_1^2 \tag{16}$$

Where

$$g(\varphi_2) = g_0 + g_1\varphi_2 + g_2\varphi_2^2 + \ldots$$
$$\chi(\varphi_2) = \chi_0 + \chi_1\varphi_2 + \chi_2\varphi_2^2 + \ldots$$
$$\chi_0 = g_0;\ \chi_1 = g_0 - g_1;\ \chi_2 = 2(g_1 - g_2);\ \ldots$$

Often, the concentration dependence of g satisfies the simple closed form

$$g = a + \frac{b_s + b_h/T}{1 - c\varphi_2} = (b_s + b_h/T)(1 + c\varphi_2 + c^2\varphi_2^2 + \ldots) \tag{17}$$

where b_s, $b_h \cdot c$ are constants.

In a network, the mixing tendency of polymer segments and solvent molecules is opposed by chain elasticity and network connectivity. As a result of solvent uptake, EANCs are stretched and an increasing force is generated which eventually counterbalances the osmotic pressure. Assuming the additivity of the Gibbs energies due to mixing and isotropic deformation of the network $\Delta F_{net} \approx \Delta G_{net} = -T\Delta S_{net}$ and ΔG_{mix}, $\Delta G_{sw} = \Delta G_{mix} + \Delta G_{net}$ and from this the change in the chemical potential of the solvent is obtained

$$\Delta\mu_1/RT = \ln(1-\varphi_2) + \varphi_2 + \chi\varphi_2^2 + V_1\nu_e(A\varphi_2^{1/3}\varphi_0^{2/3} - B\varphi_2) = \ln a_1 \approx \ln(p_1/p_1^0) \tag{18}$$

The swollen network is in equilibrium with pure solvent or with solvent vapor of partial pressure p_1. By using the Gibbs–Duhem equation $N_1\frac{\partial\Delta\mu_1}{\partial N_1} + N_2\frac{\partial\Delta\mu_2}{\partial N_2} = 0$ the chemical potential of polymer per equivalent segment $m = V_2/V_1$ is obtained

$$\Delta\mu_2/mRT = -\varphi_2 + \chi\varphi_1^2 + V_1 v_e[A\varphi_0^{2/3}(\varphi_2^{-2/3}/2 + \varphi_2^{1/3} - 3/2) + B(\ln\varphi_2 + \varphi_1)] \quad (19)$$

In equilibrium, the chemical potentials of the components in either phase are equal. According to Gibbs phase rule, three phases can coexist at constant temperature: $\mu_1' = \mu_1'' = \mu_1'''$ and $\mu_2' = \mu_2'' = \mu_2'''$. Analysis of Eqs. (18) and (19) shows that normally two phases – liquid solvent or its vapor and swollen polymer gel – are in equilibrium. Three phases can coexist only in a certain range parameters of the swelling equation [88, 89] (see below).

The swelling equilibria are shown schematically in Fig. 8.17 as interdependence of the volume fraction of polymer and temperature for a system having upper critical solution temperature (UCST), compared with an uncrosslinked (linear) polymer of increasing molecular mass.

The lower critical solution temperature (LCST) systems (increasing degree of swelling with decreasing temperature) correspond to Fig. 8.17 if it is rotated by 180° along the horizontal axis.

The degree of swelling is determined by
1. concentration of elastically active network chains, v_e,
2. polymer solvent interaction parameter χ (or g), concentration and temperature dependences
3. memory parameter φ_0 characterizing dilution at network formation and the state of coiling of network chains in the dry network
4. molar volume of the solvent, V_1
5. functionality of the crosslink (average number of bonds with infinite continuation per elastically active crosslink, f_e,
6. ordering of the swelling liquid (liquid-crystalline solvents), presence of macromolecular substances
7. strains (coating films) [90].

With increasing concentration of EANCs the degree of swelling decreases, unless the second term in Eq. (19) becomes negative in the case of small φ_0, which

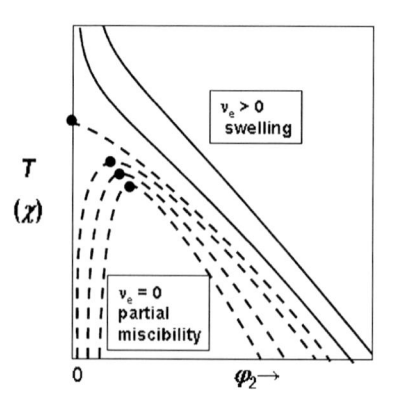

Fig. 8.17 Binodals for UCST solutions of a polymer of increasing molecular mass (dashed) and degrees of swelling as φ_2 of network of two different degrees of crosslinking levels (full).

is hardly possible even for $B>0$, because such a situation will be preceded by phase separation. The goodness of solvents increases with decreasing χ. The *temperature dependence of χ* is positive in UCST systems and negative for LCST systems; for "organic" systems usually (but not always) UCST is characteristic, aqueous systems usually show up LCST. Very important and sometimes neglected is the *concentration dependence of χ*. Concentration dependence is more frequent than concentration independence. Usually, it is obtained experimentally and expressed as a power series of φ_2 for χ or g. Concentration dependences follow from more refined theories of polymer solutions such as quasi-chemical equilibrium theory [91]. For determination of this dependence, φ_2 is usually controlled by a change in v_e, so that the concentration dependence may also reflect an increase in the concentration of crosslinks. A certain range of concentration dependence of χ gives rise to so-called "off-zero critical concentration" [89, 92] and volume phase transition [88].

Very important is the effect of *dilution at network formation* characterized by φ_0. Dilution diminishes the interchain interactions and number of entanglements. Attainment of a certain value of φ_0 is necessary for the onset of liquid–liquid and liquid–gel phase separation during network formation and facilitates volume phase transition. Increasing the *molar volume* of the swelling liquid, V_1, makes the second term in Eq. (19) more positive and φ_2 is higher.

Ionized or *polyelectrolyte networks* carry covalently attached ionized or ionizable groups (e.g., –COOH, –SO$_3$H, –N(Alk)$_2$, –N(Alk)$_3$OH). The degree of swelling of such networks can be very high. Because of the condition of electroneutrality, the charge of the ion fixed to the network is counterbalanced by a mobile ion of opposite sign (counter ion) (e.g., –COO$^-$X$^+$, or (Alk)$_3$N$^+$Y$^-$). The ions X$^+$ or Y$^-$ can be exchanged for another ion. Such swollen networks are used as ion exchangers (cation exchanger and anion exchanger) for water treatment and other applications. Hydrogels carrying ionizable groups often exhibit volume phase transition (see below). The reason for a high swelling capacity of polyelectrolyte networks is hydration of ions, especially the counterions. In a theoretical description of swelling, not only the mixing of network chain segments with solvent molecules and the elastic response of the network must be taken into account but also the effect of charges. The effect of charges is twofold: the hydration of counterions, the concentration of which is controlled by Donnan equilibrium, and repulsive electrostatic interactions of fixed charges that contribute to chain extension. There exist several models to describe the equilibrium swelling of polyelectrolyte networks. Usually, the additivity of Gibbs energies is assumed [93–95]

$$\Delta G_{sw} = \Delta G_{mix} + \Delta G_{net} + \Delta G_{ion} + \Delta G_{elst} \qquad (20)$$

although the additivity assumption has been questioned [96]. ΔG_{mix} should include all non-electrostatic interaction, ΔG_{net} should respect the chain extension limit of chains due to the high degrees of swelling encountered for polyelectrolyte gels (Langevin function or its expansion instead of Gaussian function). Of two last terms, $\Delta G_{\text{ion}} \propto \sum_j(c_j^{\text{gel}} - c_j^{\text{ext}})$ (the summation extends over all mobile ions) is more important than ΔG_{elst} (electrostatic repulsion of fixed charges). In equilibrium with external salt solution, the gel phase may also contain co-ions.

8.6.5
Phase Separation

Phase separation in polymer networks can take place *after network formation* or can start *during network formation*; in both cases, morphologically, as *macrosyneresis* (two bulk phases) or *microsyneresis* (phases dispersed or interdispersed with one another). Thermodynamical instability of the system is the reason for phase separation.

In a swollen extracted network, liquid–gel phase separation takes place when the change in external conditions (e.g., temperature) increases the solvent activity in the gel to unity. In equilibrium, the volume fraction of solvent separated, φ_s, is given by the difference between the original state characterized by $(\varphi_2)_0$ and the new state characterized by φ_2.

$$\varphi_s = \frac{V_0 - V}{V_0 - V_P} = \frac{1 - (\varphi_2)_0/\varphi_2}{1 - (\varphi_2)_0} \tag{21}$$

The driving force for phase separation is the increase in the interaction parameter from $(\chi)_0$ to χ. The volume fractions are calculated from Eq. (19). Experimentally, one often observes the formation of turbidity (microsyneresis). This is, however, not the equilibrium state. In time, the microseparated liquid (under pressure by network deformation) is expelled outside and a bulk liquid phase and clear gels are formed (Fig. 8.18).

The speed of transition to the macrophase separated state depends on crosslink density; the speed increases with increasing v_e and gels with high v_e may bypass the microsyneretic (turbid) state. If the quench in temperature is fast and deep, phase separation can also occur by spinodal decomposition and the morphology then differs from that shown in Fig. 8.18.

If diluents or additive are present during network formation, the system can phase separate in one of the ways shown in Fig. 8.19. Again, thermodynamic instability of the system is the reason for phase separation. The situation differs from phase separation in a finished gel in several respects: the instantaneously formed morphologies are partially fixed by simultaneous crosslinking, the sol component exists and participates in phase separation, the diluent, monomers and sol components can diffuse between the crosslinked and uncrosslinked phases. A bulk phase of the phase separated diluent (Fig. 8.19) can be formed if phase separation occurs well beyond the gel point when the network connectivity and level of EANCs do not allow formation of an interdispersed phase. The two other cases are typical of phase separation starting close before or close after the gel point. Whether the morphology resembles more the droplets interdispersed in crosslinked polymers or the uncrosslinked phase interpenetrates

8.6 *Properties of Polymer Networks* | 1719

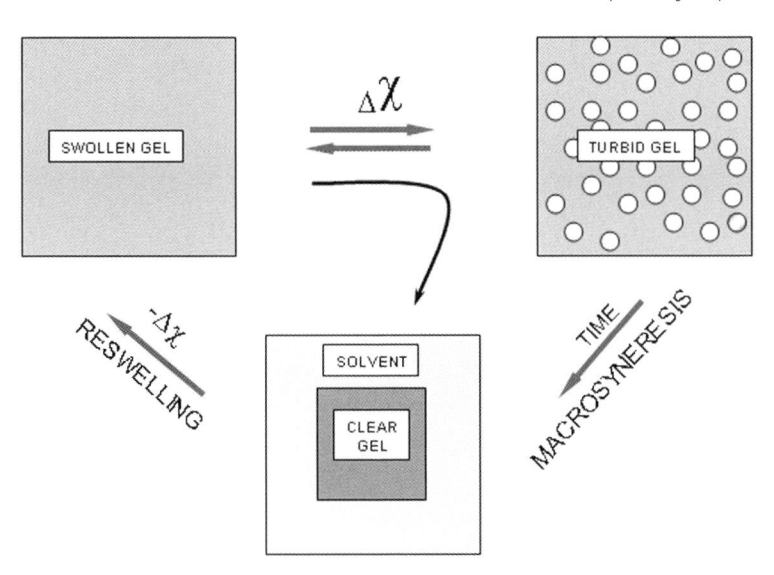

Fig. 8.18 Microsyneresis and macrosyneresis induced by a change in $\chi(T)$ in a swollen gel.

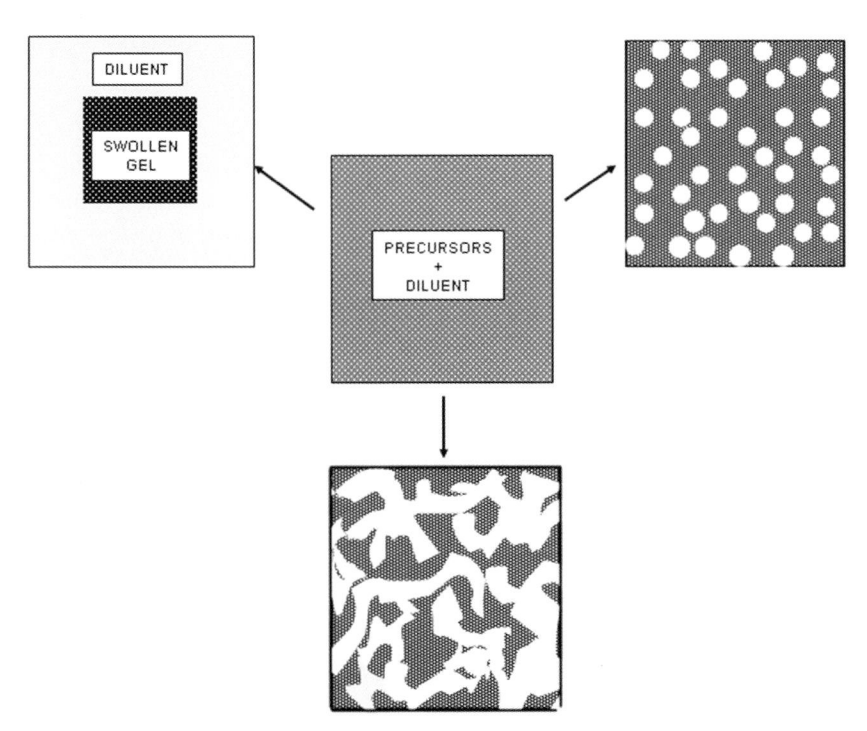

Fig. 8.19 Phase separation during network formation.

the crosslinked phase depends on several factors such as concentration of the diluent (additive), its molecular size, thermodynamic interaction, and interfacial tension (important factor). The sol also plays an important role. Considering the crosslinking system as purely binary, the condition for incipient phase separation was formulated a long time ago [97] (see also Ref. [76]): this was simply obtained by substituting $\varphi_0 = \varphi_2$ into Eq. (18) and solving for φ_0. This gives the maximum dilution for a given v_e which is changing (possibly other parameters, too). In other word, the system cannot tolerate more solvent than is its maximum swelling. To make the prediction more quantitative is extremely difficult, not only because of the effect of so many factors but especially because of fixation of non-equilibrium states during crosslinking. However, even semiquantitative considerations are helpful [2]. The thermodynamic treatment of "sweating out" of excess solvent during coating film formation [98] is more successful because it occurs well beyond the gel point by the bulk phase separation mechanism when the sol fraction is very small.

The phenomenon of phase separation during network formation is widely utilized in technology: in the preparation of porous sorbents, after the phase separated diluent or additive is removed by drying, possibly preceded by extraction; toughening of thermosets, when the finely phase separated polymer domains capture the spreading crack; formation of LC displays by phase separation of a nematic compound; polymer membranes and drug delivery hydrogel devices where the communicating pores accelerate diffusion in and out.

8.6.6
Volume Phase Transition

So far, we have been considering phase equilibria in binary systems and two coexisting phases, one of which was liquid solvent or solvent vapor (Eq. (18)). In the process of phase separation during network formation, one should consider the systems as pseudobinary (solvent + distribution of branched polymers before the gel point), pseudoternary beyond the gel point (solvent + distribution of sol molecules + infinite network polymer), or ternary containing network polymer and two additives, etc. However, theoretical analysis has shown that *two network phases* can coexist in equilibrium with liquid solvent at a certain temperature for a certain set of network parameters. For such systems, the calculated dependence of the chemical potential on φ_2 exhibited two maxima and the condition for phase equilibrium, equality of chemical potentials of a component in each phase, gave a real solution in the range $\varphi_2 \langle 0, 1 \rangle$ [88]. Since, for simple unionized networks, it was difficult to achieve experimentally the proper combination of parameters (sufficiently high at sufficiently high dilution), the experimental discovery of these transitions, characterized by a *jump in the degree of swelling*, was made 10 years later on hydrogels carrying *ionized groups* [99]. Also, it was found that un-ionized systems with a complex concentration dependence of the interaction parameter (systems exhibiting "off-zero critical concentration") exhibit this transition [89, 92]. The best known of these systems is poly(N-isopropylacry-

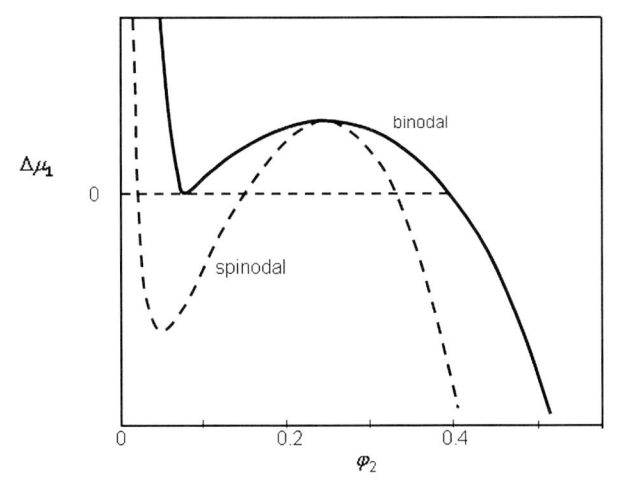

Fig. 8.20 Phase diagram for a gel exhibiting volume phase transition (corresponding to lightly crosslinked PNIPA in water). The three-phase equilibrium is shown by a dashed horizontal line (see also Ref. [89]).

lamide)-water [100] studied in hundreds of papers. This kind of transition is shown schematically in Fig. 8.20.

This volume phase transition can be induced by a number of stimuli, such as change in temperature, degree of ionization (pH), addition of co-ions, change in solvent composition, irradiation, and application of electric field. The phenomenon of phase transition is widely utilized in controlled drug delivery. For instance, a gel conjugate when administered *per os* must pass through a hostile environment before it reaches its target characterized by a certain value of pH. Then, a sudden expansion in volume exposes the gel interior to agents that split off the drug and rapid delivery of the drug is guaranteed. Other applications include concentration of dilute solutions of higher-molecular-mass, for instance, in the technology of pharmaceuticals; for some superabsorbent gels, and various control devices where a fast collapse or expansion of gel-like materials is essential. There exist a number of monographs and reviews on this subject, such as Refs. [101, 102].

Gels exhibiting swelling transition can also be obtained by incorporating into a disordered hydrogel structure a specific precursor element showing an order–disorder transition, usually a protein motif. The connective structure of the gel amplifies the response of many of the elements into a macroscopic response. The stimuli inducing the transition can react very specifically to certain metabolites or antibodies. Coiled-coil association and dissociation is an example of current development [34]. Also, interactions of the disordered hydrophilic matrix can transform the external stimulus into a change in swelling pressure which acts on the incorporated ordered motif and forces it to unfold [103]. Unfolding can result in a swelling transition of collapse type. In application, transition speed is very important and it is controlled by diffusion. The length of the diffusion path is essential, and miniaturization of constructs is important.

8.7
More Complex Polymer Networks

For many systems of practical interest inhomogeneity is a nuisance, impairing their functions, such as optical clarity, mechanical response and its sharpness, or separation selectivity. Other inhomogeneities are introduced intentionally in the form of (nano)fillers, (nano)clusters, crystalline domains, porosity, etc. In this section, we will briefly discuss chemical reasons for the formation of inhomogeneities.

An important case of inhomogeneous network formation is *chain crosslinking (co)polymerization* induced by cyclization in the early stages of polymerization. The probability that a growing macroradical meets a pendant double bond located on another chain is, at the beginning of polymerization, much lower than that of meeting the double bond pendant on the same chain. This is because, at the beginning, there are very few chains in the system and hence internally crosslinked molecules are formed. Only the pendant C=C double bonds at the microgel-like molecules' periphery can take part in the network buildup. Favorable conditions for that are: higher concentration of the polyvinyl monomer (crosslinker), not too long bridges connecting the double bonds, not too short primary chains. There exist several pieces of experimental evidence of this mechanism; the shift of the gel point conversion by a factor up to 10^2 is one of the most striking. Macrogelation then resembles more a coagulation process in which the peripheral pendant double bonds copolymerize with the monomers. When the conversion moves from the gel point to high values, the signs of inhomogeneity get weaker as the space between the particles gets filled with polymerizing monomer [104, 105]. The special features of the crosslinking chain (co)polymerization affect reaction-induced phase separation and the formation of porous networks [106, 107].

Another type of inhomogeneity, *topological cluster*, appears in multicomponent crosslinking systems if the constituent units exhibit some different property, for instance they are declared *hard* and *soft*, In the network structure, they are less or more mobile. Polyurethanes are often formed from a hard (aromatic) diisocyanate, a (hard) low-molecular-mass triol, and a soft macrodiol (Fig. 8.21). They remind one of segmented (linear) polyurethanes, but there the hard units are sequenced on linear chains and form crystalline or amorphous hard domains by physical forces. The domains can melt at higher temperatures; it is a kind of thermoplastic rubber.

To consider the network as composed of trifunctional junctions is incorrect. It should be rather viewed as being composed of polyfunctional cluster crosslinks instead of point-like crosslinks (Fig. 8.21), Their functionality and molecular mass distribution can be calculated using the branching theories [108, 109]. Un-

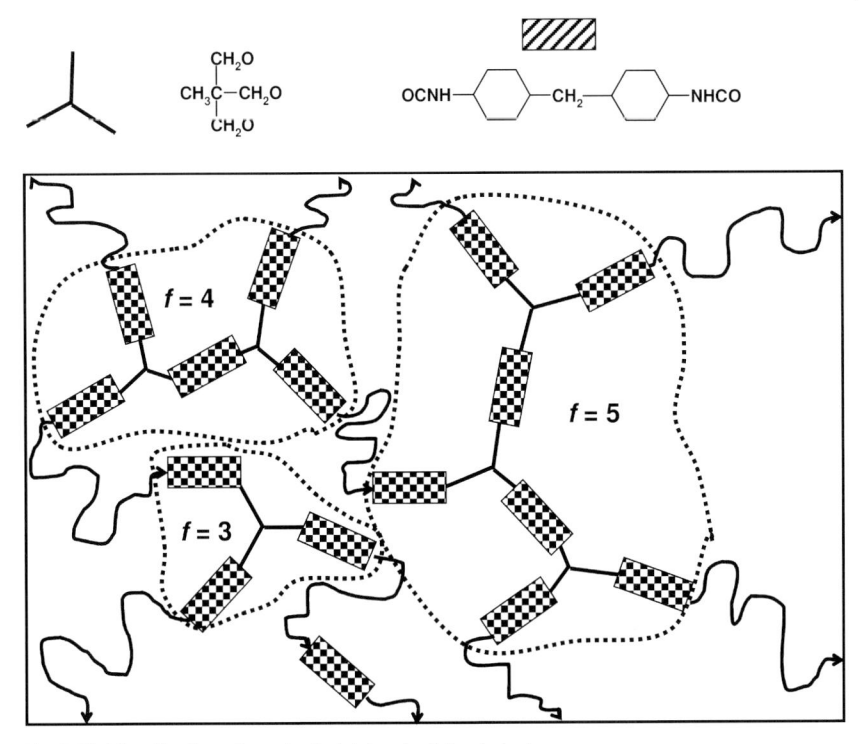

Fig. 8.21 Visualization of topological (chemical) hard clusters in a polyurethane network composed of the constituent shown above; the wavy curve is soft macrodiol. The dotted curve delimits a cluster crosslink.

der certain conditions, the size and functionality of cluster crosslinks can grow until they form a continuous hard structure: the percolation threshold of the hard structure is reached. The size distribution of topological clusters depends not only on the composition and conversion but also on precursor functionality, group reactivities, and network formation history. There is experimental evidence (SAXS) for the existence of topological clusters.

The clusters discussed so far have been three-dimensional because they are formed from at least one polyfunctional component (trimethylolpropane as shown in Fig. 8.21). However, in some networks prepared from end-linked flexible precursor chains, the end-linking process can give rise to the formation of relatively immobilized chains of another kind connecting the soft chains. Soft-chain diacrylates or polyether diamines (Jeffamine® D) crosslinked with (aromatic) diepoxides [110, 111] are examples of such systems. These chains, formed as a result of the end-linking reaction, can be viewed also as topological (but linear) clusters (Fig. 8.22), but the equilibrium elasticity is expected to be quite different from that of the systems with spherical clusters. During deformation, due to severe obstruction of conformational changes, all segments in

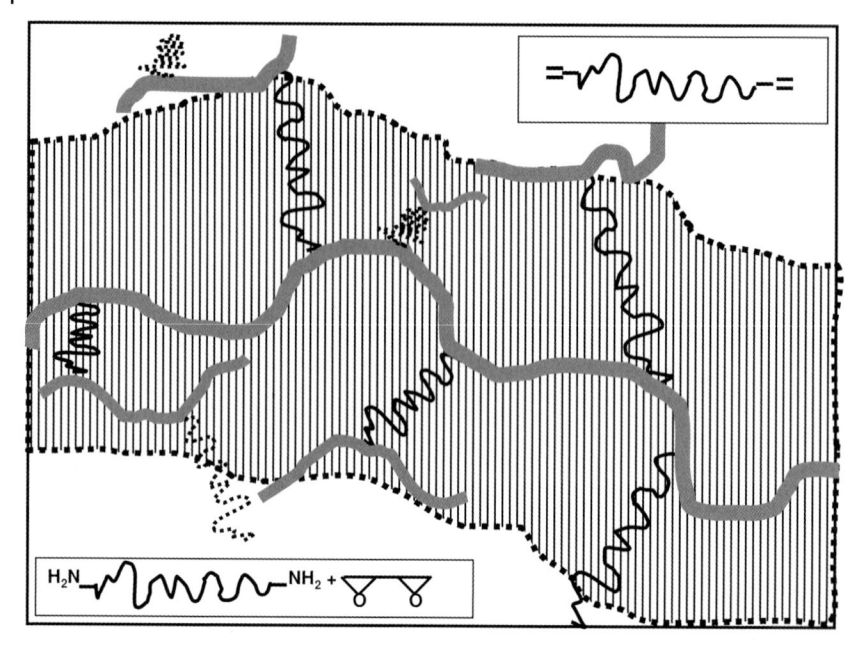

Fig. 8.22 Special networks from flexible chains end-linked by a linear polymerization or step addition process showing certain order.

these chains are expected to be displaced affinely with macroscopic deformation. When the flexible chains have a narrow degree-of-polymerization distribution, a certain degree of order is seen in these networks, for instance by SAXS, with a correlation distance corresponding to the mean end-to-end distance between the flexible chain ends.

8.7.1
Interpenetrating Polymer Networks (IPN)

Historically, interpenetrating polymer networks (IPN) represent one of the oldest rationally designed composite networks, in which the two kinds of networks interpenetrate each other on the molecular level or as separated phases. Mixing of different polymer chains, even incompatible ones, through interdispersion by crosslinking has offered a means to prepare materials of interesting mechanical, sorption, and permeation properties [112–116]. The two kinds of IPNs are shown in Fig. 8.23.

The first picture shows a truly interpenetrating network on the molecular level, the second shows a phase-separated IPN with dual phase continuity. The IPNs are formed either *simultaneously* or *sequentially*. The simultaneous method employs two chemistries (for instance, free radical polymerization and step polyaddition) and all monomers are mixed at once. In the case of the *simulta-*

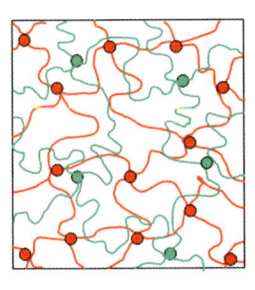

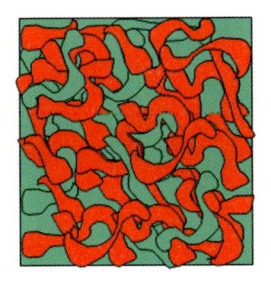

Fig. 8.23 Homogenous and phase separated interpenetrating networks.

homogeneous IPN structure   phase-separated IPN structure

neous method, it happens rarely that the two networks are formed with approximately the same speed and surpass the gel point at approximately the same time. Usually, one network is formed first (e.g., the polyurethane network at lower temperature) and the other follows when the polymerization conditions change (e.g., a vinyl-divinyl polymerization when the temperature is increased). It is important that the volume remains constant, disregarding the polymerization shrinkage. In the *sequential* method, one network is formed first and then the monomers of the second network are introduced and polymerized. The introduction of monomers for the second network is usually carried out by swelling of the first network when the chains of the first network are stretched. This confinement affects the elastic and swelling properties of the first network. Another IPN type can be prepared sequentially from a nano- or microporous polymer network when the pores are filled by the second monomer and polymerized. If the monomer of the second network does not swell the first network, the volume of the IPN does not change.

Most IPNs are phase separated. Phase separation can occur by a nucleation and growth mechanism but often by spinodal decomposition. The morphology of phase separated IPNs depends not only on the compatibility of network 1–network 2 chains, but also on the crosslink density of network 1 and on the polymerization regime. Generally, the morphology is dependent on the polymerization kinetics, as in the case of reaction induced phase separation in the course of network formation. In two-phase IPNs, there are indications of the existence of an interphase. The two glass transition temperatures of the components are moved inwards, and the transition is less sharp. The IPNs are, to a certain extent, similar to block copolymers, where the chemical connections between two blocks are substituted by interpenetration of network chains or interlocking of the two nanosized phases. However, they are very different in some respects. For instance, IPNs cannot undergo order–disorder transitions.

Some of the IPNs are homogeneous when the chemical nature of the two networks is similar. The network composition can even be identical, and still an IPN can be made. For instance, the network made from a monovinyl and divinyl monomer is swollen in the same monomer mixture and polymerized. The resulting network swells less, although the crosslink density is the same, because the network chain of the first network has been pre-stretched by addition of the second monomer portion ($\varphi_0 > 1$ for the first network!!). The importance of thermodynamic factors was documented by the formation of a porous network from components of almost identical composition. First a semi-interpenetrating network was prepared by dissolving polystyrene in styrene and divinylbenzene. After polymerization, above a certain limit of molecular mass and concentration of polystyrene and crosslinker concentration, phase separation took place in the early stage of polymerization, mainly due to the high molar volume of polystyrene. When polystyrene was extracted, the pores remained fixed in the network [117].

The variation of IPN structure by changing the composition and preparation regime is widely used to control IPN properties. The changes in structure are mainly based on empirical correlations. Theoretical quantification is difficult, especially for phase-separated IPNs, due to the dependence of phase separation on formation history. It is quite certain that phase separation is driven by attainment of thermodynamic instability, and models have been devised, based on thermodynamics, to treat structure development during IPN formation. The problem was most deeply analyzed by Binder and Frisch [114] and Schulz and Frisch [115]. For homogeneous IPNs, both networks contribute by their EANCs and there is an extra contribution by entanglements between EANCs of network 1 and network 2 [114] because the lower amount of network swelling constrains the network swelling more in its expansion [116]. However, in the pure state network 1 and network 2 have their trapped entanglements which in IPN are affected by the presence of the other network. A simple additivity rule does not apply. Moreover, in many homogeneous IPNs, grafting exists and then the elastic behavior is similar to that of the "two-networks hypothesis" [118], sum of EANCs, however, in different states of coiling (different φ_0). For phase separating networks, Schulz and Frisch [115] are aware of the very limited possibility of attaining equilibrium and treat the case of "chemical quenching" when the rate of chemical reaction is much greater than the structure relaxation rate. The microphase separation occurs by spinodal decomposition when the system enters the thermodynamically absolutely unstable region. Moreover, in IPNs phase separation of the liquid–liquid type can be coupled with the volume phase transition (see Section 8.6.4). IPNs gels can exist in four states: (i) homogeneous and collapsed, (ii) homogenous and swollen, (iii) inhomogeneous and collapsed, and (iv) inhomogeneous swollen.

During swelling of a two-phase network, expansion of one network is constrained by the other. In the ideal state, when the phase states can be characterized solely by the memory factors φ_{01} and φ_{02}, the networks are not free and one network affects the other by a pressure equal in magnitude but opposite in sign, $p_1 V_1 = -p_2 V_2$. Also, the deformation ratios of network chains due to spatial constraints are the same.

$$(\Delta\mu_1)_{mix1} + (\Delta\mu_1)_{net1} + (\Delta\mu_1)_{str1} = 0$$
$$(\Delta\mu_1)_{mix2} + (\Delta\mu_1)_{net2} + (\Delta\mu_1)_{str2} = 0$$
$$(\Delta\mu_1)_{str1} = (\Delta\mu_1)_{str2} \; ; \; \Lambda_1 = \Lambda_2 \tag{22}$$

The most important property in the application of IPNs is the fact that mostly incompatible polymers can be kept interdispersed on a more or less fine level (engineering materials, coatings). Also, IPN hydrogels can show up the two transitions which can be utilized in control devices and in drug delivery.

The possibilities for the control of material properties by physical interchain interactions are very wide. For instance, covalent attachment of special side groups to network chains such as discotics [119] exhibiting liquid-crystalline order or polyhedral organic silsesquioxanes (POSS) [120] improves the thermomechanical properties of conventional networks.

8.8
Concluding Remarks

Polymer networks have been considered here mainly from the point of view of formation and some basic physical properties. The processing and application aspects are mentioned only marginally. The reader can obtain deeper information in monographs such as Refs. [2, 82, 121, 122]. A question frequently asked by colleagues, referees, or managers is: where will polymer networks go? The development is closely tied to development in other areas of polymer science and applications. Continuing development is expected in traditional areas of crosslinked elastomers, thermosets, and composites with a crosslinked matrix by improving and further extending the chemistry and processing. Fast growth is expected in the area of biorelated gels and gel-based devices. Also, the functional character of various constructs and devices will require a controlled formation of local variations in composition and introduction of gradients of composition and properties.

References

1 *Physical Properties of Polymeric Gels*, J. P. Cohen-Addad (Ed.), J. Wiley, New York 1996.

2 J.-P. Pascault, H. Sautereau, J. Verdu, R. J. J. Williams, *Thermosetting Polymers*, Marcel Dekker, New York 2002.

3 *Polymer Networks – Principles of Their Formation, Structure and Properties*, R. F. T. Stepto (Ed.), Blackie Academics and Professional, London 1998.

4 K. Dušek, M. Dušková-Smrčková, *Progr. Polym. Sci.* **2000**, 25, 1215.

5 *Encyclopedia of Polymer Science and Technology*, J. Wiley, New York **2005**, respective entries.

6 B. Erman, J. E. Mark, *Structure and Properties of Rubberlike Networks*, Oxford University Press, Oxford **1997**.

7 G. Odian, *Principles of Polymerization*, J. Wiley, New York **1991**.

8 D. Stoye, W. Freitag, *Resins for Coatings*, Hanser Publishers, Munich **1996**.

9 *New Methods of Polymer Synthesis*, W. J. Mijs (Ed.), Springer, Berlin **1992**

10 C. A. May, *Epoxy Resins. Chemistry and Technology*, Marcel Dekker, New York **1988**.

11 *Epoxy Resins and Composites*, K. Dušek (Ed.), Part I, Adv. Polym. Sci., Vol. 72, 1985; Part II, Adv. Polym. Sci., Vol. 75, 1986; Part III, Adv. Polym. Sci., Vol. 78, 1986, Part IV, Adv. Polym. Sci., vol. 80, 1986, Springer, Berlin.

12 C. Hepburn, *Polyurethane Elastomers*, Elsevier Applied Sciences, Amsterdam **1992**.

13 Z. Wirpsza, *Polyurethanes*, Ellis Horwood, New York **1993**.
14 A. Oseiowusu, G.C. Martin, J.T. Gotro, *Polym. Eng. Sci.* **1991**, *31*, 1604.
15 *Sol-Gel Science and Technology*, Vols. 1–4, S. Sakka (Ed.), Kluwer Academic Publishers, Norwell **2003**.
16 A. Knop. W. Scheib, *Chemistry and Application of Phenolic Resins*, Springer, Berlin **1979**.
17 C.E. Hoyle, T.Y. Lee, T. Roper, *J. Polym. Sci., Part A., Polym. Chem.* **2004**, *42*, 5301.
18 K. Dušek, *Adv. Polym. Sci.* **1986**, *78*, 1.
19 K. Dušek, M. Ilavsky, S. Lunak, *J. Polym. Sci., Part C*, **1975**, *53*, 29.
20 S. Lunak, K. Dušek, *J. Polym. Sci., Part C*, **1975**, *53*, 45.
21 A. Vazquez, L. Matejka, P. Spacek, K. Dušek, *J. Polym. Sci.: Part A, Polym. Chem.*, **1990**, *28*, 2305.
22 K. Dušek, L. Matejka, *ACS Adv. Chem. Ser.* **1984**, *208*, 27.
23 L. Matejka, J. Lovy, S. Pokorny, K. Bouchal, K. Dušek, *J. Polym. Sci., Polym. Chem.***1983**, *21*, 2873.
24 T. Nishikubo, A. Kameyama, K. Kashiwagi, *Polym. J.* **1994**, *26*, 864.
25 K. Tsuchida, J.P. Bell, *J. Appl. Polym. Sci.* **2001**, *79*, 1359.
26 J.H. Saunders, K.C. Frisch, *Polyurethanes – Chemistry and Technology*, Interscience, New York **1974**.
27 *Advances in Urethane Science and Technology*, K.C. Frisch, D. Klempner (Eds.), Technomic Press, Lancaster; since 2005 published by RAPRA.
28 K. Dušek, M. Spirkova, I. Havlicek, *Macromolecules* **1990**, *23*, 1774.
29 K. Dušek, M. Dušková-Smrčková, L.A. Lewin, J. Huybrechts, R.J. Barsotti, *Surf. Coat. Int., Coat. Trans.*, **2006**, *89*, 123.
30 D. Joel, A. Hanser, *Angew. Macromol. Chem.* **1994**, *217*, 191.
31 K. Dušek, Networks from Telechelic Polymers: Theory and Application to Polyurethanes, in *Telechelic Polymers: Synthesis and Applications*, E.J. Goethals (Ed.), CRC Press, Boca Raton, **1989**, pp. 289–360.
32 K. Dušek, *Trends Polym. Sci.* **1997**, *5*, 268.
33 K. Dušek, M. Dušková-Smrčková, *Macromolecules* **2003**, *36*, 2915.
34 C. Xu, V. Breedvelds, J. Kopecek, *Biomacromolecules* **2005**, *6*, 1739.
35 G. Wenz, B.-H. Han, A. Müller, *Chem. Rev.* **2006**, *106*, 782.
36 G. Fleury, G. Schlatter, C. Brochon, G. Hadziioannou, *Polymer* **2005**, *46*, 8494
37 E. Rabinowitch, *Trans Faraday Soc.* **1937**, *33*, 1225.
38 C.D. Han, D.-S. Lee, *J. Appl. Polym. Sci.* **1987**, *33*, 2859.
39 M. Adam, *Makromol. Chem., Macromol. Symp.* **1991**, *45*, 1.
40 K. Dušek, *Polym. Gels Networks* **1996**, *4*, 383.
41 K. Dušek, I. Havlicek, *Prog. Org. Coat.* **1993**, *22*, 145.
42 I. Mita, K. Horie, *J. Macromol. Sci., Rev. Macromol. Chem. Phys.***1987**, *C27*, 91.
43 K. Dušek, *Rec. Trav. Chim. Pays-Bas* **1991**, *110*, 507.
44 M. Gordon, *Proc. R. Soc. London, Ser. A* **1962**, *268*, 240; M. Gordon, G.N. Malcolm; *Proc. R. Soc. London, Ser. A* **1966**, *295*, 29.
45 M. Gordon, G.R. Dobson, *J. Chem. Phys.* **1965**, *43*, 705.
46 K. Dušek, *Macromolecules* **1984**, *17*, 716.
47 K. Dušek, K.M. Dušková-Smrčková, J.J. Fedderly, G.F. Lee, J.D. Lee, B. Hartmann, *Macromol. Chem. Phys.* **2002**, *203*, 1936–1948.
48 K. Dušek, M. Dušková-Smrčková, J. Huybrechts, *J. Nanostructured Polym. Nanocomposites* **2005**, *1*, 45.
49 K. Dušek, B.J.R. Scholtens, G.P.J.M. Tiemersma-Thoone, *Polym. Bull.* **1987**, *17*, 239.
50 C.W. Macosko, D.R. Miller, *Macromolecules* **1976**, *9*, 199; D.R. Miller, C.W. Macosko, *Macromolecules* **1976**, *9*, 206.
51 D.R. Miller, C.W. Macosko, *J. Polym. Sci., Part B: Polym. Phys.* **1987**, *25*, 2441; D.R. Miller, C.W. Macosko, *J. Polym. Sci., Part B: Polym. Phys.* **1988**, *26*, 1.
52 K. Dušek, *Polym. Bull.* **1979**, *1*, 523.
53 S. Kuchanov, H. Slot, A. Stroeks, *Prog. Polym. Sci.* **2004**, *29*, 563.
54 J. Mikes, K. Dušek, *Macromolecules* **1982**, *15*, 93.
55 J. Somvarsky, K. Dušek, *Polym. Bull.* **1994**, *33*, 369; J. Somvarsky, K. Dušek, *Polym. Bull.* **1994**, *33*, 377.

56 K. Dušek, J. Somvarsky, *Polym. Bull.* **1985**, *13*, 313; K. Dušek, *Polym. Bull.* **1985**, *13*, 321.
57 K. Dušek, J. Somvarsky, *Polym. Int.* **1997**, *44*, 225.
58 H. Galina, J. B. Lechowicz, *Adv. Polym. Sci.* **1998**, *137*, 135.
59 J. Somvarsky, K. Dušek, M. Smrčková, *Comput. Theor. Polym. Sci.* **1998**, *8*, 201.
60 D. Stauffer, A. Coniglio, M. Adam, *Adv. Polym. Sci.* **1980**, *44*, 103.
61 H. M. J. Boots, J. G. Kloosterboer, M. M. van de Hei, R. B. Pandey, *Br. Polym. J.* **1985**, *17*, 219.
62 Y. K. Leung, B. E. Eichinger, *J. Chem. Phys.* **1984**, *80*, 3877; Y. K. Leung, B. E. Eichinger, *J. Chem. Phys.* **1984**, *80*, 3885.
63 E. R. Duering, K. Kremer, G. S. Grest, *J. Chem. Phys.* **1994**, *101*, 8169; G. Everaers, K. Kremer, G. S. Grest, *Macromol. Symp.* **1995**, *93*, 53; G. Everaers, K. Kremer, *Macromolecules* **1995**, *28*, 7291; G. Everaers, K. Kremer, *Phys. Rev. E* **1996**, *53*, R37.
64 R. F. T. Stepto, D. J. R. Taylor, in *Cyclic Molecules*, J. A. Semlyen (Ed.), Kluwer Academic Publishers, Dordrecht, **2000**.
65 R. F. T. Stepto, in *Polymer Networks – Principles of Their Formation, Structure and Properties*, R. F. T. Stepto (Ed.), Blackie Academics and Professional, London **1998**, Ch. 2.
66 J. I. Cail, R. F. T. Stepto, D. J. R. Taylor, *Macromol. Symp.* **2001**, *171*, 19.
67 R. F. T. Stepto, J. I. Cail, D. J. R. Taylor, I. M. Ward, R. A. Jones, *Macromol. Symp.* **2003**, *195*, 1.
68 M. Lang, D. Goeritz, S. Kreitmeier, *Macromolecules* **2005**, *38*, 2515.
69 K. Dušek, M. Gordon, S. B. Ross-Murphy, *Macromolecules* **1978**, *11*, 236.
70 K. Dušek, V. Vojta, *Br. Polym. J.* **1977**, *9*, 164.
71 K. Dušek, M. Dušková-Smrčková, B. Voit, *Polymer* **2005**, *46*, 4265.
72 C. Rouf-George, J.-P. Munch, F. Schosseler, A. Pouchelon, G. Beinert, F. Boue, J. Bastide, *Macromolecules* **1997**, *30*, 8344.
73 J. Nie, B. Du, W. Oppermann, *Macromolecules* **2004**, *37*, 6558.
74 S. Panyukov, Y. Rabin, *Macromolecules* **1996**, *20*, 7960.
75 E. Geissler, A.-M. Hecht, C. Rochas, F. Horkay, P. J. Basser, *Macromol. Symp.* **2005**, *227*, 27–37.
76 K. Dušek, W. Prins, *Adv. Polym. Sci.* **1969**, *6*, 1.
77 K. Dušek, *Faraday Discuss. Chem. Soc.* **1974**, *57*, 101.
78 G. Heinrich, E. Straube, G. Helmis, *Adv. Polym. Sci.* **1988**, *85*, 33
79 D. S. Pearson, W. W. Graessley, *Macromolecules* **1980**, *13*, 101.
80 M. Gottlieb, C. W. Macosko, G. S. Benjamin, K. O. Meyers, E. W. Merril, *Macromolecules* **1981**, *14*, 1039.
81 M. Ilavsky, K. Dušek, *Polymer* **1983**, *24*, 981.
82 C. W. Macosko, *RIM Fundamentals of Reaction Injection Molding*, Hanser Publ., Munich **1989**.
83 H. H. Winter, M. Mourns, *Adv. Polym. Sci.* **1997**, *134*, 165.
84 M. E. de Rosa, Dissertation, University of Massachusetts at Amherst, **1994**.
85 M. Ilavsky, in Polymer Networks – Principles of Their Formation, Structure and Properties, R. F. T. Stepto (Ed.), Blackie Academics and Professional, London **1998**, Ch. 8.
86 P. J. Flory, *Principles of Polymer Chemistry*, Cornell University Press, Ithaca 1951.
87 R. Koningsveld, W. H. Stockmayer, E. Nies, *Polymer Phase Diagrams. A Textbook*, Oxford University Press, Oxford **2001**.
88 K. Dušek, D. Patterson, *J. Polym. Sci., Part A-2*, **1968**, *6*, 1209.
89 R. Moerkerke, R. Koningsveld, H. Berghmans, K. Dušek, K. Solc, *Macromolecules* **1995**, *28*, 11103; H. Schafer-Soenen, R. Moerkerke, H. Berghmans, R. Koningsveld, K. Dušek, K. Solc, *Macromolecules* **1997**, *30*, 410; R. Moerkerke, F. Meeussen, R. Koningsveld, H. Berghmans, W. Mondelaers, E. Schacht, K. Dušek, K. Solc, *Macromolecules* **1998**, *31*, 2223.
90 K. Dušek, M. Dušková-Smrčková, *Polym. Bull.* **2000**, *45*, 83.
91 H. Tompa, *Polymer Solutions*, Butterworth, London **1956**.
92 K. Solc, K. Dušek, R. Koningsveld, H. Berghmans, *Collect. Czech. Chem. Commun.* **1995**, *60*, 1661.

93 J. Hasa, M. Ilavsky, K. Dušek, *J. Polym. Sci, Polym. Phys. Ed.* **1975**, *13*, 253.
94 M. M. Prange, H. H. Hooper, J. M. Prausnitz, *AIChE J.* **1989**, *35*, 803
95 A. I. Victorov, C. J. Radke, J. M. Prausnitz, *Mol. Phys.* **2002**, *100*, 2277; A. I. Victorov, C. J. Radke, J. M. Prausnitz, *Mol. Phys.* **2002**, *103*, 1431.
96 T. A. Vilgis, J. Wilder, *Comput. Theor. Polym. Sci.* **1998**, *8*, 61.
97 K. Dušek, *J. Polym. Sci., Part C*, **1967**, *16*, 1289.
98 M. Dušková-Smrčková, K. Dušek, P. Vlasak, *Macromol. Symp.* **2003**, *198*, 259.
99 T. Tanaka, *Phys. Rev. Lett.* **1978**, *40*, 820.
100 H. G. Schild, *Prog. Polym. Sci.* **1992**, *17*, 163.
101 *Responsive Gels. Volume Phase Transition*, K. Dušek (Ed.), Vol. I, Adv. Polym. Sci. 1993, Vol. 109; **1993**, Vol. 110, Springer, Berlin.
102 T. Okano, *Biorelated Polymers and Gels: Controlled Release and Applications in Biomedical Engineering*, Academic Press, **1998**.
103 K. Dušek, M. Dušková-Smrčková, M. Ilavsky, R. Stewart, J. Kopecek, *Biomacromolecules* **2003**, *4*, 1818.
104 K. Dušek, Network Formation by Chain Crosslinking (Co)polymerization, in *Development in Polymerization. 3*, R. N. Haward (Ed.), Applied Science Publ., Barking 1982,143–206; K. Dušek, H. Galina, J. Mikes, *Polym. Bull.* **1980**, *3*, 19.
105 K. Dušek, *Collect. Czech. Chem. Commun.* **1993**, *58*, 2245.
106 O. Okay, *Polymer* **1999**, *40*, 4117.
107 O. Okay, *Progr. Polym. Sci.* **2000**, *25*, 711.
108 B. Nabeth, J.-P. Pascault, K. Dušek, *J. Polym. Sci., Polym. Phys. Ed.* **1996**, *34*, 1031.
109 K. Dušek, J. Somvarsky, *Faraday Discuss. Chem. Soc.* **1995**, *101*, 147; K. Dušek, J.·Somvarsky, *Macromol. Symp.* **1996**, *106*, 119.
110 W. L. Wu, B. J. Bauer, *Macromolecules* **1988**, *21*, 457; W. L. Wu, B. J. Bauer, W. Su; *Polymer* **1989**, *30*, 457.
111 N. C. Beck Tan, B. J. Bauer, J. Plestil J., J. D. Barnes, D. Liu, L. Matejka, K. Dušek, W. L. Wu, *Polymer* **1999**, *40*, 1384.
112 *Interpenetrating Polymer Networks*, D. Klempner, L. H. Sperling, L. A. Utracki (Eds.), Adv. Chem. Ser. 239, American Chemical Society, Washington, D. C., 1994.
113 Y. S. Lipatov, *Phase-Separated Interpenetrating Polymer Networks*, USChTU Dnepropetrovsk **2001**.
114 K. Binder, H. L. Frisch, *J. Chem. Phys.* **1984**, *81*, 2126–2136.
115 M. Schulz, H. L. Frisch, *J. Chem. Phys.* **1997**, *107*, 2673–2682.
116 J.-M. Chenal, J. M. Widmaier; Polymer **2005**, *46*,671.
117 B. D. Andrews, A. V. Tobolsky, E. E. Hanson, *J. Appl. Phys.* **1946**, *17*, 352.
118 J. Seidl, J. Malinsky, K. Dušek, W. Heitz, *Adv. Polym. Sci.* **1967**, *5*, 113.
119 M. W. C. P. Franse, K. te Nijenhuis, J. Groenewold, S. J. Picken, *Macromolecules* **2004**, *37*, 7839.
120 A. Tsuchida, C. Bolin, F. G. Sernetz, H. Frey, R. Mulhaupt, *Macromolecules* **1997**, *30*, 2818.
121 J. I. Stanford, M. J. A. Elwell, A. J. Ryan, in *Processing of Polymers*, H. E. H. Meijer (Ed.), Wiley-VCH, Weinheim **1997**. Ch. 25. pp. 465–512.
122 A. J. Kinloch, R. J. Young, *Fracture Behaviour of Polymers*, Applied Science Publishers, London **1985**.

9
Block Copolymers for Adhesive Applications

Costantino Creton

9.1
Introduction

One of the areas of polymer science where the controlled synthesis of polymers has had the most practical impact is that of block copolymers. The possibility of incorporating, in the same molecule, two or more different monomers, no longer in a random fashion but with well-defined sequences, has opened up many new application areas that can be divided into two general categories: those where the block copolymer is essentially an additive, clearly used as a minority component, and those where the block copolymer is the majority component.

An important example of an application of the latter type is that of pressure-sensitive adhesives (PSAs) [1–4]. For this particular application, the order–disorder transition (ODT) that block copolymers typically undergo at a given temperature is highly useful. The block copolymers behave as fluids in the disordered state (usually above the ODT) but become rubbery solids in the ordered state (usually below the ODT). This is a key advantage for this application since above the ODT, they are easily processable in the melt state without need for additional solvents. This processing technique, generally called hot melt processing, is both economically and environmentally attractive [3].

In these applications, the block copolymers used are of the type ABA or AB, with A being a glassy block and B a rubbery block at the usage temperature. Typically [5], the mass ratio between the glassy block(s) and the elastomeric block can vary between 0.15 and 0.3. If the elastomeric block is a polyisoprene or polybutadiene, the ordered structures are very entangled and their elastic modulus at room temperature is too high to form a bond upon simple contact [6, 7]. Therefore the block copolymers are often blended with a low molecular weight but high-T_g molecule, called a "tackifying resin", which is miscible with the rubbery domains but should be immiscible with the glassy domains [8, 9]. The total polymer weight fraction in the final adhesive blend is typically between 30 and 50%. Although the T_g of the tackifying resin is above room temperature in its pure state, the resin/polyisoprene phase of the adhesives has a T_g around –15 °C and is therefore

Macromolecular Engineering. Precise Synthesis, Materials Properties, Applications.
Edited by K. Matyjaszewski, Y. Gnanou, and L. Leibler
Copyright © 2007 WILEY-VCH Verlag GmbH & Co. KGaA, Weinheim
ISBN: 978-3-527-31446-1

rubbery at room temperature. As a result, the weight fraction of the final adhesive blend which is in the glassy state ranges from 5 and 12% and the resulting microstructure is always that of hard spheres dispersed in a soft matrix. In principle, the same morphology can be obtained for any type of immiscible glassy/elastomeric block copolymers. However, the only systems reported for commercial applications are polystyrene–polyisoprene (S–I), polystyrene–polybutadiene (S–B) and polystyrene–poly(ethylene–butylene) (S–EB) block copolymer systems. Because of the poor oxidation resistance of the diene polymers, attempts have been made to develop equivalent block copolymers with acrylic blocks and in particular poly(methyl methacrylate)–poly(n-butyl acrylate) block copolymers [10, 11]. Although atomic force microscopy (AFM) observations show that the materials are indeed microphase separated [12] and reasonable adhesive properties have been reported [13, 14], they have not yet been used, to our knowledge, for adhesive applications, despite the fact that acrylic block copolymers based on PnBA are tacky at room temperature without the need for a tackifying resin. In the rest of this chapter we will therefore focus on the polystyrene–polyisoprene adhesive blends and comment whenever possible on how this might apply to acrylic block copolymer systems.

For the case of the S–I system, at room temperature the blend has the properties of a physically cross-linked viscoelastic gel and is very sticky to the touch. The diblock:triblock ratio fine tunes the nonlinear viscoelastic properties and the creep properties, while the tackifying resin adjusts independently the glass transition temperature of the gel and the entanglement and the physical cross-link density (number of elastic strands per unit volume). The advantage of using block copolymers for this application is that the density of elastic strands can be both tunable and homogeneous, resulting in very different properties to those of the randomly cross-linked chemical gels which are also used for similar applications [15–17].

The key property that needs to be controlled in a pressure-sensitive-adhesive is the network of cross-links, which gives the resistance to creep and the strength, and the dissipative mechanisms, which are controlled by the monomer friction coefficient (the T_g, essentially) and by the defects in the architecture (pendant chains). The use of block copolymers with their superb control of the molecular structure gives new tools to the synthetic chemist to fine tune the properties and it is the purpose of this chapter to review the current state of the art in terms of structure–property relationships.

9.2
Rheological Properties and Structure

As mentioned above, pure S–I copolymers used as base materials for PSA applications can show some self-tack (by interdiffusion), but are typically not tacky at all on solid surfaces. In order to obtain PSA properties, the entanglements of the rubbery domains must be diluted to lower the elastic modulus of the physically cross-linked gel and the material needs to be more dissipative to resist crack propagation at the interface [8]. Both modifications of properties are achieved through blending the copolymer(s) with a low molecular weight but high-T_g molecule called a tackifying resin. This low molecular weight molecule is miscible with the rubbery domains but immiscible with the glassy domains. Structures of the polymers and resins typically used for adhesive applications are shown in Fig. 9.1.

The effect of the tackifying resin on the rheological and adhesive properties is well documented [9, 18–21] and amounts to a dilution of the entanglement or cross-link density combined with an adjustment of the glass transition temperature (and therefore of the range of usage temperatures) of the PSA. We will therefore summarize here only the main findings. Figure 9.2 shows the linear viscoelastic properties, at a reference temperature of 20 °C, of a styrene–isoprene–styrene (SIS) triblock copolymer with and without tackifying resin. As one can readily see, the master curve displays the typical behavior of a cross-linked rubber for the pure block copolymer. When the resin is added (in this case 60 wt% resin and 40 wt% polymer), the plateau modulus shifts to lower values while the characteristic increase in modulus due to the glass transition shifts to lower frequencies, corresponding effectively to an increase in the T_g at fixed frequency.

When a low molecular weight tackifying resin is added to a triblock copolymer, the plateau modulus, clearly visible at low frequency and due to the entangled polyisoprene blocks, decreases with the dilution level as

$$G^0_{n,\text{adhesive}} = G^0_{n,\text{copolmyer}} \Phi^n \qquad (1)$$

Hydrogenated C5 resin

$A_x - B_y - A_x$ Triblock copolymer

$A_x - B_y$ Diblock copolymer

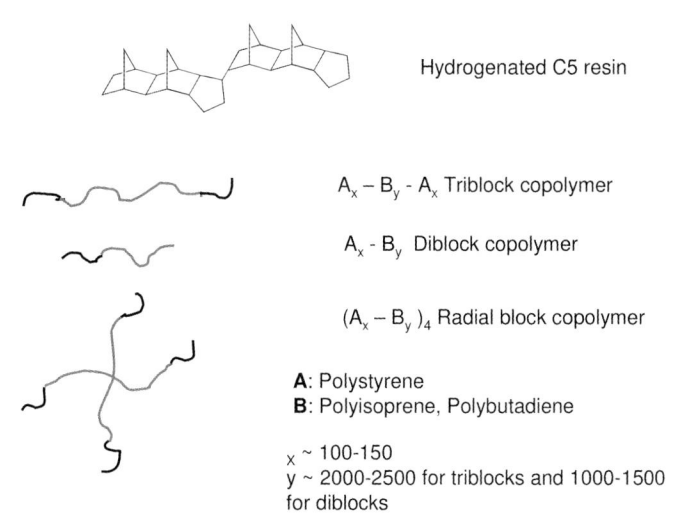

$(A_x - B_y)_4$ Radial block copolymer

A: Polystyrene
B: Polyisoprene, Polybutadiene

x ~ 100-150
y ~ 2000-2500 for triblocks and 1000-1500 for diblocks

Fig. 9.1 Molecular structures of the block copolymers and tackifying resins typically used for adhesive applications.

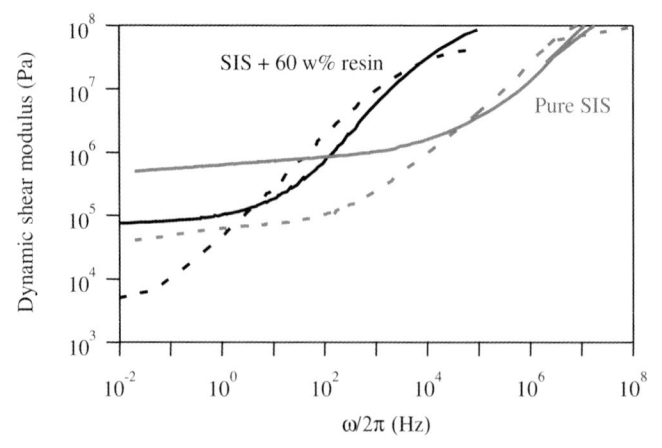

Fig. 9.2 Time–temperature shifted master curves showing the effect of the addition of 60 wt.% of C5 tackifying resin on G' (continuous lines) and G'' (dashed lines) of the SIS block copolymer. Reference temperature, 20 °C; no measurement was taken above this temperature. The black lines refer to the SIS+resin material and the gray lines to the pure SIS. Data from [22].

where is the polymer volume fraction in the adhesive and n has been reported by several workers [9,21] to be between 2.2 and 2.5. This decrease in elastic modulus is essential for the adhesive properties since it allows the material to fulfill the so-called Dahlquist criterion [6, 7], stating that a PSA should have an elastic modulus E lower than 0.1 MPa at 1 Hz.

The relative proportion of di- and triblock copolymers in the base polymer also modifies the linear viscoelastic properties but in a different way to the tackifying resin. The replacement of a triblock copolymer by two equivalent diblock copolymers each having the same styrene content and half the molecular mass of the triblock can be seen as cutting the triblock chain in half. This creates pendant chains which can relax differently to a chain bound by each end to a glassy block [8, 23].

This relaxation process is analogous to the relaxation of an arm of a star-like polymer [24] and causes G' to drop to a lower plateau modulus, the level of which is controlled only by the density of triblock chains actually bridging two styrene domains [9]. The presence of this second relaxation process is clearly apparent in Fig. 9.3a, showing the viscoelastic properties of two adhesive blends containing either SIS+resin or SIS+SI+resin. Above a frequency of 1 Hz, both blends behave identically in the linear regime. However, at lower frequency, the relaxation of the arm star of the copolymer introduces an additional relaxation process [9]. Finally, at very low frequency, G' stabilizes at a value predicted [9] by Eq. 1, which is barely seen in Fig. 9.3a for this example.

Another peculiar feature of ordered block copolymer structures is that they do not necessarily obey the time–temperature superposition principle even at moderate temperatures above ambient [9, 25]. The solid-like behavior of the ordered

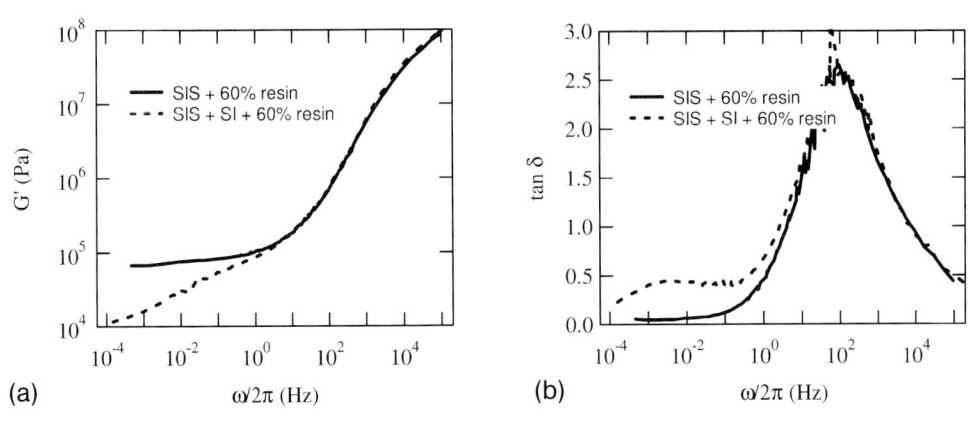

Fig. 9.3 Master curves of the linear viscoelastic properties of a triblock copolymer blend and of a di-and triblock copolymer blend. (a) G' and (b) tan. The reference temperature is 22 °C; no measurement was taken above this temperature. Data kindly provided by C. Derail.

structures comes from the presence of styrene nodules, which act as physical cross-links. In the pure copolymers they can be readily seen by AFM (Fig. 9.4a) in tapping mode or by SAXS with the distinct presence of a scattering peak representing the average domain distance (Fig. 9.4b). Of course, the degree of long-range order which can be achieved in these systems will depend on the annealing conditions. The samples shown in Fig. 9.4 have been slowly dried in air

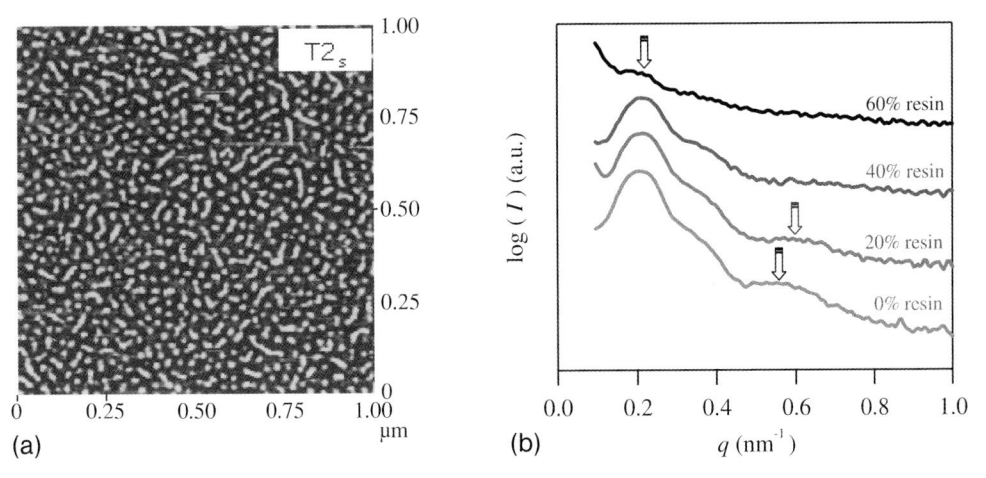

Fig. 9.4 (a) AFM of pure SIS copolymer and (b) SAXS spectra of the SIS copolymer with various amounts of tackifying resin. The intensity logarithm has been shifted for clarity. Data from [27].

and annealed at 40 °C for 24 h under vacuum: ordered structures with much better long-range order can be achieved with longer annealing times at higher temperatures [26]. As shown in Fig. 9.4b, one of the additional effects of adding the resin is a considerable reduction in the degree of long-range order for the same annealing conditions which is clearly observed by small-angle X-ray scattering (SAXS) and which was also reproduced by "quenching" mean field simulations [27]. The main scattering peak almost disappears when an amount of resin typical of adhesive applications is added to the blend and the secondary peak vanishes for even lower resin contents. However, the rheological properties confirm the presence of the physical cross-links. The structure is therefore that of randomly dispersed styrene domains in a polyisoprene–resin matrix. This loss of long-range order implies a reduction of the driving force for ordering due to the presence of the resin. It should be noted that it is likely to be even more pronounced for the adhesives processed in the melt, which are clearly out of equilibrium.

9.3
Large Strain and Adhesive Properties

For application purposes, block copolymer-based PSAs are often formulated from base polymer blends of tri- and diblock copolymers in various proportions. Setting aside cost considerations, the reasons for using a certain blend or even a pure triblock copolymer are typically based on performance in standardized PSA tests such as loop tack, peel or shear tests. Although several studies have been performed on adhesive properties of block copolymer-based adhesives [5, 20], few have focused on the details of the deformation mechanisms taking place during the debonding process [14, 17, 28]. It is worth noting that styrene–isoprene–styrene is not the only system to have been investigated: two studies have focused on all-acrylic systems where the glassy block is poly(methyl methacrylate) and the soft block is poly(n-butyl acrylate) [14, 28]. The key difference in those systems is that the relatively high average molecular weight between entanglements of the poly(n-butyl acrylate) does not require the addition of a tackifying resin to be adhesive, so that adhesive properties can be tested on the neat block copolymers. However, despite the advantage of a better temperature resistance, fully acrylic block copolymers have not been yet used in PSA commercial applications in the adhesives area.

As an illustration of the adhesive behavior of block copolymer-based adhesives, we now examine in more detail a recent study investigating the effect of the triblock:diblock ratio on the tensile and adhesive properties of styrene–isoprene-based PSA [29–31].

The study focused on four model PSA blends and their molecular characteristics and the composition of the blends are summarized in Table 9.1. The base polymers were di- and triblock copolymers provided by ExxonMobil Chemical and synthesized by Dexco. The tackifying resin was a hydrogenated C5-based re-

Table 9.1 Composition of PSA blends.

Blend	PS–PI (wt%)	PS–PI–PS M_w (kg mol^{-1})	PS in PS–PI–PS (wt.%)	PS–PI M_w (kg mol^{-1})	PS in PS–PI (wt%)
Vector 4100 D	0	154	15.1	–	–
Vector 4113	19	154	15 1	72	15
Vector 4114	42	156	15 1	72	15
DPX 565	54	176	16.1	72	16

sin, Escorez 5380, with a T_g of 48 °C, also provided by ExxonMobil Chemical. This type of resin is designed to be miscible with the polyisoprene but not with the polystyrene domains. All adhesive blends contained 40 wt% of base polymer and 60 wt% of resin and, within the base polymer, the proportion of diblock was varied from 0 to 54 wt%.

A probe test method [32, 33] was used to characterize the adhesive properties of the blends (Fig. 9.5). The output of a probe test is a force and displacement measurement as a function of time. In order to compare easily curves obtained at different probe debonding velocities and for different layer thickness, it is customary to convert these curves into nominal stress vs. nominal strain curves by normalizing the force by the maximum contact area in the compression stage (often, but not always, the complete probe surface) and by normalizing the displacement by the initial thickness of the film.

The question of the interpretation of the stress–strain curve also needs to be addressed. The simultaneous video acquisition of the data makes it possible to interpret the general features of the curve in terms of specific deformation mechanisms. The first peak in stress is due to the nucleation of cavities at the interface between the probe and the layer. These cavities nucleate on pre-existing defects due to the mismatch in surface topography between probe and layer [34, 35]. Once the cavities have been formed in sufficient numbers from the defects, the compliance of the layer increases dramatically and the force drops to a lower level before remaining roughly constant at a level which is independent of the initial population of defects. At this stage, the layer is no longer confined and the test proceeds by the progressive extension of the cavity walls in the tensile direction. The pressure inside the cavities is close to zero while the outside pressure is about 1 atm. Hence the work done by the tensile force against the atmospheric pressure (~ 0.1 MPa) should be subtracted from the measured plateau value to obtain a force representative of the viscoelastic properties of the adhesive blend.

The majority of studies focusing on the deformation of these walls started from the assumption that the observation of a filament structure is characteristic of the behavior of a liquid and requires flow. In our case, Figs. 9.2 and 9.3 show that for block copolymer-based PSA, the material is more a soft solid than a viscous liquid and cannot flow. These measurements, however, were per-

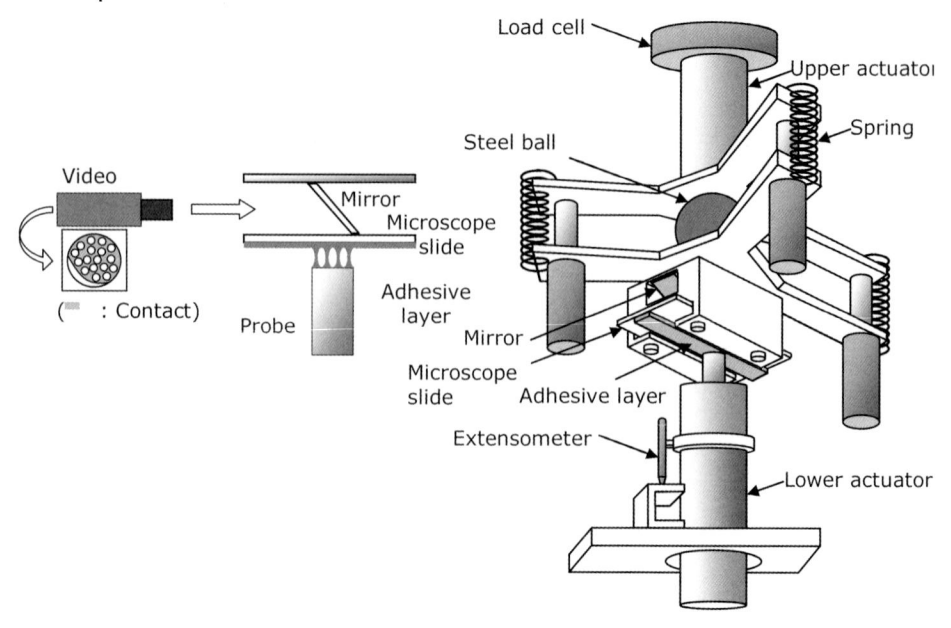

Fig. 9.5 Schematic of the experimental design and setup used for the probe tests.

formed at small strains in the linear viscoelastic regime and one could argue that at large strains, flow is likely to occur, analogously to what is observed for yield stress fluids. In order to check this hypothesis, the authors performed an interesting experiment that can be done with a probe test setup [22, 36]: instead of fully debonding the layer from the probe, the test can be stopped after the peak stress and during the cavity wall extension stage [37]. The motors are stopped and the stress is allowed to relax, giving an idea of the elastic or viscous nature of the adhesive material at large strains [22]. The results are presented as curves of force as a function of time in Fig. 9.6a and the relaxation stage, normalized by the stress at the beginning of the relaxation, is highlighted in Fig. 9.6b. After 2 min of relaxation the stress has decreased by only about 25–30%, independently of extension, showing the predominantly elastic nature of the adhesive foam formed upon debonding.

9.3.1
Effect of the Diblock Content on Adhesive and Deformation Properties

The effect of the presence of diblocks on the linear viscoelastic behavior within a frequency range of 0.01–100 Hz is shown in Fig. 9.7 and demonstrates that all adhesives have a very similar behavior in terms of elasticity above 1 Hz. As discussed earlier, the differences appear at low frequency, a regime where the free isoprene end of the diblock chain is able to relax. If we now turn to Fig. 9.8a and b, which show probe test curves for the four adhesive blends at two differ-

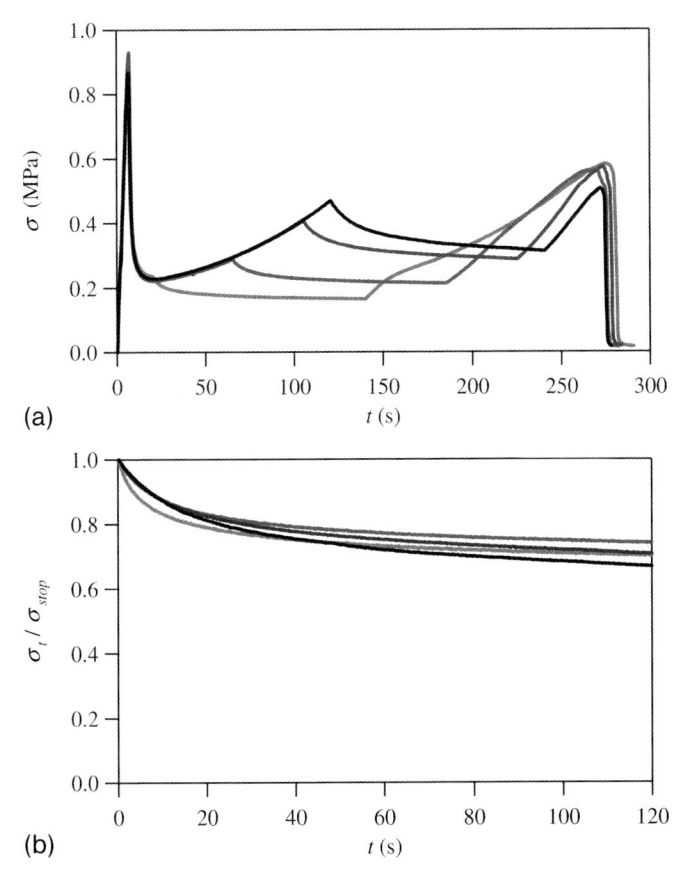

Fig. 9.6 Relaxation of the fibrils in a probe test for the SIS adhesive. (a) Force vs. time curves for consecutive tests where the displacement was stopped at various levels of deformation of the fibrils and left to relax for 120 s. (b) Relaxation curves for different stops normalized by the stress at the beginning of the stop.

ent probe debonding velocities, the differences between the four adhesive blends are striking and, although measured stresses are systematically lower for the low probe velocity, the plateau stress is clearly lowered by the presence of diblock while the maximum extension is increased. Despite the two orders of magnitude difference in strain rate applied to the adhesive layer, the effect of adding the diblock remains the same. Given the range of strain rates (typically 0.01 Hz for 1 µm s^{-1} and 1 Hz for the 100 µm s^{-1} tests), these differences cannot be simply predicted by the linear viscoelastic properties.

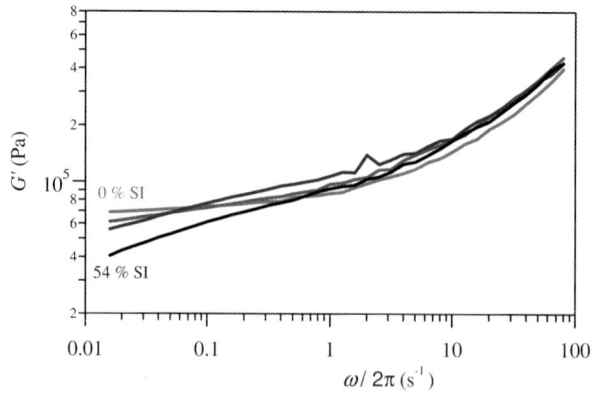

Fig. 9.7 Elastic component of the complex modulus at 22 °C for the four adhesive blends.

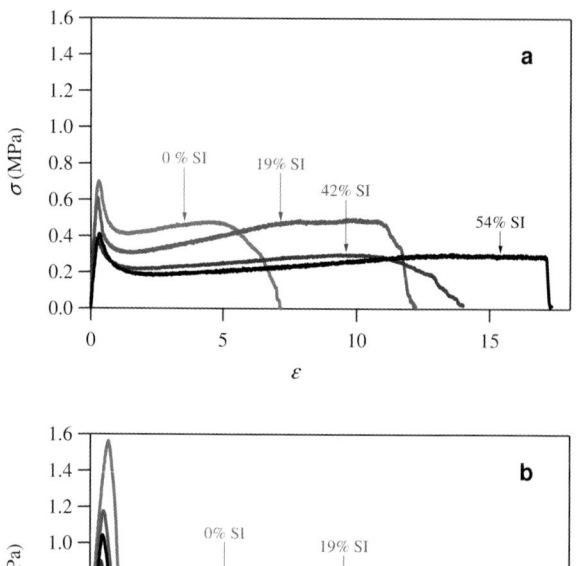

Fig. 9.8 Probe test curves for four tackified SIS+SI blends at two probe pullout rates: V_{deb} = (a) 1 and (b) 100 μm s^{-1}.

9.3.2
Understanding the Structure of the Extended Foam

Given the large strains which are imposed to the adhesive layers in this wall extension regime, a systematic characterization of the nonlinear elastic properties of the adhesive was performed with uniaxial tensile tests of adhesive layers [29] up to large strains. The main results are summarized in Fig. 9.9a and b, showing the nominal stress vs. extension of the four adhesive blends at two different cross-head velocities chosen in order to be equivalent to the strain rate applied to the materials in the probe tests at the beginning of the plateau region.

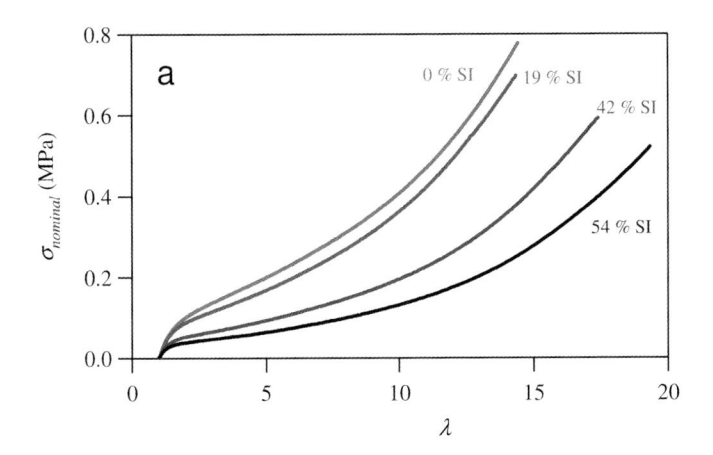

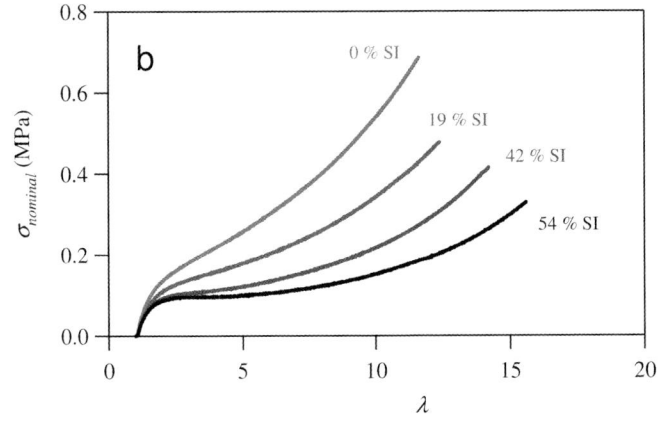

Fig. 9.9 Tensile stress–strain curves for the four adhesives at two strain rates: (a) crosshead velocity 5 mm min^{-1}, corresponding to an initial strain rate of 0.005 s^{-1}; (b) crosshead velocity 500 mm min^{-1}, corresponding to an initial strain rate of 0.5 s^{-1}.

Clearly, the differences in the nonlinear elastic properties of the adhesives, in the large-strain regime, are very pronounced whereas the initial portion of the curve does not show very marked differences. However, in this representation, a more quantitative comparison between the different adhesives is not straightforward. For those readers working with rubbers, the stress–strain curves shown in Fig. 9.9 should be familiar and display a fairly typical type of nonlinear elastic behavior observed for cross-linked rubbers. This similarity is logical since we have a structure of physically cross-linked styrene domains and a soft deformable isoprene + resin matrix. It is also very useful for the purpose of analyzing the data since rubber elasticity has been a very active field for decades and several models, both phenomenological [38, 39] and molecularly based [40, 41], are available in the literature [42].

The simplest model is the statistical theory of rubber-like elasticity, also called the affine model or neo-Hookean in the solids mechanics community [42]. It predicts the non-linear behavior at large strains of a rubber in uniaxial extension with the following relationship:

$$\sigma_N = G\left(\lambda - \frac{1}{\lambda^2}\right) \tag{2}$$

where σ_N is the nominal stress, defined as F/A_0, with F the tensile force and A_0 the initial cross-section of the sample, λ is the extension ratio and G is the shear modulus. Realizing the shortcomings of this simple model, extensive work has subsequently been performed to refine the picture of the nonlinear elastic behavior of cross-linked rubbers. The reader is referred to textbooks [42, 43] and recent reviews [40] for further information.

Among the various models that have been proposed to describe experiments, the simplest one to compare with our experimental data is a phenomenological model first proposed by Mooney [38] and further developed by Rivlin [39], which is based on the incompressibility condition and introduces a λ-dependent term in the modulus.

In uniaxial extension, the Mooney–Rivlin model predicts

$$\sigma_N = 2\left(C_1 + \frac{C_2}{\lambda}\right)\left(\lambda - \frac{1}{\lambda^2}\right) \tag{3}$$

where C_1 and C_2 are two material constants. Note that, with this model, the reduced stress defined by

$$\sigma_R = \frac{\sigma_N}{\left(\lambda - \frac{1}{\lambda^2}\right)} \tag{4}$$

depends on the deformation λ whereas it did not for the simple affine model. If the curves in Fig. 9.9b are plotted in terms of reduced stress as a function of $1/\lambda$, they appear as a set of parallel lines within the range of λ^{-1} extending from

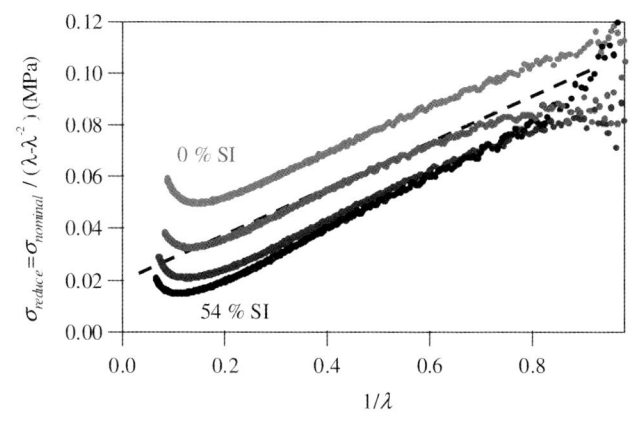

Fig. 9.10 Reduced stress representation of the stress–strain curves for the four adhesives. The dashed line is an illustrative fit of the data with the Mooney–Rivlin model.

0.2 to 0.6 (Fig. 9.10). The slope of these lines gives directly the C_2 constant while the intercept gives C_1. It then immediately becomes apparent from the graph and from Table 9.2 that the different adhesive blends have very similar values of C_2 but very different values of C_1. The upturn in reduced stress at high values of λ is due, of course, to the strain hardening of the material which is not captured by the Mooney–Rivlin model.

An example of the quality of the fit in the low to intermediate strain regime can be seen in Fig. 9.11, showing the experimental data and the fits of Fig. 9.10. The relevance of Fig. 9.11 becomes obvious when comparing its data with the portion after the peak of Fig. 9.8.

A striking feature of the probe test results is the large difference in the level of stress of the plateau region, which cannot be attributed to a variation in shear modulus G' or even in the complex modulus $G^* = (G'^2 + G''^2)^{1/2}$ taken at an equivalent strain rate (0.5 Hz in this case). On the other hand, these differences are much better correlated with the large-strain behavior of the adhesive in extension. Figure 9.12 shows the ratio between the stress level at the beginning of the plateau extracted from the probe test curves (at $\varepsilon = 2$, for a probe velocity of

Table 9.2 Mooney-Rivlin parameters obtained from the data fits.

SI (%)	$2C_1$		$2C_2$	
	5 mm min^{-1}	500 mm min^{-1}	5 mm min^{-1}	500 mm min^{-1}
0	0	0.037	0.059	0.089
19	0.019	0.019	0.064	0.101
42	0.005	0.004	0.051	0.100
54	0	0	0.050	0.109

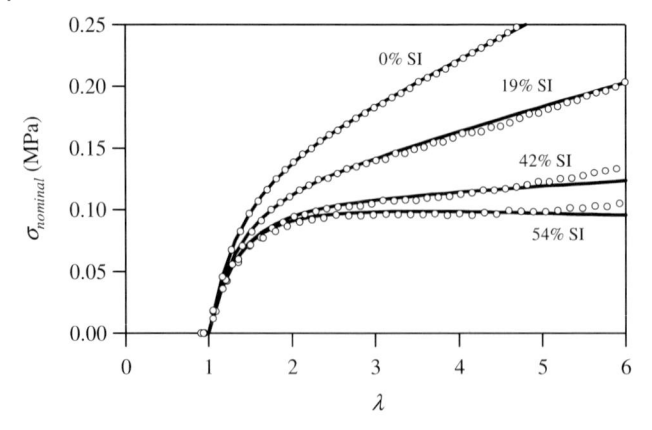

Fig. 9.11 Experimental data (circles) and best fit (full lines) with the Mooney–Rivlin model in uniaxial extension for tests performed at 500 mm min^{-1}.

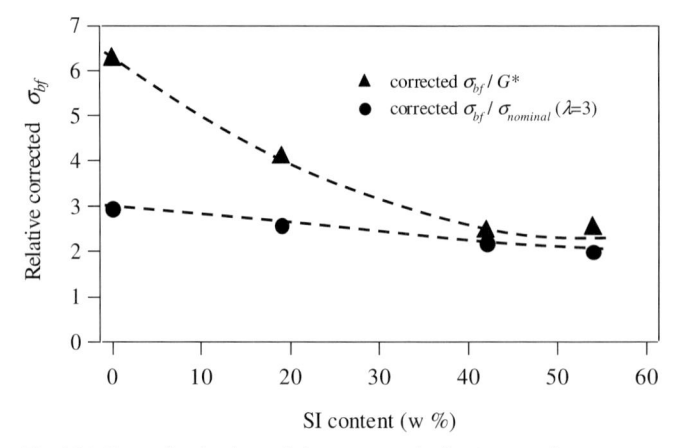

Fig. 9.12 Normalized values of the stress at the beginning of the plateau corrected by 0.1 MPa of atmospheric pressure. Values of corrected σ_{bf}/G^* (triangles) and of corrected $\sigma_{bf}/\sigma(\lambda=3)$ (circles).

100 µm s^{-1} and corrected for the work done against atmospheric pressure) and the complex small-strain modulus at 0.5 Hz and between the same corrected σ_{bf} and the stress in uniaxial extension at $\lambda=3$ for a tensile test at 500 mm/min corresponding to the position $\varepsilon=2$ in the probe test ($\lambda=\varepsilon+1$). Clearly, the stress level in the plateau region of the probe test curve is directly related to the behavior of the adhesive in uniaxial extension in large strain, which is quite different for the three adhesives, but is not proportional to the value of the shear modulus in small strains. This is a significant finding of the study, showing conclusively that the cross-linking network of the adhesive will be essential in

controlling the stress level in the plateau and therefore the work done to detach the adhesive.

In a more molecular sense, the results can be interpreted as follows [40]: the parameter C_1, which is directly related to the volume density of fixed cross-link points, varies significantly between the pure triblock adhesive and the high diblock content adhesives, where the fit gives a value close to zero. On the other hand, the parameter C_2, related to entanglements, is much higher and nearly independent of the diblock content. Since C_2 is always larger than C_1, this implies that the small-strain modulus is mainly controlled by the entanglement network of the soft domains (polyisoprene+resin) and does not change much with increasing amount of diblock, whereas the large-strain behavior is increasingly controlled by the apparent cross-link density, which decreases dramatically with increasing diblock content. In a more general way, the relative importance of physical cross-links and entanglements will depend on the ratio between average molecular weight between entanglements and average molecular weight between physical cross-links. Since acrylic-based block copolymers are much less entangled than polyisoprene or polybutadiene, the softening observed in Fig. 9.9 at intermediate strain will be less pronounced for identical molecular weights between physical cross-links, and this will make it more difficult to deform the material in the bulk for a given small-strain modulus E. This may be a reason why acrylic block copolymers have not yet found commercial applications and have inferior properties to their SIS-based counterparts.

A simple molecular picture would have the triblock chains forming bridges between styrene domains and providing the fixed cross-links. However a more detailed analysis [27] shows that the variation in bridging triblock chains can account for only a small fraction of the decrease in effective cross-link points. It is likely that permanently trapped entanglements and slowly relaxing entanglements also provide a significant contribution to the fitted value of C_1 and the role of the presence of diblock on this fraction of trapped entanglements is currently unknown. Since the block copolymer blends are not strictly rubbers (PS spheres occupy some volume and have a high functionality relative to chemical cross-link points), any quantitative comparison between the data and a molecularly based model should be treated with some caution. We feel, however, that the insight provided by the molecularly based model is essential for the understanding of the mechanical properties of these systems.

In conclusion, when such an adhesive is debonded from a high-energy surface such as steel, the large-strain properties of the adhesive control the formation and extension of the fibrillar structure which provides the bulk of the work necessary to detach the adhesive from the surface and hence the major part of the peel force. We have seen that the level of the plateau stress can be predicted quantitatively by a simple tensile test. From the studies on cavitation, we know that the nominal stress at the plateau corresponds also to the cavity growth stress for large initial defects.

Although the fibril extension stress can be predicted from the nonlinear elastic properties of the adhesive, in practice the important property that one wishes to predict is the adhesion energy rather than simply the plateau stress of the fibrillar zone. This prediction would require a better understanding of which molecular features control the detachment of the fibril from the surface, once it is highly extended. This problem is currently still open since it involves not only knowing the nonlinear elastic properties of the material at very large strains (the strain hardening) but also a microscopic criterion for fibril detachment [44].

9.3.3
Interfacial Fracture

This brings us back to more interfacial properties. It is interesting to investigate what becomes of the effect of adding diblock in the adhesive when the PSA is no longer detached from steel but from a low-adhesion surface such as a polyolefin, a silicone layer or any other release surface. A comprehensive investigation of the debonding mechanisms of the adhesives in Table 9.1 from a surface of ethylene–propylene (EP) copolymer [22] was recently carried out.

Two representative probe test curves for the detachment of an SIS adhesive from steel and from EP surfaces are shown in Fig. 9.13 and, while the initial portion of the curve is identical, the force drops rapidly to zero for the EP surface and never forms the characteristic fibrillar plateau observed on steel surfaces. How does this happen?

The initial part of the stress–strain curve is nearly identical for both surfaces and video observations of the debonding mechanism show that in both cases the peak in force is due to the appearance of cavities which grow from pre-existing small contact defects [34, 35] to a size of the order of the thickness of the adhesive film. However, the growth mechanism of these cavities then becomes

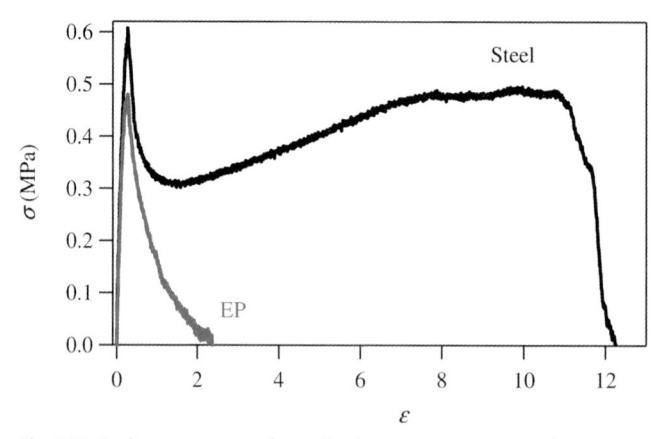

Fig. 9.13 Probe test curves of an adhesive containing 19 wt.% of diblock debonded at $V_{deb} = 1$ μm s^{-1} on a steel surface and on an EP surface.

very different. Whereas for the steel surface the projected area of the cavities in the plane of the film does not increase much and the growth occurs in the tensile direction (cigar-like cavities), for the EP surface the cavities grow mainly in the plane of the film and resemble interfacial cracks. These multiple crack are disc-like and eventually coalesce, causing the complete failure of the film. In Fig. 9.13, this complete coalescence occurs for a macroscopic deformation of $\varepsilon = 2$. Essentially, for the same applied tensile force an existing cavity at the interface between the adhesive and EP can propagate along the interface much faster than the same cavity at the interface between steel and the adhesive.

The propagation of an interfacial crack between a rubbery material and a solid surface is usually described in terms of a relation ship between critical energy release rate G_c and crack velocity. One can then write

$$G_c = G_0 \left(1 + \frac{v}{v^*}\right)^n \quad (5)$$

where G_0 is the energy per unit area to propagate a crack at zero velocity and v^* and n are material- and interface-dependent parameters. Usually for a given material, n does not vary much with different surfaces, so for a given level of dissipated energy per unit area (corresponding to a given G_c), cavities propagate much faster at the interface between the probe and the adhesive if G_0 is low and v^* is high and eventually coalesce [44]. Therefore, the difference in behavior of the same adhesive on two surfaces in Fig. 9.13 is due to a much lower resistance to crack propagation on EP than on steel.

This simple argument can be described more quantitatively by a balance between resistance to interfacial propagation G_c and resistance to bulk deformation (controlled by the elastic modulus E) for linear elastic materials [44, 45]. From the detailed description of the model, which is beyond the scope of this chapter, the key parameter is the ratio G_c/E, which is proportional to the distance over which an elastic layer needs to be deformed before being fully detached from the hard surface. This model has been experimentally verified for elastic gels [46].

For PSA layers, we need to introduce two modifications which complicate the analysis: the adhesives are both viscoelastic and strained in the nonlinear elastic regime. In other words, the term G_c will not strictly be an interfacial property and the term E should be replaced at least with a frequency-dependent equivalent and better with a strain- and frequency-dependent equivalent. With these conditions, the general principles of the elastic model should still be applicable.

The results of probe tests with the four adhesive blends on the EP surfaces are completely different from the results obtained on steel surfaces, as shown on Fig. 9.14. The high diblock content adhesives show a pronounced plateau, whereas the low or no diblock content adhesives show only a shoulder after the peak. Although the magnitude of the effect depends on the probe velocity, the tendency is clear: the presence of diblock makes it easier to extend the cavity walls in the tensile direction.

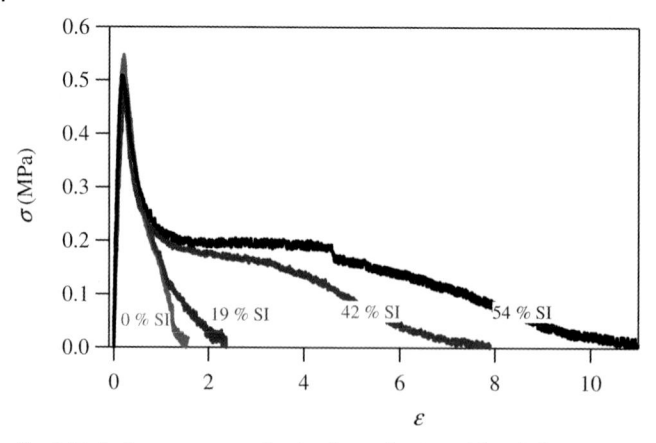

Fig. 9.14 Probe tests curves for the four adhesives debonded from an EP surface. Probe velocity 1 μm s^{-1}.

If these results are now interpreted in terms of the control parameter G_c/E, the increase in maximum extension of the fibrils observed can be due to a combination of a slight decrease in elastic modulus E or/and to an increase in G_c T. The energy release rate of an interfacial crack between a viscoelastic material and a solid surface is not quantitatively related to the linear viscoelastic properties of the adhesive [47]. but a more dissipative material will also dissipate more energy near an interfacial crack front [48–50]. The best approximation we can make here, therefore, is to consider that the dissipative properties of the adhesive near the crack tip are well represented by its viscoelastic properties in the linear regime. The value of the dissipative factor tan $\delta = G''/G'$ for the four adhesive blends is shown in Fig. 9.15. Clearly, the addition of diblock in the blend

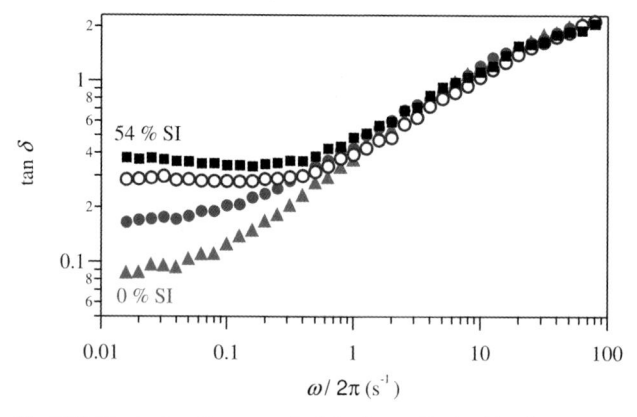

Fig. 9.15 Dissipation factor tan δ of the four adhesive blends at 22 °C (from pure SIS to 54 wt% SI).

has no effect on the dissipative properties of the adhesives at high frequency, as expected and shown also in Fig. 9.3, but has a significant effect at low frequencies. We can therefore propose, at least qualitatively, that the reduction in interfacial crack velocity observed for certain adhesives on FP surfaces is due to their more dissipative character, which slows crack propagation at the interface considerably, therefore avoiding early coalescence between separate cavities and favoring the formation of a foamy fibrillar structure with the cavity walls.

An explanation such as this remains very simplistic, since a range of strains and strain rates coexist at the crack tip and it is unlikely that a property measured in the linear regime such as G' and G'' can account alone for the behavior of the adhesive layer at the crack tip. An extensive study of the dissipative properties of the adhesive materials as a function of strain could reveal a more complex role of the presence of the diblock.

Of course, another way to influence the dissipative properties of the adhesive is to modify the T_g of the rubbery domains by using different types or different amounts of resin. This will shift the curves if Fig. 9.15 towards lower or higher frequencies, extending or modifying the useful range of temperatures for the PSA. On the other hand, the introduction of resin will not affect the large-strain behavior of the adhesive and will have very different effects on the debonding mechanisms than the addition of diblock copolymer to triblock.

9.4 Conclusions

We have reviewed the application of block copolymers with a controlled molecular structure in the adhesives field. Clearly, precise control of the molecular structure is essential in this application to obtain the desired ordered structure and therefore the optimized adhesive properties. Summarizing the main points made in this chapter:

- In the linear viscoelastic regime, the presence of diblock introduces an additional relaxation time, which is apparent at low frequency and is due to the relaxation of the free end of the diblock in the elastomeric matrix. The more diblock in the adhesive, the more pronounced is the decrease in low-frequency modulus.
- The nonlinear elastic properties of the adhesives can be reasonably described by the Mooney–Rivlin model and the fitting parameters show that the small-strain modulus of the adhesives is mainly controlled by the entanglement structure of the polyisoprene block + resin domains, while the large-strain behavior becomes increasingly controlled by the physically cross-linked structure formed by the polystyrene block domains. The incorporation of diblocks in the adhesive reduces the density of physical cross-links and causes a more pronounced softening and a delayed hardening in the large-strain part of the stress–strain curve.

Both of these material properties have direct consequences on the adhesive properties:
- On high-energy surfaces where a fibrillar structure can be formed, the cavitation stress for large defects and the level of plateau stress in the fibrillar regime are significantly higher for the pure triblock systems and are directly related to the density of physical cross-link points or trapped entanglements. The implication of this result is that pure triblock systems will perform best in PSA applications where initiation of failure is the limiting factor such as long-term resistance to shear. However, for applications where a specific peel force is required, it is not clear that the pure SIS formulations will perform better since the maximum fibril extension is systematically reduced for the triblock-based adhesives, relative to the adhesives containing diblocks.
- On low-adhesion surfaces, the addition of diblock should be much more beneficial since it provides two essential features which will slow crack propagation: a more pronounced softening at intermediate strains and a more dissipative character. When failure is initiated by the formation of cavities, the softening behavior favors crack blunting and hence the formation of fibrils, while the more dissipative character slows crack propagation and again makes it easier to form the foam structure. In terms of the parameter G_c/E, controlling the behavior of fully elastic systems, the more dissipative character increases G_c whereas the softening decreases the effective large strain E. Hence for applications where the PSA is applied to low-adhesion surfaces such as polyolefins, a reasonable percentage of diblock should increase the practical work of adhesion.

References

1 Zosel, A. *Adv. Pressure Sensitive Adhes. Technol.* **1992**, *1*, 92–127.
2 Creton, C. *MRS Bull.* **2003**, *28*, 434–439.
3 Ewins, E. E., St.Clair, D. J., Erickson, J. R., Korcz, W. H. In *Handbook of Pressure-Sensitive-Adhesive technology*, 2nd edn., Satas, D. (Ed.). Van Nostrand Reinhold: New York, **1989**, Vol. 1, pp. 317–373.
4 Creton, C., Fabre, P. In *The Mechanics of Adhesion*, Dillard, D. A., Pocius, A. V. (Eds.). Elsevier: Amsterdam, **2002**, Vol. 1, pp. 535–576.
5 Tse, M. F., Jacob, L. *J. Adhesion* **1996**, *56*, 79–95.
6 Dahlquist, C. A. In *Treatise on Adhesion and Adhesives*, Patrick, R. L. (Ed.). Marcel Dekker: New York, **1969**, Vol. 2, pp. 219–260.
7 Creton, C., Leibler, L. *J. Polym. Sci., Part B: Polym. Phys.* **1996**, *34*, 545–554.
8 Nakajima, N., Babrowicz, R., Harrell, E. R. *J. Appl. Polym. Sci.* **1992**, *44*, 1437–1456.
9 Gibert, F. X., Marin, G., Derail, C., Allal, A., Lechat, J. *J. Adhesion* **2003**, *79*, 825–852.
10 Tong, J. D., Jérme, R. *Polymer* **2000**, *41*, 2499–2510.
11 Tong, J. D., Moineau, G., Leclere, P., Brédas, J. L., Lazzaroni, R., Jerome, R. *Macromolecules* **2000**, *33*, 470–479.
12 Tong, J. D., Leclère, P., Doneux, C., Brédas, J. L., Lazzaroni, R., Jérme, R. *Polymer* **2001**, *42*, 3503–3514.
13 Flanigan, C. M., Crosby, A. J., Shull, K. R. *Macromolecules* **1999**, *32*, 7251–7262.
14 Drzal, P. L., Shull, K. R. *J. Adhesion* **2005**, *81*, 397–415.
15 Asahara, J., Hori, N., Takemura, A., Ono, H. *J. Appl. Polym. Sci.* **2003**, *87*, 1493–1499.

16 Lindner, A., Lestriez, B., S., M., Brummer, R., Maevis, T., Lühmann, B., Creton, C. *J. Adhesion* **2006**, *82*, 267–310.
17 Brown, K., Hooker, J. C., Creton, C. *Macromol. Mater. Eng.* **2002**, *287*, 163–179.
18 Kim, J., Han, C. D., Chu, S. G. *J. Polym. Sci., Part B: Polym. Phys.* **1988**, *26*, 677–701.
19 Sherriff, M., Knibbs, R. W., Langley, P. G. *J. Appl. Polym. Sci.* **1973**, *17*, 3423–3438.
20 Tse, M. F. *J. Adhesion Sci. Technol.* **1989**, *3*, 551–570.
21 Kraus, G., Rollmann, K. W., Gray, R. A. *J. Adhesion* **1979**, *10*, 221–236.
22 Roos, A., *Thesis*, Université Paris VI, **2004**, p. 350.
23 Berglund, C. A., McKay, K. W. *Polym. Eng. Sci.* **1993**, *33*, 1195–1203.
24 Frischknecht, A. L., Milner, S. T., Pryke, A., Young, R. N., Hawkins, R., McLeish, T. C. B. *Macromolecules* **2002**, *35*, 4801–4820.
25 Lim, C. K., Cohen, R. E., Tschoegl, N. W. *Adv. Chem. Ser.* **1971**, *99*, 397.
26 Hashimoto, T., Nagatoshi, K., Todo, A., Hasegawa, H., Kawai, H. *Macromolecules* **1974**, *7*, 364–373.
27 Daoulas, K., Theodorou, D. N., Roos, A., Creton, C. *Macromolecules* **2004**, *37*, 5093–5109.
28 Crosby, A. J., Shull, K. R. *J. Polym. Sci., Part B: Polym. Phys.* **1999**, *37*, 3455–3472.
29 Roos, A., Creton, C. *Macromolecules* **2005**, *38*, 7807–7818.
30 Roos, A., Creton, C. *Macromol. Symp.* **2004**, *214*, 147–156.
31 Creton, C., Roos, A., Chiche, A. In *Adhesion: Current Research and Applications*, Possart, W. G. (Ed.). Wiley-VCH: Weinheim, **2005**, pp. 337–364.
32 Lakrout, H., Sergot, P., Creton, C. *J. Adhesion* **1999**, *69*, 307–359.
33 Zosel, A. *Colloid Polym. Sci.* **1985**, *263*, 541–553.
34 Chiche, A., Pareige, P., Creton, C., *C. R. Acad. Sci. Paris, IV* **2000**, *1*, 1197–1204.
35 Chiche, A., Dollhofer, J., Creton, C. *Eur. Phys. J. E* **2005**, *17*, 389–401.
36 Josse, G., Sergot, P., Dorget, M., Creton, C. *J. Adhesion* **2004**, *80*, 87–118.
37 Lindner, A., Maevis, T., Brummer, R., Lühmann, B., Creton, C. *Langmuir* **2004**, *20*, 9156–9169.
38 Mooney, M. *J. Appl. Phys.* **1940**, *11*, 582.
39 Rivlin, R. S. *Philos. Trans. R. Soc. London, Ser. A* **1948**, *241*, 379–397.
40 Rubinstein, M., Panyukov, S. *Macromolecules* **2002**, *35*, 6670–6886.
41 Edwards, S. F., Vilgis, T. A. *Rep. Prog. Phys.* **1988**, *51*, 243–297.
42 Treloar, L. R. G. *The Physics of Rubber Elasticity*. Clarendon Press: Oxford, **1975**, pp. 210–310.
43 Gent, A. N. In *Engineering with Rubber*, Gent, A. N. (Ed.). Munich: Hanser, **1992**, pp. 33–66.
44 Shull, K. R., Creton, C. *J. Polym. Sci., Part B: Polym. Phys.* **2004**, *42*, 4023–4043.
45 Crosby, A. J., Shull, K. R., Lakrout, H., Creton, C. *J. Appl. Phys.* **2000**, *88*, 2956–2966.
46 Webber, R. E., Shull, K. R., Roos, A., Creton, C. *Phys. Rev. E* **2003**, *68*, 021805.
47 Shull, K. R. *Mater. Sci. Eng. R, Rep.* **2002**, *36*, 1–45.
48 Maugis, D., Barquins, M. *J. Phys. D* **1978**, *11*, 1989–2023.
49 Saulnier, F., Ondarcuhu, T., Aradian, A., Raphael, E. *Macromolecules* **2004**, *37*, 1067–1075.
50 Ahn, D., Shull, K. R. *Langmuir* **1998**, *14*, 3637–3645.

10
Reactive Blending

Robert Jerome

10.1
Polymer Blending: Benefits, Challenges and Compatibilization

Synthetic polymers are typically the materials of the 20th century, that witnessed their birth, growth and the remarkable innovations they gave rise to. The success of polymers has to be found in their capacity to meet the material needs of society in terms of health, clothing, housing, mobility and communication.

Since the historical publication by Herman Staudinger in 1920 on the concept of giant molecules made by the covalent linking of a large number of monomer units, the worldwide production of synthetic polymers has grown exceptionally to reach some 200 million metric tons at the end of the last century.

By the 1970s, most of the monomers used nowadays were known and exploited, and two main directions evolved in polymer production, (i) improvement of the existing polymerization processes and development of new ones, (ii) economical melt blending of existing polymers.

The commercial interest in the blending of immiscible polymers with the formation of multiphase materials lies in the opportunity to combine in an additive way attractive features of two or more synthetic polymers into one new or at least improved material. Although polymer blending is thought to be a more rapid and straightforward strategy to meet the steadily more demanding needs of customers than revisiting polymerization chemistry, major issues have to be addressed. Indeed, the polymer immiscibility, which is the rule and the condition for preserving the characteristic properties of the blended polymers, leads to phase-separated materials, that often exhibit inherent problems related to the phase morphology and the interfaces: (i) As a result of thermodynamic immiscibility, the interfacial tension is high, gross phase separation occurs spontaneously, and mechanical load-bearing properties are poor because of non-concerted deformation of the phases. (ii) The interfacial adhesion is poor as a result of a narrow interfacial width and lack of penetration of polymer chains from one phase to the other. The interface is then the "Achilles tendon" of the blend, that more likely fails much earlier than the constitutive (even brittle) polymers. (iii) The phase morphology is unstable and spontaneously evolves towards maximum separation whenever the chains are mobile enough.

The best process or technique to tackle these problems is known as compatibilization and it is inspired by colloidal science and technology. Indeed, water–oil dispersions, which are also thermodynamically unstable, are commonly converted into stable latexes by addition of a surfactant, which is a low molecular weight compound with a dual structure, i.e., a hydrophilic head and a hydrophobic tail. By analogy with traditional surfactants, suitably designed block or graft copolymers can virtually play a comparable role in immiscible polymer blends [1, 2, and references therein]. They are then designated as "compatibilizers", "emulsifiers" or "interfacial agents". For this type of copolymer with a dual structure to be located at the interfaces, each constitutive polymeric component must have a specific affinity for one phase of the polyblend (being identical to or miscible with it). Moreover, it must penetrate as deeply as possible the parent phase and form chain entanglements. Whenever these conditions are fulfilled, compatibilization is effective, thus the interfacial tension is reduced, the interfacial adhesion is enhanced, and the phase morphology is stabilized against coalescence.

Basically, there are two major strategies of compatibilization:
- the non-reactive compatibilization, that consists in adding a non-reactive compatibilizer to the immiscible polymers, during blending;
- the reactive compatibilization, that relies on the coupling reaction of mutually reactive chains with formation of the desired compatibilizer *in situ*, during blending.

There are three main routes for the reactive compatibilization of two immiscible polymers PA and PB:
- The PA and PB chains are mutually reactive, and their reaction at the interface leads to direct compatibilizer formation. Polycondensates are typical examples of chains that inherently bear functional end-groups, for instance, a carboxylic acid end-group in polyamides and polyesters.
- If one of the polymers (e.g., polypropylene) is not reactive, it must be functionalized by groups reactive towards the other polymer (such as amine end-capped polyamide). For instance, functionalization of polypropylene with maleic anhydride makes reactive compatibilization with polyamide possible.
- None of the immiscible polymers is naturally reactive. Each of them must be functionalized by appropriate reactive groups, able to react with each other and to form the desired compatibilizer. An alternative, although not general route, might consist in adding to the PA/PB blend two mutually reactive polymers (PC and PD), selected for miscibility with PA and PB, respectively. The limited number of miscible polymers is the reason for the limited application of this approach. A representative example is the interfacial reaction of polystyrene end-capped by a carboxylic acid (PS-COOH) and poly(methylmethacrylate) end-capped by an epoxide (PMMA-epoxy) in poly(phenylene oxide) (PPO misci-

ble with PS) and poly(styrene-*co*-acrylonitrile) (SAN miscible with PMMA) blends [3]. It must be noted that the exothermic interaction of each block of the *in situ* formed compatibilizer and the miscible constituent provides an additional driving force for mutual anchoring and interfacial stability.

At present, reactive compatibilization dominates the commercial exploitation of polymer blending because of a more attractive advantages/disadvantages balance and higher cost-effectiveness [4]. Briefly, non-reactive blending has the advantage of using compatibilizers with predetermined and well-characterized molecular characteristic features, which is essential to draw a fundamental molecular structure–compatibilization efficiency relationship. However, the need for a specific compatibilizer for each immiscible polymer pair is a limitation. Commercial production of limited amounts of a large variety of compatibilizers is not viable. Moreover, an excess of compatibilizer must be used to saturate the interfaces, because part of it may not have time enough to reach the interfaces within the short residence time in the processing equipment. This excess is useless and has a negative impact on the compatibilization cost. In contrast, the compatibilizer is formed where it has to be, thus at the interfaces, in reactive compatibilization. This process is also cost-effective, because the compatibilizer is formed in the bulk, i.e. a solvent-free technique that, moreover, does not require separation and purification. It is also well-suited to poorly soluble precursors. In some cases, an inherently non-reactive polymer could be functionalized in the melt as a pre-step to the reactive compatibilization itself. The lower melt viscosity of the reactive precursors compared to the pre-made compatibilizers may also be a processing advantage. Moreover, when the interface is saturated, the compatibilizer is no longer formed, so that the chance that the critical micelle concentration in the polymer phases is exceeded is low compared to the use of a pre-made compatibilizer, even though the *in situ* formed copolymer might leave the interface for thermodynamic reasons after formation. The major drawback of the reactive compatibilization is the uncertainty of the amount and composition/structure of the *in situ* formed compatibilizer.

10.2
Preparation of Reactive Polymers

The choice of the reactive groups attached to at least part of the blended chains is of prime importance for the reactive compatibilization to be successful. The functional groups must indeed be reactive enough for the interfacial reaction to occur as extensively as possible within the short residence time in the processing equipment (few seconds to few minutes). Moreover, the organic functions and the interchain bonds must be stable enough to survive the processing conditions.

There are four general routes for making reactive polymers available.

- In the most advantageous situation, the reactive groups are incorporated into the chains as inner, pendant or terminal groups as a result of the mechanism of chain formation. Polycondensates are typical examples of chains inherently end-capped by one or each of the two reactive groups involved in the step-growth polymerization, depending on the reaction stoichiometry. Furthermore, they often consist of potentially reactive repeating groups in the backbone. Traditional examples are polyamides with carboxylic acid and amine end-groups and amide inner groups and polyesters with carboxylic acid and hydroxyl end-groups and ester inner groups. As a rule, the constitutive repeating units of polyamides, polyesters such as poly(ethylene terephthalate) and poly(butylene terephthalate), and polycarbonates are prone to interchange reactions, including ester, amide, amide/ester interchanges, transesterification, aminolysis, and acidolysis, as reviewed elsewhere [5]. A large set of structures from random to block copolymers can accordingly be prepared, possibly under processing conditions, which is a substantial advantage. The major drawback might be the limited control of the final structure (average number and length of the constitutive blocks). Moreover, a series of addition polymers contain inherently reactive pendant groups, such as carboxylic acid in poly(meth)acrylic acid, epoxide in poly(glycidyl methacrylate), a double bond in 1,2-poly(butadiene) and 3,4-poly(isoprene).
- Copolymerization of the monomers of interest with a comonomer that contains the desired reactive group is a very versatile and direct route to randomly functionalized polymers. Radical copolymerization of methylmethacrylate with reactive methacrylates including methacrylic acid (MAA), glycidyl methacrylate (GMA), 2-hydroxy ethyl methacrylate (HEMA) is a representative example of this approach. The content of the reactive groups can be adjusted by the comonomer feed composition, and the reactive groups distribution is directly imposed by the reactivity ratios. These ratios change with the polymerization mechanism, which is an additional flexibility. For instance, methacrylic copolymers can be prepared by an anionic rather than by a radical route, provided that protic substituents are protected. In this respect, *tert*-butyl acrylate is a traditional precursor of carboxylic acid, whereas the hydroxyl group of 2-hydroxy ethyl methacrylate can be protected by a silyloxy group.
- Chemical modification of preformed polymers by a variety of chemical reactions is the third approach, which may be very attractive if carried out in the melt, for obvious simplicity and cost effectiveness. Indeed, extra costs associated with solution processes (solvent elimination and recycling) are suppressed, and the final cost is systematically lower when several distinct operations (polymer functionalization and blend compatibilization) can be carried out within a single processing device [6–8]. Polyolefins, which are important players in polymer blending, are synthetic polymers with a very low chemical reactivity. Free radical-induced addition of unsaturated monomers, mainly maleic anhydride, is the modification reaction more extensively carried out in the melt [9]. M. van Duin et al. investigated this grafting reaction along the

screw axis of a running extruder used as a chemical reactor [10]. The major drawback of this radical grafting is a competition between grafting and homopolymerization of maleic anhydride and β-scission of polypropylene chains or crosslinking of polyethylene ones. In addition to free radical grafting of functional (meth)acrylates (MAA, GMA, HEMA) onto a variety of polymers, substitution reactions (e.g., sulfonation and chlorination), and derivatization of terminal groups of preformed polymers are other techniques of post-chemical modification [5].

- Coming back to polymer synthesis, end-functionalization of living propagating chains offers the opportunity of attaching a variety of reactive groups, protected or not, at one or both ends of many addition polymers. Either the initiator contains the (protected) desired function or the living chains are deactivated by a properly selected/designed molecule. The two techniques can however be combined in the same process. When the deactivation approach is used, chains are end-capped at one or both end(s), depending on whether the initiator is mono- or difunctional. Living cationic polymerization of vinyl methyl ether allows poly(vinyl methyl ether) chains to be end-capped by a carboxylic acid or a primary amine. The Na salt of diethyl malonate or an imide derivative of it are the appropriate nucleophilic deactivators [11].

Reaction of living polyanions derived from styrene, α-methylstyrene and 1,3-dienes with appropriate electrophiles, such as carbon dioxide and oxirane, leads to carboxylic acid and hydroxyl terminated chains, respectively.

Macosko et al. proposed an interesting methodology based on the termination of living anionic polymers by ω-functional haloalkanes that contain protected functional groups followed by deprotection.

$$P^\ominus M_e^\oplus + X-(CH_2)_n-M \rightarrow P-(CH_2)_n-M \rightarrow P-(CH_2)_n-F$$

X: halogen
M: masked group
F: functional group

Polymers end-capped by a series of reactive groups (CHO, COOH, epoxy, NH_2, OH, SH) have been prepared with high functionality [12].

With polyolefins, model chains with terminal amine and anhydride functionality have been prepared by living anionic polymerization of 1,3-butadiene, the polydiene being ultimately hydrogenated into poly(ethyl ethylene) and poly(ethylene), depending on the microstructure (1,2 vs. 1,4 addition) of the polydiene chains. The amine and anhydride groups are protected as *tert*-butyl carbamate and *tert*-butyl ester, respectively, in this process [13].

Similarly, it is possible to end-cap poly[alkyl(meth)acrylate] chains by reaction with electrophiles, such as aldehyde, chlorosilane and aromatic acyl chloride derivatives [14]. Nowadays, controlled radical polymerization contributes to expanding the end-functionalization methods of poly(vinylaromatic), poly(meth)acrylate and poly(vinyl acetate) chains [15].

10.3
Compatibilization Reactions

As a rule, any chemical reaction known for low molecular weight compounds can be used in reactive compatibilization. Although the intrinsic reactivity of organic functions is independent of the molecular size, the reaction kinetics can be lowered in the case of macromolecular reagents for a series of reasons, including steric hindrance to the reaction sites, restricted diffusional mobility and low concentration of the reactive groups.

Because the interfacial reactions responsible for the *in situ* formation of the compatibilizer, are conducted at high temperature for a short period of time, they must be fast, selective and irreversible. It is thus mandatory to use highly reactive groups and to minimize mass transfer limitations to reaction. On the basis of these kinetic criteria, the reactive groups that more often participate in reactive compatibilization are anhydride, carboxylic acid, primary amine, alcohol, isocyanate and heterocycles (mainly epoxide and oxazoline). They are involved in a few organic reactions, the most representative of which are listed in Table 10.1, that is the imidization reaction (anhydride/primary amine), ring-opening of oxazoline (with primary amine, carboxylic acid and anhydride), ring-opening of epoxide (with protic reagents), interchange reactions (cf. *supra*). Urea and urethane formation from isocyanates, amidation (carboxylic acid and primary amine) and esterification (carboxylic acid and anhydride with alcohol) are less commonly used. Among all these possibilities, the imidization reaction is by far preferred because of unequally high reactivity and stability of the imide. Table 10.2 is a list of polymer blends reactively compatibilized by the imidization reaction. For sake of completeness, the reader is referred to Liu and Huang [5], who classified the reactive polymers according to the functional group they contain (anhydride, carboxylic acid, proton-containing amine, hydroxyl group, heterocyclic group, group prone to interchange reaction, group prone to ionic bonding, and others), and listed compatibilization reactions in which these groups are involved.

10.4
Molecular Architecture of the Compatibilizer [16]

Attention must be paid to the architecture of the compatibilizer because of possible impact on the compatibilization efficiency in relation to entanglement of the constitutive blocks with the chains to be compatibilized. This characteristic feature depends directly on the content and distribution of the reactive groups attached (on)to the precursors. Three major architectures have to be distinguished:
- Block copolymers. They result from the reaction of two end-functional linear chains.

Table 10.1 Non-exhaustive list of mutually reactive groups X and Y used in reactive blending.

X	Y	X'–Y'
Anhydride (cyclic structure with two C=O and O)	H$_2$N– O=C=N–	(succinimide-type structure with C=O, N, C=O)
Oxazoline (ring with O, N, C)	H$_2$N– HOOC– (cyclic anhydride structure)	–C(=O)–N(H)–CH$_2$CH$_2$–NH– –C(=O)–N(H)–CH$_2$CH$_2$–O–C(=O)– –C(=O)–O–CH$_2$CH$_2$–N (oxazoline ring)
Epoxide	H$_2$N– HOOC–	–CH(OH)–CH$_2$–N(H)– –CH(OH)–CH$_2$–O–C(=O)–
–N=C=O Isocyanate	H$_2$N– HOOC–	–N(H)–C(=O)–N(H)– –N(H)–C(=O)–

Table 10.2 Examples of imidization reactions used in reactive compatibilization.

Co-reactive groups	Reactive groups	Blends	Ref.[a]
Amine/anhydride	EPR-g-MA	PA6, PA6, 6/EPR	22
	EPDM-g-MA	PA6/EPDM	23
	PE-g-MA	PA6/PE	24
	PP-g-MA	PA6/PP	1
	PPE-g-MA	PA6/PPE	25
	ABS-g-MA	PA6/ABS	26
	SEBS-g-MA	PA6/SEBS	27
	S-co-MA(SMA)	PA6/ SAN	28
		PA6/ABS	29
	HIPS-g-MA	PA6/S-co-MA	30
	PMMA-MA	PA10,10/HIPS	31
		PS-NH$_2$/PMMA-MA	32

a) Reprinted with permission of Hanser from [19] and references therein.

- Graft copolymers. They are formed either by copolymerization of the monomer precursor of the main backbone and a macronomer precursor of the grafts, or by grafting of one population of chains onto (or from) reactive groups distributed along the chains of a second population. Toughening of polyamides is a typical example of the second approach, in which the primary amine end-group of polyamides is reacted with maleic anhydride groups attached to, for instance, ethylene–propylene rubbers (EPM, EPDM) [17].
- Branched copolymers. This architecture is usually ill-defined because the two reactive precursors are multifunctional, with the risk of (at least partial) crosslinking, depending on the average number of reactive groups per chain, the reaction stoichiometry and yield.

A representative example of branch/graft compatibilizer has been reported by Datta et al. [18], who prepared blends of primary amine-containing EPM (up to 6 amine groups per chain) and styrene (S), maleic anhydride (M), acrylonitrile (A) terpolymer (SMA containing 8 or 14 wt% of M).

10.5
Phase Morphology Generation and Stability [19]

Because the phase morphology has a direct impact on the mechanical performances of polyblends, it is essential to understand how this morphology develops from pellet-sized or powder-sized particles to sub-micron droplets in non-compatibilized blends. According to Favis [20], the most significant particle size reduction in polycarbonate (PC)/polypropylene (PP) blends, where PC is the dispersed phase, takes place within the first 2 min of mixing, thus when the components are melted and softened. Macosko et al. confirmed that the phase mor-

10.5 Phase Morphology Generation and Stability [19]

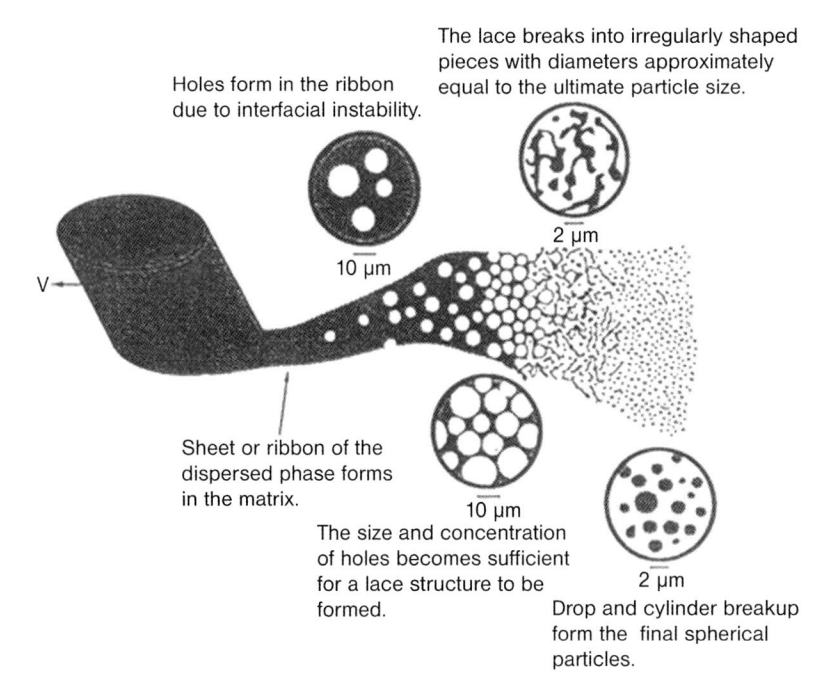

Fig. 10.1 Schematic representation of the morphology development in binary immiscible polymer blends. Reproduced with permission of Hanser from [16].

phology is developed at short mixing times and proposed the following mechanism [21]: Upon shearing, the dispersed phases form sheets or ribbons, in which holes appear as a result of interfacial instability. An increase in the size and number of holes leads to a lace structure, which then breaks down in irregularly shaped drops and cylinders that are ultimately transformed into spherical droplets (Fig. 10.1). At longer mixing times, the largest particles are converted into smaller ones, and an equilibrium between phase break-up and phase coalescence takes place and accounts for a quasi-invariant morphology. As a rule, a gross phase dispersion is observed when the matrix is less viscous than the dispersed phase, because then the dispersive forces are lower and coalescence is more favorable.

Although the phase morphology is generated during the melting/softening of the blended polymers, the interfacial reaction has several effects. The break-up of the largest particles is faster, and coalescence is severely restricted, which leads to smaller average size and narrower particle size distribution. The compatibilizer formed at the interface decreases the interfacial tension, which is favorable to phase break-up, whereas it forms a steric barrier around the dispersed particles and makes the interface less mobile, which results in strong retardation, if not inhibition, of coalescence [22–24].

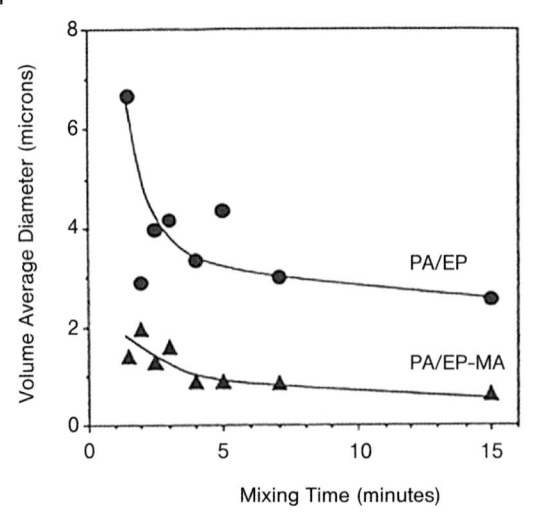

Fig. 10.2 Dependence of the volume average diameter of the dispersed rubber particles on the mixing time for PA/EP and PA/EP-MA blends. Reproduced with permission of Elsevier from C. E. Scott, C. W. Macosko, *Polymer* 1994, *35*, 5422–5433.

Figure 10.2 illustrates the change in the volume average diameter of rubbery particles of a poly(ethylene-co-propylene) copolymer (EP) dispersed within a polyamide matrix (PA) as a function of mixing time. The beneficial effect of reactive compatibilization (use of maleic anhydride grafted EP, thus EP-MA) is obvious, particularly at very short mixing times, keeping in mind that the original size of the rubber pellets was 4 mm. This size has been reduced by more than three orders of magnitude within the first 2 min of mixing, which falls in the softening process.

It may however happen that a phase coarsening is observed upon prolonged mixing of reactively compatibilized blends, as shown in Fig. 10.3 in the case of PA/EPM (70/30) blends. This deleterious effect has been accounted for by the removal of the graft copolymer from the interface by the applied shear forces. This interfacial instability is related to the thermodynamics of the system, in the sense that the graft copolymer has at least a limited miscibility with one of the phases and has a tendency to form micelles in that phase [25]. This issue is of prime importance when polymer blends are annealed or re-processed for fabrication of end-use products. The key concern must be that the structure and composition of the compatibilizer is such that the interactions of each constitutive chain with the parent phase of the blend are properly balanced for this compatibilizer to stay at the interface, whatever the post-treatments of the polyblend.

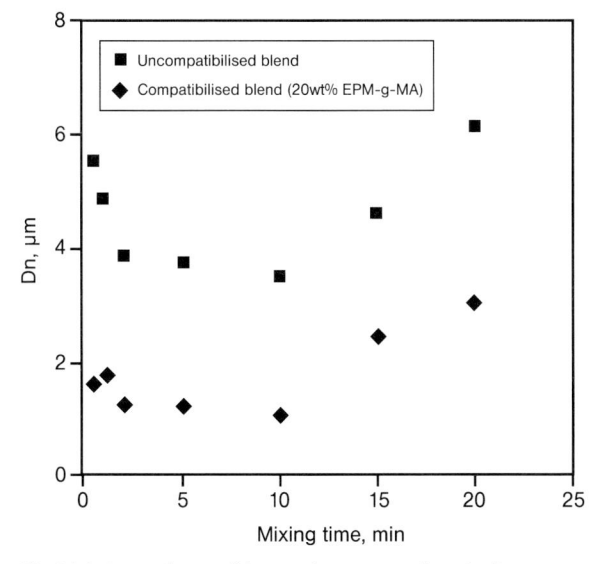

Fig. 10.3 Dependence of the number average domain diameter for uncompatibilized and reactively compatibilized 30/70 EPM/Nylon 6. Reproduced with permission of Elsevier from [35].

10.6
Kinetics of the Interfacial Reaction

Because the phase morphology and high interfacial area are generated in the very early stage of mixing, the compatibilization reaction must be as complete as possible during the short period of melting/softening of the blended components, which emphasizes the critical role of the kinetics of the interfacial reaction.

First, this problem refers not only to the intrinsic reactivity of the functional groups attached to the precursors of the compatibilizer to be formed at the interface but also to the experimental conditions under which the reaction is conducted. Indeed, attachment of the reactive groups to polymer chains, position of these groups along the chains and type of the reaction medium (solution vs. melt, homogeneous vs. heterogeneous melt, static vs. mixed heterogeneous melt) have a decisive effect on the reaction kinetics. This issue has been addressed by Macosko et al. [26], who compared the ratio of rate constants for the competitive reaction of amine end-capped PMMA and PS (PMMA-NH$_2$ and PS-NH$_2$) with PMMA bearing a phthalic anhydride in either a terminal (PMMA-eAn: k_E) or a mid-position (PMMA-mAn: k_M). In solution at 25 °C, $k_E/k_M = 2.8$ for the specific system under consideration, which indicates that functional groups at the center of polymer chains experience higher hindrance than the parent end-groups. This observation is in line with the kinetic excluded-volume effect, that predicts the scaling of the rate constant k as $(N_1+N_2)^{-\nu}$, where N_1

and N_2 are the degrees of polymerization of polymers 1 and 2, respectively [27]. The predicted value of the exponent v is 0.16 for reaction of two end-groups in solution. On this basis, k_E/k_M should have been 2.1, somewhat smaller than observed, which may be accounted for by steric effects. In the homogeneous melt of reactive PMMA chains at 180 °C, the kinetic excluded volume effect is screened (θ condition) and $k_E/k_M=1$, although it was 1.7 in this study. The shielding of the reactive groups by polymer segments might be at the origin of this discrepancy. The k_E/k_M ratio is by far the higher when the coupling reaction occurs at a static flat interface at 180 °C ($k_E/k_M > 10$), in agreement with an interface which is initially more densely populated by chain ends and which produces mainly block copolymer first. Finally, the interfacial area is continuously changed under mixing flow, which decreases the partition of the chain ends at the interface compared to that at equilibrium. Accordingly, k_E/k_M at 180 °C is decreased (2.6–3.2) although it remains higher than in the homogeneous melt. The beneficial kinetic effect of flow was emphasized by a reaction rate constant that was more than 1000 times higher for the interfacial reaction under mixing than that in the static bilayer assembly at 180 °C.

In the case of insufficient reactivity of the functional groups, their concentration should be increased or catalysts or small-size molecules should be added [28]. The beneficial effect of the reactive groups' concentration and the catalysis of the interfacial reaction was illustrated by Nakahama et al. [29], in a series of coupling reactions based on the carboxylic acid (COOH)/epoxide (E) pair. Compared to the homogeneous coupling in the melt of two populations of polystyrene chains end-capped by an acid (PS–COOH) and an epoxide (PS–E), respectively, the interfacial reaction is approximately two times faster when PMMA–E chains are substituted for the PS–E chains in a heterogeneous system, consistent with the occurrence of the reaction in a thin interfacial layer where the local concentration of the reactive functions is higher. Moreover, the addition of a catalyst insoluble in the two phases increases the kinetics of the interfacial reaction, although that of the homogeneous coupling remains unchanged.

It is also worth noting that when the compatibilization reaction is slower than the development of the phase morphology, the morphology of the uncompatibilized blend is, as expected, stabilized by the interfacial reaction [30]. This was demonstrated by experiments conducted with a low content of dispersed phase, such that coalescence was negligible, even in non-compatibilized blends [31, 32]. Another example can be found in the melt blending of 25 wt% of a rubber (EPM containing 50 wt% of maleic anhydride grafted EP chains) and 75 wt% of a styrene/acrylonitrile copolymer in which 0.028 mol/wt% of either a primary amine (SAN-NH$_2$) or the protected version, that is a carbamate (SAN-carb), was randomly distributed [33]. Although SAN-NH$_2$ is directly available to reaction with the anhydride, SAN-carb needs thermolysis at the processing temperature to release the reactive amine. Two quite different kinetics can accordingly be compared in the same compatibilization reaction. Figure 10.4 shows that 30% of SAN-NH$_2$ has reacted during the melting/softening step (~ 3 min) and that a very fine phase morphology has developed which does not change importantly

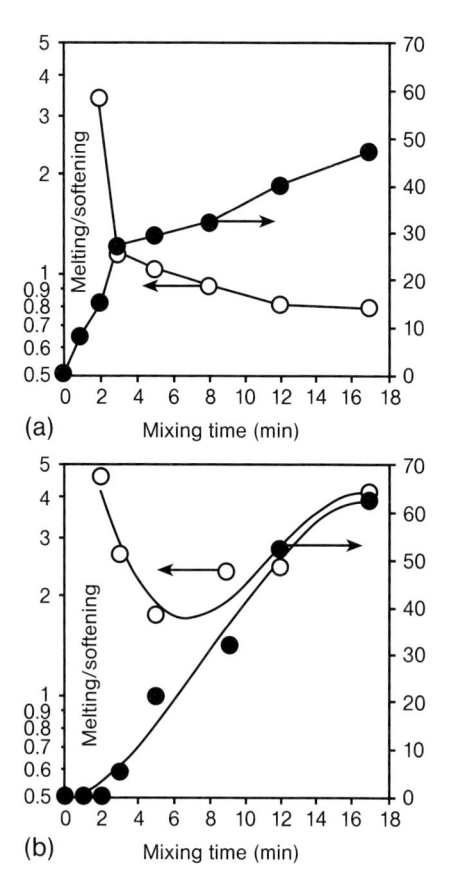

Fig. 10.4 Mixing time dependence of both the progress of the interfacial reaction and the development of phase morphology when EP-g-MA is melt blended with (a) SAN-NH$_2$ and (b) SAN-carb. Reproduced with permission of the American Chemical Society from [33].

upon further mixing. When thermolysis of carbamate into amine is the rate-determining step of the interfacial reaction, only 5 wt% of SAN has reacted at the end of the melting/softening process, and the phase dispersion is clearly challenged by coalescence, as shown by the average particle size that goes through a minimum with increasing mixing time. Furthermore, part of SAN is occluded in the rubber particles on the occasion of their coalescence, which does not occur in the rapidly compatibilized polyblend.

10.7
Molecular Weight of the Reactive Chains

The length of the reactive chains has a multifold impact on the blending process. Indeed, it has a direct influence on the blend viscosity, on the kinetics of the interfacial reaction, on the interfacial stability of the compatibilizer and on the capacity of this compatibilizer to increase the interfacial adhesion.

Briefly, the relative melt viscosity of the phases plays a leading role in the phase morphology, because, for the phase break-up and the interface generation to be effective, the matrix must be more viscous than the dispersed phase. Depending on the blend composition, this requirement might not be fulfilled because of the marked tendency of the constituent of the lower viscosity to encapsulate the second constituent in binary blends. Actually, melt viscosity and volume fraction of each constituent decide which polymer phase is continuous in two-phase polyblends according to the Jordhamo's equation [34] that expresses the condition for phase inversion:

$$\frac{\eta_1 \phi_2}{\eta_2 \phi_1} = 1$$

where η is the melt viscosity and ϕ the volume fraction of components 1 and 2.

The effect of reactive compatibilization on phase inversion and thus on phase co-continuity is not yet well understood. Experiments have concluded that the location of the phase inversion region is not perturbed by the addition of a reactive compatibilizer [35–37], although reactive compatibilization of PA6/PMMA blends by poly(styrene-co-methacrylic acid) showed that the region of phase co-continuity was smaller and shifted in the composition scale compared to the non-reactive blending [38].

Any decrease in the molecular weight of the reacting polymers is favorable to the interfacial reaction. At constant weight fraction, the concentration of the functional end-group of linear chains increases indeed with decreasing molecular weight with a positive impact on the reaction kinetics and completeness [39]. This favorable effect is amplified by the faster diffusion of lower molecular weight chains to the interface and the smaller entropic penalty that results from their location at the interface. In this respect, an interesting observation was reported by Maréchal et al. [40], who observed that the molecular weight of polyamide-6 chains (PA6) grafted onto EP-MA rubber was approximately half that of the unreacted PA6.

There is, however, a lower limit for the length of the reactive chains, below which the reactive compatibilizer can leave the interface and form micelles in one phase of the polyblend [41]. Moreover, the length of the reactive chains is also decisive for effective entanglements to be formed within the phases to be compatibilized and, thus, for the interfacial adhesion to be strong enough and to prevent cracks initiated at the interface from growing until catastrophic failure [42, 43].

The effect of the molecular weight of the reactive chains has also been illustrated by blending PMMA chains end-capped by a phthalic anhydride group (PMMA-PA) with an equimolar mixture of two populations of polystyrene chains end-capped by a primary amine (PS-NH$_2$) with $M_n = 26\,000$ and $72\,000$, respectively. The NH$_2$/anhydride molar ratio was 1.5. Again, the initial reaction rate is very high, with a conversion of approximately 25% of the PMMA-PA chains within less than 1 min. A fine phase dispersion is established at quite a

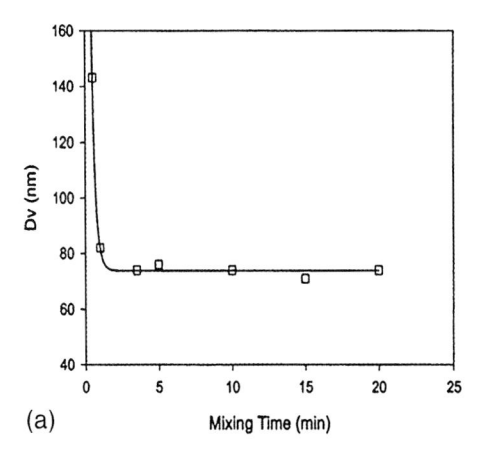

Fig. 10.5 Time dependence of the volume average diameter (D_v) (a), and conversion of reactive PMMA into diblock copolymers (b). Reproduced with permission of the American Chemical Society from [44].

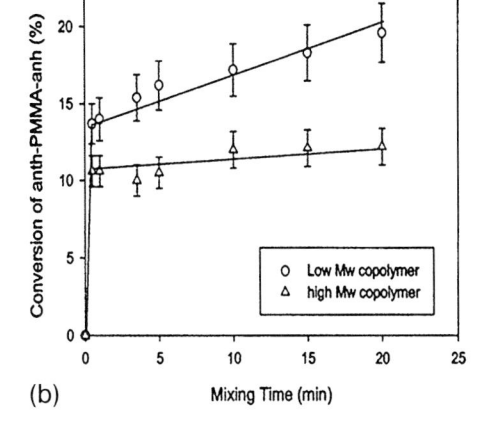

comparable rate, which remains unchanged at longer mixing times. Quite interestingly, the progress of the reaction at the interface of a constant surface area is then dramatically slowed down, consistent with the diffusion of the reactive chains to the interface, thus through the brush formed by the constitutive blocks of the diblocks. Figure 10.5 shows that the diffusion of the longer PS chains ($M_n = 72\,000$) is so slow that the apparent reaction rate is close to zero, in contrast to the shorter PS chains ($M_n = 26\,000$) that react further, although slowly (~5 wt% of the PMMA-PA chains converted within 20 min).

As a rule, the relative kinetics of two major events, that is the interfacial reaction and the renewal of the interface, has a decisive effect on the development of the phase morphology and, ultimately, on the production of block or graft copolymer by reactive blending. Two conditions must be fulfilled for the interfacial reaction to be quasi-complete and to be an effective technique for the production of block or graft copolymers in the melt: (i) the copolymer must escape from the interface as rapidly as it is formed, otherwise the progress of this reaction is slowed down, if not blocked. (ii) The reaction at the interface must be fast, in order to compensate for the migration of the copolymer away from the interface and to prevent microphase separation from occurring. These prerequisites are met whenever the reactive chains bear highly reactive groups and are short enough not to be involved in stable chain entanglements. This statement has been illustrated by melt blending at 170 °C for 10 min of primary amine containing PS chains with $M_n = 10\,000$ and 35 000, respectively, with PMMA of the same molecular weights end-capped by either an isocyanate (slow reaction) or an anhydride (fast reaction). A nanophase morphology and a quasi-complete reaction are observed in the specific case of the anhydride/amine reaction implemented with the lower M_n chains (10 000 for both the PMMA and PS) [45]. This general behavior has been confirmed by a static experiment based on the annealing at 200 °C for 2 h of a sandwich assembly that consists of a layer of PMMA chains end-capped by an anhydride ($M_n = 15\,000$) and a layer of PS end-capped by a primary amine with $M_n = 17\,000$ and 27 000, respectively [46]. When the shorter chains are allowed to react, the interfacial reaction is faster and still in progress after 2 h. Simultaneously, a kind of polymer emulsion is developed at the interface. In contrast, when the longer chains are concerned, the conversion levels off after ~40 min and only a faint roughening of the interface is observed, which is consistent with the blocking of the interface by a diblock copolymer of a higher molecular weight (Fig. 10.6).

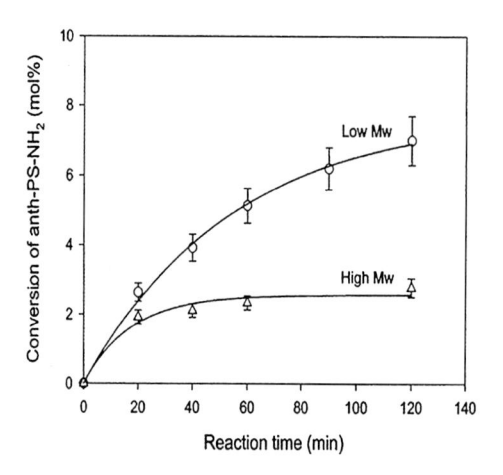

Fig. 10.6 Conversion of low and high molecular weight PS-NH$_2$ into diblock copolymer. Reproduced with permission of the American Chemical Society from Ref. [46].

10.8
Content of Reactive Group of the Reactive Polymers

The reactive group content (RGC) has to be considered when at least one of the reacting polymers is multifunctional, because it dictates the molecular architecture and composition of the compatibilizer formed at the interface and, as a result, the compatibilization efficiency, the phase morphology and the ultimate mechanical properties of the polyblends.

The effect of the molecular architecture of the compatibilizer has been studied in the specific case of polyethylene/polystyrene blends added to premade hydrogenated polybutadiene/polystyrene copolymers of various architectures [47]. As a rule, the graft copolymer was the less efficient compatibilizer based on the ultimate mechanical properties of polyblends of extreme compositions (80/20 and 20/80 wt/wt). Triblock copolymers with a linear or radial structure and a major central hydrogenated block are either more efficient or less efficient than the graft copolymer, depending on the blend composition. Diblock copolymers are far more efficient compatibilizers over the whole composition range, which may be explained by less drastic conformational restraints at the interface and better penetration and chain entanglement of each block into the parent polymer phase. Last but not least, the internal structure of diblock copolymers, all other conditions being the same, is also important because a copolymer with a progressive transition from one block to the other, known as a tapered diblock, is superior to the parent pure diblock in which this transition is sharp. It might be assumed that the interfacial region in the blends compatibilized by a tapered diblock would be more diffuse and flexible with a favorable impact on the mechanical properties.

The more complex situation of multigraft copolymers prepared by reactive blending has been investigated by Pagnoulle et al. [48] who compatibilized 75/25 wt/wt SAN/EPR blends by using EP-g-MA and SAN containing different amounts of either a primary amine or its carbamate precursor (type a: 0.004, type b: 0.028 and type c: 0.049 mol/wt%). The effect of both the kinetics of the interfacial reaction and the RGC of SAN has thus been analyzed in the same investigation. The development of the phase morphology at constant mixing time depends only on the extent of the interfacial reaction in relation to the amine (carbamate)/anhydride molar ratio and is independent of the RGC of SAN. A unique curve of the rubber particle diameter on the SAN reactive group/anhydride molar ratio is observed provided that 3.3 more carbamates than amines are added to the polyblend, which indicates that the amine is 3.3 times more reactive than the carbamate towards maleic anhydride (Fig. 10.7a). At a constant particle size, the fracture toughness of the blends however depends on the RGC of the reactive SAN chains in a way that changes with the type of reactive groups (amine vs. carbamate) as shown in Fig. 10.7b. The impact resistance is indeed independent of the carbamate content, whereas it decreases when the amine content of the SAN chains is increased. The high mutual reactivity of the amine/anhydride pair results in the multigrafting of the individual SAN

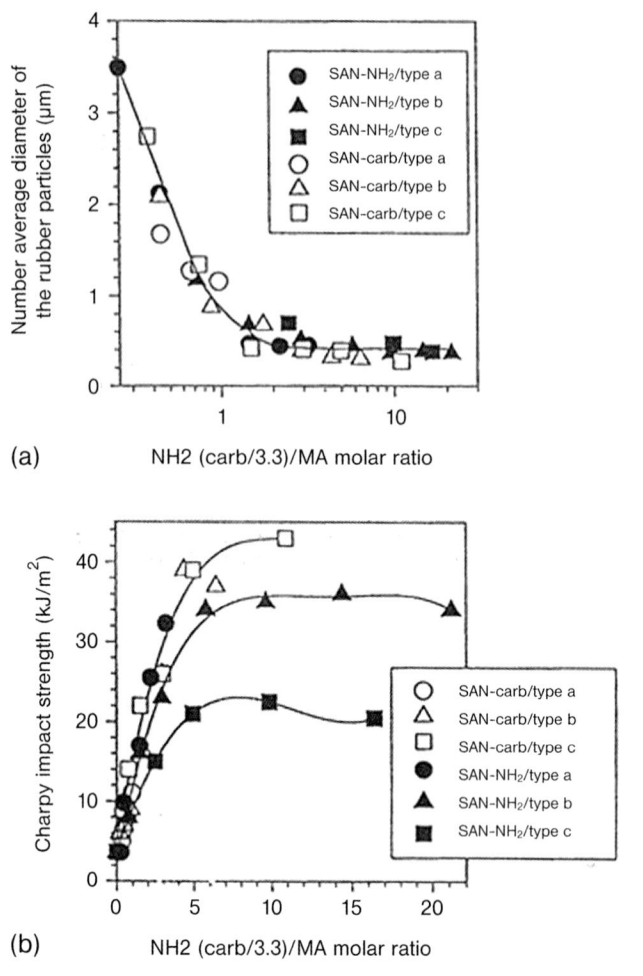

Fig. 10.7 (a) Dependence of the number average diameter of the rubber particles on the reduced carb(NH_2)/MA molar ratio for the 75/25 SAN/EPR polyblend modified by SAN-carb(NH_2) of various contents of reactive groups (type a: 0.004, type b: 0.028 and type c: 0.049 mol/wt.% of either NH_2 or carbamate) in the SAN phase and 50 wt% EP-g-MA in the EPR phase. (b) Dependence of the notched Charpy impact strength on the reduced carb(NH_2)/MA molar ratio for the 75/25 SAN/EPR polyblends modified by SAN-carb(NH_2) of various contents of reactive groups (type a: 0.004, type b: 0.028 and type c: 0.049 mol/wt% of either NH_2 or carbamate) in the SAN phase and 50 wt% EP-g-MA in the EPR phase. Reproduced with permission of the American Chemical Society from [48].

chains, the average number of grafting events per SAN chain increasing with their RGC. As a result, the SAN backbone lies as flat at the interface as the grafting degree is high, and the opportunity of becoming entangled with the SAN matrix decreases accordingly. In contrast, because of the slow thermolysis of the carbamate groups into amines, the SAN-carb chains are essentially tethered at the interface whatever their RGC, which allows them to be intermingled within the matrix to the benefit of the impact properties.

10.9
Nanostructured Polyblends

Until recently, reactive blending has been used to compatibilize immiscible polymer blends, thus to form *in situ* enough block or graft copolymer to control coalescence and to improve adhesion between the phases. Although the idea to produce nearly pure block or graft copolymers in the melt is very attractive, it is effective only in the case of short enough chains that react quickly and never saturate the interface (cf. *supra*). Between these two extreme situations, reactive blending can be exploited as a low-cost method to produce nanostructured thermodynamically stable materials with a specific morphology [49, 50]. Indeed, at concentrations higher than those used for compatibilization, block and graft copolymers self-assemble spontaneously and are nothing but nanostructuring agents for the parent polymer blends. During the last few years, special attention has been paid to this unique opportunity. For instance, Leibler et al. focused on the self-organization of graft copolymers prepared by reactive blending [51, 52]. Three major conditions have emerged from these studies for the polyblend nanostructuring to be successful and stable against annealing. Not only must enough graft copolymer be formed by the interfacial reaction, but the unreacted and/or non-reactive chains must be accommodated in the nanophases of the self-organized structures, which is directly related to the radius of curvature of these nanostructures and to the relative size of the guest chains and the host copolymer blocks. Furthermore, polydispersity of both the backbone and grafts of the copolymer and randomness in graft-attachment positions result in disorder, which is favorable to local fluctuations of the interface curvature. This seems to be a mechanism for reducing the chain stretching in the graft copolymer nanostructures and to facilitate the accommodation of the ungrafted chains. The key problem is indeed that the copolymer self-assembly must take place without expulsion and phase separation of the unreacted/non-reactive chains. In this respect, three regimes have been identified in blends of the AB symmetric copolymer/A homopolymer type [53, 54]: (i) the wet-brush regime is established when the length of the homopolymer (N_{Ah}) is shorter than the miscible block of the copolymer (N_{Ac}). Then, the brush of the A blocks is swollen by the homopolymer A and solubilization is uniform. However, in order to minimize the compression of the non-swollen B blocks, the originally flat interface exhibits an increasing curvature with the content of homopolymer to be accommodated; (ii) the dry-brush regime takes place when N_{Ah} and N_{Ac} are approximately the same. Solubilization is then localized in the copolymer domains without swelling of the brushes. The original lamellar phases persist, although the interplanar distance increases with the homopolymer content and ultimately vesicles can be formed; (iii) macrophase separation is observed when $N_{Ah} > N_{Ac}$ because of a too high energetic penalty for the homopolymer accommodation.

The grafting density plays a decisive role in the ability of graft copolymers to accommodate parent homopolymer. Let us consider the grafting of a backbone PB by arms PA. Figure 10.8 shows schematically the interface curvature in relation to the grafting density [52]. When the fraction of grafting sites is small, the interface formed by this asymmetric copolymer is strongly curved, with the grafts PA on the concave side, and a fraction of unreacted A chains should be expelled from the copolymer mesophases [52]. An interface with a large curvature radius is formed in the case of a symmetric graft copolymer (the average distance between grafts would be approximately twice the average graft length). Co-continuous structures are then stabilized and well adapted to accommodate unreacted PA and PB chains, such that the overall composition can be tuned over a wide range. The curvature of the interface can be reversed (concave towards PB) by a too small distance between the grafts together with a problem of expulsion of unreacted PB chains. In this general scenario, the possible effect of polydispersity on the self-assembly of the copolymer chains can be found in theoretical predictions that mixtures of block copolymers of different lengths should help to stabilize nanostructures in blends with homopolymers through relaxation of the stretching energy in the copolymer layers [55–57].

In a representative example, 80 wt% polyethylene containing 1.0 wt% of maleic anhydride ($M_n = 9300$) was reactively grafted by polyamide-6 (PA6; $M_n = 2500$) with one terminal amine group (anhydride/amine = 1/1) [51]. The entanglement limit of PA6 is estimated at 2000 g mol^{-1} [58]. Whatever the mixing and annealing conditions, interconnected sheets of PA6 are formed with a characteristic thickness of 20 nm. As expected, when the anhydride content of poly-

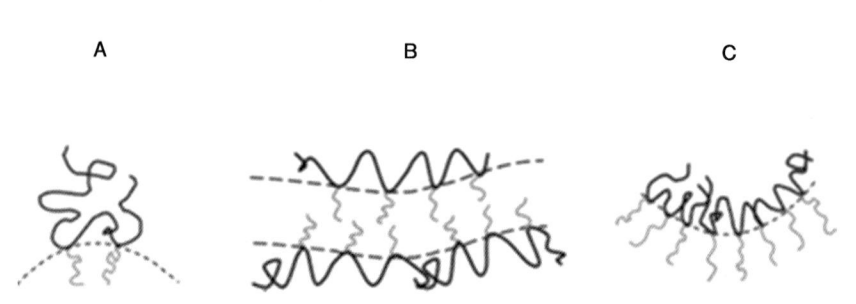

Fig. 10.8 Schematic representation of the self-assembly of PB-g-PA as a function of PB graft density, which increases from A to C. Chains PB are in black and chains PA are in grey. Reproduced with permission of Elsevier from [52].

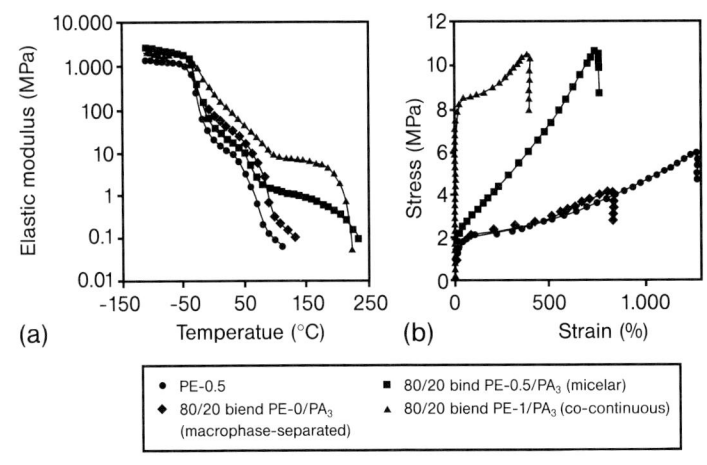

Fig. 10.9 Elastic modulus measured by dynamic mechanical analysis at 1 Hz, 0.1% strain and 3 °C min^{-1} (a). High-strain tensile properties (b). Reproduced with permission of Nature Publishing Group from [51].

ethylene is decreased by a factor of two (anhydride/amine = 1/2), the *in situ* formed graft copolymer is asymmetric, a disorganized spherical morphology is observed in which all the unreacted PA6 chains are not solubilized. The co-continuous nanophases triggered by the symmetric graft copolymer provide polyethylene with unique properties of transparency, creep and solvent resistance, together with increase in stress at break while keeping ductility. That the nanostructured material does not creep when heated above the melting temperature of polyethylene is shown in Fig. 10.9. The elastic modulus indeed remains constant at 10 Mpa between 100 and 220 °C, thus until the melting of PA6. The remarkable strengthening of the polyolefin at high temperature results from the continuity of PA6 (20 wt%) which is able to crystallize even at the nanometer scale. Clearly, both components are able to crystallize in confined nanodomains that remain stable during the timescale of crystallization.

Quite similar behavior has been observed upon substitution of glutaric anhydride containing PMMA for maleic anhydride containing polyethylene [52]. Figure 10.10 illustrates the co-continuous nanostructures exhibited by PMMA/PA6 blends with a 80/20 and 70/30 wt composition, respectively. PMMA contained approximately 24 anhydride units per chain of $M_n = 4200$, and the M_n of PA6 was 2500. The *in situ* formed copolymer thus self-organizes into nanostructures with a flat interface, that incorporate the unreacted homopolymers. Nevertheless, blending with longer PA6 chains (15000 g mol^{-1}) results in coarser and unstable dispersions. The entropic penalty associated with the localization of polymer brushes at the interface indeed increases with chain length [59], whereas pull-out of copolymer from the interface is rapidly a problem when the length of the grafted chains exceeds their entanglement limit [45, 46].

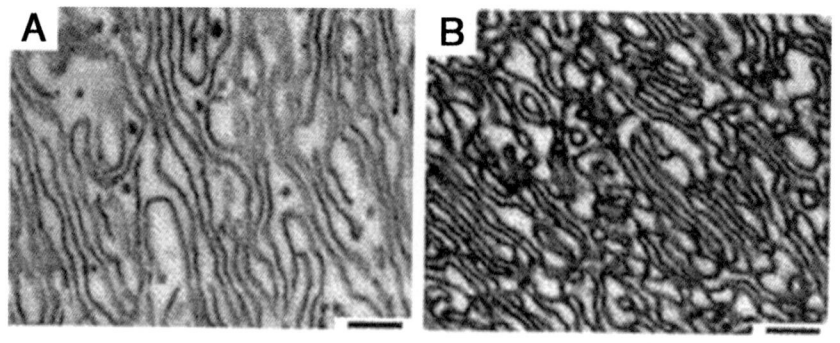

Fig. 10.10 TEM micrograph of PMMA (M_n=42000; 5.7 mol% anhydride; 2.2 mol% carboxylic acid) and PA6 (M_n=2500) blends after annealing at 225 °C for 20 h: (A) 80/20 and (B) 70/30. Reproduced with permission of Elsevier from [52].

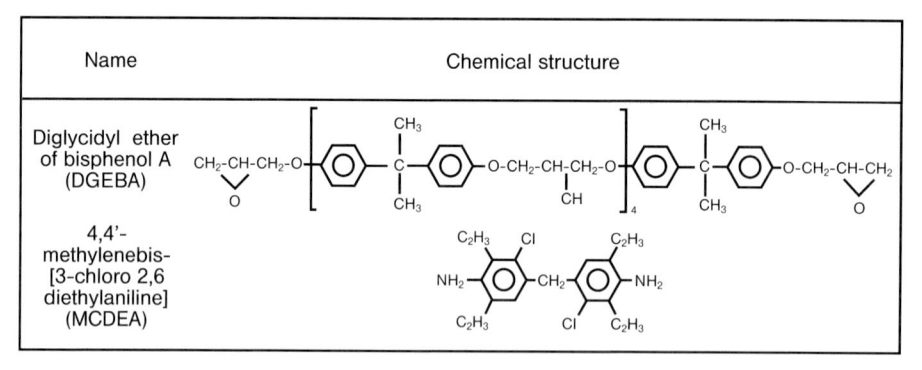

Name	Chemical structure
Diglycidyl ether of bisphenol A (DGEBA)	
4,4'-methylenebis-[3-chloro 2,6 diethylaniline] (MCDEA)	

Fig. 10.11 Chemical structure of the liquid precursors of the epoxy resin.

Because ABC triblock copolymers display a broadened spectrum of phase morphologies [60–62], their blends with AB and AC diblock copolymers have been investigated as a way to prepare new nanostructured materials [61, 63]. An interesting example of nanostructuring by SBM triblock copolymers can be found in reactive systems, such as liquid reactive mixtures used as precursors for rigid epoxy thermosets (S stands for polystyrene, B for polybutadiene and M for poly(methylmethacrylate)). Because the M block is soluble in the liquid epoxy precursors (DGEBA + MCDEA; Fig. 10.11), in contrast to the S and B blocks, SBM copolymers spontaneously form nanostructures by self-assembly, as supported by the thermal dependence of both the storage modulus (G') and loss modulus (G'') of SBM solutions of different concentrations (Fig. 10.12). Below 10 wt% SBM in the liquid diepoxide, a viscous behavior is observed. At higher SBM content, a gel-like behavior is reported, which is the direct evidence for an ordered microphase-separated structure. Quite importantly, the self-assembled nanostructures are preserved during curing and they change with the

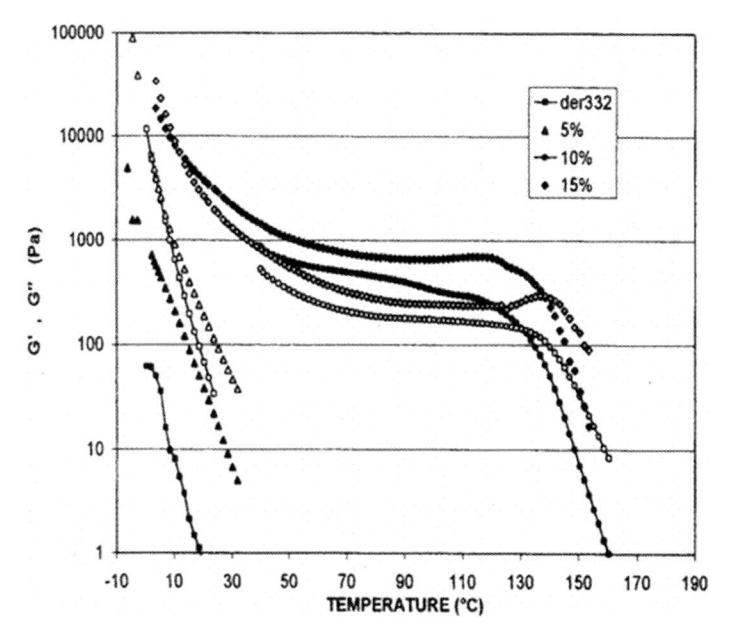

Fig. 10.12 Temperature dependence of the storage (G') and loss moduli (G'') for solutions of SBM in the epoxy precursor at different concentrations (filled symbols stand for G'). Reproduced with permission of Wiley from [68].

SBM composition and content, as illustrated in Fig. 10.13. This nanostructuring of epoxy resins by polybutadiene containing nanophases results in improved toughness, as supported by at least a twofold increase in the critical stress intensity factor (K_{IC}). It must be noted that the SBM triblocks are contaminated by SB diblocks which are completely accommodated in the nanophases, although they macrophase separate within the resin when dispersed alone.

Coming back to reactive blending of thermoplastics, 80 wt% polyamide 12 (PA12) which is intrinsically reactive (terminal primary amine), has been melt blended with polystyrene-b-polyisoprene (SI) diblocks bearing a phthalic anhydride group at the end of the polyisoprene block (SI-anh). The purpose is the thermodynamically stable nanostructuring of PA12 by an *in situ* formed triblock copolymer [50, 70]. The choice of a rubbery block (polyisoprene) aims to improve the toughness of PA12. Polyamide 12 was first melt blended with a symmetric SI diblock, such that the M_n of the individual blocks is close to the M_n of PA12. Mixing for a few minutes at 220 °C results in vesicular type nanostructures with an average size of 50 nm [71, 72]. The continuous lamellar morphology, which is expected for a symmetric triblock, cannot persist in a dilute regime under shear but evolves into more stable vesicular bilayers of triblock lamellae with a low curvature compared to the wall thickness [73, 74]. These nanophases shown in Fig. 10.14 have been observed by TEM in the direction

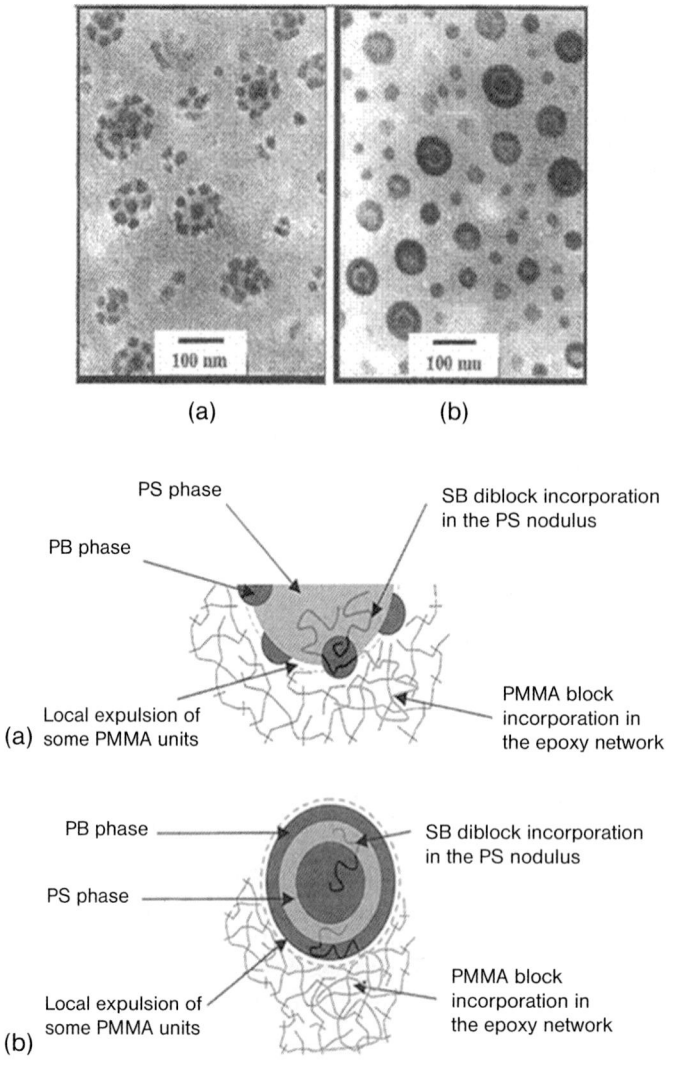

Fig. 10.13 TEM micrographs of 30 wt% (SBM + SB) in the cured epoxy resin and schematic representation of the nanodomains. (a) $S_{22}^{27}B_9M_{69}$ + 21% SB. (b) $S_{12}^{14}B_{18}M_{70}$ + 10% SB. The superscript numbers refer to $M_n \times 10^{-3}$ and the subscript numbers to the weight content of each block in the copolymer. Reproduced with permission of the American Chemical Society from [67].

perpendicular to the extrusion flow. However, vesicular cylinders are observed in the parallel direction, which indicates that these nanostructures are easily oriented by a flow field [50]. Crystallinity of PA12 is not at all perturbed by the *in situ* formed nanostructures. The vesicular morphology has the capacity to accommodate unreacted SI diblock, as illustrated by an experiment in which PA12

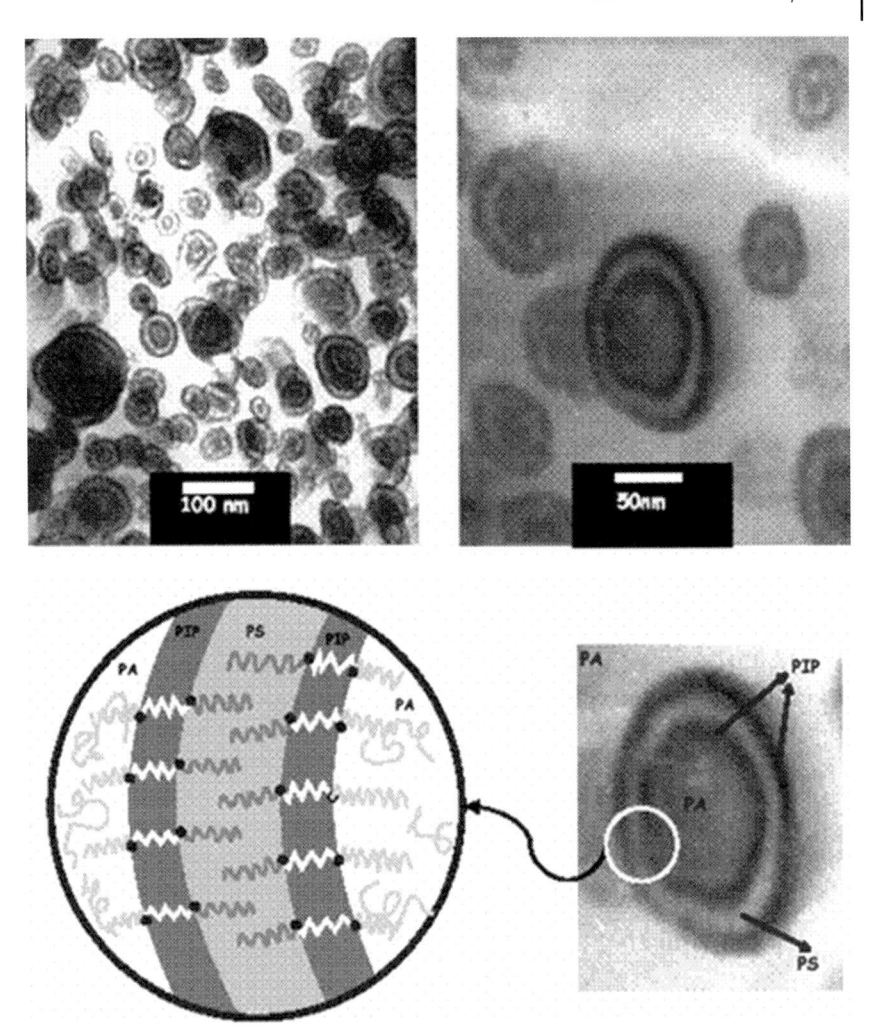

Fig. 10.14 TEM micrographs for the PA12/S$_{47}$I$_{53}$-anh (80/20 wt/wt) reactive blend, and schematic representation of the constitutive vesicles (47 and 53 refer to the weight content of the blocks). Reproduced with permission of the American Chemical Society from Ref. [72].

(80 wt%) was melt blended with an equimolar mixture of reactive SI and the non-reactive counterpart. A transition from bilayered vesicles to onion-like morphology is observed, which emphasizes the easy tuning of the vesicular morphology. Moreover, when the anhydride group is attached at the end of the S block rather than the I block, the nanostructuring remains of the vesicular type.

While keeping the total molecular weight of the diblock constant (35 to 40 kg mol^{-1}), the volume fraction of PS has been increased, which results in a

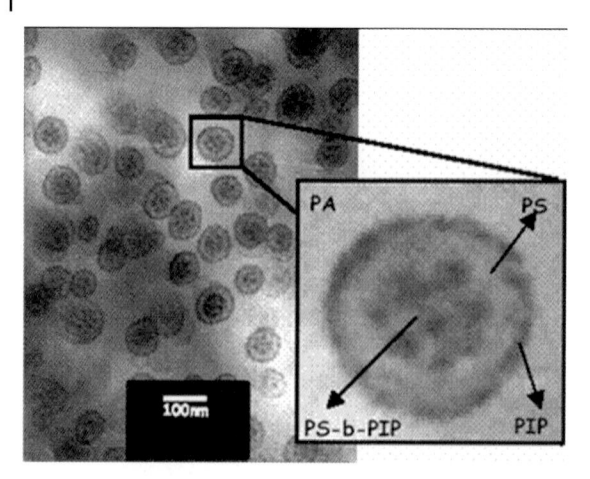

Fig. 10.15 TEM micrographs for the PA12/$S_{82}I_{18}$-anh (80/20 wt/wt) reactive blend (82 and 18 refer to the weight content of the blocks). Reproduced with permission of the American Chemical Society from [72].

transition from lamellae to cylinder (or spheres) for the neat copolymers. Accordingly, the *in situ* formed nanostructures change from an anisotropic vesicular to an isotropic core (S) and shell (I)-like morphology [50, 72]. Substructuring of the core is also observed, which is reminiscent of a slice of cucumber, and has been designated as a cucumber-like morphology (Fig. 10.15). Quite interestingly, this type of complex morphology has been theoretically predicted for surfactant nanodroplets formed by an AB diblock with $N_A/N_B = 3.0$ ($N_S/N_I = 3.0$ in this work), self-organized in a weakly selective solvent (PA12 in the polyblend) in a mild segregation regime [75]. Another key observation is that cucumber-like core–shell nanoobjects can be formed in PA12 by reactive blending of 20 wt% of an additive that consists of a blend of symmetric reactive SI diblock (precursor of the bilayered vesicles) and homo S in such an amount that the uneven volume fractions of S and I responsible for the cucumber-like morphology are restored. M_n of homo S was chosen smaller than M_n of the S block of the diblock in order to fall in the wet-brush regime of the interface [76].

The mechanical properties of PA12 (80 wt%) modified by reactive SI diblocks of constant M_n but different compositions have been measured [50]. Figure 10.16 compares fracture toughness and tensile modulus for neat PA12, and for PA12 containing vesicular nanostructures and cucumber-like core–shell nanostructures, respectively. The best and remarkable toughness–stiffness compromise is observed for the cucumber-like morphology in close relation to a very low rubber content (4 wt%) compared to the nanovesicular morphology (10 wt%). It is essential to know whether this unique toughening strategy of PA12 by a nanostructuring triblock reactively formed *in situ*, is general or not. For this purpose, two fragile thermoplastics, PMMA and SAN, have been end-

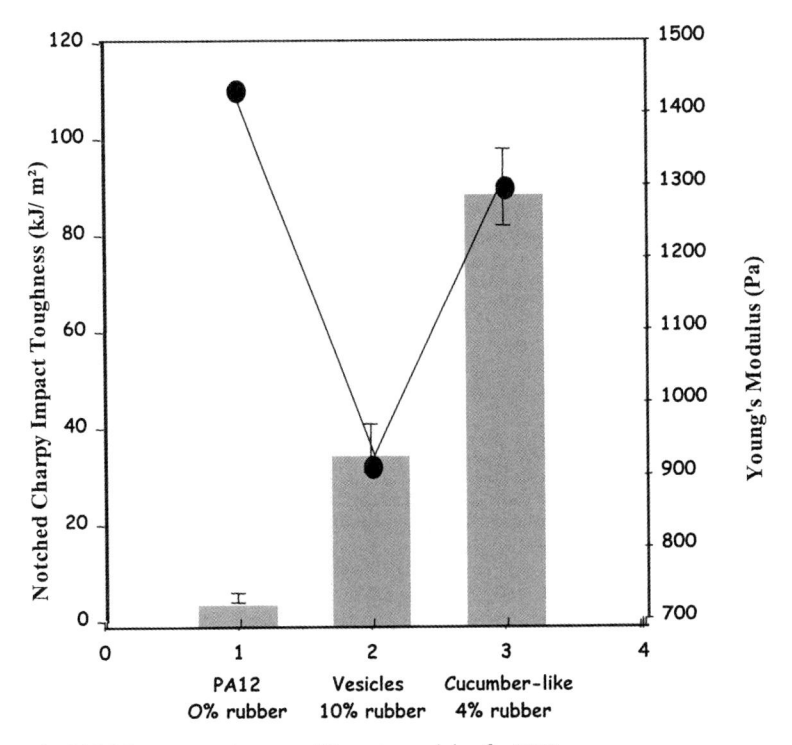

Fig. 10.16 Impact toughness and Young's modulus for PA12 modified by 20 wt% of reactive SI diblock.

capped by a primary amine and melt blended with the same reactive SI diblock. For instance, the typical cucumber-like core–shell morphology has been systematically reproduced, which confirms that the molecular architecture and composition of the ABC triblock have a decisive effect on the nanostructures that are spontaneously formed in a variety of matrices ranging from semi-crystalline PA12 to glassy SAN and PMMA [72, 77, 78]. Reactive blending of a diblock precursor with a matrix is thus quite a general strategy to form *in situ* a nanodispersed rubbery component with tunable shape and size.

10.10
Conclusions

When the polymer chemists started to synthesize block copolymers some forty years ago, they were pioneering the tremendously active research conducted nowadays at the nanometer scale. Indeed, a remarkable polymeric nanomaterial was launched on the market place very early, the well-known thermoplastic elastomers which are self-organized SBS triblock copolymers with a highly periodic spherical morphology. The unique spontaneous and thermoreversible elasticity of these rubbers results from the connection of the major rubbery central blocks B into a tridimensional network by glassy nanodomains S. Block copolymers are actually polymeric analogues of amphiphilic surfactants by the dual properties of their constitutive blocks. Their self-assemblies, which are guided by concentration, molecular structure and molecular composition, endow them with unique optical, mechanical, rheological and transport properties. In addition to a large variety of phases and nanostructures, block and graft copolymers are superior to low molecular weight surfactants by their capacity to bridge distinct domains as illustrated not only by the thermoplastic elastomers but also by the herein discussed compatibilized immiscible polymer blends. In this time honoured activity of polymer blending, major progress has been made in two directions. First, reactive blending has emerged as a useful and straightforward technique of compatibilization, as a result of the steadily increasing capacity of making chains (end-) reactive and having them reacted quickly in the melt. This solventless chemistry is of the utmost importance for economic and environmental reasons. Moreover, traditional processing equipments are now considered as a unique tool for yielding remarkable combinations of properties by self-assembly in commercially viable nanostructured block copolymer/homopolymer(s) blends.

Undoubtedly, the future of reactive blending is in the "direct" synthesis of block and graft copolymers and the nanostructuring of commodity or engineering polymers, making them players in new high-value-added applications.

Acknowledgments

The author is very much indebted to the "Belgian Science Policy" for funding in the frame of the "Pôles d'Attraction Interuniversitaire" programmes. He is very grateful to his collaborators at the University of Liège: Ph. Teyssié, R. Fayt, C. Pagnoulle, Z. Yin and C. Koulic.

References

1 C. E. Koning, M. van Duin, C. Pagnoulle, R. Jérôme, *Prog. Polym. Sci.* **1998**, *23*, 707–757.
2 L. A. Utracki, *Polymer Alloys and Blends. Thermodynamics and Rheology*, Hanser Publishers, Munich, Vienna, New York, **1989**.
3 C. Auschra, R. Stadler, I. G. Voigt-Martin, *Polymer* **1993**, *34*, 2081–2093 and 2094–2110.
4 W. Baker, C. Scott, G. H. Hu, *Reactive Blending*, Hanser Publishers, Munich, **2001**.
5 N. C. Liu, H. Huang, in *Reactive Blending*, W. Baker, C. Scott, G. H. Hu (Eds.), Hanser Publishers, Munich, **2001**, Ch. 2.
6 A. Bouilloux, J. Druz, M. Lambla, *Polym. Eng. Sci.* **1987**, *27*, 1221–1228.
7 Z. N. Frund Jr., *Plast. Compound* **1986**, Sept./Oct., 24–38.
8 S. B. Brown, C. M. Orlando, *Encycl. Polym. Sci. Eng.* **1988**, *14*, 169–189.
9 D. Braun, G. P. Hellmann, *Macromol. Symp.* **1998**, *129*, 43–52.
10 M. van Duin, A. V. Machado, J. Covas, *Macromol. Symp.* **2001**, *170*, 29–39.

11 T. Higashimura, S. Aoshima, M. Sawamoto, *Makromol. Chem., Makromol. Symp.* **1988**, *13/14*, 457–471.
12 A. Hirao, S. Nakahama, C. W. Macosko, *Polym. Prepr. (Am. Chem. Soc., Div. Polym. Chem.)* **1996**, *37*, 722–723.
13 T. D. Jones, C. W. Macosko, B. Moon, T. R. Hoye, *Polymer* **2004**, *45*, 4189–4201.
14 S. K. Varshney, Ph. Bayard, C. Jacobs, R. Jérôme, R. Fayt, Ph. Teyssié, *Macromolecules* **1992**, *25*, 5578–5584.
15 (a) S. Perrier, P. Takolpuckdee, C. A. Mars, *Macromolecules* **2005**, *38*, 2033–2036; (b) K. Matyjaszewski, J. Xia, *Chem. Rev.* **2001**, *101*, 2921–2990; (c) E. Harth, C. J. Hawker, W. Fan, R. M. Waymouth, *Macromolecules* **2001**, *34*, 3856–3862.
16 R. Jérôme, C. Pagnoulle, in *Reactive Blending*, W. Baker, C. Scott, G. H. Hu (Eds.), Hanser Publishers, Munich, **2001**, Ch. 4.
17 R. J. M. Borggreve, R. J. Gaymans, J. Schnijer, J. F. Ingen Hausz, *Polymer* **1988**, *28*, 1489–1496.
18 S. Datta, N. Dhrmarajan, G. Ver Strate, L. Ban, *Polym. Eng. Sci.* **1993**, *33*, 721–735.
19 G. Groeninckx, C. Harrats, S. Thomas, in *Reactive Blending*, W. Baker, C. Scott, G. H. Hu (Eds.), Hanser Publishers, Munich, **2001**, Ch. 3.
20 B. D. Favis, *J. Appl. Polym. Sci.* **1990**, *39*, 285–300.
21 C. W. Macosko, Ph. Guégan, K. Khandpur, A. Nakayama, Ph. Maréchal, T. Inoue, *Macromolecules* **1996**, *29*, 5590–5598.
22 B. D. Favis, *Polymer* **1994**, *35*, 1552–1555.
23 B. D. Favis, J. M. Willis, *J. Polym. Sci., Part B: Polym. Phys.* **1990**, *28*, 2259–2269.
24 J. C. Lepers, B. D. Favis, R. J. Tabar, *J. Polym. Sci., Polym. Phys.* **1997**, *35*, 2271–2280.
25 K. Dedecker, G. Groeninckx, *J. Appl. Polym. Sci.* **1999**, *73*, 889–898.
26 H. K. Jeon, C. W. Macosko, B. Moon, T. R. Hoye, Z. Yin, *Macromolecules* **2004**, *37*, 2563–2571.
27 A. R. Khokhlov, *Makromol. Chem., Rapid Commun.* **1981**, *2*, 633–636.
28 M. Seadan, PhD Thesis, Ecole d'Application des Polymères, Strasbourg, France, **1992**.
29 Ph. Guégan, C. W. Macosko, T. Ishizone, A. Hirao, S. Nakahama, *Macromolecules* **1994**, *27*, 4993–4997.
30 U. Sundararaj, PhD Thesis, University of Minnesota, USA, **1994**.
31 J. J. Elmendorp, PhD Thesis, Technical University of Delft, The Netherlands, **1986**.
32 B. Majumdar, H. Keskkula, D. R. Paul, *Polymer* **1994**, *35*, 1386–1398.
33 C. Pagnoulle, C. E. Koning, L. Leemans, R. Jérôme, *Macromolecules* **2000**, *33*, 6275–6283.
34 G. M. Jordhamo, J. A. Manson, L. H. Sperling, *Polym. Eng. Sci.* **1986**, *26*, 517–524.
35 S. Thomas, G. Groeninckx, *Polymer* **1999**, *40*, 5799–5819.
36 P. T. Hietaoja, R. M. Holsti-Miettinen, J. Seppälä, O. T. Ikkala, *J. Appl. Polym. Sci.* **1994**, *54*, 1613–1623.
37 B. De Roover, PhD Thesis, Université Catholique de Louvain, Belgium, **1994**.
38 K. Dedecker, G. Groeninckx, *Polymer* **1998**, *39*, 4993–5000.
39 C. E. Koning, A. Ikker, R. Borggreve, L. Leemans, M. Möller, *Polymer* **1993**, *34*, 4410–4416.
40 Ph. Maréchal, PhD Thesis, Université Catholique de Louvain, Belgium, **1993**.
41 A. Nakayama, Master Thesis, Tokyo Institute of Technology, Japan, **1994**.
42 J. Washiyama, E. J. Kramer, C. Y. Hui, *Macromolecules* **1993**, *26*, 2928–2934.
43 J. Washiyama, E. J. Kramer, C. Creton, C. Y. Hui, *Macromolecules* **1994**, *27*, 2019–2024.
44 Z. Yin, C. Koulic, H. K. Jeon, C. Pagnoulle, C. W. Macosko, R. Jérôme, *Macromolecules* **2002**, *35*, 8917–8919.
45 Z. Yin, C. Koulic, C. Pagnoulle, R. Jérôme, *Macromolecules* **2001**, *34*, 5132–5139.
46 Z. Yin, C. Koulic, C. Pagnoulle, R. Jérôme, *Langmuir* **2003**, *19*, 453–457.
47 R. Fayt, R. Jérôme, Ph. Teyssié, *J. Polym. Sci., Polym. Phys. Ed.* **1989**, *27*, 775–793.
48 C. Pagnoulle, R. Jérôme, *Macromolecules* **2001**, *34*, 965–975.
49 A. V. Ruzette, L. Leibler, *Nat. Mater.* **2005**, *4*, 19–31.
50 C. Koulic, R. Jérôme, J. G. P. Goossens, in *Phase Morphology and Interfaces in*

Micro- and Nano-Structured Multiphase Polymer Systems, C. Harrats, S. Thomas, G. Groeninckx (Eds.), CRC Press, Boca Raton, FL **2006**, Ch. 10.

51 H. Pernot, M. Baumert, F. Court, L. Leibler, *Nat. Mater.* **2002**, *1*, 54–58.

52 M. Freluche, I. Iliopoulos, J. J. Flat, A. V. Ruzette, L. Leibler, *Polymer* **2005**, *46*, 6554–6562.

53 L. Leibler, *Macromolecules* **1980**, *13*, 1602–1617.

54 L. Leibler, *Makromol. Chem., Macromol. Symp.* **1988**, *16*, 1–17.

55 S. Milner, T. A. Witten, M. E. Cates, *Macromolecules* **1989**, *22*, 853–861.

56 N. Dan, S. A. Safran, *Macromolecules* **1994**, *27*, 5766–5772.

57 R. B. Thompson, M. W. Matsen, *Phys. Rev. Lett.* **2000**, *85*, 670–673.

58 L. J. Fetters, D. J. Lohse, D. Richter, T. A. Witten, A. Zirkel, *Macromolecules* **1994**, *27*, 4639–4647.

59 B. O'Shaugnessy, U. Sawhney, *Macromolecules* **1996**, *29*, 7230–7239.

60 R. Stadler, C. Auschra, J. Beckmann, U. Krappe, I. Voight-Martin, L. Leibler, *Macromolecules* **1995**, *28*, 3080–3097.

61 V. Abetz, T. Goldacker, *Macromol. Rapid Commun.* **2000**, *21*, 16–34.

62 W. Zheng, Z.-G. Wang, *Macromolecules* **1995**, *28*, 7215–7223.

63 T. Goldacher, V. Abetz, R. Stadler, I. Erukhimovich, L. Leibler, *Nature* **1999**, *398*, 137–139.

64 M. A. Hillmeyer, P. M. Lipic, D. A. Hadjuk, K. Almdal, F. S. Bates, *J. Am. Chem. Soc.* **1997**, *119*, 2749–2750.

65 J. M. Dean, R. B. Grubbs, W. Saad, R. Cook, F. S. Bates, *J. Polym. Sci., Polym. Phys.* **2003**, *41*, 2444–2456.

66 S. Ritzenthaler, F. Court, L. David, E. Girard-Reydet, L. Leibler, J. P. Pascault, *Macromolecules* **2002**, *35*, 6245–6254.

67 S. Ritzenthaler, F. Court, L. David, E. Girard-Reydet, L. Leibler, J. P. Pascault, *Macromolecules* **2003**, *36*, 118–126.

68 E. Girard-Reydet, J. P. Pascault, A. Bonnet, F. Court, L. Leibler, *Macromol. Symp.* **2003**, *198*, 309–322.

69 V. Rebizant, A. S. Venet, F. Tournilhac, E. Girard-Reydet, C. Navarro, J. P. Pascault, L. Leibler, *Macromolecules* **2004**, *37*, 8017–8027.

70 C. Koulic, Z. Yin, C. Pagnoulle, R. Jérôme, *Angew. Chem. Int. Ed.* **2002**, *41*, 2154–2156.

71 C. Koulic, R. Jérôme, *Macromolecules* **2004**, *37*, 888–893.

72 C. Koulic, R. Jérôme, *Macromolecules* **2004**, *37*, 3459–3469.

73 D. J. Kinning, E. Thomas, *J. Chem. Phys.* **1990**, *10*, 5806–5825.

74 D. J. Kinning, K. Winey, E. Thomas, *Macromolecules* **1998**, *21*, 3502–3506.

75 J. G. E. M. Fraaije, G. J. A. Sevink, *Macromolecules* **2003**, *36*, 7891–7893.

76 C. Koulic, R. Jérôme, *Prog. Colloid Polym. Sci.* **2004**, *129*, 70–75.

77 C. Koulic, PhD Thesis, Université de Liège, Belgium, **2004**.

78 C. Koulic, G. François, R. Jérôme, *Macromolecules* **2004**, *37*, 5317–5322.

11
Predicting Mechanical Performance of Polymers*

Han E. H. Meijer, Leon E. Govaert, and Tom A. P. Engels

11.1
Introduction

Not unlike humans beings, freshly made polymers are ductile and tough but in time become fragile and brittle. This highly undesirable phenomenon is often due to gradual changes that occur in the structure of the material that lead to more or less severe forms of strain localization, which in turn are triggered by strain softening resulting from a gradual increase in yield stress. This process, which generally is referred to as [physical] "ageing", inevitably already commences during shaping of polymers into artifacts and, to complicate matters, progresses differently for different polymers. Here, we present a set of constitutive equations that is designed to capture the intrinsic mechanical response of polymers as a function of the kinetics of the very changes that occur from the time of shaping them into useful objects. With these equations we are able to quantitatively predict both short- and long-term mechanical properties, directly from the processing conditions, without performing even a single mechanical test. Unfortunately, until now, such analyses can be conducted only for the simple case of amorphous polymers. Semi-crystalline polymers, by their very nature, are more complex. Partly because they comprise two phases, i.e. ordered (crystalline) and unordered (amorphous), and additionally because the crystalline phase is prone to become anisotropic (oriented) due to flow, an omnipresent event in manufacturing processes. This structure causes anisotropy of the properties of a finished product. Suggestions will be made on how to extend the demonstrated, successful analyses of amorphous polymers to the more challenging case of semi-crystalline polymers.

In concert with the philosophy of this book's editors, we wish this chapter to be as accessible as possible for non-specialists in the area of the mechanics of polymers, who we seek to address. They usually are polymer chemists and ma-

---

* A list of mechanical and structural jargon and a list of nomenclature can be found at the end of this chapter.

Macromolecular Engineering. Precise Synthesis, Materials Properties, Applications.
Edited by K. Matyjaszewski, Y. Gnanou, and L. Leibler
Copyright © 2007 WILEY-VCH Verlag GmbH & Co. KGaA, Weinheim
ISBN: 978-3-527-31446-1

terials scientists with the major task of developing polymeric materials with improved properties. We realize that it is, if not impossible, difficult to discuss mechanical performance of polymers without the use of some *jargon* that easily could discourage readers from examining sections that could be of interest to them. Therefore, we add – perhaps in an unusual, pedantic manner – not only a *list of nomenclature* at the end of the chapter, but also a compilation of *mechanical and structural jargon* as used in this text. The definitions are kept as simple as possible and, hence, are not totally accurate or fully descriptive, but they may help readers to overcome difficulties that could be encountered in the first pages of the chapter. We hope that this will encourage readers to continue to learn that the mechanical behavior of polymers is, in fact, rather universal in many aspects.

Prediction of the mechanical performance of polymeric materials directly from their molecular structure, and their microstructure in the case of heterogeneous polymer systems, is hampered by limitations in computational time and memory if the *ab initio* approach is adopted [1–5]. Consequently, we are forced to coarse grain the analyses and, in the averaging steps necessary for this process, we finally lose the details that distinguish polymers from each other and also from other construction materials, like metals, ceramics, stone and glasses. A useful approach is to introduce an intermediate step, that of their *intrinsic properties*, which is the response to load during homogeneous deformation without localization of strain [6–8]. The latter can be conveniently and sufficiently circumvented in lubricated compression tests, in which small cylindrical samples, with typical dimensions diameter and height 6 mm, are compressed using Teflon© films and water–soap mixtures to induce slip between sample and plates [9]. The results of these tests can be captured in an appropriate set of constitutive equations. In finite element analyses, these equations allow prediction of the polymer's mechanical response in notched or un-notched tensile tests, under static or dynamic loading conditions in the short- and the long-term, therewith rendering the modelling quantitative. The intrinsic properties of all polymers look rather similar and reflect their initial non-linear elastic response up to yield, strain softening after yield, followed by strain hardening until the stretched network fractures [6, 7, 10]. In order to delocalize strain, the breaking stress should be higher than the yield stress and the first cause of brittleness of polymeric materials is a lack of sufficient length of the chains in the network. Hence, in order to be load bearing, the polymer's length, i.e. molecular weight, should be above a minimum value [11–15]. This is especially true, of course, for physically entangled networks of amorphous polymers, since we need to load the chains in the network, and avoid chain slippage resulting in disentangling of the network. In practice it proves that a number averaged molecular weight M_n of 8–10 times the molecular weight between entanglements M_e is sufficient to reach the maximum (breaking) stress in the chains.

In modelling the intrinsic properties of polymers, it was found to be rather useful to define their completely young, fresh, state or – alternatively – their completely *rejuvenated* state, as the reference state [8, 16]. In this state, the poly-

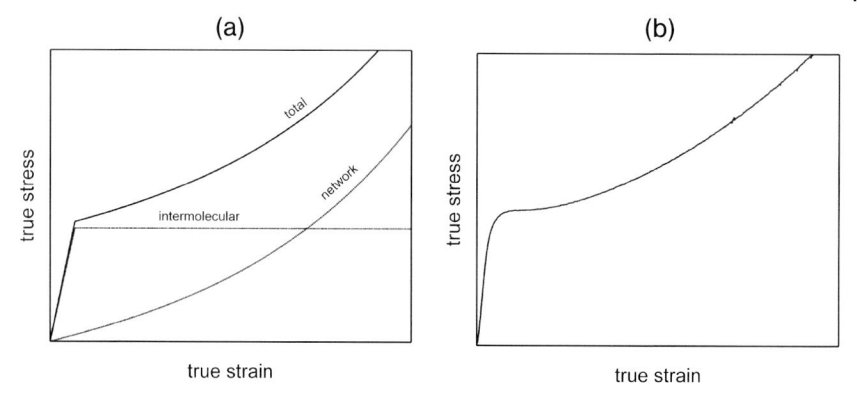

Fig. 11.1 Stress–strain relation and intrinsic response of young (or completely rejuvenated) polymers (a); schematic decomposition of the signal into two parallel contributions (b).

mer's response to loading is quite straightforward. First we find a (usually nonlinear) elastic contribution up to and including yield, determined by the polymer's chain stiffness and the *secondary* (weak) interactions (usually Van der Waals or hydrogen bonds) between the chains. Second, at larger strains the parallel contribution of the entangled macromolecular network becomes active, originating from physical or chemical crosslinks that connect the constituting long molecules, which themselves are mostly linear assemblies of atoms linked by strong *primary* interactions (usually covalent bonds) (see Fig. 11.1).

This figure shows that in the reference state all polymers are ductile, since everywhere within the sample where strain localization starts, strain hardening sets in that delocalizes the strain [17, 18] (see Section 11.6 for a more critical and precise analysis). Interesting now is that, while the completely young, fresh state of polymers is difficult to obtain (mainly because polymers usually experience a history already during their creation and subsequent processing – an often necessary step to obtain a product), the *rejuvenated* state that approaches the ideal situation indeed can be reached sometimes by thermal treatments [19], but more often by mechanical rejuvenation, for instance, through plastic pre-deformation [17, 18, 20, 21].

Not surprisingly, rejuvenated polymers are ductile. An extreme example is a mechanically rejuvenated, normally brittle, polystyrene (PS) strip that can be twisted and curled without showing any onset of crazes, and without failure [22, 23] (see Fig. 11.2 b).

Ductile polycarbonate (PC), after mechanical rejuvenation, no longer displays necking (Fig. 11.2 d) and, while crazes can be considered as an extreme form of strain localization, a neck is a mild form of the same phenomenon. Localizations are induced by intrinsic strain softening, which is absent in rejuvenated samples [7, 10]. However, in time the yield stress of polymers increases due to *physical ageing* [24, 25], a process that induces strain softening in the material's

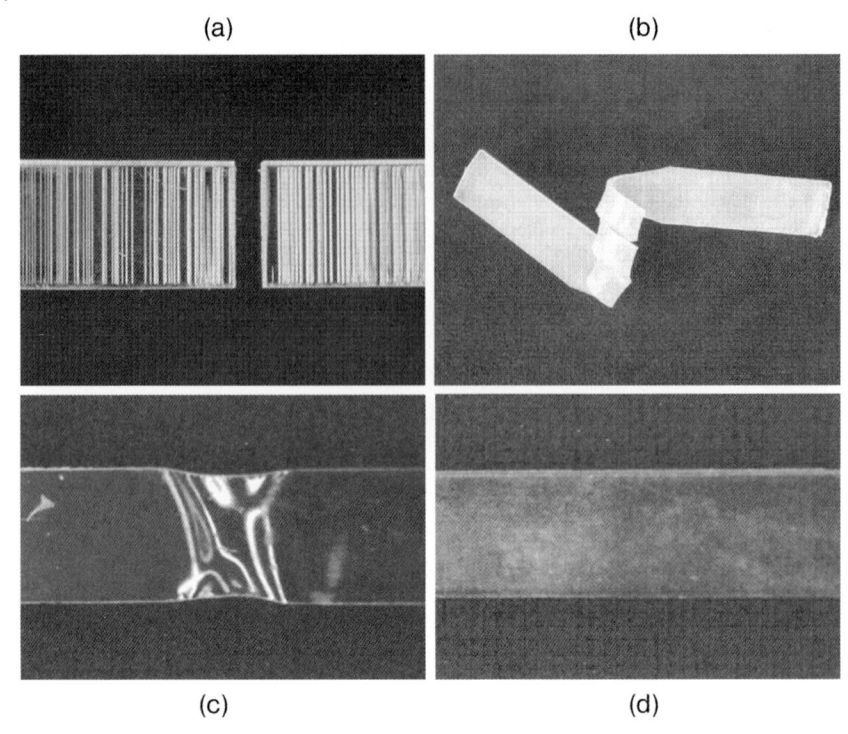

Fig. 11.2 Macroscopic deformation of (a) aged PS, (b) mechanically rejuvenated PS, (c) aged PC and (d) mechanically rejuvenated PC. Loading direction is horizontal.

intrinsic response and that brings back the localization phenomena in tensile loading. A single state parameter S in the constitutive equation, which evolves in time, proves to be able to uniquely define and determine the state of the polymer relative to the fresh, young or rejuvenated reference state [8, 16]. Polymers differ in ageing kinetics, which is one of the main causes of the different mechanical response, brittle or ductile, of products made from them. Moreover, the yield stress of polymers – essentially extremely viscous fluids – depends on external factors like testing temperature and speed.

Apart from localization induced by strain softening after yield, the second cause of brittleness is the extent to which strain hardening, originating in the network-response, can stabilize localization. Comparing PS and PC, we find not only much faster ageing kinetics in PS, but also a much less pronounced strain hardening modulus [7, 8, 10] (see Fig. 11.3).

This can be rationalized by considering the difference in entanglement density in the two polymers [9], that is low for PS which is a relatively stiff polymer due to the bulky side groups on the backbone, with a value of $M_e \approx 20\,000$, and high for PC which is a flexible polymer with a 10 times lower $M_e \approx 2000$. Since

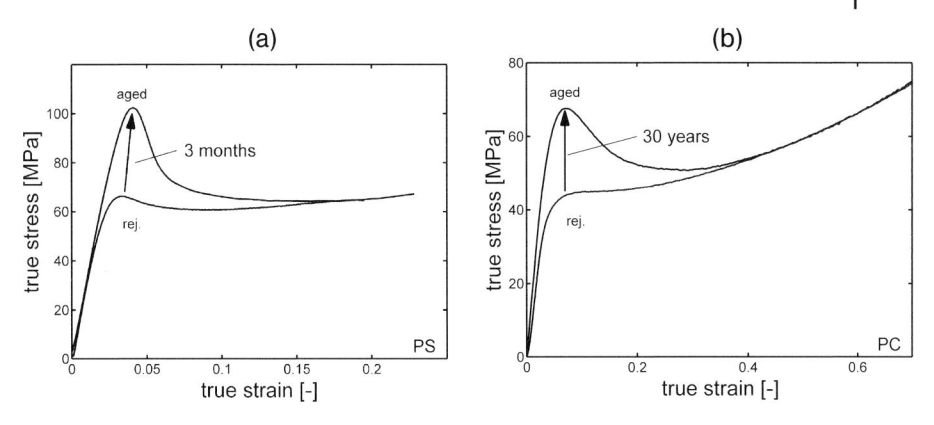

Fig. 11.3 Intrinsic deformation, recorded at room temperature, of aged and mechanically rejuvenated PS (a) and PC (b).

most other amorphous polymers exhibit a chain flexibility causing their M_e to be usually somewhere in between these two values, PS and PC can be considered as two examples of this class of polymers that illustrate the extremes.

A special phenomenon to be discussed, that influences the mechanical performance of polymer solids, is embrittlement of even ductile polymers, like PC, once a sharp notch is introduced. The principal cause of this problem is the development of a critical triaxial stress state under the notch that the material cannot sustain [7]. In unidirectional loading of relatively thin samples, like fibers, material can flow from both the *width and thickness* directions to allow elongation in the axial direction. Bidirectional loading of e.g. films allows the material to flow from the thickness direction into the two directions of biaxial elongation: width and length. In triaxial loading, by contrast, material cannot flow from anywhere and only a change in density can allow deformation (see Fig. 11.4). Loading of a notched sample introduces a triaxial stress state inside the material, under the notch, where a first cavity is formed once the critical cavitation stress is locally surpassed [26, 27]. This occurs invariably under the surface, since at the surface the stress state is always biaxial at most, since material can flow from the surface inwards.

The cavity acts as a new stress "concentrator" and, immediately after the first, a second cavity develops, followed by a third and so on, until rather rapidly a craze is formed in a direction perpendicular to that of the first principle stress, which is – in notched geometries – usually perpendicular to the direction of the external load applied. The craze ultimately fails, once the maximum stress in the stretched fibrils bridging the craze is reached. Although inside the fibrils large-scale deformation takes place (up to failure of the macromolecular network), macroscopically the craze represents a catastrophic localization of strain with overall brittle behavior as a result. The craze-initiation stress depends on the network density [26–30]; hence, also in this respect, PS is the weaker polymer and PC the more resistant one. The depth under the surface where the first cavity develops depends on the details of the polymer's intrinsic behavior,

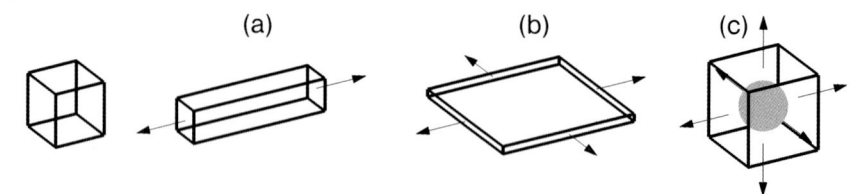

Fig. 11.4 Schematic representation of uniaxial (a), biaxial (b) and triaxial (c) loading of a material and the resulting deformation.

as captured by the constitutive equation. In particular, it is the ratio between strain softening and strain hardening that directly determines the spatial limit of localization. The critical distance for PS is thus much smaller than that for PC, causing PS to be not only notch-sensitive but even sensitive to simple scratches [31]. It follows that, in order to make polymers resistant against such geometrical imperfections, they need to be made heterogeneous by introducing a dispersed phase in a manner that the distance between two dispersed particles is sufficiently low so that the critical stress for craze initiation cannot develop [32–34]. From the above it can be extrapolated that rendering PS tough requires a much finer morphology (related to defect depth and curvature) than for PC.

This chapter summarizes the most important aspects of our present understanding of the mechanical performance of amorphous polymers, emphasizing the important interplay between modelling and experiment, and attempts to extrapolate to what should be done for semi-crystalline polymers. The text is organized as follows. We will commence in Section 11.2 with the basic constitutive equation which consists of a compressible version of a 3D Maxwell element that models the pre-yield behavior of the contributing secondary interactions, in parallel with a neo-Hookean spring representing the network contribution that provides the post-yield response at large strains. Next, in Section 11.3, we capture, with a single state parameter S, the development (read *increase*) of the yield stress with time, as a measure of physical ageing as it occurs during use under the influence of, and accelerated by, temperature and stress. In addition, this section is concerned with mechanical rejuvenation which occurs during segmental flow in the post-yield regime and results in a *decrease*, rather than increase, in the value of S, emphasizing the time and deformation reversible physical character of the state of the material. In Section 11.4, the applicability of the model is further extended by introducing kinetics already occurring during forming and shaping of a polymer product. The conclusion is that – after determination of the temperature and stress shift factors for the ageing kinetics for a specific class of materials – quantitative predictions of a polymer material's short- and long-term performance now prove possible directly from the precise thermal history during processing, without having to perform even a single mechanical experiment on the product at hand. In Section 11.5 we discuss the need for introducing heterogeneity to polymer systems by examining the craze-initiation and cavitation processes. With the combined knowledge of the constitutive equations describing the polymer's intrinsic behavior and of a craze-initiation criterion, we are able to analyse a notched specimen. In Section 11.6 we investigate the principal features that distinguish semi-crystalline from amorphous polymers, focussing on the influence of crystallinity on the intrinsic response and on the phenomenon of geometrical softening in the absence of intrinsic softening. Section 11.7 focuses on the critical influence of flow on structures formed in semi-crystalline polymers. Section 11.8 discusses possible ways to extend the framework developed for amorphous polymers to include both the heterogeneous structures present and the anisotropic character of structures developed in flow in order to predict macroscopic mechanical behavior. The chapter ends with some conclusions in Section 11.9.

11.2
Modelling Intrinsic Behavior: the Rejuvenated State

The mechanical analogue of the intrinsic response of young, fresh polymers, sketched in Fig. 11.1, comprises a constitutive equation that consists of two basic elements placed in parallel [35] (see Fig. 11.5 a). First, we find a Maxwell (spring-dashpot-in-series) fluid element that captures the small-strain, pre-yield behavior of polymers [36]. This represents the contribution of the secondary interactions. Because these interactions determine their small-strain character, the element illustrates that polymers are essentially fluids featuring their typical strain-rate dependent response (cf. Fig. 11.5 b). Flow at yield can be considered as a stress-induced passage of the glass transition temperature T_g, i.e. an increase in segmental mobility caused by stress, rather than temperature. The non-linear viscosity, η, of the dashpot represents the changes at yield. It starts to reveal its "magic" character, being the macroscopic volume-averaged representation of the stress-induced changes in intermolecular interactions, usually represented as an effective energy landscape [37, 38], that results in segmental mobility which, in the end, allows the polymer to flow. Increasing the stress from zero to – only just – 40 MPa, causes the viscosity of, for instance, PC to drop by no less than 12 decades [16, 39]. (N.B. The influences of temperature and pressure on the non-linear viscosity function as well as the effects of physical ageing and mechanical rejuvenation will be incorporated in Section 11.3).

The compressible version of the 3D objective and frame-indifferent Maxwell element containing a non-linear stress-dependent viscosity, is represented by Eq. (1):

$$\boldsymbol{\sigma}_s = K(J-1)\mathbf{I} + G\tilde{\mathbf{B}}_e^d$$

$$\overset{0}{\tilde{\mathbf{B}}}_e = (\mathbf{D}^d - \mathbf{D}_p) \cdot \tilde{\mathbf{B}}_e + \tilde{\mathbf{B}}_e \cdot (\mathbf{D}^d - \mathbf{D}_p)$$

$$\dot{J} = J \, \text{tr}\,(\mathbf{D}) \, ; \; \mathbf{D}_p = \frac{\boldsymbol{\sigma}_s^d}{2\eta(\bar{\tau})} \, ; \tag{1}$$

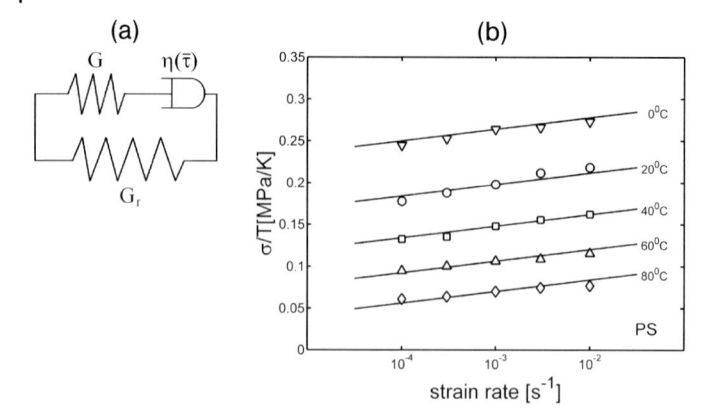

Fig. 11.5 Mechanical analogue of the intrinsic response of polymers (a) and the fluid character of polymers illustrated by a plot of yield stress of a PS versus strain rate at different temperatures (b).

(For a list of symbols used see the end of the chapter). Parallel to the Maxwell element we find the entropic, neo-Hookean, spring representing the network contribution that constitutes the post-yield response at large strains [9, 16–18]. The total response of the element of Fig. 11.5a is the sum of the contributions of the Maxwell element and the entropic spring, see Eq. (2):

$$\sigma = \sigma_s + \sigma_r$$

$$\sigma_r = G_r \tilde{B}^d \tag{2}$$

The notion that this spring renders the total element a solid is incorrect, in that the value of the spring's modulus G_r is two orders of magnitude smaller than that of the Maxwell spring G. Therefore, its contribution is noticeable only at small strains if the temperature is high, above the glass transition temperature T_g, where the contribution of the Maxwell element is substantially decreased (relaxation time $\gamma = \eta/G$) and where the entropic spring G_r captures relaxation of frozen-in orientation and the changes in dimensions of the product caused by this relaxation. The spring G_r is, however, also active at high strains after yield, where it causes strain hardening of the solid, which is of critical relevance here. In the post-yield regime, the contribution of the Maxwell element is still important, since flow is only possible in the presence of stress. The influence of the temperature on the value of the modulus of the spring that is surrounded by this "strange" fluid which instantly turns into a solid once the stress is relieved, is large [9] (see Fig. 11.6a), but its relation to the molecular entanglement or crosslink density is also obvious (see Fig. 11.6b).

During ageing the material strives towards equilibrium and local densification occurs via local segmental rearrangements [37, 38]. This changes the energy

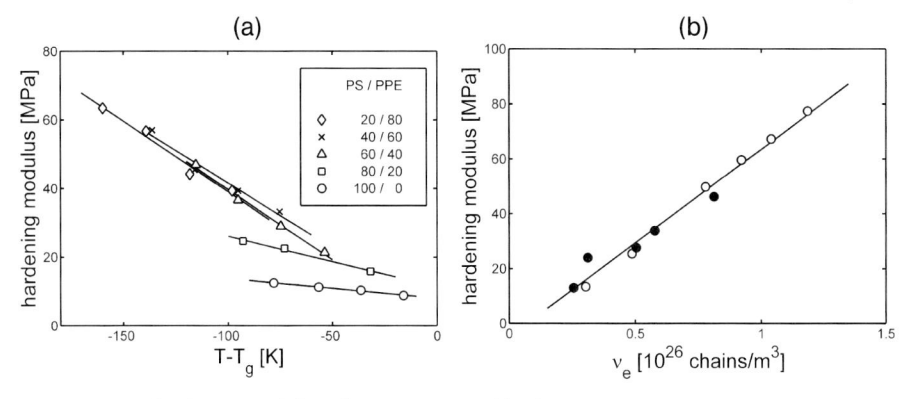

Fig. 11.6 Strain-hardening modulus of various PS/PPE blends as a function of the relative temperature $T-T_g$ (a) [9] and, at room temperature, of the entanglement or crosslink density ν_e (open symbols, PS/PPE blends) and ν_c (closed symbols, crosslinked PS), respectively [9] (b).

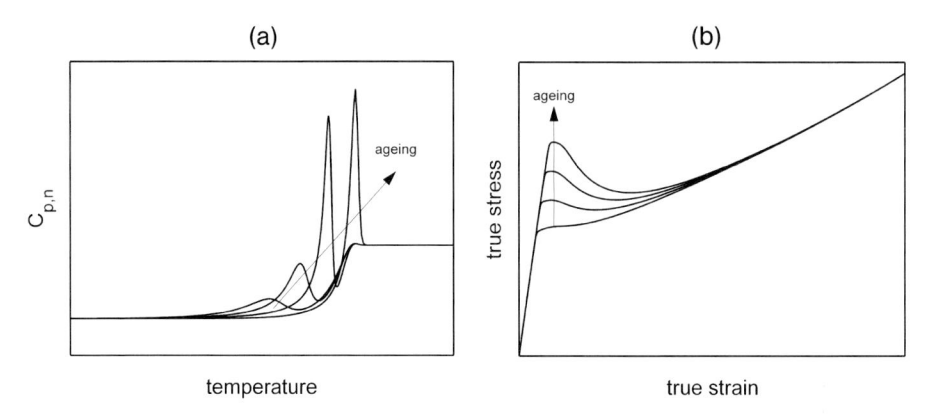

Fig. 11.7 Differential scanning calorimetric (DSC) measurements of all amorphous polymers show a developing endotherm at T_g with increasing age (a), as well as a growing "peak" at yield in intrinsic stress–strain measurements (b).

landscape which is reflected both in the temperature-induced passage of T_g (see Fig. 11.7 a) and the stress-induced passage of T_g at the yield point (cf. Fig. 11.7 b) [6, 40, 41].

Note that the additional enthalpy required to compensate for the increase in local interactions due to changes in the energy landscape is small compared to the enthalpy necessary to melt e.g. polyethylene or a simple wax (three orders of magnitude smaller). By contrast, the additional force that is needed to mechanically eliminate the local densification in a yielding process, which is a stress- rather than temperature-induced passage of T_g, is not that small; even on an absolute scale the increase in yield stress upon ageing is not negligible. Once the material has experienced flow on a segmental scale, the local minima in the energy landscape are "pulled out" [38], and we return to the origin. The mechanically rejuvenated polymer can now flow further at a lower (constant) stress.

Both thermal and mechanical rejuvenation remove the consequences of ageing and illustrate the reversible character of the processes involved – but are not identical [38, 42]. More interesting and counterintuitive, though, is that upon applying a most simple and versatile mechanical rejuvenation process based on a thickness reduction of a sample at room temperature in a two-roll mill (30% for PS, 10% proves sufficient for PC [10, 22, 23]), the overall density of the polymer *increases*. This is a remarkable result and opposite to classical free volume theories developed to describe physical ageing. In the mechanical rejuvenation experiments the yield stress decreases with increasing density, i.e. with *decreasing* free volume.

11.3
The State Parameter S that Captures Physical Ageing

Physical ageing, which conspicuously manifests itself in an increase in yield stress with time, is accelerated at increased temperature and, interestingly, also by applied stress. (N.B. Too much stress leads to the opposite: yield and mechanical rejuvenation during subsequent segmental flow that decreases the stress necessary to induce and maintain flow.) Using mechanical rejuvenated samples as starting materials, ageing can be monitored in relatively short time scales by recording changes in yield stress as a function of time [16, 23] because that dependence was found to be linear in a semi-logarithmic plot (see Fig. 11.8).

Remarkably, all curves can be shifted to yield a master curve. Accordingly, we only need to know the shift factors for the temperature- and stress-dependence of the yield stress to determine the influence of both ageing accelerating conditions. This implies that we can capture the state of a polymeric material with only one state parameter S, which is a simple combination of the two processes involved: the ageing contribution, S_a, and the mechanical rejuvenation contribution R_y [16]. Equation (3) summarizes how the state parameter S unites the two independent counteractive contributions of ageing and rejuvenation:

$$S = S_a \cdot R_y$$

$$R_y(\bar{\dot{\gamma}}_p) = (1 + (r_0 \cdot \exp(\bar{\dot{\gamma}}_p))^{r_1})^{\frac{r_2-1}{r_1}} / (1 + r_0^{r_1})^{\frac{r_2-1}{r_1}}$$

$$S_a(t) = c_0 + c_1 \cdot \log(t_{\text{eff}})$$

$$t_{\text{eff}} = \int_0^1 a_\sigma^{-1}(\bar{\tau}(t')) dt' \tag{3}$$

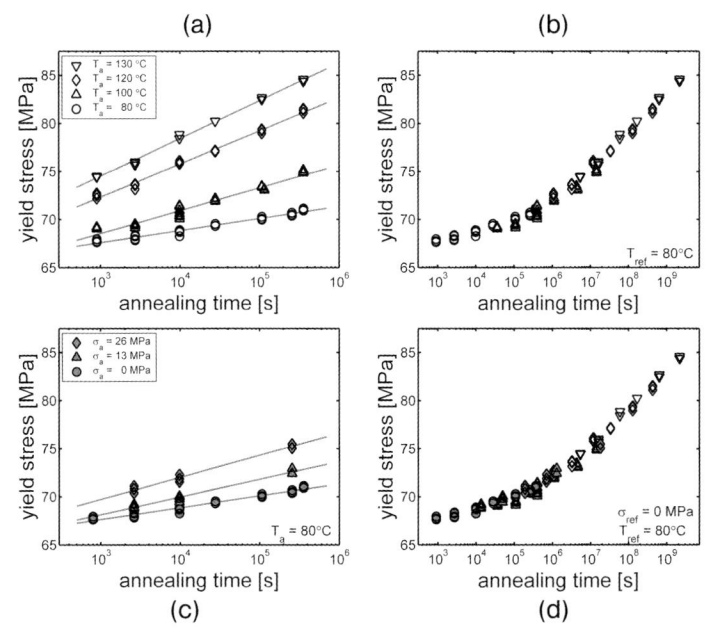

Fig. 11.8 Physical ageing of PC: yield stress as a function of time at different temperatures (a) and the master curve constructed at a reference temperature of 80 °C (b) and yield stress as a function of time at different stresses (c) and the master curve constructed at a reference stress of 0 MPa (d).

Figure 11.9 shows the change in yield stress upon physical ageing of PC and, schematically, the associated development of the ageing parameter S_a.

The fact that S_a initially remains constant – at a level that depends on the processing conditions – is due to ageing that occurred already during processing (see Section 11.4, Fig. 11.12 and text below that figure).

Table 11.1 summarizes the full set of equations that represent the 3D constitutive model for amorphous polymers. The left side of the table comprises the 3D tensorial compressible viscoelastic constitutive model of the mechanical analogue sketched in Fig. 11.5 (Eqs. (1) and (2) in Table 11.1). The neo-Hookean spring contribution of the network is found in Eq. (3), and the compressibility in Eq. (4). The strain in the Maxwell element is determined by the move of the dashpot and its evolution equation is given in Eq. (5). The dashpot contribution to the stress, comprising the rather non-linear viscosity dependence on stress, temperature, pressure and the state of the material, is found in Eq. (6). All those dependences are presented on the right side of the table that needs the averaged values for the (equivalent) stress Eq. (7) and for the plastic strain Eq. (8). The right side exhibits the expressions derived for the viscosity function that depends on temperature and pressure, as given in Eqs. (9) and (10), and the state parameter S. S itself depends on ageing and the opposite phenomenon:

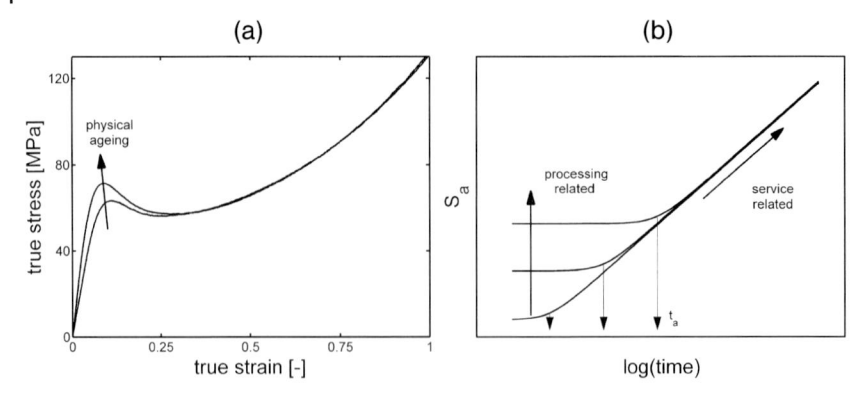

Fig. 11.9 Change in yield stress upon physical ageing of PC (a) and the development of the ageing parameter S_a (b).

Table 11.1 Summary of the set of constitutive equations used to predict mechanical properties of amorphous polymers [16].

$\sigma = \sigma_s + \sigma_r$	(1)	$\eta(\bar{\tau}, T, p, S) = \eta_{0,r}(T) \dfrac{\bar{\tau}/\tau_0}{\sinh(\bar{\tau}/\tau_0)} \exp\left(\dfrac{\mu p}{\tau_0}\right) \exp(S(t, \bar{\dot{\gamma}}_p))$	(9)
$\sigma_s = K(J-1)\mathbf{I} + G\tilde{\mathbf{B}}_e^d$	(2)	$\eta_{0,r}(T) = \eta_{0,r,\text{ref}} \cdot \exp\left[\dfrac{\Delta U}{R}\left(\dfrac{1}{T} - \dfrac{1}{T_{\text{ref}}}\right)\right]; \; \tau_0 = \dfrac{kT}{v^*}$	(10)
$\sigma_r = G_r \tilde{\mathbf{B}}^d$	(3)	$S(t_{\text{eff}}(t,T,\bar{\tau}), \bar{\dot{\gamma}}_p) = S_a(t_{\text{eff}}(t,T,\bar{\tau})) \cdot R_\gamma(\bar{\dot{\gamma}}_p)$	(11)
$\dot{J} = J \, \text{tr}(\mathbf{D})$	(4)	$R_\gamma(\bar{\dot{\gamma}}_p) = (1 + (r_0 \cdot \exp(\bar{\dot{\gamma}}_p))^{r_1})^{\frac{r_2-1}{r_1}}/(1+r_0^{r_1})^{\frac{r_2-1}{r_1}}$	(12)
$\overset{\circ}{\tilde{\mathbf{B}}}_e = (\mathbf{D}^d - \mathbf{D}_p) \cdot \tilde{\mathbf{B}}_e + \tilde{\mathbf{B}}_e \cdot (\mathbf{D}^d - \mathbf{D}_p)$	(5)	$S_a(t_{\text{eff}}(t,T,\bar{\tau})) = c_0 + c_1 \cdot \log((t_{\text{eff}}(t,T,\bar{\tau})) + t_a)$	(13)
$\mathbf{D}_p = \dfrac{\sigma_s^d}{2\eta(\bar{\tau}, T, p, S)}$	(6)	$t_{\text{eff}} = \int_0^t a_T^{-1}(T(t'))a_\sigma^{-1}(\bar{\tau}(t'))dt'$	(14)
$\bar{\tau} = \sqrt{\frac{1}{2}\text{tr}(\sigma_s^d \cdot \sigma_s^d)}$	(7)	$a_T(T) = \exp\left[\dfrac{\Delta U_a}{R}\left(\dfrac{1}{T} - \dfrac{1}{T_{\text{ref}}}\right)\right]$	(15)
$\bar{\dot{\gamma}}_p = \sqrt{2 \cdot \text{tr}(\mathbf{D}_p \cdot \mathbf{D}_p)}$	(8)	$a_\sigma(\bar{\tau}) = \dfrac{\bar{\tau}/\tau_a}{\sinh(\bar{\tau}/\tau_a)}; \; \tau_a = \dfrac{kT}{v_a^*}$	(16)

mechanical rejuvenation Eq. (11); the expression used for rejuvenation is given in Eq. (12). Ageing kinetics are logarithmic in "effective" time Eq. (13) that depends on temperature and stress, Eq. (14), for which the shift factors are defined in Eqs. (15) and (16). Table 11.2 gives the parameter values for PC.

In order to verify the predictive capabilities of the model outlined above, long term creep loading tests were performed with PC samples at different stresses [43]. The material was characterized only by determining its initial age t_a, by determining the original yield stress and computing S. Failure is defined as the onset of sudden plastic flow, as illustrated in Fig. 11.10.

11.3 The State Parameter S that Captures Physical Ageing

Table 11.2 Parameter values used for polycarbonate in the equations in Table 11.1.

K (MPa)	G (MPa)	G_r (MPa)	$\eta_{0,r}$ (MPa s)	τ_0 (MPa)
3750	308	26	2.1×10^{11}	0.7

τ_a (MPa)	μ	ΔU (kJ mol^{-1})	ΔU_a (kJ mol^{-1})	T_{ref} (°C)
1.85	0.08	289	205	23

r_0	r_1	r_2	c_0	c_1
0.965	50	−5	−4.41	3.3

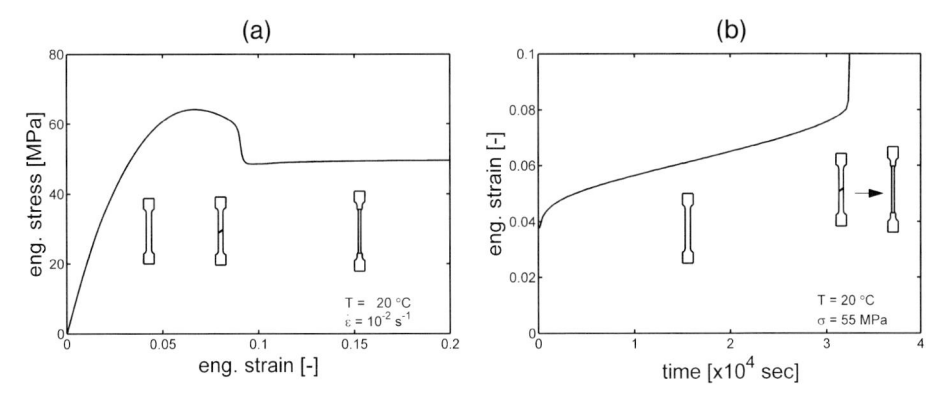

Fig. 11.10 Constant strain rate loading of a PC sample resulting in an elastic response, followed by yield and flow under reduced stress, due to strain softening (a), compared with constant stress loading (b), characterized by elastic deformation and accompanying stationary slow creep, followed by the sudden onset of plastic flow.

In modelling the creep experiments, stress-enhanced physical ageing is included. As a consequence, not only the accurate time-to-failure is predicted under all loading conditions, but also the so-called "endurance limit", that is the limit of stress below which no failure occurs [43] (see Fig. 11.11 a). The explanation is that ageing occurs more rapidly than the accumulation of plastic strain and, hence, dominates. Interesting to note is that progressive ageing is, in fact, beneficial to resist long term loading below yield. Unfortunately, it nevertheless embrittles the polymer once it flows, since once the yield point is surpassed, localization sets in via strain softening. The latter, of course, occurs only after the yield stress has increased due to ageing [8, 10].

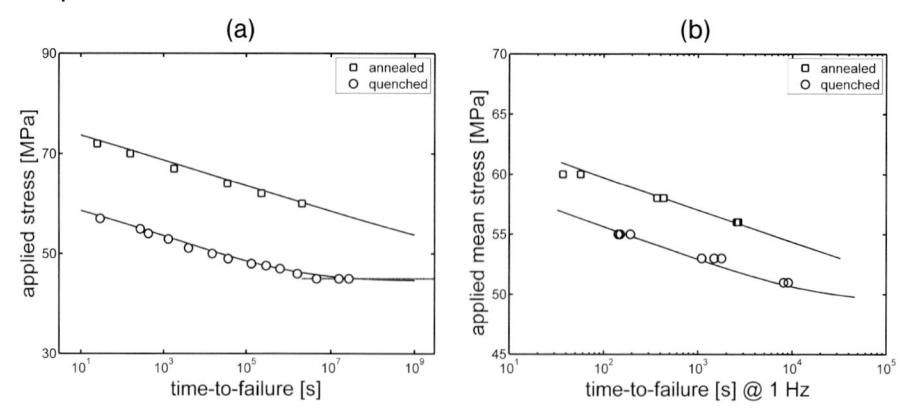

Fig. 11.11 Predicted long term failure of PC in creep loading at different loads (a) and during fatigue loading (b). Symbols are the experimental data, curves are the predictions with the model after determination of the starting age of the samples via measuring S only.

Gratifyingly, not only failure under creep, but also during fatigue loading is predicted accurately (see Fig. 11.11 b). These are remarkable results indeed and illustrate the validity of the concepts behind the constitutive modelling and provide confidence in the predictive power of the associated set of constitutive equations.

11.4
Evolution of S, Physical Ageing During Processing

The success of the model, demonstrated in Section 11.3, is now further extended by applying its kinetic components already during forming and shaping of the final polymer product [44, 45]. We use the same parameter set and shift factors as determined via accelerated ageing after mechanical rejuvenation and as applied in the simulations during use. The initial condition is the completely rejuvenated state at temperatures above the glass transition temperature T_g. Injection molding is chosen as an example of a basic polymer processing step since – in contrast to continuous production processes like extrusion, film blowing or casting, or fiber spinning – in this manufacturing technology the thermal and flow history is different at every point. A simple square object with two significantly different thicknesses (1 and 4 mm) manufactured from PC is chosen as a sample product (Fig. 11.12).

The thermal history from T_g to room temperature needed at any point of the product is easily calculated during the injection, packing and cooling stages of the process using, e.g., a Moldflow© simulation. Figure 11.12 shows (a) the geometry, (b) the results of transient temperature profiles during cooling of the thick and thin samples, (c) the yield stress profiles over the thickness as a func-

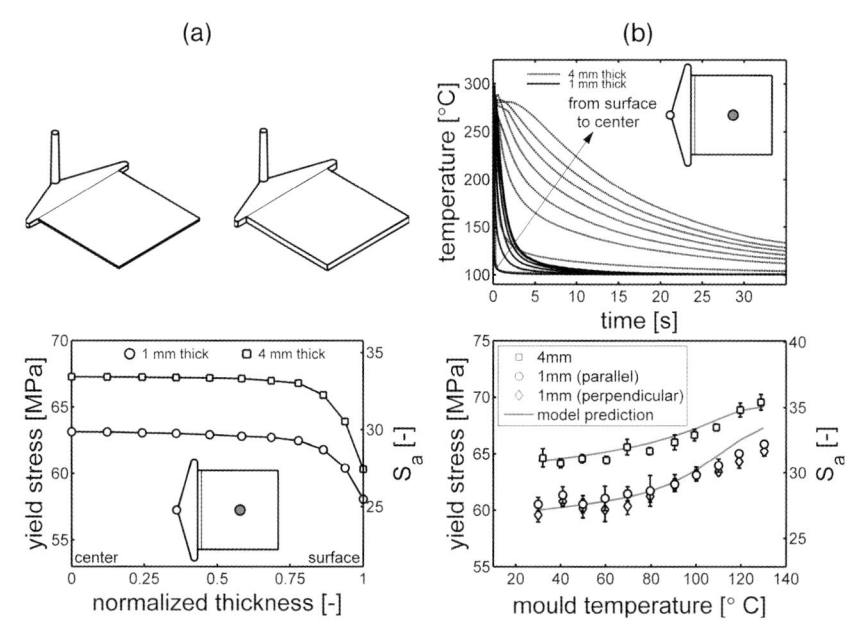

Fig. 11.12 Injection molded thin and thick square sample plaques of PC (a), the computed temperature profiles during cooling (b), computed yield stress distributions (c), and thickness-averaged yield stresses upon changing the mold wall temperature during processing (d). Symbols are experimental data, curves the predictions.

tion of time for the two samples, computed via computation of S_a during molding, and, finally, (d) the measured and predicted across-the-thickness-averaged value of the yield stress as a function of the wall temperature employed during molding [44, 45].

Using low mold temperatures, close to the mold walls and for thin samples, cooling is rapid and, generally, results in "young" polymers. By contrast, high mold temperatures and manufacturing of thick samples is associated with low cooling rates and, thus, yields samples that already are considerably aged. This is directly reflected in the values of their yield stress. Agreement between calculated and experimental values of the yield stress distribution is excellent (cf. Fig. 11.12) [44].

Having considered and understood this part of the process, for a moment we return to the observation that during some time the yield stress of polymers while in use does not change, and neither does S_a (cf. Fig. 11.9) and address the question why the level of the yield stress and this apparent induction time depends on processing conditions, such as the mold wall temperature. The reason is obvious. We tend to think that when we deal with a newly produced object we count its age to be zero to monitor changes, for instance, in its yield stress. However, the shaped polymer already is not young since it experienced a temperature history during which temperature-enhanced ageing took place. Compared to the (effective) age

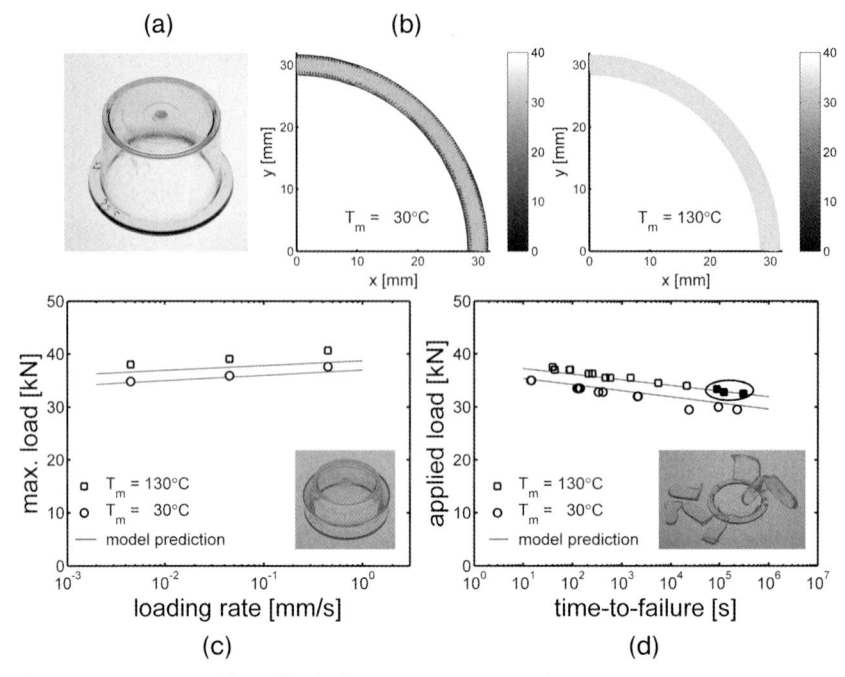

Fig. 11.13 Injection molding of a thick PC cup (a), computed S_a distributions (b), results of compression tests up to failure for different constant displacement rates (c) and for different constant loads (d). Symbols are experimental data, curves the predictions. Open symbols represent ductile failure, see photo in (c). Closed symbols are brittle failure, see photo in (d).

of the polymer in the product, initially the use time is negligible and only once the time of employment becomes of the same order as the polymer's effective age will changes become noticeable and measurable.

In order to, finally, examine the validity of the method adopted and the model developed, we end this section with an investigation of the mechanical performance of an actual injection-molded product [46]. For this we chose a thick-walled cup, fabricated from polycarbonate at two different mold temperatures. Figure 11.13 summarizes the results: (a) shows the geometry of the cup; (b) details the computed temperature histories, and from that the distributions in S_a (thus the yield stress); (c) and (d) provide the experimental and predicted results of compression tests up to failure upon loading at constant displacement rates and failure at constant applied loads.

Clearly, the cup can withstand large loads, typically that of a small truck including a characteristic driver, but for a short time; a 35.5 kN load results in a time-to-failure of 10.5 s. A slightly lower load (with the driver out on a break: 34 kN) already significantly increases the time-to-failure to 110 s. Further lowering of the load naturally results in long survival times. The influence of processing conditions is striking: a 100 °C higher mold wall temperature yields a 300-

fold increase in life time, at the same load. The predictions of such tests are surprisingly accurate indeed. Also intriguing is that both ductile failure at high loads, as well as brittle failure at lower loads (but at longer times) are accurately predicted using the same model that ("only") is concerned with ageing kinetics during loading, without the need to introduce additional failure criteria.

11.5 Why Heterogeneity is Always Needed: Craze-initiation by Surpassing a Critical Cavitation Stress and Its Consequences for the Microstructure

Tough, bullet-proof PC that normally deforms via shear yielding commences to craze like brittle polymers once a sharp notch is introduced into a sample or a product. Except in its (re-)juvenated form, PS always crazes already due to tiny scratches on the surface, or other defects, that invariably are present. In both cases the cause is the localization of strain below the surface, where a geometrically triggered triaxial stress state develops, controlled by the strain softening of the polymer. After surpassing the network strength that resists multi-axial stresses up to a limit that uniquely depends on the molecular weight between entanglements, M_e, inevitably cavitation starts [47]. The cavity formed acts as a second stress concentrator. The process repeats and, in the end, a craze is formed in the material, in the most damaging direction perpendicular to the loading direction [26, 27]. If the fibrils that bridge the two surfaces of the craze break, because the maximum strength of the completely stretched entangled network is reached, or if stress-induced disentangling of the constituent macromolecules takes place, catastrophic failure occurs. To understand the problems involved, finite element calculations are performed using the multi-level finite element method, which allows analyses also of complex structures and on a very local scale. The solution to the problem proves to be control over the localization processes that occur inside the polymer by directing deformation and fibrillation out of the plane perpendicular to the load. The analyses require an independent craze initiation criterion which is what we here consider first. Different authors have attempted to identify craze initiation criteria for amorphous polymers. Narisawa and coworkers investigated a range of notched tensile and bending tests, and determined values for critical dilative stresses [48, 49]. However, considering its scratch sensitivity and, consequently, initiation of crazes just below the surface of the scratch, these tests are inappropriate for polystyrene. This was recognized by Camwell and Hull [50], who compressed a polystyrene sheet between flat dies, generating both shear bands and crazes in the same specimen. Narisawa et al. [51] slightly modified this experiment into a plane-strain indentation between parallel plates using a flat-punch. Crazes nucleate at the tip of the plastic deformation zone where the hydrostatic stress component – caused by the plastic constraint – is highest. From the load at failure and the location of the craze origin, the critical hydrostatic stress for craze nucleation was determined using a modified slip-field theory.

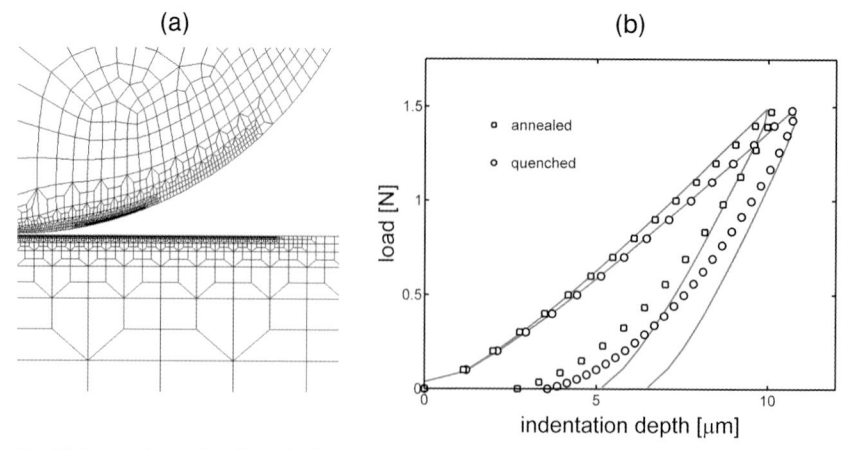

Fig. 11.14 Mesh used in the calculations of the indentation test (a) and resulting force–indentation curves for two PS samples of different age (b). Symbols are the experimental data, curves are the predictions [47].

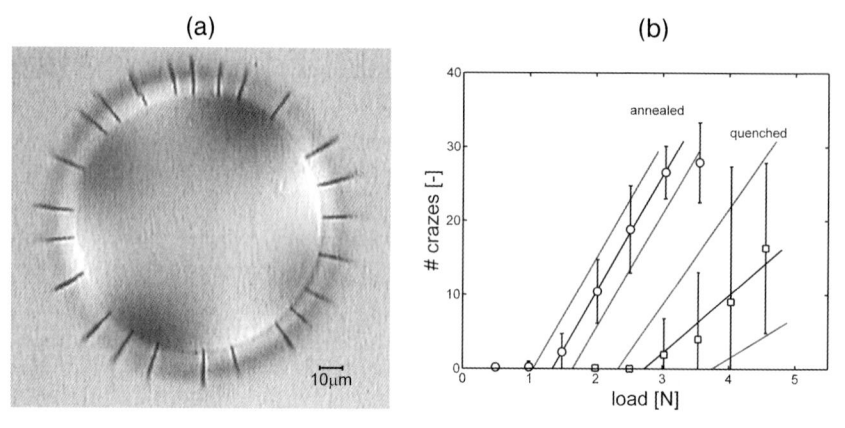

Fig. 11.15 Example of the radial crazes found after indenting a PS sample (a) and results for different normal loads (b).

The large size of the samples needed in the above experiments results in uncertainty about the exact thermodynamic state of the material, making a clear distinction between quenched and annealed material virtually impossible. Therefore we scaled the experiment down and used micro-indentation with a spherical indenter (diameter 250 µm) [47]. The results are illustrated with a young and an aged polystyrene sample. Force–displacement curves are accurately modelled in both cases (see Fig. 11.14b).

In a subsequent experiment, the samples were loaded with different normal forces and, after releasing the load, the number of radial crazes was counted (see Fig. 11.15).

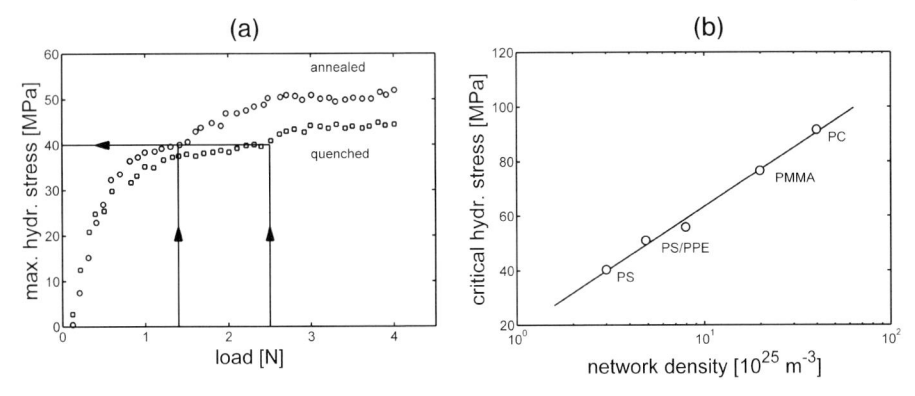

Fig. 11.16 Maximum triaxial stress in the two PS samples of Figs. 11.14 and 11.15 as a function of indentation force (a) and the resulting cavitation stress found for different polymers versus M_e (b).

Extrapolation to zero crazes results in two different values for the apparent critical normal force: around 2.5 N for the young sample and 1.4 N for the aged sample (see Fig. 11.15 b).

The largest triaxial stress in the sample is then computed for both cases (see Fig. 11.16 a). Interestingly, despite the fact that the samples responded rather differently in the indentation test, we found a critical triaxial stress of 40 MPa for both specimens [47]. Apparently, that value represents the critical cavitation stress of PS, which should indeed be independent of the age of the sample used. For different polymers that are characterized by their network density represented by the molecular weight between entanglements M_e, a linear relation between the cavitation stress and log (M_e) is observed in a semi-logarithmic plot (see Fig. 11.16 b).

Knowledge of the critical, age-*independent* cavitation stress of PS $\approx$ 40 MPa and of PC $\approx$ 90 MPa permits us to apply these values as craze-initiation criteria in finite element analyses, for instance of notched Izod bars. Under the notch, a scratch is introduced that mimics the fresh, chilled razor blade cut needed to experimentally obtain reproducible impact energy results (see Fig. 11.17 a).

The localization of stresses in PS is very pronounced. This could be expected given its large strain softening and low degree of strain hardening. Just underneath the scratch, craze-initiation commences as the critical cavitation stress of 40 MPa is surpassed (see Fig. 11.17 b). This already occurs at low macroscopic strains of $\varepsilon < 0.2\%$, which explains the brittleness of PS. Given its less prominent strain softening and its sufficient strain hardening, PC easily survives the scratch defect and never approaches its critical cavitation stress of 90 MPa, which is more than a factor of two higher than for PS. Nevertheless, upon further straining of the sample, up to a macroscopic strain of $\varepsilon \approx 1\%$, PC's critical cavitation stress is also surpassed, at some distance under the notch (see Fig. 11.17 c). As a consequence, eventually both polymers fail by crazing. Con-

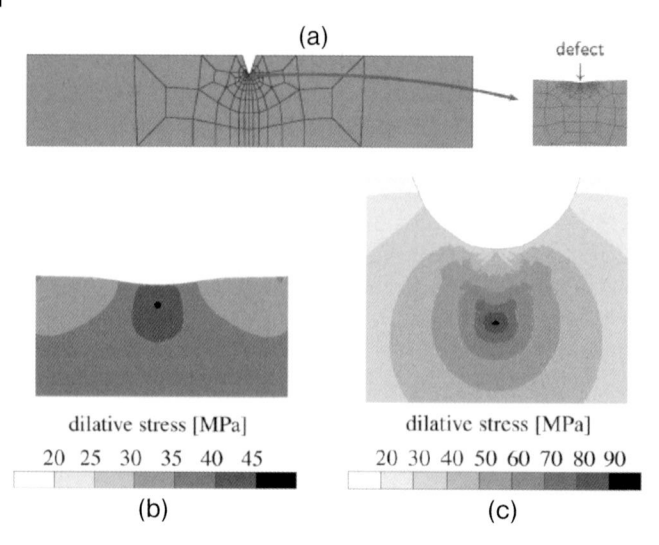

Fig. 11.17 Mesh used in the analyses of notched Izod bars with a scratch in the bottom of the notch (a); the resulting triaxial stress reaches its critical value for craze-initiation for PS of 40 MPa at a low macroscopic strain under the scratch (b); PC survives the scratch, but reaches its critical value of 90 MPa at an intermediate macroscopic strain under the notch (c). Both polymers fail by crazing [31].

sidering that both of these two materials – which represent the extremes in this class of polymers – display this behavior, it is with confidence that we state that all amorphous polymers suffer from craze initiation.

To circumvent this problem all polymers have to be made heterogeneous, which commonly is accomplished by modification with rubber. To quantitatively analyse notched, heterogeneous polymer systems we developed a multi-level finite element method that uses, in the integration points of the elements on the macro level, a finite element analysis of the structure on the micro level as a non-closed form of a constitutive equation [52, 53]. These analyses allow an optimization of the microstructure (see the reviews [7, 8, 34]). Here we present some of the results comparing homogeneous PC and PS and heterogeneous systems filled with 30% voids and with 30% special core–shell rubbers (see Fig. 11.18).

From these computations it is concluded that solving the craze initiation problem in PC is relatively straightforward. Adding just holes to induce heterogeneity is sufficient (shown in the figure is 30%, but addition of 5% holes proved to be already enough). For PS, the voids assist in delocalizing the strain (Fig. 11.18, middle right), but only the addition of rubber shells in the holes that support the deforming ligaments between them, resulting in more pronounced strain hardening of the local structure, gives rise to massive delocalization out of the plane perpendicular to the load applied [32, 34] (Fig. 11.18, bottom right).

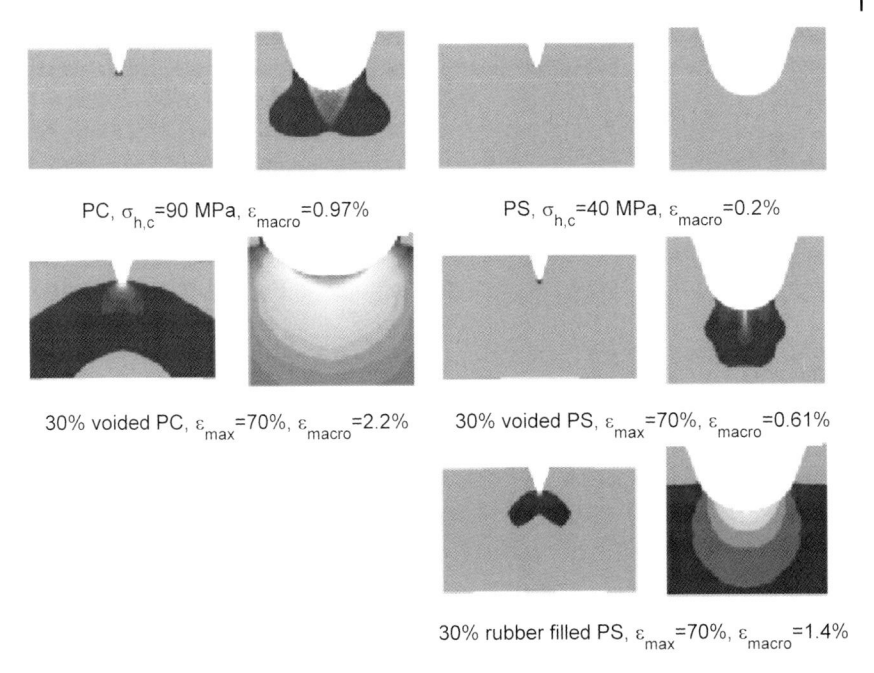

Fig. 11.18 Results of multi-level finite element analysis of loading of notched, scratched homogeneous and heterogeneous PC and PS. The strain localization is shown. Heterogeneous systems are filled with 30% holes (middle) or 30% core-shell rubbers.

11.6
Semi-crystalline Polymers, Intrinsic Behavior and Geometrical Softening

The modelling approach developed for amorphous polymers can be characterized as large-strain solid-state rheology. Given its generality, similar positive results could be expected when that same approach is applied to semi-crystalline polymers. The difficulty is, however, that semi-crystalline polymers consist of two distinct phases, of which at least the crystalline phase is anisotropic and that is, moreover, easily oriented in flow, enhanced by even minute amounts of the high-end tail of the molecular weight distribution. Hence, we need to consider not only anisotropic crystal plasticity, but also nucleation and crystallization in inhomogeneous transient high Deborah number flows.

First, we reconsider the statement made for amorphous polymers that, in the absence of strain softening, no strain localization is found. Is this really correct? Already in 1888 Considère derived that, even in the absence of intrinsic strain softening, *geometrical* softening may also lead to localization and neck formation [54, 55]. Based on Considère's analysis, Haward elegantly derived in 1993 that once the ratio between yield stress and strain hardening modulus exceeds 3, a change from homogeneous deformation to neck formation occurs [56]:

Table 11.3 Expected tensile behavior of four different amorphous (poly(styrene), poly(methylmetacrylate), poly(carbonate), and poly(vinylchloride)) and four semi-crystalline (poly(tetrafluorethylene), high-density poly(ethylene), poly(propylene), and poly(amide-6)) polymers, using σ_b=ca. 200 MPa.

Polymer	σ_y (MPa)	G_r (MPa)	σ_y/G_r	κ_t	κ_γ
PS	40	15	2.7	5	
PMMA	50	30	1.6	4	
PC	40	30	1.3	5	1.4
PVC	40	13	3	5	
PTFE	13	5	2.7	15	1
HDPE	20	4	5	7	1
PP	45	3	15	5	1.1
PA6	80	11	7.5	3	1

$$\sigma_y/G_r > 3 \tag{4}$$

This analysis has been extended to include strain softening [10], which enables us to derive a qualitative estimate of the localization strength, i.e. the draw ratio in the neck during stable cold drawing. By including the tensile strength of the necked region, $\sigma_b = \kappa_t \cdot \sigma_y$, regions of different macroscopic response in tension are predicted [10] (see Fig. 11.19).

Table 11.3 lists some typical values for the yield stress and strain-hardening modulus of some amorphous and semi-crystalline polymers and estimates their breaking stresses. From this, their tensile behavior is extrapolated (compare with Fig. 11.19).

With some caution we can conclude that amorphous polymers exhibit homogeneous deformation (region I) in the absence of softening ($\kappa_\gamma = 1$ [1]) and stable neck formation, once strain softening is present but not too pronounced (region II), while for semi-crystalline polymers, with the exception of PTFE, generally the ratio σ_y/G_r is large, with the consequence that, even in the absence of softening, geometrically induced neck formation occurs (region II, left). Sometimes σ_y/G_r is so large that brittle failure without neck formation is observed (region III). If softening occurs due to an increase in yield stress with time due to physical ageing, $\kappa_\gamma \neq 1$ no longer, and necking must always occur (see Fig. 11.19b).

Next we explore the influence of the degree of crystallinity, χ, on the intrinsic deformation behavior [57]. For that purpose we chose poly(ethyleneterephthalate), PET as an example, since quenching this polymer yields essentially the amorphous state while slow cooling and high temperature ageing results in a (semi-)crystalline solid. From Fig. 11.20a it is seen that crystallinity *increases* the yield stress and *decreases* strain softening (which is a property of the amorphous state only), while the strain-hardening modulus is unaffected, unless macromo-

---

1) Where $\kappa_\gamma = \sigma_y/\sigma_{y,\text{rej}}$ the ratio between the yield stress and the rejuvenated yield stress.

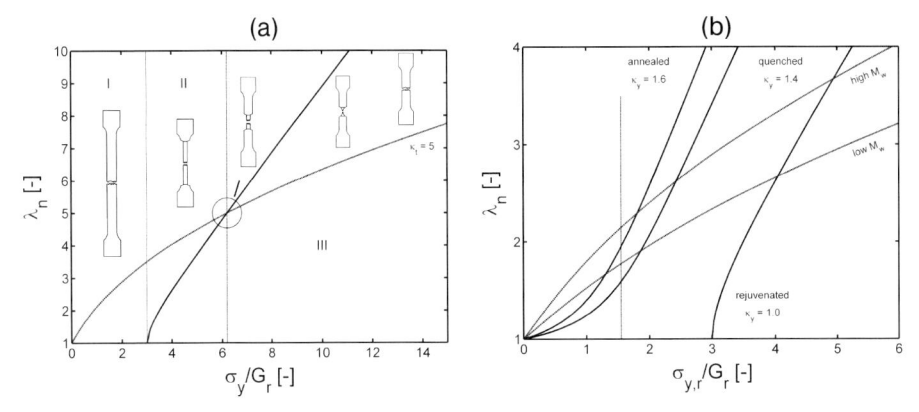

11.6 Semi-crystalline Polymers, Intrinsic Behavior and Geometrical Softening

Fig. 11.19 (a) Typical stability analysis of tensile tests on polymers [10]. Solid line: stable neck growth for polymers without strain softening. Dashed line: tensile strength limit ($\kappa_t = 5$). Three regions are distinguished: I Homogeneous deformation until the breaking stress is reached. II Stable neck formation until the breaking stress in the neck is reached. III Unstable localization and failure without neck formation. (b) Same analysis extended to strain-softening materials, $\kappa_y \neq 1$, where the first region I is absent, see text. The examples show results of quenched and annealed PC with, dotted lines, the tensile strength limit for two different molecular weights.

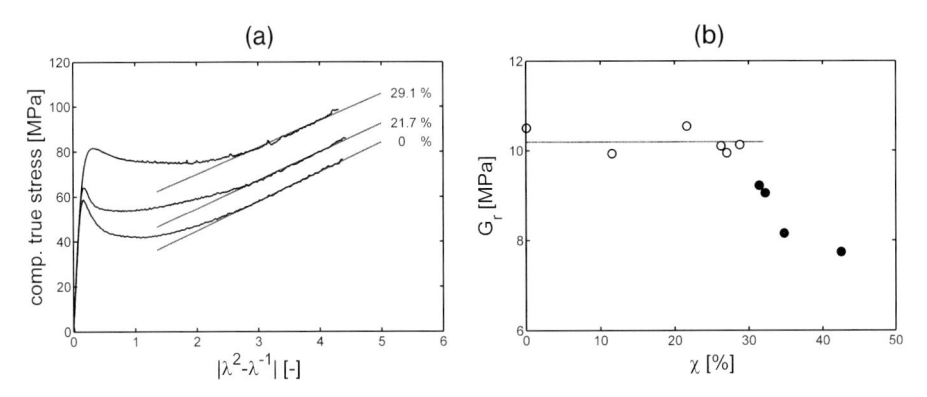

Fig. 11.20 (a) Intrinsic behavior of PET of different degrees of crystallinity [57]. (b) Strain hardening modulus of PET as a function of crystallinity. Closed symbols represent samples made by slow cooling (1 °C/min).

lecular reeling-in happens during extremely slow cooling (closed symbols in Fig. 11.20b).

This same conclusion is true for all other semi-crystalline polymers; just one other example, high-density poly(ethylene), HDPE, is given in Fig. 11.21 [57].

The yield stress is strongly dependent on the degree of order, i.e. crystallinity, and specifically on the lamellar thickness of the crystals [58, 59]. Its temperature and rate dependence appear to be related to the ease of translational motion of

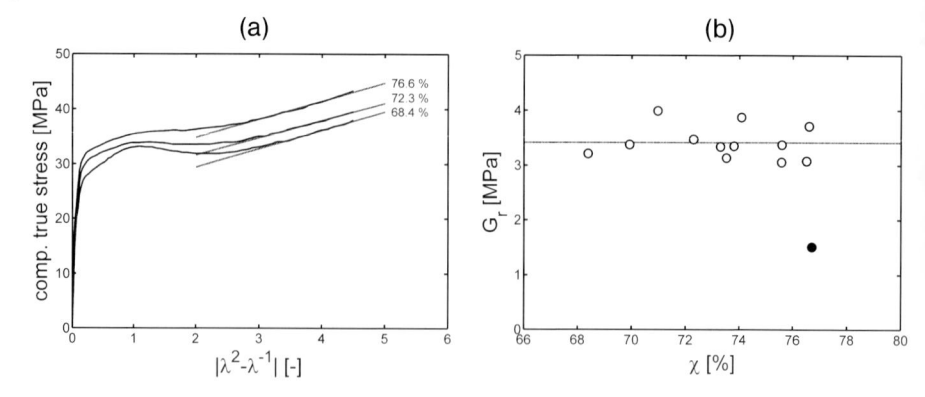

Fig. 11.21 (a) Intrinsic deformation behavior of HDPE of different degrees of crystallinity. (b) Strain-hardening modulus of HDPE as a function of crystallinity. Closed symbols represent a sample made by extremely slow cooling (0,01 °C/min).

Table 11.4 Yield stress and strain-hardening modulus measured in compression for one PET-grade, subjected to different crystallization conditions, Q quenched, IC isothermally crystallized at 110 °C and SC slowly cooled (1 °C min^{-1})

Polymer	χ (%)	l_c (nm)	σ_y (MPa)	G_r (MPa)	σ_y/G_r	λ_n	κ_y
PET-Q	0	–	57	10.3	5.5	3.2	1.3
PET-IC	42.5	4.6	82	10.3	7.9	–	1.1
PET-SC	42.6	6.9	95	7.7	12.3	–	1.05

the macromolecular chains within the crystal [57, 60, 61]. In some polymers double yield is found [61, 62] (see Fig. 11.22a). The deformation mechanisms of both yield points are generally associated as fine chain slip within the lamellae at the first yield point and a process of coarse chain slip resulting in lamellar fragmentation at the second yield point [63–66]. Nonetheless, both yield points clearly correlate with the lamellar thickness [57] (see Fig. 11.22b).

Now we investigate the influence of crystallinity on the polymer's macroscopic behavior in tensile testing, using poly(ethyleneterephthalate) as an example. For one PET-grade, which was subjected to different crystallization histories, different values for yield stress and hardening modulus are found in lubricated uniaxial compression tests (see Table 11.4).

A stability analysis of these PET samples predicts the occurrence of necking for the quenched sample and brittle fracture for the other two samples, as indeed found in tensile tests [67] (see Fig. 11.23). The fact that the draw ratio in the neck of PET-Q is rather low indicates that this sample is also only just in the stable neck forming region. Already a slight increase in tensile deformation

11.6 Semi-crystalline Polymers, Intrinsic Behavior and Geometrical Softening | 1807

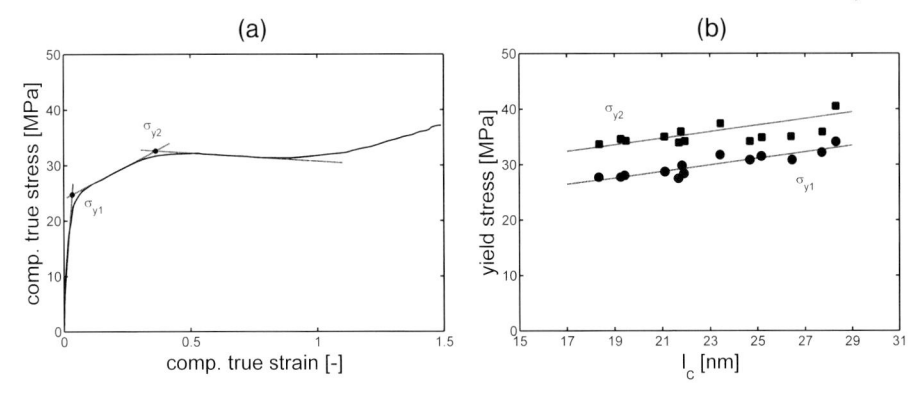

Fig. 11.22 (a) Intrinsic deformation behavior of HDPE with fine chain slip at the first yield point and course grain slip at the second yield point. (b) Yield stress at both yield points as a function of the lamellar thickness.

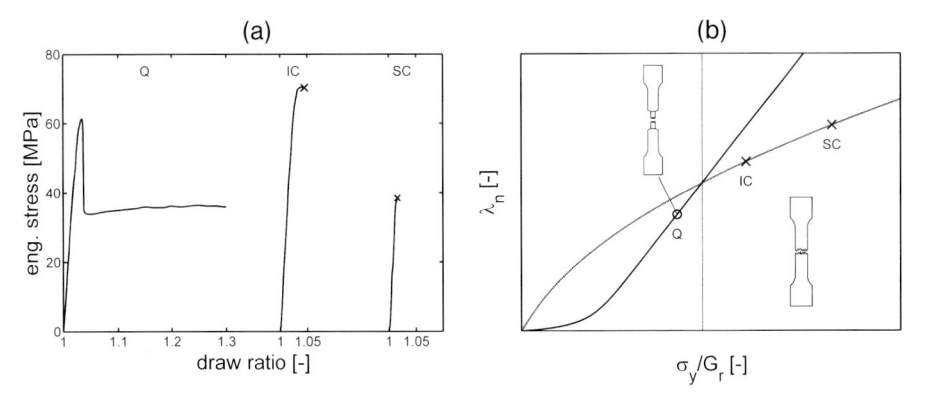

Fig. 11.23 (a) Tensile deformation behavior of PET samples produced under different crystallization conditions: quenched (Q), isothermally crystallized at 120 °C (IC120) and slowly cooled during which macromolecular reeling-in occurred (SC). (b) Stability analysis for a strain-softening material such as PET.

rate resulted in brittle fracture, due to the rate dependence of the yield stress. Because, in isothermal crystallization, strain hardening is not affected, the brittle failure of PET-IC is due to an increase in yield stress. In contrast the brittleness of PET-SC is due to a decrease in hardening modulus.

The above analysis proved to be highly useful and applicable to a variety of generally ductile and sometimes brittle samples of a broad spectrum of polymers, including HDPE, PP and PA.

11.7
Semi-crystalline Polymers: the Influence of Flow

So far, we have considered the influence of crystallinity on the polymer's intrinsic deformation behavior and neck formation or brittle failure. Life becomes considerably more complicated if flow is applied prior to crystallization. Flow causes, in particular, macromolecular chains of the high-end tail of the molecular weight distribution to extend and orient, which accelerate crystallization [68]. Flow-induced nucleation is anisotropic and the resulting crystal structures are, as a consequence, also anisotropic [67, 69]. While much is known about flow-induced orientation, much less is known about how this influences anisotropy in the mechanical properties of the resulting structures; see Fig. 11.24 for some, not only remarkable, but also typical and illustrative examples on test-direction-dependent deformation behavior.

Extruded samples produced with different draw down ratios (take-up rate/extrusion rate) display a marked effect of orientation on the strain-hardening modulus, G_r (see Fig. 11.25). The maximum degree of orientation in extrusion

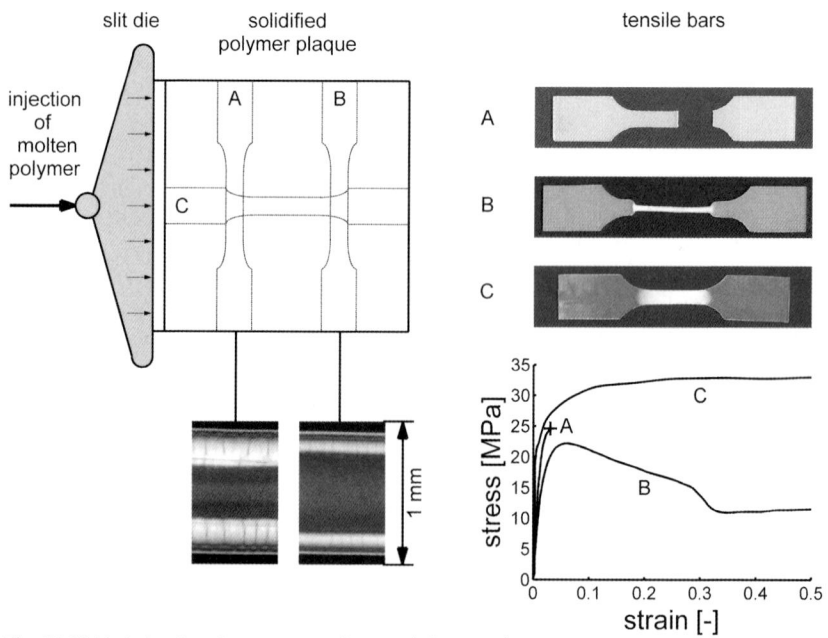

Fig. 11.24 Variation in microstructure (bottom left, optical micrograph of a cross-section, viewed between crossed polarizers) over the thickness in a simple product, and the resulting different mechanical responses (right) of samples A, B, and C, cut from different parts of a typical injection-molded plaque of high-density polyethylene (HDPE) (top left).

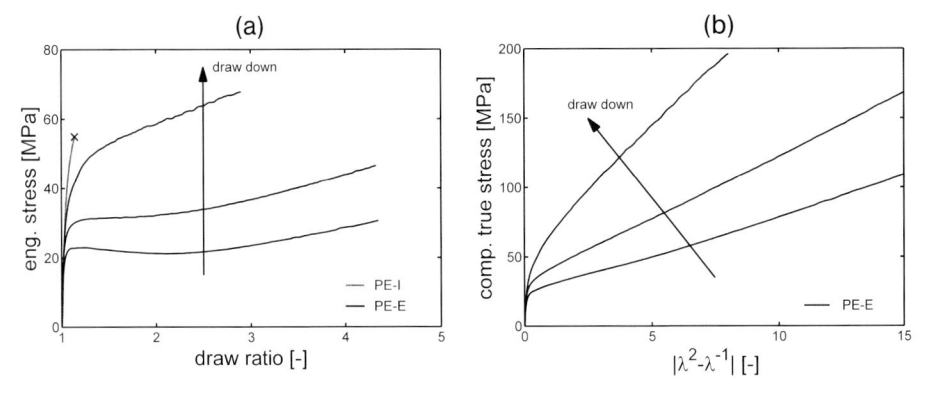

Fig. 11.25 (a) Tensile deformation behavior (engineering stress–strain) of PE injection molded (PE-I) and cast film (PE-E) extruded at three different draw down ratios: 2.5, 5.3, and 17.5, respectively. (b) Their true stress–strain response.

is considerably smaller than that observed in injection moulding, but in the former process the induced orientation is not restricted to shear layers close to the walls and is more homogeneous over the sample thickness than in the latter. Apparently, this homogeneity dominates as all extruded samples were found to deform homogeneously when tested in the extrusion direction.

When fillers are added [70–78], either soft fillers like a dispersed rubbery phase to enhance toughness, or, alternatively, hard fillers to enhance the modulus, the effect of flow on orientation-induced structures and their properties becomes even more pronounced. Depending on the process employed, compression or injection molding, and on the position in the mold (close to or far from the gate), a conspicuously different influence of the added fillers (hard in this case) on the toughness of the filled system is found [76, 79] (see Fig. 11.26a). As expected, there exists a marked effect of the mold thickness and test direction on the efficiency of flow-induced crystallization in injection molding [67, 69, 79] (see Fig. 11.26b).

WAXS patterns of injection-molded, filled and unfilled, HDPE, (4 mm thick) samples show the expected oriented layers close to the mold walls. For the filled systems an unexpected orientation is found in the mid-plane of the mold [79] (see Fig. 11.27). This orientation must be due to the flow history earlier in the process, during flow in the extruder, the runners and the gate. A similar effect of the remarkable memory of flow history on resulting crystal structures was found in melts filled with only a minor amount of nanoparticles [80–82]. The samples in Fig. 11.27 were rather thick, 4 mm, and the mid-plane orientation was detected only over a limited width. This explains the findings of the pronounced influence of the mold thickness on properties measured in the flow direction caused by anisotropic structures (see Fig. 11.26b). It also compares well with the results of extrusion processes (see Fig. 11.25), that again indicated that

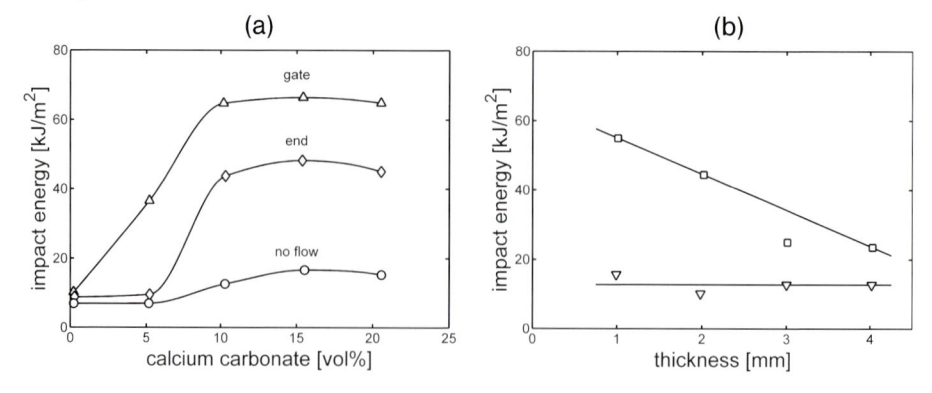

Fig. 11.26 (a) Impact strength of samples of compression-molded $CaCO_3$-filled PE taken at three different positions: gate, end and without flow. (b) Impact strength of 15% $CaCO_3$-filled PE as a function of sample thickness measured perpendicular to flow ($\triangledown$) and in the direction of flow ($\square$) [79].

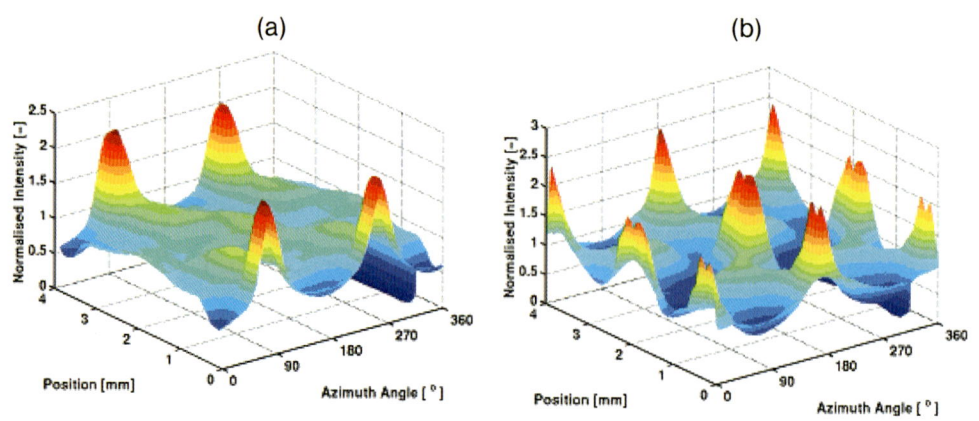

Fig. 11.27 (a) WAXS pattern of injection-molded unfilled HDPE featuring typical orientation at the mold walls due to the parabolic flow velocity. (b) HDPE filled with 15% $CaCO_3$ showing additional orientation in the mid plane due to the flow history accumulated in runners and gates.

a more uniform across-the-thickness anisotropy is rather dominant in determining the anisotropic properties of the samples produced.

The pioneering work of the Du Pont- and MIT groups [70–78] on toughness improvement of semi-crystalline polymers using hard fillers rather than soft rubber, yielded the intriguing results that they succeeded in achieving a concomitant *increase*, rather than decrease, of other important mechanical properties, such as the stiffness. These authors explained the improved toughness found,

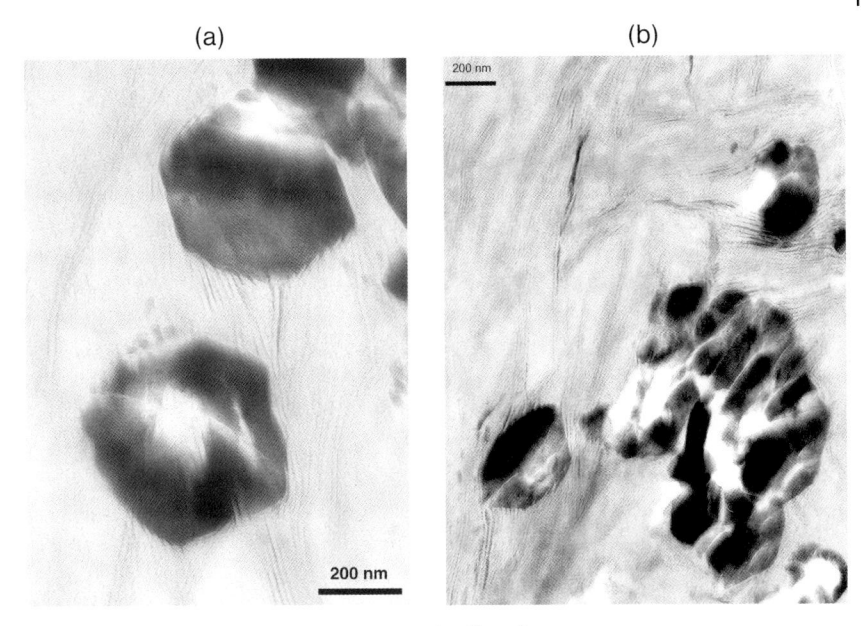

Fig. 11.28 (a) Transmission electron micrograph of lamellar orientation in injection-molded 15% CaCO$_3$-filled HDPE. Flow direction is horizontal. (b) Same in compression-molded sample [67].

advancing that at sufficiently small interparticle distances the threshold was reached for percolation of trans-crystalline layers that grow onto the surface of particles [70, 71]. The beneficial effect on the impact strength is accordingly induced by anisotropic crystal plasticity, effectively lowering the yield stress in a direction favorable for delocalization of strain [83–85]. In order to avoid triaxial stress states, the adhesion between fillers and matrix should be optimal, i.e. sufficiently strong for proper dispersion of the particles during mixing and sufficiently weak to allow debonding to occur upon mechanical loading. However, the specific crystalline arrangement associated with this explanation could not be confirmed by other researchers [67, 86, 87]. On the contrary, they observed an overall orientation in which the chains were oriented along the direction of flow, comprised in crystals arranged perpendicular to it, at least in injection-molded samples (cf. Fig. 11.28).

Compelling and direct evidence of the overruling influence of flow on the properties of such filled systems is that upon recrystallization of the samples after confined melting (no flow) all beneficial properties disappear [67, 79, 86, 87]. Indirect proof is found in simulations of the influence of flow on molecular orientation, nucleation and the subsequent crystal orientation in particle-filled systems. This is a complex task, as it involves (i) non-linear viscoelastic material flow, (ii) hydrodynamic interactions between multiple particles, and (iii) non-isothermal phase changes – issues which are strongly coupled.

Molecular rheology is needed in that the details of the molecular weight distribution determine which events occur during flow. As stated above, it is specifically the macromolecules of molecular weights at the high-end tail that orient most easily during flow. Therefore, we need a quantitative model that properly captures the prediction of molecular orientation and extension, in complex 3D flow fields with a combined shear and elongational character under transient conditions. At present, the Leonov model [88] and the extended PomPom model [89, 90] are successfully used. In addition, we require a theory that quantifies how oriented chain segments facilitate the formation of crystalline nuclei, how fast these nuclei grow, in which direction and how fast they relax. Coupled to this issue is the need for a theory for quiescent and flow-induced crystallization, since both occur simultaneously. The S_{J2} model [91] is used since molecular orientation represented by the recoverable strain in this model proves to be more important as a driving force for crystallization than shear or elongation rate or total shear [92]. Bi-periodic boundary conditions are used to compute highly filled systems with still practical mesh sizes and reasonable computing times [93, 94]. Both fully coupled and decoupled computations are performed. The decoupled approach is generally applied in simulations of injection molding, since the high Deborah or Weissenberg numbers under practical conditions (of the order of 10^4) by far exceed the numbers that can be dealt with in computational rheology (of the order of 10^1). (It should be noted though that recently a major step forward has been made in finding a route to overcome these problems that have frustrated the computational rheology community for at least a quarter of a century [95–97].) In the decoupled approach, first the kinematics of the flow and the temperature history are computed using a generalized Newtonian constitutive equation for the polymer melt. Subsequently, they are substituted in a viscoelastic constitutive model to compute molecular stresses and orientation. These are, in their turn, the driving force for nucleation and crystallization, whereby the direction of growth is the direction of the molecular orientation. Only a simplified problem has been studied consisting of two steps, i.e. first a shearing step where the melt is subject to isothermal flow, followed by isothermal crystallization after cessation of flow at a lower temperature [98]. Some characteristic results are reproduced in Fig. 11.29.

The fluid's relaxation time is $\lambda_f = 5.3$ s, the solid area fraction in this 2D computation is $\varphi = 30\%$, the shear rate imposed equals $\dot{\gamma} = 0.5$ s^{-1} and the shear time $t = 20$ s, yielding a Weissenberg number of $Wi = \lambda_f \cdot \dot{\gamma} = 3.8$ (only). Figure 11.29 comprises two parts, each containing two sets of images, in (a) and (b) we find the results of the completely coupled viscoelastic computations and in (c) and (d) those of the decoupled simulations. The two blue sets of images (a) and (c) show the trace of the deformation tensor **B**, that represents the molecular orientation induced by the flow; the two red ones (b) and (d) show the spatial distribution of the degree of space filling, ξ, due to flow-induced crystallization. Moreover, the figure is divided into three horizontal sections, that give results at different times of flow (4, 10 and 20 s after starting the flow (a) and (c)) and iso-

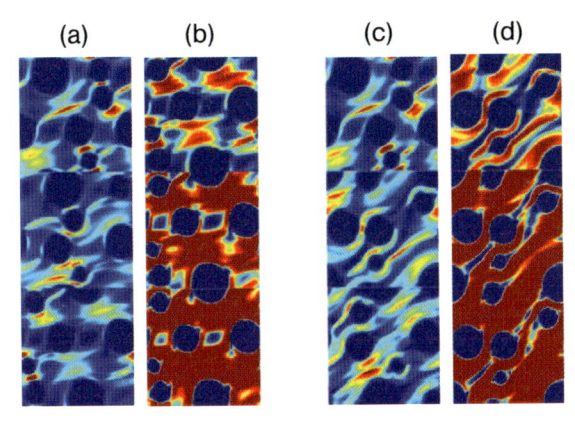

Fig. 11.29 (a) and (b) Fully coupled viscoelastic calculations of molecular orientation in flow at 4, 10 and 20 s after initiation of the flow (a, from top to bottom) and resulting anisotropic crystallization at 2500, 7500 and 15 000 s after start of isothermal crystallization (b). (c) and (d) Same in a decoupled computation. Flow direction is horizontal.

thermal crystallization times (at a high crystallization temperature, hence slow crystallization: 2500, 7500 and 15 000 s after the start, (b) and (d)).

The results are interesting. Regions of a high degree of molecular extension exist (colored yellow to red in Fig. 11.29a and c), generated on the one hand by passing and separating particles – typical for concentrated viscoelastic suspensions and not found in Newtonian suspensions – and, on the other hand, near the top and bottom of particles, especially when two particles are in close proximity. Both mechanisms for molecular extension were found to become more dominant when increasing the solid area fraction [98]. This could explain the existence of a critical interparticle distance for impact improvement, as reported for both hard and soft fillers in semi-crystalline polymers [70–76]. The highly non-homogeneous molecular conformation induced by this locally complex flow, naturally, affects the resulting morphology of the solidified filled polymer, and thus the properties and oriented crystalline structures that are created in those regions where highly extended molecules are present (b) and (d). At the same time, we find shielded regions where no crystalline structures develop (at least no oriented, flow-induced structures, since at the selected crystallization temperature, quiescent crystallization does not occur). Compared to the fully coupled computations, the results of the decoupled simulations show similarities as well as differences. The most important difference is the direction of both the molecular orientation and the resulting crystalline structures. In the results of the decoupled calculations we find them both more or less in a direction of 45° relative to the direction of flow, which is horizontal, while the full simulations yield more alignment with the flow. The main reason proved to be the rotation speed of the individual particles in the coupled simulations, which is three times larger than in the decoupled computations. This is easily understood since viscoelasticity of the matrix counteracts free rotation of particles, generated by the shear flow. The consequences for molecular and crystal orientation are important, because freely rotating particles acts as stirrers, randomizing orientation induced by flow. Since, in all experiments conducted, only crystal orientations with the chains in the direction of flow are found (and never less than 45° relative to the flow direction) we have to conclude that, in order to be able to model anisotropic flow-induced crystallization in heterogeneous, soft or hard particle-filled systems, we need to perform fully coupled viscoelastic computations. Therefore, the recently reported breakthrough to finally solve the high Wi number problem [95] could be of great practical importance.

11.8
From Heterogeneous Structures to Anisotropic Mechanical Properties

The final step to be made is to bridge the gaps in length scales between the anisotropic structure that develops during processing and the macroscopic properties of the resulting solids. Under conditions of flow the kinetics of the crystallization process, especially that of nucleation, are strongly influenced, and, more importantly, the formation of anisotropic structures is promoted. Under relatively weak flow, nucleation of (normally isotropic) spherulitic structures may already be significantly enhanced [99, 100], and fine-grained, slightly oriented spherulitic structures are formed. At increasing flow rates or times, the crystalline morphology changes and oriented crystalline structures such as stacked lamellae (row-) and fibrillar, shish-kebab-like entities are observed [92, 99, 101, 102]. In injection molding of semi-crystalline polymers, during which the flow rate often shows strong variations throughout the thickness of the mold, in general, all these structures are found [69, 103, 104].

Nucleation and crystallization theories should be focused on predicting under which processing conditions for which molecular weight distributions which structures develop in which direction, given the polymer's overall crystallization kinetics that can vary from virtually zero to extremely fast, depending on the details of the regularity of the chains and the amount of supercooling.

The next major step, given the complexity of these structures, is to determine what the local, three-dimensional, anisotropic mechanical response will be of all various microstructures that feature within a product. Therefore, an extension of the existing multi-scale, micromechanical modelling approaches [105–108] in the direction of quantitative predicting capabilities are needed. A typical result of the present capabilities of multi-scale modelling [105–108] of semi-crystalline polymers is given in Fig. 11.30.

To evaluate products of semi-crystalline polymers and predict their lifetime, as was demonstrated for their amorphous counterpart, unfortunately, the multilevel approach is not suitable and a macroscopic, closed form constitutive equation is needed. The model summarized in Table 11.1 may act as a starting point

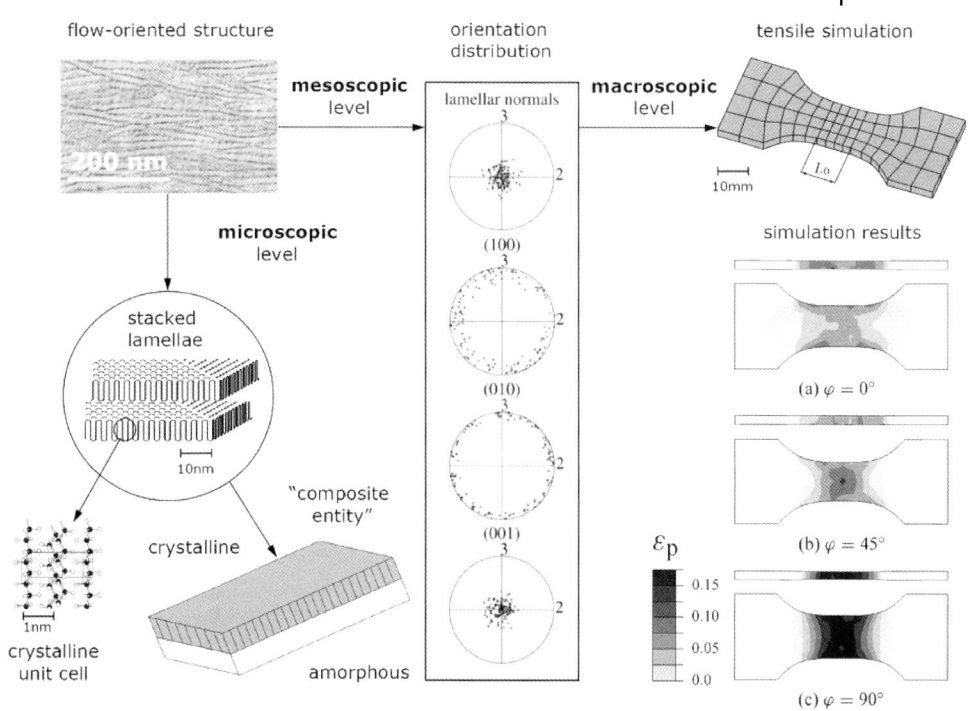

Fig. 11.30 Different length scales that need to be considered in micromechanical modelling and some preliminary results on predicted mechanical response (distribution of equivalent plastic strain, right-hand side) of an aggregate of 125 "composite entities" and their corresponding X-ray pole figures.

for such an equation. Extensions are needed, however, to incorporate anisotropy. A possible way to do that is to couple the stress dependence of the viscosity to Hill's anisotropic definition of effective stress [109], governed by the six parameters R_{ij} (see Eq. (5)).

$$\mathbf{D}_p = \frac{{}^4\boldsymbol{\Psi}(R_{ij}) : \sigma_s^d}{2\eta(\tau_{\text{Hill}}, p, S, \chi, L)} \tag{5}$$

Here the stress dependence of the rate of plastic flow is still governed by a single (scalar) viscosity function. This can be rationalized by considering that anisotropic deformation also originates from the same molecular process for all loading situations. The resulting anisotropic distribution of the strain rate is determined by a fourth order operator ${}^4\boldsymbol{\Psi}$ that is anticipated to depend solely on the six parameters R_{ij}.

Also the expression for the two independent processes, physical ageing S_a and mechanical rejuvenation R_y that uniquely determine the state S of amorphous polymers, $S = S_a \cdot R_y$, (Eq. (3), Table 11.1), needs adaptation. For instance, for semi-crystalline polymers the mechanically rejuvenated reference state is not unique as it depends on the initial crystalline morphology which, in part, persists during plastic deformation. Second, the yield stress of semi-crystalline polymers increases upon deformation due to rotation of crystalline entities (texture-induced hardening). Intrinsic strain softening occurs during plastic flow, accompanied by an initial decrease in crystallinity because of a breakdown of the crystal structure. Sometimes, when a semi-crystalline polymer is close to the glassy state (e.g. isotactic polypropylene, nylons and polyesters) softening is also partly related to rejuvenation of the amorphous state [57, 110]. In order to include texture-enhanced hardening (inducing intrinsic softening) a possible modification of the expression of the state parameter S could be envisioned (Eq. (6))

$$S = S_H \cdot R_H(\gamma_p) + S_s \cdot R_s(\gamma_p) \qquad (6)$$

The first part of the expression addresses texture hardening with time (S_H) and its development upon straining (R_H), while the second part describes all combined strain softening, where S_S relates to the reduction in stress and R_S to the evolution with plastic strain. Of course, not only do the mathematics need to be developed, as sketched above, but, more importantly, a significant number of experiments need to be designed – and conducted – to reveal the characteristics of all the features involved in the description of the mechanical behavior of anisotropic, inhomogeneous structures.

11.9
Conclusions

For amorphous polymers, it is the balance between strain softening, caused by physical ageing, and strain hardening, controlled by the test temperature relative to the polymer's glass transition temperature T_g and the entanglement network density ν_e, that is reflected in the polymer's intrinsic response and that fully determines the polymer's final performance – brittle or ductile – during tensile loading, given the maximum stress that the stretched network ultimately can bear.

Once the temperature and stress shift factors for ageing kinetics are determined, which are independent of the molecular weight distribution (as expected given that only segmental motions control ageing), fully quantitative predictions of the polymer's short- and long-term mechanical performance now prove possible directly from the thermal history during processing (especially the time spent just below the glass transition temperature T_g) without performing a single mechanical experiment on the product at hand.

In order to circumvent catastrophic triaxial stresses underneath notches and scratches, all polymers must be made heterogeneous on a length scale that prevents the first cavity from developing. The critical cavitation stress depends on the entanglement network density. For polymers with strong tendencies for localization, i.e. polymers with pronounced intrinsic softening and moderate intrinsic strain hardening, like PS, the critical distance below the surface where the first cavity is formed is extremely small, rendering these polymers highly sensitive to scratches, and modification of their structure to prevent the critical distance from being present remains a major challenge. Often local stabilizing of the structure of stretched matrix filaments by adding rubber strands is beneficial: local structure hardening. Polymers that show a less pronounced localization, because their intrinsic softening is moderate and their stabilizing hardening sufficient, like PC, are only sensitive to notches, which is easily solved, for instance by the incorporation of about only 5% of voids.

The extension of the approach that proved to be rather successful for amorphous polymers to semi-crystalline polymers is quite complex. Not only because this class of polymer solids is inherently heterogeneous, but, in addition, the heterogeneous structures are anisotropic and are characterized by highly direction-dependent mechanical properties. This anisotropy is completely determined by the details of the flow that the polymer experiences during the fabrication process, and thus the complete thermal mechanical history of every part within the product needs to be taken into account. Accordingly, modelling of the complete process requires major contributions of different disciplines. It would appear that this represents a considerable challenge.

List of Nomenclature

$\mathbf{B}$	Left Cauchy–Green strain tensor
$\mathbf{D}$	Deformation rate tensor [s^{-1}]
$\mathbf{I}$	Unity tensor
$\boldsymbol{\sigma}$	Cauchy stress tensor [MPa]
$\boldsymbol{\sigma}_s$	Driving stress tensor [MPa]
$\boldsymbol{\sigma}_r$	Hardening stress tensor [MPa]
a_σ	Stress shift factor
a_T	Temperature shift factor
$C_{p,n}$	Normalized specific heat at constant pressure
c_0	Parameter in the S_a evolution
c_1	Parameter in the S_a evolution
G	Shear modulus [MPa]
G_r	Strain hardening modulus [MPa]
K	Bulk modulus [MPa]
k	Boltzmann's constant [J K^{-1}]
J	Volume change factor
$\dot{J}$	Rate of volume change [s^{-1}]
l_c	Lamellar thickness [nm]
M_e	Molecular weight between entanglements [g mol^{-1}]
M_n	Number-average molecular weight [g mol^{-1}]

M_w	Weight-average molecular weight [g mol^{-1}]
p	Hydrostatic pressure [MPa]
R	Universal gas constant [J mol^{-1} K^{-1}]
R_H	Development of texture hardening during straining
R_s	Development of softening during straining
R_{ij}	Parameter of Hill equation
R_γ	Softening function
r_0	Parameter of the softening function R_γ
r_1	Parameter of the softening function R_γ
r_2	Parameter of the softening function R_γ
S	State-parameter accounting for the thermo-mechanical history of the glass
S_a	Momentary value of S in the undeformed state
S_H	Texture hardening parameter
S_s	Strain softening parameter
T	Temperature [K]
T_a	Annealing temperature [K]
T_{ref}	Reference temperature [K]
T_g	Glass transition temperature [K]
T_m	Mold temperature [K]
t	Time [s]
t_a	Initial age [s]
t_{eff}	Effective aging time [s]
Wi	Weissenberg number
$\bar{\gamma}$	Equivalent shear strain
$\dot{\gamma}$	Shear strain rate [s^{-1}]
ε	Strain
ε_{macro}	Macroscopic strain
ε_{max}	Maximum local strain
$\dot{\varepsilon}$	Strain rate [s^{-1}]
φ	Solid area fraction
λ	Draw ratio
λ_f	Fluid relaxation time [s]
λ_n	Draw ratio in the neck
η	Viscosity [Pa s]
$\eta_{0,r}$	Zero viscosity of the rejuvenated state [Pa s]
$\eta_{0,r,ref}$	Zero viscosity of the rejuvenated state at a reference temperature [Pa s]
κ_t	Ratio of tensile strength to yield stress
κ_y	Ratio of yield stress to rejuvenated yield stress
σ	Stress [MPa]
σ_a	Annealing stress [MPa]
σ_b	Tensile strength [MPa]
σ_y	Yield stress [MPa]
$\sigma_{h,c}$	Critical hydrostatic stress [MPa]
σ_{ref}	Reference stress [MPa]

μ	Pressure dependence parameter
v^*	Activation volume [nm^3]
v_c	Crosslink network density [chains m^{-3}]
v_e	Entanglement network density [chains m^{-3}]
$\bar{\tau}$	Equivalent shear stress [MPa]
τ_{Hill}	Hill's effective stress [MPa]
τ_0	Characteristic stress in the stress shift factor for yield [MPa]
τ_a	Characteristic stress in the stress shift factor for aging [MPa]
ΔU	Activation energy for yield [kJ mol^{-1}]
ΔU_a	Activation energy for aging [kJ mol^{-1}]
$^4\boldsymbol{\Psi}$	Fourth order anisotropy operator
x	Crystallinity [%]
superscript d	Indicating deviatoric part of the tensor
superscript h	Indicating hydrostatic part of the tensor
subscript e	Indicating elastic part of the tensor
subscript p	Indicating plastic part of the tensor
$\sim$	Indicating isochoric tensor

List of Mechanical and Structural Jargon as Used in This Chapter

General Jargon

brittle	property of a material that is hard, but prone to break easily
ductile	capability of a material to become permanently deformed
elasticity	capability of a material to retain original dimensions after deformation
fragile	easily broken or damaged
loading long term	testing typically for months, usually even at elevated temperatures
loading short term	material's response to load typically in minutes
modulus	measure of elasticity, equal to ratio of stress applied and resulting strain
plastic flow	permanent flow/strain of a solid material; irreversible deformation
post-yield response	stress–strain during permanent flow of a solid after passing the yield point
strain	length increase divided by original length
stress	load divided by cross-sectional area
tough	difficult to break and high energy take up until failure
viscoelastic	material with both a fluid and a solid character
yield stress	stress beyond which a material ceases to be elastic and becomes plastic

yield point · strain at which yielding starts

Specific Jargon

craze · fine crack bridged by highly extended fibrillar material
geometric softening · apparent weakening as a result of the reduction of cross-sectional area in extension
heterogeneous deformation · deformation with substantial spatial variations
homogeneous deformation · deformation largely equal in the volume of interest
necking · local, non-catastrophic, sudden decrease in cross-sectional area
shear yielding · plastic deformation induced by shear, usually in defined bands at an angle of 30°–45° to the tensile loading direction
strain hardening · when slope of stress–strain curve at large deformations is positive
strain localization · presence of zone within a material where deformation accumulated
strain softening · when slope of stress–strain curve is negative (after yield)

Physical Jargon

energy landscape · potential energy function describing the distribution of free energy
intrinsic softening · material's real softening recorded in tests without strain localization
local densification · local rearrangements on segmental scale
mechanical rejuvenation · returning to un-aged state by application of plastic deformation
physical ageing · striving of a material to reach thermodynamic equilibrium structure
rejuvenated state · thermodynamic state of a polymer after strain softening; un-aged state

Testing Jargon

compression test · samples are compressed uniaxially at constant strain rate
creep loading · samples are loaded with a constant force
endurance limit · stress level below which no long-term failure is observed
fatigue loading · samples are loaded with a dynamic force or strain

engineering strain	ratio of extension and original cross-section; approximation of true strain
engineering stress	ratio of force and original cross-section; approximation of true stress
macroscopic strain	strain defined and applied at the macroscopic level
micro-indentation	hardness-test with a micrometer scale indenter
strain rate	rate at which strain changes with time
tensile test	samples are uniaxially extended at a constant rate of elongation or strain
true strain	natural logarithm of deformed length divided by undeformed length
true stress	force applied divided by momentary cross-sectional area

Geometry Jargon

craze-initiation criterion	criterion for onset of craze initiation; negative pressure
cavitation stress	critical level of negative pressure at which material starts to cavitate
critical triaxial stress	same as critical cavitation stress; critical negative pressure
notch	macroscopic geometrical (sharp) defect
scratch	microscopic surface defect

Material Structure Jargon

amorphous	material of no long-range order
anisotropy	direction dependent structure, properties
entangled network	molecular network created through entanglement of macromolecules
flow-induced nucleation	during flow, long molecules orient and act as nuclei for crystallization
heterogeneous	different structure, properties in different positions
recoverable strain	elastic part of molecular deformation
semi-crystalline	material with partial molecular order
shish-kebabs	long, extended crystalline backbone surrounded by folded chain crystals
spherulitic	spherical crystal structures
texture hardening	increase in yield stress due to rotation of broken crystalline lamellae

Modelling Jargon

constitutive equation	material-specific relation between stress and resulting deformation (solid) or deformation rate (fluid) and vice versa
Deborah number	ratio of molecular relaxation time and observation or process time
entropic spring	mechanical analogue of entropy-based elasticity
glass transition T_g	temperature of kinetic transition from the glassy to the rubbery state
intrinsic response	stress–strain curve in homogeneous deformation
Maxwell element	in-series connection of a spring and a dashpot
molecular rheology	molecular details during flow
neo-Hookean	linear elastic element; isothermal response from theory of rubber elasticity
non-linear elastic	stress not proportional to deformation, reversible deformation
non-linear viscous	non-constant viscosity changes with applied stress
shift factors	to model acceleration via temperature and stress in effective time theories
solid-state rheology	flow processes in the solid state
state parameter	parameter that uniquely determines the state of a material, irrespective of the path used to reach that state
Weissenberg number	product of molecular relaxation time and shear rate in flow

Acknowledgments

This chapter was written during a sabbatical leave of the first author at the ETH Zürich in the Polymer Technology Group of Paul Smith. The authors are indebted to Paul for his critical reading and helpful comments.

The chapter is based on the theses of a number of PhD students. All theses can be downloaded from *www.mate.tue.nl*. where you will find them ordered by year of publication. Marko van de Sanden (1993), Pieter Vosbeek (1994), Theo Tervoort (1996), Bernd Jansen (1998), Robert Smit (1998), Hans Zuidema (2000), Olaf van der Sluis (2001), Harold van Melick (2002), Ilse van Casteren (2003), Hans van Dommelen (2003), Rugi Rastogi (2003), Bernard Schrauwen (2003), Maurice van der Beek (2005), and Edwin Klompen (2005). Moreover the chapter contains contributions from the research work of our present PhD students: Lambert van Breemen, Jan Willem Housmans, Roel Janssen, and Jules Kierkels.

References

1. A. Uhlherr, D. N. Theodorou, *Curr. Opin. Solid State Mater. Sci.* **1998**, *3*, 544–551.
2. D. N. Theodorou *Mol. Phys.* **2004**, *102*, 147–166.
3. D. N. Theodorou, U. W. Suter *Macromolecules* **1986**, *19*, 379–387.
4. J. Baschnagel, K. Binder, P. Doruker, A. A. Gusev, O. Hahn, K. Kremer, W. L. Mattice, F. Müller-Plathe, M. Murat, W. Paul, S. Santos, U. W. Suter, V. Tries *Adv. Polym. Sci.* **2000**, *152*, 41–156.
5. A. Uhlherr, D. N. Theodorou *J. Chem. Phys.* **2006**, *125*, 084107/1–084107/14.
6. R. N. Haward, R. J. Young (Eds.) The Physics of Glassy Polymers, 2nd edn., Chapman & Hall, London, **1997**.
7. H. E. H. Meijer, L. E. Govaert *Macromol. Chem. Phys.* **2003**, *204*, 274–288.
8. H. E. H. Meijer, L. E. Govaert *Prog. Polym. Sci.* **2005**, *30*, 915–938.
9. H. G. H. van Melick, L. E. Govaert, H. E. H. Meijer *Polymer* **2003**, *44*, 2493–2502.
10. H. G. H. van Melick, L. E. Govaert, H. E. H. Meijer *Polymer* **2003**, *44*, 3579–3591.
11. P. J. Flory *J. Am. Chem. Soc.* **1945**, *6*, 2048–2050.
12. H. W. McCormick, F. M. Brower, L. Kin *J. Polym. Sci.* **1959**, *39*, 87–100.
13. A. N. Gent, A. G. Thomas *J. Polym. Sci., Polym. Phys. Ed.* **1972**, *10*, 571–573.
14. P. I. Vincent *Polymer* **1960**, *1*, 425–444.
15. B. H. Bersted, T. G. Anderson *J. Appl. Polym. Sci.* **1990**, *39*, 499–514.
16. E. T. J. Klompen, T. A. P. Engels, L. E. Govaert, H. E. H. Meijer *Macromolecules* **2005**, *38*, 6997–7008.
17. L. E. Govaert, P. H. M. Timmermans, W. A. M. Brekelmans *J. Eng. Mater. Technol.* **2000**, *122*, 177–185.
18. T. A. Tervoort, L. E. Govaert *J. Rheol.* **2000**, *44*, 1263–1277.
19. A. Cross, R. N. Haward, N. J. Mills *Polymer* **1979**, *20*, 288–294.
20. C. Bauwens-Crowet, J. C. Bauwens, G. Homès *J. Polym. Sci. A2* **1969**, *7*, 735–742.
21. C. G'Sell, in McQueen H. (Ed.), *Strength of Metals and Alloys*, Pergamon Press, Oxford **1986**, pp. 1943–1982.
22. L. E. Govaert, H. G. H. van Melick, H. E. H. Meijer *Polymer* **2001**, *42*, 1271–1274.
23. H. G. H. van Melick, L. E. Govaert, B. Raas, W. J. Nauta, H. E. H. Meijer *Polymer* **2003**, *44*, 1171–1179.
24. J. M. Hutchinson *Prog. Polym. Sci.* **1995**, *20*, 703–760.
25. L. C. E. Struik, *Physical Ageing of Amorphous Polymers and Other Materials*, Elsevier, Amsterdam **1978**.
26. E. J. Kramer *Adv. Polym. Sci.* **1983**, *52/53*, 1–56.
27. E. J. Kramer, L. L. Berger *Adv. Polym. Sci.* **1990**, *91/92*, 1–68.
28. C. S. Henkee, E. J. Kramer *J. Polym. Sci., Polym. Phys. Ed.* **1984**, *22*, 721–737.
29. A. M. Donald, E. J. Kramer *J. Polym. Sci., Polym. Phys. Ed.* **1982**, *20*, 899–909.
30. A. M. Donald, E. J. Kramer *Polymer* **1982**, *23*, 1183–1188.
31. R. J. M. Smit, W. A. M. Brekelmans, H. E. H. Meijer *J. Mater. Sci.* **2000**, *35*, 2855–2867.
32. R. J. M. Smit, W. A. M. Brekelmans, H. E. H. Meijer *J. Mater. Sci.* **2000**, *35*, 2869–2879.
33. R. J. M. Smit, W. A. M. Brekelmans, H. E. H. Meijer *J. Mater. Sci.* **2000**, *35*, 2881–2892.
34. H. E. H. Meijer, L. E. Govaert, R. J. M. Smit *ACS Symp. Ser.* **2000**, *759*, 50–70.
35. R. N. Haward, G. Thackray *Proc. R. Soc. London Ser. A* **1967**, *302*, 453–472.
36. T. A. Tervoort, R. J. M. Smit, W. A. M. Brekelmans, L. E. Govaert *Mech. Time-Depend. Mater.* **1998**, *1*, 269–291.
37. M. J. Osborne, D. J. Lacks *J. Phys. Chem. B* **2004**, *108*, 19619–19622.
38. D. J. Lacks, M. J. Osborne *Phys. Rev. Lett.* **2004**, *93*, 255501/1–4.
39. T. A. Tervoort, E. T. J. Klompen, L. E. Govaert *J. Rheol.* **1996**, *40*, 779–797.
40. G. A. Adam, A. Cross, R. N. Haward *J. Mater. Sci.* **1975**, *10*, 1582–1590.
41. C. Bauwens-Crowet, J. C. Bauwens *Polymer* **1982**, *23*, 1599–1604.
42. G. B. McKenna *J. Phys. Condens. Matter* **2003**, *15*, S737–S763.
43. E. T. J. Klompen, T. A. P. Engels, L. C. A. van Breemen, P. J. G. Schreurs, L. E. Go-

vaert, H. E. H. Meijer *Macromolecules* **2005**, *38*, 7009–7017.
44 L. E. Govaert, T. A. P. Engels, E. T. J. Klompen, G. W. M. Peters, H. E. H. Meijer *Int. Polym. Proc.* **2005**, *20*, 70–177.
45 T. A. P. Engels, L. E. Govaert, G. W. M. Peters, H. E. H. Meijer *J. Polym. Sci. B* **2006**, *44*, 1212–1225.
46 T. A. P. Engels, L. E. Govaert, G. W. M. Peters, H. E. H. Meijer, in *13th International Conference on Deformation, Yield and Fracture of Polymers*, L. E. Govaert, H. E. H. Meijer (Eds.), Rolduc Abbey, Kerkrade, Netherlands, **2006**, pp. 69–72.
47 H. G. H. van Melick, O. F. J. T. Bressers, J. M. J. Den Toonder, L. E. Govaert, H. E. H. Meijer *Polymer* **2003**, *44*, 2481–2491.
48 M. Ishikawa, I. Narisawa, H. Ogawa *J. Polym. Sci., Polym. Phys. Ed.* **1977**, *15*, 1791–1804.
49 M. Ishikawa, H. Ogawa, I. Narisawa *J. Macromol. Sci. B* **1981**, *19*, 421–443.
50 L. Camwell, D. Hull *Philos. Mag.* **1973**, *27*, 1135–1150.
51 I. Narisawa, M. Ishikawa, H. Ogawa *Philos. Mag. A* **1980**, *41*, 331–351.
52 R. J. M. Smit, W. A. M. Brekelmans, H. E. H. Meijer *Comput. Methods Appl. Mech. Eng.* **1998**, *155*, 181–192.
53 R. J. M. Smit, W. A. M. Brekelmans, H. E. H. Meijer *J. Mech. Phys. Solids* **1999**, *47*, 201–221.
54 M. Considère *Die Anwendung von Eisen und Stahl in Konstruktionen*, **1888**, Gerold, Wien.
55 P. I. Vincent *Polymer* **1960**, *1*, 7–19.
56 R. N. Haward *Polymer* **1987**, *28*, 1485–1488.
57 B. A. G. Schrauwen, R. P. M. Janssen, L. E. Govaert, H. E. H. Meijer *Macromolecules* **2004**, *37*, 6069–6078.
58 R. J. Young *Philos. Mag.* **1974**, *30*, 85–94.
59 R. J. Young *Mater. Forum* **1988**, *11*, 210–216.
60 V. Gaucher-Miri, R. Séguéla *Macromolecules* **1997**, *30*, 1158–1167.
61 R. Séguéla, O. Darras *J. Mater. Sci.* **1994**, *29*, 5342–5352.
62 N. W. Brooks, R. A. Duckett, I. M. Ward *Polymer* **1992**, *33*, 1872–1880.
63 M. F. Butler, A. M. Donald, A. J. Ryan *Polymer* **1998**, *39*, 39–52.
64 M. F. Butler, A. M. Donald, A. J. Ryan *Polymer* **1998**, *39*, 781–792.
65 M. F. Butler, A. M. Donald, W. Bras, G. R. Mant, G. E. Derbyshire, A. J. Ryan *Macromolecules* **1995**, *28*, 6383–6393.
66 P. B. Bowden, R. J. Young *J. Mater. Sci.* **1974**, *9*, 2034–2051.
67 B. A. G. Schrauwen *Deformation and failure of semi-crystalline polymer systems*, Ch. 3, PhD Thesis, Eindhoven University of Technology, **2003**, pp. 25–47.
68 M. Seki, D. W. Thurman, J. P. Oberhauser, J. A. Kornfield *Macromolecules* **2002**, *35*, 2583–2594.
69 B. A. G. Schrauwen, L. C. A. van Breemen, A. B. Spoelstra, L. E. Govaert, G. W. M. Peters, H. E. H. Meijer *Macromolecules* **2004**, *37*, 8618–8633.
70 O. K. Muratoglu, A. S. Argon, R. E. Cohen, M. Weinberg *Polymer* **1995**, *36*, 921–930.
71 O. K. Muratoglu, A. S. Argon, R. E. Cohen, *Polymer* **1995**, *36*, 2143–2152.
72 O. K. Muratoglu, A. S. Argon, R. E. Cohen, M. Weinberg *Polymer* **1995**, *36*, 4771–4786.
73 Z. Bartczak, A. Galeski, A. S. Argon, R. E. Cohen *Polymer* **1996**, *37*, 2113–2123.
74 Z. Bartczak, A. S. Argon, R. E. Cohen, T. Kowalewski *Polymer* **1999**, *40*, 2367–2380.
75 Z. Bartczak, A. S. Argon, R. E. Cohen, M. Weinberg *Polymer* **1999**, *40*, 2331–2346.
76 Z. Bartczak, A. S. Argon, R. E. Cohen, M. Weinberg *Polymer* **1999**, *40*, 2347–2365.
77 M. W. L. Wilbrink, A. S. Argon, R. E. Cohen, M. Weinberg *Polymer* **2001**, *42*, 10155–10180.
78 Y. S. Thio, A. S. Argon, R. E. Cohen, M. Weinberg *Polymer* **2002**, *43*, 3661–3674.
79 B. A. G. Schrauwen, L. E. Govaert, G. W. M. Peters, H. E. H. Meijer *Macromol. Symp.* **2002**, *185*, 89–102.
80 B. Yalcin, D. Valladares, M. Cakmak *Polymer* **2003**, *44*, 6913–6925.
81 B. Yalcin, M. Cakmak *Polymer* **2004**, *45*, 2691–2710.
82 Y. Konishi, M. Cakmak *Polymer* **2005**, *46*, 4811–4826.

83 J. A. W. van Dommelen, W. A. M. Brekelmans, F. P. T. Baaijens *Mech. Mater.* **2003**, *35*, 845–863.
84 J. A. W. van Dommelen, W. A. M. Brekelmans, F. P. T. Baaijens *Comput. Mater. Sci.* **2003**, *27*, 480–492.
85 J. A. W. van Dommelen, W. A. M. Brekelmans, F. P. T. Baaijens *J. Mater. Sci.* **2003**, *38*, 4393–4405.
86 L. Corté, *Renforcement de polymères semi-crystallins*, PhD Thesis, Universite Pierre et Marie Curie, **2006**.
87 L. Corté, F. Beaume, L. Leibler *Polymer* **2005**, *46*, 2748–2757.
88 F. P. T. Baaijens *Rheol. Acta* **1991**, *30*, 284–299.
89 T. C. B. McLeish, R. G. Larson *J. Rheol.* **1998**, *42*, 81–110.
90 W. M. H. Verbeeten, G. W. M. Peters, F. P. T. Baaijens *J. Rheol.* **2001**, *45*, 823–843.
91 H. Zuidema, G. W. M. Peters, H. E. H. Meijer *Macromol. Theory Simul.* **2001**, *10*, 447–460.
92 G. Eder, H. Janeschitz-Kriegl, in *Materials Science and Technology*, H. E. H. Meijer (Ed.), Verlag Chemie **1997**, Vol. 18, pp. 269–342.
93 W. R. Hwang, M. A. Hulsen, H. E. H. Meijer *J. Comput. Phys.* **2004**, *194*, 742–772.
94 W. R. Hwang, M. A. Hulsen, H. E. H. Meijer *J. Non-Newtonian Fluid Mech.* **2004**, *121*, 15–33.
95 M. A. Hulsen, R. Fattal, R. Kupferman *J. Non-Newtonian Fluid Mech.* **2005**, *127*, 27–39.
96 R. Fattal, R. Kupferman *J. Non-Newton. Fluid Mech.* **2004**, *123*, 281–285.
97 R. Fattal, R. Kupferman *J. Non-Newton. Fluid Mech.* **2005**, *126*, 23–37.
98 W. R. Hwang, G. W. M. Peters, M. A. Hulsen, H. E. H. Meijer *Macromolecules* **2006**, *39*, 8389–8398.
99 H. Janeschitz-Kriegl, E. Ratajski, M. Stadlbauer *Rheol. Acta* **2003**, *42*, 355–367.
100 C. Tribout, B. Monasse, J. M. Haudin *Colloid Polym. Sci.* **1996**, *274*, 197–208.
101 R. H. Somani, L. Yang, B. S. Hsiao, P. K. Agarwal, H. A. Fruitwala, A. H. Tsou *Macromolecules* **2002**, *35*, 9096–9104.
102 L. Yang, R. H. Somani, I. Sics, B. S. Hsiao, R. Kolb, H. Fruitwala, C. Ong *Macromolecules* **2004**, *37*, 4845–4859.
103 C. M. Hsiung, M. Cakmak *J. Appl. Polym. Sci.* **1993**, *47*, 125–147.
104 Y. Ulcer, M. Cakmak, J. Miao, C. M. Hsiung *J. Appl. Polym. Sci.* **1996**, *60*, 669–691.
105 J. A. W. van Dommelen, D. M. Parks, M. C. Boyce, W. A. M. Brekelmans, F. P. T. Baaijens *J. Mech. Phys. Solids* **2003**, *51*, 519–541.
106 J. A. W. van Dommelen, D. M. Parks, M. C. Boyce, W. A. M. Brekelmans, F. P. T. Baaijens *Polymer* **2003**, *44*, 6089–6101.
107 J. A. W. van Dommelen, B. A. G. Schrauwen, L. C. A. van Breemen, L. E. Govaert *J. Polym. Sci. B* **2004**, *42*, 2983–2994.
108 J. A. W. van Dommelen, H. E. H. Meijer, in *Mechanical Properties of Polymers based on Nanostructure and Morphology*, G. H. Michler, F. J. Balta-Calleja (Eds.) CRC Press, Boca Raton, FL **2005**, pp. 317–318.
109 R. Hill *The Mathematical Theory of Plasticity*, Oxford University Press, London **1950**.
110 R. Rastogi, W. P. Vellinga, S. Rastogi, C. Schick, H. E. H. Meijer *J. Polym. Sci. B* **2004**, *42*, 2092–2106.

12
Scanning Calorimetry

René Androsch and Bernhard Wunderlich

12.1
Introduction

This chapter focuses on the specific heat capacity for the analysis of the thermodynamic stability and phase structure of polymeric materials. Polymer materials which consist of linear and flexible macromolecules may exist in a large variety of metastable phases of differing size down to nanometers due to kinetic restrictions of the crystallization process. These restrictions may result in the coexistence of multiple phases, which, according to the Gibbs phase rule, are not in equilibrium [1]. The structure and stability of these coexisting phases need to be analyzed since they control the macroscopic properties of materials. In particular, the thermodynamic stability of phases can be accessed with the specific heat capacity since this thermal property permits the calculation of the thermodynamic functions of state: free enthalpy, G, enthalpy, H, and entropy, S. These functions need to be determined for each possible phase and comparisons allow the prediction of the equilibrium phase with minimum G, which is a consequence of the combination of the first and second laws of thermodynamics. The knowledge of the thermodynamic functions of state of the various phases furthermore permits a discussion of the nature of phase transitions. The transition from the liquid phase to the solid crystalline phase occurs, if viewed on a global scale, on an irreversible path with entropy production due to the need for nucleation [2, 3]. The recent introduction of temperature-modulated calorimetry [4–7], unexpectedly, identified also reversible crystallization and melting for most polymers, which is a process located at the surface of crystals and which apparently does not need nucleation. Analysis of the heat capacity is a valuable tool to quantify irreversible and reversible changes of structure on temperature perturbation which ultimately gives information about structure on the nanoscale [8].

This chapter starts with a brief summary of the link between the experimentally accessible thermal property *specific heat capacity* and the integral thermodynamic functions of state G, H and S and a discussion of the nature of phase transitions in polymers from a thermodynamics point of view.

Macromolecular Engineering. Precise Synthesis, Materials Properties, Applications.
Edited by K. Matyjaszewski, Y. Gnanou, and L. Leibler
Copyright © 2007 WILEY-VCH Verlag GmbH & Co. KGaA, Weinheim
ISBN: 978-3-527-31446-1

The next part is an explanation of the experimental determination of the heat capacity of polymers using temperature-modulated calorimetry. The analysis of the heat capacity by adiabatic calorimetry and standard differential scanning calorimetry is not described in detail since it is outlined in numerous textbooks and has been state of the art for many decades [9–11].

The raw data for heat capacity from a typical calorimetric experiment, next, need to be separated carefully into different contributions of structural response to temperature perturbation since they are a superposition of several thermal effects on largely different time-scales. These contributions include as extremes the vibrational heat capacity on a time-scale of picoseconds to femtoseconds and long-term reorganization at a time-scale up to many years resulting from life-time determinations.

Subsequently, selected polymer structure–thermal property relations are discussed, focusing on current research activity including the analysis of the glass transition for describing the rigid amorphous phase in semicrystalline polymers and the effect of crystal morphology on reversible melting. The effect of chemical and physical structure on thermal behavior, in general, is summarized for the sake of completeness.

Finally, it is explained that application of additional tools for thermal analysis of polymers is necessary for the complete establishment of structure–property relations in polymers. A brief overview of important experimental techniques in the field of thermal analysis is given.

Before turning to the specific discussion of the heat capacity for the evaluation of thermodynamic functions and phase structure of materials, a broader view of the importance of thermal properties is given below. A property in material science is the characteristics of a material which describe the kind and the magnitude of a response to a specific imposed stimulus [12]. In the case of thermal properties, it is the response of a material to a change of temperature, T. Thermal properties, which describe the macroscopic behavior of a material on temperature variation, include, in addition to the specific heat capacity at constant pressure and composition, c_p, the linear and volume expansivity, α and β, respectively, and the thermal conductivity, λ, which are defined in Eqs. (1–3), where V and A represent the appropriate volume and area:

Heat capacity at constant pressure (C_p):

$$C_p(T) = \left.\frac{dH(T)}{dT}\right|_p \tag{1}$$

Volume expansivity (β):

$$\beta(T) = \frac{1}{V_0}\left.\frac{dV(T)}{dT}\right|_p \tag{2}$$

Thermal conductivity (λ):

$$\frac{dQ}{dt} = -\lambda A \operatorname{grad} T \qquad (3)$$

Furthermore, it is useful to include in this listing the glass transition temperature, T_g, and melting temperature, T_m, since all properties of a material change with temperature variation. The above properties or material characteristics are not only of fundamental importance for the estimation of the stability and phase structure of a system or material but also need to be known for many practical purposes.

The heat capacity C_p in J K^{-1} is the absolute amount of energy H in J which is needed to raise the temperature of a substance by 1 K. The specific heat capacity c_p is obtained by normalizing with the mass, typically in units of grams, or with the number of moles. This simple translation of Eq. (1) explains why the knowledge of c_p is necessary to analyze the thermodynamics of a system. Knowledge of the specific heat capacity is absolutely indispensable for processing and application of polymeric materials. Regarding the processing of polymeric materials, information about the specific heat capacity is required to design, for instance, the flow of thermal energy. Processing of thermoplastics typically involves heating to a temperature where viscous flow of macromolecules is possible. Subsequently, the polymers are shaped, using a die in extrusion processes or a mold in an injection-molding machine, and cooled to allow solidification. Heating and cooling characteristics are controlled by, among others, the specific heat capacity and latent heat of first-order transitions such as melting, which, if known, allow computer-aided design of machinery or processes, respectively. The specific heat capacity of polymers is usually larger than the specific heat capacity of metals or ceramics by a factor of 2–10 because of their lower density. The heat capacity of polyethylene at ambient temperature is ≈ 2 J g^{-1} K^{-1}, which is about 2.5 times the heat capacity of aluminum oxide ceramic (≈ 0.75 J g^{-1} K^{-1}) and about 4 times the heat capacity of steel (≈ 0.5 J g^{-1} K^{-1}).

The volume expansivity β describes the change in volume V on varying the temperature. The quantity V_0 in the Eq. (2) is the initial volume. Replacing volume by length yields the linear expansion coefficient α with the same dimension as β, namely K^{-1}. The knowledge of the thermal expansion characteristics of a polymeric material is of similarly extreme importance as the knowledge of the heat capacity, in particular since in case of polymers the thermal expansion may be increased by 1–2 orders of magnitude compared with metals or ceramics. Typical values for polymers are in the range of about $(1–2) \times 10^{-4}$ K^{-1}, which should be compared with 2×10^{-5} K^{-1} for copper and 1×10^{-6} K^{-1} for quartz. The reason for the relatively large thermal expansion of polymers are the weak intermolecular forces between the macromolecules. The anisotropy of the chemical bonding, that is, strong covalent bonding in the chain direction and weak secondary bonding laterally, implies a further polymer-specific phenomenon, which is the anisotropy of properties. This, in particular, includes the thermal expansion, which can even be negative in the chain direction. For design of processing, the thermal expansion of polymeric materials must be known, just as the heat capacity. The thermal shrinkage of a shaped polymeric part in the mold of an injection-molding machine due to cooling to ambient temperature may easily cause distortions of the geometry of the final product and must therefore be exactly compensated by continued filling of the mold during the cooling process until the temperature is lower than the so-called freezing temperature, that is, the glass transition temperature, T_gg",4,1>. From the point of view of application of polymeric materials, the thermal expansion coefficient must be known for estimation of the geometry or size in the entire temperature range of service. This temperature range can be as large as 100 K, as is the case for parts in the engine compartment of automobiles. Serious distortion of the geometry due to hindered thermal expansion on heating may lead to thermally induced stress, viscous flow and failure of the functionality.

The thermal conductivity λ of polymers is a further important property of polymers since it describes the capability to transfer thermal energy. In Eq. (3), dQ/dt is the flux of heat through a defined area A in the presence of a temperature gradient grad T. In the case of polymers the heat conductivity is about 10^{-1} W m^{-1} K^{-1}. This is considerably less than the heat conductivity of metals, which is about 10^2 W m^{-1} K^{-1}, or ceramics, with $\lambda \approx 10^0$–10^1 W m^{-1} K^{-1}. The increased heat conductivity in the case of metals is easy to understand since the energy transfer is not hindered by low energy of secondary bonding, as is evident in the cross-chain direction of macromolecules.

The heat capacity, thermal expansivity and thermal conductivity depend on temperature and show distinct discontinuities at the transition temperatures. Furthermore, these properties are strongly dependent on the phase structure, that is, they are controlled by the degree of crystallinity in semicrystalline polymers. A detailed discussion of the heat capacity with respect to temperature and phase composition is given below.

12.2
Heat Capacity and Thermodynamic Functions of State

Figure 12.1 shows the specific heat capacity of liquid (thin line) and solid polyethylene (thick line) as function of temperature. These data are typical of linear macromolecular structures and are listed for numerous polymers in the ATHAS Database [13].

The specific heat capacity of the solid, that is, of the crystalline phase or the amorphous glass, increases continuously with increasing temperature. The heat capacity is caused by the molecular motion of the segments of the molecules. The frequencies of the vibrations are a function of temperature and were identified for polyethylene using model calculations fitted to the normal mode analysis of the experimental frequency spectrum obtained from infrared and Raman spectroscopy [14]. At 0 K all thermal motion is frozen and the heat capacity is zero. Between 0 and 200 K, the heat capacity of solid polyethylene is caused by

12.2 Heat Capacity and Thermodynamic Functions of State

Fig. 12.1 Specific heat capacity of solid and liquid polyethylene as function of temperature.

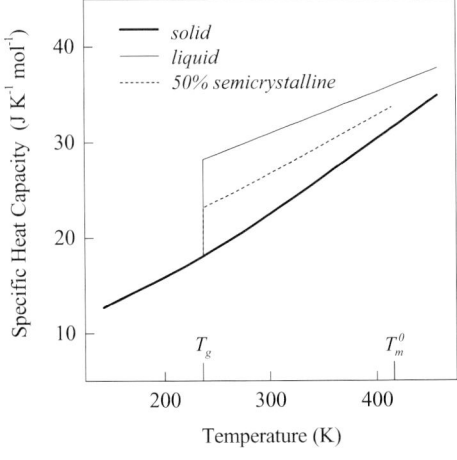

specific heat capacity of semicrystalline polymers

$$c_p(T) = m_c\, c_{p,c} + (1 - m_c)\, c_{p,l} \quad (4)$$

m_c mass fraction crystalline phase
$c_{p,c}$ crystal heat capacity
$c_{p,l}$ liquid heat capacity

torsion- and accordion-like motion of the backbone of the macromolecule, which is called the skeletal vibrations. The skeletal vibrations approach full excitation at about 200 K. The further increase of the heat capacity of the solid with increasing temperature is due to the onset of group vibrations. These are vibrations of small groups of atoms along the backbone, including bending, rocking, wagging and twisting motion of the hydrogen atoms and stretching of C–C bonds. The C–H stretching vibration finally starts to contribute at about 400 K and would be fully excited at about 700–800 K, but melting intervenes and changes the polymer to the liquid state (equilibrium melting temperature = 414.6 K). At low temperature the heat capacity of glassy polyethylene is somewhat larger than that of solid polyethylene. Devitrification of the amorphous phase on heating occurs at 237 K and is connected with a step-like increase in the heat capacity by about 50% from 18 to 28 J K^{-1} mol^{-1}. This is mainly caused by a change in the torsional vibrations to large-amplitude, cooperative rotations of molecule segments. The large-amplitude cooperative motion of molecule segments is coupled above the glass transition with an increased free volume, that is, it is directly related to the macroscopically measurable thermal expansivity, which also increases step-like on heating at the glass transition temperature.

At temperatures between the glass transition of the liquid, l, that is, of the amorphous phase and the melting of the crystalline phase, c, the heat capacity of semicrystalline polymers is, according to Eq. (4), given by the corresponding mass average of the two phases. The heat capacities of the crystalline (solid) and amorphous (liquid) phases are $c_{p,c}$ and $c_{p,l}$, respectively, and m_c is the mass fraction of the crystalline phase, the crystallinity. In other words, the heat capacity of semicrystalline polymers is intermediate between the heat capacities of the liquid and solid phases, as is indicated for a 50% crystalline preparation by the dotted line in

Fig. 12.1. This line is the so-called heat-capacity baseline, which must be known for the given crystallinity for any quantitative evaluation of additional thermal effects in calorimetry which can produce a latent heat. The latent heat, L, is defined as $(\partial H/\partial n)_T$ with ∂n representing the change in composition at constant temperature. Note that for a detailed thermal analysis with changing temperature, dn/dT and the accompanying heat capacity (due to the concentration change) must be known. For a melting on heating or crystallization on cooling, which may occur over a narrow temperature range, it is sufficient to integrate the measured apparent heat capacity over the transition range, that is, integrate the calorimeter response due to the heat capacity *and* latent heat, $dH/dT = c_p(T) + (\partial L/\partial T)_p dn$. By subtracting the proper baseline integral $[\int c_p(T)dT]$, adjusted for the change in concentration, the latent heat can be evaluated.

The specific heat capacity at constant pressure can deliver valuable information about the thermodynamics of a system or material, since it permits the evaluation of the enthalpy H, the entropy S and the free enthalpy G as a function of temperature [Fig. 12.2, Eqs. (5–7)] [9].

Figure 12.2 shows as a typical example of the integral thermodynamic functions of state G and H of polyethylene as a function of temperature at constant atmospheric pressure. Note that G and H are not available as absolute values since the enthalpy of the crystalline phase at $T_0 = 0$ K, H_0, is unknown and larger than zero. This is accounted for in Eq. (6) by the offset H_0 for calculation of the enthalpy H. The data in Fig. 12.2 must therefore be considered as being vertically shifted downwards by H_0. Such data are listed in the ATHAS Databank and were calculated from critically reviewed experimental and calculated heat capacity data and heat of fusion data available in the literature. The gray-shaded area represents the entropy multiplied by temperature, which is a convenient translation of Eq. (5) $[TS(T) = H(T) - G(T)]$. The subscripts c and l in Fig. 12.2 de-

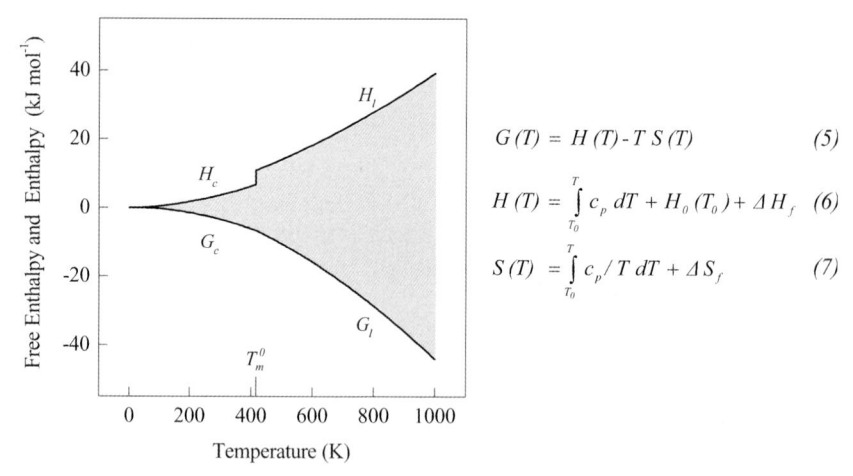

$$G(T) = H(T) - TS(T) \qquad (5)$$

$$H(T) = \int_{T_0}^{T} c_p \, dT + H_0(T_0) + \Delta H_f \qquad (6)$$

$$S(T) = \int_{T_0}^{T} c_p / T \, dT + \Delta S_f \qquad (7)$$

Fig. 12.2 Free enthalpy G and enthalpy H of polyethylene as function of temperature.

note the solid crystalline phase and liquid phase, respectively. The total enthalpy H in Eq. (6) includes the enthalpy change ΔH_f due to melting at the equilibrium melting temperature, as a prominent example of a first-order phase transition. In a pure one-component system as the chosen polyethylene, the transition must occur at constant temperature, according to the Gibbs phase rule, and, therefore, causes a step-like increase in the enthalpy at the equilibrium melting temperature in Fig. 12.2. The bulk specific heat of fusion of polymer crystals, ΔH_f, in the simplest case could be measured directly if fully crystalline samples were available. Incomplete crystallization, however, requires the preparation of sets of samples with different amounts of crystalline phase and extrapolation to the fully crystalline state, as demonstrated on the example of polyamide in [15–17]. The determination of the degree of crystallinity in these samples is possible by measurement of the density ρ. The density of the amorphous phase, ρ_a, of polymers is generally known, either by preparation of fully amorphous polymers or by extrapolation from the melt, and the density of the crystalline phase, ρ_c, can be derived by X-ray analysis of the crystal structure. The mass fraction crystallinity, X_c, is then calculated by assumption of a two-phase model $[X_c = (\rho_c/\rho)(\rho-\rho_a)/(\rho_c-\rho_a)]$ [18]. Often such a two-phase model is a first approximation only and additional phases need to be introduced, such as the common rigid amorphous phase, RAF, caused by the coupling of the amorphous phase to the crystal, to be discussed below.

The entropy change on melting, ΔS_f, in Eq. (7) cannot be measured and must be calculated using Eq. (5) with the absolute values of G, H and S replaced by the corresponding differences between the crystalline and liquid phase $[\Delta G(T) = \Delta H(T) - T\Delta S(T)]$. The value of ΔG between the crystalline and liquid phases is zero at $T = T_m^0$, which then allows one to calculate the entropy of equilibrium melting at the equilibrium melting temperature. Note that the entropy production during equilibrium melting and crystallization is zero, that is, the entropy increase on melting of crystals and crystallization of the melt at T_m^0 is fully compensated by the measured latent heats of fusion and crystallization that accompany these processes.

The thermodynamic functions of state of Fig. 12.2 represent thermodynamic equilibrium and allow the prediction of the thermodynamic stability of phases as a function of temperature. Outside of T_m^0, in a one-component system, such as monodisperse polyethylene, there can only be one stable phase at a given temperature, which is also expressed by the Gibbs phase rule. At temperatures below 414.6 K and atmospheric pressure the crystalline orthorhombic phase is stable and above 414.6 K the liquid phase. At the equilibrium melting temperature of orthorhombic polyethylene, both phases coexist and can convert reversibly into each other by absorption or evolution of latent heat. As soon as one of the phases is exhausted, further changes in enthalpy must change the temperature away from T_m^0. In other words, the overall phase structure can only change at the melting temperature. The liquid phase at temperatures $T < T_m^0$ or the crystalline phase at $T > T_m^0$ must crystallize or melt, respectively, in order to attain equilibrium. The non-equilibrium state, however, may be arrested, for ex-

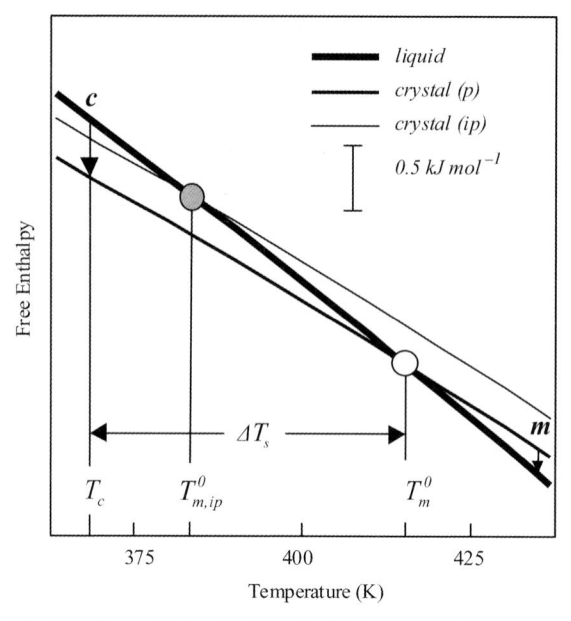

Fig. 12.3 Free enthalpy G of the liquid phase, perfect orthorhombic crystalline phase and imperfect crystalline phase of polyethylene as function of temperature. Data representing the liquid phase and perfect orthorhombic crystalline phase were taken from [13].

ample by kinetic restrictions due to nucleation barriers or very slow melting mechanisms. The two mentioned examples lead to supercooling and superheating, respectively. The directionality of the phase transition at $T \neq T_m^0$ can easily be explained by comparison of the free enthalpies of the crystalline and liquid phases, as is shown in Fig. 12.3.

The combination of the first and second laws of thermodynamics predict thermodynamic stability of a phase only if the free enthalpy is a minimum, that is, a spontaneous phase transition will only occur if the change in free enthalpy is negative. This is shown with the downwards arrows in Fig. 12.3, the left arrow (c) indicating crystallization and the right arrow (m) indicating melting. The decrease of G at constant temperature classifies a transition as being irreversible, that is, the transition can only be reversed by further variation of the thermodynamic variables of state temperature, pressure or concentration. Irreversible crystallization requires the above mentioned supercooling (ΔT_s) to overcome diverse energy barriers, necessary for formation of crystal surfaces (crystal nucleation) [19], and for decreasing the local entropy of molecule segments in the liquid phase before becoming attached to an existing crystal (molecular nucleation) [19–21]. Irreversible melting, in contrast, often only needs a minor activation energy since melt nuclei exist in any semicrystalline structure at the surface of the crystals, that is, superheating of polymer crystals is less often observed than supercooling of the liquid [22].

For kinetic reasons, the completion of the crystallization process of macromolecules is strongly hindered, which typically results in semicrystalline polymers and the coexistence of crystalline and non-crystalline phases, apparently violating the Gibbs phase rule. The resulting semicrystalline polymers are therefore metastable states since further progress towards equilibrium is arrested. Additionally, not only the extent of crystallization may be affected but also the quality of the crystalline phase. There is evidence of a large number of intermediate phases (ip), with the free enthalpies between those of the perfect extrapolated liquid and the perfect crystal (p). The free enthalpy of such an imperfect phase is even more difficult to establish from experimental data and is indicated in Fig. 12.3 by the thin line. This curve is shifted upwards compared with that of the perfect crystal and indicates by the intersection with the free enthalpy curve of the liquid an equilibrium melting point depression ($T_m^0 - T_{m,ip}^0$). A common reason for a substantial increase in free enthalpy is chain folding. Internal crystal defects usually cause a smaller increase in G. For these two imperfections the overall entropy of the phase is little affected, so that their free enthalpy is parallel to that of the crystal, as drawn in Fig. 12.3. Other in-between phases may have a different structure such as seen for liquid crystalline or conformationally disordered (condis) phases. In these cases the entropy is larger than that of the crystal, so that the slope of G is larger. It is even possible that such a free enthalpy curve intersects both the G of the crystal and the liquid and yields a range of stability for the intermediate phase, also called a mesophase. A similar discussion is possible regarding an upward shift of the free enthalpy curve of the perfect liquid, for example due to partial ordering by orientation or strain, which causes an increase in the melting point or a downward shift of the free enthalpy curve of the perfect liquid, for example due to thermodynamic miscibility with a second component, which leads to a decrease in the melting point [23]. Note that melting at $T_{m,ip}^0$ still is zero entropy production melting, since at this temperature the imperfect crystal is as metastable as the liquid [24].

The analysis of the thermodynamic heat capacity of the various phases that are possible in polymers, in combination with the evaluation of the characteristics of transitions, that is, the equilibrium temperature of the transition, bulk specific heat and entropy of transition, is a necessary prerequisite for the analysis of the thermal behavior and the properties of materials.

12.3
Measurement of the Heat Capacity and Separation of Thermal Events by temperature-modulated DSC

Measurement of the specific heat capacity at constant pressure by standard differential scanning calorimetry (DSC) is well described in the literature [9] and is only briefly explained here for easier understanding of the recently introduced temperature-modulated differential scanning calorimetry (TMDSC). Instrumentation includes a sample calorimeter and a reference calorimeter which either are placed

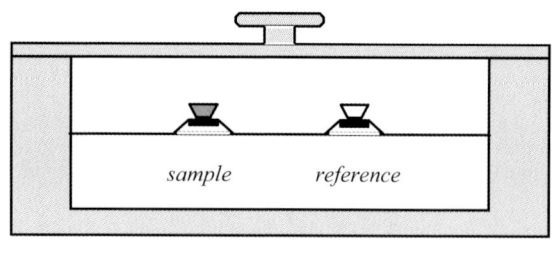

- controlled change of temperature -
- exact temperature measurement -
- defined conduction of heat -

Fig. 12.4 Schematic of the measuring cell in a differential scanning calorimeter (DSC).

in a single furnace in the case of heat flux calorimetry or which are separated in the case of power compensation calorimetry [9, 11]. The schematic in Fig. 12.4 shows the sample and reference calorimeters in a single furnace for heat flux DSC. All technical details, common in state of the art instruments for improved accuracy, are removed for the sake of clarity. The calorimeters consist of metal pans to increase the conductivity in the sample and are usually made of Al, Au, Ag or Pt or their alloys, depending on the temperature limits and chemical inertness required. The calorimeters are exposed to controlled temperature changes using the furnace and well-defined heat conduction pathways. The path from the furnace wall is chosen such that the temperature gradient within the sample can be neglected, making the heat flow-rate into the sample proportional to the temperature difference between reference and sample, $\Delta T_{r,s} = (T_r - T_s)$. The temperature of each calorimeter is recorded separately as output of the experiment. In standard experiments, the pan volume and sample mass are of the order of 10–50 µL and 1–20 mg, respectively. The furnace can be purged by gases at a rate ensuring laminar flow. Typically dry nitrogen at flow-rates of 10–50 mL min^{-1} are used to minimize thermal-oxidative degradation of the polymer.

The measurement of the specific heat capacity of the sample requires heating or cooling at a controlled rate q, which is a consequence of the definition of the heat capacity being the amount of energy required to change the temperature by 1 K. Heat capacity cannot be measured isothermally. The left part of Fig. 12.5 shows the dependence of temperature on time, as programmed (dotted line) and as measured underneath the reference calorimeter (thin line) and the sample calorimeter (thick line). The measured temperature lags behind the program due to the gradient between the furnace and the calorimeters. As outlined above, it then depends approximately on the heat capacity of the sample, the heating rate q and heat conduction and geometry of the apparatus, which is expressed by Newton's law constant K. The latter two parameters are approximately identical for both calorimeters. The lag is larger for the sample calorimeter since heat must be absorbed. The constant K converts the temperature difference between sample and reference into a differential heat flow-rate, HF,

temperature versus time

- .... program
- —— reference
- —— sample

$\Delta T_{s,r}$

$$\Delta T_{s,r} \propto \Delta c_{p_{s,r}}\, m\, q/K \quad (8)$$

heat-flow rate versus time

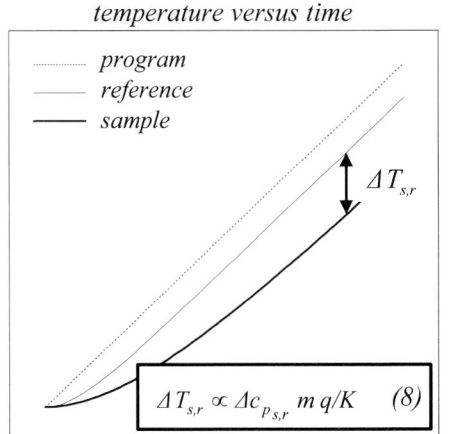

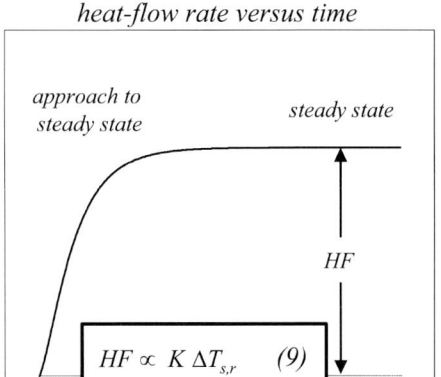

approach to steady state

steady state

HF

$$HF \propto K\, \Delta T_{s,r} \quad (9)$$
$$c_p \propto HF/q \quad (10)$$

Fig. 12.5 Program, sample and reference temperature (a) and the heat flow-rate into the sample (b) as a function of time.

with the prerequisite of a proper calibration of the instrument. In addition to the heat flow-rate calibration, the temperature must also be calibrated. Furthermore, it is necessary to correct the instrumental asymmetry. This correction compensates differential heat flow-rate into sample and reference calorimeters in the absence of any sample and is done by measurement of an instrumental baseline under identical conditions as the sample run, that is, using the same temperature–time program, purging condition and mass of aluminum pans. Within limits, measurement of the heat capacity using standard DSC is more accurate when using a large sample mass, high heating rate, optimum heat conduction between pan and sample and no cross-flow of heat between sample and reference calorimeter. The error of heat capacity measurements is typically about 3%, which is mainly due non-reproducible heat conduction between furnace, pan and sample. Note that heat capacity measurement by standard DSC requires steady-state conditions, as indicated by the arrow in Fig. 12.5. Any change in heating or cooling rates in a DSC experiment requires instrumental equilibration, which can be quantified by a time constant, typically being in the range of seconds to minutes, depending on the design of the instrument.

TMDSC has been commercially available since 1993 and offers several advantages compared with standard DSC [4–7]. The basic idea of TMDSC is the simultaneous use of two rates of temperature change, in order to resolve thermal events of different kinetics in the sample. A typical example is the separation of the heat capacity, which contributes to the heat flow-rate signal even at the fastest rates of temperature change and slow processes such as enthalpy-relaxation, crystallization, melting or chemical reactions, which may be sufficiently slow not to follow fast temperature perturbation. Note that enthalpy changes measured with scanning calorimetry may originate not only from true heat capaci-

program temperature versus time

sum

linear component (slow)

modulated component (fast)

linear component

$\langle q \rangle \approx 10^0$ K min^{-1}

modulated component

sinusoidal, sawtooth-like

rate $\quad q \quad 10^1$ K min^{-1}

amplitude $\quad A^T \approx 0.1...2$ K

period $\quad p \approx 10...200$ s

frequency $\quad \omega = 2\pi/p$

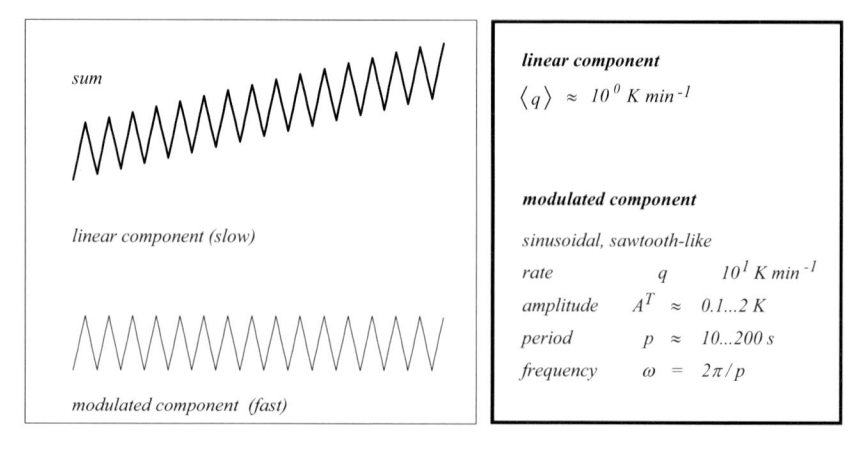

Fig. 12.6 Program temperature as a function of time in a modulated DSC.

ties, as defined in Eq. (1), but also from latent heats which change the enthalpy without change in temperature, but with changes in composition, such as in phase transitions like melting (with heats of fusion), evaporation (with heats of evaporation) and chemical changes (with heats of reactions). Processes with latent heats often follow slow kinetics. Frequently they are also of large magnitude and cause instrument lags which must be considered separately.

A typical temperature–time program for TMDSC is shown in Fig. 12.6. It is a superposition of a linear underlying component, with a rate of temperature change between 0 (isothermal mode) and 2 K min^{-1} and a modulated component with an instantaneous rate of temperature change between 1 and a maximum of 20 K min^{-1}. The profile of modulation may be sinusoidal or sawtooth-like, as is predetermined by the instrument provider. In general, the larger the difference between linear and modulated rates of temperature-change, the higher is the probability of complete separation of thermal effects. The maximum available difference of about one decade, unfortunately, often is not sufficient, which permits in this case only a qualitative discussion of results. A zero underlying rate of temperature change is called the quasi-isothermal TMDSC mode and permits the measurement of heat capacity with only the small changes in temperature caused by the modulation. This mode is used to separate the effects caused by slow latent heat development [25].

The simultaneous application of two rates of temperature change for the purpose of separation of thermal events of different kinetics requires advanced data evaluation techniques. Separation of the raw data of the modulated heat flow into total, reversing and non-reversing components is done by discrete Fourier transformation [26] and is summarized in Eqs. (11–19). The total component represents the response to the linear underlying rate of temperature change and the reversing component is the response to the modulated part of the temperature program. The Fourier transformation of the modulated heat flow-rate $HF(t)$ and temperature $T(t)$ in the time domain [Eqs. (11) and (12), respectively] yield the average or total component $\langle HF \rangle$ [Eq. (13)] and with the amplitudes of the first harmonic ($v=1$), per definition, the reversing components, A_v^{HF} and A_v^T [Eqs. (16) and (17), respectively]. The total component of the heat flow-rate, $\langle HF \rangle$, is equivalent to the standard DSC signal and can be converted into a total heat capacity by division with the underlying rate of temperature change $\langle q \rangle$ [Eq. (18)]. The reversing heat capacity is similarly calculated using the amplitude of the reversing heat flow-rate and amplitude of modulated heating rate [$= A_v^T \times v\omega$, with the frequency of the first harmonic $\omega = 2/p$; see Eq. (19)].

$$HF(t) = \langle HF \rangle + \sum_{v=1}^{\infty} \left[a_v^{HF} \sin(v\omega t) + b_v^{HF} \cos(v\omega t) \right] \tag{11}$$

$$T(t) = \langle T \rangle + \sum_{v=1}^{\infty} \left[a_v^T \sin(v\omega t) + b_v^T \cos(v\omega t) \right] \tag{12}$$

$$\langle HF \rangle = \frac{1}{n} \sum_{t=-p/2}^{t=+p/2} HF(t) \tag{13}$$

$$a_v^{HF} = \frac{2}{n} \sum_{t=-p/2}^{t=+p/2} HF(t) \sin(v\omega t) \tag{14}$$

$$b_v^{HF} = \frac{2}{n} \sum_{t=-p/2}^{t=+p/2} HF(t) \cos(v\omega t) \tag{15}$$

$$A_v^{HF} = \sqrt{\left(a_v^{HF}\right)^2 + \left(b_v^{HF}\right)^2} \tag{16}$$

$$A_v^T = \sqrt{\left(a_v^T\right)^2 + \left(b_v^T\right)^2} \tag{17}$$

Total heat capacity:

$$\langle c_p \rangle = \frac{\langle HF \rangle}{\langle q \rangle} \tag{18}$$

Reversing heat capacity:

$$c_{p,\text{rev}} = \frac{A_v^{HF}}{A_v^T v\omega} K(v\omega) \tag{19}$$

Non-reversing = Total − Reversing

Note that Eqs. (11) and (12) are only available at $p/2$ because of the need to evaluate $\langle HF \rangle$ in Eq. (13) and $\langle T \rangle$ in an equivalent averaging of $T(t)$. These two averages accomplish the deconvolution of the reversing and total functions. The sums in Eqs. (14) and (15), which again extend over one modulation period, increase the delay to p. Furthermore, it is common practice to smooth the final, deconvoluted data once more over a full modulation period, which increases the delay to $1.5p$.

The reversing component filters thermal events in the sample which can follow the temperature modulation. The remaining part (total–reversing) is therefore the non-reversing component. The meaning of n in Eqs. (13–15) is the number of data points per period, which is the calculation window for observation of the Fourier coefficients at a given time t. Note that hardware and instrument calibration in TMDSC are identical with those in standard DSC, but additional precautions must be taken regarding stationarity and linear response, baseline subtraction, smoothing or instrumental frequency correction. These points are addressed in the literature [27–32]. The validity of linear response between temperature perturbation and heat flow-rate permits the use of higher harmonics v of the Fourier transform for data evaluation. Higher harmonics $v > 1$ are evident if the temperature modulation is non-sinusoidal. In the case of the frequently applied sawtooth modulation, which is shown in Fig. 12.6, the amplitudes of even-numbered harmonics ($v = 2, 4, 6, \ldots$) are zero and the amplitudes of the odd-numbered harmonics ($v = 1, 3, 5, \ldots$) are non-zero due to the symmetry of the temperature profile about the underlying, linear component. If this symmetry in the sample-temperature modulation is lost in the heat flow-rate response, the even-numbered harmonics also are non-zero, which, if analyzed, gives valuable information about the kinetics of thermal events in the sample.

The calculation of the reversing heat capacity needs an additional correction by a frequency-dependent factor $K(v\omega)$, accounting the different equilibration behaviors of sample and reference calorimeters. In contrast to standard DSC, the transient heat flow-rate after changing the heating rate must be included in the Fourier transformation, that is, a steady state is not required (see also Fig. 12.5). This leads to the advantage that rather short modulation periods, down to 15 s in typical present-day equipment, can be applied, which is the limit for ensuring stationarity and linear response of the sample. Unfortunately, the correction function $K(v\omega)$ depends on the heat flow path from the furnace to the sample and within the sample, that is, $K(v\omega)$ cannot be calibrated with standards such as sapphire. Determination of $K(v\omega)$ requires analysis of the instrumental frequency dependence of the uncorrected reversing heat capacity, as is shown in Fig. 12.7 for poly(ethylene-co-1-octene) and sapphire. The calculation of the uncorrected reversing heat capacity is done by assuming initially that $K(v\omega) = 1$. Data were collected quasi-isothermally ($\langle q \rangle = 0$ K min^{-1}) at 299 K on a DSC 820 heat-flux calorimeter (Mettler-Toledo), using a sawtooth modulation of the temperature and a modulation amplitude of 1 K. The figure includes data from three experiments with modulation periods of 240, 120 and 60 s. The heat

12.3 Measurement of the Heat Capacity and Separation of Thermal Events

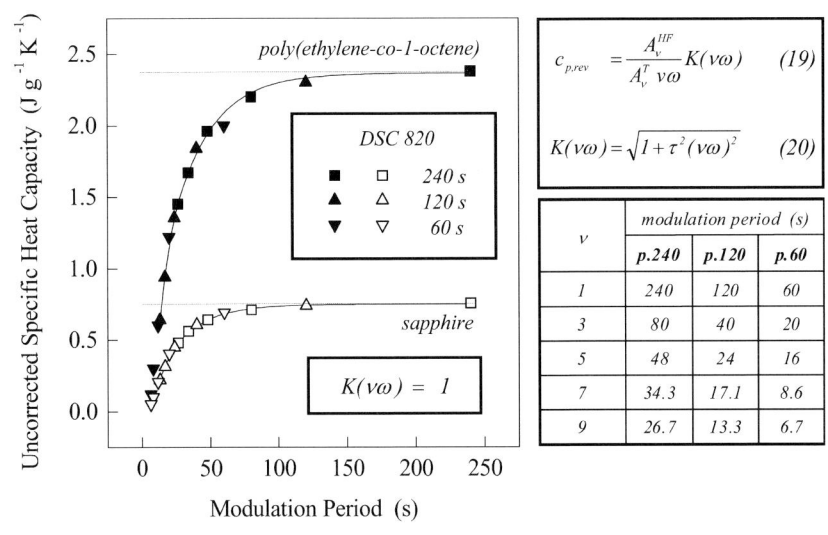

$$c_{p,rev} = \frac{A_v^{HF}}{A_v^T \, v\omega} K(v\omega) \quad (19)$$

$$K(v\omega) = \sqrt{1 + \tau^2 (v\omega)^2} \quad (20)$$

v	modulation period (s)		
	p.240	p.120	p.60
1	240	120	60
3	80	40	20
5	48	24	16
7	34.3	17.1	8.6
9	26.7	13.3	6.7

Fig. 12.7 Uncorrected, reversing heat capacity of poly(ethylene-co-1-octene) and sapphire as function of modulation period. Data were taken from [32].

capacities were calculated as suggested before from the amplitudes of the first harmonic ($v=1$) of the Fourier series of the heat flow-rate and temperature, respectively and additionally from higher harmonics with a frequency $v \times \omega$, that is, with a reduced modulation period as shown in the table in Fig. 12.7. The data in Fig. 12.7 fit a single function which originally was derived for the special case of the heat-flux calorimeter of TA Instruments [33], and which was modified later by introduction of a time constant τ, depending not only on the instrument parameters but also sample parameters such as mass and heat conductivity [Eq. (20)] and must be calibrated for each sample, calibrant and calorimeter configuration [30–32]. Particularly at short periods, the uncorrected heat capacity deviates strongly from the expected value, which is given with the horizontal line in Fig. 12.7. In order to obtain precise data, a correction is necessary using the square-root expression in Eq. (20) or the measurement must be performed at a sufficiently large modulation period of about 60–100 s, depending on the type of instrument or τ.

For correction, the reversing heat capacity can be obtained by plotting the inverse of the squared, uncorrected heat capacity as a function of the square of the frequency according to Eq. (22), which is illustrated in Fig. 12.8 for the data of Fig. 12.7. The intercept at $(v\omega)^2 = 0$ yields the inverse of the squared, correct specific heat capacity and in addition the slope delivers information about the time constant τ. The data points at high frequency or short modulation periods of less than ≈ 20 s do not fit the predicted correction function and indicate loss of linear response and can only be used by treating τ as an empirical function. Details about the frequency correction for several types of instruments and ex-

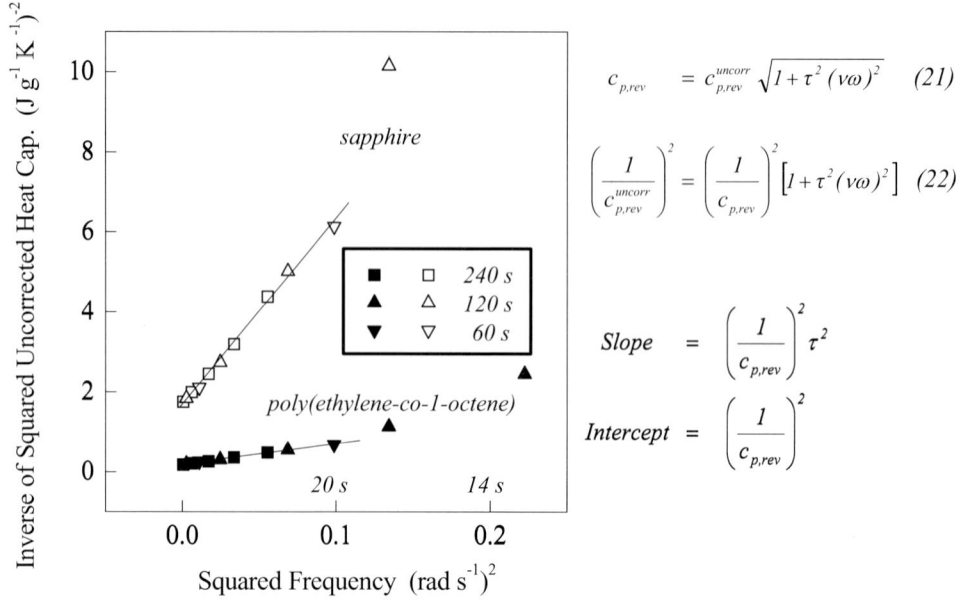

Fig. 12.8 Determination of the corrected value of the reversing heat capacity by extrapolation to zero frequency. Data were taken from [32].

$$c_{p,rev} = c_{p,rev}^{uncorr}\sqrt{1+\tau^2(v\omega)^2} \quad (21)$$

$$\left(\frac{1}{c_{p,rev}^{uncorr}}\right)^2 = \left(\frac{1}{c_{p,rev}}\right)^2 \left[1+\tau^2(v\omega)^2\right] \quad (22)$$

$$\text{Slope} = \left(\frac{1}{c_{p,rev}}\right)^2 \tau^2$$

$$\text{Intercept} = \left(\frac{1}{c_{p,rev}}\right)^2$$

perimental setups are available in the literature [30–32]. The advantage of the use of higher harmonics v for determination of the reversing heat capacity is not only the tremendous reduction in the time required for calibration but also the determination of the frequency dependence of thermal events in the sample within a single experiment and under identical conditions. The disadvantage of the lower amplitude of the higher harmonics can be compensated by tailoring the profile of the temperature modulation [34, 35].

Application of standard TMDSC ($\langle q \rangle \neq 0$) for the separation of thermal effects is extensively described in the literature. Regardless of the advantages of non-isothermal TMDSC, which mainly is the relatively fast gain of information about overlapping thermal events of different time-scales, its major disadvantage is that it is not quantitative, due to permanent loss of equilibrium by triggering new irreversible thermal processes by changing the average temperature. In other words, the amplitude of the first harmonic of the heat flow-rate may include irreversible thermal events. This shortcoming can only be avoided by quasi-isothermal TMDSC, which may exclude permanent triggering of irreversible events if the amplitude of modulation is sufficiently low. Figure 12.9 is a schematic overview of combinations of thermal events, which are different in kinetics and reversibility. The left column shows the modulated heat flow-rate as a function of temperature on modulation with an amplitude $amp.T$ about the average temperature. The right column is the corresponding amplitude of the

12.3 Measurement of the Heat Capacity and Separation of Thermal Events

heat-flow rate versus temperature

reversing heat capacity versus time

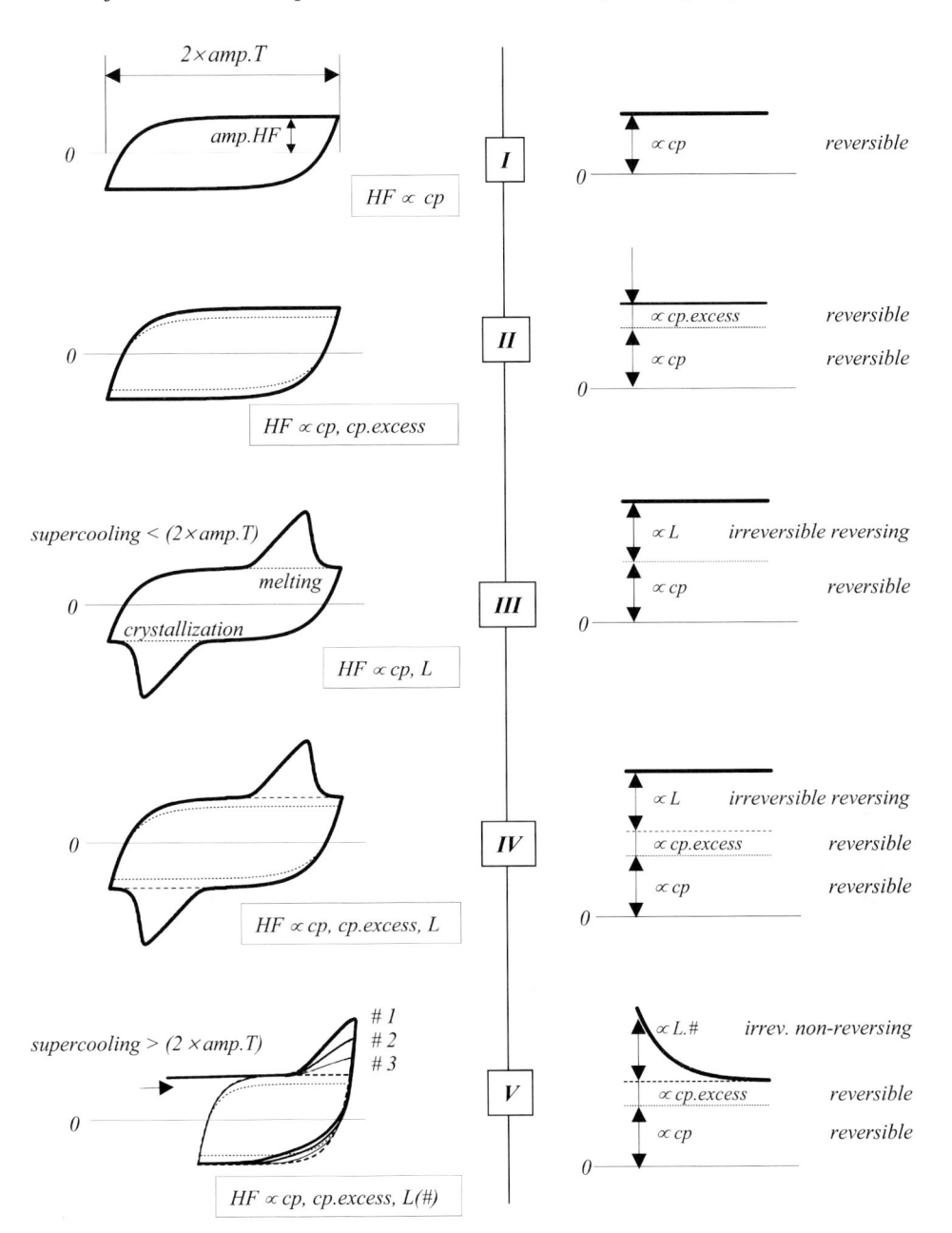

Fig. 12.9 Heat flow-rate as a function of temperature on quasi-isothermal temperature modulation for different combinations of thermal events (a) and resulting reversing heat capacity as function of time or cycle number (b).

heat flow-rate, which is proportional to the reversing heat capacity as a function of time or number of modulation periods. Parts I–V are discussed below.

I: The heat flow-rate is caused by the heat capacity c_p only. Each change in the heating rate results in a new approach of instrumental steady state and afterwards the amplitude remains nearly constant. Note that the temperature dependence of the heat capacity can be neglected at typical amplitudes of modulation of less than 2 K. The average amplitude during one period is not a function of time or number of cycles. The heat capacity is classified as a reversible event. Note that the time interval where a steady state has been reached can also be evaluated by the standard DSC method of Fig. 12.5, while Fourier analysis of the TMDSC of the full cycle needs the evaluation of Eqs. (11–19) and Figs. 12.7 and 12.8.

II: In this example, additional reversible melting and crystallization are illustrated, which increase *amp.HF*. The measured apparent heat capacity contains the heat capacity c_p and an excess heat capacity $c_{p,excess}$ due to reversible melting and crystallization. Again, a standard DSC could be applied to the parsed segments of the plot which refer to the steady state.

III: This experiment shows a superposition of the heat capacity c_p with irreversible melting and crystallization, that is, with an additional exchange of latent heat, L. The classification of melting and crystallization as an irreversible event is indicated by the supercooling. The range of temperature modulation ($2 \times amp.T$), however, more than covers the temperature difference between crystallization and melting. This results in complete reversal of the phase transition. The measured apparent reversing heat capacity is therefore not a function of the number of repetitions. Although melting and crystallization contribute to the reversing heat capacity, these are irreversible processes from the point of view of thermodynamics. One should note that the terms *reversing* and *thermodynamically reversible* have distinctly different meanings. A reversible process can only be identified after experimentation with decreasing modulation amplitudes and, even then, it may only be possible to suggest limits of the reversibility. For sharp melting substances, such as indium, it was possible by typical quasi-isothermal TMDSC to set the limit of reversibility to $< \pm 0.05$ K [36, 37].

IV: In addition to the situation in **III**, the reversing heat capacity may be increased by reversible melting and crystallization (see also **II**) and give the result shown in the right part of Fig. 12.9. The separation of the various effects becomes more complicated in this case and it may not always be possible to resolve the effects without additional experimentation with different heating rates and frequencies. In the shown case of reaching a short-time steady state before melting and crystallization, a standard DSC evaluation of the data using Fig. 12.5 could resolve the various effects. Otherwise, adjusting the frequency and amplitude of modulation may allow the various effects to be separated.

V: In contrast to the situation in **III**, in V crystallization and melting are still irreversible, but initially there is a reversing contribution. The amplitude of temperature modulation is small enough to avoid crystallization, but melting continues in each subsequent modulation cycle. The melting peak extends to the maximum temperature in each modulation cycle and even on subsequent cooling melting continues. In this case, the average amplitude of the heat flow-rate decreases as a function of time or number of cycles and approaches a constant value when all crystals are melted. Naturally, intermediate cases are possible when both melting and crystallization are incomplete. Since melting and crystallization rates are only exactly equal at the melting temperature and the melting rate usually increases more with increasing temperature than the crystallization rate increases with decreasing temperature [38], it should be a rare occurrence that incomplete phase transitions could produce a constant $\langle amp.HF \rangle$ with cycle number and even then the result would not be a symmetric event.

In addition to the separation of the thermal events, the application of quasi-isothermal TMDSC implies another advantage compared with standard DSC, which is the ability to measure heat capacity under quasi-isothermal conditions. In standard DSC, a temperature range of at least about 10–20 K must be analyzed for reliable determination of the heat capacity, whereas in TMDSC an amplitude of <1 K is sufficient. The precision in TMDSC is even higher in TMDSC since long-term drift of experimental conditions show up only in the non-reversing component.

12.4
Contributions to the Apparent Heat Capacity of Semicrystalline Polymers

The heat flow-rate into the sample in DSC, in general, is a superposition of a heat capacity contribution (c_p) and various latent-heat contributions (L) due to phase transitions [Eq. (23)]. The heat capacity depends on temperature, T, and the structure, as outlined with Eq. (4). The latent heat, evolved by the sample during exothermic processes or absorbed in endothermic processes as a result of a phase transition or chemical reaction, additionally depends on time, t. Consequently, the heat capacity also depends on time if L is non-zero, since a phase transition changes the overall composition of the phase. Formal translation of the heat flow-rate into units of heat capacity using Eqs. (10), (18) or (19) therefore yields an apparent heat capacity, $c_p^{\#}$, which also depends on time as expressed by Eq. (25). Separation of the heat capacity contribution from the latent heat in the integral heat flow-rate signal is possible with knowledge of the heat capacity baseline (see Section 12.2) and is a prerequisite for further analysis of structure and structure formation in polymers. Separation of the superimposed different latent heat contributions, in addition, may be possible by evaluation of the time dependence, using, for example, TMDSC.

Figure 12.10 shows a typical standard DSC experiment on poly(ethylene-co-1-octene) with a 1-octene content of 38 wt% and a density of about 0.87 g cm^{-3} [39, 40]. The polymer was heated to 403 K and subsequently melt-crystallized by cooling at a rate of 10 K min^{-1}. The bold line shows the apparent heat capacity of the sample as a function of temperature. The symbols indicate the heat capacity of the liquid and crystalline phases of polyethylene and the thin line is

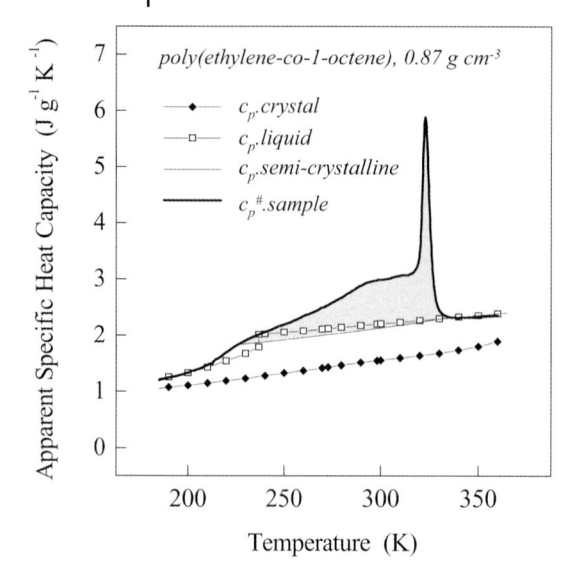

$$HF(T,t) = c_p[T,L(t)]\,q + \sum_i f\,[L_i(T,t)] \quad (23)$$

$$\frac{HF(T,t)}{q} = c_p[T,L(t)] + \frac{\sum_i f\,[L_i(T,t)]}{q} \quad (24)$$

$$\frac{HF(T,t)}{q} = c_p^\#(T,t) \quad (25)$$

Fig. 12.10 Apparent specific heat capacity of poly(ethylene-co-1-octene) of density 0.87 g cm^{-3} on cooling from the melt at 10 K min^{-1}, plotted as a function of temperature [39, 40].

the heat capacity baseline. The gray-shaded area represents latent heat as evolved in the sample by the crystallization. The data clearly reveal that crystallization occurs over a wide temperature range, starting at about 330 K and continuing down to the glass transition temperature of the amorphous phase, controlled by the ethylene sequence length distribution. The maximum enthalpy-based crystallinity is obtained by integration of the gray-shaded area and is about 29%.

The particular copolymer is representative of a series of linear low-density polyethylenes with inter- and intramolecular homogeneous distribution of copolymers units, that is, branches. These low-density ethylene copolymers attracted renewed interest with the invention of new synthesis routes to specifically structured polymers and were recently investigated in detail regarding the relationship between thermal behavior and structure, including the DSC analysis of the various contributions to the apparent heat capacity [39–43].

A major step in analyzing the various contributions to the apparent heat capacity during the crystallization was the application of quasi-isothermal TMDSC. Figure 12.11 shows the reversing apparent heat capacity of the polymer discussed in Fig. 12.10 as a function of time during isothermal annealing or secondary crystallization at 298 K. Before annealing, the sample was melt-crystallized at a rate of 10 K min^{-1}. The continuous cooling process was interrupted at 298 K in order to perform the TMDSC annealing experiment. The reversing apparent heat capacity decreases as a function of time to reach a metastable state after 500–1000 min, indicated by constancy of the apparent heat ca-

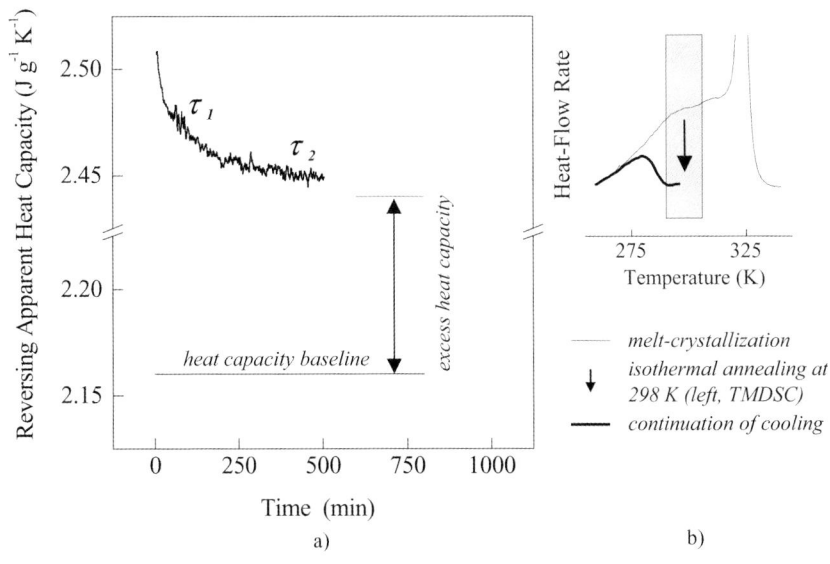

Fig. 12.11 (a) Reversing apparent heat capacity of poly(ethylene-co-1-octene) of density 0.87 g cm^{-3} as a function of time during isothermal secondary crystallization and annealing at 298 K, after cooling from the melt [40]. (b) The heat flow-rate of a complete melt crystallization on cooling at 10 K min^{-1} (thin line) and the continuation to the glass transition after annealing at 298 K (thick line).

pacity. The decrease in the heat flow-rate is shown in the temperature-domain of Fig. 12.11 as the downwards arrow. Subsequently, the sample was also cooled to sub-ambient temperature (thick line), in order to compare the continuation of the crystallization process with that of a continuously cooled sample (thin line). The isothermal decrease in the apparent heat capacity on the left can be fitted using two exponential functions with time constants τ_1 and τ_2, which are in the order of 10 and 100 min, respectively. Visual inspection of the modulated, raw heat-flow data revealed that the fast process is due to continued crystallization of the appropriate ethylene sequence length and that the slow process is due to annealing and reorganization. At present there exists in the literature a number of efforts further discuss the structural reasons of the isothermal decay of the apparent heat capacity by using different mathematical models to fit the experimental data or by analyzing of activation energies [40, 44–48]. A further important observation in the experiment in Fig. 12.11 is the unexpected proof of an excess heat capacity due to reversible melting. In the particular example, the equilibrium heat capacity at infinity time is 2.44 J g^{-1} K^{-1}, which is larger than the expected baseline heat capacity of polyethylene of 15% crystallinity of 2.16 J g^{-1} K^{-1}. The excess of 0.28 J g^{-1} K^{-1} was attributed to reversible melting and crystallization, that is, to zero entropy production crystallization and melting at $T^0_{m,ip}$ (see Fig. 12.3). The width of the temperature range of metastability after isothermal secondary crystallization/annealing was estimated by variation of the amplitude of temperature modulation and is shown in the right part of Fig. 12.11 by the rectangular bar. The apparent reversing heat capacity remains constant if the amplitude of modulation is less than about 2–5 K, that is, reversible melting is evidenced only within this temperature range. Analysis of subsequent cooling (bold line in the right part of Fig. 12.11) allowed the decrease in the apparent heat capacity to be associated with a specific population of ethylene sequences since the crystallization at lower temperature was not affected.

The experiments in Figs. 12.10 and 12.11 reveal six different thermodynamic contributions to the apparent heat capacity in the melting and crystallization region of the linear low-density copolymer of ethylene and 1-octene [48], as follows:

1. skeletal and group vibrations;
2. large-amplitude motion of molecule segments;
3. reversible crystallization and melting;
4. crystal perfection;
5. secondary crystallization;
6. initial crystallization.

The first and largest contribution to the apparent heat capacity in any polymer is due to skeletal and group vibrations. This contribution is visualized with the filled symbols in Fig. 12.10 and the bold line in Fig. 12.1, and is additionally described in Section 12.2.

The second contribution is due to large-amplitude motion of molecule segments. Large-amplitude motion in the amorphous structure is possible above the glass transition temperature where the structure contains sufficient free volume to permit an increase in the dynamic ratio between gauche and trans conformations. As the temperature is decreased to the glass transition temperature, this conformational motion becomes increasingly cooperative, that is, several neighboring conformations must adjust in concert to accomplish the motion. Ultimately, at the low-temperature limit of the glass transition, the cooperative motion has changed completely to a torsional oscillation. The heat capacity increases as the ratio of gauche-to-trans conformation increases with temperature since the potential energy of the gauche conformation is higher than that of the trans conformation. At sufficiently high temperature, equipartition between the conformations is reached at a gauche-to-trans ratio of 2:1 and no additional heat capacity contribution is seen. The time for conformational changes is in the time-scale for molecular motion, in the picosecond range (10^{-12} s), and increases from minutes to years (10^2–10^7 s) and beyond when cooling through the glass transition temperature due to the cooperativity. This change in conformational contribution is the main cause of the difference between the heat capacities of the solid and the liquid, which is 0.73 J g^{-1} K^{-1} at 250 K.

The third contribution is reversible melting and crystallization. These are zero entropy production, first-order phase transitions (see Fig. 12.3) which cannot be monitored under isothermal conditions. Rather, it is possible to detect reversible crystallization by the absence of supercooling of the liquid state during decreasing temperature in the modulation cycle. Reversible crystallization is due to a local thermodynamic equilibrium at the crystal surfaces involving molecule sequences decoupled from the crystal. In polyethylene, local equilibria were identified at the fold surface and at the lateral surface of crystals, both depending on the surface-specific coupling of crystalline and amorphous phases. The contribution to the apparent heat capacity in the particular example shown in Figs. 12.10 and 12.11 is 0.28 J g^{-1} K^{-1} at 298 K, but depends strongly on temperature, crystallinity, crystal size and crystal perfection. Reversible melting seems typical for semicrystalline polymers in the temperature range of irreversible crystallization and melting since it has been detected in numerous polymers including polyolefins [48–50], polyesters [51–53] and polyamides [54].

The remaining contributions to the apparent heat capacity in the above list are irreversible and include crystal perfection, secondary crystallization and primary crystallization. Primary initial crystallization and secondary crystallization are direct transformations from the liquid to the crystal. Primary crystallization can be considered as undisturbed crystallization which follows the primary nucleation, whereas secondary crystallization can be connected with crystallization in the confined environment created by primary crystallization. For example, any crystallization between already existing lamellae would be considered secondary. Primary and secondary crystallization can be identified by the kinetics of growth under isothermal conditions. Secondary crystallization is slower than primary crystallization. In non-isothermal crystallization, these processes overlap and are then difficult to separate. Crystal perfection, in contrast, is not a direct transformation from the liquid to the crystal. The internal order, surface structure and size of existing crystals may improve as a result of localized diffusion processes or as a result of complete or incomplete melting and recrystallization, where even after complete melting the crystallization kinetics is faster than on primary crystallization. Crystal perfection is commonly associated with an annealing peak in calorimetric analysis, a small endotherm which occurs at somewhat higher temperature than the annealing, clearly indicating the irreversibility of the process.

Contributions 3–6 involve the exchange of latent heat which can be quantified with the knowledge of the baseline heat capacity, that is, the contributions 1 and 2. The invention of TMDSC was therefore a milestone in the history of calorimetry which (a) helps to identify truly reversible exchange of latent heat (3) and (b) allows the continuous detection of irreversible changes of structures as in crystal perfection, even if the process involves only small amount of heat (4).

12.5
Recent Advances in Thermal Analysis of Polymers

Since the invention of modern thermal analysis some 100 years ago, a large part of its application has always been the qualitative check for chemical and phase transitions as documented in the major compendium dealing with the application to macromolecules [55]. Over the last 50 years, quantitative thermal analysis as described in Sections 12.2–12.4 has become more common, but by far not as widely accepted as the enormous proliferation and improvement in precision of the DSC equipment would allow. As an example of the growth of information on going from qualitative analyses, which may be better called differential thermal analysis (DTA) since it yields no calorimetric information, to the quantitative DSC one can look at Fig. 12.10. Without a calibration, the data can only be plotted as a relative heat flow-rate, that is, the ordinate disappears. This still allows a good assessment of the beginning of crystallization and its maximum at the peak. However, it is already difficult to find the end of crystallization and the possible glass transition. Only an analysis of the thermodynamic functions and the calibration to heat capacity allows the introduction of the reference solid and liquid baselines, basic for the discussion given above. The recent advances discussed in this section give a picture of how TMDSC has begun to revolutionize the understanding of semicrystalline macromolecular structure and morphology by studying the reversible melting and the rigid-amorphous nanophase. In the last part of this section, a short preview is given of what is expected for the future when quantitative DSC may be possible at superfast heating and cooling rates by the application of calorimetry to nanogram samples on an electronic chip [super-fast chip calorimetry (SFCC)].

12.5.1
Reversible Melting of Polymers

Primary crystallization of polyethylene from a homogeneous, nucleus-free melt requires supercooling to ≈ 10 K and is therefore thermodynamically irreversible. Many other polymers need supercooling of 10–50 K. The supercooling is required to form a crystal nucleus to initiate growth with a reduced free energy barrier and for local reduction of the entropy of the molecular segments before becoming attached to the crystal after molecular nucleation. Molecular nucleation is specific for crystallization of macromolecules and is not required for crystallization of metals or low molecular mass substances. Figure 12.12 summarizes the dependence of supercooling on the chain length or number of carbon atoms in the backbone for alkanes and oligo- and polyethylene and gives information about the nucleation mechanism in polymers [49, 56]. The data were collected in the presence of primary crystal nuclei and show, in this case, that supercooling is required only if the number of carbon atoms exceeds ≈ 75. The presence of primary nuclei avoids the crystal nucleation and reveals that supercooling is then due only to molecular nucleation. The crystal morphology, apparently, has no major effect on the requirement of molecular nucleation since supercooling is necessary for both extended-chain and folded-chain crystals. The commonly observed chain folding arises at a molecular length of $x \approx 300$.

Thermodynamically, irreversible crystallization results in the formation of a *globally* metastable structure consisting of an amorphous phase and a crystalline phase. Crystallization of macromolecules is rarely complete for kinetic reasons.

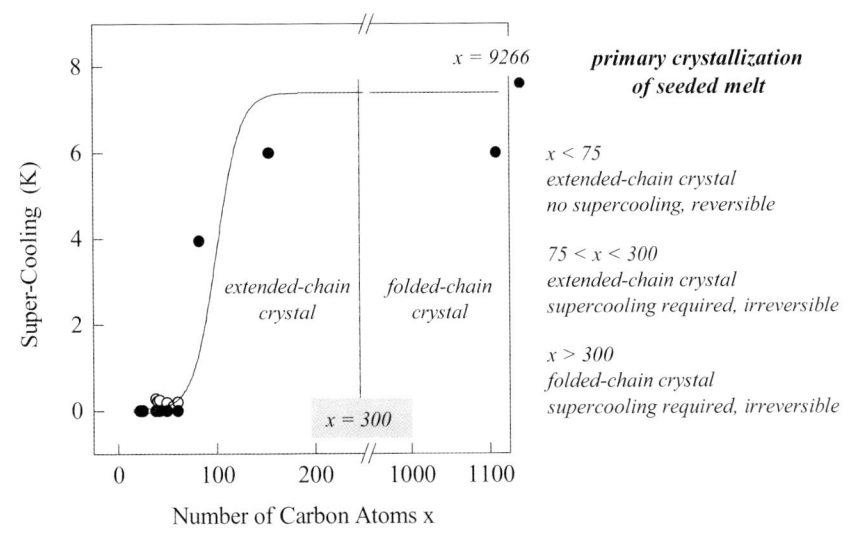

Fig. 12.12 Supercooling of alkanes and oligo- and polyethylene as a function of the number of backbone carbon atoms. Data were taken from [49].

Plot labels:
- $x = 9266$ — *primary crystallization of seeded melt*
- $x < 75$ extended-chain crystal, no supercooling, reversible
- $75 < x < 300$ extended-chain crystal, supercooling required, irreversible
- $x > 300$ folded-chain crystal, supercooling required, irreversible
- $x = 300$ boundary between extended-chain crystal and folded-chain crystal

The thermodynamic metastability of such a semicrystalline structure leads to a violation of the Gibbs phase rule, which must hold in the case of equilibrium. The phase rule requires a one-component system at constant pressure to be present in a single-phase system, except at the equilibrium melting temperature, where the crystal transforms into the melt or *vice versa*. The *local* thermodynamic equilibria at shorter length scale, which are indicated by the reversible crystallization and melting seen by TMDSC, are superimposed on the *globally* metastable crystalline and amorphous phases. Early investigations of reversible crystallization and melting due to local equilibria were proposed, based on temperature-resolved small-angle X-ray scattering experiments on semicrystalline polyethylene, and led to specific models of reversible insertion–crystallization at the lateral crystal surface in branched low-density polyethylene and reversible fold-surface melting in high-density polyethylene. The recent introduction of TMDSC revealed truly reversible melting in numerous semicrystalline polymers and allowed the development of a better understanding of the properties of semicrystalline polymers. Present research on reversible melting proves that there is a link between decoupling of molecule segments, preservation of a molecular nucleus and the degree of reversible melting.

Figure 12.13 shows schematically the morphology of an extended-chain crystal (a) and a typical folded-chain crystal (b). This scheme permits a straightforward explanation of the widely differing degree of reversibility of melting in these crystals. In the case of an extended-chain crystal, a selected macromolecule is part of a single crystal only and melts completely at the melting temperature. This process cannot be reversed without supercooling, necessary for formation

extended-chain crystal
- entire molecule is part of a single crystal
- molecule melts at a single temperature
- molecular nucleus is not preserved

- no reversible melting is expected

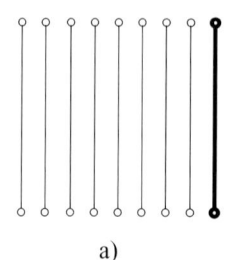

folded-chain crystal
- molecule may be part of several crystals of different melting temperature, or of regions of different thermodynamic stability within a crystal
- molecule melts not at a single temperature
- molecular nucleus can be preserved

- reversible melting is expected

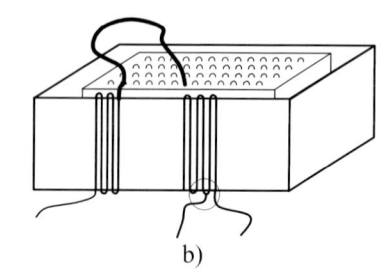

a)   b)

Fig. 12.13 Sketch of the morphology of an extended-chain crystal (a) and a folded-chain crystal (b) to explain the dependence of reversible melting on morphology.

of a molecular nucleus, that is, the melting process is irreversible. In the case of a folded-chain crystal, there may be a situation where a selected macromolecule is part of several crystals or, as is shown in Fig. 12.13, part of regions of a single crystal of different thermodynamic stability. The latter situation would imply an intracrystalline and intramolecular distribution of melting points, which are reasonable to assume since at the beginning of the crystallization the kinetic restrictions are less than at the end of the crystallization. It would therefore be possible that surface regions melt at slightly lower temperatures than interior regions and melting of a selected macromolecule may occur in part at different temperatures. The partial melting of a molecular segment at the crystal surface allows the preservation of a molecular nucleus which can be considered as an entropy-reducing residual of the partly molten chain and reversible crystallization on cooling becomes possible. This interpretation of reversible melting is also supported by the observation that poorer crystals which have more irregular folds show a larger amount of locally reversible melting. Recent results revealed, furthermore, that sharp folding in crystals of short-chain poly(oxyethylene) also show no locally reversible melting [57].

Figure 12.14 shows the reversing apparent specific heat capacity of extended-chain crystals (filled squares) and folded-chain crystals (open squares) of polyethylene (PE) and polytetrafluoroethylene (PTFE) as a function of temperature in the *globally* metastable state. The broken lines are the corresponding heat capacities of the liquid and crystal and serve for calculation of the excess heat capacity due to reversible melting. Extended-chain crystals of PE were obtained by

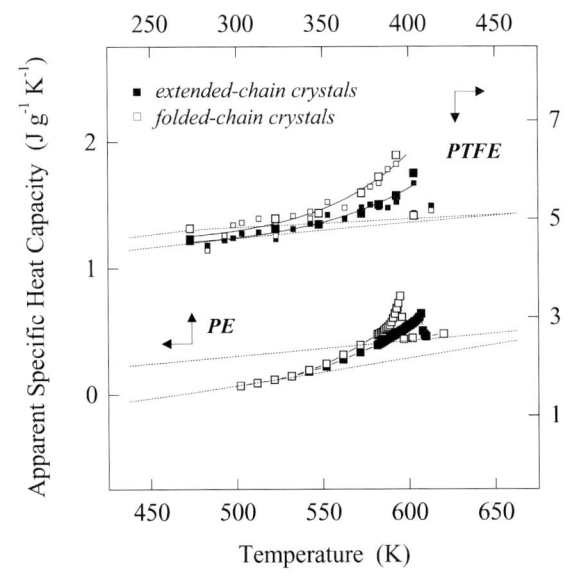

Fig. 12.14 Apparent reversing specific heat capacity of extended-chain and folded-chain crystals of polyethylene (PE) (left and top scales) and polytetrafluoroethylene (PTFE) (right and bottom scales) in the globally metastable state as a function of temperature [58]. Reproduced with kind permission of Springer Science and Business Media.

high-pressure crystallization, whereas in the case of the investigated as-synthesized PTFE, quasi-extended-chain, folded-ribbon morphology is known to exist for the crystals. In both cases the crystals are almost perfect, the crystallinity is high and the samples melt irreversibly, close to their respective equilibrium melting temperatures. Non-isothermal recrystallization on cooling from the melt results in the formation of less crystalline samples with folded-chain crystals, which melt at lower temperature than the extended-chain crystals. The data were obtained stepwise at increasing temperatures by quasi-isothermal TMDSC and clearly evidence a higher apparent specific heat capacity for the folded-chain crystals, despite the much reduced crystalline content. Normalization of the excess heat capacity by the actual crystallinity, which yields an average specific degree of reversibility of melting, would even enhance the difference between extended-chain and folded-chain crystals [58].

Further details about the effect of crystal morphology on reversible crystallization were collected by controlled variation of the condition of melt crystallization on isotactic polypropylene (iPP) [59, 60] and low-density polyethylene [48]. In general, one observes an increased excess heat capacity and degree of reversibility in samples with small and defective crystals. Figure 12.15 shows atomic force microscopy (AFM) images of slowly cooled (a) and quenched iPP (b). Both preparations were subsequently annealed at 423 K in order to enhance the differences in the crystal morphology. Note that annealing results in an increase in the crystal size, but does not significantly change the crystal shape. In the

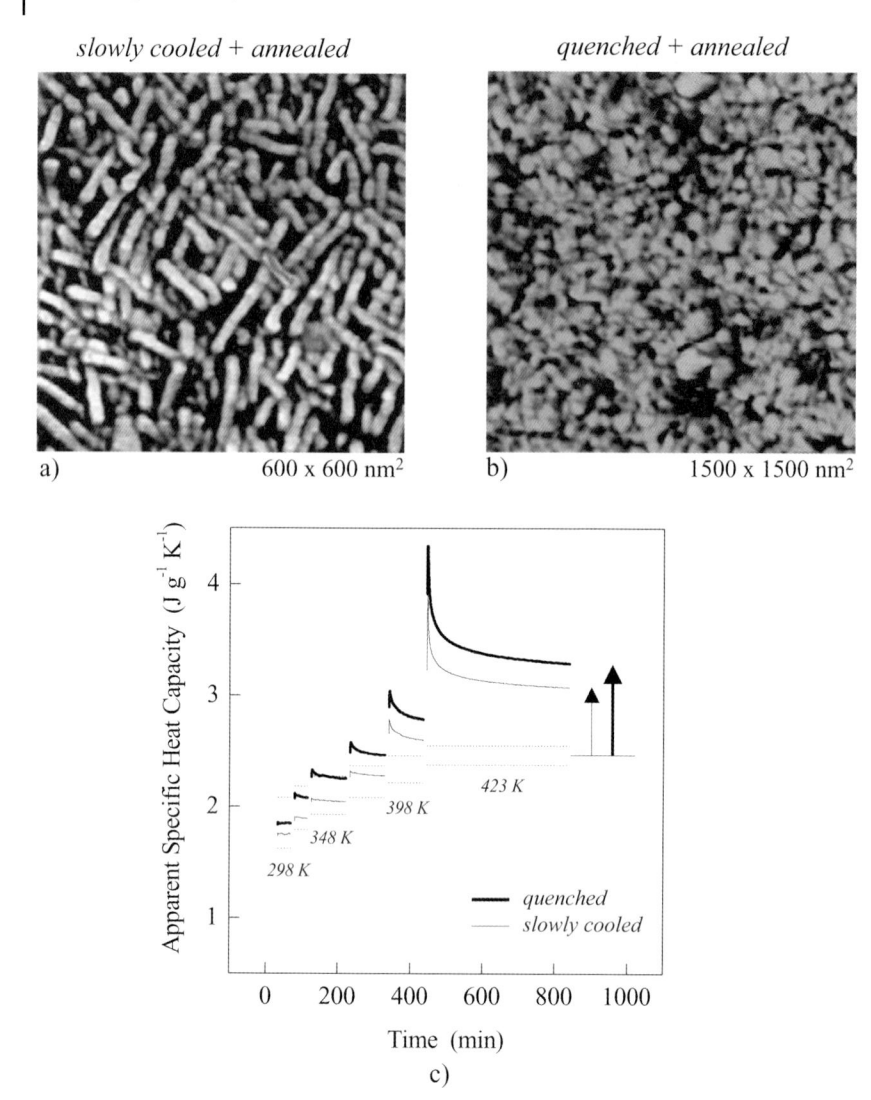

Fig. 12.15 Atomic force microscopy images of initially slowly cooled isotactic polypropylene (iPP) (a) and quenched iPP (b), both after annealing at 423 K. The reversing, apparent specific heat capacity is plotted in (c) as a function of time obtained on isothermal annealing on stepwise heating to the indicated temperatures. The broken lines represent the heat capacities of liquid and crystalline iPP. The arrows visualize the excess reversing heat capacity after annealing at 423 K. Reprinted with permission from [59]. Copyright 2001 American Chemical Society.

case of the slowly cooled and annealed sample, well-organized lamellae were detected, whereas in quenched and annealed iPP, at best, short lamellae with nearly globular shape were identified. The difference in the crystal morphology is directly mirrored by the apparent reversing specific heat capacity, which is shown as a function of time in Fig. 12.15c. The heat capacity of both preparations was analyzed by quasi-isothermal TMDSC at stepwise increased temperatures from 298 to 423 K. The data show in the beginning of each isothermal annealing step an irreversible approach to a metastable state and a final distinct difference in reversing apparent heat capacity between the slowly cooled and quenched samples. The degree of the reversibility of melting is about 30% higher in the case of the initially quenched samples, as is exemplified in Fig. 12.15 by the arrows indicating the excess heat capacity, that is, the difference between expected and measured heat capacity. An increased ratio between the lateral crystal surface area and the crystal volume is identified as major cause of the increased reversibility of melting of the initially quenched iPP.

Quantification of reversible melting is possible with the newly developed concept of specific reversibility of crystallization [61]. The main idea of this concept is the assignment of the excess heat capacity to a well-defined crystal population. It is assumed that at a given temperature the crystals which are present in typical polycrystalline, semicrystalline polymers do not contribute equally to the excess heat capacity caused by reversible melting. Figure 12.16 summarizes the

total reversibility $R(T)$

$$R_{total}(T) = \frac{c_{p,excess,\infty}(T)}{\Delta h_{m,0}(T)} \quad (26)$$

average specific reversibility
$R(T) \sim crystallinity\ (T)$

$$\langle R_{specific}(T) \rangle = \frac{R_{total}(T)}{X_c(T)} \quad (27)$$

specific reversibility
$R(T) \sim [d(crystallinity)/dT]\ (T)$

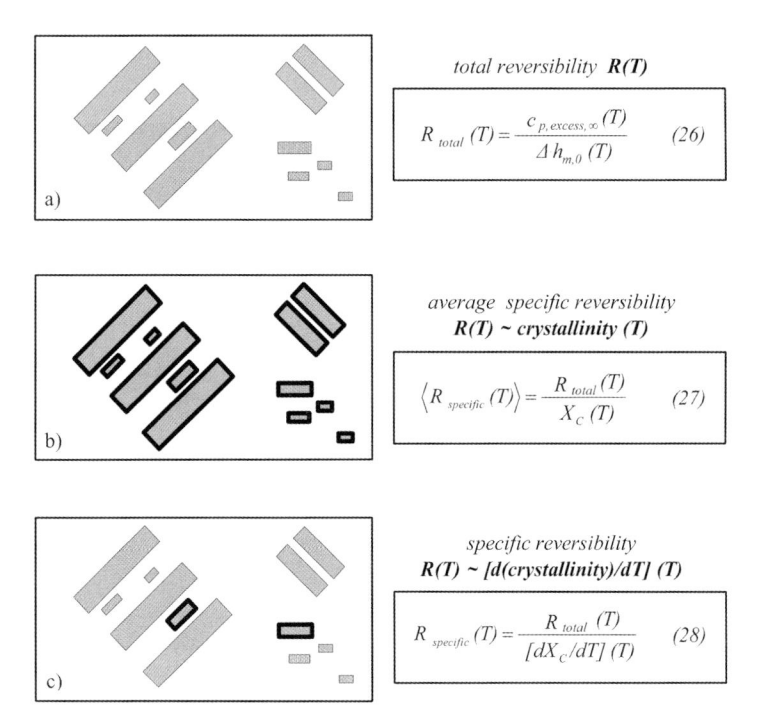

$$R_{specific}(T) = \frac{R_{total}(T)}{[dX_c/dT](T)} \quad (28)$$

Fig. 12.16 Processing of the equilibrium excess heat capacity into total reversibility (a), average specific reversibility (b) and specific reversibility of melting (c). Graphics illustrating the corresponding structures are given on the left [61].

options for the processing of the experimental heat capacity to obtain further information about melting in semicrystalline polymers.

The total reversibility of melting, $R(T)$, is obtained by normalizing the equilibrium excess heat capacity $c_{p,excess,\infty}$ with the bulk specific heat of fusion $\Delta h_{m,0}$, which yields the total change in crystallinity per change in temperature by 1 K. The total reversibility is related to the change in structure of the entire sample, crystalline and non-crystalline polymer, which is delimited in Fig. 12.16a by the overall, bold border. The amount of $R(T)$ typically is less than 1% K^{-1}. In the example in Fig. 12.11, an excess heat capacity of 0.28 J g^{-1} K^{-1} is observed, which can be recalculated into a total reversible melting of 0.097% K^{-1} [$\approx$ (0.28 J g^{-1} K^{-1}/290 J g^{-1} K^{-1})$\times$100%)], that is, there is about a 0.1% reversible change in crystallinity per 1 K temperature change. The calculation of the average specific reversibility of melting relates the excess heat capacity to the weight fraction of crystallinity, X_c. This is a reasonable approach since reversible melting occurs at the crystal surfaces only. It is an average of the specific reversibility in so far as in its calculation it is assumed that all crystals of different thermodynamic stability contribute equally to the reversible melting at a given temperature. In the example of linear low-density polyethylene in Fig. 12.11, the average specific reversibility with an approximate crystallinity of 0.15 is 0.65% K^{-1} (=0.097% K^{-1}/0.15) at the temperature of analysis. In other words, each crystal, as indicated by the bold lines in Fig. 12.16b, is assumed to change its crystallinity reversibly by 0.65% per 1 K temperature change. The calculation of specific reversibility, finally, assigns the excess heat capacity to a small, specific population of crystals as outlined in Fig. 12.16c. It is assumed that only those crystals show reversible melting at the temperature of analysis which grow or melt at the same or only negligibly higher temperature. The specific reversibility is then calculated by normalizing the total reversible melting to the fraction of irreversibility at the given temperature, that is, with the actual change of crystallinity, $dX_c/dT(T)$. The specific reversibility of crystallization is much larger than the total or average specific reversibility. First estimates lead to values of 10–50% [61, 62]. The physical meaning of such specific reversibility is that 10–50% of the crystals with a thermodynamic stability such that they grow in the immediate vicinity of the particular temperature can melt reversibly.

The concept of the specific reversibility of crystallization and melting, that is, the idea of assigning the measured excess heat capacity to a well-defined crystal population, was developed from numerous experimental data which proved a direct relation between the irreversible change of crystallinity, $dX_c/dT(T)$ and the total reversibility of crystallization, $R_{total}(T)$, at a given temperature. Figure 12.17 illustrates this relation with a fairly simple experiment on initially melt-crystallized isotactic polypropylene with a crystallinity of about 50% after cooling to ambient temperature [61]. The bold line in Fig. 12.17a is the decrease in crystallinity on continuous heating, which reaches zero close to 440 K. Interruption of the heating process at discrete temperatures, followed by isothermal annealing until a structural, globally metastable state is reached, causes an irreversible increase of the crystallinity, as is indicated by the arrows. The length of the arrow

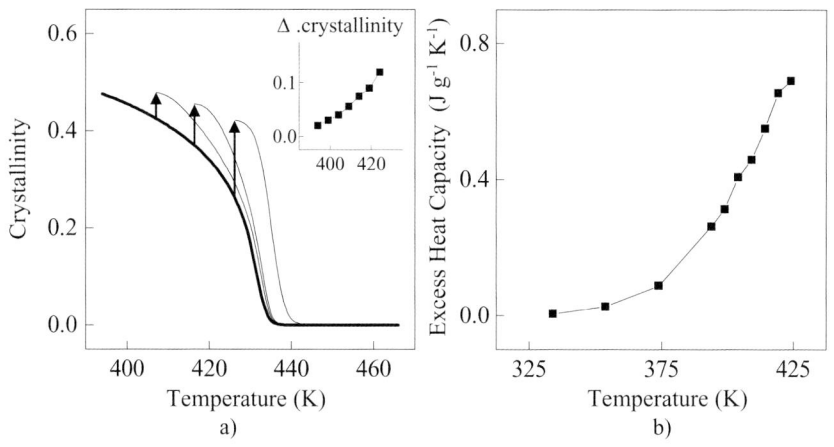

Fig. 12.17 Enthalpy-based crystallinity of isotactic polypropylene (iPP) as a function of temperature (a). The samples were continuously heated (bold line) or isothermally annealed at discrete temperatures before final melting (thin lines). The arrows show the irreversible increase of crystallinity on iso- thermal annealing. The inset summarizes the change of crystallinity as a function of annealing temperature. (b) The excess heat capacity in metastable equilibrium after annealing as function of the annealing temperature [61].

reflects the magnitude of the irreversible change in crystallinity, which is plotted in the inset as a function of the annealing temperature. Figure 12.17b shows the excess specific heat capacity measured at the end of the annealing process as a function of the annealing temperature. The data clearly prove proportionality between the excess heat capacity and the irreversible increase in crystallinity, measured at the given temperatures. The larger the irreversible increase in crystallinity, the higher is the excess heat capacity.

A further indication of the correctness of the concept of the specific reversibility of melting is the fact that an excess heat capacity without prior equilibration of the structure due to reversible melting has not been detected. Segments of macromolecules convert reversibly between crystalline structure and liquid only at temperatures close to their zero entropy production melting temperature (see Fig. 12.3). From this point of view, reversible melting is different from the well-documented reversible lamellar thickening in folded-chain crystals of polyethylene, which may be better described as a defect diffusion from the crystal to the amorphous phase. Reversible lamellar thickening is reported to occur even at temperatures lower by 100 K than the temperature of irreversible formation of crystals and points to a qualitatively different phenomenon.

The experiment-based model of reversible melting of specific crystal segments in semicrystalline polymers over a limited range of temperature may be based on an intracrystalline distribution of the melting temperature. It is assumed that the local defect concentration in a selected crystal increases during the growth process since the kinetic restrictions increase with the lateral crystal size due to the confinement by neighboring crystals, increasingly limited capability for diffusion of chain segments or increasing concentration of chemical and physical irregularities (such as entanglements and copolymer units, respectively). These restrictions may have the result that sharp folding of macromolecules to achieve a low surface free energy is not possible. Rather, molecules may occasionally leave the crystal at the basal planes without adjacent re-entry. The non-folded macroconformation has a lower local melting temperature than a folded-chain macro-conformation if the stem lengths for the two situations are identical. Figure 12.18a shows for polyethylene the experiment-based Broadhurst equation for estimation of the melting temperature, T_m, of chains in an extended conformation (left) and the Gibbs–Thomson equation fitted to the measured melting temperature of chains with a folded conformation as a function of the chain length, x, that is, the number of CH_2 segments uninterrupted by folds. The end-groups for the extended-chain conformation and the conformation of the folded-chain cause depressions of the melting temperature, but lead to close-to-identical equilibrium melting temperatures for both types of crystals at $x = \infty$. In Fig. 12.18b, an example calculation is shown to illustrate the effect of the conformation of the basal plane on the melting at the growth face. We assume a molecule of length $x = 90$ segments, which folds twice in order to achieve a crystal thickness of 30 segments, then has a melting temperature of almost 344 K. The calculation considers the equilibrium melting temperature of the molecule, which is, according to the Broadhurst equation, 385.95 K, and the proper melting point depression of a molecule with two end-groups per two folds with a combined total of six chain cross-sections for the two basal surfaces. An extended-chain crystal of identical height of 30 segments would have a melting point of only 337.35 K. Figure 12.18c is a graph of the dependence of the melting temperature on the crystal thickness for an equilibrium extended-chain crystal (Broadhurst equation, thin line), for a folded-chain crystal of infinite molar mass (Gibbs–Thomson equation, thick line) and for a part of a molecule with three stems and two end-groups in a crystal patch (symbols). The calculation shows that the melting temperature of a decoupled sequence of a molecule within the crystal decreases with its lateral size.

Summarizing the recent research on reversible melting using TMDSC, the following novel conclusions about thermal properties of semicrystalline polymers can be drawn:

1. Local equilibria on the crystal surfaces are superimposed on the globally metastable thermodynamic equilibrium.
2. The local equilibria are controlled by temperature and affect the degree of crystallinity as a function of temperature. The degree of crystallinity controls the properties of materials including, for example, the modulus and yield strength. For this reason, reversible melting needs to be quantified for better sample characterization.
3. Reversible melting depends on the crystal morphology. Extended-chain and probably also sharply folded polymer crystals show no reversible melting, in contrast to the common folded-chain and fringed-micellar crystals. The last

12.5 Recent Advances in Thermal Analysis of Polymers

extended – chain conformation
- Broadhurst equation -

T_m for $x = \infty$ ⟵ depression due to end-groups

$T_m = 414.3 \, [(x - 1.5)/(x + 5.0)]$

folded – chain conformation
- Gibbs-Thomson equation -

T_m for $x = \infty$ ⟵ depression due to fold-surface

$T_m = 414.2 \, [1 - (6.27/(1.27 \, x))]$

a)

example-calculation for a crystal of height x=30 (3.7 nm) with an extended-chain or folded-chain macro-conformation with two folds

T_m for $x = 90$ includes depression by end-groups

depression due to fold-surface, corrected for proper fold-number

$T_{m, 2/3 \, FCC} = 385.95 \, [1 - 2/3 \, (6.27/(1.27 \, x))]$

$T_{m, ECC} = \mathbf{337.35 \, K}$ $T_{m, 2/3 \, FCC} = \mathbf{343.61 \, K}$

$T_m(x = 90) = 385.95 \, K$
equilibrium melting point of the extended-chain conformation of a segment of length x=90

b)

— folded - chain
— extended - chain
—□— two folds

Melting Temperature (K) vs Crystal Thickness (nm)

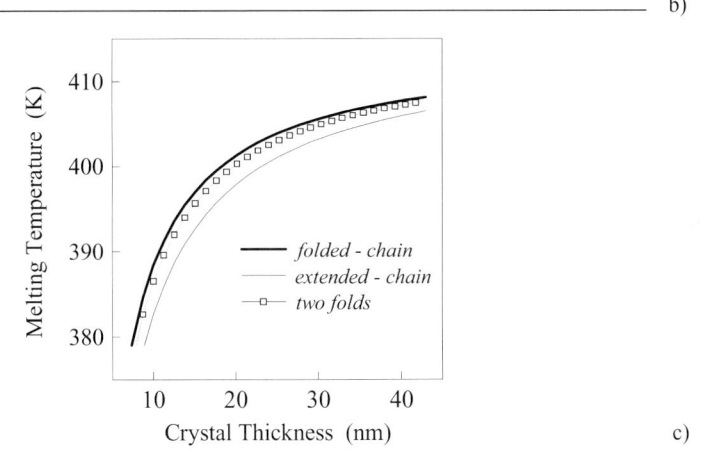

c)

Fig. 12.18 (a) Melting temperature of extended-chain and folded-chain crystals as a function of chain length x, where x is the number of CH_2 groups. (b) Example-calculation of the melting temperature of a folded-chain of length x=90 CH_2 groups with two folds to achieve a crystal height of 30 segments (right) and a comparison with the melting temperature of an extended chain of 90 CH_2 groups. (c) Graphical representation of the melting temperature of an extended-chain crystal (thin line), folded-chain crystal of infinite lateral dimension (thick line) and a crystal of lateral dimension of three stems containing two folds (symbols) as a function of crystal thickness [62].

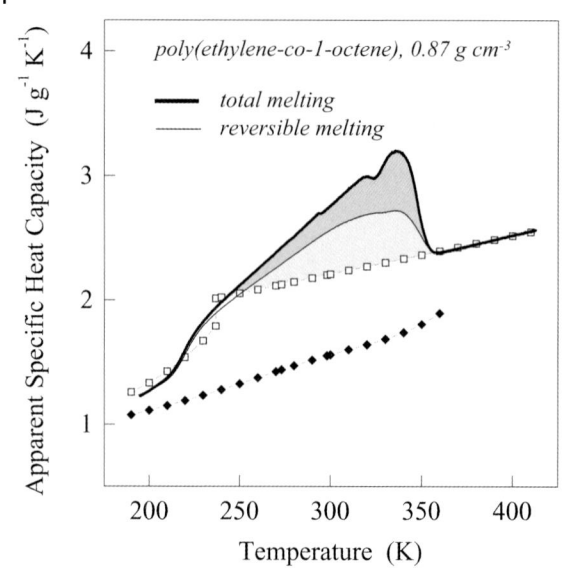

Fig. 12.19 Apparent specific heat capacity of poly(ethylene-co-1-octene) of density 0.87 g cm^{-3} as a function of temperature on heating. The bold line is the total apparent heat capacity and the thin line is a measure of the reversible fraction of melting. The shading between the curves and the heat capacity baseline, which is approximated by heat capacity of the liquid (open symbols), visualizes the related heat of irreversible and reversible melting [39].

two allow partial melting of decoupled molecule segments. These decoupled segments may maintain their molecular nucleus, which melts at higher temperature and serves as an entropy reduction of the melted segment. Points of decoupling may be chemical irregularities such as branches in linear low-density polyethylene discussed above, or conformational defects introduced by the crystallization history. The study of reversible melting therefore permits conclusions about the coupling between crystalline and amorphous phases, as will also be pointed out below in the discussion of the rigid amorphous phase by DSC and TMDSC.

4. Reversible melting of polymer crystals must be the step preceding irreversible melting. Figure 12.19 shows the apparent heat capacity on heating poly(ethylene-co-1-octene) of density 0.87 g cm^{-3}. The bold line is the total signal and the thin line the reversing signal. The ratio between reversing and total excess heat capacity, that is, the specific reversibility of melting, can reach values above 50% in an extended range of temperature. The dominance of reversible melting in this particular example reveals that polymer crystals can change their size to a large extent without total loss of thermodynamic stability. On crystallization, the initial reversible growth must be involved in the rejection of species of lower molar mass below their melting temperatures as extended-chain or folded-chain crystals. The irreversibly growing, longer species should provide a perfect surface for crystal growth [63].

The analysis of the specific reversibility of a crystal is therefore expected to provide valuable information about the local crystallization, melting behavior and thermodynamic stability of a selected crystal in a polycrystalline amorphous superstructure, not yet available by alternative methods.

12.5.2
The Rigid Amorphous Fraction

Semicrystalline, linear polymers contain crystalline and amorphous phases, as indicated in Fig. 12.20b. If these two phases were as independent as drawn in the bottom sketch, their phase properties would have to be identical to separate crystals and the bulk melt or glass. The two phases, however, are usually inter-

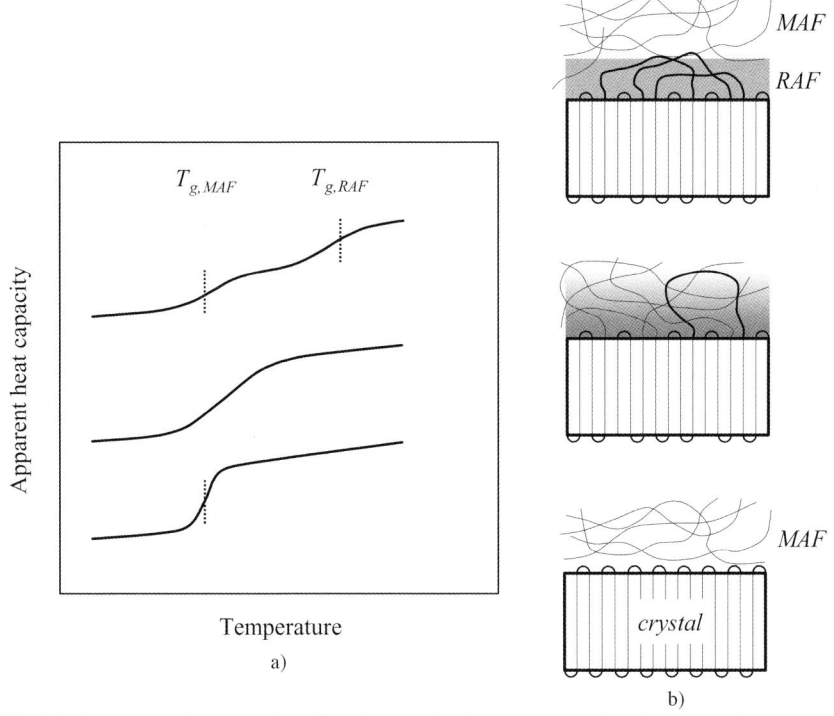

Fig. 12.20 Apparent heat capacity of semicrystalline polymers as a function of temperature (a), illustrating the effect of coupling of crystalline and amorphous phases on the glass transition. Completely decoupled crystalline and amorphous phases [(b) bottom sketch] show only one glass transition as would be observed in a fully amorphous structure. Coupling of the crystalline and amorphous phases either leads to a broadening glass transition (center curve) due to a gradient of the restriction of the mobility of the amorphous segments [(b) center sketch], or the occurrence of two glass transitions (top curve), if there is a distinct rigid amorphous fraction [(b) top sketch]. The gray shading in the schematics of structure indicates the degree of restriction of the mobility of amorphous segments.

connected by tie molecules. Tie molecules continue from the crystalline to the amorphous phase and are decoupled at the interface, but are still able to transmit stress from one phase to the other. The decoupled segments are characterized by different functions of state and properties. The thickness of the interface is controlled by the structure of the crystal surface and the resulting conformational strain within the tie molecules. Figure 12.20b illustrates the decoupling due to the presence of tie molecules with the top and center sketches. The amount of stress transmitted across the interface can be judged from the change of the glass transition temperature as shown in Fig. 12.20a for differently coupled crystalline and amorphous phases.

Complete separation of the crystalline and amorphous phases causes an interface of negligible thickness, that is, the interface is not thick enough to be considered as a separate phase with different thermodynamic functions of state and properties. Decoupled tie molecules can result in an increase in the thickness of the interface and may reach nanometer dimensions, so that the interface must be considered as a separate nanophase [64, 65]. The degree of decoupling of the molecules traversing the crystalline and amorphous phases is caused by the kinetic/conformational restrictions developing during the crystallization process and leave a signature in the glass transition range, assessable by thermal analysis (Fig. 12.20a). In general, the interphase at the boundary between the crystalline and amorphous structures is characterized by a local entropy which is less than the entropy of a fully amorphous structure, measurable by the difference in heat capacity caused by a higher glass transition. The microscopic reason is the locked microconformation of decoupled segments when entering the crystalline register. Glass transition analyses suggest that the amorphous mobility is reached only by a small fraction of the non-crystalline molecule segments, a reasonable observation, since larger phase areas with truly amorphous behavior, naturally, would crystallize in accord with the Gibbs phase rule. The changed structure of the crystal–amorphous transition layer, compared with the truly amorphous structure, in combination with the locking of chains in the crystal lattice, reduces the local chain mobility and cooperative mobility of chain segments. The interphase is therefore denoted a rigid amorphous phase (RAF) [66, 67]. The term *rigid amorphous* is not necessarily related to the actual state of vitrification; rather, it points to the existence of a restraint in the amorphous structure with an increased glass transition temperature. The bottom curve in Fig. 12.20a reveals a single glass transition at a temperature identical with the glass transition of a fully amorphous polymer, labeled in Fig. 12.20 as mobile amorphous phase or fraction (MAF). The center curve is related to the center structure in Fig. 12.20b and illustrates that the glass transition, in general in semicrystalline polymers, is broadened towards higher temperatures due to the restriction of the mobility at the crystal surface. The restriction decreases exponentially with increasing distance from the crystal surface, which is indicated by the fading gray shade in the schematic in Fig. 12.20b. The top curve shows two separate glass transitions, as is seen, for example, in poly(ethylene terephthalate) or poly(butylene terephthalate). The occurrence of separate glass transitions indicates the existence of distinctly different mobile amorphous phases with a rather sharp boundary. Across this boundary, again, strains exist, so that both glass transitions are broadened relative to the bulk-amorphous phase.

The need for characterization of the RAF in semicrystalline polymers, among others, is due to its direct impact on the local glass transition and melting behavior [68, 69], which is linked to the macroscopic mechanical behavior [70, 71].

The detection of different mobile-amorphous fractions in semicrystalline polymers by thermal analysis is straightforward. In the absence of processes connected with latent heat effects, such as structural changes, crystallization, melting, evaporation and chemical changes, DSC yields the thermodynamic specific heat capacity, c_p, as a function of temperature, T. In the case of multiphasic structures, the measured c_p is well represented by the sum of the c_p values of the various phases multiplied by the respective weight fractions (m), as shown in Fig. 12.21 with Eq. (29) for the special case of a semicrystalline polymer consisting of a crystalline phase, labeled 'cry', rigid amorphous phase, labeled 'am, rigid' and mobile amorphous phase, labeled 'am, mobile'. The temperature dependences of the heat capacity of the fully crystalline and amorphous structures are known and listed in the ATHAS Database. The heat capacity of the crystalline phase increases steadily with increasing temperature, whereas the temperature dependence of the heat capacity of the amorphous phase shows on heating a step-like increase at the glass transition temperature, T_g, due to the onset of the large-amplitude, cooperative motion characteristic of the liquid. The heat capacity of the glass is identical with or close-to that of the crystal and, above T_g, the heat capacity of the amorphous structure is that of the liquid. The increase in c_p at T_g in the absence of crystals is approximately 11 J K^{-1} mol^{-1} of small mobile unit in the liquid and should be linearly reduced with increasing crystallinity if the devitrification occurs as a single event (Eq. 30). The magnitude of Δc_p at T_g is also called the relaxation strength of the amorphous phase. The expected c_p at T_g can therefore be estimated, if the crystallinity is known. For illustration of the detection of a RAF by DSC, the apparent specific heat capacity of cold-crystallized poly(ethylene terephthalate) (PET) is shown in Fig. 12.21 as a function of temperature on heating. The upper plot shows the complete run and the bottom plot is an enlargement of the temperature range where the glass transition of the MAF occurs. The two dotted lines are the heat capacities of the liquid and solid and the dashed line is the heat-capacity-baseline of the sample which is 24% crystalline. The arrow labeled '1' is the expected heat capacity increment if the two-phase model were valid according to Eq. (30), that is, in absence of RAF, and the arrow labeled '2' is the heat capacity decrement due to the existence of the RAF.

First experimental observations of the RAF by thermal analysis were made more than 20 years ago on poly(oxymethylene) [66] and poly(ethylene terephthalate) [67] and has by today been evidenced for numerous semicrystalline polymers such as poly(trimethylene terephthalate) [52], poly(butylene terephthalate) [72, 73], poly(ether ketone) [74], polycarbonate [75, 76], poly(3-hydroxybutyrate)

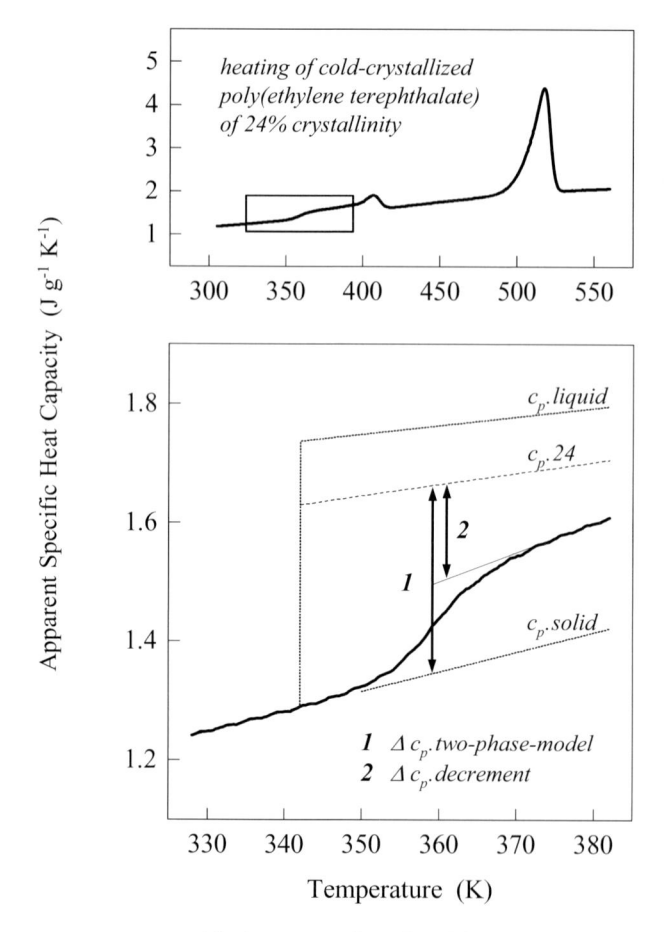

Fig. 12.21 Apparent specific heat capacity of cold-crystallized poly(ethylene terephthalate) as a function of temperature on heating. The top graph is an overview and the bottom graph is an enlargement showing the glass transition. Details of the analysis of the glass transition for detection of the RAF using Eqs. (29) and (30) are described in the text.

$$c_p(T) = m_{cry}(T)\, c_{p,cry}(T) + m_{am,rigid}(T)\, c_{p,am,rigid}(T) + m_{am,mobile}(T)\, c_{p,am,mobile}(T) \tag{29}$$

$$\Delta c_p(T = T_g) \approx (c_{p,liquid} - c_{p,solid}) \times (1 - \text{crystallinity}) \tag{30}$$

[75], poly(oxy-2,6-dimethyl-1,4-phenylene) [77, 78], poly(phenylene sulfide) [79, 80], polypropylene [75, 81] and polyethylene [82].

Research in the field of the RAF must address, before anything else, the quantitative evaluation of the total amount within the semicrystalline structure. The RAF depends strongly on the molecular architecture and the thermal and mechanical history. Information about the RAF therefore cannot be given with-

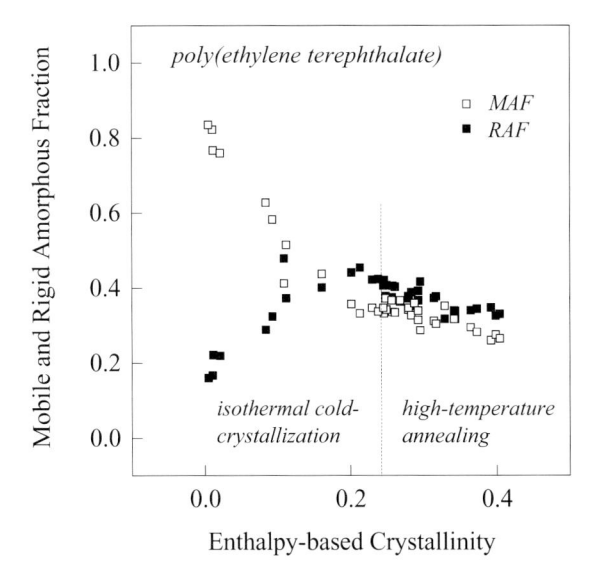

Fig. 12.22 Rigid amorphous and mobile amorphous fractions of poly(ethylene terephthalate) as a function of crystallinity. Data were observed by isothermal cold crystallization at 390 K (data left of the vertical line) and by subsequent annealing at elevated temperatures up to 513 K (data right of the vertical line). Reprinted from [83], with permission from Elsevier.

out considering the processing of the sample. A distinct effect on the RAF arises from the condition of crystallization. The condition of crystallization controls crystallinity, the crystal size and shape, specific surface area and surface structure, that is, the main effects which arise from the coupling of crystalline and amorphous phases. Figure 12.22 shows as an example the crystallinity dependence of the rigid and mobile amorphous fractions of cold-crystallized PET which was subsequently annealed at elevated temperature [83]. The initial crystallinity was adjusted by interruption of the isothermal cold crystallization process at 390 K after different times, which ultimately yield a maximum crystallinity of 24% (data left of the vertical line). This was followed by subsequent annealing at elevated temperature (data right of the vertical line). The total fraction of rigid amorphous phase increases with increasing crystallinity during the isothermal cold crystallization and yields a maximum of 44%. Subsequent annealing at elevated temperature up to 513 K increases the crystallinity from 24 to 44%, with the RAF decreasing to 33% due to changing of the coupling of the crystalline and amorphous phase by reorganization and perfection of crystals.

More information about the interplay between crystal morphology and RAF was gained by calculation of the specific RAF, which is the RAF per unit crystal or per 1% crystallinity. The specific RAF is largest at the beginning of the crystallization process when the crystals have a large surface-to-volume ratio. After completion of cold crystallization the specific RAF is about 2 and after perfection of the crystals the specific RAF is reduced to <1. Early investigations of the RAF of PET [84–86] revealed the thickness of the transition layer, which was estimated based on X-ray investigations, to be about 2 nm. It was proposed that the layer thickness is independent of the thickness of the lamellar crystals, which also implied the conclusion that with increasing crystallinity the specific RAF decreases. Qualitatively similar observations were collected by the analysis of poly(butylene terephthalate) (PBT). Melt crystallization resulted in a crystallinity of 36.3% and led to a reduction in the RAF to 21.3%, that is, a specific RAF of <1. On ultra-quenching in liquid nitrogen-cooled Al_2O_3 powder, followed by cold crystallization, the specific RAF could be increased to 3.75 at a crystallinity of 12% [87].

The formation of the RAF is coupled to the crystal formation. For melt-crystallized polycarbonate and poly(3-hydroxybutyrate), it is suggested that the vitrification occurs parallel to the crystal formation. The exact temperature of vitrification on cooling and devitrification on heating, however, is often difficult to determine, since the glass transition of the RAF is expected to be gradual, according to the gradual change in structure between crystal and unrestrained amorphous phase. The analysis of the RAF has, furthermore, shown that the RAF may only broaden the glass transition of the amorphous phase. Devitrification then starts at the same temperature as for the MAF, but extends to 1–50 K higher temperatures and moves the midpoint of the increase in c_p to higher temperature. This midpoint is usually chosen as the glass transition temperature. Typical examples of such behavior are found in polyethylene and poly(oxyethylene). For polymers such as poly(ethylene terephthalate) and poly(butylene terephthalate), a separate glass transition is seen for the MAF and the RAF, with the T_g of the RAF still being below the melting transition. In this case, the semicrystalline state is truly represented by three phases, separately characterized by two separate glass transitions and the melting transition. Other polymers such as poly(oxymethylene) and poly(oxy-2,6-dimethyl-1,4-phenylene) seem to have the RAF devitrify together with melting. For poly(oxy-2,6-dimethyl-1,4-phenylene) it was even shown that the RAF devitrifies on heating at a different rate than the melting of crystals occurs. The specific RAF decreased with ongoing melting, showing that crystals with a higher melting point exhibit less coupling to the surrounding amorphous phase.

The application of modern instrumentation pointed to complete formation and vitrification of the RAF during isothermal crystallization of a polycarbonate [poly(4,4'-isopropylidene diphenylene carbonate) (PC)] and poly(3-hydroxybutyrate) (PHB) [75]. Figure 12.23 shows the reversing heat capacity as a function of time during isothermal crystallization of PC at 456.8 K. The heat capacity of the amorphous sample at the beginning of the crystallization experiment agrees with the heat capacity of the liquid (top dotted line) and decreases with ongoing crystallization towards the heat capacity of the crystal which is indicated with the bottom dotted line. The heat capacity baseline was calculated using Eq. (4), assuming a two-phase structure with 22% crystallinity. Alternatively, the heat capacity baseline was calculated from the experimentally observed heat capacity deficiency at the glass transition of the mobile amorphous structure using Eq. (30). Figure 12.23 clearly shows coincidence of the measured reversing heat capacity at 456.8 K after

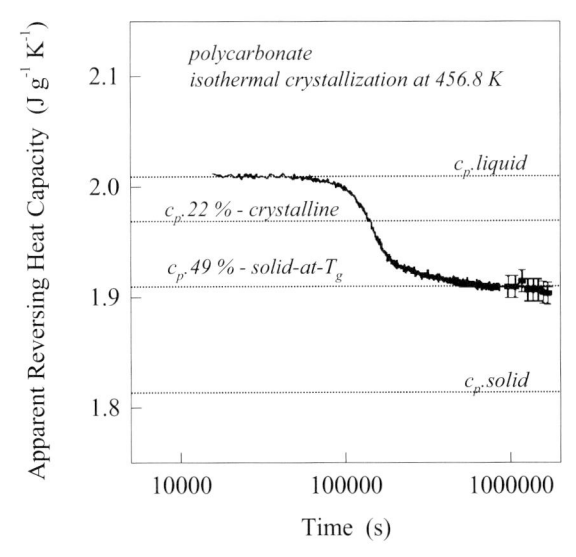

Fig. 12.23 Apparent reversing heat capacity as function of time during isothermal crystallization of polycarbonate at 456.8 K. The dotted lines represent, from top to bottom, the heat-capacity of liquid PC, the heat-capacity baseline of semi-crystalline PC of 22% crystallinity, calculated assuming the two-phase model by Eq. (4), the heat capacity baseline calculated from the heat-capacity deficiency at T_g, using Eq. (30) and the heat capacity of the solid. Reprinted from [75], with permission from Elsevier.

completed isothermal crystallization and the heat capacity baseline calculated from the three-phase model. This experiment shows that the RAF of this particular polymer vitrifies entirely at the temperature of crystallization.

Summarizing the research on RAF using DSC and TMDSC, the following novel conclusions about the thermal properties of semicrystalline polymers can be drawn:

1. Semicrystalline polymers may contain three instead of only two phases, which, based on thermal analysis, have been called crystalline, mobile amorphous and rigid amorphous (crystal, MAF and RAF).
2. The glass transition, even of the MAF, is broadened as a consequence of the stress developed at the points of decoupling from the crystal, which decreases the mobility of the non-crystalline segments. Since the low-temperature glass transition usually begins at the same temperature as the glass transition of the purely amorphous polymer, one does not expect any appreciable number of polymer segments in the metastable, semicrystalline polymer which are as mobile as in the melt at the same temperature. This is an expected result, since otherwise the semicrystalline polymer would not be an arrested, metastable state, but continue to crystallize.
3. The RAF with a distinctly different, higher glass transition can be considered as a separate phase. Close to the interface, one expects orientation and lower entropy relative to the bulk amorphous polymer.

4. Generating crystals out of an oriented amorphous phase or by deformation of a semicrystalline bulk sample, the RAF may attain added orientation and, as a result, show a substantial decrease in enthalpy and entropy relative to the melt, measurable by thermal analysis and analyzed for gel-spun, ultra-high molar mass polyethylene and poly(ethylene terephthalate) [88].
5. Knowledge about the RAF is critical when discussing the properties of semicrystalline polymers since it can affect crystallization, melting, glass transition temperatures and therefore the mechanical and chemical properties. Similarly, the processing is affected. At the glass transition of the RAF and MAF properties such as drawability, heat distortion, modulus and dye diffusion change. Also, annealing of a sample with a high fraction of RAF will soften the material, whereas under analogous conditions the annealing of crystals will harden the material. Without knowledge about the transition temperatures of all phases present in the globally metastable, semicrystalline polymers, much more trial and error optimization is needed and drastically improved materials might elude detection.

12.5.3
High-speed Calorimetry

High-speed calorimetry is of interest in materials analysis not to speed up the work, but to study metastable and unstable polymer systems quickly enough to avoid thermodynamic stabilization that would lead to changes of the sample within the calorimeter [89]. Furthermore, many industrial processes are very fast. Spinning of fibers, for example, may be done at rates of 100–10 000 m min^{-1}. A temperature change of 60 K over a length of 1 m in the thread-line, then, causes rates of temperature change of 6000–600 000 K min^{-1} (100–10 000 K s^{-1}). It is one of the goals of thermal analysis to model such fast processes.

The main advance in thermal analysis over the last 100 years was the decrease in the mass to be analyzed from kilograms to grams and to milligrams and to point to methods capable of assessing micrograms to picograms. With such decreases in mass, the rate of temperature change can be increased drastically; however, the sampling of the material may become atypical, which leads to the need to couple thermal analysis with microscopy to check the material to be analyzed.

The higher heating rates obviously cause larger temperature gradients within the sample. Figure 12.24 may be used to judge the limits of useable heating rates without becoming qualitative. The temperature inside a solid polymer cylinder of radius R is shown at a distance r from the center [9]. The thermal diffusivity, k_s, chosen was 10^{-7} m^2 s^{-1}, typical for most polymers. A sample of $R=0.004$ m and 0.02 m length, to avoid end-effects, would need about 1 g of material. In order to keep the temperature gradient negligible, one should keep the heating rate below 1.0 K min^{-1}. With a sample of 4×10^{-4} m radius and 0.002 m length, the sample mass is of the order of 1 mg and the ordinate must

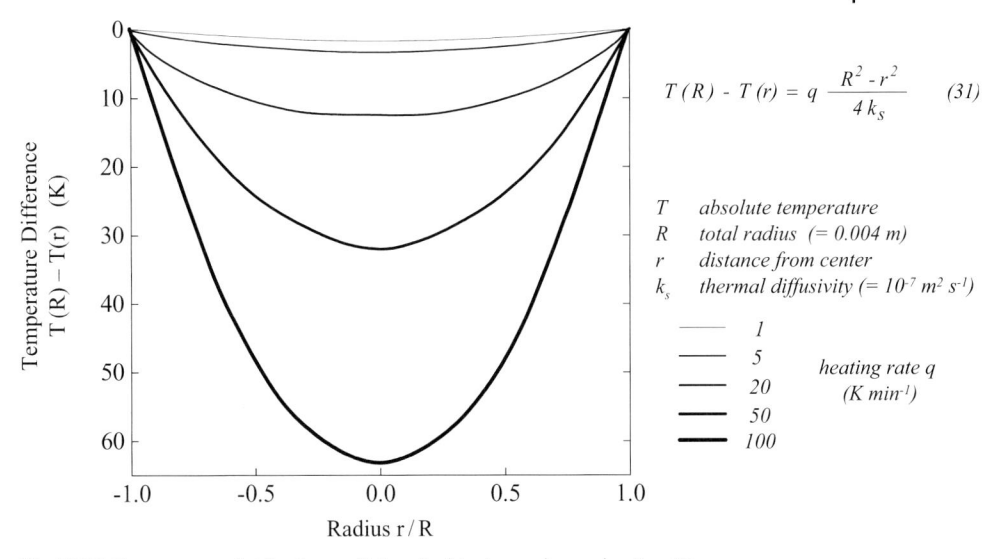

Fig. 12.24 Temperature distributions within cylindrical samples on heating [9].

$$T(R) - T(r) = q \, \frac{R^2 - r^2}{4 k_s} \quad (31)$$

- T absolute temperature
- R total radius ($= 0.004$ m)
- r distance from center
- k_s thermal diffusivity ($= 10^{-7}$ m^2 s^{-1})

heating rate q (K min^{-1}): 1, 5, 20, 50, 100

be multiplied by 0.01. This means that for the same temperature lag as before, the analysis can now be performed at rates of 100 K min^{-1}. Such dimensions (and heating rates) can be realized in most present-day DSCs. Furthermore, one can extrapolate to designs suitable for microgram, nanogram and even picogram quantities. Heating rates as high as 10^2, 10^4 and 10^6 K s^{-1}, respectively, should then be possible. Decreasing the sample mass is therefore the route for increases in the heating rate.

With DTA using glass capillaries as sample holders with a 0.0008 m radius with an inside thermocouple, it was possible to reach heating rates up to 600 K min^{-1} [90] and, on direct insertion of special, thin-walled glass tubes into hot and cold baths of widely differing temperatures, one could reach rates as high as 9000 K min^{-1} for the study of intermediates in the crystallization [91].

Microgram and smaller samples could be studied with a thin-film hot-stage heated at 3200 K min^{-1} mounted on an optical interference microscope used to assess their lamellar thicknesses to a precision of 1 nm [92]. A recent step to speed up DSC was taken with a commercial power-compensation DSC instrument (high-performance DSC) [93, 94]. It is claimed to reach 500 K min^{-1} in quantitative analysis.

A unique solution to super-fast calorimetry was the foil calorimeter [95]. A copper foil as heater is folded in such a way that two sheets of the sample (also very thin, so that the mass remains small) separate the foil. Into the center of the stack of folded copper foil and sample, a copper–constantan thermocouple was placed. The instrument was then heated fast enough so that it reacted practically adiabatically. Heating rates of up to 30 000 K min^{-1} have been achieved in measuring heat capacities by recording temperature, time and the rate of change of temperature for a given heat input. This solution for fast calorimetry of up to 500 K s^{-1} has seen little application because of the rather complicated arrangement of the sample, heater and thin thermocouple, which, in practice, require building a new calorimeter for each sample. To achieve such fast calorimetry, modulated calorimetry such as AC calorimetry or the 3ω-method was also applied [96].

Modern versions of fast-reacting calorimeters are based on silicon chip technology for measuring heat flow and temperature using thermopiles and resistors for heating, all integrated in the chip, such as the super-fast chip calorimeter (SFCC). Figures 12.25 and 12.26 illustrate the construction [97] and first results on polymers of such an SFCC [98, 99]. It consists of a 500-nm thick Si_3N_x membrane of 0.1×0.05 mm^2 dimensions with a thin-film thermopile and a resistive film heater placed at the center of the membrane. All electrical connections are covered by an additional 700-nm SiO_2 layer for electrical insulation and protection. The six hot junctions of the thermopile are placed around the central heated area. The cold junctions are placed at the silicon support, 0.1 mm from the center. The cooling is achieved by overcompensating the heater power with a cooled purge gas. Figure 12.25 summarizes a broad range of cooling experiments on polyethylene. The slowest experiments are possible by standard DSC as described in Figs. 12.4–12.6. The next series use a high-performance DSC, as mentioned above [93, 94]. The fastest runs are made with the SFCC [98, 99]. Clearly, polyethylene crystallizes so fast that an amorphous sample is not obtained by cooling at 5000 K s^{-1}. A similar experiment with 100 ng of poly(butylene terephthalate) on cooling and heating at 6600 K s^{-1} is depicted in Fig. 12.26, covering the temperature range 450–320 K within about 20 ms

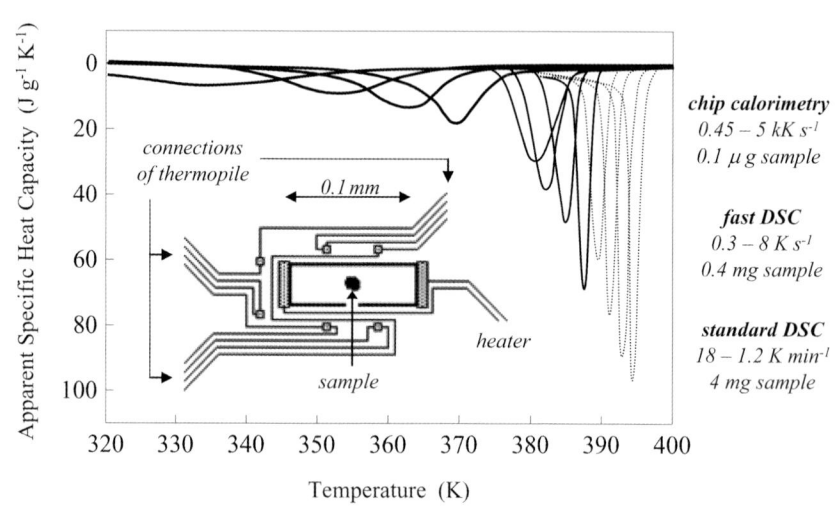

Fig. 12.25 Change in the rate of crystallization of polyethylene with cooling rate. Sample masses are 4 mg for standard DSC, 0.4 mg in a fast DSC and 0.1 μg in the SFCC. A top view of the calorimeter cell is shown [97]. Data were taken from [98].

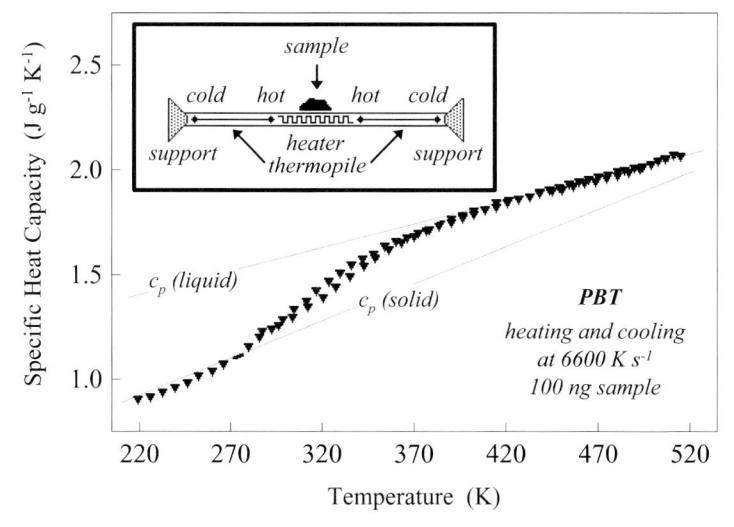

Fig. 12.26 The specific heat capacity, c_p, on superfast cooling and heating of poly(butylene terephthalate) at 6600 K s^{-1} in an SFCC. The schematic side view of the calorimeter is given [97]. The overall radius is about 1.0 mm. Data were taken from [87].

[87]. In this temperature range, isothermal crystallization is possible and reaches a maximum within 0.2–1.0 s, that is, the heating and cooling rates of the SFCC are sufficient to avoid crystallization on cooling and also on heating, avoiding cold-crystallization. The application of the SFCC as shown in Figs. 12.25 and 12.26 has also been used for AC calorimetry [100, 101] in the frequency range 1 mHz–100 Hz, which compares with a 0.1 Hz limit typical for TMDSC. It may be possible to analyze nanogram samples by evaporation from solution on the sample position.

Summarizing SFCC, one can draw the following conclusions about the future of thermal analysis for the analysis of the properties of the nanophase structure of semicrystalline polymers:

1. The rates of analysis can be extended to rates as high as 10^6 K s^{-1} to model many of the fast industrial processes quantitatively, and also to assess the properties of metastable phases as they become unstable on heating or cooling.
2. By coupling super-fast calorimetry with AFM [102], the nanophases within the globally metastable structure of semicrystalline polymers may be assessed with nanometer-scale resolution and thermally analyzed simultaneously without losing the metastability.
3. These developments in high-speed calorimetry point to a continuation of the expansion of the DSC method for the characterization of polymers, despite the many intricacies involved in data extraction and evaluation. Besides providing the basic thermodynamic functions, DSC has developed into an impor-

tant technique in the evaluation of thermal, macroscopic material properties and enabled them to be linked to microscopic structure and motion. All this allows a better understanding of the applications of new, precisely synthesized macromolecules [103].

12.6
Experimental Tools for Further Thermal Analysis of Polymers

Despite the many advances in the scanning calorimetry of polymers, from the point of view of instrumentation and also data interpretation, a complete thermal analysis must always involve additional experimentation to gain complementary information. The calorimetric information only provides the macroscopic response caused by molecular motion, which, in turn, is strongly influenced by the microscopic phase structure. Direct information about the molecular motion is obtained by infrared and Raman spectroscopy, and also nuclear magnetic resonance (NMR) spectroscopy. Direct structure and morphology information is obtained by light, X-ray and neutron scattering, and also the different types of microscopy. A demonstration of the requirements for the application of additional techniques is given in Fig. 12.27. It shows the standard DSC heating scan of polytetrafluoroethylene (PTFE) in the temperature range 275–325 K [104]. The two peaks with maxima at about 295 and 305 K provide unambiguous evidence of first-order transitions, connected with an enthalpy of transition. DSC allows the exact determination of the temperature and enthalpy of these transitions and even the kinetics, but cannot provide further information

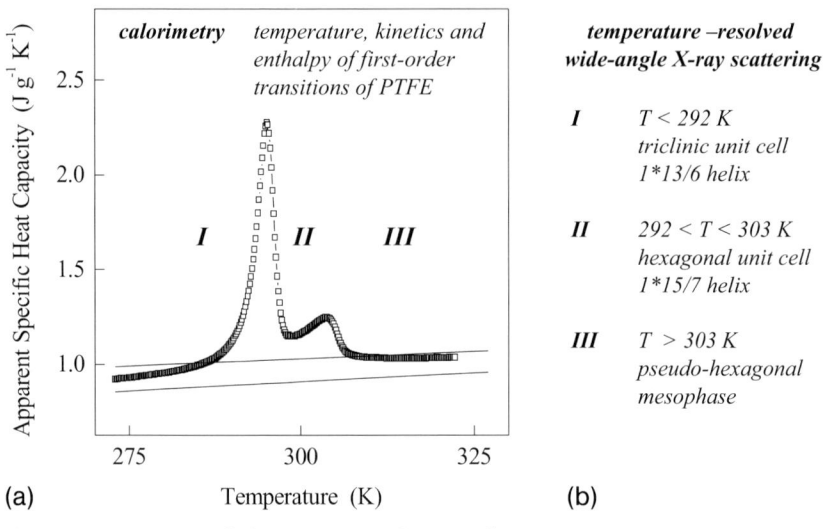

Fig. 12.27 Apparent specific heat capacity as function of temperature on heating polytetrafluoroethylene from 275 to 325 K.

about the structural changes within the sample. This information can be obtained by temperature-resolved wide-angle X-ray scattering (WAXS), summarized in Fig. 12.27b. Analysis of X-ray data yields the information that below 292 K the PTFE crystals are triclinic with the molecules adopting a 1*13/6 helical twist. Heating above this temperature causes a strong increase of the amplitude of torsional oscillations of the molecule, which results in a 1*15/7 helical structure and a change of the crystal symmetry from triclinic to hexagonal, with increased disorder along the chain axis. In the X-ray fiber pattern the equatorial spots remain sharp and point to maintained lateral order of the lattice. At 303 K, the longitudinal order is lost completely and therefore the three-dimensional character of the lattice is changed by further untwisting to an irregular conformation and the symmetry is now best described as pseudo-hexagonal [105–107].

A further example of the need for combination of different analysis tools is shown in Fig. 12.28. Figure 12.28a shows the DSC heating scan of poly(ethylene-co-1-octene) of 0.87 g cm^{-3} density, which has already been discussed with Figs. 12.10 and 12.19. DSC allowed the identification of the exact temperature range of melting and evaluation of the corresponding heat of fusion. These data, however, cannot be further interpreted without structural information. Temperature-resolved Fourier-transform infrared (FTIR) spectroscopy [108], in combination with X-ray analyses [109], provided evidence that this particular copolymer, as a representative of the class of linear low-density polyethylenes, not only crystallizes in the orthorhombic form, as is typical of less-branched high-density polyethylene, but also in a mesophase with a pseudo-hexagonal arrangement of the molecules [108, 109]. The X-ray fiber pattern in Fig. 12.28b shows in addition to the 110 and 200 reflections from the orthorhombic crystalline structure a sharp and intense reflection at a spacing of about 0.45 nm, due to the existence of an additional mesophase. The relative ratio between these two metastable polymorphs depends on the branch concentration and the condition of crystallization. Temperature-resolved FTIR allowed, furthermore, the analysis of the specific melting and crystallization behaviors of these two phases, which cause absorbance at specific wavenumbers. Figure 12.28c shows absorbance indexes being proportional to the absorbance at 720 and 716 cm^{-1} and the corresponding phase fractions of orthorhombic and mesomorphic structure, both as a function of temperature. The data show that at low temperatures melting of the mesophase occurs, whereas the orthorhombic structure melts at higher temperatures, indicated by the filled and hollow arrows in the DSC scan in Fig. 12.28a. Furthermore, by comparison of the enthalpy-based crystallinity, obtained by DSC, with the true phase composition, obtained by WAXS, it was estimated that the mesophase has a considerably lower bulk specific heat of fusion than the orthorhombic phase [109].

Thermal analysis, by definition, includes all methods which allow the analysis of the physical properties of a substance as a function of temperature with the limitation that the temperature–time profile must be well defined and controlled [11]. Scanning calorimetry is therefore just one prominent example of

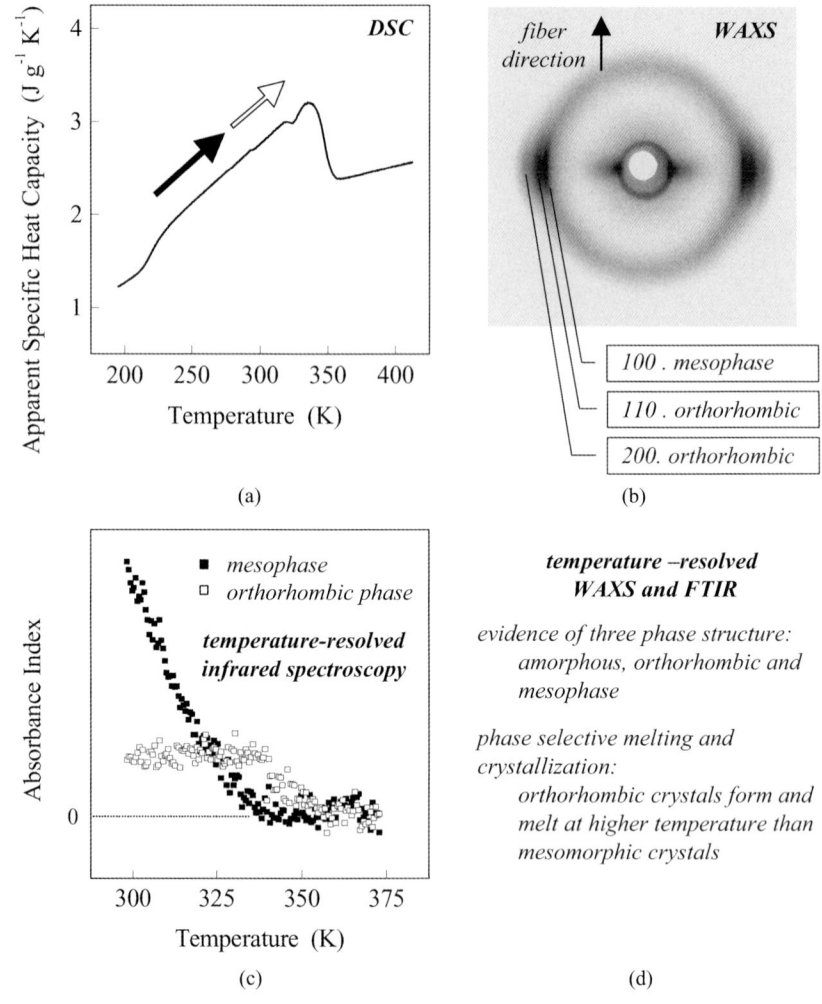

Fig. 12.28 (a) Apparent specific heat capacity of poly(ethylene-co-1-octene) of density 0.87 g cm^{-3} as a function of temperature on heating. (b) Wide-angle X-ray fiber pattern of oriented poly(ethylene-co-1-octene) of density 0.87 g cm^{-3}, providing evidence of the coexistence of amorphous, orthorhombic crystalline and pseudo-hexagonal mesomorphic phases. (c) FTIR absorbance indexes, representing relative fractions of orthorhombic and mesomorphic structure as a function of temperature (data were taken from [108]).

numerous techniques for the thermal characterization of polymers. Others include thermogravimetry (TGA), dynamic mechanical analysis (DMA), dilatometry and thermomechanical analysis (TMA), which all can be considered as standard methods in thermal analysis laboratories. Table 12.1 gives a brief summary of the major information which can be obtained by these techniques, without

Table 12.1 List of selected thermal analysis techniques.

Thermal analysis technique	Perturbation	Answer measured as function of temperature and/or time	Major information and some analyzed phenomena
Differential scanning calorimetry (DSC)	Temperature Atmosphere	Differential heat flow-rate into sample and reference	Heat capacity Temperature, enthalpy and kinetics of transitions and chemical reactions Glass transition and enthalpy relaxation Melting and crystallization Cross-linking and degradation Oxidative induction time Temperature and enthalpy of mixing/demixing
Thermogravimetry (TGA)	Temperature Atmosphere	Mass	Temperature, kinetics and mechanisms of degradation Moisture content Filler content Carbon black content in rubber
Dynamic mechanical analysis (DMA)	Temperature Dynamic force (stress) Dynamic deformation	Deformation Force (stress)	Dynamic modulus of elasticity Viscoelasticity Glass transition Secondary relaxation Crystal α-relaxation Creep and relaxation
Dilatometry	Temperature Hydrostatic pressure	Length Volume	Linear or volume thermal expansion coefficient of solids and liquids Glass transition Temperature and specific change of volume of first-order transitions
Thermal mechanical analysis (TMA)	Temperature Force	Length Penetration	See dilatometry Heat distortion temperature

claiming to be complete, due to the continuing advances in all fields of instrumentation. In addition, this table could be extended by more sophisticated instrumentation which, however, is reserved for solving special scientific problems. These include temperature-resolved wide-angle and small-angle X-ray scattering and neutron scattering, temperature-resolved FTIR and solid-state NMR, thermo-optical analysis and temperature-resolved AFM.

References

1. Gibbs JW (1876) On the equilibrium of heterogeneous substances, *Trans Conn Acad 3*: 108–248 and 343–524, 1878. Reprinted in *The Scientific Papers of J. Willard Gibbs*, pp. 55–371, Longmans, Green, London, 1906.
2. Zettlemoyer AC (1969) *Nucleation*, Marcel Dekker, New York.
3. Volmer M (1938) *Kinetik der Phasenbildung*, Theodor Steinkopff, Dresden.
4. Reading M, Hahn BK, Crowe BS (1993) Method and apparatus for modulated differential analysis, *US Patent 5 224 775*.
5. Reading M (1993) Modulated differential scanning calorimetry – a new way forward in materials characterization, *Trends Polym Sci 8*: 248–253.
6. Gill PS, Sauerbrunn SR, Reading M (1993) Modulated differential scanning calorimetry, *J Thermal Anal 40*: 931–939.
7. Reading M, Elliott D, Hill VL (1993) A new approach to the calorimetric investigation of physical and chemical transitions, *J Thermal Anal 40*: 949–955.
8. Wunderlich B (2003) Reversible crystallization and the rigid-amorphous phase in semi-crystalline macromolecules, *Prog Polym Sci 28*: 383–450.
9. Wunderlich B (2005) *Thermal Analysis of Polymeric Materials*, Springer, Berlin.
10. Mathot VBF (1994) *Calorimetry and Thermal Analysis of Polymers*, Carl Hanser, Munich.
11. Hemminger WF, Cammenga HK (1989) *Methoden der Thermischen Analyse*, Springer, Berlin.
12. Callister WD (2003) *Materials Science and Engineering: an Introduction*, Wiley, New York.
13. Gaur U, Shu H-C, Mehta A, Wunderlich B (1981) Heat capacity and other thermodynamic properties of linear macromolecules, *J Phys Chem Ref Data 10*: 89–117; Gaur U, Wunderlich B (1981) II. Polyethylene, *J Phys Chem Ref Data 10*: 119–152; III. Polyoxides, *J Phys Chem Ref Data* (1981) *10*: 1001–1049; IV. Polypropylene, *J Phys Chem Ref Data* (1981) *10*: 1051–1064; V. Polystyrene, *J Phys Chem Ref Data* (1982) *11*: 313–325; Gaur U, Lau S-F, Wunderlich BB, Wunderlich B (1982) VI. Acrylic Polymers, *J Phys Chem Ref Data 11*: 1065–1089; Gaur U, Wunderlich BB, Wunderlich B (1983) VII. Other Carbon Backbone Polymers, *J Phys Chem Ref Data 12*: 29–63; Gaur U, Lau S-F, Wunderlich BB, Wunderlich B (1983) VIII. Polyesters and Polyamides, *J Phys Chem Ref Data 12*: 65–89; Gaur U, Lau S-F, Wunderlich B (1983) IX. Aromatic and Inorganic Polymers, *J Phys Chem Ref Data 12*: 91–108; Varma-Nair M, Wunderlich B (1991) X. Update of the ATHAS 1980 Data Bank, *J Phys Chem Ref Data 20*: 349–404.
14. Baur H (1999) *Thermophysics of Polymers*, Springer, Berlin; Godovsky YK (1992) *Thermophysical Properties of Polymers*, Springer, Berlin; Wunderlich B, Baur H (1970) Heat capacities of linear high polymers, *Fortschr Hochpolym Forsch 7*: 151–368.
15. Illers KH, Haberkorn H (1971) Schmelzverhalten, Struktur und Kristallinität von 6-Polyamid, *Makromol Chem 142*: 31–67.
16. Illers KH, Haberkorn H (1971) Spezifisches Volumen, Schmelzwärme und Kristallinität von 6.6- und 8-Polyamid, *Makromol Chem 146*: 267–270.
17. Illers KH (1978) Polymorphie, Kristallinität und Schmelzwärme von Poly(ε-caprolactam), 2. Kalorimetrische Untersuchungen, *Makromol Chem 179*: 497–507.
18. Alexander LE (1969) *X-Ray Diffraction Methods in Polymer Science*, Wiley, New York.
19. Wunderlich B (1976) *Macromolecular Physics, Vol. 2, Crystal Nucleation, Growth, Annealing*, Academic Press, New York.
20. Wunderlich B, Mehta A (1974) Macromolecular nucleation, *J Polym Sci, Polym Phys 12*: 255–263.
21. Mehta A, Wunderlich B (1975) A study of molecular fractionation during the crystallization process, *Colloid Polym Sci 253*: 193–205.
22. Wunderlich B (1980) *Macromolecular Physics, Vol. 3, Crystal Melting*, Academic Press, New York.

23. Flory PJ (**1953**) *Principles of Polymer Chemistry*, Cornell University Press, Ithaca, NY.
24. Wunderlich B (**1997**) Metastable mesophases, *Macromol Symp 113*: 51–65.
25. Boller A, Jin Y, Wunderlich B (**1994**) Heat capacity measurement by modulated DSC at constant temperature, *J Thermal Anal 42*: 307–329.
26. Wunderlich B (**1997**) Modeling the heat flow and heat capacity of modulated differential scanning calorimetry, *J Thermal Anal 48*: 207–224.
27. Merzlyakov M, Schick C (**1999**) Complex heat capacity measurements by TMDSC. Part 1. Influence of the non-linear thermal response, *Thermochim Acta 330*: 55–64.
28. Merzlyakov M, Schick C (**1999**) Complex heat capacity measurements by TMDSC. Part 2. Algorithm for amplitude and phase angle correction, *Thermochim Acta 330*: 65–73.
29. Schawe JEK, Winter W (**1997**) The influence of heat transfer on temperature-modulated DSC measurements, *Thermochim Acta 298*: 9–16.
30. Androsch R, Moon I, Kreitmeier S, Wunderlich B (**2000**) Determination of heat capacity with a sawtooth-type, power-compensated temperature-modulated DSC, *Thermochim Acta 357/358*: 267–278.
31. Androsch R, Wunderlich B (**1999**) Temperature-modulated DSC using higher harmonics of the Fourier transform, *Thermochim Acta 333*: 27–32.
32. Androsch R (**2000**) Heat-capacity measurements using temperature-modulated heat flux DSC with close control of the heater temperature, *J Thermal Anal Cal 61*: 75–89.
33. Wunderlich B, Jin Y, Boller A (**1994**) Mathematical description of differential scanning calorimetry based on periodic temperature modulation, *Thermochim Acta 238*: 277–293.
34. Wunderlich B, Androsch R, Pyda M, Kwon YK (**2000**) Heat capacity by multi-frequency sawtooth modulation, *Thermochim Acta 348*: 181–190.
35. Kwon YK, Androsch R, Pyda M, Wunderlich B (**2001**) Multi-frequency sawtooth modulation of a power-compensation differential scanning calorimeter, *Thermochim Acta 367/368*: 203–215.
36. Androsch R, Wunderlich B (**2000**) Reversing melting and crystallization of indium as a function of temperature modulation, *Thermochim Acta 364*: 181–191.
37. Androsch R, Wunderlich B (**2001**) Reversibility of melting and crystallization of indium as a function of the heat conduction path, *Thermochim Acta 369*: 67–78.
38. Cheng SZD, Wunderlich B (**1986**) Molecular segregation and nucleation of poly(ethylene oxide) crystallized from the melt. II. Kinetic study, *J Polym Sci, Polym Phys 24*: 595–617.
39. Androsch R (**1999**) Melting and crystallization of poly(ethylene-co-octene) measured by modulated d.s.c and temperature-resolved X-ray diffraction, *Polymer 40*: 2805–2812.
40. Androsch R, Wunderlich B (**1999**) A study of annealing of poly(ethylene-co-octene) by temperature-modulated and standard differential scanning calorimetry, *Macromolecules 32*: 7238–7247.
41. Mathot VBF, Scherrenberg RL, Pijpers MFJ, Bras W (**1996**) Dynamic DSC, SAXS and WAXS on homogeneous ethylene–propylene and ethylene–octene copolymers with high comonomer contents, *J Thermal Anal 46*: 681–718.
42. Mathot VBF, Scherrenberg RL, Pijpers TFJ, Engelen YMT (**1996**) Structure, crystallization and morphology of homogeneous ethylene–propylene, ethylene–1-butene and ethylene–1-octene copolymers with high comonomer contents, in *New Trends in Polyolefin Science and Technology*, ed. S. Hosoda, Research Signpost, Trivandrum (India), p. 71.
43. Mathot VBF, Scherrenberg RL, Pijpers TFJ (**1998**) Metastability and order in linear, branched and copolymerized polyethylenes, *Polymer 39*: 4541–4559.
44. Schick C, Merzlyakov M, Wunderlich B (**1998**) Analysis of the reorganization of poly(ethylene terephthalate) in the melting range by temperature-modulated calorimetry, *Polymer Bull 40*: 297–303.

45 Höhne GWH, Kurelec L, Rastogi S, Lemstra PJ (2003) Temperature-modulated differential scanning calorimetric measurements on pre-melting behavior of nascent ultrahigh molecular mass polyethylene, *Thermochim Acta* 396: 97–108.

46 Huang Z, Marand H, Cheung WY, Guest M (2004) Study of crystallization processes in ethylene–styrene copolymers by conventional DSC and temperature-modulated calorimetry: linear polyethylene and low styrene content copolymers. *Macromolecules* 37: 9922–9932.

47 Pyda M, Di Lorenzo ML, Pak J, Kamasa P, Buzin A, Grebowicz J, Wunderlich B (2001) Reversible and irreversible heat capacity of poly[carbonyl(ethylene-*co*-propylene)] by temperature-modulated calorimetry, *J Polym Sci, Polym Phys* 39: 1565–1577.

48 Androsch R, Wunderlich B (2000) Analysis of the degree of reversibility of crystallization and melting in poly(ethylene-*co*-1-octene), *Macromolecules* 33: 9076–9089.

49 Pak J, Wunderlich B (2001) Melting and crystallization of polyethylene of different molar mass by calorimetry, *Macromolecules* 34: 4492–4503.

50 Tanabe Y, Strobl GR, Fischer EW (1986) Surface melting in melt-crystallized linear polyethylene, *Polymer* 27: 1147–1153.

51 Okazaki I, Wunderlich B (1997) Reversible melting in polymer crystals detected by temperature-modulated differential scanning calorimetry, *Macromolecules* 30: 1758–1764.

52 Pyda M, Wunderlich B (2000) Reversible and irreversible heat capacity of poly(trimethylene terephthalate) analyzed by temperature-modulated differential scanning calorimetry, *J Polym Sci, Polym Phys* 38: 622–631.

53 Righetti MC (1999) Reversible melting in poly(butylene terephthalate), *Thermochim Acta* 333: 131–135.

54 Di Lorenzo ML, Pyda M, Wunderlich B (2001) Reversible melting in nanophase-separated poly(oligo-*alt*-oligoether)s and its dependence on sequence length, crystal perfection and molecular mobility, *J Polym Sci, Polym Phys* 39: 2969–2981.

55 Turi E (1997) *Thermal Characterization of Polymeric Materials*, revised 2nd edn, Academic Press, New York,

56 Pak J, Wunderlich B (2000) Thermal analysis of paraffins as model compounds for polyethylene, *J Polym Sci, Polym Phys* 38: 2810–2822.

57 Qiu W, Pyda M, Nowak-Pyda E, Habenschuss A, Wunderlich B (2005) Reversibility between glass and melting transitions of poly(oxyethylene), *Macromolecules* 38: 8454–8567.

58 Androsch R, Wunderlich B, Radusch HJ (2005) Analysis of reversible melting in polytetrafluoroethylene, *J Thermal Anal Cal* 79: 615–622.

59 Androsch R, Wunderlich B (2001) Reversible crystallization and melting at the lateral surface of isotactic polypropylene crystals, *Macromolecules* 34: 5950–5960.

60 Androsch R, Wunderlich B (2001) Heat of fusion of the local equilibrium of melting of isotactic polypropylene, *Macromolecules* 34: 8384–8387.

61 Androsch R, Wunderlich B (2003) Specific reversible melting of polymers, *J Polym Sci, Polym Phys* 41: 2039–2051.

62 Androsch R, Wunderlich B (2003) Specific reversible melting of polyethylene, *J Polym Sci, Polym Phys* 41: 2157–2173.

63 Wunderlich B (1979) Molecular nucleation and segregation, *Discuss Faraday Soc* 68: 239–243.

64 Wunderlich B (1997) Detection of multiple nanophases by DSC, *J Thermal Anal* 49: 513–520.

65 Wunderlich B (2003) The thermal properties of complex, nanophase-separated macromolecules as revealed by temperature-modulated calorimetry, *Thermochim Acta* 403: 1–13.

66 Suzuki H, Grebowicz J, Wunderlich B (1985) Glass transition of poly(oxymethylene), *Br Polym J* 17: 1–3.

67 Menczel J, Wunderlich B (1981) Heat capacity hysteresis of semicrystalline macromolecular glasses, *J Polym Sci, Polym Lett* 19: 261–264.

68 Zachmann HG (1967) Der Einfluss der Konfigurationsentropie auf das Kristallisations- und Schmelzverhalten von hochpolymeren Stoffen, *Kolloid-Z Z Polym* 216/217: 180–191.

69 Wunderlich B (2003) Reversible crystallization and the rigid-amorphous phase in semicrystalline macromolecules, *Prog Polym Sci* 28: 383–450.

70 Rastogi R, Vellinga WP, Rastogi S, Schick C, Meijer HEH (2004) The three-phase structure and mechanical properties of poly(ethylene terephthalate), *J Polym Sci, Polym Phys* 42: 2092–2106.

71 Fu Y, Annis B, Boller A, Jin Y, Wunderlich B (1994) Analysis of structure and properties of poly(ethylene terephthalate) fibers, *J Polym Sci, Polym Phys* 32: 2289–2306.

72 Cheng SZD, Pan R, Wunderlich B (1988) Thermal analysis of poly(butylene terephthalate), its heat capacity, rigid amorphous fraction and transition behavior, *Makromol Chem* 189: 2443–2458.

73 Pyda M, Nowak-Pyda E, Mays J, Wunderlich B (2004) Heat capacity of poly(butylene terephthalate) *J Polym Sci, Polym Phys* 42: 4401–4411.

74 Cheng SZD, Cao MY, Wunderlich B (1986) Glass transition and melting behavior of poly(oxy-1,4-phenyleneoxy-1,4-phenylenecarbonyl-1,4-phenylene), *Macromolecules* 19: 1868–1876.

75 Schick C, Wurm A, Mohamed A (2002) Dynamics of reversible melting revealed from frequency dependent heat capacity, *Thermochim Acta* 392/393: 303–313.

76 Schick C, Wurm A, Merzlyakov M, Minakov A, Marand H (2001) Crystallization and melting of polycarbonate studied by temperature-modulated DSC (TMDSC), *J Thermal Anal Cal* 64: 549–555.

77 Cheng SZD, Wunderlich B (1987) Glass transition and melting behavior of poly(oxy-2,6-dimethyl-1,4-phenylene), *Macromolecules* 20: 1630–1637.

78 Pak J, Pyda M, Wunderlich B (2003) Rigid amorphous fractions and glass transitions in poly(oxy-2,6-dimethyl-1,4-phenylene), *Macromolecules* 36: 495–499.

79 Cheng SZD, Wu ZQ, Wunderlich B (1987) Glass transition and melting behavior of poly(thio-1,4-phenylene), *Macromolecules* 20: 2802–2810.

80 Huo P, Cebe P (1992) Dielectric relaxation of poly(phenylene sulfide) containing a fraction of rigid amorphous phase, *J Polym Sci, Polym Phys* 30: 239–250.

81 Grebowicz J, Lau SF, Wunderlich B (1984) The thermal properties of polypropylene, *J Polym Sci, Polym Symp* 71: 19–37.

82 Kolesov IS, Androsch R, Radusch HJ (2005) Effect of crystal morphology and crystallinity on the mechanical α- and β-relaxation processes of short-chain branched polyethylene, *Macromolecules* 38: 445–453.

83 Androsch R, Wunderlich B (2005) The link between rigid amorphous fraction and crystal perfection in cold-crystallized poly(ethylene terephthalate), *Polymer* 46: 12556–12566.

84 Schick C, Krämer L, Mischok W (1985) Der Einfluss struktureller Veränderungen auf den Glasübergang in teilkristallinem Polyethylenterephthalat, 1. Isotherme Kristallisation, *Acta Polym* 36: 47–53.

85 Schick C, Fabry F, Schnell U, Stoll G, Deutschbein L, Mischok W (1988) Der Einfluss struktureller Veränderungen auf den Glasübergang in teilkristallinem Polyethylenterephthalat, 2. Charakterisierung der übermolekularen Struktur, *Acta Polym* 39: 705–710.

86 Schick C, Wigger J, Mischok W (1990) Der Einfluss struktureller Veränderungen auf den Glasübergang in teilkristallinem Polyethylenterephthalat, 3. Der Glasübergang in den zwischenlamellaren Bereichen, *Acta Polym* 41: 137–142.

87 Pyda M, Nowak-Pyda E, Heeg J, Huth H, Minakov AA, Di Lorenzo ML, Schick C, Wunderlich B (2006), Melting and crystallization of poly(butylene terephthalate) by temperature-modulated and superfast calorimetry, *J Polym Sci, Polym Phys* 44: 1364–1377.

88 Kwon YK, Boller A, Pyda M, Wunderlich B (2000) Melting and heat capacity of gel-spun, ultra-high molar mass polyethylene fibers, *Polymer* 41: 6237–6249.

89 Wunderlich B (2004) Fast and super-fast DTA and calorimetry, in *Proceedings of the NATAS Annual Conference on Thermal Analysis and Applications*, p. 21, ed. Rich MJ, printed by Omnipress, Madison.

90 Wolpert SM, Weitz A, Wunderlich B (1971) Time-dependent heat capacity in

the glass transition region, *J Polym Sci, Polym Phys* 9: 1889–1895.
91. Wu ZQ, Dann VL, Cheng SZD, Wunderlich B (1988) Fast DSC applied to the crystallization of polypropylene, *J Thermal Anal* 34: 105–114.
92. Hellmuth E, Wunderlich B (1965) Superheating of linear high-polymer polyethylene crystals, *J Appl Phys* 36: 3039–3044.
93. Pijpers TFJ, Mathot VBF, Goderis B, Scherrenberg RL, van der Vegte EW (2002) High-speed calorimetry for the study of the kinetics of (de)vitrification, crystallization and melting of macromolecules, *Macromolecules* 35: 3601–3613.
94. Kolesov I, Androsch R, Radusch HJ (2004) Non-isothermal crystallization of polyethylenes as function of cooling rate and concentration of short chain branches, *J Thermal Anal Cal* 78: 885–895.
95. Hager NE (1972) High speed thermal analysis with thin-foil calorimeter, *Rev Sci Instrum* 43: 1116–1122.
96. Jeong YH (2001) Modern calorimetry: going beyond tradition, *Thermochim Acta* 377: 1–7; see also Jung DH, Moon IK, Jeong YH, Lee SH (2003) Differential 3ω calorimeter, *Thermochim Acta* 403: 83–88.
97. van Herwaarden AW (2005) Overview of calorimeter chips for various applications, *Thermochim Acta* 432: 192–201; see also: http://www.xensor.nl.
98. Adamovsky SA, Minakov AA, Schick C (2003) Scanning microcalorimetry at high cooling rate, *Thermochim Acta* 403: 55–63.
99. Minakov AA, Mordvintsev DA, Schick C (2005) Isothermal reorganization of poly(ethylene terephthalate) revealed by fast calorimetry (1000 K s^{-1}; 5 ms), *Faraday Discuss* 128: 261–270.
100. Minakov AA, Mordvintsev DA, Schick C (2004) Melting and reorganization of poly(ethylene terephthalate) on fast heating (1000 K/s), *Polymer* 45: 3755–3763.
101. Merzlyakov M (2003) Integrated circuit thermopile as a new type of temperature modulated calorimeter, *Thermochim Acta* 403: 65–81.
102. Moon I, Androsch R, Chen W, Wunderlich B (2000) The principles of microthermal analysis and its application to the study of macromolecules, *J Thermal Anal Cal* 59: 187–203.
103. Qiu W, Sworen J, Pyda M, Nowak-Pyda E, Habenschuss A, Wagener KB, Wunderlich B (2006) Effect of the precise branching of polyethylene at each 21st CH$_2$ group on its transition behavior, *Macromolecules* 39: 204–217.
104. Androsch R (2001) Reversibility of the low-temperature transitions of polytetrafluoroethylene as revealed by temperature-modulated differential scanning calorimetry, *J Polym Sci, Polym Phys* 39: 750–756.
105. Bunn CW, Howells ER (1954) Structures of molecules and crystals of fluorocarbons, *Nature* 174: 549–551.
106. Clark ES (1962) Partial disordering and crystal transitions in poly(tetrafluoroethylene), *Z Kristallogr* 117: 119–127.
107. Sperati CA, Starkweather HW Jr (1961) Fluorine-containing polymers, II. Polytetrafluoroethylene, *Fortschr Hochpolym-Forsch* 2: 465–495.
108. Androsch R, Kolesov I, Radusch HJ (2003) Temperature-resolved derivative FTIR: melting and formation of mesomorphic poly(ethylene), *J Thermal Anal Cal* 73: 59–70.
109. Androsch R, Blackwell J, Chvalun SN, Wunderlich B (1999) Wide- and small-angle X-ray analysis of poly(ethylene-co-octene), *Macromolecules* 32: 3735–3740.

13
Chromatography of Polymers

Wolfgang Radke

13.1
Introduction

In the last decade, considerable improvements have been made in the field of macromolecular synthesis. Novel polymerization techniques, especially those termed controlled or living polymerizations, make it possible to synthesize new macromolecular structures such as special copolymers or topologies. The possibility of obtaining well-defined model structures in combination with improvements in polymer physics and polymer theory resulted in an increasing understanding of structure–property relationships. Nowadays materials can be tailored to fulfill the requirements of a particular application. The macroscopic features required for a particular application originate from the molecular structure due to the self-organization of the macromolecules. Since small deviations at the molecular level might result in severe changes to the bulk properties of the materials, characterization techniques at the molecular level are of fundamental importance. Among these, chromatography is especially important. In contrast to most other analytical techniques, chromatography not only yields an average value for a particular property, such as average molar mass or composition, it can also separate the different molecular structures. This separation allows quantification and therefore the determination of the distribution functions for the different structural parameters. The ability to characterize the distribution function of a property instead of yielding only average properties makes polymer chromatography a unique analytical tool.

13.1.1
Heterogeneity of Polymers

Polymers can be heterogeneous in different ways. The most obvious heterogeneity is that with respect to molar mass. Since every polymerization involves statistical processes of chain initiation, growth, termination and transfer, the different chains formed in a polymerization are never of equal length, but show a molar

Macromolecular Engineering. Precise Synthesis, Materials Properties, Applications.
Edited by K. Matyjaszewski, Y. Gnanou, and L. Leibler
Copyright © 2007 WILEY-VCH Verlag GmbH & Co. KGaA, Weinheim
ISBN: 978-3-527-31446-1

mass or weight distribution (MMD, MWD). In copolymers, individual chains might vary in their comonomer compositions, giving rise to a chemical composition distribution (CCD). Another structural feature is functional groups attached to the chain ends. This results in a functionality-type distribution (FTD). If the polymer mixture contains molecules of different topology, e.g. branched in addition to linear polymers, a topology distribution exists. In general, different types of heterogeneity will coexist. A copolymer, being heterogeneous with respect to chemical composition, will at the same time be polydisperse with respect to molar mass. In order to understand the dependences of the macroscopic properties on molecular structure, it is necessary to separate and identify the different entities. The identification can be performed by suitable spectroscopic or spectrometric methods. However, without prior separation, these methods yield only an average of a certain property, which can result from different compositions of the polymer components. For example, an average composition of 50% component A and 50% component B might result from a homopolymer blend, from a block or statistical copolymer or a proper mixture of homo- and copolymers. Clearly, although the compositions are identical, the material properties will be very different. Chromatographic separation and the subsequent identification and quantification of the different components can result in a comprehensive characterization of such complex polymer samples. This makes chromatographic methods unique.

This chapter provides an overview of the different chromatographic separation techniques for polymers, the detection methods allowing structural characterization of the components and their capabilities. It is not the aim to give a comprehensive review of the field, but rather to provide a general understanding of the ideas, strategies and type of information that can be extracted from a particular chromatographic experiment. In order to do so, selected applications will be presented and discussed.

A number of excellent reviews are available for further reading [1–3].

13.2
Modes of Liquid Chromatography of Polymers

High-performance liquid chromatography (HPLC) is a separation method taught to probably every student of chemistry. The separation is based on the differences in the retention times or retention volumes of the different analyte molecules as they pass in a flowing liquid through a column. The retention volume, V_R, at which an analyte elutes from a chromatographic column can be described by the following general chromatographic equation [4, 5]:

$$V_R = V_0 + KV_{st} \tag{1}$$

where V_0 and V_{st} are the volumes of the mobile and stationary phase, respectively, and K is the distribution coefficient, defined as the ratio of the analyte concentration in the stationary phase to that in the mobile phase. There is still a debate concerning the exact definition of the volume of the stationary phase for polymer molecules [4]. However, for the purpose of this chapter, it will be sufficient to assume the volume of the stationary phase to be the entire pore volume, $V_{st} = V_P$, and to identify the volume of the mobile phase with the interparticle volume, $V_0 = V_i$.

It should be mentioned that in HPLC of low molar mass compounds, the void or hold-up volume is usually defined differently as the sum of the pore and interstitial volume, $V_M = V_i + V_P$ [6, 7].

Due to their large size, the macromolecules cannot penetrate the complete pore volume. Entering the pore from the free mobile phase causes a loss of entropy. Certain conformations of the polymer molecule simply do not fit into the pore. In addition, for a given fixed polymer conformation, the center of gravity cannot access certain regions of the pore volume, due to steric exclusion of parts of the molecule with the pore wall [8]. If no enthalpic interaction exists between the polymer molecule and the pore wall, the above steric exclusion will result in a distribution coefficient $K < 1$ and therefore in a retention volume $V_R < V_i + V_P$. This mode of chromatography, where the distribution coefficient is determined purely by the steric exclusion of the polymer molecule from the pore, is called size-exclusion chromatography (SEC) or gel permeation chromatography (GPC). The size of a linear polymer molecule increases with its molar mass. Hence larger polymer molecules will be more strongly excluded from the pores than smaller molecules, resulting in an increase in the distribution coefficient and therefore in an increase in retention volume with decrease in molar mass (Fig. 13.1).

Decreasing eluent strength results in an attractive interaction between the pore wall and the repeating units of the polymer molecule. If this interaction is

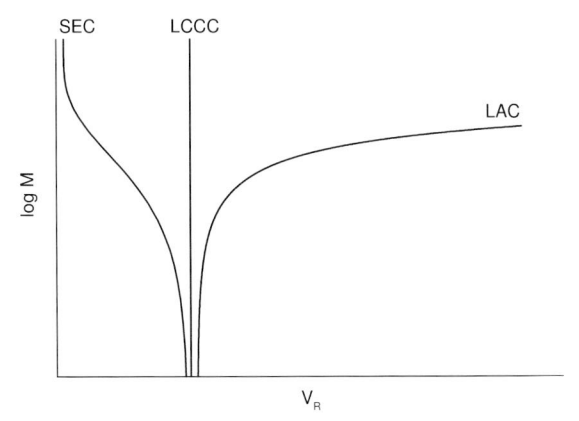

Fig. 13.1 Schematic representation of the dependences of the distribution coefficient, K, on molar mass for size exclusion, adsorption and chromatography under critical conditions of adsorption.

sufficiently strong, retention of the macromolecule will occur and the molecule will elute later than the solvent band ($K > 1$). Since the number of repeating units increases with the molar mass of the polymer molecule, the total interaction energy increases in the same direction. Consequently, the macromolecule will be adsorbed the stronger the higher is its molar mass. Therefore, the retention volume will increase with the molar mass of the macromolecule. This molar mass dependence of retention volume characterizes the adsorption mode of chromatography [liquid adsorption chromatography (LAC)].

Finally, it is possible to adjust exactly the eluent strength such that adsorption and size-exclusion effects compensate each other. Under these conditions, homopolymers of a given type elute irrespective of their molar mass at the same retention volume, $V_R = V_i + V_P$. This elution behavior is termed liquid chromatography under critical conditions of adsorption (LCCC). The different molar mass dependences in polymer chromatography are depicted schematically in Fig. 13.1.

It should be noted that the picture given above is an idealized one. If, for example, weak adsorptive interaction exists in addition to steric exclusion, this might still result in a decreasing retention volume with increasing molar mass. We shall term such a situation SEC-like elution in order to distinguish it from true, purely size-based separation in SEC.

The loss of entropy upon entering a pore is much higher for a polymer than for a small molecule. Therefore, the entropy changes due to the confinement of the pore are of minor importance in the chromatography of small molecules. As a result, SEC and LCCC have no counterparts in the chromatography of small molecules. In the chromatography of small molecules, the term high-performance liquid chromatography (HPLC) is therefore used interchangeably with adsorption chromatography, whereas for macromolecules HPLC is used in a broader sense, covering all the modes of chromatography, namely SEC, LAC and LCCC.

The three modes of polymer chromatography and their combinations provide the tools for successful polymer separations according to certain molecular parameters. A basic understanding of the different chromatographic modes is necessary to choose the right chromatographic approach for a given separation problem. The following sections will provide a more detailed insight into the different chromatographic modes and will give typical examples for separations in these.

13.2.1
Size-Exclusion Chromatography (SEC, GPC)

13.2.1.1 Basic Size-Exclusion Chromatography
Size-Exclusion chromatography (SEC), also referred to as gel permeation chromatography (GPC), is certainly the chromatographic mode most often used by polymer scientists. SEC is used in research and in quality control laboratories. It is the simplicity of SEC and the detailed information that it provides that

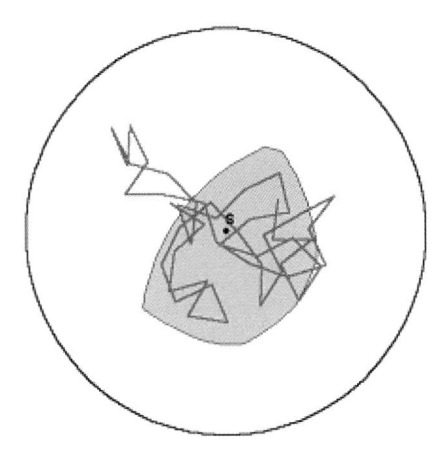

Fig. 13.2 Schematic representation of the accessible pore volume (shaded area) for the center of gravity (S) of a rigid polymer conformation in a spherical pore. Reprinted from [8], copyright 2001, with permission from Wiley-VCH.

make SEC a work horse for polymer characterization. SEC provides not only molar mass averages, like a variety of other methods, but also the complete molar mass distribution in addition. Therefore, small differences between samples can be easily detected. The separation principle of SEC is simple. However, the very simple picture describing SEC often results in its uncritical use and erroneous interpretation of the results.

In SEC, the chromatographic separation takes place due to the different sizes of macromolecules in solution. Larger molecules have less access to the pore volume than smaller ones. This is illustrated in Fig. 13.2, where the accessible pore volume of a rigid polymer conformation is shown schematically. The center of gravity of the molecules is placed at the center of the spherical pore. The molecule can be shifted in different directions until a polymer segment intersects with the pore wall. The calculation of the maximum displacement length in different directions results in the volume fraction accessible to the center of gravity of the polymer molecule (shaded area in Fig. 13.2). This volume fraction is identical with the SEC distribution coefficient of the macromolecule in the pore.

The accessible volume decreases with increasing size, resulting in a decreasing retention volume with increasing molar mass. In order to assign an retention volume to a particular molar mass, a calibration curve has to be established. This is usually done by running several samples having a narrow molar mass distribution and known molar mass. Assigning the volume at the peak maximum of the chromatogram to the molar mass yields the desired calibration curve. Figure 13.3 shows a schematic SEC calibration curve. At low retention volume, a steep increase in molar mass is observed. The size of the macromolecule is too large to fit into the pore. As a result, all molecules above a certain

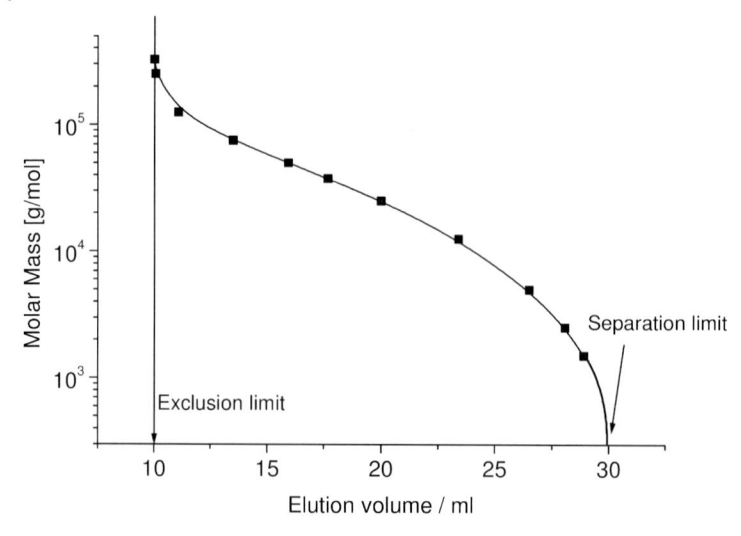

Fig. 13.3 Typical SEC calibration curve.

size elute at the same retention volume irrespective of their molar mass. The corresponding volume is termed the exclusion limit of the column or column set. If the size of the molecule becomes comparable to the pore size, a nearly linear dependence of the logarithm of molar mass on retention volume is observed. This more or less linear region of the calibration curve provides the best molar mass selectivity. At high retention volumes, the calibration curve becomes steeper again, indicating a less effective separation. In this region of the calibration curve the low molar mass molecules are much smaller than the pores and can penetrate the complete pore volume without any restriction. The separation limit of the column is observed.

From the above, it follows that each pore size provides only a separation within a certain range of molar sizes. Therefore, several columns of different pore size have to be combined to analyze samples having a broad molar mass distribution, resulting in a larger separation range at the cost of longer retention times. As an alternative, a single column containing particles with a distribution of pore sizes (sometimes termed mixed-bed or linear columns) can be used, resulting in shorter analysis times, at the expense of a decrease in resolution.

There is no theoretical equation relating the molecular size to the retention volume for a given experimental setup. Therefore, the column has to be calibrated. Since the size of the molecule differs for different chemical structures of the same molar mass, the calibration has to be performed using standards having the same chemical structure as the sample to be analyzed. SEC calibration standards are available for a variety of polymers. However, often for a particular material standards are not available. In such cases, calibration is most commonly performed using polystyrene standards. The resulting molar masses are therefore reported as polystyrene equivalents. For quality control or simple data comparison, reporting of polystyrene equivalents might be sufficient. However, if absolute molar masses are required, calibration has to be performed using polymer standards of the same chemical nature as the sample to be analyzed. To avoid confusion when comparing results from different laboratories, it is essential to provide the information on the type of the calibration and on the eluent used.

After calibrating the column set, the molar mass distribution is obtained from the signal of a concentration detector, typically a differential refractometer [refractive index (RI) detector]. In order to obtain the correct distribution function, care has to be taken to ensure that the detector response is independent of molar mass or chemical composition. The conversion of the elugram, $S(V)$, into the molar mass distribution, $w(M)$, is performed using

$$w(M) = \frac{S(V)}{\sigma M} \qquad (2)$$

where M is the molar mass at the retention volume under investigation and $\sigma = \mathrm{d}\log M/\mathrm{d}V$ is the slope of the calibration curve. Equation (2) accounts for the fact that volume slices of identical width at the higher and lower mass ends of the chromatogram correspond to different molar mass intervals. Hence the corresponding mass has to be distributed over a larger number of molecular species.

The average molar masses can be easily obtained from the molar mass distribution (Eq. 2) via

$$M_n = \frac{\int w(M)\mathrm{d}M}{\int w(M)/M \mathrm{d}M} \qquad (3)$$

$$M_w = \frac{\int w(M)M\mathrm{d}M}{\int w(M)\mathrm{d}M} \qquad (4)$$

$$M_z = \frac{\int w(M)M^2\mathrm{d}M}{\int w(M)M\mathrm{d}M} \qquad (5)$$

$$M_\eta = \left[\frac{\int w(M)M^a \mathrm{d}M}{\int w(M)\mathrm{d}M}\right]^{\frac{1}{a}} \qquad (6)$$

where M_n, M_w, M_z and M_η are the number-, weight-, z- and viscosity-average molar mass, respectively. The calculation of the viscosity-average molar mass requires a knowledge of the Mark–Houwink exponent, a, of the polymer in the respective solvent at the given temperature.

There are three crucial requirements in SEC which determine the correct interpretation of SEC results.

1. Most important: SEC is a relative method. Correct molar masses will be obtained only if the calibration is performed using standards of the same chemical identity and topology as the sample under investigation.
2. SEC works correctly only if no adsorptive interaction between the stationary phase and the analyte is observed.
3. Correct transformation of the signal, S, into the molar mass distribution requires a signal that is proportional to the mass concentration and does not show a dependence on molar mass or chemical composition.

13.2.1.2 Calibration in SEC

As indicated above, SEC is a relative method and the molar masses obtained depend crucially on the calibration curve used. The simplest and therefore usual way of calibration is to use polymer standards of known molar mass, having a narrow molar mass distribution. However, the synthesis of polymers with a narrow molar mass distribution is restricted to the relatively small number of monomers that can be polymerized in a living or at least controlled manner. This results in the fact that molar masses are often reported as polystyrene equivalents. However, there are methods to overcome the calibration problem for those polymers where no narrow-distribution standards are commercially available. Some of these methods can be easily implemented in any laboratory.

Universal calibration Since the introduction of SEC as a method of molar mass characterization, there have been numerous attempts to relate the distribution coefficient and therefore the retention volume to a parameter which is independent of the chemical and topological structure of the polymer. Benoît and co-workers assumed that the parameter determining the elution behavior of a polymer molecule in SEC is the hydrodynamic volume and not the molar mass of the polymer [9, 10]. They suggested that the hydrodynamic volume, V_h, of a polymer is proportional to the product of the intrinsic viscosity, $[\eta]$, and the molar mass, $V_h \propto [\eta]M$. By plotting $[\eta]M$ as a function of retention volume for a large variety of chemically and topologically different polymers, they showed that indeed all polymers investigated follow a common calibration curve. Today, the concept of universal calibration is widely accepted.

Applying the concept of universal calibration, it is easy to convert the calibration curve of one polymer into that of another. If both polymers obey the universal calibration principle, the following equation is valid at any retention volume:

$$[\eta]_1 M_1 = [\eta]_2 M_2 \tag{7}$$

where 1 and 2 denote the two different molecular species. The well-known Mark–Houwink equation relates the intrinsic viscosity to the molar mass of a polymer in a particular solvent:

$$[\eta] = KM^a \tag{8}$$

Introducing Eq. (8) into Eq. (7) and solving for M_2, we obtain

$$M_2 = \left(\frac{K_1 M_1^{a_1+1}}{K_2}\right)^{\frac{1}{a_2+1}} \qquad (9)$$

Equation (9) allows the calculation of the calibration curve for the polymer under investigation (2) based on the calibration curve of a different polymer (1), provided that the Mark–Houwink parameters are given for the two polymers. The Mark–Houwink coefficients of each polymer have to be taken for the particular eluent at the temperature of the experiment. It should be noted that the use of Eq. (9) implies constant Mark–Houwink parameters. This is usually the case if the molar mass of the polymer is sufficiently high. For typical vinyl polymers a good choice is to have a degree of polymerization of at least 50–100 repeating units.

SEC calibration using samples of broad molar mass distribution   Sometimes only samples having a broad molar mass distribution are available. These samples might be well characterized in terms of their average molar masses, e.g. by light scattering, osmometry or other methods. Due to the broad molar mass distribution, a broad chromatogram will result from an SEC experiment. A correct assignment of the peak maximum to a molar mass is difficult in such cases, since the peak maximum does not represent any molar mass average. However, samples having a broad molar mass distribution can be used for calibration purposes.

Two different methods for calibration using samples of broad molar mass distribution are usually applied. The first approach assumes a linear dependence of $\log M$ on retention volume:

$$\log M = A - \sigma V_R \Leftrightarrow M = 10^{A-\sigma V_R} \qquad (10)$$

Introducing Eqs. (10) and (2) into the definitions of the average molar masses, M_n or M_w (Eqs. 3 and 4), results in

$$M_n = \frac{\int S \, dV_R}{\int S/10^{A-\sigma V_R} \, dV_R} \qquad (11)$$

$$M_w = \frac{\int S \times 10^{A-\sigma V_R} \, dV_R}{\int S \, dV_R} \qquad (12)$$

$$D = \frac{M_w}{M_n} = \frac{\int S \times 10^{-\sigma V_R} \, dV_R \int S \times 10^{\sigma V_R} \, dV_R}{\left(\int S \, dV_R\right)^2} \qquad (13)$$

Equations (11) and (12) calculate the average molar masses of the broadly distributed polymer sample from the chromatogram as a function of the parameters A and σ. It can be seen from Eq. (13) that the polydispersity of the sample is influ-

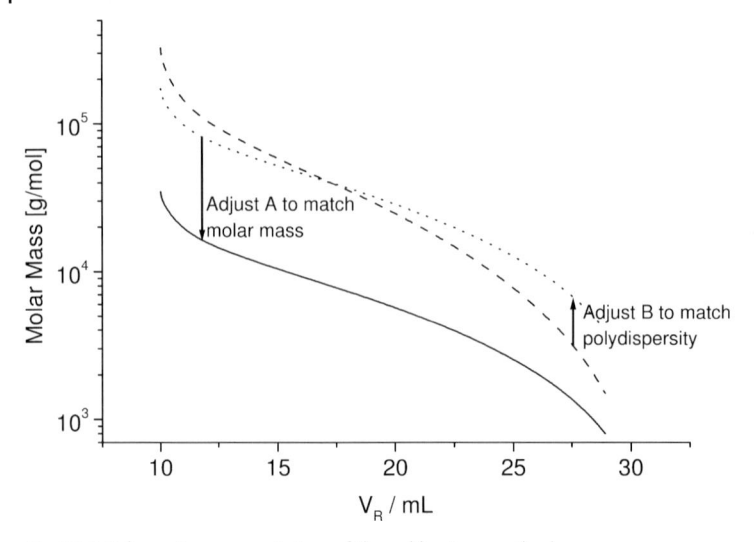

Fig. 13.4 Schematic representation of the calibration method according to Eqs. (16) and (17). The approach according to Eqs. (11) and (12) is obtained from the figure by selecting a linear calibration curve.

enced only by the slope of the calibration curve (σ). Thus, for a broad sample the slope of the calibration curve is first selected to fit the polydispersity of the sample (Fig. 13.4). In a second step, the offset of the calibration curve is adjusted by selecting a suitable value for A, which fits the molar masses at the given value of σ. Using the parameters obtained from the fitting procedure the calibration curve can be calculated from Eq. (10). If several samples are available, which are characterized by only one average molar mass, a similar approach is possible. However, in such cases non-linear fitting procedures have to be applied.

The basic disadvantage of the above calibration method is the assumption of a linear calibration curve. Since the true shape of the calibration curve might be non-linear, a different approach to calibration based on samples of broad molar mass distribution uses a true calibration curve of an arbitrary calibrant:

$$\log M_1 = f(V_R) \Leftrightarrow M_1 = 10^{f(V_R)} \tag{14}$$

Combining Eq. (14) and the universal calibration approach (Eq. 9) results in

$$M_2 = \left[\frac{K_1 \times 10^{(a_1+1)f(V_R)}}{K_2}\right]^{\frac{1}{a_2+1}} = A \times 10^{Bf(V_R)} \tag{15}$$

with

$$A = \left(\frac{K_1}{K_2}\right)^{\frac{1}{a_2+1}}; \quad B = \frac{a_1+1}{a_2+1}$$

The parameters A and B are unknown. However, it is possible to calculate the molar mass averages from the chromatogram [$S(V_R)$], the calibration curve of the arbitrary calibrant $f(V_R)$ and an arbitrary set of parameters A and B:

$$M_n = \frac{\int S \, dV_R}{\int S \, dV_R / A \times 10^{Bf(V_R)}} \tag{16}$$

$$M_w = \frac{\int SA \times 10^{Bf(V_R)} dV_R}{\int S \, dV_R} \tag{17}$$

$$D = \int S \times 10^{Bf(V_R)} dV_R \times \int S \times 10^{-Bf(V_R)} dV_R \tag{18}$$

The unknown parameters A and B have to be optimized in order to match the given average molar masses. Similarly to the approach described above, the polydispersity is dependent on only one parameter (B) and for a single sample with known M_n and M_w the same procedure as above can be applied. However, if more than a single sample is used, e.g. several samples with M_w, the parameters A and B have to be determined using non-linear fitting procedures. In these procedures the parameters A and B will be varied until the best fit between the calculated and the experimentally determined average molar masses is obtained. The knowledge of the parameters A and B allows the calculation of the calibration curve using Eq. (15). Adjusting the parameters A and B in Eq. (15) is identical with changing the slope and height of the calibration curve, similar to the calibration method described above. However, the present method preserves the non-linear shape, the exclusion volume and separation limit of the base calibration curve [11–13].

Integral calibration A further calibration method applicable without additional financial effort is integral calibration. This calibration method utilizes at least a single sample of broad molar mass distribution but with known integral or cumulative molar mass distribution, $I(M) = \int_0^M w(M') dM'$. The cumulative or integral molar mass distribution at a given molar mass, M, is identical with the weight fraction of material having molar masses smaller than M. A single SEC run of the sample on the column to be calibrated has to be performed to obtain the chromatogram $S(V_R)$. The value of the integral mass distribution $I(M)$, i.e. the mass fraction having lower molar masses than the selected molar mass, has to be identical with the area fraction of the normalized signal for retention volumes larger than $V_R(M)$:

$$I(M) = \int_0^M w(M') dM' = \frac{\int_{V(M)_e}^{\infty} S(V_R) dV_R}{\int_0^{\infty} S(V_R) dV_R} = I(V_R) \tag{19}$$

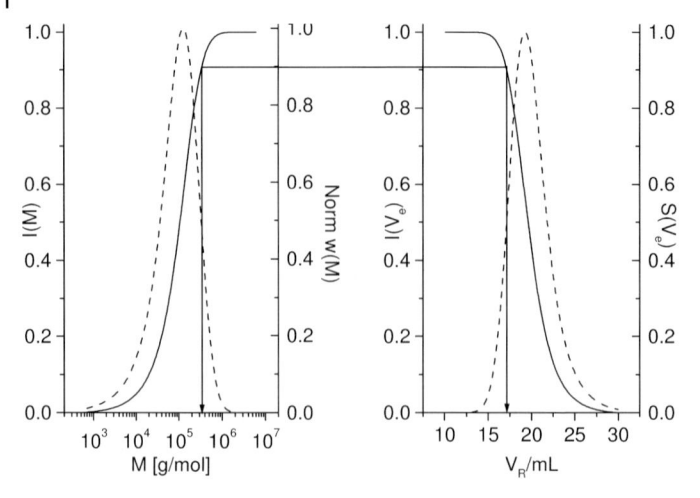

Fig. 13.5 Schematic representation of integral calibration method according to Eq. (19).

According to Eq. (19), each molar mass value of the cumulative molar mass distribution is assigned to a unique retention volume of the SEC trace, resulting in the desired calibration curve. A schematic representation of this procedure is given in Fig. 13.5.

As has been shown above, SEC calibration can be performed using samples having a broad molar mass distribution, despite the usual assumption that samples of narrow molar mass distribution are required. Nowadays modern SEC software packages support the use of the calibration methods described above, improving the determination of true molar masses by SEC. Despite this fact, in some cases, e.g. for branched samples, no calibration standards, either broad or narrow, exist. In such cases, the use of molar mass-sensitive detectors might be the method of choice.

13.2.1.3 SEC with Molar Mass-sensitive Detection

SEC/light scattering The drawbacks of SEC as a relative method, for which sometimes useful polymer standards are not easily available, resulted in the introduction of new chromatographic detectors. Since static light scattering is a method for the determination of absolute molar masses, there has been considerable effort to develop light scattering detectors for on-line use in SEC. The theoretical basis for the evaluation of static light scattering experiments is given by the following equation [14, 15]:

$$R(\theta) = \frac{I(\theta) r^2}{I_0} = K c M P(\theta)(1 - A_2 c M + \ldots) \tag{20}$$

where

$$K = \frac{4\pi^2 n_0^2}{\lambda_0^4 N_A} \left(\frac{dn}{dc}\right)^2$$

$R(\theta)$ is the excess scattering intensity of the polymer sample of concentration c and molar mass M at a scattering angle θ and A_2 is the second virial coefficient, which describes the thermodynamic solvent quality for the polymer. K is an optical constant which depends on the solvent refractive index, n_0, the wavelength of the incident light in vacuum, λ_0, the refractive index increment dn/dc of the polymer and Avogadro's number, N_A. The refractive index increment dn/dc describes the contrast between the solvent and the scattering molecule. For molecules of a size comparable to the wavelength of the scattered light, destructive interferences of the scattered radiation from different segments within the same molecule result in a decreasing scattering intensity with increasing angle of observation. The particle scattering factor $P(\theta)$ accounts for that. Irrespective of the shape of the scattering molecule, the particle scattering factor can be approximated for small scattering angles by

$$\lim_{\theta \to 0} P(\theta) = 1 - \frac{\mu^2}{3} \langle R_g^2 \rangle \tag{21}$$

$$\mu = \frac{4\pi}{\lambda} \sin\left(\frac{\theta}{2}\right) \tag{22}$$

with $\langle R_g^2 \rangle$ being the mean square radius of gyration or more precisely the square distance of the segments from the center of gravity [16]. In order to determine the molar masses by static light scattering experiments, the scattering intensity at different angles has to be measured for different concentrations. According to Eqs. (20–22), the molar mass of the polymer molecule is obtained by performing a double extrapolation to zero angle and zero concentration. The mean square radius is obtained from the slope of the angular dependence at low angles, while the second virial coefficient is obtained from the slope of the concentration dependence. This double extrapolation is often performed using a Zimm plot (Fig. 13.6). In a Zimm plot, the angular dependences of the scattering intensities for different concentrations are shifted linearly with concentration. By this procedure a grid-like plot is obtained (Fig. 13.6).

The above equations assume monodisperse samples. For polydisperse samples, evaluation of light scattering data using these equations results in the weight-average molar mass, M_w, and the z-average of the mean square radius of gyration, $\langle R_g^2 \rangle_z = \sum \langle R_g^2 \rangle_i w_i M_i / \sum w_i M_i$.

Until now we have summarized the basic equations for the evaluation of static light scattering experiments up to a level allowing us to follow the principles of SEC light scattering. A good overview on SEC/light scattering can be found in [16]. An on-line light scattering setup consists of a conventional SEC setup of

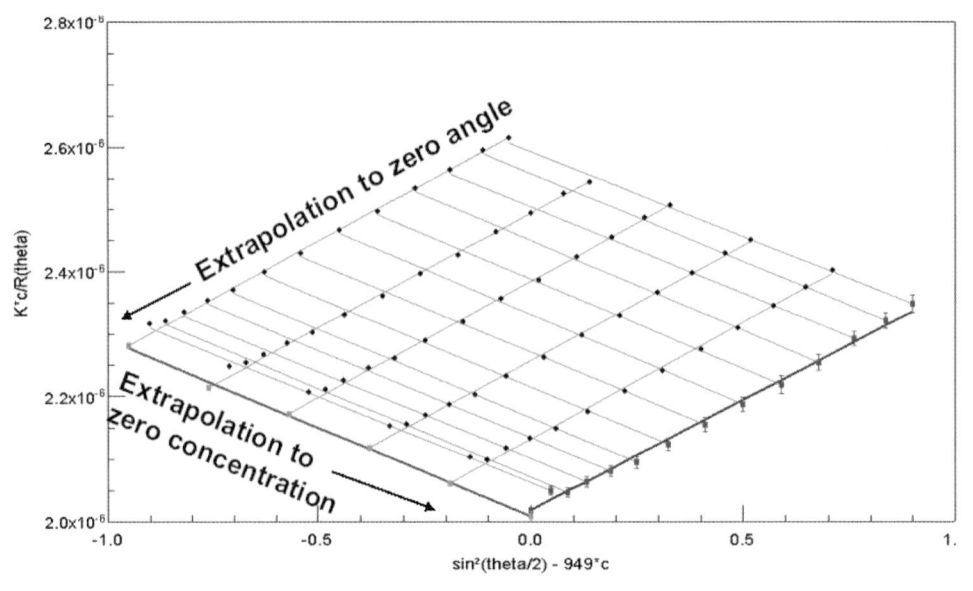

Fig. 13.6 Example of a Zimm plot to evaluate light scattering data. Sample: dextran 5×10^5 g mol^{-1}. Concentrations: $c=2\times10^{-4}$, 4×10^{-4}, 6×10^{-4}, 8×10^{-4}, 1×10^{-3} g mL^{-1}. Courtesy of Wyatt Technology Europe GmbH.

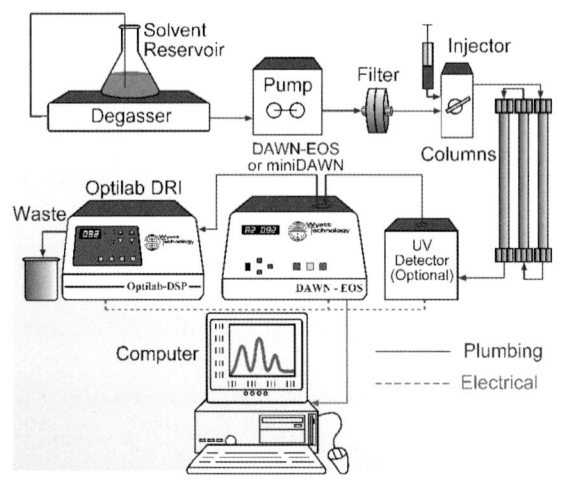

Fig. 13.7 Schematic setup of an SEC light scattering system. Courtesy of Wyatt Technology Europe GmbH.

pump, injector and columns to which a light scattering detector and a concentration detector (usually an RI detector) are attached (Fig. 13.7). The light scattering instrument contains a flow cell through which a laser beam is focused. The eluting molecules scatter the light of the incident beam in different directions. The light scattering intensities at one or more angles of observation are

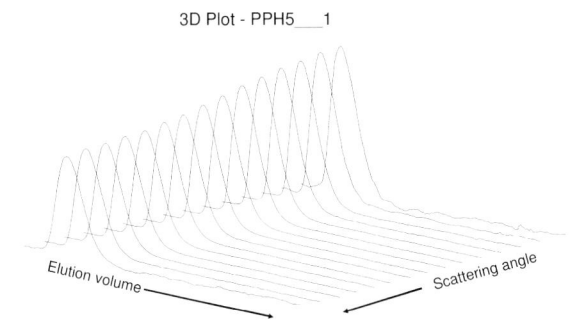

Fig. 13.8 MALS detector signals at different angles.

monitored as a function of retention volume by a number of photodiodes around the flow cell (Fig. 13.8). Thus, for each retention volume the angular dependence of the scattered radiation is obtained. The determination of the polymer concentration required to evaluate the light scattering data is performed by using the signal of the RI detector. The concentrations are obtained from the signal of a differential refractometer, RI_i, via

$$c_i = \frac{RI_i}{k_{inst} dn/dc} \qquad (23)$$

where k_{inst} is an instrument constant.

In contrast to an off-line light scattering experiment, only a single concentration is available for each fraction. However, the actual concentrations at the detectors are orders of magnitude smaller as compared with conventional light scattering experiments. Thus, term $A_2 cM$ in Eq. (20) is negligible compared with unity. For a typical value of A_2 of the order of 10^{-4} mol mL g^{-2}, a molar mass of 10^6 g mol^{-1} and a rather high concentration at the detector of 10^{-1} g L^{-1}, the term $A_2 cM$ is calculated as 10^{-2}, which is much lower than unity. Therefore, ignoring this term is suitable for most conditions encountered experimentally. As a consequence, the extrapolation to zero concentration, i.e. the second term in Eq. (20), can be neglected. Hence Eq. (20) can be approximated by

$$R(\theta) = KcMP(\theta) \qquad (24)$$

where

$$M = \frac{R(\theta)}{KcP(\theta)}$$

After determination of the scattering intensity at different angles [multi-angle light scattering (MALS)], it is possible to extrapolate the angular variation of the scattering intensity to zero angle. The slope at $\theta=0$ allows the calculation of the mean square radius for each SEC retention volume (Fig. 13.9). The value

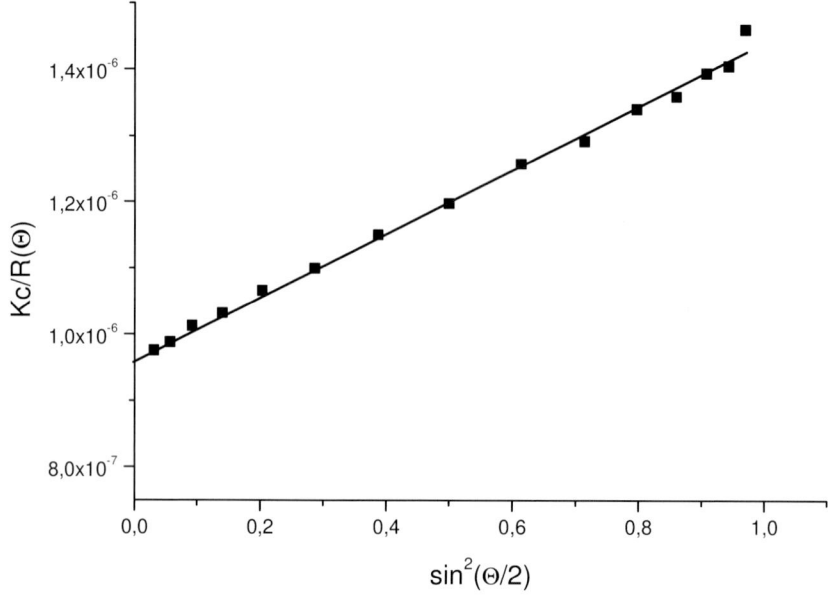

Fig. 13.9 Zimm graph for a single slice of the elugram of poly(p-methylstyrene) having a concentration of $c = 6.3 \times 10^{-5}$ g mL^{-1}. The intercept corresponds to a molar mass of $M = 1.04 \times 10^6$ g mol^{-1} and the slope results in $R_g = 43$ nm.

$R(\theta=0)/Kc$ yields the molar mass of the polymer at the given retention volume from Eq. (24). This procedure is identical with using only one extremely low concentration of the Zimm plot (Fig. 13.6). Having determined the particle scattering factor from the angular variation, the molar mass for each slice can be calculated. Comparison of the concentrations in Figs. 13.6 and 13.9 clearly shows that neglecting extrapolation to zero concentrations usually does not produce a significant error in the molar masses. Some SEC light scattering detectors provide only a single detector at a very low angle. This angle is sufficiently low to allow the approximation of $P(\theta)$ in Eq. (22) by unity [low-angle light scattering (LALS)]. Since only one angle is used, no information on the angular dependence of the light scattering intensity is obtained. LALS therefore cannot yield size information of the eluting molecules.

The extrapolation of the angular dependence of the scattering intensity can be performed for every SEC retention volume, resulting in the molar mass and radius of gyration for each retention volume, allowing the construction of the calibration curve from a single sample (Fig. 13.10). Using this self-calibrating system, the different molar mass averages can be calculated from the molar masses of the different SEC slices and the corresponding concentration of the RI signal. Figure 13.10 shows the calibration curve for poly(p-methylstyrene) obtained by SEC/light scattering. It can be observed that in the central region a

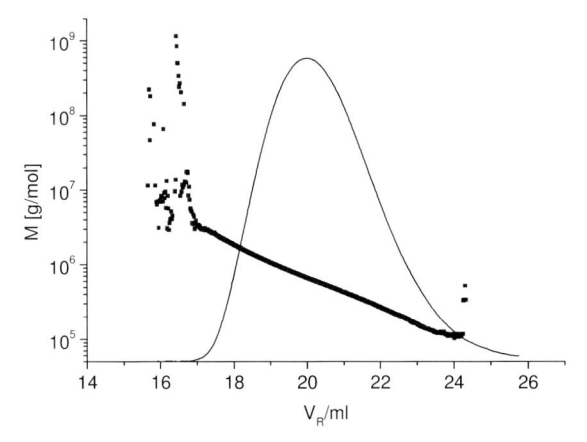

Fig. 13.10 Dependence of molar mass on retention volume determined by SEC/light scattering for a poly(p-methylstyrene) having a broad molar mass distribution. The solid line represents the RI chromatogram.

smooth calibration curve is obtained. At the edges of the chromatogram the calibration curve shows substantial scattering. This is due to the different responses of the RI and light scattering detectors to molar mass. In the low molar mass end the light scattering signal loses intensity much faster than the RI signal, whereas at the high molar mass end the opposite is true. Whereas the weight-average molar mass is not be influenced by these effects, the number-average molar mass determined by SEC/light scattering typically is higher than that obtained from standard SEC experiments, i.e. the polydispersity will be too low. This problem is usually overcome by fitting the accurate region of the calibration curve obtained by SEC/light scattering with an appropriate function and extrapolation across the complete chromatogram.

If a MALS system is used, valuable additional information can be extracted from the SEC/light scattering experiment. Since MALS allows the determination of the mean square radius in addition to the molar mass for each retention volume, it is possible to obtain information on the shape of the eluting molecules. The structure of the polymer molecules influences the scaling exponent, a_S, of the mean square radius with respect to molar mass, $\langle R_g^2 \rangle^{\frac{1}{2}} \propto M^{a_s}$. The exponent a_s can be easily obtained from the slope of a plot of log R_g versus log M for the different SEC slices of a single sample. Typical exponents a_s are given in Table 13.1.

In addition, the combination of size information (mean square radius) and molar mass allows information to be obtained on branching in polymers. It is well known that branches in polymers result in a decreasing coil size relative to the corresponding linear polymer having the same molar mass. This reduction of the size of the molecule is described by the branching ratio $g = \langle R_{g,\mathrm{br}}^2 \rangle / \langle R_{g,\mathrm{l}}^2 \rangle$ introduced by Zimm and Stockmayer in 1949 [17]. The g-ra-

Table 13.1 Exponents of the conformation plots $\langle R_g^2 \rangle^{\frac{1}{2}} \propto M^{a_s}$ and the Mark–Houwink exponents, a.

Structure	a_s	a
Random coil (coil under θ-conditions)	0.5	0.5
Coil in good solvent	0.5–0.6	0.5–0.8
Solid sphere	$1/3$	0
Rigid rod	1	2

tios have been calculated for a variety of different structures. However, interpretation of the experimental results of static light scattering is complicated by the polydispersity of the samples. SEC with light scattering detection allows this problem to be overcome, since radii and molar masses are obtained on nearly monodisperse fractions. Now it becomes possible to compare the radii of the branched sample for a given molar mass with the corresponding radius of a strictly linear sample, such that branching information can be extracted.

Having explained the basic principals of SEC with light scattering detection, the application to poly(p-methylstyrene) can serve for illustration. Figure 13.8 shows the angular dependence of the scattered light at different scattering angles for the SEC/light scattering experiment on a broad molar mass distribution poly(p-methylstyrene). It can be observed that the scattering intensity increases at low scattering angles (lower detector numbers) as a result of the large size of the macromolecules.

Figure 13.11 shows the comparison of the normalized 90° light scattering signal and the RI signal for the same sample. It can clearly be seen that the peak maximum of the light scattering trace is shifted towards lower retention volume

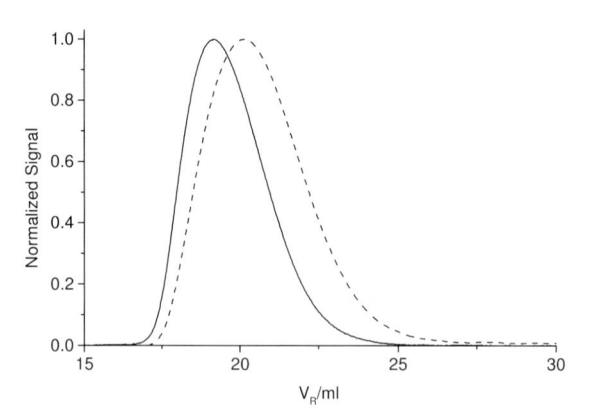

Fig. 13.11 Normalized 90° light scattering (solid line) and RI trace (broken line) for a linear poly(p-methylstyrene) having a broad molar mass distribution.

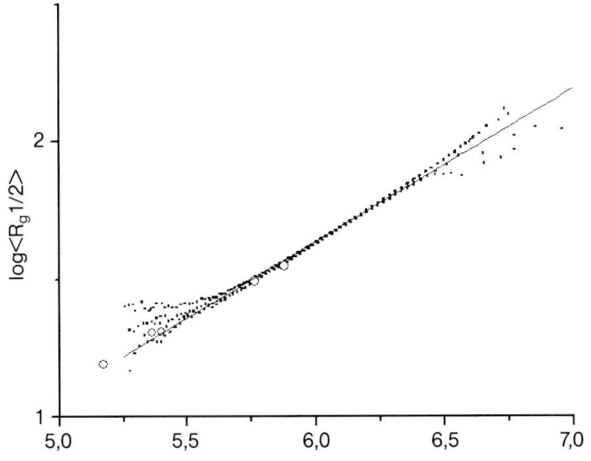

Fig. 13.12 Dependence of root mean square radius of gyration on molar mass for poly(p-methylstyrene)s in THF having narrow (○) and broad (■) molar mass distribution. The solid line corresponds to the relationship $R_g = 18.4 \times 10^{-3}$ nm $M^{0.57}$. Reprinted with permission from [18]. Copyright (2005) American Chemical Society.

compared with the RI signal. This is due to the fact that the light scattering signal is proportional to the product Mc, whereas the RI trace is proportional to c only. At lower retention volume the increase in the molar mass therefore enhances the light scattering signal stronger than at the lower molar mass end of the chromatogram. This results in the observed shift of the peak maximum. Figure 13.9 shows the angular dependence of the light scattering intensity for a single SEC slice. The concentration is assumed to be close to zero. Therefore, the intercept with the y-axis yields the molar mass, whereas the slope allows the calculation of the mean square radius (Eqs. 20–22).

Having determined the molar mass and root mean square radius for each retention volume, the dependence of root mean square radius on molar mass can be determined. This is shown in Fig. 13.12, where the small symbols represent data from the individual slices of broad molar mass samples, whereas the larger symbols correspond to data obtained on samples having a narrow molar mass distribution. It can clearly be seen that the results obtained from samples having a broad and narrow molar mass distribution closely coincide. The double logarithmic plot allows the determination of the exponent a_s of the scaling behavior of the poly(p-methylstyrene) (Fig. 13.12). According to Table 13.1, the determined value of $a_s = 0.57$ indicates a coil structure in a good solvent.

SEC with light scattering detection certainly has distinct advantages over conventional SEC. However, some drawbacks and peculiarities exist, which should at least be mentioned at this point.

Separation mechanism: Since the use of light scattering for the determination of molar masses does not require calibration of the SEC column, the molar masses obtained by SEC with light scattering should be correct, even if attractive interactions between the sample and the column material occur. This is true as long as the sample is completely eluted from the column and a separation according to molar mass can be achieved.

Copolymers: Light scattering of copolymers is complicated by the fact that the molar mass obtained for chemically heterogeneous samples is an apparent molar mass only. Due to the different chemical compositions of the molecules, each chain has its own refractive index increment, dn/dc, which must be used for correct data evaluation. However, only an average refractive index increment can be determined from bulk dn/dc measurements. Therefore, without going into the details, the true molar mass for a chemically heterogeneous polymer can only be obtained from light scattering experiments after determination of the apparent molar masses in at least three solvents of different refractive index. More information can be found in textbooks on light scattering [14, 15]. Although this problem is often ignored in practice, chemically heterogeneous polymers are encountered fairly often. They arise from high conversion statistical copolymers, if the difference in the reactivity ratios results in a change in the monomer composition with conversion, which in turn produces changes in the composition of the instantaneously produced copolymer chains. Grafting reactions usually yield mixtures of homo- and graft copolymers. Block copolymers might contain homopolymer impurities due to unavoidable side-reactions.

Number-average molar mass: As a peculiarity SEC/light scattering allows the true number-average molar mass, M_n, of a copolymer to be obtained, provided that the individual slices can be assumed to be homogeneous with respect to chemical composition and molar mass [19]. Usually extrapolation of the calibration curve is necessary, due to the loss in signal intensity at the ends of the chromatograms. The determination of the true M_n of a copolymer by light scattering seems to be odd, since light scattering usually yields the weight-average molar mass, M_w. The described peculiarity results from a fortunate cancellation of the errors made in the determination in the molar masses and concentrations for the individual SEC slices.

Mixed solvents and salt-containing solvents: Light scattering in mixed or salt-containing solvents is complicated by preferential adsorption. One component of the mixed solvent (an added salt is simply regarded as a second solvent component) is preferentially adsorbed on the polymer chains, resulting in a different solvent composition in the vicinity of the polymer molecule as compared with the free solvent region. The bulk dn/dc determination results in a dn/dc value that is not appropriate to use, since it does not adequately describe the scattering contrast between the scattering polymer segment and the surrounding solvent. Without going into details, the correct dn/dc-determination in mixed solvents usually requires tedious and time-consuming dialysis of the sample. More detailed information can be found in textbooks on light scattering [14, 15]. Since dn/dc enters into the light scattering equation to a power of two, small errors in the dn/dc determination result in pronounced errors in the absolute molar mass. A peculiarity of SEC with light scattering is that the SEC separation has the same effect as a dialysis. As a result, the refractive index increment that is be determined using the RI detector is the value which is needed for the correct treatment of the light scattering data in the mixed solvent [20, 21].

SEC/viscometric detection As has already been mentioned, SEC separates according to hydrodynamic volume, allowing the construction of a universal calibration curve by plotting [η]M as a function of retention volume [9, 10]. Universal calibration is valid as long as a pure SEC mechanism without enthalpic interaction is operating. Once the universal calibration curve has been constructed, it is possible to calculate the true molar mass if the intrinsic viscosity of the eluting polymer in the given SEC slice is known.

Therefore, different attempts have been made to design SEC detectors allowing the determination of the intrinsic viscosity of the eluting highly diluted polymer solution. Simple capillary viscosimeters do not provide sufficient sensitivity for this task, due to the extreme dilute solutions in chromatography. The breakthrough was the introduction of differential viscosimeters, responding to the difference in the viscosities of the eluting polymer solution and the pure solvent. Today's most commonly applied setup uses four capillaries of identical flow resistance and two pressure transducers in a design that looks like a Wheatstone bridge (Fig. 13.13) [22–24]. The flow of the polymer solution coming from the SEC is split into two branches of identical flow resistance. In one branch of the bridge a hold-up reservoir is incorporated. When the polymer solution enters this reservoir, the polymer is retained, allowing pure solvent to enter the following capillary. At the same time, the polymer solution passes through the second branch of the bridge. The differences in the viscosities result in a pressure difference between the two branches, which is detected by the differential pressure transducer, DP. The total pressure drop of the branches is

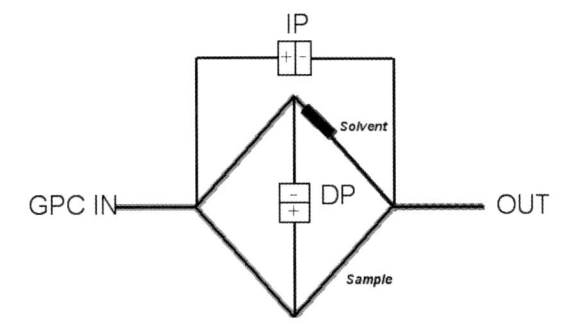

Fig. 13.13 Schematic representation of the four-capillary design of an SEC viscosity detector. Courtesy of Viscotek Corporation.

proportional to the flow-rate which also influences DP. Therefore, the pressure drop across the bridge is monitored by the inlet pressure transducer, IP. It can be shown that the specific viscosity, η_{sp}, can be directly calculated from the two pressure readings. A benefit of the design is that pressure fluctuations arising from pump pulsations are effectively cancelled out, resulting in a remarkable sensitivity of such viscosimeters.

However, for universal calibration we need to know the intrinsic viscosity, $[\eta]$, which is defined as the limiting value of η_{sp}/c:

$$[\eta] = \lim_{c \to 0} \frac{\eta_{sp}}{c} \tag{25}$$

The concentration in each SEC slice is calculated from a concentration detector, similar to SEC/light scattering experiments Eq. (23). The concentrations at the detector are sufficiently low that the extrapolation to zero concentration in Eq. (25) can be neglected. Therefore, by taking the ratio of the specific viscosity from the viscosity detector and the concentration determined by the RI detector, the intrinsic viscosity is obtained. In order to use an on-line viscometer for the determination of molar masses, a universal calibration curve has to be constructed using a series of arbitrary calibrants. Dividing the value $[\eta]M$ of the previously established universal calibration curve by the intrinsic viscosity of the unknown sample at a particular retention volume, the true molar mass of the unknown sample at the given retention volume is obtained. By this procedure, the molar mass is calculated slice by slice, resulting in the calibration curve for the particular sample. Having determined the calibration curve, the average molar masses are calculated from the concentration signal. It should be emphasized that the samples used to establish the universal calibration curve do not need to be of a chemical structures identical with that of the sample to be analyzed. However, the standards and the sample to be analyzed have to follow the universal calibration principle, i.e. a pure size-exclusion mechanism is required.

SEC with viscosity detection allows the determination of the molar mass and intrinsic viscosity of a particular sample for a given retention volume. From these data, additional information on the polymer structure can be obtained.

The Mark–Houwink relation

$$[\eta] = KM^a \tag{26}$$

describes the scaling behavior of the polymer's intrinsic viscosity with respect to molar mass. Similarly to the scaling exponent a_s of the mean square radius, the exponent a is characteristic of the structure of a polymer in solution. Typical Mark–Houwink exponents, a, are given in Table 13.1.

Since the intrinsic viscosity and the molar mass of the polymer are known for each retention volume, a Mark–Houwink plot can easily be constructed from a single sample. As an example, the Mark–Houwink plot obtained by SEC/viscometry in THF for a poly(p-methylstyrene) sample having a broad molar mass

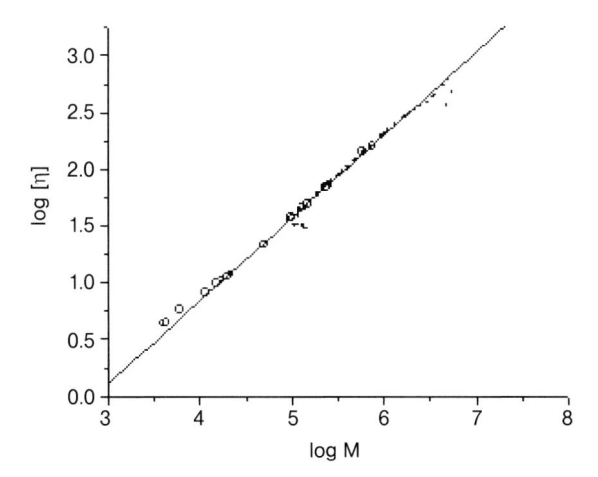

Fig. 13.14 Mark–Houwink plot of poly(p-methylstyrene) in THF obtained from SEC/viscometry on a sample having broad molar mass distribution (■) in comparison with the results obtained on narrow distributed samples (○). Reprinted with permission from [18]. Copyright (2005) American Chemical Society.

distribution is shown in Fig. 13.14 in comparison with the Mark–Houwink relation obtained on narrowly distributed poly(p-methylstyrene)s. As can be observed, both datasets coincide. The Mark–Houwink parameters were determined as $K = 8.3 \times 10^{-3}$ mL g^{-1} and $a = 0.73$. The value for a is typical for a coil in a good solvent, similar to the results obtained by SEC/light scattering (Fig. 13.12).

As has already been mentioned in the section on SEC with light scattering detection, the size of a molecule having a given molar mass is reduced by the introduction of branches. Since the intrinsic viscosity is a measure of the specific volume of the polymer molecule in solution, the intrinsic viscosity of a branched molecule is also reduced relative to that of the corresponding linear polymer having the same molar mass. The comparison of the intrinsic viscosities for a given molar mass can therefore be used to obtain branching information. Figure 13.15 shows the Mark–Houwink plots of comb-shaped poly(p-methylstyrene)s in comparison with the linear polymer. The intrinsic viscosities of the combs are significantly reduced compared with the linear polymer of the same molar mass, showing the size reduction introduced by branching. In addition, the Mark–Houwink plots have similar slopes and hence similar Mark–Houwink exponents, a. This indicates that the coil structure is preserved despite the introduction of branches.

Similarly to SEC with light scattering detection, some drawbacks and peculiarities exist also with viscometric detection.

Mechanism: The correct determination of the molar masses using SEC with viscosity detection requires a pure SEC mechanism.

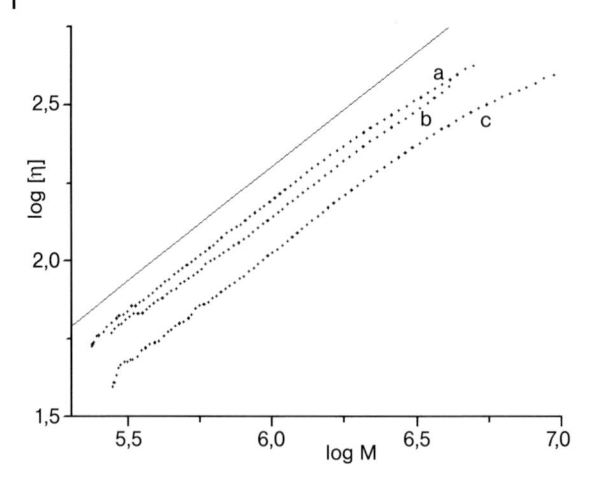

Fig. 13.15 Comparison of Mark–Houwink relationships for comb polymers of varying branching frequencies, λ, at constant side-chain length, $M_{n,sc} = 1.11 \times 10^4$ g mol^{-1}. Fraction of grafted backbone units (λ): $\lambda =$ (a) 0.38%, (b) 0.72% and (c) 1.55%. The solid curve shows the Mark–Houwink relation for linear poly(p-methylstyrene) in THF. Reprinted with permission from [18]. Copyright (2005) American Chemical Society.

Copolymers: For chemically heterogeneous copolymers, the refractive index increment might vary with retention volume, resulting in an erroneous concentration determination from Eq. (23). Consequently, the determined intrinsic viscosity and therefore the molar mass calculated from the universal calibration curve will also be erroneous. However, as has been shown by Goldwasser, the determination of the number average molar mass of the sample is not affected by these errors [25, 26]. Hence SEC with viscosity detection provides a method for the determination of the true M_n of a copolymer. Since SEC with light scattering detection also yields the number-average molar mass of copolymers, but with different requirements [19], either method can be used to validate the results of the other.

For further information on the use of SEC with molar mass-sensitive detectors, the review by Yau [27] is recommended.

SEC with triple detection The above sections have identified the strengths and weaknesses of SEC with molar mass-sensitive detection. Both light scattering and viscosity detectors have their advantages and disadvantages. While light scattering has the advantage of being independent of the separation mechanism, it has the disadvantage that size information can only be obtained for structures that are sufficiently large to be comparable with the wavelength of the incident light. This means for typical light sources that only mean square radii above 10–20 nm can be determined. These sizes correspond to molar masses of polystyrenes of the order of 100–200 kg mol^{-1}. In contrast, size information on much smaller structures can be determined using viscosity detection.

However, the determination of molar mass by viscosity detection depends crucially on the separation mechanism. Hence a combination of both light scattering and viscosity detection seems to be of advantage. However, the high cost of the instrumentation prevents the use of both detectors in many laboratories.

In recent years, the so called triple detection has become popular. Unfortunately, there is no well-defined definition of what is meant by "triple detection". This term is used in some cases to describe the determination of intrinsic viscosity using viscosity detection and molar mass determination using light scattering at a single angle. This angle can be sufficiently small that the particle scattering factor even for large molecules can be approximated by unity. Alternatively, the 90° scattering angle, which usually provides the best signal-to-noise ratio, is used for small molecules. Under these circumstances, the molecules have to be sufficiently small for the angular variation of the scattered light to be ignored. Hence molar masses are obtained from the combination of light scattering and concentration detection, whereas intrinsic viscosities are calculated from viscosity and concentration detection. In other cases, the term "triple detection" is used for a system composed of a viscosity, a single-angle light scattering and a concentration detector, the signals of which are processed using a special algorithm [28]. This algorithm, which includes some assumptions and therefore results in severe limitations, is often unknown. This is the motivation to give a short description of the triple detection principle, its assumptions and its limitations.

We will assume that the light scattering detector acquires the data at 90°, although any other scattering angle can also be used. To a first approximation, the evaluation of the data ignores the angular variation of the scattering intensity. An approximate molar mass, M_1, is determined from Eq. (24) using

$$M_1 = \frac{R(\theta)}{Kc} \tag{27}$$

The Flory–Fox equation

$$[\eta] = \frac{6^{\frac{3}{2}} \Phi \langle R_g^2 \rangle^{\frac{3}{2}}}{M} \tag{28}$$

correlates the intrinsic viscosity and molar mass with the mean square radius, via the Flory constant $\Phi = 2.86 \times 10^{23}$ mol^{-1}. Using the above estimate of the molar mass, M_i, the mean square radius is approximated from Eq. (28) using

$$\langle R_g^2 \rangle_i = \left(\frac{[\eta] M_i}{6^{\frac{3}{2}} \Phi^{\frac{3}{2}}} \right)^{\frac{2}{3}} \tag{29}$$

Using this estimate of $\langle R_g^2 \rangle$ the particle scattering factor of a corresponding Gaussian coil at the angle of observation is calculated:

$$P_i(\theta) = 2\left[\frac{\exp(-x) - (1-x)}{x^2}\right] \qquad (30)$$

where

$$x = \frac{8}{3}\langle R_g^2\rangle_i^{\frac{1}{2}}\left[\frac{n\pi}{\lambda}\sin\left(\frac{\theta}{2}\right)\right]^2$$

Using this particle scattering factor, a new estimate for the molar mass is obtained:

$$M_{i+1} = \frac{R(\theta)}{KcP_i(\theta)} \qquad i \geq 1 \qquad (31)$$

and the calculations according to Eqs. (28–31) can be repeated until the molar mass and mean square radius approach limiting values.

The advantage of this procedure is that the molar mass and size information obtained are independent of the separation mechanism. The results, however, depend crucially on the approximations involved. For structures other than coiling molecules, the approximation of the particle scattering factor according to Eq. (30) has to be considered. The Flory constant, Φ, depends on solvent quality. In addition, it has been shown that Φ varies with branching density. Hence the uncritical use of triple detection might result in severe misinterpretation of experimental results.

The advantages of molar mass-sensitive detectors described above for the determination of molar masses and the additional information that can be obtained by these techniques resulted in the extensive use today of molar mass SEC detectors. Both systems have their advantages and disadvantages and the decision as to which of the two systems to apply has to be made after careful consideration of the type and molar mass of the samples to be analyzed.

13.2.2
Interaction Chromatography of Polymers

SEC separations are based purely on the size of the macromolecules, avoiding any enthalpic interaction between the stationary phase and the analyte molecule. Most polymer separations performed today still use SEC. This is astonishing since almost every student of chemistry is taught about the use of HPLC as the preferred separation method for small molecules. Separations by interaction chromatography make use of the interactions between the stationary phase and the analyte, which have to be avoided in SEC. The reason why interaction chromatography is used less frequently in polymer science might be that polymer separations show features which are different to those of low molar mass molecules. The difference in the chromatographic behavior of macromolecules and low molar mass molecules is due to their large size difference. Chromatographic separations are performed in columns packed with porous materials. Low molar mass molecules penetrate the pore volume almost without any restriction, since even for columns of a small pore size of 100 Å the diameter of the pore is much larger than the size of the molecule. As a consequence, for small molecules the loss of entropy upon entering the pore is negligible. For low molar mass compounds the effect of the porous structure of the column material is mainly to provide a high surface area for the interaction between the molecule and the stationary phase, in order to improve the separation efficiency. This is different in polymer chromatography, where the size of the molecule is comparable to the pore size. As already discussed, entering the pores is accomplished by a significant loss of entropy for macromolecules. Hence the change in free energy has a considerable entropic contribution. Depending on the relative size of the enthalpic and entropic contributions to the change in free energy, different correlations between the retention volume (SEC, LAC, LCCC) and molar mass are observed. In ideal SEC, enthalpic contributions (interactions) between the stationary phase and the analyte molecules are avoided by the proper selection of the stationary and mobile phases. In LAC the enthalpic contributions dominate and in LCCC enthalpic and entropic contributions compensate exactly. Therefore, both modes, LAC and LCCC, are modes of interaction chromatography and will be treated in the following sections.

13.2.2.1 Liquid Adsorption Chromatography (LAC) of Polymers

With decreasing eluent strength, interaction between the repeating units and the stationary phase will occur. If the adsorptive interaction is strong enough to exceed the entropic loss of the macromolecule due to the restrictions of the pores, the macromolecule will elute at a retention volume larger than the total void volume of the column ($V_i + V_P$), i.e. it is retarded. Since the number of the repeating units increases with molar mass and so does the net interaction of the macromolecule, the molecules will elute in order of increasing molar mass.

At weak interaction with the stationary phase, the retention volume increases approximately exponentially with molar mass [5, 7]. This can be easily understood on the basis of the multiple attachment mechanism [29]. Let us assume that a repeating unit might exist in two states, an adsorbed and a non-adsorbed state. The probability of finding a repeating unit adsorbed will be denoted p_M. Hence the probability of the repeating unit not being adsorbed will be $1-p_M$. For the time when even a single repeating unit is adsorbed, the molecule will be fixed on the stationary phase, unable to travel through the column. Hence a polymer molecule of degree of polymerization P can only move through the column if all repeating units are desorbed at the same time. Assuming that each repeating unit interacts independently of the others, the probability of all repeating units being desorbed is given by $p=(1-p_M)^P$. Therefore, in order to exit the column at the retention time t_R, the molecule has to be desorbed for the time $t_0 = p t_R$. The retention time is therefore given by $t_R = t_0/p$. Hence even a low probability of the repeating unit being adsorbed will result in a long retention time for a macromolecule. For example, if a repeating unit is adsorbed only 5% of the time, this would result for a corresponding low molar mass compound in a retention coefficient as low as 0.05. In contrast, the probability of finding the corresponding macromolecule of degree of polymerization $P=100$ desorbed is $1-p=0.95^{100} \approx 0.006$. The corresponding retention is therefore 170 times higher than that of the repeating unit. For a typical HPLC column having a void volume of 2.5 mL this would correspond to a retention time of 7 h at a flow-rate of 1 mL min^{-1}. Since already a very low interaction strength of a repeating unit suffices to result in extremely high retention times for high molar mass polymers, even a small change in solvent composition often results in a sudden transition from infinite retention to complete desorption. Therefore, the elution behavior of polymers seems to occur by an on–off mechanism, unlike the elution of low molar mass compounds.

The extreme dependence of retention time on molar mass for high molar mass polymers tends to partial irreversible adsorption even for narrow distributed samples, resulting in a limited recovery. In order to understand this, we recall the definition of the non-uniformity $U = D - 1$:

$$U = D - 1 = \frac{M_w^2 - M_n^2}{M_n^2} = \frac{\sigma^2}{P_n^2} \tag{32}$$

where P_n is the number-average degree of polymerization and σ is the variance of the distribution, which for a Gaussian distribution is related to the peak width at half-height by $w_{\frac{1}{2}} = 2.355\sigma$. For a Gaussian distribution, approximately 99% of all chains will have a degree of polymerization in the range $P = P_n \pm 3\sigma$. For low interaction energies, the distribution coefficient increases approximately exponentially with the degree of polymerization, P:

$$K \propto \exp(AP) \tag{33}$$

Combining Eqs. (32) and (33) results in

$$\frac{K(P_n \pm 3\sigma)}{K(P_n)} = K(P_n)^{\pm 3\sqrt{U}} \tag{34}$$

Equation (34) allows the estimation of the retention volume range of a chromatographic peak. Let us assume we have a narrow distributed sample with a number-average degree of polymerization $P_n = 100$ and a polydispersity $D = 1 + U = 1.05$, which is a typical value for polymer standards prepared by anionic polymerization. About 99% of all chains have a degree of polymerization in the range $P = P_n \pm 3\sigma = P_n(1 \pm 3\sqrt{U})$, i.e. $P = 33$–167. Suppose that isocratic conditions have been selected such that the peak maximum elutes at $K = 3$. This corresponds to a retention volume of approximately 5.5 mL for a typical HPLC column (25 cm length, 0.46 cm inner diameter, assuming $V_i = 1$, $V_P = 1.5$). Under these conditions 99% of all chains will have a distribution coef-

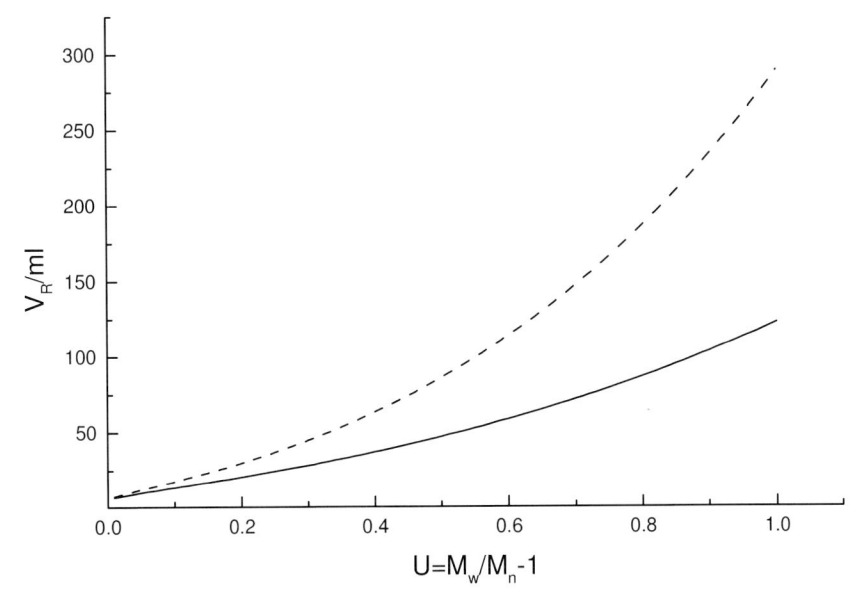

Fig. 13.16 Dependence of the retention volume for 99% polymer elution on sample polydispersity for Gaussian peaks. Solid line, $K(P=100)=3$; dotted line, $K(P=100)=4$.

ficient in the range $K=1.4$–6.3, corresponding to a retention volume between 3.1 and 10.4 mL, with the peak maximum at 5.5 mL. Hence, although the sample has a very narrow molar mass distribution, the eluting peak will be fairly broad. The dependence of the high molar mass retention volume on sample polydispersity is depicted in Fig. 13.16 for two different eluent strengths, corresponding to K at peak maxima of 3 and 4, respectively. It can be observed that the retention volume of the high molar mass end of the peak increases rapidly with increasing polydispersity or increasing K at peak maximum. Therefore, even moderate retention allows only for isocratic elution of narrow distributed samples. For samples of higher polydispersity, the K value at the peak maximum has to be close to unity to allow complete elution within reasonable times. This statement is identical with that isocratic elution of high molar mass polymers is possible only very close to critical conditions.

An experimental example of isocratic polymer elution is given in Fig. 13.17. Here the chromatogram for the isocratic elution of a low molar mass poly(ethylene oxide) ($M_w=960$ g mol^{-1}, $D=1.04$) is shown. Under the experimental conditions the sample can be separated into individual oligomers. Despite its low polydispersity the sample elutes between 5 and 35 mL with its center around 12 mL. The above crude estimation results in a calculated retention volume range between 4.3 and 34.7 mL, in good agreement with the estimations given above.

An additional drawback of isocratic adsorption chromatography is the limited resolution of macromolecular species. For high molar mass polymers even of

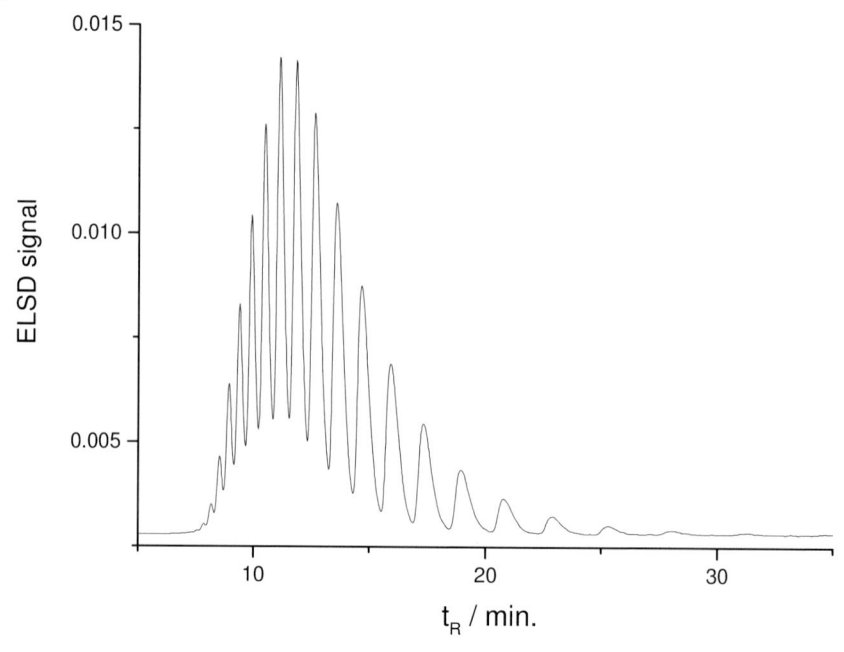

Fig. 13.17 Chromatogram of poly(ethylene oxide) ($M_w = 960$ g mol^{-1}, $D = 1.04$) for isocratic elution in MeOH–water (46:54, v/v) on a reversed-phase column.

low polydispersity, a large number of individual species have to be resolved in a limited retention volume range. Due to peak broadening, these species will not be observed as individual peaks, but as a single broad peak. As a consequence, peaks of molecular species differing, e.g., in end-groups or other structural parameters will not be separated, but hidden due to the polydispersity of the sample. Finally, the peak broadening due to sample polydispersity results in the fact that for a given injected sample mass the signal intensity at a given retention volume decreases, resulting in detection problems.

As a consequence of these drawbacks, we can state that useful separations in the isocratic adsorption mode can be performed only for oligomers of rather narrow molar mass distribution. In these cases the individual peaks might be sufficiently separated from each other, allowing identification of peaks having different functional groups or stereochemistry. For high molar masses the peak width due to the molar mass distribution will hide minor differences in the polymer structure. The same holds true if samples are eluted under SEC conditions. In order to detect such small variations in the macromolecular structure, chromatographic conditions have to be chosen which allow for molar mass-independent chromatographic elution.

13.2.2.2 Chromatography at the Critical Point of Adsorption (LCCC)

As already explained in the preceding sections, the size of the molecule determines the retention volume in SEC, resulting in a decreasing retention volume with increasing molar mass. SEC is performed in strong eluents in order to suppress interaction between the stationary phase and the analyte. In contrast, adsorption chromatography corresponds to weak eluents, allowing enthalpic interactions between the stationary phase and the macromolecule. If these interactions are of sufficient strength, the retention volume increases with molar mass (Fig. 13.1). It is therefore interesting to explore the elution behavior of the macromolecule in mixtures of strong and weak eluents. This allows for a systematic variation of the interaction strength between the macromolecule and the stationary phase. It has been found that at a given solvent strength (relating to a specific solvent composition), all macromolecules of a given chemical structure elute at the same retention volume, irrespective of their molar mass. Non-functionalized homopolymers elute at the column's void volume [5, 7]. Since the repeating units no longer contribute to the chromatographic retention, this observation is often described as the repeating units becoming chromatographically "invisible". Chromatography performed under such conditions is referred to as chromatography under critical conditions, critical chromatography or chromatography at the exclusion adsorption transition point. The critical conditions for a given type of polymer depend crucially on the stationary phase and the temperature. Even small deviations from the critical conditions result in a molar mass-dependent elution. For illustration, the dependence of the retention volume on solvent composition and temperature is depicted in Fig. 13.18 for the elution of polystyrenes of various molar masses in CH_2Cl_2–CH_3CN mixtures. In Fig. 13.18a the critical conditions were found at constant temperature by variation of the solvent composition, whereas in Fig. 13.18b the temperature was varied at a constant solvent composition. It can be observed in Fig. 13.18a that at a CH_2Cl_2 content of less than 57% the retention volume increases with increase in molar mass, indicating the typical LAC behavior. If the amount of CH_2Cl_2 is raised above 57%, the retention volume decreases with increase in molar mass, indicating SEC-like elution order. Sometimes such behavior is referred to as the SEC mode, which is not accurate. As already explained, SEC corresponds to a pure size-determined elution behavior without any additional interaction between the stationary phase and the macromolecule. However, as can be seen in Fig. 13.18a, the retention volume can be varied by changing the eluent composition despite an SEC-like elution order. Hence interaction between the macromolecule and the stationary phase exists, which will be influenced by variations of the solvent composition. At a CH_2Cl_2 content of 57%, all polystyrene standards elute at the same retention volume irrespective of molar mass, indicating critical conditions.

In Fig. 13.18b, the eluent composition is kept fixed at 57% CH_2Cl_2. An increase in retention volume with increase in molar mass is found at temperatures below 30.5 °C, whereas an SEC-like elution order is found at higher temperatures. At 30.5 °C, critical behavior is observed. The effect of changes in eluent composition is usually much stronger than that of variations in tempera-

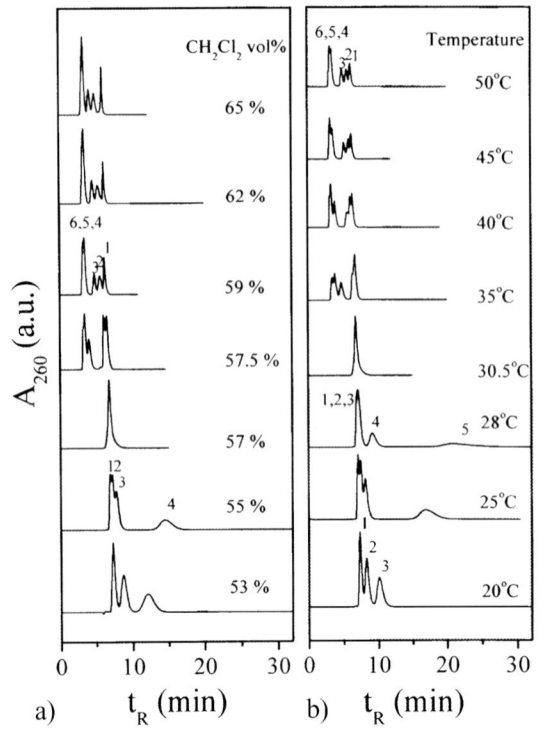

Fig. 13.18 Effects of (a) solvent composition and (b) temperature on the separation of mixtures of polystyrene standards [$M_w =$ (1) 2.5, (2) 12, (3) 29, (4) 165, (5) 502 and (6) 1800 kg mol^{-1}] ($T = 30.5$ °C). Eluent, CH_2Cl_2–CH_3CN; reversed-phase column. Reprinted from [32], copyright 1999, with permission from Wiley-VCH.

ture. Therefore, eluent variations are usually used to approach critical conditions. Fine tuning might then be done by variation of temperature.

The usual experimental procedure to determine the critical eluent composition involves a number of isocratic experiments using polymer standards of different molar masses. Starting from a strong eluent, which results in an SEC-like elution order, the eluent strength is systematically reduced by adding a weak eluent. For each new eluent composition the dependence of retention volume on molar mass has to be determined after carefully equilibrating the column. The addition of too much of the weak eluent changes the elution behavior to the LAC mode, which for high molar mass polymers might result in irreversible adsorption. As a result, time consuming flushing and re-equilibration of the column become unavoidable. By this trial and error approach, an eluent composition is finally found where the molar mass dependence of the retention volume vanishes. A simple way to estimate a reasonable starting composition to establish the critical eluent composition has been proposed recently [30]. In this approach, a linear eluent gradient is applied to the highest molar mass standard available. From the retention time, dwell and void volume, the composition at the time of elution is calculated. As will be discussed below, the composition of the eluent should be close to the critical composition and can therefore be used as the starting composition for subsequent isocratic runs for the determination of critical conditions. Recently, a collection of critical conditions has been published [31].

The simple argument that the enthalpic interactions have to compensate the entropic loss of the molecule entering the pore often results in the assumption that critical chromatography requires a pore size which is larger than the size of the macromolecule. This is not necessary, however. In fact, it might be detrimental in some cases. According to theory, critical conditions exist also for pores smaller than the size of the molecule [5]. However, the transition from adsorption- to SEC-like behavior occurs over a much more narrow solvent range for small than for wide pores, making adjustment of critical conditions more difficult. This again seems to favor large pores. However, the choice of the correct pore size depends on the separation that is aimed at. In certain cases, small pores should be used, as will be explained further below.

As has been mentioned above, at the critical conditions the repeating units of a certain type of polymer become chromatographically invisible. The elution behavior of the molecule is then determined by other chromatographically active parts of the polymer as there are end-groups, topology [33–46] or segments of different chemistry as in block [47–53] or graft copolymers [54, 55]. Separations of blends [56–59] and even separations based on tacticity have been reported [60]. The ability to separate according to small changes in molecular structure makes critical chromatography a valuable tool for the characterization of macromolecules by parameters other than molecular size. A few examples will illustrate the capabilities of chromatography under critical conditions of adsorption.

If an end-functionalized polymer is subjected to critical conditions, the elution behavior of the whole polymer is governed by the end-groups. Since the end-groups will not give rise to substantial exclusion from the column, the selection of the column has to be performed in such a way that the end-groups should increase retention. An example of a separation based on the type of end-group is given in Fig. 13.19 for poly(ethylene oxide)s initiated by various alkoxylates. A mixture of different fatty alcohol ethoxylates was subjected to a separation under critical conditions for PEG on a reversed-phase column. The separation is achieved according to the non-polar character of the different alkyl chains. A variety of peaks can be resolved. The peak at the lowest retention volume elutes at the same retention time as PEG. The peaks correspond to a polyethoxylates initiated by the different alcohols. The longer the alkyl chain of the alcohol, the longer the polymer is retained on the column. The peaks show no broadening due to the polydispersity of the polyethoxylate block.

Instead of being regarded as end-functionalized macromolecules, the fatty alcohol ethoxylates can also be regarded as block copolymers, due to the lengths of their alkyl chains. Trathnigg and coworkers extensively used fatty alcohol ethoxylates to compare experimental data with theoretical expectations on polymer chromatography [61–64].

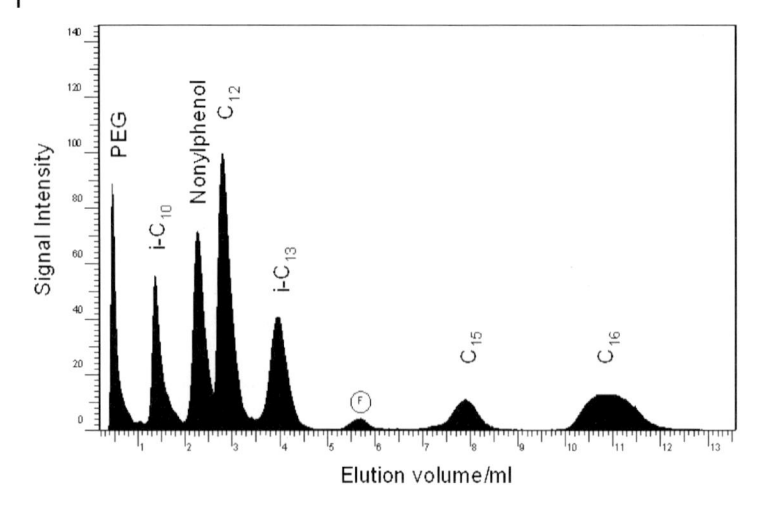

Fig. 13.19 Separation of poly(ethylene oxide)s according to the type of end-group.

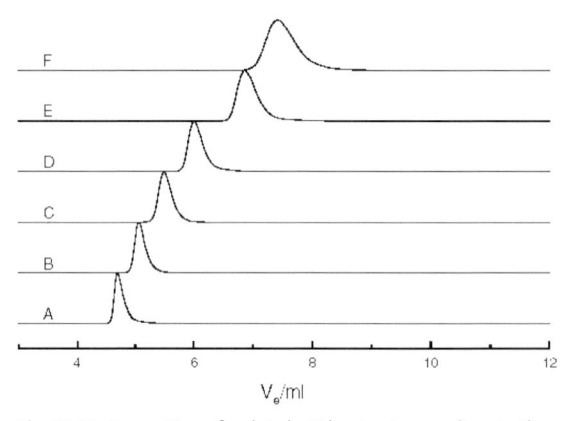

Fig. 13.20 Separation of poly(L-lactide stars) according to the number of terminal OH groups. Traces A–F = 1–6 OH-terminated arms. Reprinted from [37], copyright (2005), with permission from Elsevier.

Another example of a separation according to functionality by LCCC is given in Fig. 13.20. Here poly(L-lactide) (PLA) stars were investigated at critical conditions of PLA on a normal-phase column. Each arm of the PLA stars is end-functionalized by an OH functionality. Under the chromatographic conditions the OH groups interact with the stationary phase more strongly than the repeating units, giving rise to additional retention. To a first approximation, each additional OH functionality results in an additional constant shift to higher retention volumes, allowing separation according to the number of arms. End-group analysis by NMR or titration will yield the average number of OH functionalities per molecule but no information about arm number distributions. The evaluation of g-factors by light scattering or of viscosity data to obtain information on the number of arms would be difficult, since the molar mass distribution and the arm-number distribution might overlap. However, the arm number distribution could be easily obtained by LCCC due to the good separation according to the number of arms.

Liquid chromatography at the critical conditions of adsorption allows the effective separation of linear and cyclic molecules having the same molar mass. Whereas linear chains and rings of the same molar mass are separated only slightly in SEC, good separations result under critical conditions. Hence LCCC allows pure cyclic fractions to be obtained [33, 35].

In addition to separations according to end-groups, LCCC provides an effective tool to characterize polymer blends and segmented copolymers. If a blend is chromatographically investigated at critical conditions of one blend component, the second component will show either an SEC- or an adsorption-like behavior, depending on whether the interaction of this component is weaker or stronger than that of the other component. Hence critical chromatography allows the quantification of the amounts of the blend components. In addition, the molar mass and the molar mass distribution of the non-critical component can be determined. In practice, the situation where one component is subjected to critical conditions whereas the other shows SEC-like elution behavior is preferred, due to the difficulties associated with isocratic adsorption chromatography. In order to achieve a sufficient separation under such conditions between the component which is at critical conditions and a second blend component having a low molar mass, a column with small pores has to be selected despite the problem that critical conditions are more difficult to establish. The reason for using a small pore size column is that the critical component is expected to elute at $V_i + V_P$, whereas the other component elutes at $V_i + KV_P$. Therefore, in order to achieve a good separation, K should be as low as possible, meaning that the pore size should be as small as possible.

An example of the separation of polymer blends by LCCC is shown in Fig. 13.21, which is taken from work of Pasch and Rode [58]. Blends of poly(decyl methacrylate) (PDMA) and PMMA were separated at critical conditions of PDMA on a reversed-phase column. As can be seen, both components are clearly separated, allowing quantification of the amounts of the components. PDMA elutes at the same retention volume irrespective of the molar mass, whereas PMMA, as the more polar component, clearly shows SEC-like elution behavior, allowing the determination of the average molar mass of the PMMA.

If an AB diblock copolymer is subjected to critical conditions of block A, the elution should be completely determined by block B. If under these conditions the type B monomer shows a weaker adsorption than the type A monomer, the homopolymer B will elute before the void volume in a SEC-like elution order. Since block A is chromatographically invisible and does not contribute to the retention volume, the block copolymer will behave chromatographically identically with a homopolymer B having the same molar mass as the B block. Therefore,

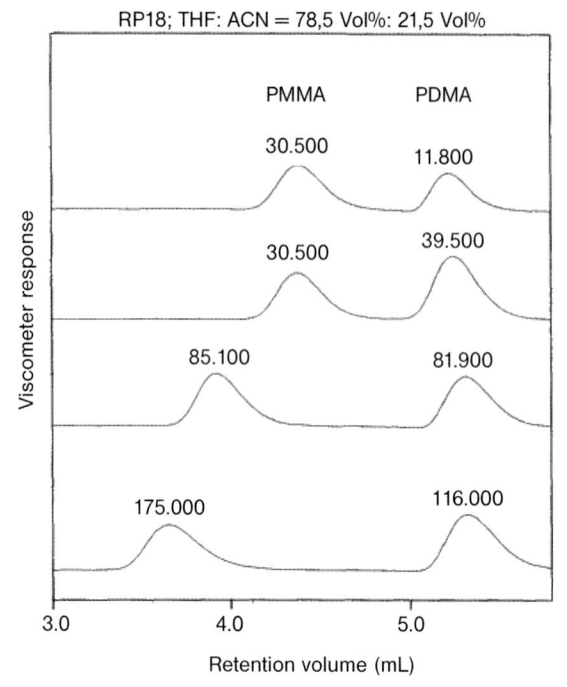

Fig. 13.21 Separation of blends of poly(methyl methacrylate) and poly(decyl methacrylate) at critical conditions of poly-(decyl methacrylate). Reprinted from [58], copyright (1998), with permission from Elsevier.

homopolymer B and the block copolymer will elute before homopolymer A and its elution is determined solely by the molar mass of block B. Similar, if the interaction between the stationary phase and monomer B is stronger than that of monomer A, the block copolymer and the homopolymer of type B will elute later than the homopolymer of type A at a retention volume determined by the molar mass of the B block. In both cases homopolymer A, which is subjected to critical conditions, is separated from the other components and its amount can be determined. In order to determine the amount of homopolymer B, the sample has to be analyzed at critical conditions for homopolymer B. Under these conditions, homopolymer A shows the same retention behavior as the block copolymer and both are separated from homopolymer B.

Therefore, in order to determine the molar mass distributions of both homopolymer components, the first investigation might be performed on a non-polar reversed-phase column, such that the less polar component is investigated at critical conditions, whereas the more polar homopolymer elutes in the SEC mode. The analysis of the more polar component should then be performed on a polar column, such that the less polar component shows less adsorption and elutes before the void volume. By similar procedures it is possible to separate homopolymer fractions in graft copolymers.

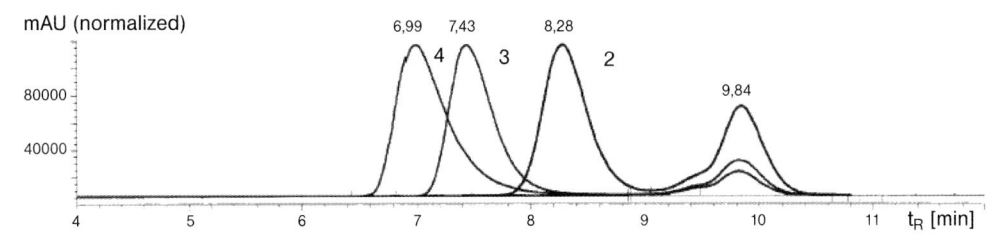

Fig. 13.22 LCCC chromatograms of P*t*BMA-*b*-PMMA (precursor P*t*BMA, $M_n = 73\,000$ g mol^{-1}) with different PMMA block lengths at critical conditions of P*t*BMA. Reprinted with permission from [47]. Copyright (2000) American Chemical Society.

The example in Fig. 13.22 illustrates the determination of residual homopolymer in block copolymers composed of poly(*tert*-butyl methacrylate) (P*t*BMA) and PMMA. The conditions correspond to the critical conditions of P*t*BMA on a reversed-phase column. As is shown in Fig. 13.22, all samples show two clearly separated peaks. The peak at 9.84 min can be assigned to the P*t*BMA precursor. Since PMMA elutes in an SEC-like order, the retention volume of a block copolymer is lower than that of the P*t*BMA and decreases with increasing PMMA block length. Since the P*t*BMA block is chromatographically invisible, the molar mass distribution and the average molar mass of the PMMA block can be determined after establishing a PMMA calibration curve.

Based on experimental evidence, there is still a contradiction as to whether or to what extent the "invisible" block in an AB block copolymer has an influence on retention behavior [47, 65]. However, it can be stated that for diblock copolymers the approach described above gives at least a good estimate for the molar mass distribution of the chromatographically active block.

It should be mentioned that for ABA block copolymers or graft copolymers the situation is more complicated. According to theory [66, 67], if the segment which is at its critical point (B segment in ABA) is attached at both ends to chromatographically active segments (A segments in an ABA block copolymer or backbone in graft copolymers), the molar mass of the "invisible" segment will still influence the retention time of the polymer. If the invisible segments are attached to a chromatographically active segment at one end only (A segments in an ABA block copolymer or side-chains in graft copolymers), the invisible block should not contribute to the retention time and the polymer molecule behaves chromatographically like pure B block or backbone, respectively.

The previous examples have shown the possibilities of chromatography at the critical point of adsorption. However, as an isocratic method it has its difficulties when sample components with very different polarities have to be analyzed. In such cases, gradient chromatography might be the method of choice.

13.2.2.3 Gradient Chromatography of Polymers

In gradient chromatography, the eluent strength is systematically increased during the duration of the chromatographic experiment. This is usually done by systematically changing the eluent composition during the chromatographic run. However, as has been shown by Chang and coworkers, a temperature gradient can also be applied, resulting in high-resolution separations [32, 68–73].

Usually a very weak eluent is used at the beginning of the gradient, resulting in strong adsorption of the polymer molecules on the stationary phase. Desorption occurs with increasing eluent strength. Due to the strong molar mass dependence of the distribution coefficient, desorption occurs only in a narrow region of eluent compositions close to the critical composition. Lower molar mass polymer molecules, being relatively weakly adsorbed, desorb at an eluent composition well below the critical one whereas stronger eluents are required to desorb higher molar masses. As the polymer becomes desorbed it will be surrounded by an eluent below the critical composition, i.e. the polymer moves with a velocity lower than that of the eluent. As a consequence, it will be overtaken by compositions of increasing eluent strength, resulting in an acceleration of the polymer molecules. However, this acceleration can only result as long as the eluent moves faster than the polymer. The condition of polymer velocity being equal to the eluent velocity corresponds to the critical conditions, $K=1$. Therefore, the polymer molecule will elute at a composition below or equal to the critical value [4, 30, 74]. This is the reason why, in gradient chromatography, often no molar mass dependence of the retention volume is observed. However, for sufficiently shallow gradients a molar mass-dependent elution behavior can be found. Whether a molar mass-dependent or -independent elution close to the critical composition is observed depends crucially on the product of molar mass and gradient slope [4, 30, 74]. High molar mass polymers will therefore elute in order of their critical solvent composition. Since the critical composition depends strongly on the chemical nature of the polymer molecule, a separation according to chemical composition can be achieved. Therefore, gradient chromatography is often applied for separations of polymer blends or copolymers according to chemical composition [75–80].

Figure 13.23 shows the separation of a polymer blend composed of P*n*BA, P*t*BA and two PMMAs of different molar masses using a multi-step toluene–THF gradient on pure silica. As can be seen, the two poly(butyl acrylate)s can be baseline separated despite the small difference in structure. The more polar PMMA elutes at a significantly higher THF content. The flat gradient during PMMA elution results in a molar mass-dependent elution of the PMMA.

The second example deals with the separation of statistical copolymers composed of *n*BA and styrene (Fig. 13.24). The samples were synthesized by performing low-conversion copolymerizations in order to obtain samples having a low chemical heterogeneity. Samples of different composition can be separated using a gradient on a normal-phase column. As can be seen, the retention increases with increasing acrylate content. The established dependence of the retention volume on chemical composition can be used to determine the chemical

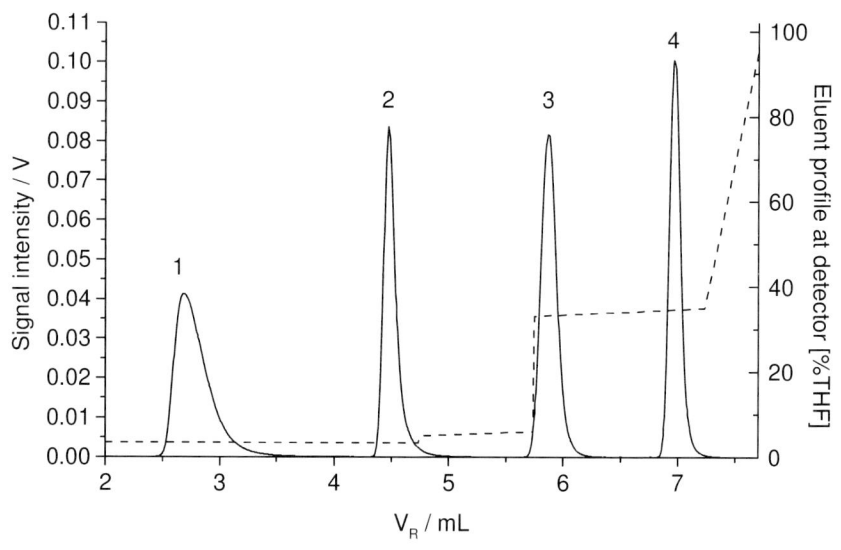

Fig. 13.23 Separation of a blend of (1) P*n*BA 22.6 kg mol^{-1}, (2) P*t*BA 210 kg mol^{-1}, (3) PMMA 24.4 kg mol^{-1} and (4) PMMA 263 kg mol^{-1} by multi-step gradient chromatography. Stationary phase, Nucleosil pure silica, 1000 Å pore size, 7 μm particle size; eluent, toluene–THF gradient.

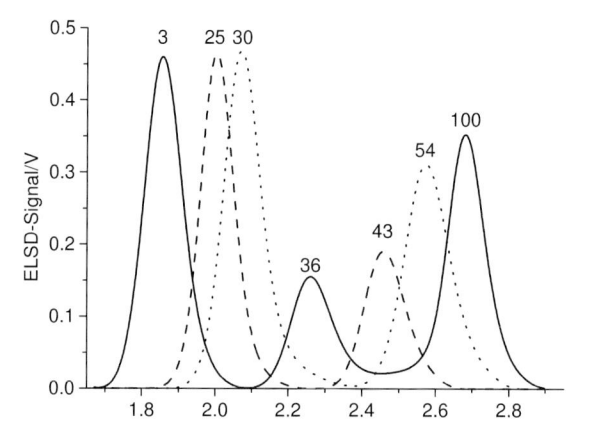

Fig. 13.24 Separation of stat. poly(styrene–*n*-butyl acrylate) copolymers according to chemical composition by gradient chromatography. The numbers at the peaks indicate the weight fraction of styrene in the copolymer. Reprinted from [81], copyright 2005, with permission from *e-Polymers*.

composition distribution of samples at high conversions, which will show a considerable chemical heterogeneity distribution due to variations in the comonomer ratio associated with the different reactivities of the comonomers.

A change in solvent composition provides a rather abrupt change in the interaction between the polymer molecule and the stationary phase. A change in interaction energy can also result from changes in temperature. Changing the temperature provides a less abrupt variation of solvent strength than changing the eluent composition, allowing very delicate separations to be performed. Hence temperature gradient interaction chromatography (TGIC) has been successfully applied for high-resolution separations of homopolymers according to molar mass [82]. The resolution achieved is higher than that which can be obtained by SEC. The high sensitivity to small differences in molar mass allows star polymers to be separated according to the number of arms. This allows the linking reaction to be followed during star polymer synthesis. Separations of polymer blends and graft copolymers have also been reported [83]. However, the range of interaction strength that can be covered by TGIC is limited. Therefore, TGIC can only be applied if the interaction strengths of the different components in the given solvent are not too different. From that point of view, TGIC provides a means to fill the gap between isocratic adsorption chromatography and eluent gradient chromatography. However, there is a pronounced difference between eluent gradients and temperature gradients. As explained above, a homopolymer in an eluent gradient will elute in an eluent composition slightly lower than or equal to the critical composition. That is, for principal reasons an increasing retention time with increasing molar mass or a molar mass- independent elution is expected during eluent gradients. In TGIC, however, the situation is different. If the temperature variation relative to the flow-rate is high, the temperature which would provide critical conditions for isocratic experiments might be reached while the molecules are still within the column. A further increase in temperature will then allow SEC-like elution conditions to be developed. Hence TGIC should allow the chromatographic experiment to be started under adsorbing conditions and ended under SEC conditions. Depending on the ratio of temperature rise to the flow-rate, either increasing or decreasing retention times with increasing molar mass can be observed. The capabilities of this approach have not yet been fully explored, however.

Figure 13.25 shows the separation by TGIC of 10 polystyrene standards ranging in molar mass from 3.2 to 1530 kg mol^{-1}. As can be observed, all standards can be separated with nearly baseline resolution. If the conditions are selected such that at the initial temperature one component of a blend elutes in an SEC-like elution order, whereas the second component can be eluted by TGIC, the first component will elute in order of decreasing molar mass whereas the second component elutes in order of increasing molar mass. Such an example is depicted in Fig. 13.26 for the separation of PMMA and polystyrene standards having different molar masses. The five PMMA standards having molar masses ranging from 1500 to 2 kg mol^{-1} elute in the SEC mode at the beginning of the temperature gradient. The 11 polystyrene standards in the molar mass range

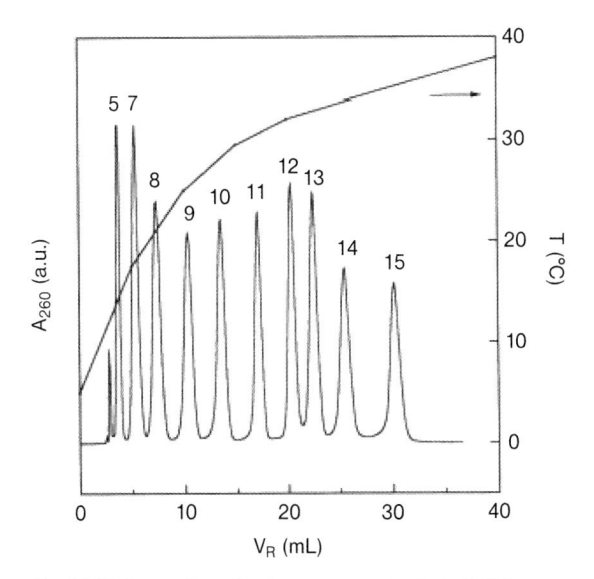

Fig. 13.25 Separation of polystyrene standards by TGIC. Reprinted from [82], copyright (1999), with permission from Elsevier.

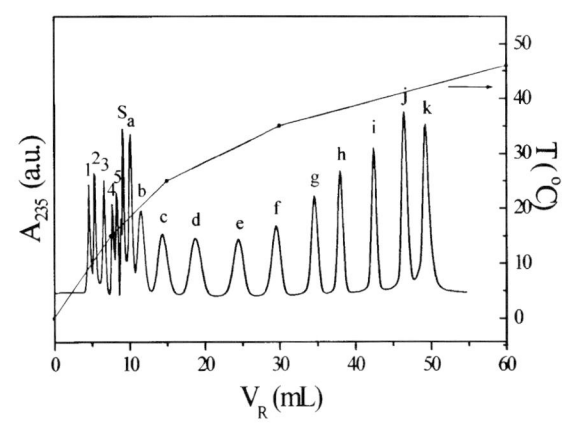

Fig. 13.26 Separation of a blend of PMMA standards (1–5: M_w = 1500, 501, 77.5, 8.5, 2 kg mol^{-1}) and PS standards (a–k: M_w = 1.7, 5.1, 11.6, 22, 37.3, 68, 114, 208, 502, 1090, 2890 kg mol^{-1}) by TGIC. Reprinted with permission from [83]. Copyright (1996) American Chemical Society.

1.7–2890 kg mol^{-1} elute in order of increasing molar mass. Despite the similar molar mass ranges, the 11 polystyrene standards show a better separation than the PMMAs, indicating the much better separation that can be achieved by TGIC as compared with SEC.

The preceding sections have given an overview of the separations that can be obtained with the different modes of polymer chromatography. However, if dealing with a complex polymer having more than one type of heterogeneity, none of these separation techniques will fully resolve the complex structure. Therefore, two-dimensional separations might have to be used.

13.2.3
Two-dimensional Chromatography

Application of a single chromatographic separation technique will generally fail to resolve fully the structural details of complex polymers. A polymer blend subjected to SEC will be separated in terms of hydrodynamic volume, but polymers of different chemical composition and different molar mass might coelute due to their identical hydrodynamic volumes. A functionalized polymer subjected to critical conditions might be separated according to the type of end-groups, but this separation will lack the information on the molar masses of the fractions having different functionality. In order to obtain information on the complete complex distribution function, a separation according to one structural parameter only should first be performed (e.g. chemical composition). The resulting fractions should then be analyzed according to a second structural feature, e.g. molar mass.

The first two-dimensional chromatographic experiments were tediously performed by manually taking fractions from SEC and performing their subsequent analysis by gradient chromatography. An excellent overview is given by Glöckner [84]. Today, fully automated two-dimensional chromatography is commercially available [85–88]. The system consists of two coupled liquid chromatographs (Fig. 13.27). The heart of the setup is the transfer valve, which allows the injection of fractions from the first chromatographic separation into the second chromatograph. While the first loop of the transfer valve is filled with the eluent of the first dimension, the content of the second loop is analyzed in the second dimension (SEC). After performing the SEC analysis, the valve position is changed, resulting in filling of the second and SEC analysis of the content of the first loop. By this procedure, a series of SEC chromatograms is produced. Knowing the flow-rate of the first dimension and times of valve actuation, the corresponding retention volume of the first dimension can be assigned to each SEC chromatogram. Valve control, data acquisition and the subsequent data analysis are done using a commercial software package. The switching times, loop sizes and flow-rates of the different chromatographic dimensions have to be carefully adjusted in order to obtain a complete transfer of the sample from the first to the second dimension, without too much loss of resolution of the first dimension. The software uses this information to create two-dimensional chromatograms for further analysis. Areas of the two-dimensional chromatogram can be integrated to yield the molar mass distribution or to quantify the amount of the corresponding fraction.

Two-dimensional chromatography has been used for the determination of the combined functionality type (FTD) and molar mass distribution (MWD) or the

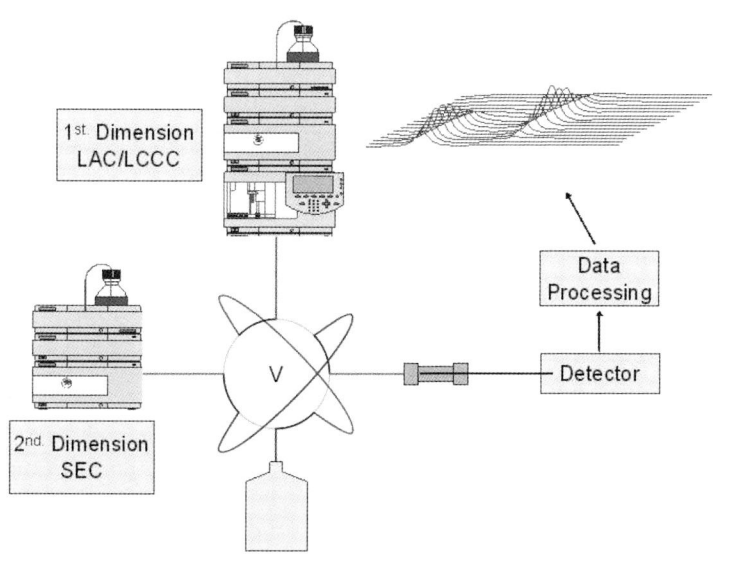

Fig. 13.27 Schematic setup for online two-dimensional experiments. Reprinted from [55], copyright 2005, with permission from Wiley-VCH.

joint distribution of chemical composition (CCD) and molar mass, which usually cannot be obtained by one chromatographic mode alone [43, 52, 55, 85–92]. Also, topology-based separations have been reported [91, 93, 94], allowing the separation of linear and star-shaped molecules having identical hydrodynamic sizes. Examples of different types of separations are discussed below.

The first example, which is taken from the work of Schoenmakers and coworkers [90], shows a two-dimensional separation of a mixture of non-functionalized, mono-OH- and bis-OH-terminated PMMAs according to functionality and molar mass (Fig. 13.28). Near critical conditions were selected for the separation in the first dimension. The fractions were subjected on-line to SEC analysis in a setup similar to that in Fig. 13.27. From the two-dimensional representation it becomes clear that fractions of similar functionality elute at (near) the same retention volume in the first dimension (LCCC, ordinate). The separation of the fractions having the same functionality occurs in the second dimension (SEC, abscissa). Two-dimensional chromatography therefore provides a detailed insight into the correlation of functionality and molar mass.

The second example shows the separation of a complex graft copolymer. The sample was synthesized by grafting n-butyl acrylate on to a styrene–butadiene diblock copolymer [54]. The separation in the first dimension was performed at critical conditions for PnBA. Under the conditions chosen, the polybutadiene shows SEC-like elution behavior, whereas polystyrene elutes in the adsorption mode. The two-dimensional chromatogram shows a very complex pattern (Fig. 13.29). This is due to the fact that in addition to the desired product,

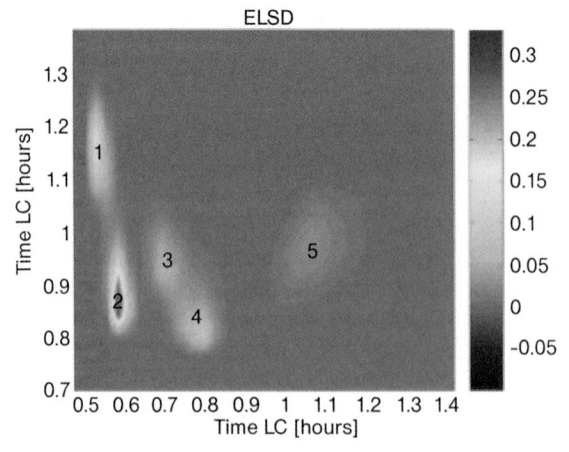

Fig. 13.28 Two-dimensional separation of a mixture of PMMA standards. Peaks 1 and 2: non-functional; peaks 3 and 4: PMMA-OH (monofunctional); peak 5: PMMA-bis-OH. Reprinted from [90], copyright (2005), with permission from Elsevier.

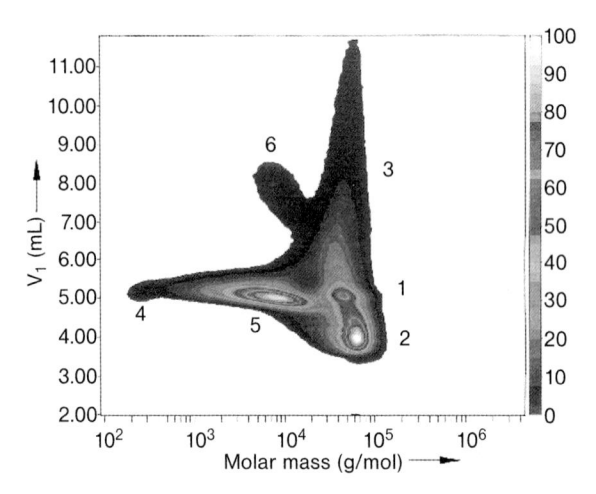

Fig. 13.29 Two-dimensional separation of a graft copolymer of (PS-b-PB)-g-PnBA. 1, ungrafted PS-b-PB; 2, graft polymer; 3, graft and block copolymer with degraded PB blocks; 4, additive; 5, PnBA; 6, maleic acid copolymer. First dimension, LCCC for PnBA; second dimension, SEC. Reprinted from [54], copyright (2000), with permission from Elsevier.

homo-PnBA is formed and the complete block copolymer is not grafted during the grafting reaction. In addition, the block copolymer itself contains residual polystyrene as a side-product of its synthesis by anionic polymerization. Finally, during the grafting reaction the block copolymer is partially degraded, resulting in products which also form graft products. The traces for the SEC and the LCCC experiment at critical conditions of P*n*BA can be reconstructed from the projections on to the abscissa and ordinate, respectively. Thus, as can be seen from Fig. 13.29, in the SEC experiment peaks 1, 2 and 3 would coelute as well as peaks 5 and 6, since these peak pairs have identical elution ranges on the abscissa. In the LCCC experiment, peaks 1, 4 and 5 would coelute as well as peaks 3 and 4, due to their identical retention volumes on the ordinate. Therefore, only two-dimensional chromatography allows complete information of the complex product to be obtained.

The last example shows a topology-based separation of star-shaped from linear polystyrenes [93]. In gradient chromatography, the star-shaped and the linear polymer elute according to molar mass nearly independent of their architecture. Hence a separation of linear and star-shaped molecules having identical molar masses is not possible. In SEC, however, coelution of molecules having identical hydrodynamic volumes occurs. None of the chromatographic experiments alone is able to separate a mixture of linear and star-shaped molecules. Combination of both modes of chromatography results in a separation of linear and star-shaped molecules, as shown in Fig. 13.30. A fraction of the first dimension con-

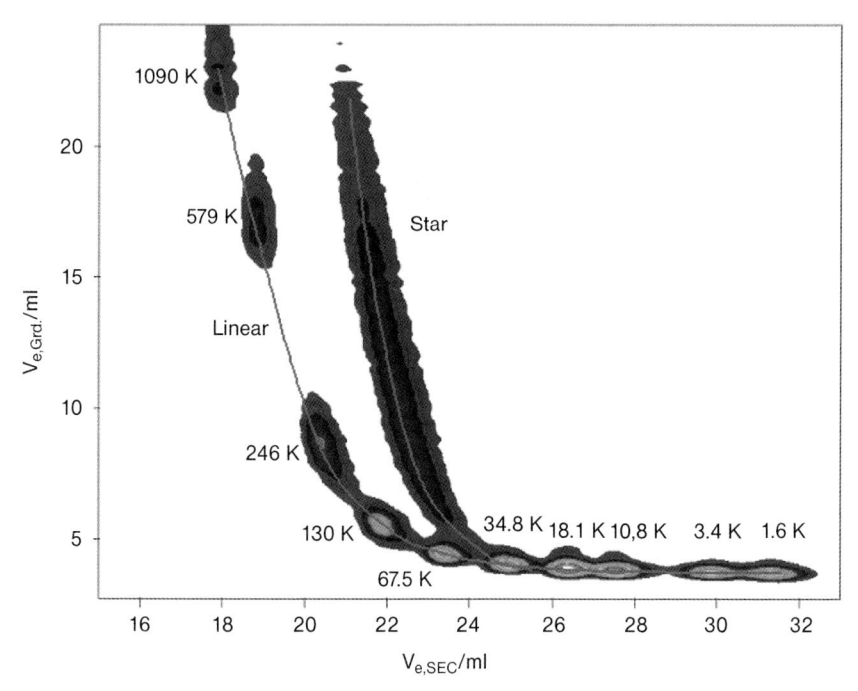

Fig. 13.30 Two-dimensional separation of linear and star-shaped polystyrenes. First dimension: gradient chromatography; second dimension SEC. Reprinted from [93], copyright (2005), with permission from Elsevier.

tains molecules of different topology but similar molar mass. If this fraction is injected into the SEC system, two peaks can be separated, due to the difference in the hydrodynamic volume resulting from the different topologies.

This section explained the use and the potential of the different chromatographic modes existing for polymer chromatography and their combination in two-dimensional separations to analyze complex polymer mixtures. However, in addition to the separation into individual components, the identification of the separated species is also an important issue.

13.3
Coupling Liquid Chromatography with Spectrometric and Spectroscopic Methods

Spectrometric and spectroscopic methods provide the major tools for structural characterization. However, the value of their information is sometimes limited since these methods provide only average properties such as average composition or functionality. In combination with chromatographic separations, however, their usefulness is enhanced.

13.3.1
Coupling LC with FTIR Spectroscopy

Fourier transform infrared (FTIR) spectroscopy is known to every chemist. The FTIR spectrum contains information on the vibrations of the different structural elements in a molecule. Apart from the detection of absorptions which can be directly assigned to structural elements (group vibrations), e.g. the C=O or C–H absorptions, the region between 1000 and 1500 cm^{-1} is very sensitive to structural changes (fingerprint region), allowing identification by comparison with FTIR databases. FTIR spectroscopy is a very sensitive method; only small amounts of the material are necessary to obtain FTIR spectra. The small amounts of material required in conjunction with the information that can be extracted from FTIR experiments seem to make FTIR spectroscopy an ideal detection method for chromatographic separation. However, the typical eluents used in chromatography are strong IR absorbers themselves, hiding the absorbances of low-concentration analytes. The use of an IR flow cell is therefore possible only in rare cases, where the absorbances of the analyte match absorbance-free windows of the solvent spectrum. The drawback of solvent bands can be overcome if the solvent is removed from the eluting fractions. This could be done by taking fractions and evaporating the solvent off-line. A more effective way uses a commercial interface [95–97]. The main element of the interface is a heated nozzle positioned above a rotating germanium disk (Fig. 13.31). The eluent is heated upon flowing through the nozzle. At the same time, the back-pressure of the nozzle decreases towards the end of the nozzle, resulting in solvent vaporization at the nozzle end. Non-volatile substrates, e.g. polymers, will not evaporate but are deposited on the Ge disk. Due to the disk rotation, fractions

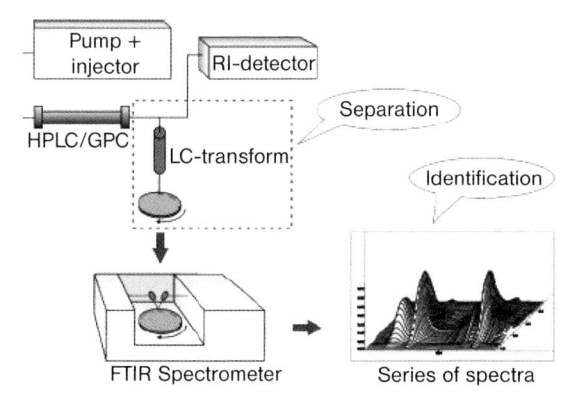

Fig. 13.31 Schematic representation of LC/FTIR using the LabConnection FTIR Interface. Reprinted from [55], copyright 2005, with permission from Wiley-VCH.

corresponding to different chromatographic elution times will appear at different positions on the disk. After ending the chromatographic experiment, the disk is placed on a motor-driven optical device within a conventional FTIR spectrometer. The optical device consists of a number of mirrors redirecting the FTIR beam on to the Ge disk and further to the detector of the FTIR instrument. The motor allows the rotation of the disk, such that the fractions at the different positions can be positioned in the FTIR beam. This allows solvent-free FTIR spectra to be obtained for every position of the deposited track. Each spectrum corresponds to a different elution time of the chromatographic separation.

From the IR spectra, the different separated components can be identified. By following specific absorption bands of a particular monomer unit, the elution profile of individual components in a polymer can be followed. This information can be used to obtain information on the change in chemical composition with retention volume.

An example of the use of LC/FTIR taken from the work of Pasch et al. [55] is given in Fig. 13.32. The example shows a separation of poly(ethylene oxide)-g-poly(methacrylic acid) at critical conditions of PEG. Spectra were taken along the chromatogram. The Gram–Schmidt reconstruction corresponds to the chromatogram and approximates the total polymer concentration. By plotting the intensities of the IR bands due to the methacrylic acid (MAA) and ethylene oxide (EO) units, respectively, the change in the composition can be clearly detected. At low retention volumes the bands due to EO units are lower than those of the MAA units. Thus, an PMAA-rich product is detected. With increasing retention volume, the relative intensity of the band due to the EO units increases. Therefore, at ~2.9 mL a PMAA-poor product is found followed by pure PEG at a retention volume slightly above 3 mL.

The example shows the potential of LC/FTIR. Spectroscopic methods alone cannot yield such detailed information. It is the combination of separation and

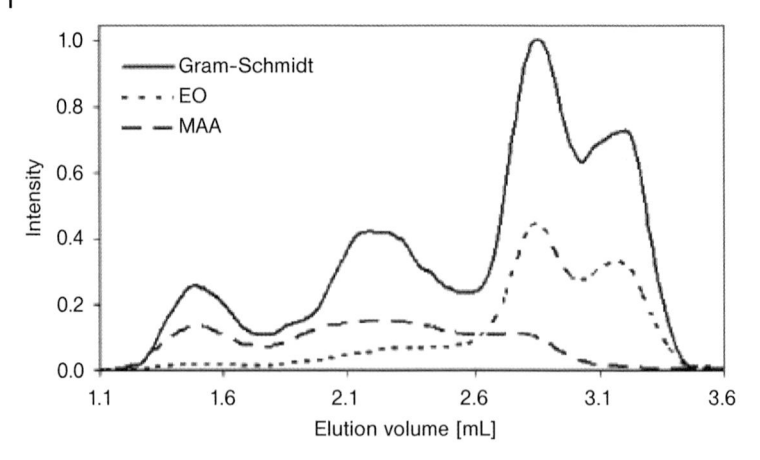

Fig. 13.32 LC/FTIR analysis of PEG-g-PMAA. Reprinted from [55], copyright 2005, with permission from Wiley-VCH.

identification which provides a deep insight into the polymer structure. Despite the large amount of information that can be gathered from LC/FTIR experiments, it also has some drawbacks. Since the width of the deposited track depends on a variety of parameters, e.g. the viscosity of deposited sample, quantification in terms of absolute concentrations is not possible. If the chromatographic method requires the use of buffers, care has to be taken to use volatile buffers in order to avoid disturbances of the analyte spectra. In some cases, e.g. in acrylic copolymers, the differences in the FTIR spectra are not sufficient to distinguish between the different components. In such cases, more structure-sensitive methods have to be used.

13.3.2
Coupling LC with NMR Spectroscopy

Nuclear magnetic resonance (NMR) spectroscopy is one of the most informative methods for the structural characterization of polymers. It provides much more detailed information than FTIR spectroscopy. Information on chemical composition, end-groups, sequence distribution, tacticity, branch points and sometimes the number-average molar mass can be extracted from NMR experiments. Similarly to conventional FTIR, NMR provides only average information of the sample. For complex polymers, the coupling of liquid chromatography and NMR spectroscopy is highly informative. However, two problems exist in coupling NMR to chromatography. NMR is usually performed in deuterated solvents. The use of such expensive eluents would prevent the use of NMR together with chromatography. Fortunately, modern solvent suppression techniques allow the solvent signals to be reduced significantly. Although complete elimination of the solvent signals is not possible NMR can be used even in gradient chromatography. The second problem is sensitivity. NMR is not a very sensitive method, which at low concentrations restricts the use of LC/NMR to proton NMR only. The lack of sensitivity can be partially compensated using high concentrations, resulting in a decreased resolution due to column overloading. An enhanced signal-to-noise ratio can also be obtained by performing more scans. Therefore, either lower flow-rates or even stop-flow experiments can be performed if necessary. As an alternative, the fractions can be stored in a loop collector and be automatically analyzed off-line after the chromatographic experiment. This allows even ^{13}C NMR experiments to be performed for structural characterization.

Two examples will serve to illustrate the potential of LC/NMR. The first example is devoted to the structural characterization of poly(ethylene oxide)s (PEOs) (Figs. 13.33 and 13.34). A mixture of PEGs having different non-polar end-groups was separated using a mixture of acetonitrile and water. On-line NMR experiments were performed using solvent suppression techniques. The intense signals between 2.1 and 2.6 ppm and at 1.4 ppm are due to acetonitrile and its impurities, which could not be suppressed. However, there is still enough information available for structural identification (Fig. 13.33). The contour plot with the chemical shift on the abscissa and the retention time on the ordinate clearly shows the resonances of the PEO CH$_2$O groups at $\sim$3.9 ppm. In addition, aromatic resonances between 7 and 8 ppm can be detected for peaks 2 and 4. Finally, a detailed investigation of the spectrum of peak 4 reveals additional peaks at 4–4.28 ppm and in the aliphatic region (Fig. 13.34). The analysis shows that peak 4 is due to PEOs of an average degree of polymerization $n=5.5$ started by an isooctylphenoxy group. The analysis of the ^{1}H resonances of the other peaks identify peak 1 as PEG started presumably by impurities. Peak 2 corresponds to a *tert*-butylphenoxy-

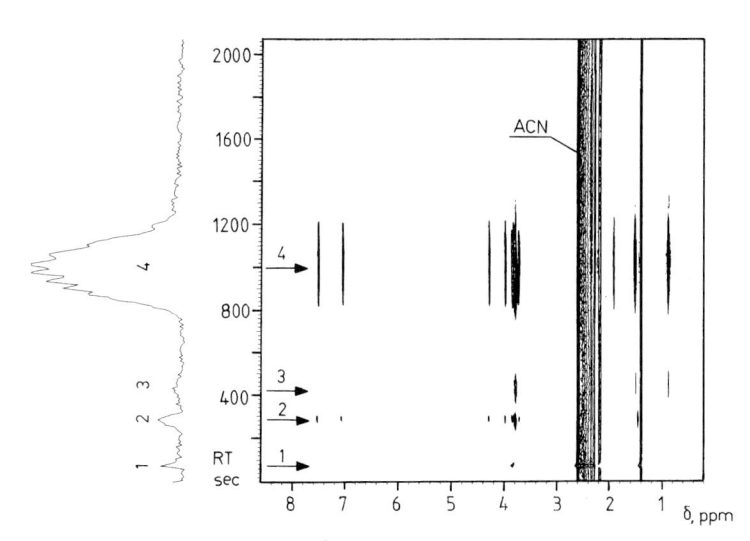

Fig. 13.33 Resonances of on-flow ^{1}H NMR spectra of poly-(ethylene oxide)s. Reprinted with permission from [98]. Copyright (1996) American Chemical Society.

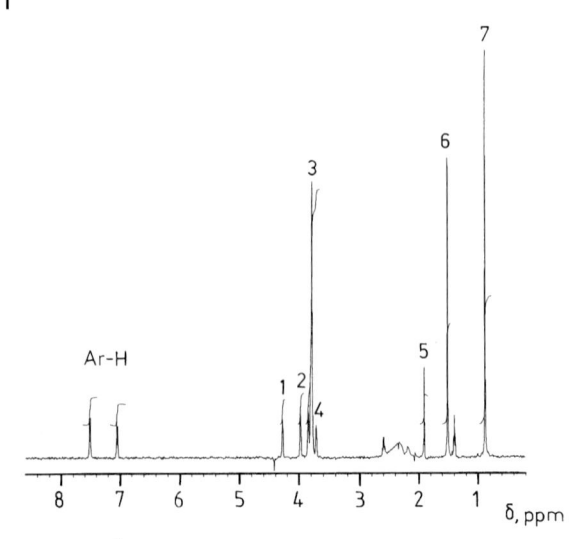

Fig. 13.34 ^{1}H NMR spectrum of peak 4 in Fig. 13.33. Reprinted with permission from [98]. Copyright (1996) American Chemical Society..

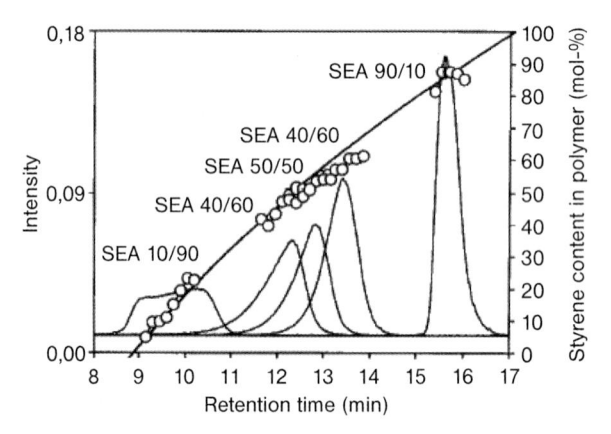

Fig. 13.35 HPLC of styrene–ethyl acrylate copolymers of different compositions with high conversion. Styrene content obtained by (——) HPLC calibration and (circles) on-line HPLC/^{1}H NMR experiments. Reprinted from [99], copyright 2000, with permission from Wiley-VCH.

PEO of average degree of polymerization $n=6$. The retention times of the different peaks confirm this structural assignment, since the increase in the non-polar part of the molecule is accompanied by a longer retention time.

The second example of LC/NMR deals with the determination of the chemical composition distribution of poly(styrene–ethyl acrylate) copolymers. A similar example has already been treated in the section on gradient chromatography. Gradient chromatography permits separation according to chemical composition. However, calibrating the relationship between composition and retention volume requires samples of narrow chemical composition distribution, which have to be synthesized by low-conversion copolymerization. Coupling ^{1}H NMR to gradient chromatography allows the relation between chemical composition and retention volume to be established without prior calibration. Figure 13.35 shows the chemical compositions that were obtained by on-line ^{1}H NMR on four samples showing some chemical heterogeneity. The chemical composition at each retention volume can be obtained using the ratio of the five aromatic protons of the styrene units to the methylene protons adjacent to the ester group of the side-chain. The analysis shows that the acrylate content increases with increasing retention volume. The calibration curve of composition versus retention volume obtained by LC/NMR closely follows that obtained from the conventional calibration using low-conversion samples having a narrow chemical composition distribution. The time and effort required to obtain the result are significantly lower, however.

13.3.3
Coupling LC with MALDI-TOF-MS

Valuable information on polymer structure can be extracted from matrix-assisted laser desorption/ionization time-of-flight mass spectrometry (MALDI-TOF-MS). MALDI-TOF-MS uses a solid matrix in which the polymer is incorporated usually by mixing and drying a polymer solution with a matrix solution. The matrix–polymer mixture is irradiated with a UV laser. The uptake of the laser energy by the matrix results in an explosion-like evaporation of the matrix, which in turn transfers the intact polymer molecules into the gas phase, where they become ionized and accelerated, allowing mass determination by a TOF analyzer. More details on the basics of MALDI-TOF-MS can be found elsewhere in this book. The high sensitivity of mass spectrometry in conjunction with its high precision in mass determination allows valuable information to be obtained on the fractions separated by chromatographic techniques. Unfortunately, the use of high vacuum in the mass spectrometer does not allow for direct coupling of MALDI-TOF-MS with chromatography. However, automated mixing of the fractions with the matrix and spotting on to the MALDI target has been performed using different instruments. In a variety of experiments, MALDI-TOF-MS was performed on fractions obtained from chromatographic experiments [100–103]. Often the molar mass in an SEC fraction was determined and assigned to the corresponding retention volume, allowing the calibration problem in SEC to be overcome.

However, MALDI-TOF-MS can also be used to identify fractions obtained by interaction chromatography. Figure 13.36 shows the chromatogram obtained for a mixture of C_{16} and C_{18} fatty alcohol alkoxylates obtained under critical conditions of PEG. It becomes clear that more than the two expected molecular spe-

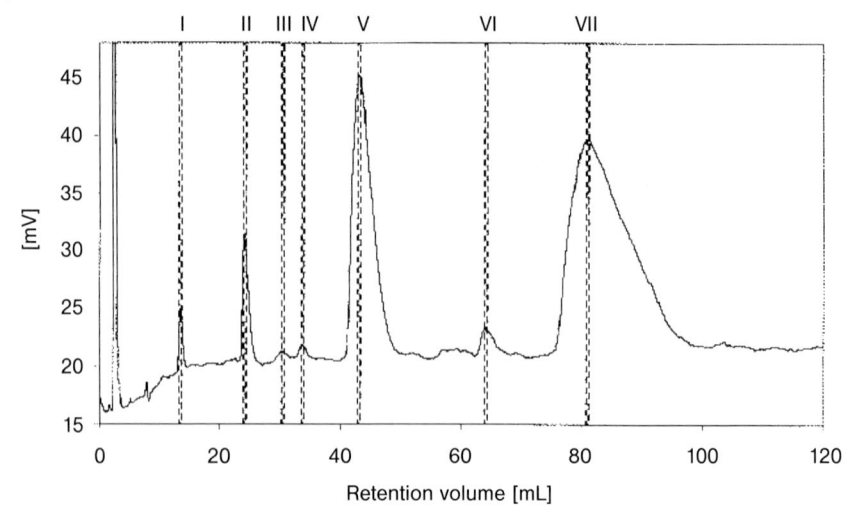

Fig. 13.36 Chromatogram of a separation of glycerol started poly(propylene oxide)s indicating the positions of fractions used for MALDI-TOF-MS analysis. Reprinted from [103], copyright 2003, with permission from Wiley-VCH.

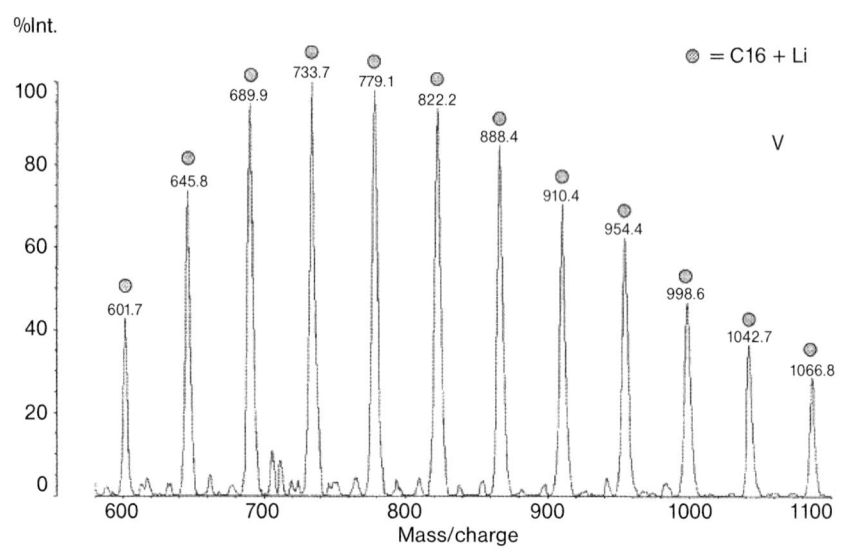

Fig. 13.37 MALDI-TOF mass spectrum of fraction V. Reprinted from [103], copyright 2003, with permission from Wiley-VCH.

cies are present in the product. The origin of the additional peaks is unclear. Fractions were taken at the positions indicated in the chromatogram. The fractions were analyzed by MALDI-TOF-MS. The spectra of fractions V and VII are shown in Figs. 13.37 and 13.38. In both cases mainly series of $[M+Li]^+$ molecu-

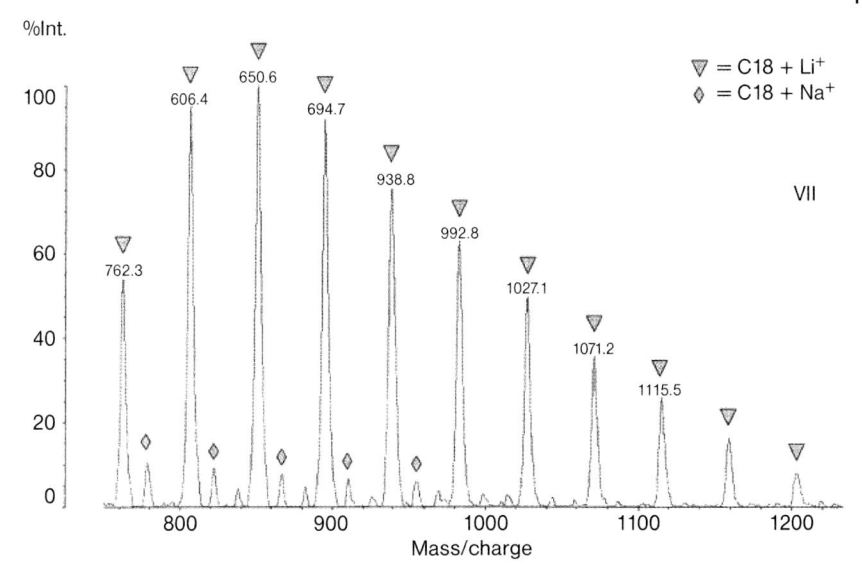

Fig. 13.38 MALDI-TOF mass spectrum of fraction VII. Reprinted from [103], copyright 2003, with permission from Wiley-VCH.

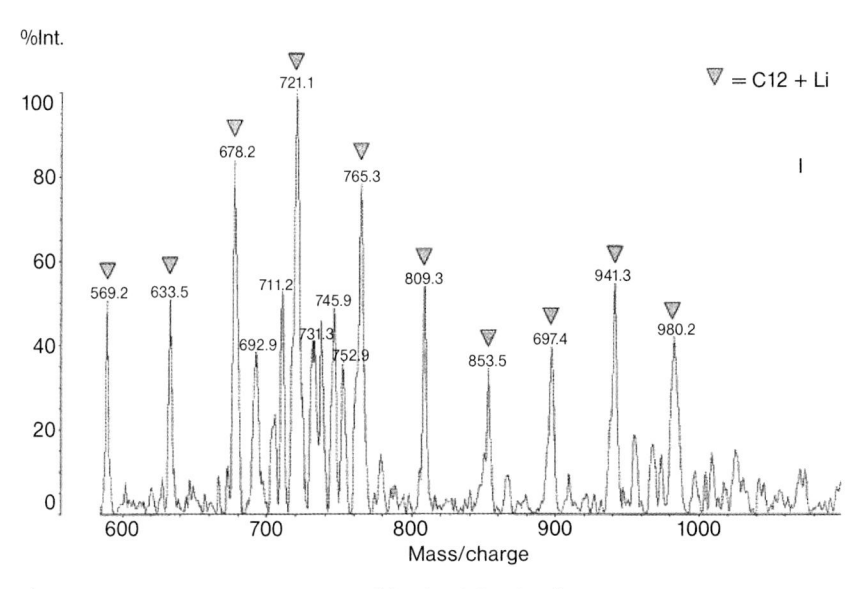

Fig. 13.39 MALDI-TOF mass spectrum of fraction I. Reprinted from [103], copyright 2003, with permission from Wiley-VCH.

lar ions were found and series having a mass differences of 44 g mol^{-1} could be identified. The difference of 44 mass units corresponds exactly to the molar mass increase due to the addition of a single repeating unit. This indicates that both series originate from PEOs having different degrees of polymerization. It is interesting that a complete PEO series is obtained despite the very narrow fractions taken. This result is in accordance with a separation at critical conditions of PEO. Since the elution does not depend on molar mass, all species having identical end-groups elute at the same retention volume, giving rise to the observed series for the fractions. The residual masses calculated from the spectra correspond to the end-groups of C_{16} and C_{18} fatty alcohol alkoxylates, respectively. The MALDI-TOF mass spectrum of fraction I is presented in Fig. 13.39. Again, a series having a mass separation of 44 g mol^{-1} is observed. The residual mass is calculated to be 169.3 g mol^{-1}. This mass equals the expected residual mass for a C_{12} fatty alcohol alkoxylate. Similarly, the other peaks were assigned to C_{14} (II), C_{14}/C_{15} (III, IV) and C_{16}/C_{17} (VI) fatty alcohol alkoxylates. The combination of chromatography and MALDI-TOF-MS therefore provides an additional powerful tool for the characterization of polymeric samples, especially if information on the functionality type distribution is required.

13.4
Conclusions

It has been shown that separations of polymers can be performed according to their different structural parameters. However, a single chromatographic technique often is not sufficient for a complete separation of a complex polymer sample. In addition, it is necessary to identify the structures of the different fractions. Chromatographic separations, especially if combined with spectroscopic or spectrometric methods, provide powerful tools for such detailed structural characterization of polymers. The examples given should serve to select the most suitable approach for a given characterization problem.

References

1 H. Pasch, *Adv. Polym. Sci.*, **2000**, *150*, 1.
2 T. Chang, *Adv. Polym. Sci.*, **2003**, *163*, 1.
3 H. Pasch, B. Trathnigg, *HPLC of Polymers*. Springer, Berlin, **1998**.
4 Y. Brun, Y. Alden, *J. Chromatogr. A*, **2002**, *699*, 25.
5 A. A. Gorbunov, A. M. Skvortsov, *Adv. Colloid Interface Sci.*, **1995**, *62*, 31.
6 L. R. Snyder, J. J. Kirland, J. L. Glajch, *Practical HPLC Method Development*. Wiley, New York, **1977**.
7 S. G. Entelis, V. V. Evreinov, A. V. Gorshkov, *Adv. Polym. Sci.*, **1986**, *76*, 129.
8 W. Radke, *Macromol. Theory Simul.*, **2001**, *10*, 668.
9 Z. Grubisic, P. Remp, H. Benot, *Polym. Lett.*, **1967**, *5*, 753.
10 H. Benoît, Z. Grubisic, P. Rempp, D. Decker, V. J. Zilliox, *Chem. Phys.*, **1966**, *63*, 1507.
11 H. K. Mahabadi, K. F. Driscoll, *J. Appl. Polym. Sci.*, **1977**, *21*, 1283.

12 A. R. Weiss, E. Cohn-Ginsberg, *J. Polym. Sci., Part B*, **1969**, *7*, 379.
13 S. Mori, *Anal. Chem.*, **1981**, *53*, 1813.
14 M. B. Huglin, *Light Scattering from Polymer Solutions*. Academic Press, London, **1972**.
15 P. Kratochvil, *Classical Light Scattering from Polymer Solutions*. Elsevier, Amsterdam, **1987**.
16 P. J. Wyatt, *Anal. Chim. Acta*, **1993**, *272*, 1.
17 B. H. Zimm, W. H. Stockmayer, *J. Chem. Phys.*, **1949**, *17*, 1301.
18 W. Radke, A. H. E. Müller, *Macromolecules*, **2005**, *38*, 3949.
19 W. Radke, P. Simon, A. H. E. Müller, *Macromolecules*, **1996**, *29*, 4926.
20 W. Radke, *GIT Lab. Fachz.*, **2005**, *5*, 428.
21 S. A. Berkowitz, *J. Liq. Chromatogr.*, **1983**, *6*, 1359.
22 M. Haney, *J. Appl. Polym. Sci.*, **1985**, *30*, 3023.
23 M. Haney, *J. Appl. Polym. Sci.*, **1985**, *30*, 3037.
24 M. A. Haney, *Am. Lab.*, **1985**, *17*, 116.
25 J. M. Goldwasser, *International CPC Symposium Proceedings*, **1989**, p. 150.
26 R. A. Sanayei, K. G. Suddaby, A. Rudi, *Makromol. Chem.*, **1993**, *194*, 1953.
27 W. W. Yau, *Chemtracts Macromol. Chem.*, **1990**, *1*, 1.
28 M. A. Haney, C. Jackson, W. W. Yau, *International GPC Symposium Proceedings*, **1991**, p. 49.
29 G. Glöckner, *Polymercharakterisierung durch Flüssigkeitschromatographie*. VEB Deutscher Verlag der Wissenschaften, Berlin, **1980**.
30 M. A. Bashir, A. Brüll, W. Radke, *Polymer*, **2005**, *46*, 3223.
31 T. Macko, D. Hunkeler, *Adv. Polym. Sci.*, **2003**, *163*, 62.
32 T. Chang, H. C. Lee, W. Lee, S. Park, C. Ko, *Macromol. Chem. Phys.*, **1999**, *200*, 2188.
33 W. Lee, H. Lee, H. C. Lee, D. Cho, T. Chang, A. A. Gorbunov, J. Roovers, *Macromolecules*, **2002**, *35*, 529.
34 H. Pasch, A. Deffieux, I. Henze, M. Schappacher, L. Rique-Lurbet, *Macromolecules*, **1996**, *29*, 8776.
35 H. C. Lee, H. Lee, W. Lee, T. Chang, J. Roovers, *Macromolecules*, **2000**, *33*, 8119.

36 T. Biela, A. Duda, K. Rode, H. Pasch, *Polymer*, **2003**, *44*, 1851.
37 W. Radke, K. Rode, A. V. Gorshkov, T. Biela, *Polymer*, **2005**, *46*, 5456.
38 H. Kukula, H. Schlaad, J. Falkenhagen, R.-P. Krueger, *Macromolecules*, **2002**, *35*, 7157.
39 X. Jiang, P. J. Schoenmakers, J. L. J. van Dongen, X. Lou, V. Lima, J. Brokken-Zijp, *Anal. Chem.*, **2003**, *75*, 5517.
40 X. Jiang, P. J. Schoenmakers, X. Lou, V. Lima, J. L. J. van Dongen, J. Brokken-Zijp, *J. Chromatogr. A*, **2004**, *1055*, 123.
41 X. Jiang, V. Lima, P. J. Schoenmakers, *J. Chromatogr. A*, **2003**, *1018*, 19.
42 A. V. Gorshkov, H. Much, H. Becker, H. Pasch, V. V. Evreinov, S. G. Entelis, *J. Chromatogr.*, **1990**, *523*, 91.
43 J. Adrian, D. Braun, H. Pasch, *Angew. Makromol. Chem.*, **1999**, *267*, 82.
44 J. Adrian, D. Braun, K. Rode, H. Pasch, *Angew. Makromol. Chem.*, **1999**, *267*, 73.
45 A. V. Gorshkov, V. V. Evreinov, B. Lausecker, H. Pasch, H. Becker, G. Wagner, *Acta Polym.*, **1986**, *37*, 740.
46 H. Pasch, I. Zammert, *J. Liq. Chromatogr.*, **1994**, *17*, 3091.
47 J. Falkenhagen, H. Much, W. Stauf, A. H. E. Müller, *Macromolecules*, **2000**, *33*, 3687.
48 H. Pasch, Y. Gallot, B. Trathnigg, *Polymer*, **1993**, *34*, 4986.
49 H. Lee, T. Chang, D. Lee, M. S. Shim, H. Ji, W. K. Nonidez, J. W. Mays, *Anal. Chem.*, **2001**, *73*, 1726.
50 W. Lee, D. Cho, T. Chang, K. J. Hanley, T. P. Lodge, *Macromolecules*, **2001**, *34*, 2353.
51 H. Pasch, M. Augenstein, *Makromol. Chem.*, **1993**, *194*, 2533.
52 H. Pasch, C. Brinkmann, H. Much, U. Just, *J. Chromatogr.*, **1992**, *623*, 315.
53 H. Pasch, C. Brinkmann, Y. Gallot, *Polymer*, **1993**, *34*, 4100.
54 J. Adrian, E. Esser, G. Hellmann, H. Pasch, *Polymer*, **2000**, *41*, 2439.
55 H. Pasch, M. Adler, F. Rittig, S. Becker, *Macromol. Rapid Commun.*, **2005**, *26*, 438.
56 E. Esser, D. Braun, H. Pasch, *Angew. Makromol. Chem.*, **1999**, *271*, 61.
57 H. Pasch, *Polymer*, **1993**, *34*, 4095.
58 H. Pasch, K. Rode, *Polymer*, **1998**, *39*, 6377.

59 H. Pasch, K. Rode, N. Chaumien, *Polymer*, **1996**, *37*, 4079.
60 D. Berek, M. Janco, K. Hatada, T. Kitayama, N. Fujimoto, *Polym. J.*, **1997**, *29*, 1029.
61 A. Gorbunov, A. Skvortsov, B. Trathnigg, M. Kollrose, M. Parth, *J. Chromatogr. A*, **1998**, *798*, 187.
62 B. Trathnigg, B. Maier, A. Gorbunov, A. Skvortsov, *J. Chromatogr. A*, **1997**, *791*, 21.
63 A. Gorbunov, B. Trathnigg, *J. Chromatogr. A*, **2002**, *955*, 9.
64 C. Rappel, B. Trathnigg, A. Gorbunov, *J. Chromatogr. A*, **2003**, *984*, 29.
65 T. Chang, W. Lee, S. Park, D. Cho, *Am. Lab.*, **2001**, *33*, 24.
66 A. M. Skvortsov, A. A. Gorbunov, *J. Chromatogr.*, **1990**, *507*, 487.
67 A. A. Gorbunov, A. M. Skvortsov, *Int. Lab.*, **1995**, *25*, 8J.
68 J. Ryu, S. Park, T. Chang, T, *J. Chromatogr. A*, **2005**, *1075*, 145.
69 D. Cho, S. Park, T. Chang, A. Avgeropoulos, N. Hadjichristidis, *Eur. Polym. J.*, **2003**, *39*, 2155.
70 T. Chang, W. Lee, H. C. Lee, D. Cho, S. Park, *Am. Lab.*, **2002**, *34*, 39.
71 W. Lee, H. C. Lee, S. Park, T. Chang, K. H. Chae, *Macromol. Chem. Phys.*, **2000**, *201*, 320.
72 H. C. Lee, W. Lee, T. Chang, J. S. Yoon, D. J. Frater, J. W. Mays, *Macromolecules*, **1998**, *31*, 4114.
73 H. C. Lee, T. Chang, S. Harville, J. W. Mays, *Macromolecules*, **1998**, *31*, 690.
74 Y. Brun, *J. Liq. Chromatogr. Relat. Technol.*, **1999**, *22*, 3067.
75 M. Augenstein, M. Stickler, *Macromol. Chem.*, **1990**, *191*, 415.
76 T. Kawai, M. Akashima, S. Teramachi, *Polymer*, **1995**, *36*, 2851.
77 S. Teramachi, A. Hasegawa, T. Matsumoto, K. Kitahara, Y. Tsukahara, Y. Yamashita, *Macromolecules*, **1992**, *25*, 4025.
78 S. Teramachi, S. Sato, H. Shimura, S. Watanabe, Y. Tsukahara, *Macromolecules*, **1995**, *28*, 6183.
79 D. Braun, I. Krämer, H. Pasch, S. Mori, *Macromol. Chem. Phys.*, **1999**, *200*, 949.
80 D. Braun, I. Henze, H. Pasch, *Macromol. Chem. Phys.*, **1997**, *198*, 3365.
81 I. R. Garcia, H. Pasch, *e-Polymers*, **2005**, *042*.
82 W. Lee, H. C. Lee, S. Park, T. Chang, J. Y. Chang, *Polymer*, **1999**, *40*, 7227.
83 H. C. Lee, T. Chang, *Macromolecules*, **1996**, *29*, 7294.
84 G. Glöckner, *Gradient HPLC of Copolymers and Chromatographic Cross-Fractionation*. Springer, Berlin, **1991**.
85 H. Pasch, P. Kilz, *GIT Lab. Fachz.*, **1999**, *43*, 239.
86 P. Kilz, R. P. Krüger, H. Much, G. Schulz, *ACS Adv. Chem.*, **1995**, *247*, 223.
87 P. Kilz, R. P. Krüger, H. Much, G. Schulz, *PMSE Prepr.*, **1993**, *69*, 114.
88 P. Kilz, *Lab. Praxis*, **1993**, *6*, 64.
89 T. Biela, A. Duda, K. Rode, S. Penczek, H. Pasch, *J. Polym. Sci., Part A: Polym. Chem.*, **2003**, *41*, 2884.
90 X. Jiang, A. van der Horst, V. Limac, P. J. Schoenmakers, *J. Chromatogr. A*, **2005**, *1076*, 51.
91 J. Gerber, W. Radke, *e-Polymers*, **2005**, *045*.
92 J. Falkenhagen, H. Much, W. Stauf, A. H. E. Müller, *Polym. Prepr.*, **1999**, *40*, 984.
93 J. Gerber, W. Radke, *Polymer*, **2005**, *46*, 9224.
94 K. Im, Y. Kim, T. Chang, K. Lee, N. Choi, *J. Chromatogr. A*, **2006**, *1103*, 235.
95 L. M. Wheeler, J. N. Willis, *Appl. Spectrosc.*, **1993**, *47*, 1128.
96 E. Esser, H. Pasch, P. Montag, *GIT Lab. Fachz.*, **1996**, *16*, 68.
97 J. N. Willis, J. K. Dwyer, L. M. Wheeler, *Polym. Mater. Sci.*, **1993**, *69*, 120.
98 W. Hiller, H. Pasch, *Macromolecules*, **1996**, *29*, 6556.
99 I. Krämer, W. Hiller, H. Pasch, *Macromol. Chem. Phys.*, **2000**, *210*, 1662.
100 J. Falkenhagen, R.-P. Krüger, G. Schulz, in P. Jutzi, U. Schubert (Eds.) *Silicon Chemistry*, **2003**, 406.
101 S. M. Weidner, J. Falkenhagen, H. Much, R.-P. Krüger, J. F. Friedrich, *Polym. Prepr.*, **2000**, *41*, 655.
102 R. P. Krüger, J. Falkenhagen, G. Schulz, J. Gloede, *GIT Lab. Fachz.*, **2001**, *45*, 380.
103 K.-H. Spriestersbach, K. Rode, H. Pasch, *Macromol. Symp.*, **2003**, *193*, 129.

14
NMR Spectroscopy

Hans Wolfgang Spiess

14.1
Introduction

Precise knowledge of the structure and dynamics of macromolecules of well-defined architecture is key to tailoring them for specific functions. For instance, such diverse technological challenges as efficient fuel cells, photonic sensors and devices and gene delivery systems all require transport of electrons, holes, protons or other ions. Likewise, the properties of conventional polymers can be substantially improved by controlling, e.g., their microstructure and the processing conditions. Properties of macromolecular systems critically depend on the arrangement of the building blocks of the material relative to each other and their mobility on different length- and time-scales [1]. Therefore, success in polymer science requires the development of characterization techniques that are able to provide information on these aspects. Hence scattering of light [2], X-rays [3] and neutrons [4], the various forms of microscopy [5] and mechanical [6] and dielectric [7] relaxation are well established in polymer science.

NMR spectroscopy, on the other hand, is often considered just as an analytical tool, monitoring the various synthetic steps that eventually lead to the new polymers or supramolecular systems. High-resolution NMR in one and two dimensions [8] provides a number of important parameters that are specific to macromolecules [9]. Examples are stereochemical configuration, geometric isomerism and regioregularity. The ability to identify selectively the NMR signals of end-groups offers a precise way of determining the molar mass of macromolecules and advanced multidimensional techniques allow the determination of the three-dimensional structure of biopolymers in solution [10]. It is interesting that important advances in NMR technology have always been applied to synthetic and even biological macromolecules almost immediately after they had been invented [11].

More recently, the rapid development of solid-state NMR has substantially broadened the application of NMR to polymers [12]. In particular, molecular and collective motions can now be characterized in unprecedented detail. The new NMR techniques use concepts well established in X-ray and neutron scattering, yet on much longer time-scales and provide structural information with atomic resolution even for systems that lack crystalline order in the traditional sense [13]. Moreover, solid state NMR concepts can also be applied to partially ordered mobile systems such as liquid crystals [14] and also highly viscous materials such as polymer melts in the vicinity of the glass transition temperature or elastomers.

After a brief outline of the basics of solid-state NMR, this chapter describes specific examples of the study of macromolecular and supramolecular systems by this technique.

14.2
Background

NMR spectroscopy is remarkably versatile and, therefore, is widely applied in many fields of science, in particular physics, chemistry, biology, materials science and medicine [15]. This is due to the fact that this form of spectroscopy combines a number of features that make it an almost ideal tool: NMR is non-destructive, sample preparation is easy and the possibility of observing different nuclei and isotopes provides extreme structural selectivity [8]. Moreover, dynamic features can be studied over many decades of characteristic times, from picoseconds to minutes [12], and length-scales from interatomic distances in the 100-pm range up to 1 m or so in NMR imaging [16]. In fact, despite today's dominance of the applications of high-resolution liquid-state NMR in polymer chemistry, the relation of molecular dynamics and NMR relaxation parameters was elucidated long before the discovery of the chemical shift and the J-coupling, which provided the basis of chemical applications. Moreover, the introduction of Fourier transform NMR and its extension to two and higher dimensions [8] made it possible to include low-sensitivity, yet highly informative, spectroscopy of rare nuclei, such as ^{13}C and ^{15}N.

The wealth of information accessible by NMR spectroscopy results from the fact that a variety of interactions of the nuclear spins with their surroundings can be exploited [8, 12]. The structure of polymers and other supramolecular systems can be elucidated via several routes. The chemical shift provides the basis of site selectivity and direct geometric parameters such as internuclear distances and dihedral angles are encoded in the dipole–dipole and the J-coupling. Of the great variety of molecular motions possible in polymers, rotations have the most pronounced effects on NMR spectra and relaxation parameters, because the spin interactions are angle dependent. However, conformational dynamics and translational motions can also be tackled. The anisotropic nuclear spin interactions also offer a means to probe the alignment of residues in partially ordered systems such as drawn fibers [17] and oriented liquid crystals [14]. Therefore, a brief outline of these anisotropic spin interactions is given here.

Table 14.1 Interactions of nuclei between themselves and with their surroundings as a basis for elucidating structure and dynamics of polymers.

Interaction	Electronic structure	Geometry	Nuclei	Structure	Dynamics
Chemical shift	Yes	Intrinsic and orientation	^{1}H, ^{13}C, ^{15}N, ^{19}F, ^{29}Si, ^{31}P	Conformation, through-space proximities	Conformational transitions, rotational motions
Dipole–dipole coupling	No	Internuclear distance and orientation	^{1}H, ^{13}C, ^{15}N, ^{19}F, ^{29}Si, ^{31}P	Through-space distances	Translational and rotational motions
J-coupling	Yes	Intrinsic, internuclear distance and orientation	^{1}H, ^{13}C, ^{15}N, ^{19}F, ^{29}Si, ^{31}P	Conformation and intergroup binding	Conformational transitions, rotational motions
Quadrupole coupling	Yes	Intrinsic and orientation	^{2}H, ^{14}N, ^{17}O, ^{23}Na, ^{27}Al	Symmetry of electronic environment, chemical bonding	Rotational motions

14.2.1
Anisotropic Spin Interactions

As noted above, the information that can be extracted from solid-state NMR spectra is encoded via spin interactions such as the chemical shift, the quadrupolar interaction and homo- and heteronuclear dipolar interactions, in addition to J-couplings [8–10, 12]. For convenience, these couplings and the information that is accessible through them are listed in Table 14.1.

A common characteristic of the relevant spin interactions is that they are anisotropic and can be described by second-rank tensors. The resulting orientation-dependent NMR frequency for each of the interactions is alike and of the following form [8, 12]:

$$\omega(\alpha,\beta) - \omega_\mathrm{L} = \omega_\mathrm{iso} + \frac{\delta}{2}(3\cos^2\beta - 1 - \eta\sin^2\beta\cos 2\alpha) \qquad (1)$$

where ω_L is the Larmor frequency, ω_iso is the isotropic chemical shift and the other terms reflect the deviations due to angular-dependent contributions. The strength of the anisotropic interaction is specified by δ and η ($0 \leq \eta \leq 1$) is the asymmetry parameter which describes the deviation from axial symmetry. The polar angles α and β relate the orientation of the principal axes system of the interaction tensor with the external magnetic field.

The most important interaction with respect to chemical information is the *chemical shift*, which results from the shielding of the magnetic field at the position of the nucleus by the electrons. Major advances have been achieved recently in quantum chemical calculations of this parameter in bulk at different levels of precision and calculational cost [18–20]. The NMR frequency is thus shifted by an isotropic contribution ω_{iso} and by angle-dependent contributions. Assuming an equal probability for all directions, one can calculate the powder average and a broad powder spectrum is obtained which reflects the chemical shift anisotropy. For the case of an axially symmetric chemical shift tensor, η is zero and the angle-dependent term in Eq. (1) simplifies to $\delta(3\cos^2\beta-1)/2$. Usually the chemical shift in polymers spans about 10 ppm for ^{1}H, 200 ppm for ^{13}C and 400 ppm for ^{15}N nuclei.

For abundant nuclei with spin 1/2, the spectrum is often dominated by heteronuclear or homonuclear *dipole–dipole interactions*, i.e. the interactions between the magnetic moments of two neighboring spins. In this case there is no isotropic contribution and η is zero, so that Eq. (1) simplifies correspondingly. For a two-spin system one obtains a Hamiltonian of the form

$$H_D = \left(\frac{\mu_0}{4\pi}\right) \frac{\gamma_1 \gamma_2 \hbar}{r_{12}^3} \left(\frac{3\cos^2\beta - 1}{2}\right) (3I_{1z}I_{2z} - \vec{I}_1\vec{I}_2) \tag{2}$$

where r_{12} is the magnitude of the vector connecting the two spins, θ is the angle of this vector to the magnetic field, the $\vec{I}_i$ are spin operators and the γ_i ($i=1,2$) are the magnetogyric ratios of the spins. That is, the strength of the resulting line splitting depends strongly on the distance between the two spins, so that distance information can be extracted from such spectra. Homonuclear interaction (equivalent spins with $\gamma_1=\gamma_2=\gamma$) and heteronuclear interaction (non-equivalent spins with $\gamma_1 \neq \gamma_2$) have to be distinguished. In the latter case, the flip-flop term which is part of the product $\vec{I}_1\vec{I}_2$ in Eq. (2) can be neglected. For a powder sample one again has to take into account all angles β and thus obtains the so-called Pake spectrum with a considerable anisotropic line broadening of up to 50 kHz for homonuclear ^{1}H–^{1}H and up to 25 kHz for heteronuclear ^{13}C–^{1}H or ^{15}N–^{1}H dipolar interaction. Since the dipole–dipole coupling is a through-space interaction, however, we have in principle to evaluate the sum over all possible pair interactions. This and the presence of molecular motions lead to considerable complications and are responsible for the experimental finding that in practice not a Pake spectrum but a relatively structureless lineshape is obtained.

In the case of a deuterated sample (^{2}H, spin $I=1$), the spectra are usually dominated by the *quadrupole interaction*, that is, the coupling of the nuclear quadrupole moment with the electric field gradient of the C–^{2}H bond. For deuterons in C–^{2}H bonds, this can lead to a splitting of about 250 kHz. As in the case of the dipole–dipole interaction, a Pake spectrum is obtained for a powder sample. The z-principal axis of the quadrupolar interaction is oriented along the bond axis, which makes deuteron NMR particularly useful for studies of segmental orientations and molecular dynamics (reorientation) [12].

In sufficiently mobile (i.e. liquid-like) systems, the anisotropy is averaged out by the isotropic thermal motions, leaving only the isotropic contributions.

14.2.2
Manipulation of Spin Interactions

The rich information content of solid-state spectra makes them difficult to handle, in particular if more than one of the interactions introduced above has to be taken into account. NMR methodology, however, provides the possibility of decoupling and recoupling spin interactions nearly as desired (see [21] for a comprehensive introduction). Moreover, the different sources of information can be separated and correlated using the two-dimensional techniques discussed below.

The most prominent example of a technique for decoupling or line narrowing is magic angle spinning (MAS) [8, 12, 21]. Here, the angle-dependent part of the interactions is modulated by rapidly spinning the sample around an axis inclined at an angle Θ to the magnetic field. If the spinning axes is chosen along the so-called magic angle $\Theta_m = 54.7°$, the relevant scaling factor $(3\cos^2\Theta - 1)/2$ becomes zero and the anisotropic part of the interaction vanishes (Fig. 14.1). Without MAS, the static ^{1}H NMR line for organic solids is a Gaussian, typically 30–35 kHz in width. Under 30-kHz spinning in high magnetic fields, the ^{1}H chemical shift is partially resolved. In mobile systems such as elastomers or swollen polymers, so called high-resolution MAS (HRMAS) with moderate spinning (below 10 kHz) provides even higher spectral resolution, which is still substantially lower, however, than in solution. Note that the ^{1}H chemical shift in solids can be much larger than in solution due to packing effects, indicated by different horizontal scales in Fig. 14.1. Today, MAS rotation frequencies reaching 70 kHz [22] are possible. This paves the way for easy access to high spectral resolution in solid-state ^{1}H NMR and, more important, greatly simplifies the dipolar coupled network [23] that prevents the dipole–dipole coupling between ^{1}H from being used for specific structural investigations on non-spinning samples.

In addition, it is possible to manipulate the spin interactions by pulse techniques [8, 12, 21]. These act on the spin operators in the corresponding Hamiltonians (see, for example, Eq. 2) rather than on the geometric part. Depending on the applied pulse sequence, a given spin interaction can be switched on and off in order to discriminate the different contributions to the desired information. For instance, ^{13}C chemical shift values can be determined by selectively irradiating the protons. This so-called heteronuclear dipolar decoupling removes any influence of the coupling to the protons from the ^{13}C spectrum. Heteronuclear dipole–dipole couplings are also averaged out by MAS, yet a simple train of 180° pulses properly synchronized with the rotor phase reintroduces the coupling for a well-defined period of time. This so-called rotational-echo double-resonance (REDOR) technique [24] can be seen as the basis of many more sophisticated ways of reintroducing homo- and heteronuclear dipole–dipole as well as anisotropic chemical shift interactions under MAS [25–27]. Pictorially, the mechanical rotation by MAS in real space is partially offset by counter rotations in spin space.

Magic angle spinning modulates the spin interactions periodically. This means that it generates rotational echoes and data acquisition can be performed in two ways [12]: if only the echo height is monitored (rotor-synchronized acquisition)

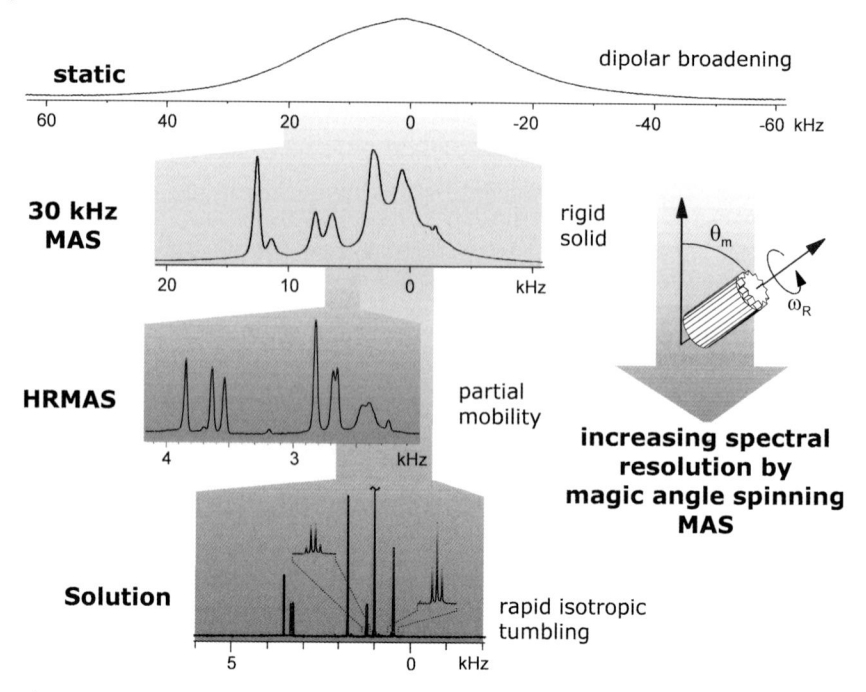

Fig. 14.1 Increase in spectral resolution of NMR spectra by magic angle spinning (MAS) at a ^{1}H NMR frequency of 700 MHz.

a single line results in the spectrum for each spectroscopically resolved site and information about the anisotropic coupling is lost; if the whole echo train is monitored, a sideband pattern results, which retains information about the anisotropic coupling, yet with spectral resolution of the different sites. This is important for precise structural elucidation, based on the dipole–dipole coupling as well as using this interaction to study molecular dynamics resulting in reduction of the coupling.

14.2.3
Double Quantum NMR

The basic concept for using distance- and angle-dependent dipole–dipole coupling for structural studies is displayed in Fig. 14.2a; details are described in extended reviews [25, 26]. In a two-dimensional experiment, double quantum (DQ) coherence is created, e.g. between two ^{1}H with like or different chemical shifts. During excitation of the DQ coherence, the dipole–dipole coupling between the two spins, which is largely reduced by MAS, is recoupled by a suitable pulse sequence. In the evolution time of DQ coherence recoupling is turned off, such that the different residues can be distinguished by their different chemical shifts. In the subsequent reconversion to single quantum coher-

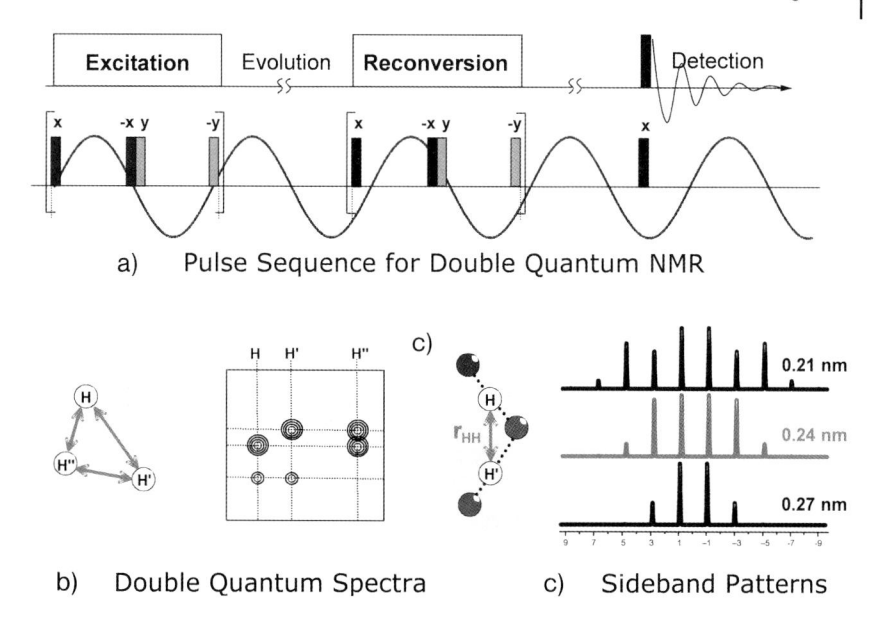

Fig. 14.2 Principle of double quantum NMR as a tool for studying structure and dynamics of supramolecular systems. (a) Pulse sequence; (b) double quantum NMR spectra; (c) sideband patterns.

ence needed for signal detection, recoupling is again applied. Thus, a 2D spectrum as shown in Fig. 14.2b is recorded, in which information about internuclear distances is, first of all, encoded in the strength of the DQ peaks.

It should be appreciated that the technique is based on analyzing signals resulting from a *coherence* originating from two spatially separated nuclei. Thus the basic physics is analogous to coherent X-ray or neutron *scattering*, where the coherent superposition of scattered waves is exploited to deduce the distance between the scattering centers. In NMR, however, no translational symmetry is necessary to determine distances in the range 100–1000 pm. In fact, proximities of ^{1}H in the same or different moieties can be probed by DQ NMR by analyzing the intensities of a so-called rotor-synchronized spectrum, which can typically recorded in about 10 min for 10 mg as-synthesized samples, without the need for isotopic labeling. If the dipole–dipole coupling needs to be determined more accurately, for more precise determination of internuclear distances or molecular dynamics (see above), DQ sideband spectra, as displayed in Fig. 14.2c, are recorded. The measurement time is then considerably longer, typically overnight.

Such DQ spectra can be recorded for both homonuclear ^{1}H–^{1}H and heteronuclear ^{1}H–^{13}C and ^{1}H–^{15}N coherences, exploiting the much higher site selectivity of ^{13}C and ^{15}N chemical shifts [25, 26, 28]. In the heteronuclear case, polarization transfer and recoupling take advantage of the popular REDOR technique introduced above [24], usually applied to isotopically labeled samples.

Moreover, the sensitivity of such heteronuclear experiments can be increased significantly by detecting the signal of the rare spin, in particular ^{15}N through ^{1}H. By this so-called indirect detection, the measurement time for heteronuclear ^{1}H–^{15}N spectra was recently be reduced [29] by a factor of about 100.

14.2.4
Two-dimensional NMR Spectroscopy

Many, if not most, of the advanced NMR techniques make use of two-dimensional (2D) spectroscopy, because of the superb increase in resolution and ease of information encoding. There is insufficient space to explain these techniques in detail, but comprehensive books are available [8, 12] and only the basics are described here. In general, a 2D NMR experiment is divided into several time periods that follow each other. In order to record a 2D NMR spectrum, a two-dimensional data set is acquired as a function of two time variables, t_1 and t_2, as shown schematically in Fig. 14.2a for the specific case of double quantum NMR. This is preceded by the so-called preparation period in which coherences are excited by a suitable pulse sequence, which in the simplest case is only one radiofrequency pulse. Unlike in conventional (1D) NMR spectroscopy, the excited signal is not directly acquired but is allowed to evolve in the so-called evolution period under the influence of the relevant spin interactions. The evolution time t_1 is incremented in subsequent experiments and provides the first time dimension of the experiment. After the evolution period (and an optional mixing time), the remaining signal is directly detected in the detection period for each time increment, thus generating a 2D data set. Two-dimensional Fourier transformation then gives the 2D spectrum. Optionally a so-called mixing period of length t_m can be inserted between the evolution and detection periods. During t_m, changes in the system can occur, for instance, by molecular motions, spin interactions, relaxation or spin manipulation (e.g. recoupling).

The different aspects of 2D NMR spectroscopy are reflected in the different variants that can be distinguished. One variant, *separation* spectroscopy, is used to separate different interactions taking advantage of the spin manipulation techniques. For instance, during the evolution period the spin manipulation can be made such that only the isotropic chemical shift is acquired whereas in the detection period the full spectrum is acquired. This leaves the anisotropy to be studied site selectively.

Other 2D NMR techniques, so-called *correlation* techniques, aim at obtaining new information by correlating different interactions. For instance, the ^{13}C chemical shift anisotropy can be correlated with the heteronuclear dipolar powder pattern in order to obtain information on the relative orientation of the two tensors. Moreover, since no signal is detected during the evolution time, it can conveniently be used to provide the basis for recording a double quantum spectrum which correlates single and double quantum coherences. Considering the manifold of spin manipulation techniques, there is a wealth of such 2D NMR techniques that can be derived for different purposes.

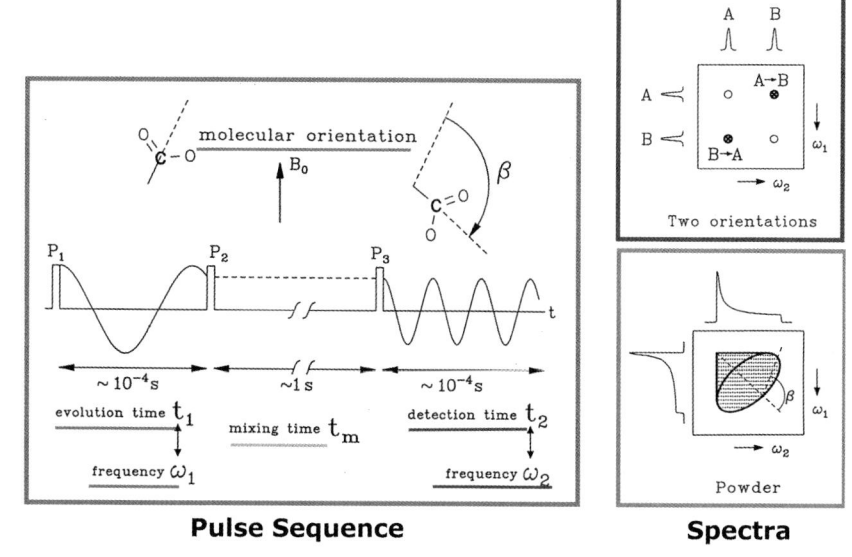

Fig. 14.3 Principle of 2D exchange NMR as a tool for studying molecular motions.

Finally, introducing a mixing time t_m, 2D *exchange* spectroscopy can be performed (Fig. 14.3). The most important application of such exchange techniques with respect to polymer investigations is the study of slow molecular dynamics. In these experiments, reorientations due to molecular dynamics are allowed to take place during the mixing time t_m and lead to characteristic off-diagonal patterns in the resulting 2D spectra. If the mixing time is increased in a series of 2D experiments, slow dynamics in the range of milliseconds to seconds can be investigated in detail. For instance, rotation of molecules by a well-defined angle leads to an elliptical exchange ridge for a powder. This can be viewed as a Lissajous figure, from which the angle through which the molecules have rotated can be read off directly using a ruler [30]. The measuring time can be dramatically reduced in a 1D variant under MAS [31]. Likewise, exchange of magnetization can also occur by cross-relaxation or mutual spin flips, the latter being designated *spin diffusion*. It can be described by a diffusion equation and provides access to domain dimensions in phase-separated systems.

14.3
Applications

Due to the versatility of NMR spectroscopy, innumerable applications exist, which cannot be dealt with fully in a single chapter. Therefore, selected examples of recent applications concentrating on the relation between chain architecture and the architecture of building blocks and the resulting macromolecular or supramolecular systems have been selected, taken mostly from the work of our own group. Other examples are readily available from reviews in series such as *Progress in NMR Spectroscopy* [32], *Annual Reports on NMR Spectroscopy* [33] and *Advances in Polymer Science* [34].

14.3.1
Chain Microstructure

The macroscopic properties of polyalkenes depend strongly on the chain microstructure. In recent years, new single-site catalysts [35] have allowed much greater synthetic control over the polydispersity, type of branch and branch content. In polyethylene, the physical properties of both the solid and the melt can be tuned by the presence of branches of various lengths in the polymer backbone. Short-chain branches (SCBs) can be introduced by copolymerization of ethylene with short alkenes or by isomerization reactions during ethylene homopolymerization. Such branches form structural defects during crystallization and thus strongly affect crystallization rates, ultimate crystallinity and other bulk mechanical properties [36]. Similarly, long-chain branches (LCBs) are

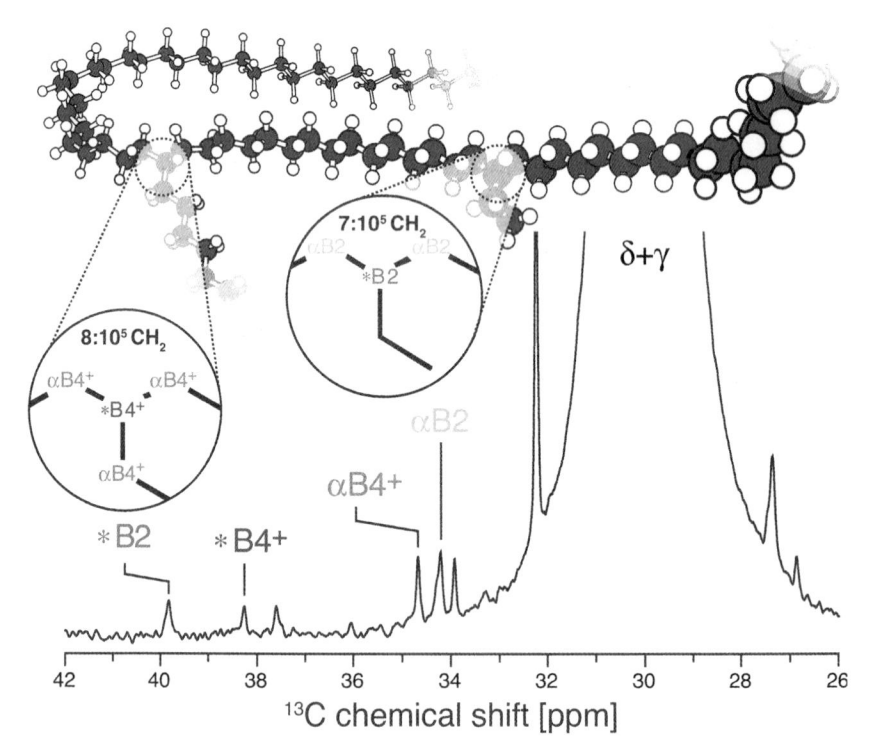

Fig. 14.4 Quantification of low branch content in optimized melt state ^{13}C NMR under MAS [39].

formed by macromonomer incorporation during polymerization. With these branches typically being longer than the entanglement molecular weight, their presence strongly affects the processability of the bulk polymer. Long-chain branching is known to influence the zero-shear viscosity even at concentrations of 2 LCBs per 100 000 CH_2 groups. Hence it is very important to quantify the degree of chain branching and ^{13}C NMR in solution seems to be the method of choice as the chemical shifts of branch points and also adjacent carbon positions can be distinguished from the backbone resonances [9].

However, when applied to polyethylene, a number of problems arise, the most important being the low solubility of polyethylene, even at high temperatures. The low concentration of ^{13}C nuclei results in low NMR signals requiring extended measurement times, especially for the quantification of very low levels of branching. In one study, up to 2×10^6 transients had to be acquired at high field (600 MHz) in order to determine branch contents of 3–8 branches per 100 000 CH_2 groups [37]. Another inherent limitation of solution-state NMR is the lack of access to cross-linked and other non-soluble fractions of polyethylene. Solid-state NMR, under MAS, on the other hand, can be used to overcome some of these limitations, despite the substantial loss of spectral resolution compared with solution NMR [38]. Under optimized conditions, time savings of about two orders of magnitude were achieved, allowing quantification of 7–9 branches in 100 000 CH_2 groups within 13 h [39] (Fig. 14.4).

14.3.2
Heterogeneous Polymer Melts

Semicrystalline polymers containing amorphous and crystalline regions usually have intimately mixed chains. The resulting topological constraints (entanglements) in the amorphous regions limit the drawability in the solid state [40, 41]. By controlled synthesis, the number of entanglements can be reduced. Ultimately, crystals composed of single chains are feasible, where the chains are fully separated from each other. If such separation can be maintained in the melt, a new melt state can be formed. This can be achieved through slow and carefully controlled melting of such polymer crystals, resulting in a heterogeneous melt with more entangled regions, where the chains are mixed, and less entangled ones, composed of individually separated chains. Advanced NMR spectroscopy was applied to probe the local chain dynamics, which is indicative of structural differences. This technique is able to detect dynamic heterogeneities and also chain order in polymer melts, even if the dynamics in different regions are very similar, that is, in cases of extremely low structural contrast. In an early example reduced four-dimensional spectroscopy was combined with spin diffusion at temperatures close to the glass transition [42]. Another approach uses double quantum methods [43] to study chain organization in the melt and is applicable at higher temperatures. As shown in Fig. 14.5, marked differences in double quantum filtered ^{1}H spectra of polyethylene were detected between a heterogeneous melt obtained after slowly heating a sample composed

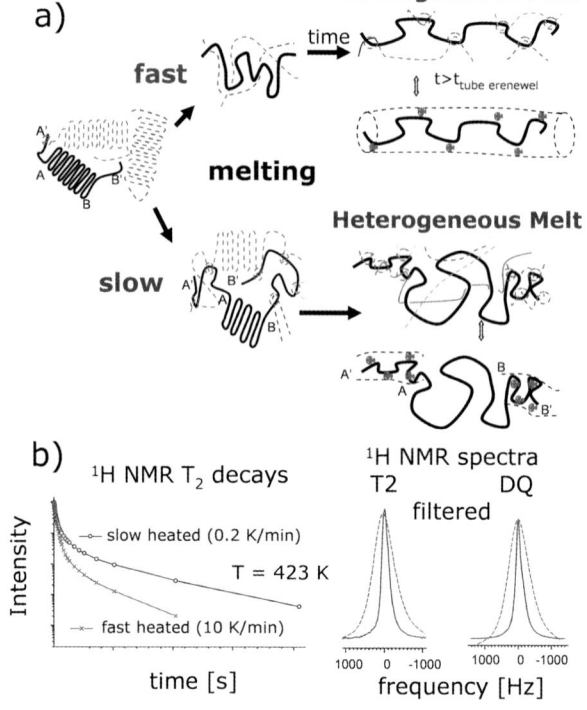

Fig. 14.5 (a) Scheme of a heterogeneous melt generated by slowly heating crystals composed of single chains. The entanglements are at the chain ends; the middle part of a chain remains disentangled. (b) ^{1}H NMR transverse relaxation curves and NMR spectra after filtering for the mobile parts. The heterogeneity shows up in the slowly heated sample through a much narrower line [44].

of single-chain single crystals and a homogeneous melt of the same system, where, remarkably, the heterogeneous melt contains a fraction of chain segments with higher mobility attributed to unentangled chains [44]. The long-lived heterogeneous melt shows decreased melt viscosity and provides enhanced drawability on crystallization. Chain reptation, required for the homogenization of the entanglement distribution, was found to be considerably hindered.

14.3.3
Micellar Aggregates from Block Copolymers

A central objective in micellar science is the synthesis of new materials to increase control of micellar architecture and functionality. However, the scope of conventional micelles composed of low molecular weight surfactants or block copolymers is inherently limited by the use of only one type of amphiphilic molecule generating only one type of micellar surface or core environment in solution. In nature, many of the problems associated with molecular recogni-

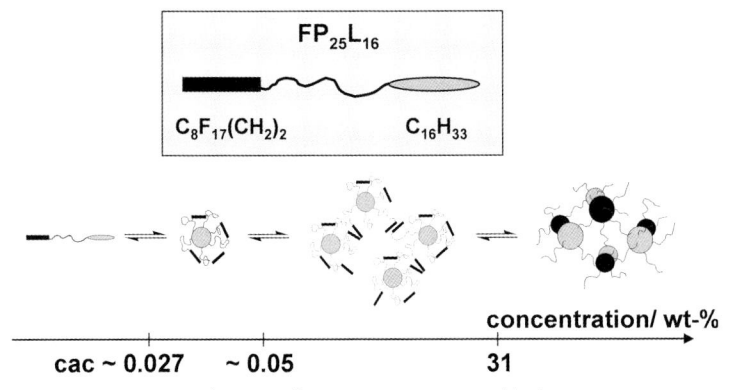

Fig. 14.6 Aggregation diagram of an asymmetric ABC triblock copolymer in water with a hydrophilic polymer end-tagged with a fluorocarbon and a hydrocarbon groups [46].

tion, catalysis and molecular engineering have been solved by compartmentation of highly complex biological systems into well-separated functional and structural units [45]. A straightforward approach for generating two-compartment micellar systems is based on the self-association behavior of asymmetric ABC triblock copolymers. As an example, we studied hydrophilic poly(N-acylethylenimine) polymers, end-tagged with hydrophobic fluorocarbon and hydrocarbon groups of variable length [46]. The aggregation behavior was probed by steady-state fluorescence spectroscopy with pyrene as the probe molecule and ^{19}F NMR spectroscopy. Transverse (T_2) relaxation for all positions of the fluorocarbon group is bimodal, where the shorter relaxation is ascribed to the micelles. Moreover, the aggregation also manifests itself in the ^{19}F chemical shift.

The aggregation behavior of these systems is summarized in Fig. 14.6. Below the critical aggregate concentration (cac), only unimers, 1.4 nm in dimension, exist in aqueous solution. Above the critical micellar concentration (cmc), only the hydrocarbon chain ends form aggregates which are detected by steady-state fluorescence. These micellar aggregates have a diameter of around 5.5 nm. At higher concentrations, in addition aggregates formed by fluorinated end groups are identified in ^{19}F NMR relaxation experiments. Because of the steric hindrance of the polymer chain in the inner block the fluorocarbon end-groups form dimers, trimers, etc., with increasing polymer concentration. The aggregates are star-like micelles with a fluorocarbon surface which form a network at higher concentrations. The infinite network defined by the complete aggregation of all fluorocarbon and hydrocarbon end-groups is reached at 31 wt%. Thus, the hydrocarbon and fluorocarbon aggregates are formed at different concentrations, because the stability of such aggregates is different. Overall, the formation of the infinite network is similar to that in symmetric ABA triblock copolymers reported earlier [47]. In the asymmetric ABC triblock copolymer, however, micelle and network formation can clearly be separated.

14.3.4
Elastomers

From the viewpoint of NMR, elastomers and other viscoelastic polymers above their glass transition temperatures exhibit both solid-like and liquid-like features. Whereas the segmental motions give rise to the liquid-like behavior, the presence of permanent or non-permanent cross-links leads to residual dipolar couplings, which are responsible for the solid-like properties [48]. While this promises that both properties can be exploited, the application of 2D NMR techniques to viscoelastic materials has to deal with the difficulties related to both rigid and mobile samples. Compared with the wealth of applications in solution and in the solid state, NMR spectroscopy has not been widely applied to viscoelastic polymers, although it can provide information on such important fields as the chain dynamics in elastomers, the local structure, residual couplings (induced by chemical cross-links and topological constraints), dynamic order parameters, internuclear distances, intermolecular interactions (which are important, for instance, for the miscibility), the effects of fillers on molecular motions, segmental orientation under mechanical stress and others. A detailed description of the information on rubbers that can be obtained from different spectroscopic methods is available in a recent book on the subject [49].

Again, double quantum NMR proved to be particularly informative. In the data analysis, however, special care has to be taken as the chain order in networks detected by NMR bears no simple relation with the actual cross-link density [50, 51]. Although the overall scaling is found to be linear as expected (which can justify simple calibration procedures), the prefactor is not clear within a factor of the order of two in either direction. This might be due to the fact that the quantity measured in NMR is formed as a running time average of a second-order tensor. This renders studies of well-defined model networks particularly important [50]. They showed that the chain order parameter distribution broadens upon swelling accompanied by a non-affine change of this distribu-

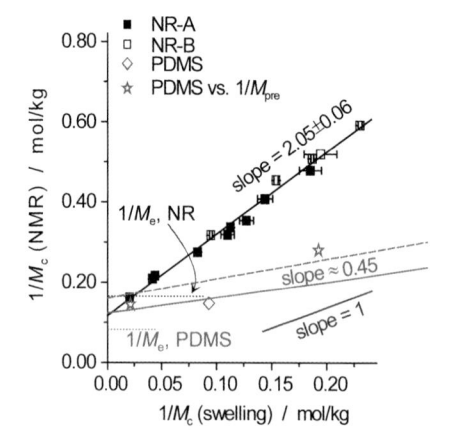

Fig. 14.7 Comparison of NMR-determined reciprocal inter-cross-link molecular weights of natural rubber (NR) with corresponding results from swelling experiments [51].

tion, indicating a heterogeneous swelling process. Comparing the simulated tensor order parameter with the autocorrelation function of segments, major deviations were observed in the swollen state but also for the long-chain fraction of the dry, bimodal network. Despite its undisputed use in comparing samples and assessing relative differences in the cross-link density with good accuracy, the chain order phenomenon as seen by NMR is far from being fully understood, with uncertainties arising at various stages of theoretical modeling. In simple cases, however, and with proper care, scaling relations between NMR observables and inter-cross-link molecular weights are observed [51], as shown in Fig. 14.7.

14.3.5
Melts Composed of Stiff Macromolecules

"hairy rod" polymers, molecular composites made of stiff macromolecules with flexible, short side-chains, represent a new class of polymers with the tendency to self-assemble into supramolecular structures when thin films are cast from solution [52] (Fig. 14.8a). The resulting films exhibit high dimensional stability since the rigid backbones act as reinforcing elements in an amorphous matrix formed by the side-groups. A hairy rod polymer composed of stiff poly(p-phenylene) backbone and short ethylene oxide (EO) side-groups (Fig. 14.8b) was synthesized and the possibility of using it as a solid polymer electrolyte when

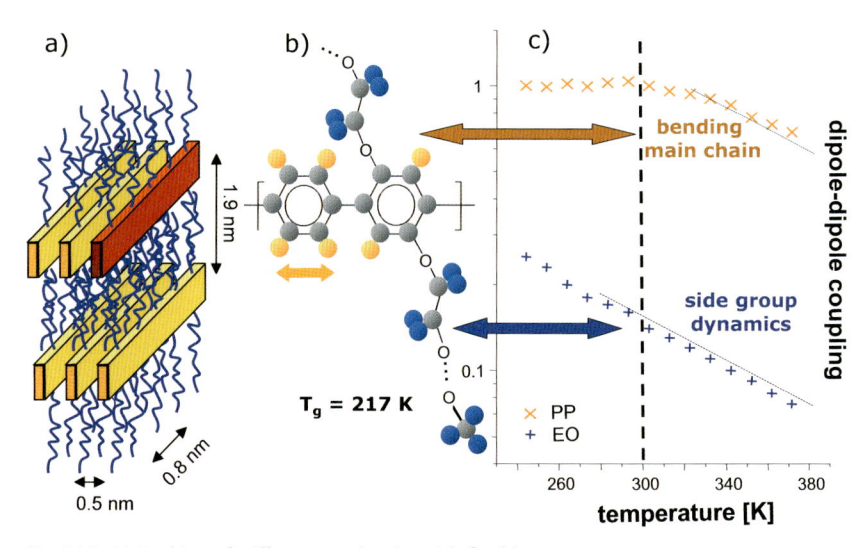

Fig. 14.8 (a) Packing of stiff macromolecules with flexible side-groups. (b) Structure of poly(p-phenylene) with ethylene oxide side-groups. (c) ^{1}H–^{1}H dipole–dipole coupling determined by double quantum NMR, separating bending motion of main-chain from side-group dynamics [54].

blended with lithium salt was explored [53]. One of the factors affecting ion mobility in such systems is the intrinsic dynamics of the EO side-chains and of the rigid backbone. Since the electric dipole moment is located on the EO side-chain, the mobility can conveniently be probed by dielectric relaxation. On the other hand, advanced NMR techniques are also sensitive to the backbone mobility. Therefore, the combination of the two techniques can help to unravel the molecular origin of dynamic processes in such a complex system [54].

In fact, dielectric relaxation detects two dynamic processes, the β-process attributed to the side-chain dynamics and the α-process associated with the glass transition. However, unlike the usual segmental process in homopolymers, the length of the short EO side-chains was found to control the glass transition temperature and the associated dynamic response. This effect originates from the confinement of the flexible side-chains to the relatively rigid backbones forming the periodic lattice. The α-process is of more complex molecular origin. At low temperatures, the main part of the relaxation originates again from the "fast" outer EO units. At intermediate temperatures, this relaxation is assisted by small-angle fluctuations of the substituted rings and, at even higher temperatures, internal modes of the backbone contribute to the slower part of the relaxation process. This is nicely confirmed by solid–state NMR, where the different processes could be distinguished via heteronuclear dipole–dipole couplings, which are sensitive to the flipping motion of the unsubstituted phenylene rings, in addition to the small angle fluctuations of the substituted rings. Bending motion of the backbone is permitted by the relatively short persistence length (13 nm), which indicates that the backbone is not completely rigid. The unfreezing of the backbone dynamics can be probed by homonuclear ^{1}H–^{1}H dipole–dipole coupling (Fig. 14.8c), which shows that it occurs only well above the glass transition temperature and does not lead to the dissolution of the structure, which occurs only at even higher temperatures. Comparison with the relaxation map from dielectric spectroscopy (see Fig. 4 in Ref. [54]) shows that at 300 K the α-relaxation is in the region of 10 MHz and the β-relaxation is 2–3 orders of magnitude faster, whereas the bending motion occurs in the region of only 50 kHz.

14.3.6
Conformational Memory in Poly(n-Alkyl Methacrylates)

Stiff macromolecules with flexible side-groups lack conformational freedom within the backbone, which leads to the formation of a layered structure even in the melt and a highly anisotropic motion as described above. The question then arises of whether in more conventional polymers extended conformations involving several repeat units can exhibit conformational memory manifesting itself in collective anisotropic motions. Randomization of conformation leading to locally isotropic reorientation would then occur as a separate process on a longer time-scale. Indeed, structurally heterogeneous poly(n-alkyl methacrylates), which consist of a polar backbone and flexible non-polar side-groups $R_nC_nH_{2n+1}$, are candidates for polymers with orientational memory and indeed exhibit unusual relaxation behavior [55]. The backbone has extended syndiotactic sequences which leads to extended chain conformations (Fig. 14.9a). The molecular dynamics of a macromolecular chain involve both conformational and rotational motions. Along these lines, the backbone dynamics of poly(n-alkyl methacrylates) have recently been clarified using advanced solid-state NMR, which enables one to probe conformational and rotational dynamics separately [56]. The former is encoded in the isotropic ^{13}C chemical shift in a highly resolved NMR spectrum, recorded under MAS. The latter is probed via the anisotropic ^{13}C chemical shift of the carboxyl group with a unique axis along the local chain axis. Randomization of conformations and isotropization of backbone orientation occur on the same time-scale, yet they are both much slower than the slowest relaxation process identified previously. This effect is attributed to extended backbone conformations, which retain conformational memory over many steps of restricted locally axial chain motion (Fig. 14.9b and c). These findings were rationalized in terms of a locally structured polymer melt, in which the polar and less flexible polymethacrylate backbones form disordered layers. This structure has recently been confirmed through temperature-dependent wide-angle X-ray scattering (WAXS) [57]. The anisotropic chain motion occurs within the layers; conformational randomization and rotational isotropization require extended chain units to translate from one structured unit to another. The variation in the molecular weight of poly(ethyl methacrylate) (PEMA) showed that a minimum chain length of 5–10 repeat units is required for this effect to occur [58].

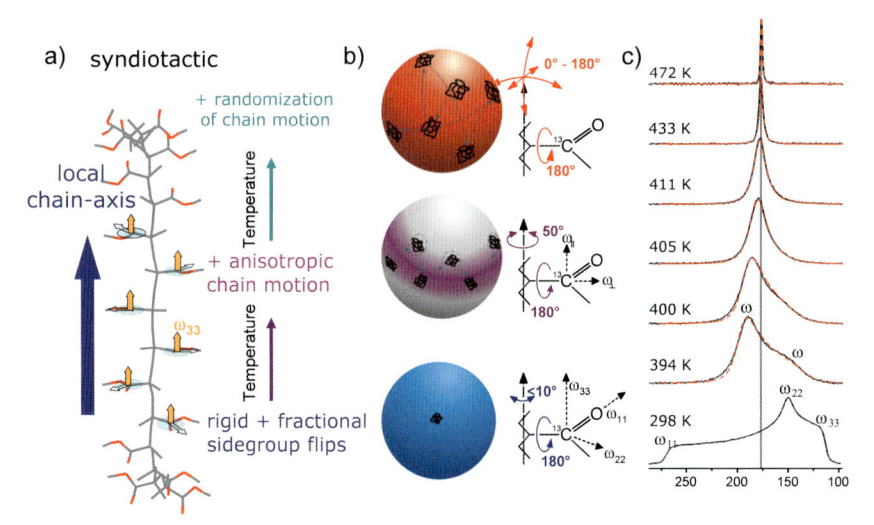

Fig. 14.9 (a) Extended chain conformation of syndiotactic poly(n-alkyl methacrylates); (b) anisotropic chain motion during glass process; (c) ^{13}C NMR spectra indicating anisotropic motion above T_g [56].

With this detailed understanding of the complex chain dynamics, the peculiarities noted previously for molecular dynamics in the series of poly(n-alkyl methacrylates) are easily resolved. The WLF parameters for the different members of the series deduced from the randomization/isotropization process exhibit minor systematic variations such as a decrease in fragility with alkyl side-group length [56]. Indeed, with increasing alkyl side-group length, the structuring of the polymer melt is predominantly driven by side-group association, the influence of the extended main chain conformations is reduced and the distinct a-relaxation and isotropization processes merge.

14.3.7
Applications in Supramolecular Chemistry

The techniques introduced above have been applied to elucidate the structure of various supramolecular systems of current interest. Early examples are included in an extended review [59]. In particular, it has been possible to locate protons in hydrogen bonds, study their dynamics and relate these findings, e.g. to thermal rearrangements of smart supramolecular polymers [29, 60] or proton conductivity [61], specify the mutual arrangements of aromatic moieties and relate them to charge carrier mobilities in photonic materials [62, 63] and also to molecular recognition in guest–host systems [64], and elucidate the organization of macromolecules and surfactants attached to surfaces in colloidal particles [65] and relate these observations to the function of such systems [66, 67].

14.3.7.1 Hydrogen Bonds in Supramolecular Polymers

As hydrogen bonds provide probably the most important secondary interactions exploited in natural and supramolecular systems, their elucidation was first on our list. Particularly challenging was the structure of the quadruple hydrogen bonds used to generate supramolecular polymers [60], as shown in Fig. 14.10. The backbone consists of oligomers linked together via quadruple hydrogen bonds. The linkers exhibit both a keto and an enol form, which had to be distinguished. The homonuclear ^1H–^1H 2D DQ NMR spectra differ markedly for the two forms (Fig. 14.10), making it possible to determine the ^1H–^1H and the ^1H–^{15}N distances (after moderate ^{15}N labeling) with high accuracy [68]. Thus, the complex hydrogen bonding of the linker group is fully elucidated. This should be particularly appreciated in view of the fact that these polymers do not crystallize. Hence our new solid-state NMR approach can provide information not available from scattering. Moreover, data acquisition in ^1H NMR is rapid enough to follow the tautomeric rearrangement in this system in real time, providing thermodynamic and kinetic parameters [29].

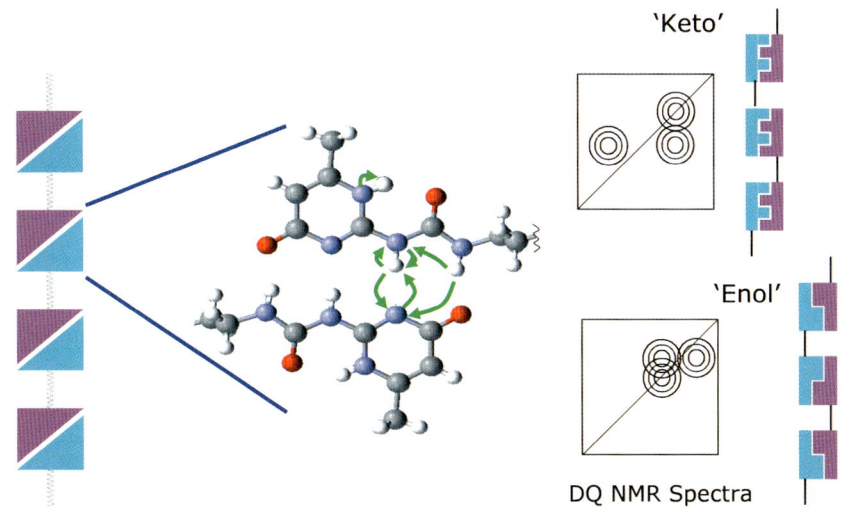

Fig. 14.10 Structure of quadruple hydrogen bonds in supramolecular polymers as deduced from DQ NMR spectroscopy [29]. The blue and red schemes indicate hydrogen bonding donor and acceptor groups linked together in different ways.

14.3.7.2 Proton Conductors

The structure and dynamics of the protons are particularly interesting in proton conductors. In a recent study [61], we applied advanced solid-state ^{1}H NMR to a new class of imidazole-based proton conductors [69], aimed at generating improved membranes for fuel cells. Here the imidazole units are linked via flexible ethylene oxide spacers, improving their mechanical properties. The main findings of the study led to a complex structural picture as displayed in Fig. 14.11. The organization in this functional supramolecular system involves at least three states of different order: strongly ordered regions in which the functional groups form dimers held together by strong H-bonds and the dimers are linked by slightly weaker H-bonds, disordered hydrogen-bonded aggregates and free molecules. Rapid exchange between the last two species is the main source of the proton conductivity. The structure of the ordered regions is obtained from a single-crystal X-ray structure of one member of the series. ^{1}H solid-state NMR in combination with computer simulation showed that this structure is present in the whole series studied, including those members which are amorphous. Moreover, the ^{1}H chemical shift in the melt proves that the complex heterogeneous organization prevails after melting, providing an example of a structured amorphous melt.

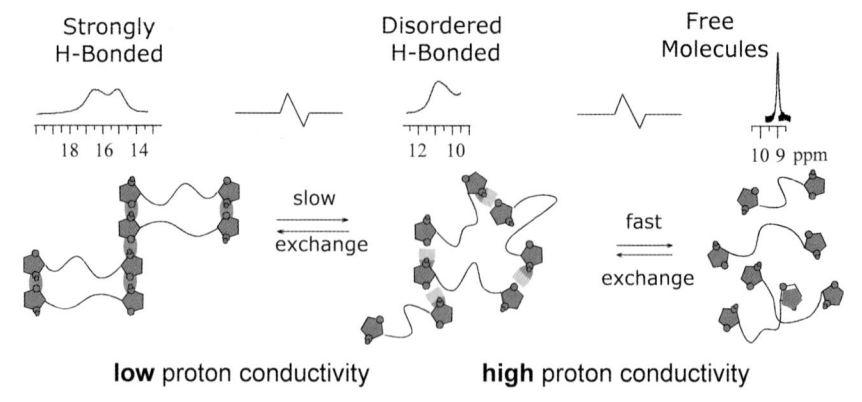

Fig. 14.11 Structure of imidazole proton conductors as deduced from solid-state NMR and X-ray scattering [61].

14.3.7.3 Supramolecular Assembly of Dendritic Polymers

The most complex example studied so far by the new advanced solid-state NMR methods are supramolecular systems with controlled shape and size [70]. They consist of a polymer backbone with dendritic side-groups and self-assemble into a columnar structure [71]. The NMR experiments are performed on as-synthesized samples. The analysis of ^{1}H NMR chemical shift effects and dipolar ^{1}H–^{1}H and ^{1}H–^{13}C couplings provides site-specific insight into the local structure within the dendrons. Relative changes of ^{1}H chemical shifts serve as distance constraints and allow protons to be positioned relative to aromatic rings

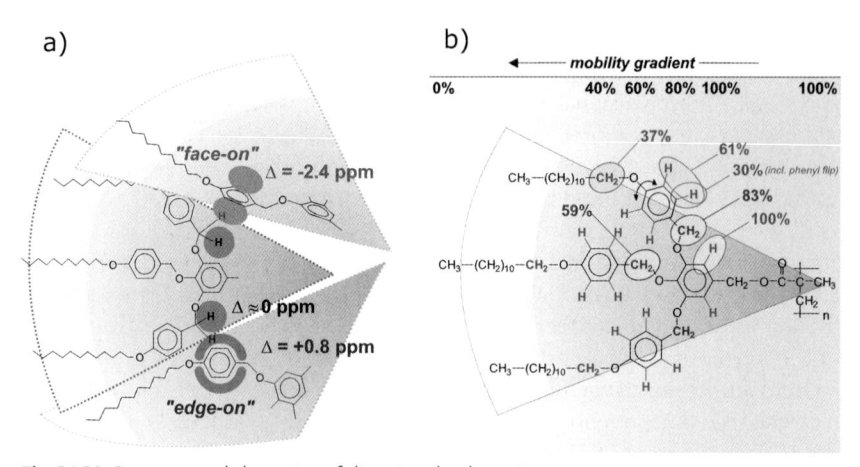

Fig. 14.12 Structure and dynamics of directing dendrons in cylindrical supramolecular macromolecules [70]. (a) Local packing allowing the formation of helices; (b) restricted motion as indicated by high dynamic order parameters; see axis on the top, with mobility gradient inside/out.

(Fig. 14.12 a). Intergroup π-packing phenomena were identified, allowing the identification of the dendron cores as the structure-directing moieties within the supramolecular architecture. The study was carried out over a representative selection of systems which reflect characteristic differences, such as different polymer backbones, sizes of dendritic side-groups and length and flexibility of linking units. Whereas the polymer backbone was found to have virtually no effect on the overall structure and properties, the systems are sensitively affected by changes to the generation or the linkage of the dendrons. This structural information provided by NMR was augmented by site-selective measurements of the ^{13}C–^{1}H heteronuclear dipole–dipole couplings (dynamic order parameters) of the different groups, as shown in Fig. 14.12 b. The dendritic groups themselves are highly flexible due to the presence of the ethylene oxide linkers. In the supramolecular assembly, however, they exhibit dynamic order parameters as high as 40–80%, displaying a gradient of mobility, decreasing from inside out. This significant immobilization nicely demonstrates their role as structure-directing moieties. The results help in understanding the self-assembly process of dendritic moieties and aid in the chemical design of self-organizing molecular structures. Indeed, this concept fostered the design of supramolecular helical dendrimers as promising electronic materials [63].

14.3.7.4 Discotic Photoconductors Based on Hexabenzocoronene (HBC)

Another important intermolecular aspect which governs the mutual arrangement of molecules in condensed phases is the π–π interaction, in particular that between aromatic moieties. There is currently much interest in polycyclic aromatic materials, which form columnar mesophases, on account of, for example, their potential applications as charge transport layers in devices such as field-effect transistors, photovoltaic cells and light-emitting diodes [72]. One such class of materials are the hexaalkyl-substituted hexa-*peri*-hexabenzocoronenes, which have high one-dimensional charge carrier mobility along the columns [73]. In order to control their organization and processability, π-stacking can be combined with hydrogen bonding. Such synergetic combination offers an attractive means to enhance and control supramolecular order. Here, combined studies of solid-state NMR, X-ray scattering and scanning tunneling microscopy can provide a clear and consistent picture of the effect of these two interactions on supramolecular organization [74] (Fig. 14.13). It turned out that the tether between the carboxylic group and the aromatic core needs special attention. The desired improvement of the order is achieved only for relatively short tethers. It should be noted that the arrangement of the HBC cores can either be staggered (tilted herringbone arrangement) or planar, with the ring normal along the columnar axis. These different stackings lead to very different ^{1}H–^{1}H double quantum NMR spectra, which are easily distinguished [75]. The highest order was obtained for a system with two carboxyl groups attached via short tethers in 'para' position to each core.

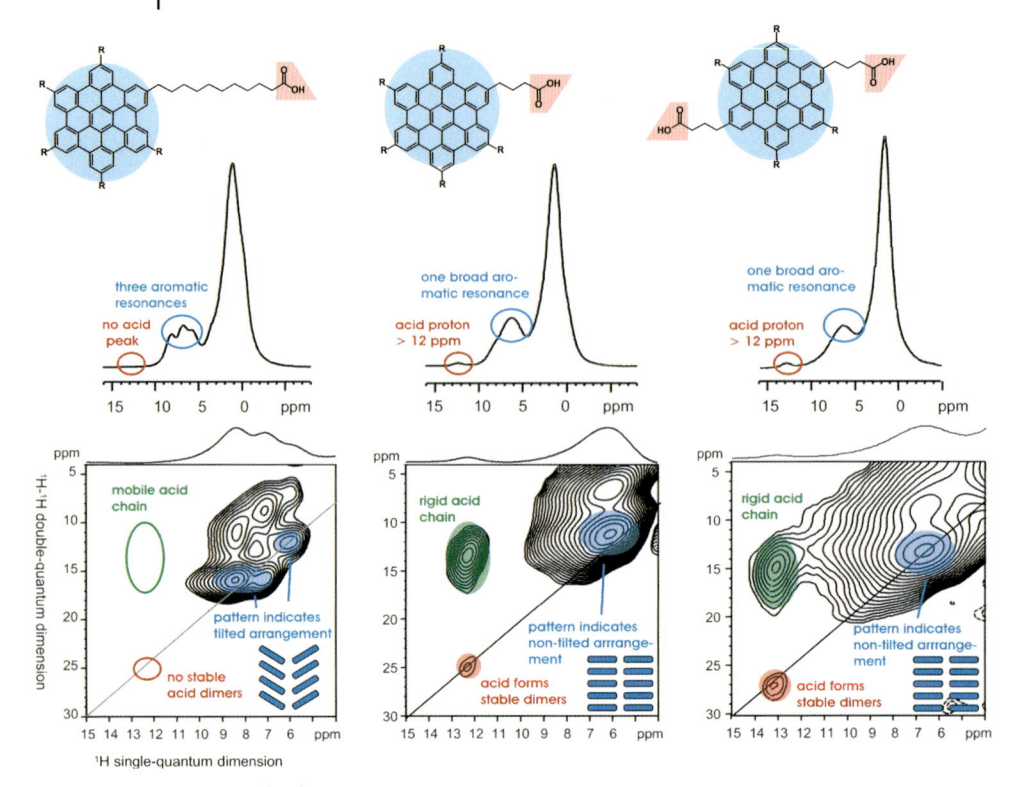

Fig. 14.13 ^{1}H–^{1}H double quantum filtered and two-dimensional ^{1}H–^{1}H double quantum NMR spectra indicating the influence of hydrogen bonds on the columnar packing of photoconducting HBCs [74].

14.3.7.5 Polyphenylene Dendrimers as Shape-persistent Nanoparticles

Polyphenylene dendrimers, macromolecules with a regularly branched structure emanating from a well-defined focal point, represent an interesting class of new materials which are expected to have dense shells and open cores (Fig. 14.14). Hence they are attractive candidates for applications as nanoreactors, since they contain molecular cavities which may be guest specific [76]. Shape persistence of these molecules has been checked via ^{1}H–^{13}C-dipolar DQ spinning sideband patterns excluding mobility on the microsecond time-scale. On the much longer 100-ms time-scale, restricted motion of the phenyl rings is observed, indicating dominance of intramolecular constraints consistent with rigid arms, proving shape persistence on this much longer time-scale also. The dependence of the correlation times of this restricted motion as a function of dendrimer generation was analyzed in terms of the local density experienced by the aromatic rings. It was concluded that these NMR results represented the first experimental evidence of a class of dendrimers in which the radial segment density distribution is caused by truly extended arms and for which the dense-shell packing

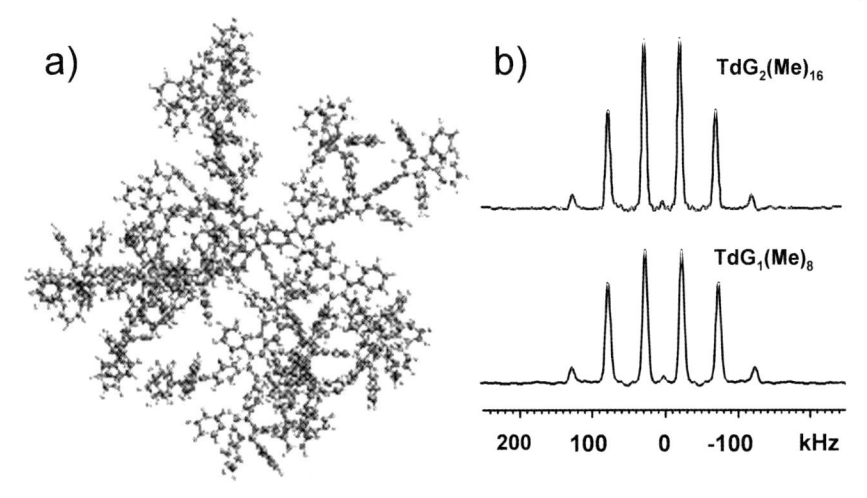

Fig. 14.14 (a) Structure of a shape-persistent polyphenylene dendrimer. (b) ^{1}H–^{13}C dipole–dipole DQ spinning sideband patterns excluding mobility on the microsecond time-scale [77].

limit is reached for generation four [77]. Notably the same conclusion was later reached by use of small-angle neutron scattering, applying contrast variation [78]. The comparison of these data with simulations demonstrated that the scaffold of the dendrimer is rigid, as expected from its chemical structure. The positions of the various units setting up consecutive shells of the dendrimer are relatively well localized and the entire structure cannot be modeled in terms of spherically symmetric models. No back-folding of the terminal groups can occur and the model calculations demonstrate that higher generations of this dendritic scaffold must exhibit a dense shell and congestion of the terminal groups.

14.3.7.6 Organic–Inorganic Hybrid Materials

Hybrid materials composed of organic and inorganic residues are at the forefront of polymer science [79]. These structures are often well defined. NMR, of course, can be applied to both the organic and the inorganic phases. As example, consider the structure of a novel molecularly ordered 2D silicate framework in a surfactant-templated mesophase (Fig. 14.15). These materials are unusual in their combination of headgroup-directed 2D crystalline framework ordering, zeolite-like ring structures within the layers and long-range mesoscopic organization without 3D atomic periodicity. The absence of registry between the silicate sheets, resulting from the liquid-like disorder of the alkyl surfactant chains, prevents the determination of framework structures in these and similar materials. Double quantum ^{29}Si NMR correlation experiments, however, established the interactions and connectivities between distinct intra-sheet silicon sites from which the structure of the molecularly ordered inorganic framework could be determined [80].

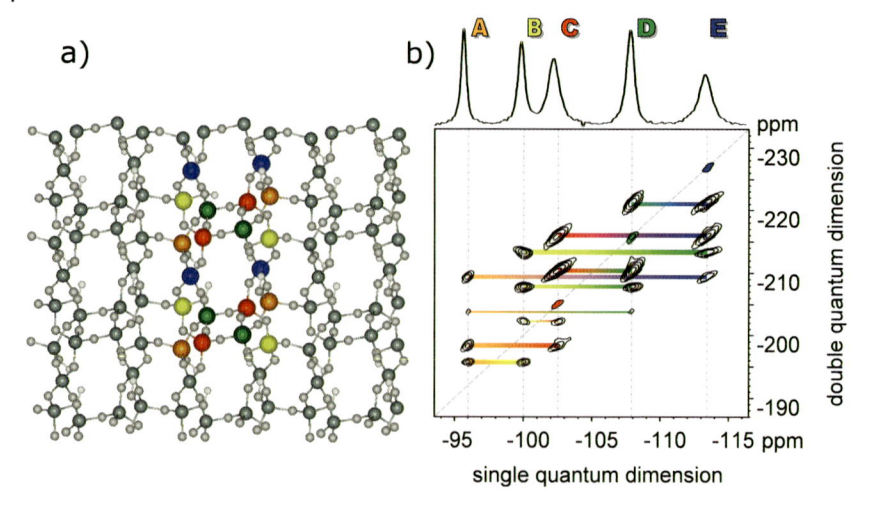

Fig. 14.15 (a) Structure of molecularly ordered silicate sheets in a lamellar silicate. (b) ^{29}Si DQ NMR spectrum, from which spatial proximities between pairs of dipole–dipole-coupled ^{29}Si sites are established [80].

In fact, control of the shape and size of organic–inorganic hybrid materials is a key feature of natural growth phenomena. For instance, in biomineralization, complex architectures on different length-scales are usually obtained through cooperative self-assembly of organic and inorganic species [81]. Despite the success in understanding the basic principles of self-assembly, it remains a challenge for scientists to mimic such natural pathways and develop simple but efficient routes to materials structured all the way down to the nanometer scale. Block copolymers can be regarded as macromolecular analogues of low molecular weight surfactants. Their self-assembly in block-selective solvents has been shown to result in a variety of nanoscale morphologies including spheres, rods, lamellae, vesicle tubules, cylinders and large compound vesicles, micelles or rod micelles [82]. Therefore, the use of this self-assembly to gain structured organic–inorganic composites on the nanometer scale is appealing and has indeed been successful (for a review, see [83]).

The phase behavior of poly(isoprene-*b*-ethylene oxide) (PI-*b*-PEO) is well understood [83] and provided the basis of aluminosilicate–PI-*b*-PEO hybrids. In this system, the organic polymer may significantly penetrate the aluminosilicate network. Therefore, the interface between the inorganic and the organic part is of special interest. Two possible models for the distribution of the aluminosilicate network in the PEO phase of PI-*b*-PEO should be distinguished (Fig. 14.16): a three-phase scenario (a) with an interfacial layer or interphase of pure PEO or a two-phase scenario (b) where the aluminosilicate network is distributed throughout the PEO phase. These two scenarios can be nicely distinguished in NMR spin-diffusion experiments, based on selection of the mobile

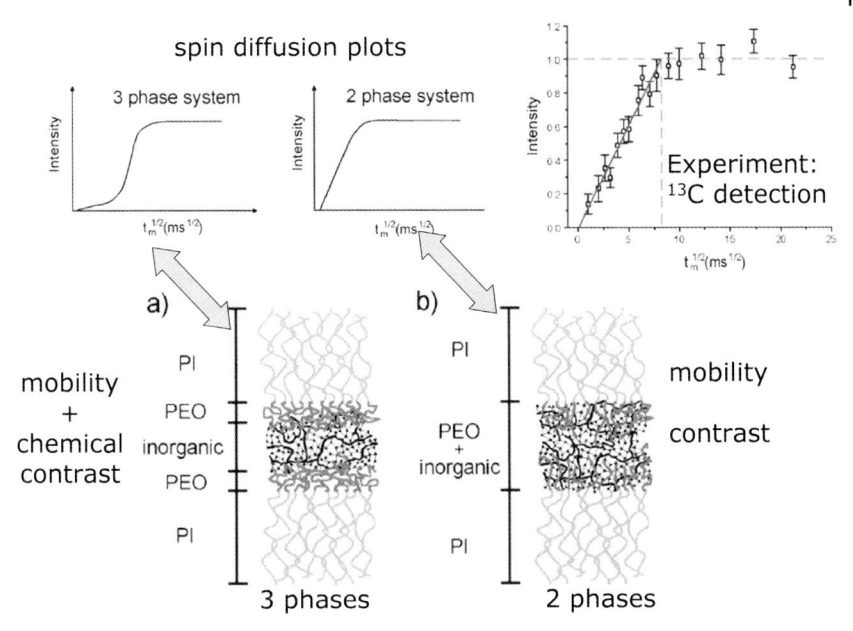

Fig. 14.16 Two possible models for the distribution of the aluminosilicate network in the PEO phase of PI-b-PEO: (a) three-phase scenario; (b) two-phase scenario. The spin diffusion curves differ as shown in the upper part. The experimental curve indicates the two-phase scenario [84].

PI phase and detection of a resolved signal characteristic of the inorganic phase. If there were a PEO interphase (scenario (a)), magnetization would initially diffuse from the isoprene to this PEO phase. Only at later times would it reach the groups bound to the aluminosilicate. Such a situation would correspond to a spin-diffusion build-up plot for the Si–CH$_2$– peak, which has an intercept on the abscissa that is greater than zero. The experimental plot in [84] does not show such a "lag time" and, therefore, indicates that the model scenario (b), where PEO and the aluminosilicate are intimately mixed, is representative of the system.

14.4
Conclusion and Outlook

The examples given above, mostly taken from our own group, were chosen to indicate the scope of using solid-state NMR as a tool for elucidating the structure and dynamics of macromolecular and supramolecular systems. With the impressive developments in synchrotron radiation, X-ray scattering has experienced a remarkable development [85]. In Table 14.2, we compare the types of information currently available through scattering and NMR. Both approaches

Table 14.2 Comparison of scattering and NMR techniques and the information provided about structure and dynamics of materials.

	Scattering		NMR	
	Incoherent	Coherent	Single quantum	Double quantum
Dynamics				
Molecular	n-Quasi-elastic	n-Quasi-elastic	2D, 3D, 4D exchange	Sidebands
Collective		n-Spin-echo	2D exchange	Decay of DQC
Structure				
Molecular		WAXS, WANS	Chemical shift, sidebands	2D pattern, sidebands
Collective (packing)		X-ray pole figures, SAXS, SANS	DECODER, chemical shift	2D signal pattern

are very versatile and powerful. Structure and dynamics are accessible and molecular and collective phenomena can be elucidated. Both approaches can be applied over large length- and time-scales, which partially overlap and nicely complement each other. As the problems we face in soft matter research are very complex, none of the techniques can provide the full picture and combining the different approaches is most fruitful.

NMR is particularly suited to tackle non-crystalline systems. As far as supramolecular organic systems are concerned, highly resolved solid-state spectra can nowadays be recorded on small amounts (10–15 mg) of as-synthesized samples and specific results are obtained overnight. Organization of organic moieties on colloidal particles can now be studied by NMR at sub-monolayer coverage [67]. The methodology, however, can be applied in general. For instance, high-resolution DQ MAS NMR spectroscopy of isotopically labeled systems of biological interest yields torsional angles in peptides [86–88] and thus important information about the chain conformation. The sensitivity of ^{1}H chemical shifts to ring currents in aromatic systems has been exploited in determining the packing of porphyrins. The full information about distances and the relative orientation of specific moieties is available from DQ NMR spectroscopy on static samples of specifically labeled materials [89] and has provided important information about the chain conformation and the packing of synthetic macromolecules [90].

The development of the technique is far from being complete and major advances are expected in the years to come. Hence solid-state NMR is expected to become as indispensable a tool in macromolecular, supramolecular and biological science as solution-state NMR is today.

Acknowledgments

The author gratefully acknowledges the important role of outstanding students and postdocs, who did the work described here and are co-authors in the references given below. Continuous financial support from the Max Planck Society, the Deutsche Forschungsgemeinschaft, the Bundesministerium für Bildung und Forschung and the European Union is gratefully acknowledged.

References

1 K.H.J. Buschow, R.W. Calm, M.C. Flemings, B. Ilschner, E.J. Kramer, S. Mahajan (Eds.), *Encyclopedia of Materials Science and Technology*, Elsevier, Amsterdam, **2001**.
2 B.J. Berne, R. Pecora, *Dynamic Light Scattering*, Dover, Mineola, **2000**.
3 B.E. Warren, *X-Ray Diffraction*, Dover, Mineola, **1990**.
4 J.S. Higgins, H.C. Benoit, *Polymers and Neutron Scattering*, Oxford University Press, Oxford, **1997**.
5 L.C. Sawyer, D.T. Grubb, *Polymer Microscopy*, Springer, Heidelberg, **1987**.
6 I.M. Ward, J. Sweeney, *An Introduction to the Mechanical Properties of Solid Polymers*, Wiley, Chichester, **2004**.
7 F. Kremer, A. Schönhals (Eds.), *Broadband Dielectric Spectroscopy*, Springer, Heidelberg, **2002**.
8 R.R. Ernst, G. Bodenhausen, A. Wokaun, *Principles of Nuclear Magnetic Resonance in One and Two Dimensions*, Clarendon Press, Oxford, **1987**.
9 F.A. Bovey, *Chain Structure and Conformation of Macromolecules*, Academic Press, New York, **1982**.
10 K. Wüthrich, *NMR of Proteins and Nucleic Acids*, Wiley, New York, **1986**.
11 H.W. Spiess, *Macromol. Chem. Phys.* **2003**, *204*, 340–346.
12 K. Schmidt-Rohr, H.W. Spiess, *Multidimensional Solid State NMR and Polymers*, Academic Press, New York, **1994**.
13 H.W. Spiess, *J. Polym. Sci. A* **2004**, *42*, 5031–5044.
14 D. Demus, J.W. Goodby, G.W. Gray, H.W. Spiess, V. Vill (Eds.), *Handbook of Liquid Crystals*, Wiley-VCH, Weinheim, **1998**.
15 D.M. Grant, R.K. Harris (Eds.), *Encyclopedia of Nuclear Magnetic Resonance*, Wiley, Chichester, **1996**.
16 B. Blümich, *NMR Imaging of Materials*, Clarendon Press, Oxford, **2000**.
17 I.M. Ward, *Structure and Properties of Oriented Polymers*, Chapman and Hall, London, **1997**.
18 M. Kaupp, M. Bühl, M. Malkin, G. Vladimir (Eds.), *Calculation of NMR and EPR Parameters*, Wiley-VCH, Weinheim, **2004**.
19 C. Ochsenfeld, J. Kussmann, F. Koziol, *Angew. Chemie Int. Ed.* **2004**, *43*, 4485–4489.
20 D. Sebastiani, *ChemPhysChem* **2006**, *7*, 164–175.
21 S. Hafner, H.W. Spiess, *Concepts Magn. Reson.* **1998**, *10*, 99–128.
22 A. Samoson, in *Encyclopedia of Nuclear Magnetic Resonance*, Vol. 9, D.M. Grant, R.K. Harris (Eds.), Wiley, Chichester, **2002**, pp. 59–64.
23 C. Filip, S. Hafner, I. Schnell, D.E. Demco, H.W. Spiess, *J. Chem. Phys.* **1999**, *110*, 423–440.
24 T. Gullion, J. Schaefer, *Adv. Magn. Reson.* **1989**, *13*, 57–83.
25 I. Schnell, H.W. Spiess, *J. Magn. Reson.* **2001**, *151*, 153–227.
26 I. Schnell, *Prog. Nucl. Magn. Reson. Spectrosc.* **2004**, *45*, 145–207.
27 I. Fischbach, F. Ebert, H.W. Spiess, I. Schnell, *ChemPhysChem* **2004**, *5*, 895–908.
28 K. Saalwächter, H.W. Spiess, *J. Chem. Phys.* **2001**, *114*, 5707–5728.
29 (a) I. Schnell, B. Langer, S.H.M. Söntjens, M.H.P. van Genderen, H.W. Spiess, *J. Magn. Reson.* **2001**, *150*, 57–

70; (b) I. Schnell, B. Langer, S. H. M. Söntjens, M. H. P. van Genderen, H. W. Spiess, *Phys. Chem. Chem. Phys.* **2002**, *4*, 3750–3758.
30 C. Schmidt, S. Wefing, B. Blümich, H. W. Spiess, *Chem. Phys. Lett.* **1986**, *130*, 84–90.
31 E. R. deAzevedo, W.-G. Hu, T. J. Bonagamba, K. Schmidt-Rohr, *J. Am. Chem. Soc.* **1999**, *121*, 8411–8412.
32 J. W. Emsley, J. Feeney, L. H. Sutcliffe (Eds.), *Prog. Nucl. Magn. Reson. Spectrosc.* **2006**, *48*, 1–160.
33 G. A. Webb (Ed.), *Annu. Rep. NMR Spectros.* **2006**, *59*, 1–280.
34 A. Abe, et al. (Eds.), *Adv. Polym. Sci.* **2005**, *182*, 1–308.
35 H. H. Brintzinger, D. Fischer, R. Mülhaupt, B. Rieger, R. M. Waymouth, *Angew. Chem. Int. Ed.*, **1995**, *34*, 1143–1170.
36 M. Zhang, S. E. Wanke, *Polym. Eng. Sci.* **2003**, *43*, 1878–1888.
37 P. M. Wood-Adams, J. M. Dealy, A. W. deGroot, O. D. Redwine, *Macromolecules* **2000**, *33*, 7489–7499.
38 M. Pollard, K. Klimke, R. Graf, H. W. Spiess, M. Wilhelm, O. Sperber, C. Piel, W. Kaminsky, *Macromolecules* **2004**, *37*, 813–825.
39 K. Klimke, M. Parkinson, C. Piel, W. Kaminsky, H. W. Spiess, M. Wilhelm, *Macromol. Chem. Phys.* **2006**, *207*, 382–395.
40 B. P. Rotzinger, H. D. Chanzy, P. Smith, *Polymer* **1989**, *30*, 1814–1819.
41 P. Smith, P. J. Lemstra, *J. Mater. Sci.* **1980**, *15*, 505–514.
42 U. Tracht, M. Wilhelm, A. Heuer, H. Feng, K. Schmidt-Rohr, H. W. Spiess, *Phys. Rev. Lett.* **1998**, *81*, 2727–2730.
43 R. Graf, A. Heuer, H. W. Spiess, *Phys. Rev. Lett.* **1998**, *80*, 5738–5741.
44 S. Rastogi, D. R. Lippits, G. W. M. Peters, R. Graf, Y. Yao, H. W. Spiess, *Nat. Mater.* **2005**, *4*, 635–641.
45 B. Alberts, D. Brass, J. Lewis, M. Raff, K. Roberts, J. D. Watson, *Molecular Biology of the Cell*, 3rd edn, Garland, New York, **1995**.
46 R. Weberskirch, J. Preuschen, H. W. Spiess, O. Nuyken, *Macromol. Chem. Phys.* **2000**, *201*, 995–1007.
47 J. Preuschen, S. Menchen, M. A. Winnik, A. Heuer, H. W. Spiess, *Macromolecules* **1999**, *32*, 2690–2695.
48 J.-P. Cohen Addad, *Prog. Nucl. Magn. Reson. Spectros.* **1994**, *25*, 1–316.
49 V. M. Litvinov, P. P. De, *Spectroscopy of Rubbers and Rubbery Materials*, RAPRA, Shawbury, **2002**.
50 K. Saalwächter, F. Kleinschmidt, J.-U. Sommer, *Macromolecules* **2004**, *37*, 8556–8568.
51 K. Saalwächter, B. Herrero, M. A. Lopez-Manchado, *Macromolecules* **2005**, *38*, 9650–9660.
52 G. Wegner, *Macromol. Chem. Phys.* **2003**, *204*, 347–357.
53 U. Lauter, W. H. Meyer, G. Wegner, *Macromolecules* **1997**, *30*, 2092–2101.
54 M. Mierzwa, G. Floudas, M. Neidhöfer, R. Graf, H. W. Spiess, W. H. Meyer, G. Wegner, *J. Chem. Phys.* **2002**, *117*, 6289–6299.
55 M. Beiner, *Macromol. Rapid Commun.* **2001**, *22*, 869–895.
56 M. Wind, R. Graf, A. Heuer, H. W. Spiess, *Phys. Rev. Lett.* **2003**, *91*, 155702, 1–4.
57 M. Wind, R. Graf, S. Renker, H. W. Spiess, W. Steffen, *J. Chem. Phys.* **2005**, *122*, 014906, 1–10.
58 M. Wind, R. Graf, S. Renker, H. W. Spiess, *Macromol. Chem. Phys.* **2005**, *206*, 142–156.
59 S. P. Brown, H. W. Spiess, *Chem. Rev.* **2001**, *101*, 4125–4155.
60 R. P. Sijbesma, F. H. Beijer, L. Brunsveld, B. J. B. Folmer, J. H. K. K. Hirschberg, R. F. M. Lange, J. K. L. Lowe, E. W. Meijer, *Science* **1997**, *278*, 1601–1604.
61 G. R. Goward, M. F. H. Schuster, D. Sebastiani, I. Schnell, H. W. Spiess, *J. Phys. Chem. B* **2002**, *106*, 9322–9334.
62 I. Fischbach, T. Pakula, P. Minkin, A. Fechtenkötter, K. Müllen, H. W. Spiess, K. Saalwächter, *J. Phys. Chem. B* **2002**, *106*, 6408–6418.
63 V. Percec, M. Glodde, T. K. Bera, Y. Miura, I. Shiyanovskaya, K. D. Singer, V. S. K. Balagurusamy, P. A. Heiney, I. Schnell, A. Rapp, H. W. Spiess, S. D. Hudson, H. Duan, *Nature* **2002**, *419*, 384–387.

64 S. P. Brown, T. Schaller, U. P. Seelbach, F. Koziol, C. Ochsenfeld, F.-G. Klärner, H. W. Spiess, *Angew. Chem. Int. Ed.* **2001**, *40*, 717–720.

65 G. Decher, *Science* **1997**, *277*, 1232–1237.

66 M. McCormick, R. N. Smith, R. Graf, C. J. Barrett, L. Reven, H. W. Spiess, *Macromolecules* **2003**, *36*, 3616–3625.

67 S. Pawsey, M. McCormick, S. De Paul, R. Graf, Y. S. Lee, L. Reven, H. W. Spiess, *J. Am Chem. Soc.* **2003**, *125*, 4174–4184.

68 M. Schulz-Dobrick, T. Metzroth, H. W. Spiess, J. Gauss, I. Schnell, *ChemPhysChem* **2005**, *6*, 315–327.

69 M. F. H. Schuster, W. H. Meyer, G. Wegner, H. G. Herz, M. Ise, M. Schuster, K. D. Kreuer, *Solid State Ionics* **2001**, *145*, 85–92.

70 A. Rapp, I. Schnell, D. Sebastiani, S. P. Brown, V. Percec, H. W. Spiess, *J. Am. Chem. Soc.* **2003**, *125*, 13284–13297.

71 V. Percec, C.-H. Ahn, G. Ungar, D. J. P. Yeardley, M. Möller, S. S. Sheiko, *Nature* **1998**, *391*, 161–164.

72 W. Pisula, A. Menon, M. Stepputat, I. Lieberwirth, U. Kolb, T. Pakuly, K. Müllen, *Adv. Mater.* **2005**, *17*, 684–689.

73 A. M. van de Craats, N. Stutzmann, O. Bunk, M. M. Nielsen, M. Watson, K. Müllen, H. D. Chanzy, H. Sirringhaus, R. H. Friend, *Adv. Mater.* **2003**, *15*, 495–499.

74 D. Wasserfallen, I. Fischbach, N. Chebotareva, M. Kastler, W. Pisula, F. Jäckel, M. D. Watson, I. Schnell, J. P. Rabe, H. W. Spiess, K. Müllen, *Adv. Funct. Mater.* **2005**, *15*, 1585–1594.

75 C. Ochsenfeld, S. P. Brown, I. Schnell, J. Gauss, H. W. Spiess, *J. Am. Chem. Soc.* **2001**, *123*, 2597–2606.

76 G. R. Newkome (Ed.), *Advances in Dendritic Macromolecules*, Elsevier, Amsterdam, **2005**.

77 M. Wind, K. Saalwächter, U.-M. Wiesler, K. Müllen, H. W. Spiess, *Macromolecules* **2002**, *35*, 10071–10086.

78 S. Rosenfeldt, N. Dingenouts, D. Potschke, M. Ballauff, A. J. Berresheim, K. Müllen, P. Lindner, K. Saalwächter, *J. Lumin.*, **2005**, *111*, 225–238.

79 J. Pyun, K. Matyjaszewski, *Chem. Mater.* **2001**, *13*, 3436–3448.

80 N. Hedin, R. Graf, S. C. Christiansen, C. Gervais, R. C. Hayward, J. Eckert, B. F. Chmelka, *J. Am. Chem. Soc.* **2004**, *126*, 9425–9432.

81 S. Mann, G. A. Ozin, *Nature* **1996**, *382*, 313–318.

82 U. Wiesner, *Macromol. Chem. Phys.* **1997**, *198*, 3319–3352.

83 P. F. W. Simon, R. Ulrich, H. W. Spiess, U. Wiesner, *Chem. Mater.* **2001**, *13*, 3464–3486.

84 S. M. De Paul, J. W. Zwanziger, R. Ulrich, U. Wiesner, H. W. Spiess, *J. Am. Chem. Soc.* **1999**, *121*, 5727–5736.

85 For current developments, see http://www.esrf.fr/.

86 M. Hong, J. D. Gross, R. G. Griffin, *J. Phys. Chem. B* **1997**, *101*, 5869–5874.

87 X. Feng, P. J. E. Verdegem, Y. K. Lee, D. Sandstrom, M. Eden, P. Bovee Gueurts, W. J. de Grip, J. Lugtenburg, H. J. M. de Groot, M. H. Levitt, *J. Am. Chem. Soc.* **1997**, *119*, 6853–6857.

88 F. Castellani, B. van Rossum, A. Diehl, M. Schubert, K. Rehbain, H. Oschkinat, *Nature* **2002**, *420*, 98–102.

89 K. Schmidt-Rohr, *Macromolecules* **1996**, *29*, 3975–3981.

90 K. Schmidt-Rohr, W. Hu, N. Zumbulyadis, *Science* **1998**, *280*, 714–717.

15
High-throughput Screening in Combinatorial Polymer Research

Michael A. R. Meier, Richard Hoogenboom, and Ulrich S. Schubert

15.1
Introduction

The concept of high-throughput screening (HTS, "a process for rapid assessment of the activity of samples from a combinatorial library or other compound collection") [1] dates back to the 1950s and was mainly developed due to the need for fast and automated analysis in clinical testing and medicine [2]. The successful implementation of these combinatorial and high-throughput methods in pharmaceutical research [3, 4] resulted in increased interest in parallel and combinatorial approaches in other fields of research, such as the synthesis and discovery of new inorganic materials [5–10], catalysts [11–14] and organic polymers. However, the large sample throughput (up to 1 million tests per week) of pharmaceutical ultra-high-throughput screening (UHTS) approaches is unlikely to be reached (and is also not necessary) for combinatorial approaches in polymer and materials chemistry. Combinatorial material research (CMR) is still a young field of research but has attracted great attention during the last few years [15–20]. It offers very promising approaches in the field of polymer chemistry since it is possible to vary and evaluate a large parameter space in a reduced amount of time. For polymers, these parameters include molecular weight, polydispersity index, polymerization kinetics, viscosity, hardness and stiffness, to mention only a few. Nevertheless, the fast preparation and screening of combinatorial material libraries does not necessarily speed up the time to market or understanding of quantitative structure–property relationships (QSPR) of new materials. Only the integration of design of experiments [21, 22] (DoE) and data handling solutions [23] can complete, optimize and close the combinatorial workflow cycle [16–24] in CMR, as depicted in Fig. 15.1. All aspects of this workflow [design–of–experiments (DoE), automated synthesis and formulation, high-throughput screening (HTS), thin film and dot preparation techniques, property investigations and data handling] are crucial for the success of a combinatorial approach in materials science.

Macromolecular Engineering. Precise Synthesis, Materials Properties, Applications.
Edited by K. Matyjaszewski, Y. Gnanou, and L. Leibler
Copyright © 2007 WILEY-VCH Verlag GmbH & Co. KGaA, Weinheim
ISBN: 978-3-527-31446-1

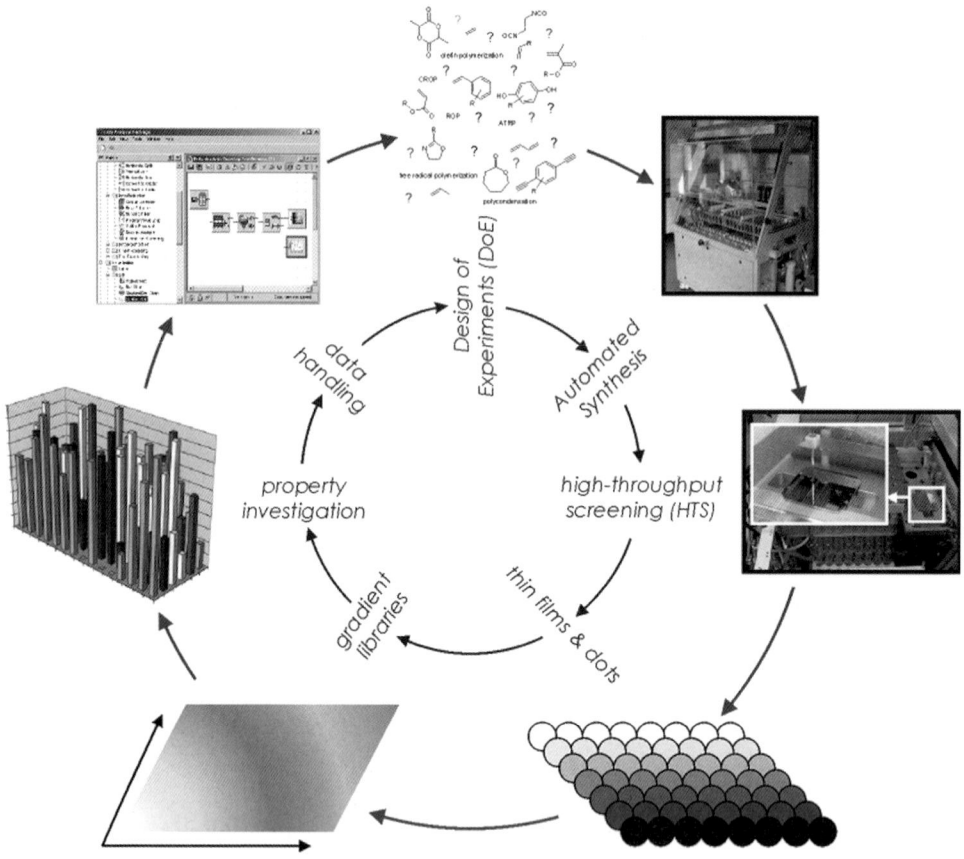

Fig. 15.1 Schematic representation of a possible workflow in combinatorial materials research covering typical stages of experimentation ranging from the design of experiments through various steps to data handling and thereby closing the combinatorial cycle. From Schubert et al. *Macromol. Rapid Commun.* **2004**, 25, 21–33 [16]. Reprinted with permission of John Wiley and Sons, Inc.

Here, we provide an overview of the progress made in the field of high-throughput screening of polymeric materials with regard to their molecular characteristics and some of their application-directed properties. We first give a very brief overview of synthetic possibilities for combinatorial approaches in polymer science and subsequently focus on the discussion of different high-throughput screening methods covering possibilities for the fast determination of molecular weights, polydispersity indices, polymerization kinetics, copolymer compositions and others. Finally, we describe several automated parallel techniques for the fast and accurate evaluation of certain polymer properties, such as surface energy and thermal properties.

15.2
Automated and Parallel Polymer Synthesis

Automated and parallel synthesis (and also formulation) of polymers can provide several advantages for the fast discovery of new materials with certain structure–property relationships. However, some drawbacks regarding the special requirements of the equipment have to be overcome [25]. Polymerizations have to be carried out over a wide range of temperatures and pressures. Furthermore, the automated acquisition of samples in order to monitor polymerization kinetics should be available. Finally, the strongly varying properties of the polymers (e.g. viscosity and composition) and the handling of solids (e.g. in blends or hybrid materials) are additional challenges. In general, the goal of automated parallel synthesis and formulation in CMR is the faster preparation and understanding of new polymeric materials and their processing conditions. Literature examples of the automated synthesis of polymers range from simple free radical polymerizations via polycondensations to various controlled and living polymerization techniques. Thereby, both simple manual parallel approaches and fully automated and optimized workflows have been applied. In general, the automated and/or parallel synthesis of polymers can be divided into three groups: (i) parallel synthesis applying standard laboratory techniques; (ii) semi-automated synthesis using parallel manual approaches; and (iii) fully automated and parallel synthesis utilizing robotic systems. Here, we only briefly describe the possibilities of fully automated robotic systems as they are utilized in our laboratories [26]. Subsequently, we shall address prominent literature examples that describe fully automated polymer synthetic approaches utilizing such or similar equipment in order to provide an introduction to the field. However, for more detailed descriptions of the synthetic possibilities within combinatorial materials research readers are referred to recent literature reviews [15–17], since the focus of this chapter is on high-throughput screening and characterization

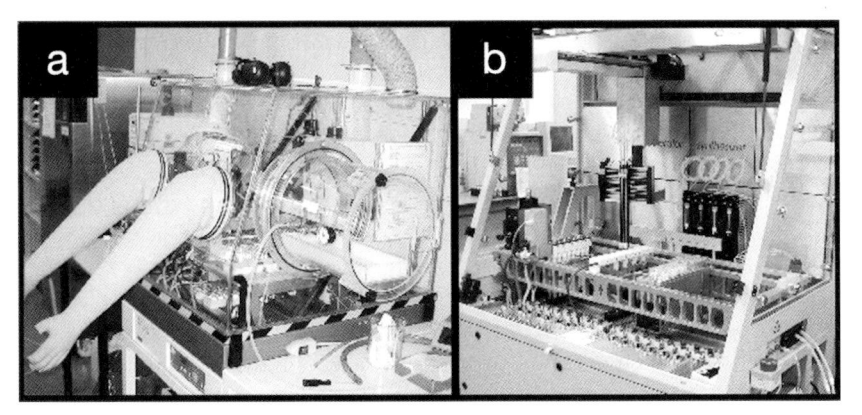

Fig. 15.2 Photographs of (a) the ASW2000 and (b) the Accelerator SLT100 automated synthesizer platforms.

of polymeric materials. Our available automated synthesis robots for combinatorial materials research can be divided into two classes: synthesizers for small-scale reactions that are applied for screening of reaction parameters and for library synthesis and a robot for scale-up and process development. The main difference of the scaling-up and process development synthesizer is its mechanical stirring and continuous feed possibility that match industrial scale polymerizations more closely. Here, we only want to describe the possibilities of the ASW2000 and Accelerator SLT100 (Fig. 15.2) platforms from Chemspeed Technologies, since they are the most frequently applied systems for typical tasks in the CMR routine, such as the monitoring of polymerization kinetics or the preparation of polymer libraries. They have in common a modular layout that allows the straightforward combination of different reactor arrays, microtiter plates and custom racks and also large vial and stock solution racks. Both synthesizers are equipped with vortex stirring (up to 1400 rpm), a needle/syringe-based liquid handling system (one needle in the case of ASW2000 and a four-needle head in the case of the SLT100), glove-box for inert atmosphere and the possibility of using different sizes of glass reactors (13, 27, 75 or 100 mL).

The reactions can be cooled or heated with a cryostat (–70 to 145 °C) that pumps its oil through the double-jacket heating mantles of the reactors. On top of the reactors an array of cold-finger reflux condensers can be placed for higher temperature reactions. The temperature of these condensers (5–50 °C) can be controlled via a second cryostat. The possibility of heating the condensers is a valuable tool for evaporating solvents from the reactors. The final part in assembling the reaction arrays is the placement of a reaction block on top of the reflux condensers. This reaction block has a ceramic drawer inside that can switch between opening the reactors, opening the reactors under argon, closing the reactors under argon or vacuum and closing the reactors independently, thereby offering possibilities similar to common Schlenk techniques in the laboratory and providing a second inert atmosphere within the reaction vessels. The SLT100 synthesizer offers additional features, such as an overhead solid dosing unit with microbalance for the accurate distribution of solids to the reactors and individually heatable reactors and pressure reactors. Therefore, together with the larger available space within the Accelerator (compared with the ASW2000), this machine offers more flexibility for certain types of chemistry due to the solid handling possibility and the possibility of working under pressure conditions. Such synthesizer platforms have proven to be very versatile tools for the automated synthesis of a large variety of different controlled/living polymerizations including anionic polymerization [27, 28], cationic ring-opening polymerization (CROP) [29, 30], atom transfer radical polymerization (ATRP) [31, 32], reversible addition–fragmentation chain transfer polymerization (RAFT) [33, 34] and controlled ring-opening polymerization [35]. One of the most prominent literature examples is a high-throughput screening methodology applied to the discovery of new polyolefin catalysts [36]. This example made use of a three-stage screening strategy with a primary (high-throughput, catalyst discovery, 384 experiments), secondary (intermediate-throughput, catalyst optimization, 96 experiments in a focused library) and tertiary (conventional-throughput, laboratory batch reactor, two reactions) screening and discovered new 1-octene polymerization catalysts. Finally, the development of a new and versatile high-temperature LLDPE (linear low-density polyethylene) catalyst was possible. Moreover, already in 1997, Kohn and coworkers described the "combinatorial approach for polymer design" to prepare a 112-membered library of strictly alternating A–B-type copolymers with predictable and systematic material property variations by the co-polycondensation of 14 tyrosine-derived diphenols and eight diacids [37]. The resulting library was investigated for structure–property correlations and it was discovered that the glass transition temperature (T_g) and the air-water contact angle of the polymers increased in a defined fashion as the number of carbon or oxygen atoms in the polymer backbone and pendent chain decreased. Subsequently, this library was utilized to derive models for protein adsorption properties [38, 39] or gene expression levels and it was attempted to correlate the outcome of these tests with the polymer structure [40, 41].

For all these polymerizations it is crucial to be able to investigate and analyze the large number of samples (e.g. for kinetic investigations) produced during automated synthesis in an as fast as possible but still accurate manner. Moreover, the synthesized polymer libraries with specific variations in polymer structure, molecular weight and/or composition need to be evaluated in terms of their structure and properties in order to establish the above-mentioned structure–property relationships. Therefore, fast, accurate and reliable analytical techniques are required to avoid bottlenecks in the workflow. For that reason, special high-throughput screening techniques were developed, adapted and integrated into the CMR workflow. Their application, evaluation and performance will be discussed in the following sections.

15.3
High-throughput Screening

High-throughput screening in combinatorial polymer chemistry cannot be directly compared with HTS in pharmaceutical research since the parameters of interest are rather different. The structural characterization of small molecule libraries can be easily accomplished utilizing gas chromatography/mass spectrometry (GC/MS) or high-performance liquid chromatography/mass spectrometry (HPLC/MS) techniques. In addition, screening these libraries for a potential drug candidate is mostly performed by optical screening methods (based on plate reader approaches), resulting in qualitative binding information of a certain compound to a certain receptor. On the other hand, one of the most important structural parameters of a (synthetic) polymer is its molecular weight and the corresponding molecular weight distribution. Moreover, a large variety of different important properties, such as the glass transition temperature (T_g), melting temperature (T_m) or application-specific parameters, need to be determined for polymer libraries in order to develop new materials. Therefore, new screening methods had to be developed and/or adapted to be suitable for HTS in combinatorial polymer research.

15.4
Screening for Molecular Weight and Polydispersity Index

The molecular weight and its distribution are essential characterization parameters for synthetic polymers. Size-exclusion chromatography (SEC) can be considered as the easiest and most commonly used method to determine them [42, 43]. Considering high-throughput experimentation, several approaches have been reported to decrease the analysis time, which is conventionally more than 30 min [44–46]. An automated sample preparation technique for high-temperature SEC utilizing robotic systems providing excellent reproducibility was described in 1997 [47]. In general the speeding up of conventional SEC systems can be achieved by several methods, e.g. decrease in the column length, increase in the flow-rate of the chromatographic system, flow injection analysis (FIA) or parallelization [42, 43, 45]. Nowadays, similar techniques can be utilized for online monitoring of polymerization reactions within automated synthesizer robots [46]. The combination of both robotic systems for SEC sample preparation and the utilization of commercially available fast SEC columns (2–5-min analysis times) is a powerful screening technology for polymeric materials in CMR. The integration of a SEC system into a high-throughput workflow for direct monitoring can be accomplished by utilizing the injection port that is available within the above described automated synthesizer platforms [46].

Therefore, the tubing of the chromatographic system was extended from this injection port to the pump of the system allowing a direct and automated injection of SEC samples within the synthesizer (Fig. 15.3). As soon as the sample is injected by the liquid handling system of the robot, a trigger signal will start the SEC measurement. To increase the speed of analysis further, a commercially available high-speed column was installed on the online SEC system. Especially for fast polymerizations it might be necessary to have a shorter analysis time (compared with the standard 15-min analysis time that is common for mixed pore-size columns), because otherwise a bottleneck might be present and not

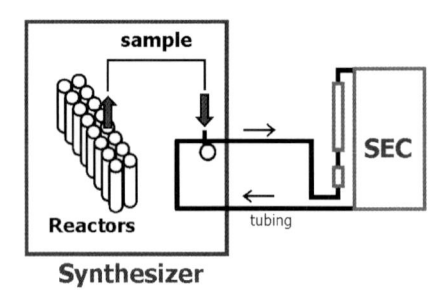

Fig. 15.3 Schematic representation of the online SEC characterization method utilized. From Schubert et al., Macromol. Rapid Commun. 2004, 25, 237–242 [46]. Reprinted with permission of John Wiley and Sons, Inc.

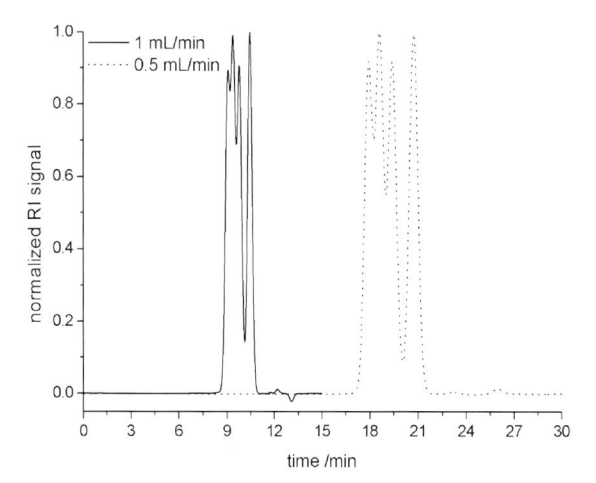

Fig. 15.4 Four polystyrene standards ($M_p = 2000–12000$ Da) investigated with SEC at different flow-rates.

enough samples can be processed. However, by the use of (shorter) high-speed SEC columns or an increase in the flow-rate of the chromatographic system, a decrease in resolution (plate count) will occur, as demonstrated in Fig. 15.4.

It is obvious that a doubling of the flow-rate reduced the separation capability of the chromatographic system but also decreased the analysis time by a factor of two. This trend was also observed even on doubling the flow-rate again to 2 mL min^{-1}, reaching analysis times of less than 7 min with a conventional SEC column, but losing accuracy. In the case of a 2 mL min^{-1} flow-rate the peak maxima of the same polystyrene standards as shown in Fig. 15.4 were no longer separated. Therefore, it is crucial to choose the column and flow-rate of the SEC system according to the needs of the investigation, e.g. if highly accurate kinetic data are required, longer columns with lower flow-rates are mandatory. However, if the amount of samples is too high or less accurate data are sufficient, high-speed columns at high flow-rates offer a very good alternative for high-throughput determinations of the molecular weight and polydispersity index in CMR. In addition to the mentioned approaches, ultra-fast SEC screening techniques were developed by Symyx (Rapid GPC system), which are capable of analyzing SEC samples in 40 s–2 min [48]. Moreover, Dow showed that this system can be adopted for their purposes [49]. Subsequently, Symyx compared data obtained from the Rapid GPC with the results obtained by conventional SEC measurements, revealing good agreement and showing that these approaches are practicable for the screening of molecular weights of polymer libraries [50]. Moreover, more advanced SEC systems were shown to be automatable. In this respect, refractive index, light scattering and viscosity techniques were coupled with an SEC system in order to obtain reproducible M_w data for polyvinylpyrrolidine (PVP) and poly(ethylene oxide) (PEO) polymers in an automated approach [51].

Matrix-assisted laser desorption/ionization time-of-flight mass spectrometry (MALDI-TOF-MS), on the other hand, is a very versatile and accurate technique for the determination of absolute molecular masses and has advanced to the perhaps most powerful MS tool for the analysis of polymers [52] since its introduction in the 1980s [53, 54]. However, until recently, no feasible technique was described for the high-throughput determination of molar masses, molar mass distributions and end-groups of synthetic polymers by MS. Moreover, it should be mentioned that MALDI in general is applicable only to rather low molecular weight synthetic polymers with a narrow molecular weight distribution. Recently, a multiple-layer spotting technique for MALDI-TOF-MS of synthetic polymers was developed [55] that is ideally suited for applications in combinatorial materials research [15–17]. This method is able to reduce significantly the time required for sample preparation [56], to improve the analytical results [55, 57] and to be automated [56, 57] and miniaturized [57]. The possibilities and application examples of this automation approach were recently discussed [58]. In general, the key factor and therefore the most important requirement for any successful MALDI experiment are an easy applicable and reproducible sample preparation technique [52]. One of the most widely applied sample preparation techniques for MALDI-TOF-MS is the dried droplet (DD) method [52, 59]. For this method solutions of analyte, matrix and ionizing salt (most likely in different solvents) are mixed by volume and an aliquot of the mixture (usually 0.5–1.0 μL) is deposited on the MALDI target and air dried. One particular problem with this method for the analysis of synthetic polymers is the solvent selection, since all three compounds should be readily soluble in a certain solvent mixture. However, salts, for instance, do not show high solubilities in common organic solvents, which makes the DD method difficult for hydrophobic synthetic polymers. This can lead to a small amount of polymer non-solvent in the final mixture, which can affect the signal reproducibility [59]. Therefore, other sample preparation techniques for improved reproducibility and analytical results were developed, including vacuum drying [60], overlayer (two-layer or seed layer) [61, 62] and fast evaporation [63]. All the methods mentioned have in common the effort to control the crystallinity of the matrix utilized and therefore the improvement of the analytical results. Concerning the high-throughput analysis of libraries by MALDI-TOF-MS, especially the seed-layer technique could be automated for the analysis of peptides and proteins [62] and of a 41-compound library of organic molecules [64]. In general, the requirements for an automated MALDI-TOF-MS analysis of combinatorial polymer libraries are similar to those discussed above. However, for combinatorial approaches a reproducible and widely applicable sample preparation technique is even more relevant, since a large variety of different compounds or, in the case of polymers, a large variety of different molecular weights have to be analyzed utilizing the same sample preparation technique. Moreover, the technique applied should be easy to automate and to integrate into the combinatorial workflow and be as fast as possible. In order to meet these challenges, a multiple-layer spotting technique was developed, evaluated [55] and later automated and integrated into the workflow

Fig. 15.5 (a) Conventional mixing sample preparation for MALDI; (b) multiple-layer sample preparation for MALDI. Reprinted with permission from *Rev. Sci. Instrum.* **2005**, *76*, 062211/15 [58]. Copyright 2005, American Institute of Physics.

of combinatorial material research [56, 57]. Generally, the multiple-layer sample preparation decouples the handling of the three main components of a MALDI-TOF-MS sample, namely the matrix, the doping salt and the analyte. These three components are spotted on top of one another in the mentioned order from different solutions, allowing the free choice of solvent for every component. As a result, difficulties in the solubility of analyte, salt and matrix in different solvents can be avoided. Moreover, it was observed that this technique is able to provide superior analytical results compared with the conventional dried droplet technique [55]. Figure 15.5 depicts schematically the difference between the multiple-layer sample preparation and the conventional mixing approach.

It should be noted that this representation is oversimplified since redissolution of previous layers can occur to a certain extent. Nevertheless, the free choice of solvent for the crystallization of the matrix is a significant advantage of the developed technique and provides clearly improved analytical results. Figure 15.6 clearly illustrates this behavior for the matrix a-cyano-4-hydroxycinnamic acid (CAHA). A poly(ethylene glycol) (PEG) standard with a molecular weight of 5000 Da was analyzed by MALDI-TOF-MS applying exactly the same instrument settings for the measurement (e.g. laser intensity, acceleration voltage) and the multiple-layer spotting technique (with NaI as salt) for the sample preparation. The only difference was the solvent utilized to prepare the first (matrix) layer of the sample: for the sample resulting in the spectrum displayed in Fig. 15.6a, CAHA was deposited from acetone, whereas in Fig. 15.6b it was deposited from methylene chloride. It is obvious that CAHA (and other matrices) is able to ionize PEG providing well-resolved MALDI spectra as displayed in Fig. 15.6a. In this case acetone was applied for the crystallization of CAHA, resulting in a continuous and crystalline layer of CHCA (see microscopy picture: inset in Fig. 15.6a).

On the other hand, if methylene chloride was used to prepare the matrix layer of CHCA, a discontinuous film showing hardly any crystalline features was obtained (see microscopy picture: inset, Fig. 15.6b). Only a baseline signal could be measured from this poorly crystallized matrix film, as shown in Fig. 15.6b. This correlation between matrix crystallinity and MALDI-TOF-MS spectral quality can also be observed for other matrices and is moreover described in the literature (see the discussion above). Investigating this effect in more detail, the crystallization behavior of six MALDI matrices [*trans*-2-[3-(4-*tert*-butylphenyl)-2-methyl-2-propenylidene]malononitrile (DCTB), a-cyano-4-hydroxycinnamic acid

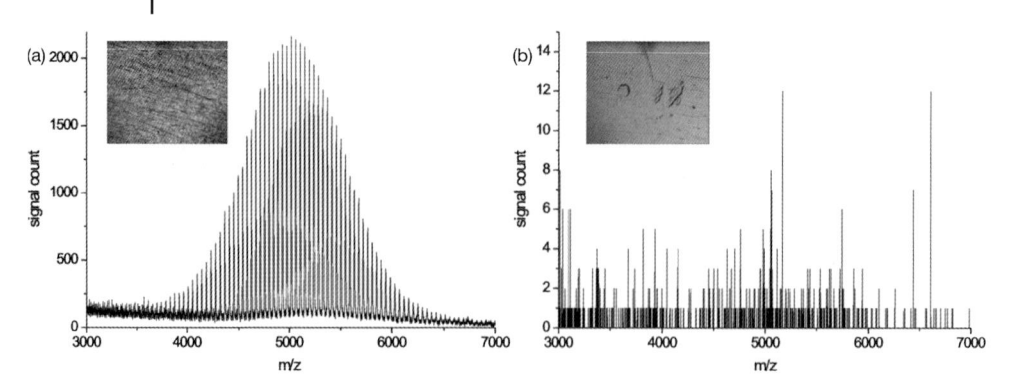

Fig. 15.6 MALDI-TOF mass spectra of a 5000-Da poly(ethylene glycol) standard applying the multiple-layer sample preparation. (a) CHCA layer prepared from acetone; (b) CHCA layer prepared from methylene chloride.

(CAHA), 2-(4-hydroxyphenylazo)benzoic acid (HABA), *trans*-3-indoleacrylic acid (IAA), 2,5-dihydroxybenzoic acid (DHB) and 1,8,9-anthracenetriol (dithranol)] from six solvents was studied [58].

Figure 15.7a shows optical microscopy pictures of the matrix layer of a 36-membered library of MALDI-TOF-MS samples that were prepared utilizing the multiple layer approach. Figure 15.7b displays the resulting relative signal intensities of this library when sodium iodide in acetone and poly(ethylene glycol) standard (M_n = 5000 Da) in dichloromethane were spotted on all 36 positions as second and third layers, respectively. For this MALDI-TOF-MS experiment, spectra of all 36 samples were taken with the same instrument settings and the corresponding signal intensities were subsequently compared. The most striking result is that DCTB is the superior matrix and is not very selective on the solvent used for crystallization. All DCTB solvent combinations, except DCTB with methanol, were able to produce an at least 10-fold higher signal intensity compared with all other sample preparations. It should also be noted that the combination of DCTB and methanol (in contrast to the other DCTB solvent combinations) was not able to produce a crystalline film as shown in Fig. 15.7a, which resulted in a poor MALDI result for reasons already discussed above. Moreover, the trend that DCTB is able to produce considerably higher signal intensities compared with other matrices is described in the literature for fullerene derivatives [65] and was observed in our laboratories also for other classes of polymers. Therefore, DCTB in combination with the discussed multiple layer spotting approach for MALDI-TOF-MS of synthetic polymers is the method of choice for the high-throughput screening of synthetic polymers since it allows easy implementation of the sample preparation and provides good to excellent analytical results for a large variety of different synthetic polymers.

The implementation of this sample preparation into the CMR workflow (cf. Fig. 15.1) was accomplished in a straightforward manner by manufacturing a

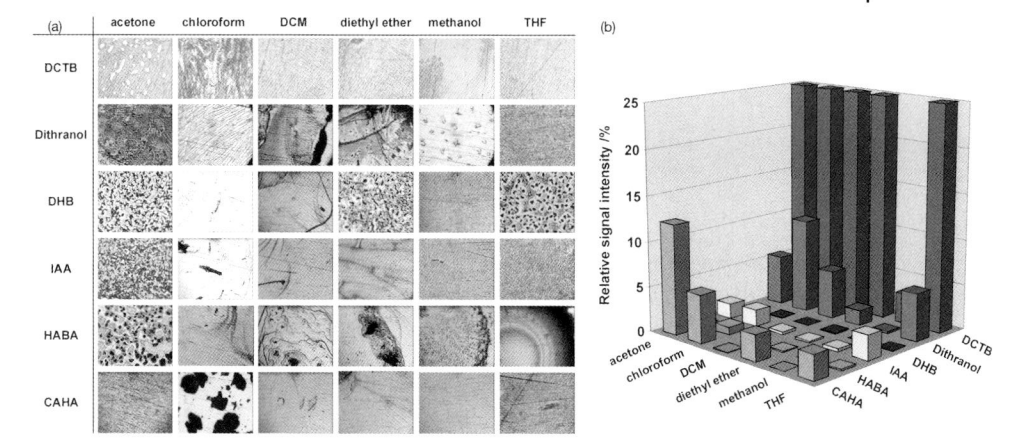

Fig. 15.7 (a) Optical microscopy pictures of layers of different matrices crystallized from different solvents. (b) The corresponding MALDI-TOF-MS signal intensities for a poly(ethylene glycol) standard ($M_n = 5000$ Da). All MALDI spectra for this experiment were recorded with the same instrument settings for comparison reasons. Reprinted with permission from *Rev. Sci. Instrum.* **2005**, *76*, 062211/1–062211/5 [58]. Copyright 2005, American Institute of Physics.

custom-made MALDI target holder in microtiter plate format. The sample positions were programmed on an *xyz* basis in the Chemspeed software (Gilson 735 Sampler Software V 2.10) of an ASW2000 synthesizer [56]. The spotting can then be carried out using the liquid handling system of the robotic synthesizer. The respective solution is aspirated and subsequently spotted at a defined position on the MALDI target. First, 1 µL of the matrix solution was spotted on the target. The time required by the robotic system to finish all sample spots was long enough to ensure complete drying of the spots before the next layers of salt and analyte were applied. Moreover, the analyte could be spotted directly from the polymerization mixtures or samples taken for SEC/GC analysis to allow online monitoring of the polymerization reactions. Figure 15.8a depicts schematically the integration approach, and Fig. 15.8b shows the needle, which is attached to the robotic arm of the ASW2000 synthetic robot (see also Fig. 15.2) spotting dithranol matrix solutions on the MALDI target in the custom-made rack.

The time required for spotting additive and matrix solutions (about 45 s per spot) is shorter than the time required for spotting the polymer sample (about 90 s per spot), because no additional rinsing steps of the robot needle are required between the drop-casting of individual spots. Moreover, the multiple-layer spotting approach not only saves time since the mixing step would take approximately an extra 90 s per sample but also saves valuable space within the automated synthesizer, as illustrated schematically in Fig. 15.9 [66]. This technique allows fast (about 3 min per sample), easy and automated sample preparation within a synthetic robot and therefore improves the workflow of combi-

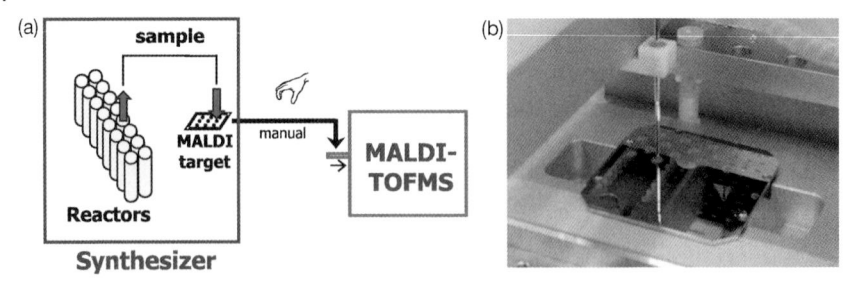

Fig. 15.8 (a) Schematic representation of online monitoring by MALDI-TOF-MS. From Schubert et al., *Macromol. Rapid Commun.* **2004**, *25*, 237–242 [46]. Reprinted with permission of John Wiley and Sons, Inc. (b) Liquid handling system of the ASW2000 synthetic robot preparing matrix layers of dithranol. Reprinted in part with permission from Schubert et al., *J. Comb. Chem.* **2003**, *5*, 369–374 [56]. Copyright 2003, American Chemical Society.

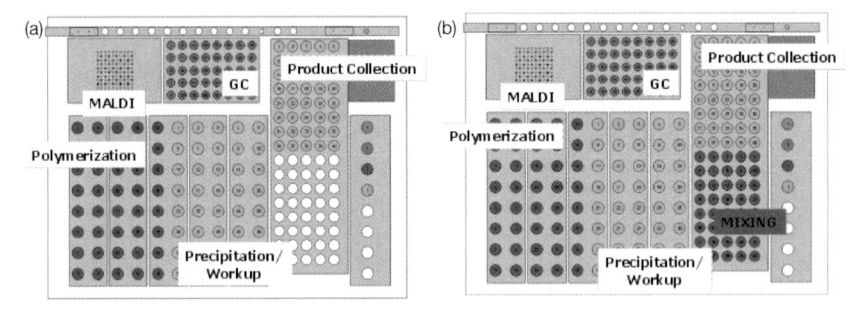

Fig. 15.9 Schematic layout of the ASW2000 automated synthesizer with 40 parallel polymerizations including workup and online monitoring. (a) Multiple-layer MALDI sample preparation. (b) Conventional MALDI sample preparation. From Schubert et al. [66]. Reprinted with permission.

natorial polymer research and offers high-throughput screening possibilities. This automated sample preparation approach was first tested and evaluated with a set of polystyrene standards of known molecular weights, clearly revealing its feasibility. As a first application of this technique, the living cationic ring-opening polymerization of 2-ethyl-2-oxazoline was monitored.

The polymers were synthesized utilizing the ASW2000 automated synthesizer and the samples were prepared from the automatically purified products (precipitation at −20 °C in diethyl ether), providing an additional proof that the polymerizations were indeed living [67]. This technique found applications in the monitoring of controlled and/or living polymerizations, such as the monitoring of cationic ring-opening polymerizations of 2-nonyl-, 2-phenyl-, 2-ethyl- and 2-methyl-2-oxazoline [68, 69] and the reversible addition–fragmentation chain transfer polymerization of methyl methacrylate [33]. Due to the optimized sam-

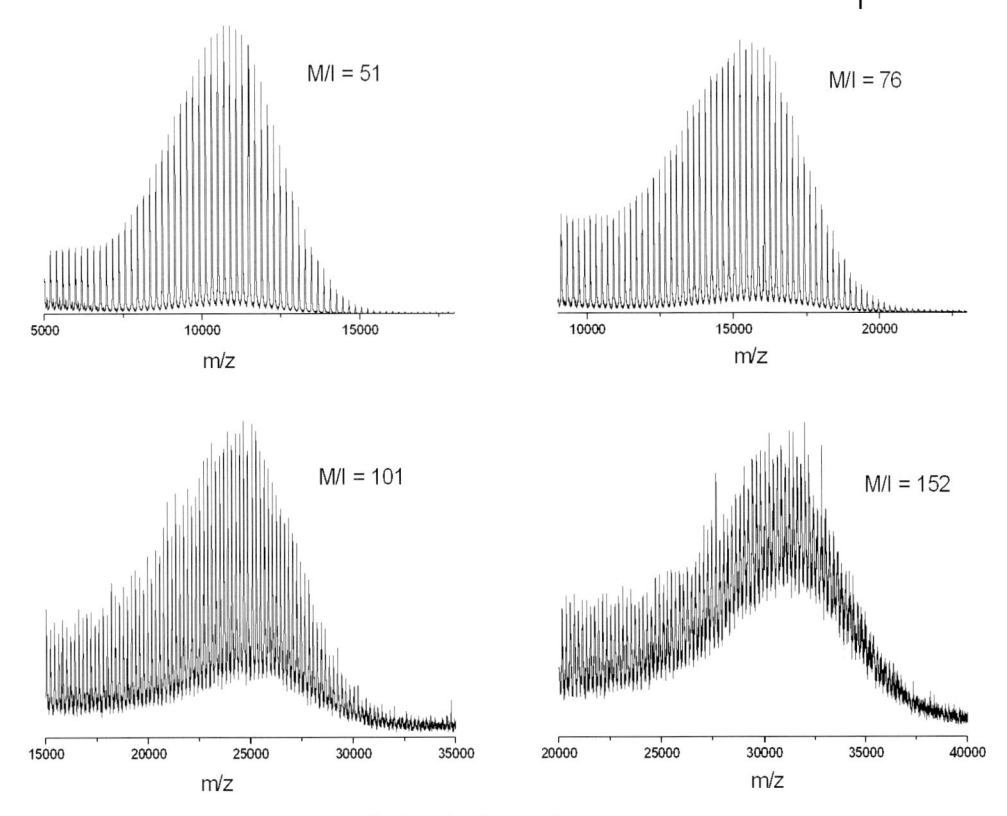

Fig. 15.10 MALDI-TOF mass spectra of high molecular weight poly(2-nonyl-2-oxazoline)s. Reprinted with permission from Schubert et al., *Macromolecules* **2005**, *38*, 5025–5034 [69]. Copyright 2005, American Chemical Society.

ple preparation, it was possible to investigate poly(2-oxazoline)s with relatively high number-average molecular weights by MALDI-TOF-MS. As an example, the MALDI-TOF-MS spectra of a series of eight poly(2-nonyl-2-oxazoline)s with molecular weights ranging from 10 000 to 30 000 Da is depicted in Fig. 15.10 [69]. The number-average molecular weights obtained by this technique showed perfect agreement between theoretical and experimental data. The samples that could be analyzed by MALDI-TOF-MS for these investigations were limited by the number-average molecular weights ($M_n < 30$ kDa) and average molecular weight distributions (PDI < 1.30), as usual for this technique. In general, all mass spectra showed the expected signal spacings for the investigated poly(2-nonyloxazoline)s (197 Da).

Therefore, the monitoring of polymerization reactions and/or the analysis of libraries of polymers by MALDI-TOF-MS is now feasible, allowing the investigation of large numbers of samples with reduced experimental effort in a shorter time. Moreover, MALDI-TOF-MS high-throughput screening is an excellent ex-

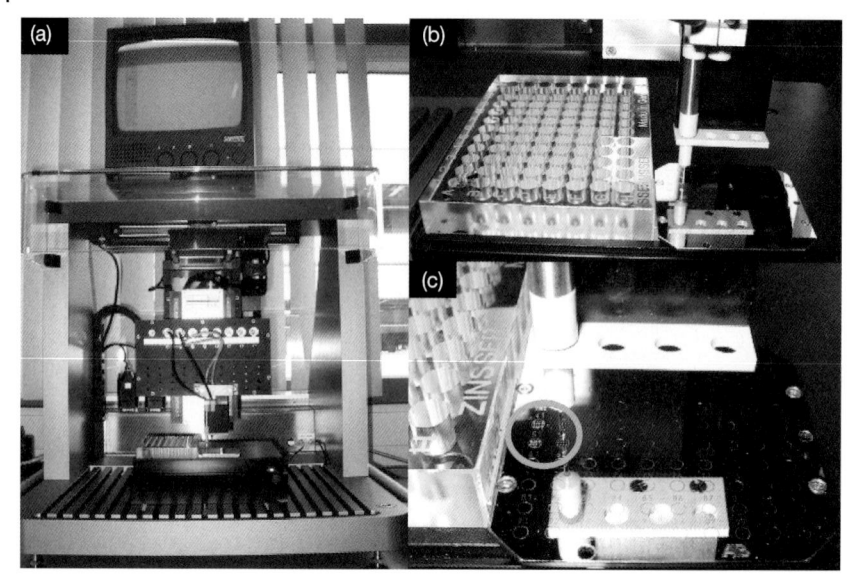

Fig. 15.11 (a) Photograph of the inkjet printer used in this study. (b) The glass microtiter plate for stock solutions and the MALDI sample target on the stage of the inkjet printer. (c) Close-up of the MALDI target. From Schubert et al., *Rapid Commun. Mass Spectrom.* **2003**, *17*, 2349–2353 [57]. Reprinted with permission of John Wiley and Sons Ltd.

ample showing that high-throughput data are not necessarily of reduced quality compared with conventional measuring approaches.

In an effort to further miniaturize and automate the described multiple-layer spotting sample preparation, the possibility of utilizing inkjet printing technology for this sample preparation was investigated [57]. The three layers of matrix, salt additive and polymer were deposited sequentially on top of each other by printing a 4×4 array of spots of each layer on the MALDI-target (Fig. 15.11).

Each spot consisted of five individual drops printed on top of each other. The spots were printed at regular intervals of 0.5 mm. The distance between the arrays was 5 mm. After drying, the matrix layer was covered with a dopant layer, which was printed in exactly the same way as the matrix. Finally, the analyte was printed on top of the two previous layers. Generally, this resulted in reproducible sample spots of size about 180–200 µm (Fig. 15.12) and in comparable analytical results for the respective polymeric analytes. The size of the inkjet printed sample spots is therefore significantly smaller than the size of hand-prepared spots (4–5 mm), allowing the deposition of at least 400 times more samples on one target plate (∼4000 samples in the case of the MALDI target utilized for the present study). The solvents used up to now for matrix crystallization, i.e. chloroform and THF, are not compatible with the inkjet printer since they evaporate too quickly. For inkjet printing experiments, the rate of evaporation of the solvent utilized should be relatively low to prevent the solution from

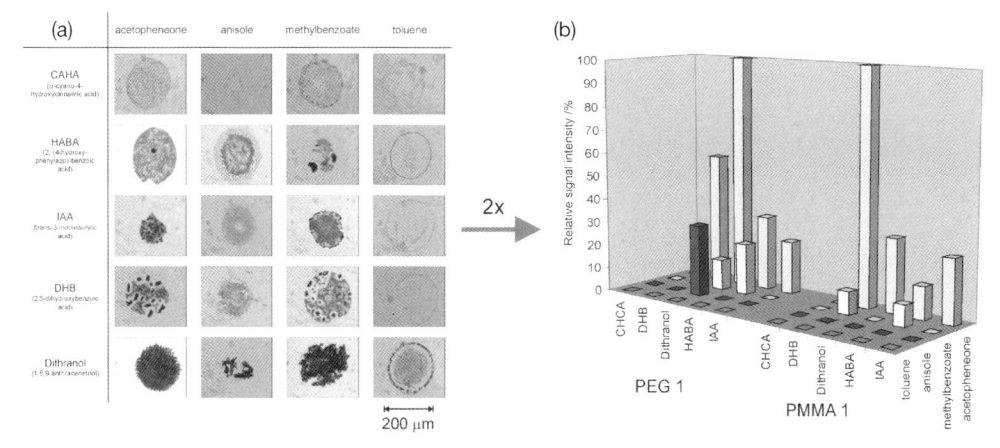

Fig. 15.12 (a) Optical microscopy pictures of inkjet printed matrix solutions. (b) Relative signal intensities obtained from MALDI-TOF mass spectra for a poly(ethylene glycol) ($M_n = 3000$ Da) and a poly(methyl methacrylate) ($M_n = 4000$ Da). From Schubert et al., *Rapid Commun. Mass Spectrom.* **2003**, *17*, 2349–2353 [57]. Reprinted with permission of John Wiley and Sons Ltd.

drying out and thereby clogging the nozzle of the inkjet printer. Therefore, solutions of five different matrices (CAHA, HABA, IAA, DHB and dithranol) in different solvents (acetophenone, anisole, methyl benzoate and toluene) were printed that can be used for the inkjet processing. After printing these matrix solutions, optical microscopy pictures were taken, revealing significant differences in spot shape and crystallinity. Figure 15.12 clearly shows that certain matrix–solvent combinations can result in very defined spots (∼180–200 µm in diameter, cf. Fig. 15.12) with high homogeneity and crystallinity. This qualitative result correlated with the observed spectrum quality for two polymer standards. Finally, both polymers were printed as the final layer on top of the matrix layer crystallized from different solvents and an NaI layer (printed from a 20 mg mL^{-1} solution in water). Subsequently, MALDI-TOF-MS spectra were measured with the same instrument settings for all 20 combinations of different sample preparations for both polymers. These spectra were evaluated by means of signal intensity and signal-to-noise ratio, resulting in a preferred matrix–solvent combination for each polymer. Some of the results are shown in Fig. 15.12b. It can be clearly seen that certain combinations of solvents and matrix show a preference for a certain analyte. For instance, toluene is a very bad solvent for the crystallization of any matrix (Fig. 15.12a) and therefore the results obtained from MALDI-TOF-MS using matrices crystallized from toluene are also poor (Fig. 15.12b).

On the other hand, acetophenone seems to be a good solvent for the matrix crystallization since both homogeneous and crystalline spots are observed for almost all matrices (Fig. 15.12a). This behavior can be very well correlated with the relative signal intensities if this solvent is utilized for matrix crystallization (Fig. 15.12b). In general, it seems that the previous conclusion that high matrix

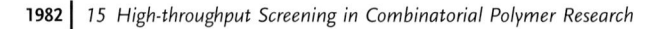

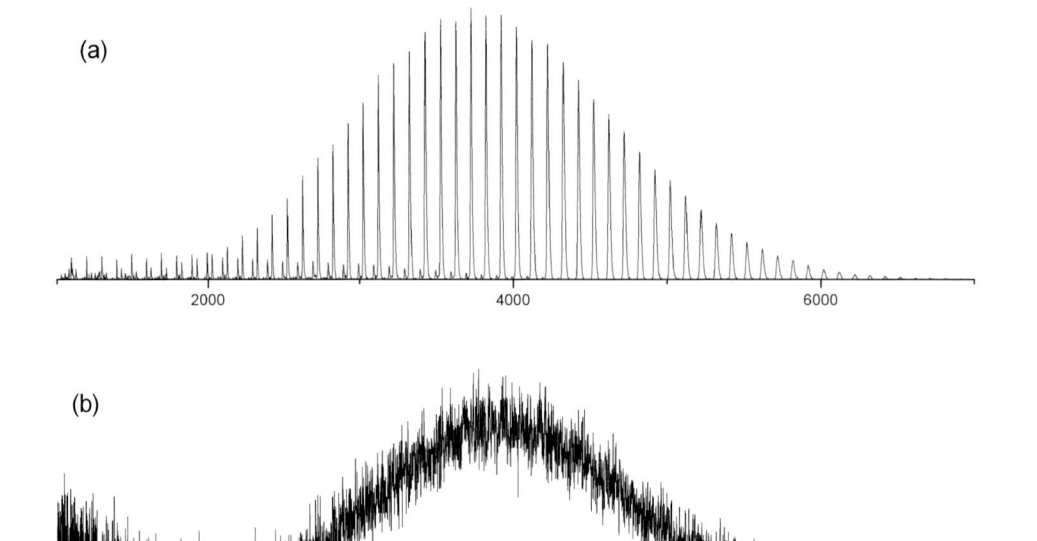

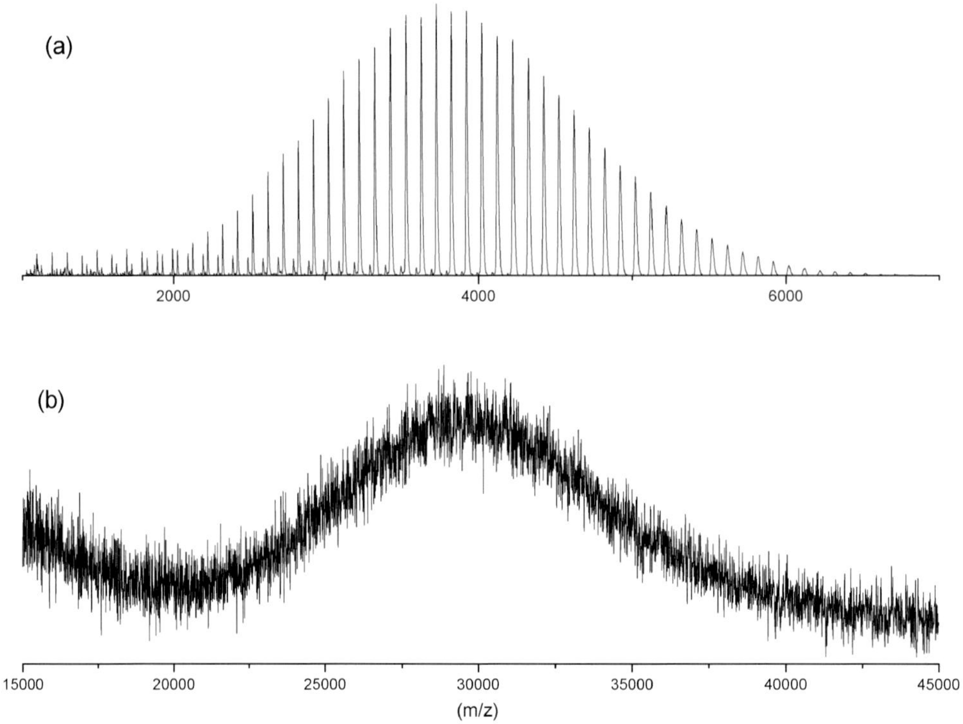

Fig. 15.13 MALDI-TOF mass spectra obtained from two PMMAs, (a) $M_n = 4000$ Da, (b) $M_n = 30\,000$ Da, with optimized settings for inkjet printing. From Schubert et al., *Rapid Commun. Mass Spectrom.* **2003**, *17*, 2349–2353 [57]. Reprinted with permission of John Wiley and Sons Ltd.

crystallinity positively influences the MALDI mass spectrum still holds if an inkjet printer is used for sample preparation. Therefore, optimized settings were obtained for MALDI-TOF-MS multiple-layer sample preparation with inkjet dispensers for the two investigated polymer classes: (i) dithranol in acetophenone, NaI in water and PEG in acetophenone; (ii) DHB in acetophenone, NaI in water and PMMA in acetophenone (layers were printed in the given order). Applying these settings, very good analytical results for both PEG and PMMA polymers could be obtained, even for high molar masses.

As an example, MALDI-TOF mass spectra of two PMMA samples with low and high molecular weights are displayed in Fig. 15.13. Inkjet printing of MALDI-TOF-MS samples not only provides good analytical results but also offers the possibility of automating the MALDI sample preparation. Moreover, the developed techniques might be interesting for ultra-high-throughput applications since it was possible to measure MALDI mass spectra from single printed spots of 200 µm size, allowing the deposition of at least 4000 samples on one target plate. Apart from the above-described techniques for an automated and/or parallel investigation of the molecular weight of polymers, it is also possible to integrate more classical techniques, such as viscosimetry, into the workflow of CMR. Capillary viscosimetry provides an easy and straightforward method for molecular weight determination by measuring the intrinsic viscosity and application of the Mark–Houwink equation. For this purpose, for example, the Processor Viscosity System from Lauda is a valuable tool since this modern and completely computer-controlled instrument allows the precise measurement of viscosity using four standardized glass capillary viscometers in parallel. Moreover, the system is equipped with an autosampler and a completely automated cleaning system, allowing up to 50 samples to be measured per day. Furthermore, automated dilution series can be performed to obtain the intrinsic viscosity. These features make the instrument perfectly suited for the combinatorial workflow and permit molecular weight determinations for polymers that are otherwise difficult to characterize due to, e.g., column interactions in SEC or too high molecular weight and/or too broad polydispersity indices for MALDI-TOF-MS.

15.5
Screening for Polymer Composition and Polymerization Kinetics

The chemical composition of polymeric materials and also polymerization kinetics can be conveniently obtained utilizing optical screening methods, such as Fourier transform infrared (FTIR) spectroscopy or chromatographic techniques, such as gas chromatography (GC) or high-performance liquid chromatography (HPLC). In general, optical methods are very well suited for high-throughput screening since the optical signal can be read out continuously for online monitoring purposes. Furthermore, by utilizing parallel approaches (e.g. plate readers or imaging setups), a large number of samples can be evaluated quasi-simultaneously. Chromatographic techniques, on the other hand, are well suited for handling a large number of samples due to the utilization of autosamplers. Moreover, these techniques can be accelerated by taking advantage of commercially available columns with reduced analysis time for HPLC, SEC and GC. For the monitoring of polymerization kinetics, both optical and chromatographic techniques can be used in a straightforward manner by measuring t_0 samples and subsequently following the process in time relative to the initial sample to obtain monomer conversions or other interesting parameters. It was shown, for instance, that a flow cell, which is positioned in the working area of the GC autosampler and connected to an injection port inside a robotic synthesizer system, can be used for high-throughput monitoring of cationic ring-opening polymerizations [46]. This setup is depicted schematically in Fig. 15.14.

The reliability and comparability of this online GC setup was investigated by monitoring the cationic ring-opening polymerization (CROP) of 2-ethyl-2-oxazoline at different concentrations [46]. Ten polymerizations with a monomer to initiator ratio [M]/[I] of 60 at different concentrations were investigated with

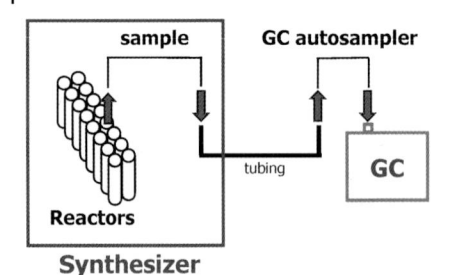

Fig. 15.14 Schematic representation of an online-monitoring setup for GC characterization. From Schubert et al., *Macromol. Rapid Commun.* **2004**, *25*, 237–242 [46]. Reprinted with permission of John Wiley and Sons, Inc.

both online and offline GC measurements. This was accomplished by taking samples from the reaction mixture and injecting them into the flow cell and subsequently into the GC column. During the GC analysis time (~8 min including cooling time), a sample was taken from the same reactor to a vial prefilled with chloroform for offline GC characterization. Figure 15.15 (inset) shows the obtained ratios of the integrals from the 2-ethyl-2-oxazoline (EtOx) monomer signal divided by the solvent [N,N-dimethylacetamide (DMAc)] signal for the different concentrations at time zero from both online and offline GC. This graph clearly shows the good correspondence between online and offline GC and, in addition, it demonstrates that, as expected, the EtOx/DMAc ratio becomes larger at higher concentrations. The CROP of oxazolines is known to follow a living mechanism with first-order kinetics [70], whereby the rate of polymerization (assuming that all initiator reacted instantaneously upon heating) can be given by the equation

$$-\frac{dM}{dt} = k_p[P^*][M] \tag{1}$$

where [M] and [P*] are the concentrations of monomer and propagating species, respectively. Integration of Eq. (1) results in Eq. (2) (assuming that [P*] is equal to the initial initiator concentration $[I]_0$):

$$\ln\left(\frac{[M]_0}{[M]_t}\right) = k_p[I]_0 t \tag{2}$$

From Fig. 15.15, it can be concluded that the polymerizations at all different concentrations follow linear first-order kinetics {ln([M]$_0$/[M$_t$]) versus time}, which is expected for a living polymerization. For the concentrated polymerizations, two slopes can be observed. The different kinetics at the beginning of the polymerization might be due to a combination of slow initiation and the different reactivities of very short chains, which are influenced by the initiating species and longer chains [71]. Moreover, for the complete kinetic investigations the same (general within 5% difference) signals and polymerization kinetics were obtained when utilizing both online and offline GC. However, at higher concentrations (>5 M), the time between samplings (60 min waiting time after

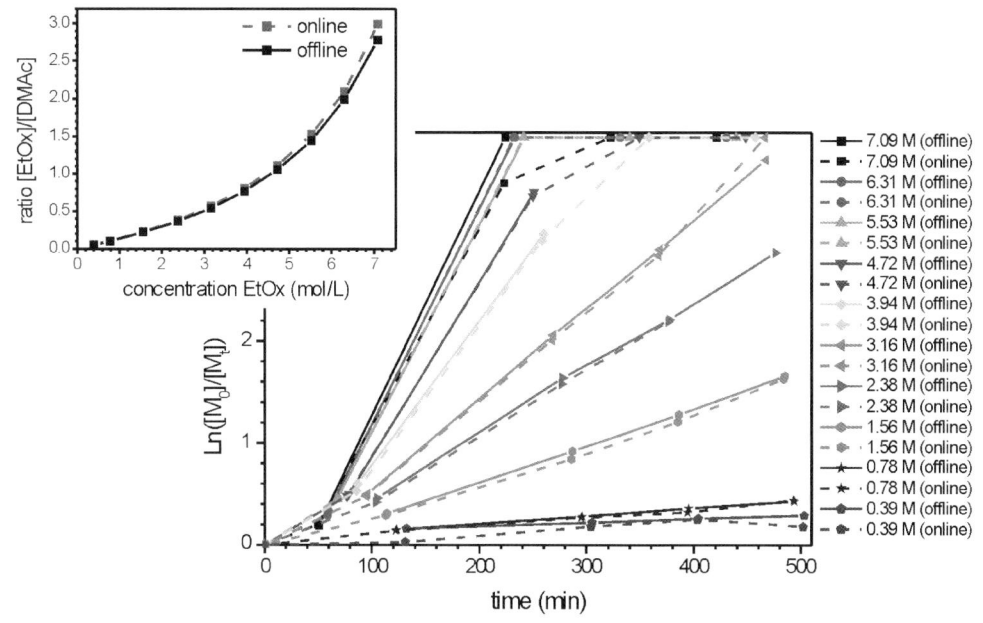

Fig. 15.15 Monitoring of the reaction kinetics by both online (dashed lines) and offline GC (solid lines) of the polymerization of 2-ethyl-2-oxazoline (EtOx) at different concentrations. The inset shows the initial EtOx/DMAc ratio obtained by GC for the different concentrations. From Schubert et al., *Macromol. Rapid Commun.* **2004**, *25*, 237–242 [46]. Reprinted with permission of John Wiley and Sons, Inc.

the second sampling) was too long to distinguish between the polymerization rates, because full conversion was already reached before the third sampling. For the evaluation of very fast polymerizations, omission of the waiting times or performing fewer reactions in parallel might still result in too slow monitoring. In that case, the monitoring speed could still be increased by installing a faster gas chromatograph or by connecting several parallel gas chromatographs to the automated synthesizer since the GC analysis time is currently the limiting factor. This example clearly demonstrates that online GC monitoring in CMR is feasible. The GC characterization results presented can, of course, also be correlated with the polymer composition obtained and, if the composition in time is analyzed, with the polymer structure (e.g. random, block-like; see, e.g., [111]).

Another analytical approach to obtain the polymer composition of polymer libraries in addition to monomer conversions of polymerization reactions in a fast and convenient fashion is the application of vibrational spectroscopy.

In general, these optical methods are very well suited for high-throughput screening since the optical signal can be read out continuously for online monitoring purposes. In that respect, fiber-optic mid-IR [72] or Fourier transform (FT) IR [73] probes were developed for the online monitoring of polymerization reactions. FTIR spectroscopy was applied for the monitoring of living isobutene and ethylene oxide polymerizations [74] and the copolymerization of ethene and 1-hexene [75]. Generally, the monitoring of monomer consumption using FTIR spectroscopy can be accomplished by following the decrease in the intensities of monomer peaks in time. It has been reported, for instance, that such IR-based systems allow the recording of one complete spectrum every 22 s [74]. Moreover, attenuated total reflection FTIR (ATR-FTIR) has been applied for the online monitoring of polymerization reactions, as shown for the terpolymerization of butyl acetate, methyl methacrylate and vinyl acetate in emulsion [76] and the carbocationic polymerization of isobutene [77]. Generally, ATR-FTIR has the advantage that direct characterization of powders and polymers without pre-processing is feasible and is therefore highly suited for the fast evaluation of large sample amounts. Mülhaupt and coworkers also reported that ATR-FTIR is applicable for the high-throughput evaluation of olefin copolymer compositions [78]. The composition of ethene–propene, ethene–1-hexene and ethene–1-octane copolymers could be calculated from the ATR-FTIR spectra with an error of less than 5% by using multivariate calibration as depicted in Fig. 15.16. Due to time savings during sample preparation, the sampling rate for these ATR-FTIR measurements can be increased to up to 40 samples per hour.

Furthermore, by utilizing parallel approaches (e.g. plate readers or imaging setups), a large number of samples can be evaluated quasi-simultaneously. Moreover, an FTIR plate reader setup was shown to be a fast tool to characterize a library of five-arm star-shaped block copolymers with a poly(ethylene glycol) (PEG) core and a poly(ε-caprolactone) (PCL) corona [35]. By first normalizing the FTIR spectra to the C=O stretch vibration at 1735 cm^{-1} (significant for the PCL block), three signals of the block copolymer spectra could, for instance, be utilized to analyze the composition of a series of six polymers with varying molecular weight of the PCL block. Furthermore, these findings could be correlated with both SEC and ^{1}H NMR spectroscopy and provided a fast and parallel evaluation of the polymer composition. Moreover, other methods to determine mono-

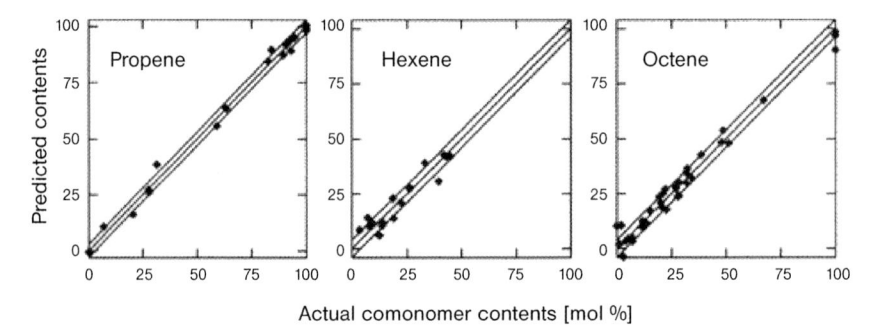

Fig. 15.16 Predicted (as determined by IR spectroscopy) vs. actual comonomer content (as determined by ^{13}C NMR spectroscopy) for ethene–propene, ethene–1-hexene and ethene–1-octene copolymers. The lines indicate the 1σ intervals. Reprinted with permission from Mülhaupt and coworkers, *J. Comb. Chem.* **2001**, *3*, 598–603 [78]. Copyright 2001, American Chemical Society.

mer conversion and/or polymer composition in a fast manner include HPLC [79]. These examples clearly show that the high-throughput screening of monomer conversion and polymer composition is feasible by utilizing well-known polymer characterization tools, such as fast GC or parallel-plate reader setups.

15.6
Polymer Property Screening

The property screening of new (polymeric) materials is challenging and special tests have to be developed and/or existing tests have to be adjusted in order to speed up the evaluation of new materials. Recently, these methods have been reviewed and discussed in the literature [15–17, 80, 81]. In particular, the screening of thin films (coatings) for certain properties of interest such as adhesion, crystallization and dewetting is of significant importance. Therefore, techniques for the preparation of continuous polymer thin-film libraries with gradients in thickness, temperature, composition and others were developed [82–87]. The advantage of such libraries is their relatively easy preparation and the possibility of obtaining material properties over a wide span of different parameters. Here, we will only discuss the screening of such libraries and refer the reader to the literature for their preparation [82–87]. Moreover, preparation techniques for libraries with variations in surface energy are available [88, 89]. Finally, the utilization of inkjet printing [90] or vapor deposition technologies [91, 92] offers advantages for the preparation of thin-film polymer libraries with discrete composition variations.

AFM was shown to be a powerful tool for the high-throughput characterization of pattern formation in symmetric polystyrene-b-poly(methyl methacrylate) (PS-b-PMMA) diblock copolymer films [84]. Thickness gradient films of copolymers with different molecular weights were prepared and their morphology was screened using AFM and optical microscopy, with the result that new morphology patterns of these materials were observed. AFM was also utilized for the evaluation of two-dimensional thickness–surface energy thin–film libraries of PS-b-PMMA. The degree of formation of islands and holes in the film was found to be dependent on differences in surface energy [93]. Moreover, optical microscopy was applied to two-dimensional composition–temperature libraries to detect phase separations and microstructures of PS–poly(vinyl methyl ether) (PS–PVME) blends [82]. An automated setup with an optical microscope was also used to evaluate the thin-film dewetting behavior of PS films on silicon [83]. By investigating PS thin-films with orthogonal, continuous variations in thickness and temperature, it was possible to obtain the temperature–thickness–time dependence of dewetting structures and kinetics. Moreover, the influence of film thickness and temperature on the crystallization behavior of isotactic polystyrene (iPS) films was studied [94]. Generally, all results were consistent with conventional experiments. It was observed, for instance, that the crystallization growth rate had a maximum at intermediate temperatures over a wide temperature range. Adhesion properties of a polydimethylsiloxane (PDMS) microlens library were obtained by "pressing" the PDMS samples against a PS-coated Si wafer and subsequently removing the wafer at constant speed [95]. An optical microscope was used to obtain digital images of this process. This led to a quantitative map of relative adhesion differences of interfaces created across the microlens array. Grunlan et al. described a method for the combinatorial screening of the moisture vapor transition rate (MVTR) [96]. A mixture of Nafion (a sulfonic acid-substituted fluoro polymer) and Crystal Violet was used as a moisture sensitive film, which was coated on to a poly(ethylene terephthalate) (PET) support film. After both sides of the PET support had been laminated with a transfer adhesive, the sensor was laminated on to a glass substrate. This substrate was then covered with barrier films of interest and the slopes of absorbance as a function of time were converted to moisture vapor transition rate values using a series of known films as references. This procedure was transferred to a combinatorial MVTR assessment. It was shown that 20 emulsion-based poly(vinylidene chloride) films (for the film/sensor preparation, see [96]) could be assessed simultaneously. In addition to the described techniques, Symyx introduced a Sensor Array Modular Measurement System (SAMMS), which has been utilized for many applications, such as the identification of phase transitions via calorimetric or thermal conductivity measurements of thin films [97]. Karim and coworkers also discussed a combinatorial approach to characterize epoxy curing in films consisting of a constant amount of fluorescent dye, curing agent and epoxy resin [98]. Therefore, FTIR microspectroscopy, confocal microscopy and axisymmetric adhesion testing were applied to study first discrete epoxy samples and subsequently a continuous temperature gradient combinatorial library. Figure 15.17 shows FTIR maps of the continuous gradient library, revealing a lower degree of curing at lower temperatures. The same trends in curing degree were found in a corresponding fluorescence map. Furthermore, the work of debonding (WOD, obtained from adhesion testing with polydimethylsiloxane lenses) could be correlated with the curing temperature and therefore the degree of curing.

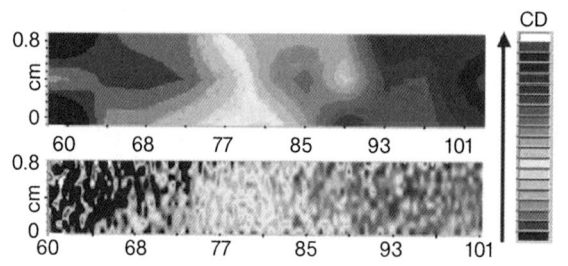

Fig. 15.17 FTIR maps of the curing degree (CD) (5×4 mm steps and 400×800 μm steps) of a thin-film epoxy library depending on temperature. From Karim et al., *Macromol. Rapid. Commun.* **2004**, *25*, 259–263 [98]. Reprinted with permission of John Wiley and Sons, Inc.

Symyx also described a generic workflow for the evaluation of synthetic polymers as selective transport agents for targeted delivery of potential drugs to human tissue [99]. After identification of an optimized tissue-mimicking substrate and a reconstructed biological liquid, the screening of 3000 diverse polymers on two selected mimics (representing two types of biological tissue) was feasible, leading to the identification of a polymer with a maximum bioactive uptake (while preserving its high tissue substantivity). The wettability of a polyurethane polymer library was evaluated by a new high-throughput screening method consisting of a liquid handing robot, a standard webcam and automated image processing software [100]. The method includes automated dispensing of a liquid on a spin-coated polymer film and subsequent image capture and automated evaluation of the spreading area. Furthermore, the results of this screening method could be correlated with conventional contact angle measurements.

Other interesting and important polymer properties include the elastic modulus, hardness and other mechanical properties. In that respect, nanoindentation is a very powerful tool for the combinatorial researcher since only small quantities of samples are necessary for a full mechanical analysis and samples can be measured in a serial and automated fashion. Moreover, these small-scale tests generally show good comparability with classical tensile tests if the necessary care is taken with sample preparation and measurement evaluation [101, 102]. Therefore, inkjet printing technology [103] and parallel (combinatorial) spin-coating [104] techniques have been investigated in detail for the preparation of defined polymer films and polymer film libraries that can subsequently be used for automated and parallel investigations of mechanical properties by nanoindentation. An example of a nanoindentation experiment of a polycarbonate polymer standard and the experimental setup is shown in Fig. 15.18. Figure 15.18a displays a load-displacement curve for a polycarbonate sample showing an elastic–plastic behavior (the plastic part is revealed by the hysteresis), whereas Fig. 15.19b displays the experi-

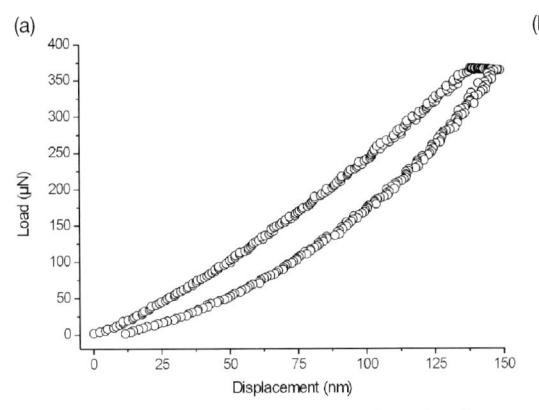

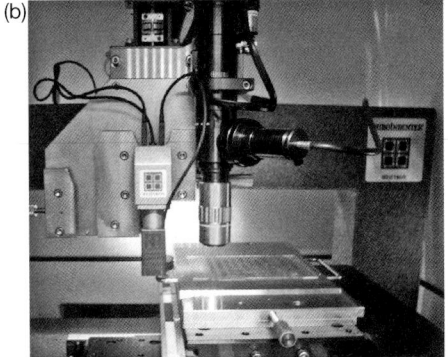

Fig. 15.18 (a) Load-displacement curve of a polycarbonate sample obtained by nanoindentation. (b) Experimental setup for nanoindentation.

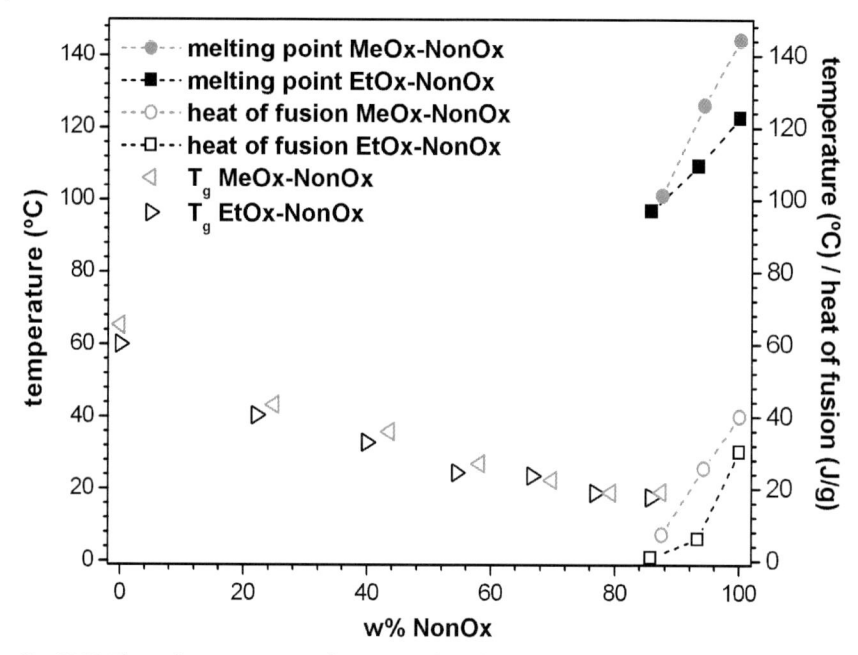

Fig. 15.19 Thermal investigations of an 18-membered poly(2-oxazoline) copolymer library. Reprinted with permission from Schubert et al., *J. Comb. Chem.* **2006**, *8*, 145–149 [78]. Copyright 2006, American Chemical Society.

mental setup with a microtiter plate-type format of the samples that allows high throughput investigations of mechanical properties.

The reduced modulus of the sample can be calculated from the initial slope of the unloading curve and by taking the indentation depth and the tip shape into account [105]. Finally, utilizing the Poisson constant, the reduced modulus can be converted to the *E*-modulus. Moreover, a recent literature example [106] showed that nanoindentation provided a general, rapid, precise and accurate mechanical characterization of the investigated discrete acrylate-based polymer library. Mechanical properties were also investigated for segmented poly(urethane urea) libraries [107, 108], resulting in structure–mechanical property relationships for libraries with continuous gradients in chain extender composition and/or cure temperature [108]. The mechanical properties, obtained from a high-throughput mechanical characterization apparatus, could subsequently be correlated with morphology, hydrogen bonding and degree of phase separation. As a result, optimum strength and percentage elongation were observed at a chain extender composition of 85 mol%. Furthermore, mechanical properties of a combinatorial library of differently thermally treated isotactic polypropylene (iPP) were investigated [109]. First, the libraries were created utilizing a temperature gradient plate apparatus in a disc-shaped form by cooling from the melt to room temperature applying different temperature gradients. Subsequently, the samples were investigated by wide-angle X-ray scattering (WAXS) to characterize the changes in crystalline morphology originating from the different thermal treatments. Finally, mechanical tensile testing was performed on specimens that were punched out from the discs. The results showed that the stress and strain at break could be correlated with the crystal morphology obtained from the WAXS experiments. Moreover, Symyx developed a mechanical thermal analyzer, a fully parallel instrument that is capable of performing 96 simultaneous measurements as a function of varying environmental conditions [110]. After validation of the new instrument, a case study was performed for tackifier and plasticizer additives to a polystyrene-b-polybutadiene-b-polystyrene triblock copolymer as a possible high-performance pressure-sensitive additive. For example, it was observed that the addition of the plasticizer dioctyl phthalate to the ABA triblock copolymer led to a decrease in the glass transition temperature of the hard styrene domain, whereas it had no effect on the soft butadiene domain. The absolute decrease could furthermore be correlated with the amount of plasticizer present in the system.

In the case of thermal property evaluation by DSC, TGA and other methods, the most convenient approach to analyze large sample sets or polymer libraries is the utilization of autosamplers that allow the automated and continuous measurement of important polymer properties, such as glass transition temperature (T_g), melting temperature (T_m) or the degradation temperature. As an example, Fig. 15.19 depicts screening results obtained by DSC with an autosampler of a systematic 2-oxazoline copolymer library prepared with an ASW2000 automated synthesizer [111]. Two copolymer series containing methyl and nonyl (MeOx:NonOx) and also ethyl and nonyl (EtOx:NonOx) side-chains were compared, because they have a gradient and random composition, respectively, and pMeOx and pEtOx homopolymers have very similar surface and thermal properties. Figure 15.19 shows the change in T_g, T_m and the heat of fusion with incorporation of NonOx.

Upon incorporation of NonOx into pMeOx or pEtOx, T_g decreases due to the higher flexibility of the nonyl side-chains. At approximately 90 wt% of NonOx, the glass transition disappears. In addition, the melting-point of pNonOx decreases on incorporation of MeOx or EtOx since the crystallinity of the pNonOx is disturbed. A more detailed discussion of these findings can be found in the literature [111]. This example clearly shows that polymer libraries can be analyzed for their thermal properties in a straightforward and automated fashion by utilizing characterization tools equipped with autosamplers that allow continuous measurements even during nights and weekends.

As already mentioned briefly, the possibility of preparing libraries of defined polymer films by inkjet printing offers possibilities for investigating these libraries by UV/visible spectroscopy or contact angle measurements in order to obtain quantitative structure–property relationships. Especially the utilization of plate reader equipment allows the full UV/visible characterization of a 96-membered library within 40 sec and the complete fluorescence spectra of this library can subsequently be obtained in less than 2 min per excitation wavelength. This

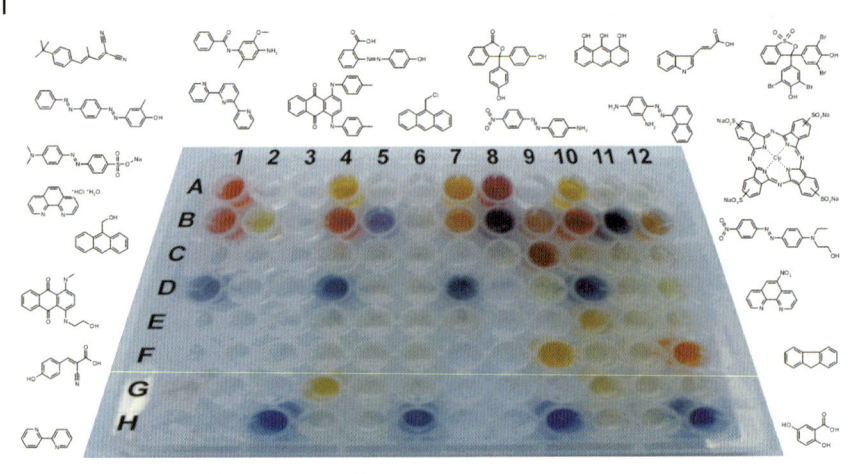

Fig. 15.20 Microtiter plate screening of host–guest interactions of star-shaped block copolymers. Reprinted with permission from Schubert et al., *J. Comb. Chem.* **2005**, *7*, 356–359 [112]. Copyright 2005, American Chemical Society.

offers the possibility of screening complete libraries of polymer thin films and of investigating certain material properties in solution. For instance, a plate reader screening was successfully applied to the evaluation of the host–guest chemistry of 24 potential guest molecules for star-shaped block copolymer host molecules [112]. This screening was performed in a microtiter plate in both basic and acidic environments and characteristic changes in a guest's UV/visible or fluorescence behavior could be interpreted as an encapsulation event. Figure 15.20 shows an optical picture of the microtiter plate as utilized for the screening [112]. This screening revealed that the investigated star-shaped block copolymers with a poly(ethylene glycol) core and a poly(ε-caprolactone) corona were able to encapsulate 22 of the 24 investigated potential guest molecules, making these polymers interesting candidates for drug delivery applications.

Moreover, photoluminescence and UV/visible spectroscopic techniques were used successfully by Muramatsu et al. [113] to characterize a library of 14 kinds of π-conjugated polymeric monolayer and 49 kinds of polymeric bilayer films. The library was screened and the bilayer with the strongest blue light emission was identified, which was the goal of the study. The authors stated that thousands of new materials can be evaluated with this setup on a weekly bases. In addition, Potyrailo and coworkers showed that it is possible to evaluate the weathering of polymeric materials [114] and the abrasion resistance of coating libraries [115] by spectroscopic methods. The abrasion resistance of UV-cured coatings was tested by automated scattered light measurements with a fiber-optic arrangement [115]. This approach was shown to be at least 10 times faster than a conventional coating development process. The UV-induced degradation of polycarbonate (PC), poly(butene terephthalate) (PBT) and their 45:55 wt% blend with two types of pigments (rutile and carbon black) was evaluated using

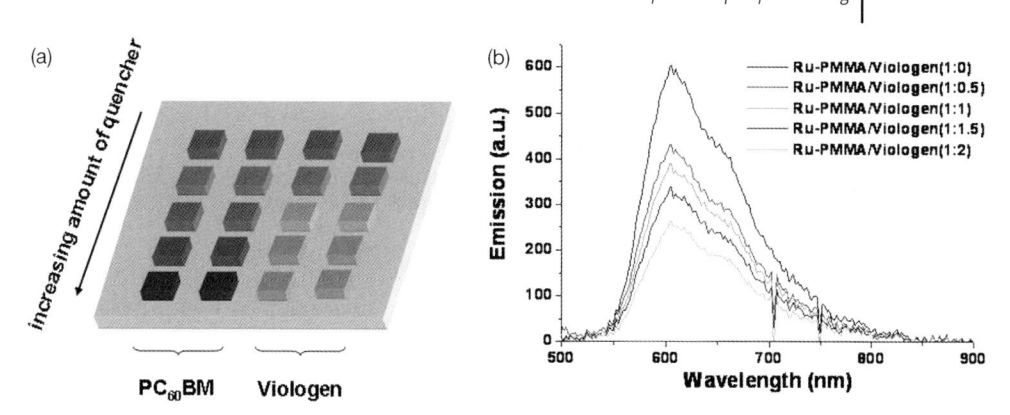

Fig. 15.21 (a) Schematic representation of a 20-array film library prepared using inkjet printing. (b) Luminescence quenching of a tris-bipyridine–Ru–PMMA copolymer caused by C7-Viologen (films inkjet printed from isopropyl alcohol–acetophenone–o-dichlorobenzene). From Schubert et al., *Macromol. Rapid Commun.* **2005**, *26*, 319–324 [116]. Reprinted with permission of John Wiley and Sons, Inc.

fluorescence imaging and spectroscopy [114]. Using this method, the screening throughput can be 800 times higher than with conventional methods, such as color change or gloss loss. For the investigation of polymer thin films that were prepared by inkjet printing, it was possible to demonstrate that UV/visible and fluorescence plate reader technology is a very versatile tool for the evaluation of material properties for potential solar cell applications. Figure 15.21 displays an example where inkjet printing was used to prepare a 20-membered library to investigate donor–acceptor combinations for bulk heterojunction solar cells [116]. Figure 15.21a shows the library layout for investigating two acceptors and one donor at different molar ratios. Figure 15.21b depicts quenching (energy transfer) results obtained from this library obtained with a fluorescence plate reader, clearly showing that it is feasible to investigate these systems in a fast and accurate manner.

Moreover, Potyrailo and coworkers demonstrated that fluorescence spectroscopy is a versatile tool in combinatorial polymer research [117, 118]. Utilizing a 96-microreactor array for the preparation of bisphenol A poly(carbonate)s, properties of interest such as molecular weight, amount of branching and catalyst selectivity could be correlated with fluorescence features. Therefore, a CCD-based spectrofluorimeter and/or a fluorescence scanning system were utilized [117]. Both fluorescence screening tools revealed good correlations with each other and with conventional characterization techniques. It was possible, for instance, to correlate the fluorescence (excitation at 280 nm, emission at 305 nm) of the solid polymer samples with their number-average molecular weight obtained by SEC. Furthermore, the selectivity of catalysts could be correlated with the ratio of fluorescence intensities (I_{400}/I_{500}) with 340-nm excitation.

Overall, optical screening techniques present very promising methods for the online monitoring or high-throughput screening of polymer properties since

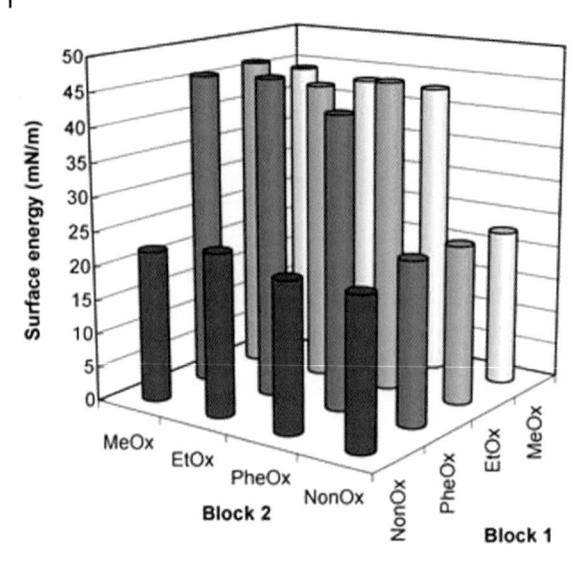

Fig. 15.22 Surface energies of block copoly-(2-oxazoline)s calculated with the equation of state method, revealing a considerably lower surface energy for the nonyl-containing polymers. From Schubert et al., *Macromol. Rapid Commun.* **2004**, 25, 1958–1962 [119]. Reprinted with permission of John Wiley and Sons, Inc.

they are fast, non-destructive and comparably cheap. However, the implementation of some of the techniques is complicated, since advanced chemometric tools are required. In addition, the identification of correlations between macroscopic polymer properties and optical screening methods will be very important in the future in order to accelerate the applied screening approaches.

Polymer film libraries can also be evaluated by means of automated contact angle measurements. If two test liquids are used to determine the contact angle on a certain surface, it is possible to calculate the surface energy of that sample, which is an important material property. Therefore, a modified commercial contact-angle measuring apparatus that is capable of automatic dispensing, analyzing and aspirating was applied for the evaluation of a 16-membered library of polyoxazoline block copolymers [119]. Utilizing diiodomethane and ethylene glycol as test liquids and Neumann's equation of state, it was possible to convert the measured contact angles into surface energies. The necessary 128 contact angle measurements could be performed within 70 min.

Figure 15.22 displays the resulting surface energies of this 16-membered polyoxazoline library, revealing that polymers containing the 2-nonyl-2-oxazoline monomer have a considerably lower surface energy. This is most likely caused by close packing (and preferential orientation) of the aliphatic nonyl chains on the surface of the samples.

15.7
Conclusion

The synthesis and screening of polymer libraries with a known and systematic variation of certain properties is one of the two possibilities for the reliable determination of quantitative structure–property relationships in combinatorial materials research. The second possibility is the application of design-of-experiments and other computational approaches for the reduction of the necessary experiments while the whole parameter space is still covered. Both approaches have their pros and cons, e.g. the coverage of the whole parameter space by the computational approaches with the disadvantage that a large number of experiments have to be performed and structure–property correlations might still not be identified because the descriptors were not chosen correctly. On the other hand, the investigation of small and defined libraries ("targeted libraries") will certainly reveal structure–property relationships, but only within a very limited parameter set and the observed correlations might not be transferable to other materials. Nevertheless, whatever approach is chosen, combinatorial materials research has already shown that these efforts can address a large variety of different problems in materials research, including the extremely large parameter space that is available.

Finally, we would like to point out the importance of the discussed screening approaches, since (i) it would not be possible to investigate polymer libraries and to reveal the desired structure–property relationships within a reasonable time frame without them and (ii) the efforts taken to optimize, miniaturize and parallelize analytical techniques generally result in a better understanding, improved analytical results and/or increased applicability of these techniques, which is not only very useful for the combinatorial researcher but also provides significant advantages for the whole scientific community.

Acknowledgments

The authors would like to thank the Dutch Polymer Institute (DPI), the Nederlandse Wetenschappelijke Organisatie (NWO) and the Fonds der Chemischen Industrie for financial support.

References

1. Glossary of Terms Used in Combinatorial Chemistry, *Pure Appl. Chem.* **1999**, *71*, 2349–2365.
2. E.W. McFarland, W.H. Weinberg, *Trends Biotechnol.* **1999**, *17*, 107–111.
3. S.P. Rohrer, E.T. Birzin, R.T. Mosley, S.C. Berk, S.M. Hutchins, D.-M. Shen, Y. Xiong, E.C. Hayes, R.M. Parmar, F. Foor, S.W. Mitra, S.J. Degrado, M. Shu, J.M. Klopp, S.-J. Cai, A. Blake, W.W.S. Chan, A. Pasternak, L. Yang, A.A. Patchett, R.G. Smith, K.T. Chapman, J.M. Schaeffer, *Science* **1998**, *282*, 737–740.
4. K.C. Nicolaou, A.J. Roecker, S. Barluenga, J.A. Pfefferkorn, G.-Q. Cao, *ChemBioChem* **2001**, *2*, 460–465.
5. S.M. Senkan, *Nature* **1998**, *394*, 350–353.
6. S.J. Taylor, J.P. Morken, *Science* **1998**, *280*, 267–270.
7. T.R. Boussie, C. Coutard, H. Turner, V. Murphy, T.S. Powers, *Angew. Chem. Int. Ed.* **1998**, *37*, 3272–3275.
8. T. Bein, *Angew. Chem. Int. Ed.* **1999**, *38*, 323–326.
9. W.F. Maier, *Angew. Chem. Int. Ed.* **1999**, *38*, 1216–1218.
10. M.T. Reetz, *Angew. Chem. Int. Ed.* **2001**, *40*, 284–310.
11. B. Jandeleit, D.J. Schaefer, T.S. Powers, H.W. Turner, W.H. Weinberg, *Angew.-Chem. Int. Ed.* **1999**, *38*, 2494–2532.
12. X.-D. Wang, X. Sun, G. Briceño, Y. Lou, K.-A. Wang, H. Chang, W.G. Wallace-Freedman, S.-W. Chen, P.G. Schultz, *Science* **1995**, *268*, 1738–1740.
13. E. Danielson, J.H. Golden, E.W. McFarland, C.M. Reaves, W.H. Weinberg, X.D. Wu, *Nature* **1997**, *389*, 944–948.
14. J. Wang, Y. Yoo, C. Gao, I. Takeuchi, X. Sun, H. Chang, X.-D. Xiang, P.G. Schultz, *Science* **1998**, *279*, 1712–1714.
15. R. Hoogenboom, M.A.R. Meier, U.S. Schubert, *Macromol. Rapid Commun.* **2003**, *24*, 15–32.
16. M.A.R. Meier, R. Hoogenboom, U.S. Schubert, *Macromol. Rapid Commun.* **2004**, *25*, 21–33.
17. M.A.R. Meier, U.S. Schubert, *J. Mater. Chem.* **2004**, *14*, 3289–3299.
18. K.C. Nicolaou, R. Hanko, W. Hartwig, *Handbook of Combinatorial Chemistry.* Wiley-VCH, Weinheim, **2002**.
19. Special Issue, *Macromol. Rapid Commun.* **2003**, *24*, 3–142.
20. Special Issue, *Macromol. Rapid Commun.* **2004**, *25*, 19–386.
21. C.H. Reynolds, *J. Comb. Chem.* **1999**, *1*, 297–306.
22. N. Adams, U.S. Schubert, *Macromol. Rapid Commun.* **2004**, *25*, 48–58.
23. N. Adams, U.S. Schubert, *J. Comb. Chem.* **2004**, *6*, 12–23.
24. R.A. Potyrailo, *Macromol. Rapid Commun.* **2004**, *25*, 77–94.
25. S. Schmatloch, M.A.R. Meier, U.S. Schubert, *Macromol. Rapid Commun.* **2003**, *24*, 33–46.
26. R. Hoogenboom, U.S. Schubert, *Rev. Sci. Instrum.* **2005**, *76*, 062202/1–7.
27. C. Guerrero-Sanchez, C. Abeln, U.S. Schubert, *J. Polym. Sci., Part A: Polym. Chem.* **2005**, *43*, 4151–4160.
28. C. Guerrero-Sanchez, D. Wouters, C.-A. Fustin, J.-F. Gohy, B.G.G. Lohmeijer, U.S. Schubert, *Macromolecules* **2005**, *38*, 10185–10191.
29. R. Hoogenboom, M.W.M. Fijten, U.S. Schubert, *Macromol. Rapid Commun.* **2004**, *25*, 339–343.
30. R. Hoogenboom, M.W.M. Fijten, U.S. Schubert, *J. Polym. Sci., Part A: Polym. Chem.* **2004**, *42*, 1830–1840.
31. H. Zhang, M.W.M. Fijten, R. Reinierkens, R. Hoogenboom, U.S. Schubert, *Macromol. Rapid Commun.* **2003**, *24*, 81–86.
32. H. Zhang, V. Marin, M.W.M. Fijten, U.S. Schubert, *J. Polym. Sci., Part A: Polym. Chem.* **2004**, *42*, 1876–1885.
33. M.W.M. Fijten, M.A.R. Meier, R. Hoogenboom, U.S. Schubert, *J. Polym. Sci., Part A: Polym. Chem.* **2004**, *42*, 5775–5783.
34. M.W.M. Fijten, R.M. Paulus, U.S. Schubert, *J. Polym. Sci., Part A: Polym. Chem.* **2005**, *43*, 3831–3839.
35. M.A.R. Meier, J.-F. Gohy, C.-A. Fustin, U.S. Schubert, *J. Am. Chem. Soc.* **2004**, *126*, 11517–11521.

36 T. R. Boussie, G. M. Diamond, C. Goh, K. A. Hall, A. M. LaPointe, M. Leclerc, C. Lund, V. Murphy, J. A. W. Shoemaker, U. Tracht, H. Turner, J. Zhang, T. Uno, R. K. Rosen, J. C. Stevens, *J. Am. Chem. Soc.* **2003**, *125*, 4306–4317.

37 S. Brocchini, K. James, V. Tangpasuthadol, J. Kohn, *J. Am. Chem. Soc.* **1997**, *119*, 4553–4554.

38 J. R. Smith, V. Kholodovych, D. Knight, W. J. Welsh, J. Kohn, *QSAR Comb. Sci.* **2005**, *24*, 99–113.

39 N. Weber, D. Bolikal, S. L. Bourke, J. Kohn, *J. Biomed. Mater. Res.* **2004**, *68A*, 496–503.

40 J. R. Smith, A. Seyda, N. Weber, D. Knight, S. Abramson, J. Kohn, *Macromol. Rapid Commun.* **2004**, *25*, 127–140.

41 S. D. Abramson, G. Alexe, P. L. Hammer, J. Kohn, *J. Biomed. Mater. Res.* **2005**, *73A*, 116–124.

42 C.-S. Wu (Ed.), *Column Handbook for Size Exclusion Chromatography*. Marcel Dekker, New York, **2002**.

43 J. Cazes (Ed.), *Encyclopedia of Chromatography*, online edition. Marcel Dekker, New York, **2002**.

44 G. Klaerner, A. L. Safir, H.-T. Chang, M. Petro, R. B. Nielson, *Polym. Prepr.* **1999**, *40*, 469–470.

45 H. Pasch, P. Kilz, *Macromol. Rapid Commun.* **2003**, *24*, 104–108.

46 R. Hoogenboom, M. M. W. Fijten, C. H. Abeln, U. S. Schubert, *Macromol. Rapid Commun.* **2004**, *25*, 237–242.

47 D. S. Poché, R. J. Brown, P. L. Morabito, R. Tamilarasan, D. J. Duke, *J. Appl. Polym. Sci.* **1997**, *64*, 1613–1623.

48 G. Klaerner, A. L. Safir, H.-T. Chang, M. Petro, R. B. Nielson, *Polym. Prepr.* **1999**, *40*, 469.

49 K. P. Peil, *DECHEMA Monogr.* **2001**, *137*, 71–78.

50 M. Petro, A. Safir, R. B. Nielson, *Polym. Prepr.* **1999**, *40*, 702.

51 R. Srelitzki, W. F. Reed, *J. Appl. Polym. Sci.* **1999**, *73*, 2359–2368.

52 S. D. Hanton, *Chem. Rev.* **2001**, *28*, 4562–4569.

53 K. Takana, H. Waki, Y. Ido, S. Akita, Y. Yoshida, T. Yoshida, *Rapid Commun. Mass Spectrom.* **1988**, *2*, 151–153.

54 M. Karas, F. Hillenkamp, *Anal. Chem.* **1988**, *60*, 2299–2301.

55 M. A. R. Meier, U. S. Schubert, *Rapid Commun. Mass Spectrom.* **2003**, *17*, 713–716.

56 M. A. R. Meier, R. Hoogenboom, M. W. M. Fijten, M. Schneider, U. S. Schubert, *J. Comb. Chem.* **2003**, *5*, 369–374.

57 M. A. R. Meier, B. J. de Gans, A. M. J. van den Berg, U. S. Schubert, *Rapid Commun. Mass Spectrom.* **2003**, *17*, 2349–2353.

58 M. A. R. Meier, U. S. Schubert, *Rev. Sci. Instrum.* **2005**, *76*, 062211/1–5.

59 M. W. F. Nielen, *Mass Spectrom. Rev.* **1999**, *18*, 309–344.

60 D. I. Papac, A. Wong, A. J. Jones, *Anal. Chem.* **1996**, *68*, 3215–3223.

61 Y. Dai, R. Whittal, L. Li, *Anal. Chem.* **1999**, *71*, 1087–1091.

62 P. Önnerfjord, S. Ekström, J. Bergquist, J. Nilsson, T. Laurell, G. Marko-Varga, *Rapid Commun. Mass Spectrom.* **1999**, *13*, 315–322.

63 O. Vorm, P. Roepstorff, M. Mann, *Anal. Chem.* **1994**, *66*, 3281–3287.

64 D. A. Lake, M. V. Johnson, C. N. McEwen, B. S. Larsen, *Rapid Commun. Mass Spectrom.* **2000**, *14*, 1008–1013.

65 L. Ulmer, H. G. Torres-Garcia, J. Mattay, H. Luftmann, *Eur. J. Mass Spectrom.* **2000**, *6*, 49–52.

66 R. Hoogenboom, M. A. R. Meier, U. S. Schubert, *Mater. Res. Soc. Symp. Proc.* **2004**, *804*, 83–94.

67 R. Hoogenboom, M. W. M. Fijten, M. A. R. Meier, U. S. Schubert, *Macromol. Rapid Commun.* **2003**, *24*, 92–97.

68 R. Hoogenboom, F. Wiesbrock, M. A. M. Leenen, M. A. R. Meier, U. S. Schubert, *J. Comb. Chem.* **2005**, *7*, 10–13.

69 F. Wiesbrock, R. Hoogenboom, M. A. M. Leenen, M. A. R. Meier, U. S. Schubert, *Macromolecules* **2005**, *38*, 5025–5034.

70 K. Aoi, M. Okada, *Prog. Polym. Sci.* **1996**, *21*, 151–208.

71 T. Saegusa, S. Kobayashi, A. Yamada, *Makromol. Chem.* **1976**, *177*, 2271–2283.

72 E. G. Chatzi, O. Kammona, C. Kiparissides, *J. Appl. Polym. Sci.* **1997**, *63*, 799–809.

73 J. E. Puskas, P. Antony, Y. Kown, C. Paulo, M. Kovar, R. R. Norton, G. Kaszas, V. Altstädt, *Macromol. Mater. Eng.* **2001**, *286*, 565–580.

74 M. Lanzendörfer, H. Schmalz, V. Abetz, A. H. E. Müller, *Polym. Prepr.* **2001**, *42*, 329–330.

75 A. Tuchbreiter, R. Mülhaupt, A. Warmbold, P. Liebertraut, B. Koppler, J. Honerkamp, *Polym. Prepr.* **2002**, *43*, 285–286.

76 H. Hua, M. A. Dubé, *J. Polym. Sci., Part A: Polym. Chem.* **2001**, *39*, 1860–1876.

77 R. F. Storey, T. L. Maggio, *Macromolecules* **2000**, *33*, 681–688.

78 A. Tuchbreiter, J. Marquardt, J. Zimmermann, P. Walter, R. Mülhaupt, *J. Comb. Chem.* **2001**, *3*, 598–603.

79 H. Pasch, M. Schrod, *Macromol. Rapid Commun.* **2004**, *25*, 224–230.

80 J. C. Meredith, *J. Mater. Sci.* **2003**, *38*, 4427–4437.

81 A. J. Crosby, *J. Mater. Sci.* **2003**, *38*, 4439–4449.

82 J. C. Meredith, A. Karim, E. J. Amis, *Macromolecules* **2000**, *33*, 5760–5762.

83 J. C. Meredith, A. P Smith, A. Karim, E. J. Amis, *Macromolecules* **2000**, *33*, 9747–9756.

84 A. P. Smith, J. F. Douglas, J. C. Meredith, E. J. Amis, A. Karim, *J. Polym. Sci., Part B: Polym. Phys.* **2001**, *39*, 2141–2158.

85 A. P. Smith, J. F. Douglas, J. C. Meredith, E. J. Amis, A. Karim, *Phys. Rev. Lett.* **2001**, *87*, 015503/1–4.

86 T. A. Dickinson, D. R. Walt, *Anal. Chem.* **1997**, *69*, 3413–3418.

87 R. A. Potyrailo, R. J. Wroczynski, J. E. Pickett, M. Rubinsztajn, *Macromol. Rapid Commun.* **2003**, *24*, 123–130.

88 K. Ashley, A. Seghal, E. J. Amis, A. Karim, *Mater. Res. Soc. Symp. Proc.* **2002**, *700*, S4.7–S4.10.

89 K. M. Ashley, J. C. Meredith, E. J. Amis, D. Raghavan, A. Karim, *Polymer* **2003**, *44*, 769–772.

90 B.-J. de Gans, E. Kazancioglu, W. Meyer, U. S. Schubert, *Macromol. Rapid Commun.* **2004**, *25*, 292–296.

91 H. Fukumoto, Y. Muramatsu, T. Yamamoto, J. Yamaguchi, K. Itaka, H. Koinuma, *Macromol. Rapid Commun.* **2004**, *25*, 196–203.

92 M. Bäte, C. Neuber, R. Giesa, H.-W. Schmidt, *Macromol. Rapid Commun.* **2004**, *25*, 371–376.

93 A. P. Smith, A. Sehgal, J. F. Douglas, A. Karim, E. J. Amis, *Macromol. Rapid Commun.* **2003**, *24*, 131–135.

94 K. L. Beers, J. F. Douglas, E. J. Amis, A. Karim, *Langmuir* **2003**, *19*, 3935–3940.

95 A. J. Crosby, A. Karim, E. J. Amis, *Polym. Prepr.* **2001**, *42*, 645–646.

96 J. C. Grunlan, A. R. Mehrabi, A. T. Chavira, A. B. Nuget, D. L. Saunders, *J. Comb. Chem.* **2003**, *5*, 362–368.

97 P. Mansky, *Polym. Prepr.* **2001**, *42*, 647–648.

98 N. Eidelman, D. Rafhavan, A. M. Forster, E. J. Amis, A. Karim, *Macromol. Rapid. Commun.* **2004**, *25*, 259–263.

99 M. Petro, S. H. Nguyen, M. Liu, O. Kolosov, *Macromol. Rapid Commun.* **2004**, *25*, 178–188.

100 J.-F. Thaburet, H. Mizomoto, M. Bradley, *Macromol. Rapid Commun.* **2004**, *25*, 366–370.

101 L. Shen, I. Y. Phang, L. Chen, T. Liu, K. Zeng, *Polymer* **2004**, *45*, 3341–3349.

102 D. Drechsler, S. Karbach, H. Fuchs, *Appl. Phys. A Mater. Sci. Process.* **1998**, *66*, S825–S829.

103 B.-J. de Gans, E. Kazancioglu, W. Meyer, U. S. Schubert, *Macromol. Rapid Commun.* **2004**, *25*, 292–296.

104 B.-J. de Gans, S. Wijnans, D. Wouters, U. S. Schubert, *J. Comb. Chem.* **2005**, *7*, 952–957.

105 W. C. Oliver, G. M. Pharr, *J. Mater. Res.* **1992**, *7*, 1564–1583.

106 C. A. Tweedie, D. G. Anderson, R. Langer, K. J. Van Vliet, *Adv. Mater.* **2005**, *17*, 2599–2604.

107 J.-L. Sormana, J. C. Meredith, *Macromol. Rapid Commun.* **2003**, *24*, 118–222.

108 J.-L. Sormana, J. C. Meredith, *Macromolecules* **2004**, *37*, 2186–2195.

109 K. Schneider, N. E. Zafeiropoulos, L. Häussler, M. Stamm, *Macromol. Rapid Commun.* **2004**, *25*, 355–356.

110 M. B. Kossuth, D. A. Hajduk, C. Freitag, J. Varni, *Macromol. Rapid. Commun.* **2004**, *25*, 243–248.

111 R. Hoogenboom, M. W. M. Fijten, S. Wijnans, A. M. J. van den Berg, H. M. L.

Thijs, U. S. Schubert, *J. Comb. Chem.* **2006**, *8*, 145–149.
112 M. A. R. Meier, U. S. Schubert, *J. Comb. Chem.* **2005**, *7*, 356–359.
113 Y. Muramatsu, T. Yamamoto, T. Hayakawa, H. Koinuma, *Appl. Surf. Sci.* **2002**, *189*, 319–326.
114 R. A. Potyrailo, J. E. Picket, *Angew. Chem.* **2002**, *114*, 4404–4407; *Angew. Chem. Int. Ed.* **2002**, *41*, 4230–4233.
115 R. A. Potyrailo, B. J. Chisholm, D. R. Olsen, M. J. Brennan, C. A. Molaison, *Anal. Chem.* **2002**, *74*, 5105–5111.
116 V. Marin, E. Holder, M. M. Wienk, E. Tekin, D. Kozodaev, U. S. Schubert, *Macromol. Rapid Commun.* **2005**, *26*, 319–324.
117 R. A. Potyrailo, R. J. Wroczynski, J. P. Lemmon, W. P. Flanagan, O. P. Siclovan, *J. Comb. Chem.* **2003**, *5*, 8–17.
118 R. A. Potyrailo, J. P. Lemmon, T. K. Leib, *Anal. Chem.* **2003**, *75*, 4676–4681.
119 S. Wijnans, B.-J. de Gans, F. Wiesbrock, R. Hoogenboom, U. S. Schubert, *Macromol. Rapid Commun.* **2004**, *25*, 1958–1962.